10th Biennial Conference on Classical and Quantum Relativistic Dynamics of Particles and Fields (IARD 2016)

Journal of Physics: Conference Series Volume 845

Ljubljana, Slovenia
6 – 9 June 2016

ISBN: 978-1-5108-4189-5
ISSN: 1742-6588

Printed from e-media with permission by:

Curran Associates, Inc.
57 Morehouse Lane
Red Hook, NY 12571

Some format issues inherent in the e-media version may also appear in this print version.

Printed by Curran Associates, Inc. (2017)

Institute of Physics
Dirac House, Temple Back
Bristol BS1 6BE UK

Phone: 44 1 17 929 7481
Fax: 44 1 17 920 0979

techtracking@iop.org

Additional copies of this publication are available from:

Curran Associates, Inc.
57 Morehouse Lane
Red Hook, NY 12571 USA
Phone: 845-758-0400
Fax: 845-758-2633
Email: curran@proceedings.com
Web: www.proceedings.com

Table of contents

Volume 845

10th Biennial Conference on Classical and Quantum Relativistic Dynamics of Particles and Fields

6–9 June 2016, Ljubljana, Slovenia

Accepted papers received: 11 May 2017
Published online: 1 June 2017

Preface

Papers

IOP Conf. Series: Journal of Physics: Conf. Series **845** (2017) 011001 doi:10.1088/1742-6596/845/1/011001

Preface

The International Association for Relativistic Dynamics was organized in February 1998 in Houston, Texas, with John R. Fanchi as president.

Although the subject of relativistic dynamics has been explored, from both classical and quantum mechanical points of view, since the work of Einstein and Dirac, its most striking development has been in the framework of quantum field theory. The very accurate calculations of spectral and scattering properties, for example, of the anomalous magnetic moment of the electron and the Lamb shift in quantum electrodynamics, and many qualitative features of the strong and electroweak interactions, demonstrate the very great power of description achieved in this framework. Yet, many fundamental questions remain to be clarified, such as the structure of classical relativistic dynamical theories on the level of Hamilton and Lagrange in Minkowski space as well as on the curved manifolds of general relativity. There, moreover, remained the important questions of the covariant classical description of systems at high energy for which particle production effects are not large, such as discussed in Synge's book, *The Relativistic Gas*, and in Balescu's book on relativistic statistical mechanics, and the development of a consistent single and many body relativistic quantum theory.

In recent years, the very high accuracy of telescopes and advanced facilities for computation have brought a high level of interest in cosmological problems such as the structure of galaxies (dark matter) and the apparently anomalous expansion of the universe (dark energy). Some of the papers reported here deal with these problems, as well as other fundamental related issues. It was for this purpose, to bring together researchers from a wide variety of fields, such as particle physics, astrophysics, cosmology, foundations of relativity theory, and mathematical physics, with a common interest in relativistic dynamics, to investigate fundamental questions of this type, that this Association was founded.

The second meeting took place, in 2000, at Bar Ilan University in Ramat Gan, Israel, the third, in 2002, at Howard University in Washington, D.C., and the fourth, on June 12-19, 2004, in Saas Fee, Switzerland. In 2006, the fifth meeting took place at the University of Connecticut campus in Storrs, Connecticut, and the sixth meeting, in Thessaloniki, Greece, in 2008, with the significant guidance of Ioannis Antoniou, the help of the Aristotle University of Thessaloniki, at the Teleglion Foundation. The seventh meeting was held in Hualien, Taiwan in 2010, organized by Da-Shin Lee with the gracious help of Bei-Lok Hu. The eighth at the Galileo Galilei Institute for Theoretical Physics (GGI) in Florence (Firenze), Italy, in 2012 and chaired by Luca Lusanna. The ninth meeting took place again at the University of Connecticut in 2014 under the guidance of Philip Mannheim and James O'Brien.

The tenth biennial meeting of IARD in 2016 was held in Ljubljana, Slovenia, organized by Matej Pavšič in cooperation with Jožef Stefan Institute. This meeting forms the basis for the Proceedings that are recorded in this issue of the Journal of Physics: Conference Series. Along with the work of some of the founding and newer but already much engaged members of the

Association, we were fortunate to have lecturers from application areas that provided strong challenges for further developments in quantum field theory, cosmological problems, and in the dynamics of systems subject to accelerations and the effects of general relativity.

We dedicate the Proceedings of IARD 2016 to the memory of Jacob Bekenstein of Hebrew University, who served on the Scientific Advisory Committee, and David R. Finkelstein of the Georgia Institute of Technology, who served on the IARD Standing Committee. We gratefully recall their support and encouragement.

Topics treated in this issue include studies in relativistic statistical mechanics, fluid mechanics and thermodynamics with relevance for the dark matter and dark energy problems, for example, in the context of Wheeler's quantum foam and its associated thermodynamics. There are also studies of the dynamics of a perfect fluid of point particles, a treatment of surface tension analog in spacetime, and advances in conformal gravity, as well as the possibility of dark matter phenomena emerging from metric modifications which change dynamically the relation between inertial and gravitational masses. The dynamics of deformed neutron stars and the effects of acceleration, expressed in a modified metric relation is studied. Work is also reported on a fundamental development of a generalization of Newtonian mechanics, and a study is made of relativistic Coulomb systems in velocity space, providing new insight into the relativistic Kepler problem. A fundamental study of the structure of spacetime is reported which provides an interpretation of time in the presence of matter, and results in an estimate for the size of the observable universe.

Discussions of electromagnetism, including a "skewon" modification (a covariant tensor quadratic term in field strengths) of the standard electrodynamics, and a study of the field equations for moving media in covariant form, a continuum dynamics in the scalar ether theory of gravitation, as well a wave equations of massless particles of any spin. There are discussions of Pizzella's experiment apparently demonstrating instantaneous Coulomb interaction.

Quantum and particle physics are discussed in a proposed spin-charge family theory as a successor to the standard model, and a study of branes and quantized fields. A geometrical model for electro-gravity, the quantum dynamics of bound states with spacetime fluctuations, and quantum models as classical cellular automata are also reported. Integrability of geodesics with the use of action angle variables is studied, and a study of the Feynman-Dyson theory, with a justification of the Feynman formulation of quantum mechanics, is reported.

Progress is reported in our understanding of the (off-shell) relativistically covariant formulation of Stueckelberg, with (classical) approaches to the stabilization of particle mass after interaction, and the relation of the 5D gauge fields resulting from the gauging of the evolution term in the Stueckelberg-Schrödinger equation to the standard Maxwell fields is discussed. In this framework, there is a new and fundamental study of neutrino oscillations. In a somewhat different direction, the possibility of a relativistic extension of Newtonian dynamics is studied for which the potential function in Hamiltonian theory can introduce curvature in spacetime.

Spinor theory is discussed in a study of free particle wave equations in curved spacetime. A generalized spin statistics theorem is given, and the transformations of spinors under automorphisms of their associated Clifford algebras are also discussed.

We thank the Scientific Advisory Committee for their invaluable guidance and advice:

Stephen Adler (Institute for Advanced Study)
Itzhak Bars (University of Southern California)
Gordon Baym (University of Illinois)
Fred Cooper (Los Alamos National Laboratory)
Bei-Lok Hu (University of Maryland)
Werner Israel (University of Victoria)
E.V. Shuryak (Brookhaven National Laboratory)
L.S. Shulman (Clarkson University)
William Unruh (University of British Columbia)
Luca Lusanna (National Institute for Nuclear Physics, INFN)
Benoit Famaey (Observatoire Astronomique de Strasbourg, CNRS)

The organizers express their gratitude to the Jožef Stefan Institute for its support in arranging excellent facilities in the charming city of Ljubljana.

We thank the participants who contributed through their lectures, personal discussions, and these papers, to the advancement of the subject and our understanding.

For the Editors and Organizing Committee,

L. P. Horwitz, Editor-in-Chief
Matej Pavšič, Chair of the Organizing Committee
Tepper Gill, IARD Treasurer
Martin Land, IARD President
James O'Brien, IARD Vice-President
Luca Lusanna, IARD SAC

IOP Conf. Series: Journal of Physics: Conf. Series **845** (2017) 011002 doi:10.1088/1742-6596/845/1/011002

Peer review statement

All papers published in this volume of *Journal of Physics: Conference Series* have been peer reviewed through processes administered by the proceedings Editors. Reviews were conducted by expert referees to the professional and scientific standards expected of a proceedings journal published by IOP Publishing.

IOP Conf. Series: Journal of Physics: Conf. Series **845** (2017) 012001 doi:10.1088/1742-6596/845/1/012001

Quantum foam, gravitational thermodynamics, and the dark sector

Y. Jack Ng

Institute of Field Physics, Department of Physics & Astronomy, University of North Carolina, Chapel Hill, NC 27599-3255, USA

E-mail: yjng@physics.unc.edu

Abstract.
Is it possible that the dark sector (dark energy in the form of an effective dynamical cosmological constant, and dark matter) has its origin in quantum gravity? This talk sketches a positive response. Here specifically quantum gravity refers to the combined effect of quantum foam (or spacetime foam due to quantum fluctuations of spacetime) and gravitational thermodynamics. We use two simple independent gedankan experiments to show that the holographic principle can be understood intuitively as having its origin in the quantum fluctuations of spacetime. Applied to cosmology, this consideration leads to a dynamical cosmological constant of the observed magnitude, a result that can also be obtained for the present and recent cosmic eras by using unimodular gravity and causal set theory. Next we generalize the concept of gravitational thermodynamics to a spacetime with positive cosmological constant (like ours) to reveal the natural emergence, in galactic dynamics, of a critical acceleration parameter related to the cosmological constant. We are then led to construct a phenomenological model of dark matter which we call "modified dark matter" (MDM) in which the dark matter density profile depends on both the cosmological constant and ordinary matter. We provide observational tests of MDM by fitting the rotation curves to a sample of 30 local spiral galaxies with a single free parameter and by showing that the dynamical and observed masses agree in a sample of 93 galactic clusters. We also give a brief discussion of the possibility that quanta of both dark energy and dark matter are non-local, obeying quantum Boltzmann statistics (also called infinite statistics) as described by a curious average of the bosonic and fermionic algebras. If such a scenario is correct, we can expect some novel particle phenomenology involving dark matter interactions. This may explain why so far no dark matter detection experiments have been able to claim convincingly to have detected dark matter.

1. Introduction

This talk is based on several loosely related pieces of work I did mostly with various collaborators. I will start with one aspect of John Wheeler's spacetime foam or quantum foam by which he was referring to a foamy structure of spacetime due to quantum fluctuations. So how large are those fluctuations? I will briefly discuss (in section 2) a gedankan experiment to measure a distance l and deduce the intrinsic limitation δl to the accuracy with which we can measure that distance, [1, 2, 3] for that distance undergoes quantum fluctuations. [1] To gain more insight I

[1] I will further show that the scaling of $\delta l \gtrsim l^{1/3} l_P^{2/3}$, as deduced from the gedankan experiment, is exactly what the holographic principle [4, 5] demands, according to which the maximum amount of information stored in a

will use (in section 2) an argument in mapping out the geometry of spacetime [7] to arrive at the same result. When I generalize the argument to the case of an expanding Universe, we will see that something akin to a (positive) cosmological constant emerges [8] — an effective dynamical cosmological constant that has its origin in the quantum fluctuations of spacetime. This dynamical cosmological constant will be shown [9, 10] (in section 3) to have the same magnitude as the one deduced by using unimodular gravity [11, 12] in combination with causal-set theory [13].

Next I switch gear to discuss (in section 4) gravitational thermodynamics /entropic gravity, inspired by the work of Ted Jacobson [14] and Eric Verlinde [15]. By generalizing their work to our spacetime with positive cosmological constant Λ, we will be led to a critical acceleration parameter a_c of the same magnitude as the one introduced by Milgrom by hand in his formulation of MOND (modified Newtonian dynamics) to explain flat galactic rotation curves. But I will argue that a_c actually is a manifestation of the existence of dark matter of a specific mass profile. My collaborators and I call that model of dark matter "modified dark matter" (MDM) to distinguish it from cold dark matter (CDM). [16, 17] Recently my collaborators and I have sucessfuly tested MDM (see Section 5) with galactic rotation curves and galactic clusters. [18]

The take-home message from this talk is this: It is possible that the dark sector (viz., dark energy and dark matter) has its origin in quantum gravity. If so, then we can perhaps understand why the dark sector is really so different from ordinary matter. And if the scenario to be sketched in Section 6 is correct, then we can expect some rather novel particle phenomenology, for the quanta of the dark sector obey not the familiar Bose-Einstein or Fermi-Dirac statistics, but an exotic statistics that goes by the name infinite statistics [19, 20, 21, 22] or quantum Boltzmann statistics. [23, 17] However, it is known that theories of particles obeying this exotic statistics are non-local — meaning that we cannot use conventional quantum field theories to describe these particles' interactions. On the positive side, this non-locality may explain why so far dark matter detection experiments have failed to definitively detect dark matter. Furthermore we expect that the extended nature of the quanta of the dark sector may connect them to certain global aspects of spacetime such as the cosmological constant and the Hubble parameter (as will be shown in Section 4).

I would like to take this opportunity to make a disclaimer on my own behalf: In a recent paper "New Constraints on Quantum Gravity from X-ray and Gamma-Ray Observations" by Perlman, Rappaport, Christiansen, Ng, DeVore, and D. Pooley [24], it was claimed that detections of quasars at TeV energies with ground-based Cherenkov telescopes seem to have ruled out the holographic spacetime foam model (with δl scaling as $l^{1/3} l_P^{2/3}$). But now I believe this conclusion is conceivably premature when correct averaging is carried out. The point is that these authors (including myself!) have considered the instantaneous fluctuations in the distance between the location of the emission and a given point on the telescope aperture. Perhaps one should average over both the huge number of Planck timescales during the time it takes light to propagate through the telescope system, and over the equally large number of Planck squares across the detector aperture. It is then possible that the net fluctuations are exceedingly small, but at the moment there is no formalism for carrying out such averages. [25]

2. Spacetime (Quantum) Foam and Effective Cosmological Constant Λ

Spacetime is foamy due to quantum fluctuations. To examine how large the fluctuations are, let us consider a gedankan experiment in which a light signal is sent from a clock to a mirror (at a

region of space scales as the area of its two-dimensional surface, like a hologram.[6]

distance l away) and back to the clock in a timing experiment to measure l. From the jiggling of the clock's position alone, the Heisenberg uncertainty principle yields $\delta l^2 \gtrsim \frac{\hbar l}{mc}$, where m is the mass of the clock. On the other hand, the clock must be large enough not to collapse into a black hole; this requires $\delta l \gtrsim \frac{Gm}{c^2}$. We conclude that the fluctuations of a distance l scales as

$$\delta l \gtrsim l^{1/3} l_P^{2/3}, \tag{1}$$

where $l_P = \sqrt{\hbar G/c^3} \sim 10^{-33}$cm is the Planck length. [2] [1, 2, 3]

One can further show that the scaling of δl given above is exactly what the holographic principle [4, 5] demands. Heuristically, this comes about because a cube with side l contains $\sim l^2/l_P^2$ number of small cubes with side δl. [27] Imagine partitioning a cubic region with side l into small cubes. The small cubes so constructed should be as small as physical laws allow so that intuitively we can associate one degree of freedom with each small cube. In other words, the number of degrees of freedom that the region can hold is given by the number of small cubes that can be put inside that region. A moment's thought tells us that each side of a small cube cannot be smaller than the accuracy δl with which we can measure each side l of the big cube. This can be easily shown by applying the method of contradiction: assume that we can construct small cubes each of which has sides less than δl. Then by lining up a row of such small cubes along a side of the big cube from end to end, and by counting the number of such small cubes, we would be able to measure that side (of length l) of the big cube to a better accuracy than δl. But, by definition, δl is the best accuracy with which we can measure l. The ensuing contradiction is evaded by the realization that each of the smallest cubes (that can be put inside the big cube) indeed measures δl by δl by δl. Thus, the number of degrees of freedom I in the region (measuring l by l by l) is given by $l^3/\delta l^3$, which, according to the holographic principle, is

$$I \lesssim l^2/l_p^2. \tag{2}$$

It follows that δl is bounded (from below) by the cube root of $l l_P^2$, the same result as found above in the gedanken experiment argument.

We can rederive the scaling of δl by another argument. Let us consider mapping out the geometry of spacetime for a spherical volume of radius l over the amount of time $2l/c$ it takes light to cross the volume.[7] One way to do this is to fill the space with clocks, exchanging signals with the other clocks and measuring the signals' times of arrival. The total number of operations, including the ticks of the clocks and the measurements of signals, is bounded by the Margolus-Levitin theorem [28] which stipulates that the rate of operations cannot exceed the amount of energy E that is available for the operation divided by $\pi\hbar/2$. This theorem, combined with the bound on the total mass of the clocks to prevent black hole formation, implies that the total number of operations that can occur in this spacetime volume is no bigger than $2(l/l_P)^2/\pi$. To maximize spatial resolution, each clock must tick only once during the entire time period. If we regard the operations as partitioning the spacetime volume into "cells", then on the average each cell occupies a spatial volume no less than $\sim l^3/(l^2/l_P^2) = l l_P^2$, yielding an

[2] Now the amount of fluctuations in the distance l can be thought of as an accumulation of the l/l_P individual fluctuations each by an amount plus or minus l_P. But note that the individual fluctuations cannot be completely random (as opposed to random-walk); actually successive fluctuations must be *entangled* and somewhat *anti-correlated* (i.e., a plus fluctuation is slightly more likely followed by a minus fluctuation and vice versa), in order that together they produce a total fluctuation less than that in a random-walk model (for which $\delta l \gtrsim l^{1/2} l_P^{1/2}$.) [26] This small amount of anti-correlation between successive fluctuations (corresponding to what statisticians call fractional Brownian motion with self-similarity parameter $\frac{1}{3}$) must be due to quantum gravity effects.

average separation between neighboring cells no less than $\sim l^{1/3}l_P^{2/3}$. [8] This spatial separation can be interpreted as the average minimum uncertainty in the measurement of a distance l, that is, $\delta l \gtrsim l^{1/3}l_P^{2/3}$, in agreement with the result found in the gedanken experiment to measure the fluctuation of a distance l.

We make two observations: [29, 23] First, maximal spatial resolution (corresponding to $\delta l \sim l^{1/3}l_P^{2/3}$) is possible only if the maximum energy density $\rho \sim (ll_P)^{-2}$ is available to map the geometry of the spacetime region, without causing a gravitational collapse. Secondly, since, on the average, each cell occupies a spatial volume of ll_P^2, a spatial region of size l can contain no more than $\sim l^3/(ll_P^2) = (l/l_P)^2$ cells. Hence, this result for spacetime fluctuations corresponds to the case of maximum number of bits of information l^2/l_P^2 in a spatial region of size l, that is allowed by the holographic principle[4, 5].

It is straightforward to generalize [29] the above discussion for a static spacetime region with low spatial curvature to the case of an expanding universe by the substitution of l by H^{-1} in the expressions for energy and entropy densities, where H is the Hubble parameter. (Henceforth we adopt $c = 1 = \hbar$ for convenience unless stated otherwise.) Thus, applied to cosmology, the above argument leads to the prediction that (1) the cosmic energy density has the critical value

$$\rho \sim (H/l_P)^2, \tag{3}$$

and (2) the universe of Hubble size R_H contains $I \sim (R_H/l_p)^2$ bits of information. (For the present cosmic epoch we have $I \sim 10^{122}$.) It follows that the average energy carried by each particle/bit is $\rho R_H^3/I \sim R_H^{-1}$. Such long-wavelength constituents of dark energy give rise to a more or less uniformly distributed cosmic energy density and act as a dynamical cosmological constant with the observed small but nonzero value

$$\Lambda \sim 3H^2. \tag{4}$$

3. Cosmological Constant Λ via Unimodular Gravity and Causal-set Theory

The dynamical cosmological constant we have just obtained will be seen to play an important role in our subsequent discussions. So let us "rederive" it by using another method based on quantum gravity. The idea makes use of the theory of unimodular gravity[11, 9] (which can be regarded as the ordinary theory of gravity except for the way the cosmological constant Λ arises in the theory). But here we will use the (generalized) version of unimodular gravity given by the Henneaux and Teitelboim action[12]

$$S_{unimod} = -\frac{1}{16\pi G}\int [\sqrt{g}(R + 2\Lambda) - 2\Lambda \partial_\mu \mathcal{T}^\mu](d^3x)dt. \tag{5}$$

One of its equations of motion is $\sqrt{g} = \partial_\mu \mathcal{T}^\mu$, the generalized unimodular condition, with g given in terms of the auxiliary field $\mathcal{T}^\mu$. Note that, in this theory, Λ/G plays the role of "momentum" conjugate to the "coordinate" $\int d^3x \mathcal{T}_0$ which can be identified, with the aid of the generalized unimodular condition, as the spacetime volume V. Hence Λ/G and V are conjugate to each other. It follows that their fluctuations obey a Heisenberg-type quantum uncertainty principle,

$$\delta V \delta\Lambda/G \sim 1. \tag{6}$$

Next we borrow an argument due to Sorkin[13], drawn from the causal-set theory, which stipulates that continous geometries in classical gravity should be replaced by "causal-sets", the discrete substratum of spacetime. In the framework of the causal-set theory, the fluctuation in the number of elements N making up the set is of the Poisson type, i.e., $\delta N \sim \sqrt{N}$. For a causal set, the spacetime volume V becomes $l_P^4 N$. It follows that

$$\delta V \sim l_P^4 \delta N \sim l_P^4 \sqrt{N} \sim l_P^2 \sqrt{V} = G\sqrt{V}. \tag{7}$$

Putting Eqs. (6) and (7) together yields a minimum uncertainty in Λ of $\delta\Lambda \sim V^{-1/2}$. This cosmological constant, like the one given by Eq. (4) from a heuristic quantum mechanical consideration, is finite and is to be identified with the fully renormalized cosmological constant from a quantum field-theoretic argument given by the path integration method, to which we turn next.

Following an argument due to Baum[30], Hawking[31], and Adler[32], one can now plausibly argue [9] that, in the framework of unimodular gravity, Λ vanishes to the lowest order of approximation and that it is positive if it is not zero. The argument goes as follows: Consider the vacuum functional for unimodular gravity given by path integrations over $\mathcal{T}^\mu$, $g_{\mu\nu}$, the matter fields (represented by ϕ), and Λ:

$$Z_{Minkowski} = \int d\mu(\Lambda) \int d[\phi]d[g_{\mu\nu}] \int d[\mathcal{T}^\mu] exp\left\{-i[S_{unimod} + S_M(\phi, g_{\mu\nu})]\right\}, \tag{8}$$

where S_M stands for the contribution from matter (including radiation) fields (and $d\mu(\Lambda)$ denotes the measure of the Λ integration). [32] The integration over $\mathcal{T}^\mu$ yields $\delta(\partial_\mu \Lambda)$, which implies that Λ is spacetime-independent (befiting its role as the cosmological constant). A Wick rotation now allows us to study the Euclidean vacuum functional Z. The integrations over $g_{\mu\nu}$ and ϕ give $exp[-S_\Lambda(\overline{g}_{\mu\nu}, \overline{\phi})]$ where $\overline{g}_{\mu\nu}$ and $\overline{\phi}$ are the background fields which minimize the effective action S_Λ. A curvature expansion for S_Λ yields a Lagrangian whose first two terms are the Einstein-Hilbert terms $\sqrt{g}(R + 2\Lambda)$. Note that (1) Λ now denotes the fully renormalized cosmological constant after integrations over all other fields have been carried out; (2) the Einstein-Hilbert terms are exactly the first two terms in Eq. (5), hence the fluctuation $\delta\Lambda \sim V^{-1/2}$ we found above now applies to the renormalized Λ. Next we can make a change of variable from the original (bare) Λ to the renormalized Λ for the integration in Eq. (8). Let us assume that for the present and recent cosmic eras, ϕ is essentially in the ground state, then it is reasonable to neglect the effects of $\overline{\phi}$. [33]. To continue, we follow Baum[30] and Hawking[31] to evaluate $S_\Lambda(\overline{g}_{\mu\nu}, 0)$. For negative Λ, S_Λ is positive; for positive Λ, one finds $S_\Lambda(\overline{g}_{\mu\nu}, 0) = -3\pi/G\Lambda$, so that

$$Z \approx \int d\mu(\Lambda) exp(3\pi/G\Lambda). \tag{9}$$

This implies that the observed cosmological constant in the present and recent eras is essentially zero (or more accurately, very small but positive). So we [9] conclude that Λ is positive and it fluctuates about zero with a magnitude of

$$\Lambda \sim V^{-1/2} \sim R_H^{-2}, \tag{10}$$

where, we recall, R_H is the Hubble radius of the Universe, contributing an energy density ρ given by: $\rho \sim + \frac{1}{l_P^2 R_H^2}$, which is of the order of the critical density as observed!

4. From Cosmological Constant Λ to Modified Dark Matter (MDM)

The dynamical cosmological constant (originated from quantum fluctuations of spacetime) can now be shown to give rise to a critical acceleration parameter in galactic dynamics. The argument [16] is based on a simple generalization of E. Verlinde's recent proposal of entropic gravity [15, 14] for $\Lambda = 0$ to the case of de-Sitter space with positive Λ. Let us first review Verlinde's derivation (or prescription, if you like) of Newton's second law $\vec{F} = m\vec{a}$. Consider a particle with mass m approaching a holographic screen at temperature T. Using the first law of thermodynamics to introduce the concept of entropic force $F = T\frac{\Delta S}{\Delta x}$, and invoking Bekenstein's original arguments [34] concerning the entropy S of black holes, $\Delta S = 2\pi k_B \frac{mc}{\hbar}\Delta x$, Verlinde gets $F = 2\pi k_B \frac{mc}{\hbar}T$. With the aid of the formula for the Unruh temperature, $k_B T = \frac{\hbar a}{2\pi c}$, associated with a uniformly accelerating (Rindler) observer, Verlinde then obtains $\vec{F} = m\vec{a}$. Now in a de-Sitter space with cosmological constant Λ, the net Unruh-Hawking temperature, [35, 36, 37] as measured by a non-inertial observer with acceleration a relative to an inertial observer, is

$$\tilde{T} = \frac{\hbar \tilde{a}}{2\pi k_B c}, \tag{11}$$

with [38]

$$\tilde{a} = \sqrt{a^2 + a_0^2} - a_0, \tag{12}$$

where $a_0 \equiv \sqrt{\Lambda/3}$. Hence the entropic force (in de-Sitter space) is given by the replacement of T and a by $\tilde{T}$ and $\tilde{a}$ respectively, leading to

$$F = m[\sqrt{a^2 + a_0^2} - a_0]. \tag{13}$$

For $a \gg a_0$, we have $F/m \approx a$ which gives $a = a_N \equiv GM/r^2$, the familiar Newtonian value for the acceleration due to the source M. But for $a \ll a_0$, $F \approx m\frac{a^2}{2a_0}$, so the terminal velocity v of the test mass m in a circular motion with radius r should be determined from $ma^2/(2a_0) = mv^2/r$. In this small acceleration regime, the observed flat galactic rotation curves (v being independent of r) now require $a \approx (2a_N a_0^3/\pi)^{\frac{1}{4}}$. But that means $F \approx m\sqrt{a_N a_c}$. This is the celebrated modified Newtonian dynamics (MoND) scaling [39, 40, 41] discovered by Milgrom who introduced the critical acceleration parameter

$$a_c = a_0/(2\pi) = cH/(2\pi) \tag{14}$$

by hand to phenomenologically explain the observed flat galactic rotation curves. Thus, we have recovered MoND with the correct magnitude for the critical galactic acceleration parameter $a_c \sim 10^{-8} cm/s^2$. From our perspective, MoND is a *classical* phenomenological consequence of *quantum* gravity (with the $\hbar$ dependence in $T \propto \hbar$ and $S \propto 1/\hbar$ cancelled out in the product TS for the entropic force). [16] As a bonus, we have also recovered the observed Tully-Fisher relation ($v^4 \propto M$).

Having generalized Newton's 2nd law, we [16] can now follow the second half of Verlinde's argument [15] to generalize Newton's law of gravity $a = GM/r^2$. Verlinde derives Newton's law of gravity by considering an imaginary quasi-local (spherical) holographic screen of area $A = 4\pi r^2$ with temperature T, and by invoking the equipartition of energy $E = \frac{1}{2}Nk_BT$ with $N = Ac^3/(G\hbar)$ being the total number of degrees of freedom (bits) on the screen, as well as the Unruh temperature formula $k_B T = \frac{\hbar a}{2\pi c}$, and the fact that $E = Mc^2$. The generalized Newton's law of gravity (for the case of de-Sitter space) is obtained by the replacement of T and M by $\tilde{T}$ and $\tilde{M}$ respectively, so that we get

$$2\pi k_B \tilde{T} = G\tilde{M}/r^2, \tag{15}$$

where

$$\tilde{M} = M + M_d \tag{16}$$

represents the *total* mass enclosed within the volume $V = 4\pi r^3/3$, with M_d being some unknown mass, i.e., dark matter. For $a \gg a_0$, consistency with the Newtonian force law $a \approx a_N$ implies $M_d \approx 0$. But for $a \ll a_0$, consistency with the condition $a \approx (2a_N\, a_0^3/\pi)^{\frac{1}{4}}$ requires [3]

$$M_d \approx \frac{1}{\pi}\left(\frac{a_0}{a}\right)^2 M \sim (\sqrt{\Lambda}/G)^{1/2} M^{1/2} r. \tag{17}$$

This yields the dark matter mass density ρ_d profile given by $\rho_d(r) \sim M^{1/2}(r_v)(\sqrt{\Lambda}/G)^{1/2}/r^2$, for an ordinary (visible) matter source of radius r_v with total mass $M(r_v)$. [4]

Thus dark matter indeed exists! And the MoNDian force law derived above, at the galactic scale, is simply a manifestation of dark matter! [16, 42] Dark matter of this kind can behave as if there is no dark matter but MoND. Therefore, we used to call it "MoNDian dark matter" which, to some people sounds like an oxymoron. Now we call it "modified dark matter". Note that the dark matter profile we have obtained relates, at the galactic scale, dark matter (M_d), dark energy (Λ) and ordinary matter (M) to one another.

5. Observational Tests of MDM

In order to test MDM with galactic rotation curves, we fit computed rotation curves to a selected sample of Ursa Major galaxies given in [43]. The sample contains both high surface brightness (HSB) and low surface brightness (LSB) galaxies. The rotation curves, predicted by MDM as given above by

$$F = m[\sqrt{a^2 + a_0^2} - a_0] = m a_N \left[1 + \frac{1}{\pi}\left(\frac{a_0}{a}\right)^2\right], \tag{18}$$

along with $F = mv^2/r$ for circular orbits, can be solved for $a(r)$ and $v(r)$. We [18] fit these to the observed rotation curves as determined in [43], using a least-squares fitting routine. As in [43], the mass-to-light ratio M/L, which is our *only* fitting parameter for MDM, is assumed constant for a given galaxy but allowed to vary between galaxies. Once we have $a(r)$, we can find the MDM density profile by using $M_d \approx \frac{1}{\pi}\left(\frac{a_0}{a}\right)^2 M$ to give $\rho_d(r) = \left(\frac{a_c}{r}\right)^2 \frac{d}{dr}\left(\frac{M}{a^2}\right)$.

Rotation curves predicted by MDM for NGC 4217, a typical HSB galaxy, and NGC 3917, a typical LSB galaxy in the sample are shown in Fig. 1 and Fig. 2 respectively. (See Ref. [18] for the rotation curves for the other 28 galaxies.)

In these figures, observed rotation curves are depicted as filled circles with error bars, and for the two curves at the bottom, the dotted and dash-dotted lines show the stellar and interstellar gas rotation curves, respectively. The solid lines and dashed lines are rotation curves predicted by MDM and the standard cold dark matter (CDM) paradigm respectively. For the CDM fits, we use the Navarro, Frenk & White (NFW) [44] density profile, employing *three* free parameters (one of which is the mass-to-light ratio.) It is fair to say that both models fit the data well;

[3] Actually the two acceleration limits have little to say about the intermediate regime; thus we expect that a more generic dark mass profile is of the form $M_d = \left[\xi\left(\frac{a_0}{a}\right) + \frac{1}{\pi}\left(\frac{a_0}{a}\right)^2\right] M$ with positive parameter $\xi \sim 1$. See discussions in the next section about the dark matter mass profile.

[4] This result can be compared with the distribution associated with an isothermal Newtonian sphere in hydrostatic equilibrium (used by some dark matter proponents): $\rho(r) = \sigma(r^2 + r_0^2)^{-1}$. Asymptotically the two expressions agree with σ identified as $\sim M^{1/2}(r_v)(\sqrt{\Lambda}/G)^{1/2}$.

IOP Conf. Series: Journal of Physics: Conf. Series **845** (2017) 012001 doi:10.1088/1742-6596/845/1/012001

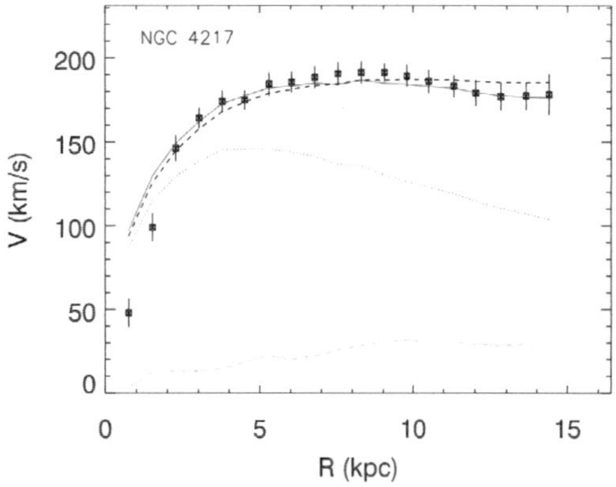

Figure 1. Galactic rotation curves for NGC 4217 (HSB).

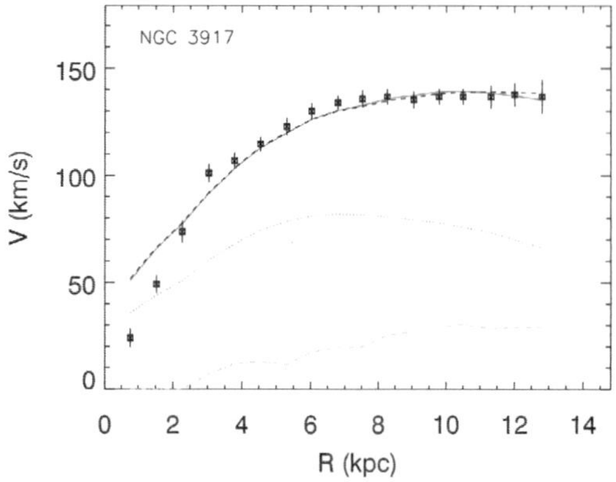

Figure 2. Galactic rotation curves for NGC 3917 (LSB).

[5] but while the MDM fits use only 1 free parameter, for the CDM fits one needs to use 3 free parameters. Thus the MDM model is a more economical model than CDM in fitting data at the galactic scale.

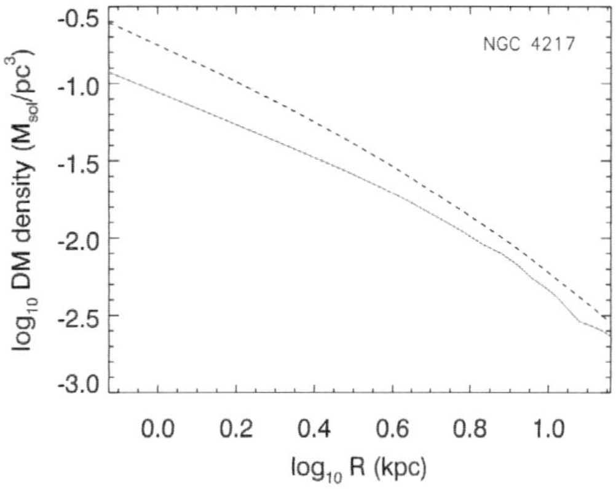

Figure 3. Dark matter density profile for NGC 4217 (HSB).

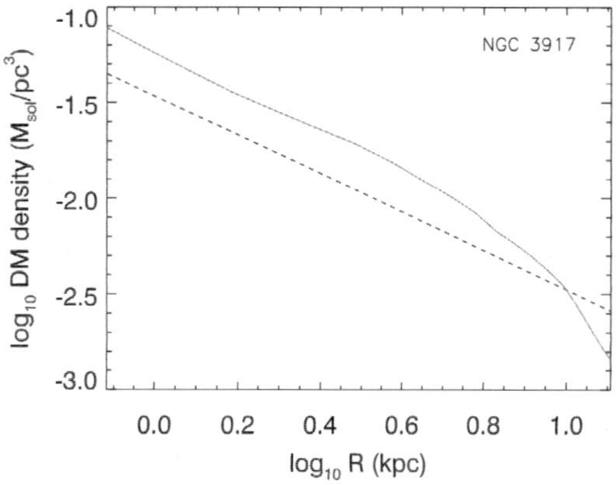

Figure 4. Dark matter density profile for NGC 3917 (LSB).

Shown in Fig. 3 and Fig. 4 are the dark matter density profiles predicted by MDM (solid lines) and CDM (dashed lines) for the HSB galaxy NGC 4217 and the LSB galaxy NGC 3917 in the sample respectively. (See Ref. [18] for details.)

To test MDM with astronomical observations at a larger scale, we compare dynamical and observed masses in a large sample of galactic clusters. [6] First, let us recall that the MDM profile $M_d = \frac{1}{\pi}\left(\frac{a_0}{a}\right)^2 M$ reproduces the flat rotation curves. But we expect that a more general profile should be of the form $M_d = \left[\xi\left(\frac{a_0}{a}\right) + \frac{1}{\pi}\left(\frac{a_0}{a}\right)^2\right] M$, with $\xi > 0$ which ensures that $M_d > 0$

[5] We should point out that the rotation curves predicted by MDM and MOND have been found [18] to be virtually indistinguishable over the range of observed radii and both employ only 1 free parameter.

[6] The comparison is made in some unpublished work by D. Edmonds et al. [18]

when $a \gg a_0$. For the more general profile, the entropic force expression is replaced by

$$F = ma_N \left[1 + \xi \left(\frac{a_0}{a} \right) + \frac{1}{\pi} \left(\frac{a_0}{a} \right)^2 \right]. \tag{19}$$

Sanders [45] studied the virial discrepancy (i.e., the discrepancy between the observed mass and the dynamical mass) in the contexts of Newtonian dynamics and MOND. We [18] have adapted his approach to the case of MDM. For his work, Sanders considered 93 X-ray-emitting clusters from the compilation by White, Jones, and Forman (WJF) [46]. He found the well-known discrepancy between the Newtonian dynamical mass ($M_{\rm N}$) and the observed mass ($M_{\rm obs}$): $\left\langle \frac{M_{\rm N}}{M_{\rm obs}} \right\rangle \approx 4.4$. And for the sample clusters, he found $\langle M_{\rm MOND}/M_{\rm obs} \rangle \approx 2.1$.

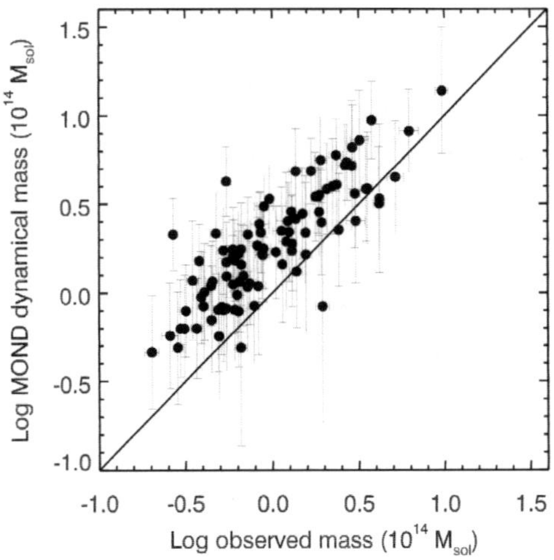

Figure 5. Fit to galactic cluster data using MONDian dynamics.

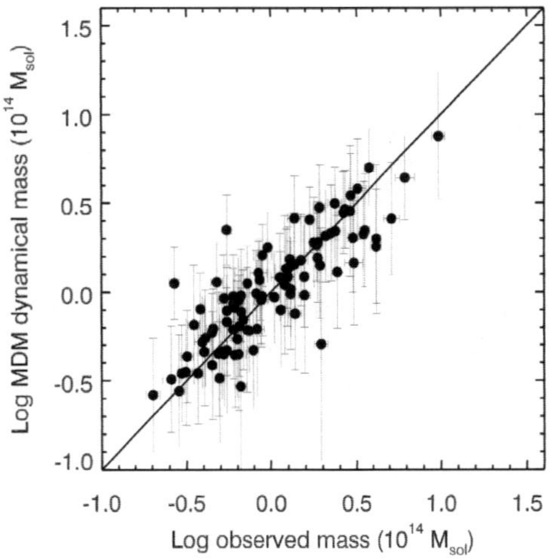

Figure 6. Fit to galactic cluster data using MDM dynamics.

For MDM, the observed (effective) acceleration is given by $a_{obs} = \sqrt{a^2 + a_0^2} - a_0$. Using the more general expression for the MDM profile, we have $a_{obs} = \frac{GM_{MDM}}{r^2}\{1 + \xi\left(\frac{a_0}{a}\right) + \frac{1}{\pi}\left(\frac{a_0}{a}\right)^2\}$. Recalling that $a_{obs} = GM_N/r^2$ for Newtonian dyanmics, we get

$$M_{\rm MDM} = \frac{M_{\rm N}}{1 + \xi\left(\frac{a_0}{a}\right) + \frac{1}{\pi}\left(\frac{a_0}{a}\right)^2}, \tag{20}$$

for the dynamical mass for MDM, using ξ as a universal fitting parameter. With $\xi \approx 0.5$, we get $\left\langle \frac{M_{\rm MDM}}{M_{\rm obs}} \right\rangle \approx 1.0$. [7] In Fig. 5 and Fig. 6, we show the MOND and MDM dynamical masses respectively against the total observed mass for the 93 sample clusters compiled by WJF. The virial discrepancy is eliminated in the context of MDM! Recalling that Sanders found $\langle M_{\rm MOND}/M_{\rm obs} \rangle \approx 2.1$, we conclude that, at the cluster scale, MDM is superior to MOND.

[7] For completeness we mention that previously we have used $\xi = 0$ when fitting galactic rotation curves. But since now the galaxy cluster sample in our current study implies $\xi \approx 0.5$, we (in unpublished work [18]) refit the galaxy rotation curves using $\xi = 0.5$ and find the fits are nearly identical with a reduction in mass-to-light ratios of about 35%.

IOP Conf. Series: Journal of Physics: Conf. Series **845** (2017) 012001 doi:10.1088/1742-6596/845/1/012001

6. The Dark Sector and Infinite Statistics

What is the essential difference between ordinary matter and dark energy from our perspective? To find that out, let us recall our discussions in Section 2, and liken the quanta of dark energy to a perfect gas of N particles obeying Boltzmann statistics at temperature T in a volume V. For the problem at hand, as the lowest-order approximation, we can neglect the contributions from matter and radiation to the cosmic energy density for the recent and present eras. Thus let us take $V \sim R_H^3$, $T \sim R_H^{-1}$, and $N \sim (R_H/l_P)^2$. A standard calculation (for the relativistic case) yields the partition function $Z_N = (N!)^{-1}(V/\lambda^3)^N$, where $\lambda = (\pi)^{2/3}/T$. With the free energy given by $F = -TlnZ_N = -NT[ln(V/N\lambda^3) + 1]$, we get, for the entropy of the system,

$$S = -(\partial F/\partial T)_{V,N} = N[ln(V/N\lambda^3) + 5/2]. \tag{21}$$

The important point to note is that, since $V \sim \lambda^3$, the entropy S in Eq. (21) becomes nonsensically negative unless $N \sim 1$ which is equally nonsensical because N should not be too different from $(R_H/l_P)^2 \gg 1$. But the solution [23] is obvious: the N inside the log in Eq. (21) somehow must be absent. Then $S \sim N \sim (R_H/l_P)^2$ without N being small (of order 1) and S is non-negative as physically required. That is the case if the "particles" are distinguishable and nonidentical! For in that case, the Gibbs $1/N!$ factor is absent from the partition function Z_N, and the entropy becomes $S = N[ln(V/\lambda^3) + 3/2]$.

Now the only known consistent statistics in greater than two space dimensions without the Gibbs factor (recall that the Fermi statistics and Bose statistics give similar results as the conventional Boltzmann statistics at high temperature) is infinite statistics (sometimes called "quantum Boltzmann statistics") [19, 20, 21]. Thus we have shown that the "particles" constituting dark energy obey infinite statistics, instead of the familiar Fermi or Bose statistics [23]. [8]

To show that the quanta of modified dark matter also obey this exotic statistics, we [17] first reformulate MoND via an effective gravitational dielectric medium, motivated by the analogy [48] between Coulomb's law in a dielectric medium and Milgrom's law for MoND. Ho, Minic and I then find that MONDian force law is recovered if the quanta of MDM obey infinite statistics.

What is infinite statistics? Succinctly, a Fock realization of infinite statistics is provided by a $q = 0$ deformation of the commutation relations of the oscillators: $a_k a_l^\dagger - q a_l^\dagger a_k = \delta_{kl}$ with q between -1 and 1 (the case $q = \pm 1$ corresponds to bosons or fermions). States are built by acting on a vacuum which is annihilated by a_k. Two states obtained by acting with the N oscillators in different orders are orthogonal. It follows that the states may be in any representation of the permutation group. The statistical mechanics of particles obeying infinite statistics can be obtained in a way similar to Boltzmann statistics, with the crucial difference that the Gibbs $1/N!$ factor is absent for the former. Infinite statistics can be thought of as corresponding to the statistics of identical particles with an infinite number of internal degrees of freedom, which is equivalent to the statistics of nonidentical particles since they are distinguishable by their internal states.

It has been shown that a theory of particles obeying infinite statistics cannot be local [22, 21]. For example, the expression for the number operator,

$$n_i = a_i^\dagger a_i + \sum_k a_k^\dagger a_i^\dagger a_i a_k + \sum_l \sum_k a_l^\dagger a_k^\dagger a_i^\dagger a_i a_k a_l + ..., \tag{22}$$

[8] Using the Matrix theory approach, Jejjala, Kavic and Minic [47] have also argued that dark energy quanta obey infinite statistics.

is both nonlocal and nonpolynomial in the field operators, and so is the Hamiltonian. The lack of locality may make it difficult to formulate a relativistic verion of the theory; but it appears that a non-relativistic theory can be developed. Lacking locality also means that the familiar spin-statistics relation is no longer valid for particles obeying infinite statistics; hence they can have any spin. Remarkably, the TCP theorem and cluster decomposition have been shown to hold despite the lack of locality [21].

According to the holographic principle, the number of degrees of freedom in a region of space is bounded not by the volume but by the surrounding surface. This suggests that the physical degrees of freedom are not independent but, considered at the Planck scale, they must be infinitely correlated, with the result that the spacetime location of an event may lose its invariant significance. Since the holographic principle is believed to be an important ingredient in the formulation of quantum gravity, the lack of locality for theories of infinite statistics may not be a defect; it can actually be a virtue. Perhaps it is this lack of locality that makes it possible to incorporate gravitational interactions in the theory. Quantum gravity and infinite statistics appear to fit together nicely, and nonlocality seems to be a common feature of both of them [23].

7. Summary and Conclusion

The dark sector in the concordant model of cosmology ΛCDM has two components: dark energy and dark matter. We have argued that quantum fluctuations of spacetime give rise to dark energy in the form of an effective cosmological constant Λ of the correct magnitude as observed – a result also expected for the present and recent cosmic eras in (generalized) unimodular gravity and causal-set theory. In a spacetime with positive Λ, gravitational thermodynamics arguments then show that dark matter (i.e., modified dark matter) necessarily exists whose mass profile is intimately related to Λ and ordinary matter, with an emergent acceleraton parameter related to Λ and the Hubble parameter H, of the magnitude required to explain flat galactic rotation curves. Thus the dark sector in our Universe may indeed have its origin in quantum gravity.

Pursuing this line of argument further, we find that the quanta of the dark sector appear to obey an unfamiliar statistics, viz, infinite statistics (or quantum Boltzmann statistics). This indicates that the dark sector is made up of extended quanta. As a result, we expect novel particle phenomenology for interactions involving dark matter, thereby "explaining" why so far dark matter detection experiments have not yet convincingly detected dark matter. The extended nature of the MDM quanta may also explain why the mass profile of MDM depends on such global aspects of spacetime as Λ and H.

MDM has passed observational tests at both the galactic and cluster scales. We can also mention that preliminary examinations have demonstrated (see Ref. [16]) that the cosmology with MDM is well described by the usual Friedmann's equations. We anticipate that this fact will allow MDM to predict the correct cosmic microwave background (CMB) spectrum shapes as well as the alternating peaks. And as briefly explained in Ref. [49], the MDM mass distribution as found appears to be consistent with the observed strong gravitational lensing.

We conclude by listing a few items on our lengthy to-do list. We plan to study concrete constraints from gravitational lensing and the bullet cluster on MDM. And we would like to answer these questions: Can we distinguish MDM from CDM? How strongly coupled is MDM to baryonic matter? How does MDM self-interact? We will also test MDM at cosmic scales by studying the acoustic peaks in the CMB and by doing simulations of structure formation. Last but not least, if the quanta of MDM indeed obey infinite statistics as we found, can quantum gravity be the origin of particle statistics and can the underlying statistics be infinite statistics such that ordinary particles obeying Bose or Fermi statistics are actually some sort of collective degrees of freedom of more fundamental entities obeying infinite statistics? And if so, what are the implications for grand unification?

Acknowledgments

This talk is partly based on work done in collaboration with (the late) H. van Dam, S. Lloyd, M. Arzano, T. Kephart, C. M. Ho, D. Minic, D. Edmonds, D. Farrah, and T. Takeuchi. I thank them all. The work reported here was supported in part by the US Department of Energy, the Bahnson Fund, and the Kenan Professorship Research Fund of UNC-CH.

References

[1] Ng Y J and van Dam H 1994 *Mod. Phys. Lett.* A **9** 335
[2] Ng Y J and van Dam H 1995 *Mod. Phys. Lett.* A **10** 2801
[3] Karolyhazy F 1966 *Il Nuovo Cimento* A **42** 390
[4] 't Hooft G 1993 Dimensional Reduction in Quantum Gravity (*Preprint* gr-qc/9310026)
[5] Susskind L 1995 *J. Math. Phys.* **36** 6377
[6] Ng Y J 2002 *Int. J. Mod. Phys.* D **11** 1585
[7] Lloyd S and Ng Y J 2004 *Scientific American* **291** #5, 52
[8] Ng Y J 2008 *Entropy* **10** 441
[9] Ng Y J and van Dam H 1990 *Phys. Rev. Lett.* **65** 1972; 2001 *Int. J. Mod. Phys.* D **10** 49
[10] Ng Y J 2003 *Mod. Phys. Lett.* A **18** 1073
[11] van der Bij J J, van Dam H and Ng Y J 1982 *Physica* A **116** 307
[12] Henneaux M and Teitelboim C 1989 *Phys. Lett.* B **222** 195
[13] Sorkin R D 1991 *Relativity and Gravitation: Classical and Quantum* ed J. C. D'Olivo *et al* (Singapore: World Scientific); 1997 *Int. J. Th. Phys.* **36** 2759
[14] Jacobson T 1995 *Phys. Rev. Lett.* **75** 1260
[15] Verlinde E 2011 *JHEP* **1104** 029
[16] Ho C M, Minic D and Ng Y J 2010 *Phys. Lett.* B **693** 567
[17] Ho C M, Minic D and Ng Y J 2012 *Phys. Rev.* D **85** 104033
[18] Edmonds D, Farrah D, Ho C M, Minic D, Ng Y J and Takeuchi T 2014 *ApJ* **793** 41; 2016 *Preprint* arXiv:1601.00662 [astro-ph.CO]; and 2015 unpublished work
[19] Doplicher S, Haag R and Roberts J 1971 *Commun. Math. Phys.* **23** 199; 1974 *Commun. Math. Phys.* **35** 49
[20] Govorkov A B 1983 *Theor. Math. Phys.* **54** 234
[21] Greenberg O W 1990 *Phys. Rev. Lett.* **64** 705
[22] Fredenhagen K 1981 *Commun. Math. Phys.* **79** 141
[23] Ng Y J 2007 *Phys. Lett.* B **657** 10
[24] Perlman E S *et al* 2015 *ApJ* **805** 10
[25] Perlman E S *et al* to appear in *Proc. 14th Marcel Grossmann Meeting on General Relativity* (July 2015 Rome) ed R Ruffini *et al* (Singapore: World Scientific)
[26] Ng Y J 2005 Quantum Foam and Quantum Gravity Phenomenology (*Preprint* arXiv: gr-qc/0405078) *Proc 40th Karpacz Winter School on Theoretical Physics* ("Planck Scale Effects in Astrophysics and Cosmology"), *Lect. Notes Phys.* **669** 321 ed. J. Kowalski-Glikman and G. Amelino-Camelia (Berlin Heidelberg: Springer)
[27] Ng Y J and van Dam H 2000 *Phys. Lett.* B **477** 429
[28] Margolus N and Levitin L B 1998 *Physica (Amsterdam)* D **120** 188
[29] Arzano M, Kephart T W and Ng Y J 2007 *Phys. Lett.* B **649** 243
[30] Baum E 1983 *Phys. Lett.* B **133** 185
[31] Hawking S W 1984 *Phys. Lett.* B **134** 403
[32] Adler S L 1982 *Rev. Mod. Phys.* **54** 729
[33] Ng Y J and van Dam H 2001 *Int. J. Mod. Phys.* D **10** 49
[34] Bekenstein J D 1973 *Phys. Rev.* D **7** 2333
[35] Unruh W G 1976 *Phys. Rev.* D **14** 870
[36] Davies P C W 1975 *J. Phys.* A **8** 609
[37] Hawking S W 1975 *Comm. Math. Phys.* **43** 199
[38] Deser S and Levin O 1997 *Class. Quant. Grav.* **14** L163

[39] Milgrom M 1983 *Astrophys. J.* **270** 365, 371, 384
[40] Famaey B and McGaugh S S 2011 Modified Newtonian Dynamics (MOND): Observational Phenomenology and Relativistic Extensions (*Preprint* arXiv:1112.3960)
[41] Milgrom M 1999 *Phys. Lett.* A **253** 273
[42] For an earlier attempt to relate MoND scaling to dark matter, see Kiplinghat M and Turner M S 2002 *Astrophys. J.* **569** L19
[43] Sanders R H and Verheijen M A W 1998 *ApJ* **503** 97
[44] Navarro J F, Frenk C S and White S D M 1996 *ApJ* **462** 563
[45] Sanders R H 1999 *ApJ Lett.* **512** L23
[46] White D A, Jones C and Forman W 1997 *MNRAS* **292** 419
[47] Jejjala V, Kavic M and Minic D 2007 *Adv. High Energy Phys.* **2007** 21586
[48] Blanchet L *Preprint* arXiv:astro-ph/0605637
[49] Ng Y J, Edmonds D, Farrah D, Minic D, Takeuchi T and Ho C M 2016 Modified Dark Matter (*Preprint* arXiv:1602.00055) to appear in *Proc. 14th Marcel Grossmann Meeting on General Relativity* (July 2015 Rome) ed R. Ruffini *et al* (Singapore: World Scientific)

A further analysis of the energy-momentum for a system of point particles and for a perfect fluid in finite volume

G. Ares de Parga[1] **and R. E. Gonzalez-Narvaez**[1]

[1]Dpto. de Física, Escuela Superior de Física y Matemáticas, Instituto Politécnico Nacional, U. P. Adolfo López Mateos, Zacatenco, C.P. 07738, México City, México.

E-mail: gadpau@hotmail.com

Abstract. A simple analysis of the total energy-momentum of a system of non-interacting point particles shows that it represents a well defined 4-vector. A generalization for interacting charged particles including the self-forces is developed. A comparison between a system of point particles in an infinite volume and the perfect fluid in a finite volume is done. The finite volume case is analyzed in order to define an energy-momentum depending on the notion of instantaneity for a perfect fluid. The relativistic transformations of thermodynamic quantities are defined in a covariant form leading to a redefined relativistic Thermodynamics.

1. Introduction

In the early times of Special Relativity, Einstein and Planck developed the theory of relativistic transformations of the thermodynamic quantities [1], [2]. However, due to the doubts that had Einstein of his own theory [3], different theories appeared in the beginning of the middle of the last century [4], [5], [6] , [7], [8], [9], [10], [11], [12], [13], [14], [15], [16]. Foregrounding the covariance of the theory, Gamba [17] improved the discussions about the correctness of the transformation laws. At the end of the last century, Sieniutycz [18] said that the controversy was still alive and the three following different proposals remained the most acceptable: the Ott [4], the Einstein-Planck [1], [2] and the Rohrlich light proposals [8]. Nevertheless, in the first years of this century, Landsberg and Matsas proposed that it is not possible to define a universal definition of the temperature [6], [19], [20] [21], [22]. Based on Fermi ideas of instantaneity and a formal definition of a 4−vector [23], [24], [25], [26], Ares de Parga et al developed a new theory which is known as the Redefined Relativistic Thermodynamics [27], [28], [29], [30], [31], [32], [33], [34].

In this paper, the concept of the 4−vector total energy-momentum is developed for a system of point particles in an infinite volume and for the perfect fluid in a finite volume [35] which permits to understand that the Einstein-Planck theory, with some corrections, represents a covariant formal theory which is called the Redefined Relativistic Thermodynamics [30]. In the course of the article, the no interaction theorem [36] is proven for classical particles and the case of a system of interacting point charged particles is analyzed including the reaction force.

IOP Conf. Series: Journal of Physics: Conf. Series **845** (2017) 012002 doi:10.1088/1742-6596/845/1/012002

2. The Energy-Momentum Vector and Tensor for a System of Point Particles

The definition of a 4−vector total energy-momentum presents many difficulties. Before going into the definition by means of an energy-momentum tensor, it is convenient to give the simplest definition (we take $c = 1$).

2.1. The total energy-momentum vector for a system of point particles

In an inertial reference frame K_1, regardless whether or not there is interaction, one would be tempted to define the total 4−vector total energy-momentum of a system of particles as

$$\underset{K_1}{P^\alpha} = \sum_{i=1}^{n} \underset{K_1}{p_i^\alpha}(x_i(t)), \tag{1}$$

where $\underset{K_1}{p_i^\alpha}$ represents the 4−vector energy-momentum of the i−particle measured in reference frame K_1. As will be shown, $\underset{K_1}{P^\alpha}$ does not represent a 4−vector in general because it is the sum of many 4−vectors in different events ($x_i(t) \neq x_j(t)$ for $i \neq j$). However, when there is no interaction, free particles, $\underset{K_1}{p_i^\alpha}$ are constant and the sum of them represents a 4−vector. Let us prove this assertion. Consider a reference frame K_2 which velocity with respect to K_1 is $\overrightarrow{u}$. Infinite volumes V_1 and V_2 are defined as 3−hypersurface in the 4−space being instantaneous in each reference frame K_1 and K_2. The times in K_1 and K_2 are denoted as t and t', respectively. If the i−particle trajectory is analyzed in reference frame K_1, it is clear that it crosses the 3−hypersurface V_2 in a time t_i (see Figure 1). Since the 4−vector energy-momentum of each particle is constant, we have

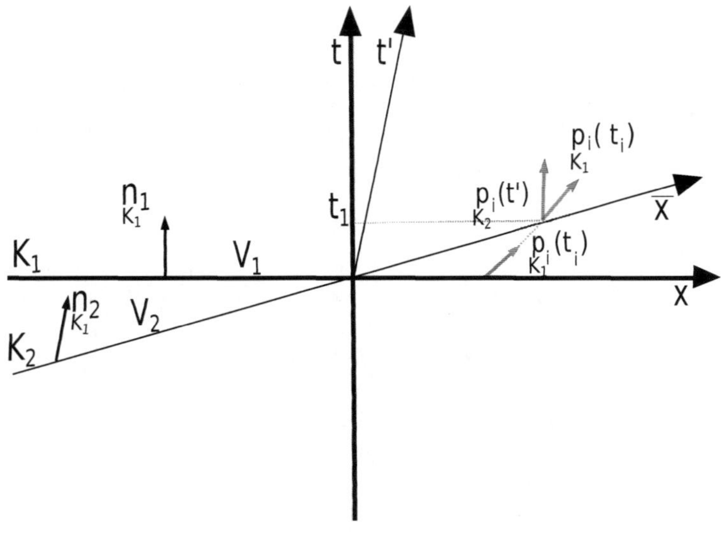

Figure 1. A system of free particles in an infinite volume. V_2 represents the infinite volume which is instantaneous at the reference frame K_2 and V_1 represents the infinite volume which is instantaneous at the reference frame K_1.

$$\underset{K_1}{p_i^\alpha}(x_i(t)) = \underset{K_1}{p_i^\alpha}(x_i(t_i)). \tag{2}$$

Therefore,

$$\begin{aligned} \underset{K_1}{P^\alpha} &= \sum_{i=1}^{n} \underset{K_1}{p_i^\alpha}(x_i(t)) \\ &= \sum_{i=1}^{n} \underset{K_1}{p_i^\alpha}(x_i(t_i)). \end{aligned} \tag{3}$$

Because each $\underset{K_1}{p_i^\alpha}(x_i(t_i))$ can be expressed as

$$\underset{K_1}{p_i^\alpha}(x_i(t_i)) = \underset{K_2 \to K_1}{\Lambda^\alpha{}_\beta}\ \underset{K_2}{p_i^\beta}(\overline{x}_i(t')), \tag{4}$$

where $\underset{K_2 \to K_1}{\Lambda^\alpha{}_\beta}$ represents the Lorentz transformation from K_2 to K_1 and $\overline{x}_i(t')$ corresponds to the position of the $i-$particle in K_2. It has to be noticed that t' is the common time for all the particles in K_2 because the time t_i corresponds to the time in K_1 where the $i-$particle crosses the 3$-$hypersurface V_2 at a common time t' in K_2. Therefore, Eq. (4) can be used in Eq. (3), obtaining

$$\begin{aligned} \underset{K_1}{P^\alpha} &= \sum_{i=1}^{n} \underset{K_2 \to K_1}{\Lambda^\alpha{}_\beta}\ \underset{K_2}{p_i^\beta}(\overline{x}_i(t')) \\ &= \underset{K_2 \to K_1}{\Lambda^\alpha{}_\beta} \sum_{i=1}^{n} \underset{K_2}{p_i^\beta}(\overline{x}_i(t')) \\ &= \underset{K_2 \to K_1}{\Lambda^\alpha{}_\beta}\ \underset{K_2}{P^\beta}, \end{aligned} \tag{5}$$

which shows that $\underset{K_1}{P^\alpha}$ is a 4$-$vector. It has to be noticed that since the 4$-$vector energy-momentum of the particles are constant, $\underset{K_1}{P^\alpha}$ and $\underset{K_2}{P^\alpha}$ are also time independent. Finally, it can be concluded that $\underset{K_1}{P^\alpha}$ is a well defined constant 4$-$vector total energy-momentum when the system of particles is composed by non interacting particles.

Let us now consider a system of interacting particles. In this case, the particles cross the 3$-$hypersurface K_2 at different times t_i^{int}. Therefore, since the momentum of each particle is not anymore constant,

$$\underset{K_1}{p_i^\alpha}(x_i(t)) \neq \underset{K_1}{p_i^\alpha}(x_i(t_i^{int})), \tag{6}$$

the above calculation cannot be repeated. Consequently, we cannot define a 4$-$vector total energy-momentum as in Eq. (1) (see Figure 2). This is equivalent to the no interaction theorem [36] for classical particles because it has been proven that a 4$-$vector total energy-momentum cannot be defined for a system of interacting particles. Indeed, this represents Pryces major idea [37] which is equivalent to the no interaction theorem. Moreover, since in the case of free particles, the 4$-$vector total energy-momentum in Eq.(1) is a 4$-$vector, it is possible to find a reference frame K_o where the 4$-$vector total energy-momentum may be written as

$$\underset{K_o}{P^\alpha} = (\underset{K_o}{P^0}, \overrightarrow{0}), \tag{7}$$

with $\underset{K_o}{P^0}$ constant. That is: a rest frame K_o exists. It has to be noticed that with this simple calculation the no interaction theorem has been proven without using Lie algebras, Lorentz groups or any sophisticated mathematical tools [36].

IOP Conf. Series: Journal of Physics: Conf. Series **845** (2017) 012002 doi:10.1088/1742-6596/845/1/012002

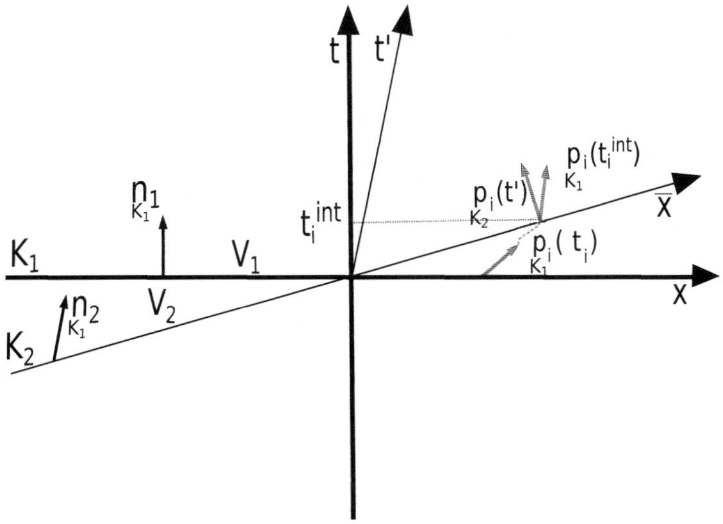

Figure 2. A system of interacting particles in an infinite volume. V_2 represents the infinite volume which is instantaneous at the reference frame K_2 and V_1 represents the infinite volume which is instantaneous at the reference frame K_1.

2.2. The energy-momentum tensor for a system of particles

The energy-momentum tensor is normally introduced for a continuous system. However, it can defined for a system of point particles in the following manner [35].

$$\underset{K}{T}^{\alpha\beta} = \sum_{i=1}^{n} \underset{K}{p_i^{\alpha}} \frac{dx_i^{\beta}}{dt} \delta^3(\overrightarrow{x} - \overrightarrow{x}_i(t)). \tag{8}$$

Apparently, these quantities do not constitute a 4−tensor of rank 2 because they are not covariantly defined and they are composed of a sum of 4−vectors that are multiplied by quantities which are not 4−vectors in different points of the space-time. However, it can be expressed in a covariant form by using the four dimension delta function;that is:

$$\underset{K}{T}^{\alpha\beta} = \sum_{i=1}^{n} \int \underset{K}{p_i^{\alpha}} \frac{dx_i^{\beta}}{d\tau_i} \delta^4(x - x_i(\tau_i)) d\tau_i. \tag{9}$$

Since $\underset{K}{T}^{\alpha\beta}$ has a covariant expression and the delta function enables it to be seen as a sum of quantities in (x, t), it can be affirmed that the energy-momentum tensor is a 4−tensor of rank 2.

Let us define a 4−quantity $\underset{K}{D}^{\alpha}$ that may coincide with the 4−vector total energy-momentum for free particles:

$$\begin{aligned}
\underset{K}{D}^{\alpha} &= \int \underset{K}{T}^{\alpha 0} dV_K \\
&= \int \left[\sum_{i=1}^{n} \underset{K}{p_i^{\alpha}} \frac{dx_i^0}{dt} \delta^3(\overrightarrow{x} - \overrightarrow{x}_i(t)) \right] dV_K \\
&= \int \left[\sum_{i=1}^{n} \underset{K}{p_i^{\alpha}} \delta^3(\overrightarrow{x} - \overrightarrow{x}_i(t)) \right] dV_K
\end{aligned}$$

$$= \sum_{i=1}^{n} \underset{K}{p_i^\alpha}(t) = \underset{K}{P^\alpha}, \tag{10}$$

where V_K represents the infinite volume in reference frame K. Therefore, $\underset{K}{D^\alpha} = \underset{K}{P^\alpha}$. However, even if we know that this 4−vector total energy-momentum just represents a 4−vector for a system of non interacting particles, it is necessary to express it in a covariant form in order to try to develop a new expression leading to a well defined 4−vector regardless of the existence or not of interaction.

2.2.1. Conservation theorem Let us consider $\Theta^{\alpha\beta}$, a 4−tensor of rank 2 with the following properties:

A- $\Theta^{\alpha\beta}$ must be of type C^1.

B- Its 4−divergence is null,

$$\frac{\partial \Theta^{\alpha\beta}}{\partial x^\beta} = 0 \tag{11}$$

C- $\Theta^{\alpha\beta}$ vanishes for large distances ($\Theta^{\alpha\beta} \to 0$ for $|\vec{x}| \to \infty$).

A 4−volume space-time V^4 is bounded by two space-like 3−hypersurface planes V_1 and V_2 and a time-like 3−hypersurface σ which will be considered at infinite large distance. Integrating the 4−divergence of $\Theta^{\alpha\beta}$ over such a 4−volume V^4 and by using the generalized Gauss theorem to four dimensions (see Figure 3), obtains

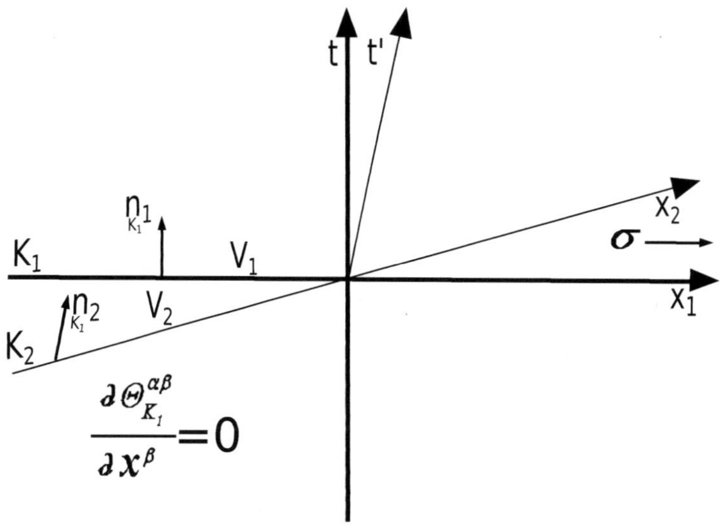

Figure 3. Conservation integral. V_2 represents the infinite volume which is instantaneous at the reference frame K_2 and V_1 represents the infinite volume which is instantaneous at the reference frame K_1.

$$\begin{aligned} 0 &= \int_{V^4} \frac{\partial \underset{K_1}{\Theta^{\alpha\beta}}}{\partial x^\beta} d^4x \\ &= \int_{V_1} \epsilon_1 \underset{K_1}{\Theta^{\alpha\beta}} \underset{K_1}{n_{1\beta}} dV_1 + \int_\sigma \epsilon_\sigma \underset{K_1}{\Theta^{\alpha\beta}} \underset{K_1}{n_{\sigma\beta}} d\sigma + \int_{V_2} \epsilon_2 \underset{K_1}{\Theta^{\alpha\beta}} \underset{K_1}{n_{2\beta}} dV_2. \end{aligned} \tag{12}$$

The unitary 4−vectors $\underset{K_1}{n_{1\beta}}$ y $\underset{K_1}{n_{2\beta}}$ are measured in K_1 and are perpendicular to the hypersurfaces V_1 and V_2, respectively. $\underset{K_1}{n_{\sigma\beta}}$ is perpendicular to the hypersurface σ and ϵ_1 , ϵ_σ and ϵ_2 take the values 1 or −1 depending on the hypersurface orientations. It is necessary to remember that all the quantities are measured in K_1 and when $\sigma \to \infty$ the tensor $\Theta^{\alpha\beta}$ vanishes. Consequently, we arrive at

$$\int_{V_1} \underset{K_1}{\Theta^{\alpha\beta}} \underset{K_1}{n_{1\beta}} \, dV_1 = \int_{V_2} \underset{K_1}{\Theta^{\alpha\beta}} \underset{K_1}{n_{2\beta}} \, dV_2. \tag{13}$$

Notice that all the events that belong to V_1 are instantaneous in K_1 and all the events that belong to V_2 are instantaneous in K_2. We have

$$\underset{K_1}{n_{1\beta}} = (1, 0, 0, 0) \quad and \quad \underset{K_1}{n_{2\beta}} = (\gamma, \gamma\vec{u}), \tag{14}$$

where $\gamma = \left(1 - u^2\right)^{-1/2}$ and $\vec{u}$ represents the relative motion between the frame K_1 and K_2. However, in the right side of Eq. (13), $\underset{K_1}{\Theta^{\alpha\beta}} \underset{K_1}{n_{2\beta}}$ is measured in K_1 and the integral must be done at the 3−hypersurface V_2. Then, it is convenient to express all the quantities in K_2. Let us call $\underset{K_i}{S^\alpha}$ as

$$\underset{K_i}{S^\alpha} = \int_{V_i} \underset{K_i}{\Theta^{\alpha\beta}} \underset{K_i}{n_{i\beta}} \, dV_i, \tag{15}$$

where $\underset{K_i}{n_{i\beta}} = (1, 0, 0, 0)$ in K_i. That is,

$$\underset{K_i}{S^\alpha} = \int_{V_i} \underset{K_i}{\Theta^{\alpha 0}} dV_i. \tag{16}$$

It must be noticed that $\underset{K_i}{S^\alpha}$ is expressed in a covariant form by using Eq. (15) and in a not apparently covariant form by using Eq. (16). Finally, it is easy to see from Eq. (13) that

$$\begin{aligned} \underset{K_1}{S^\alpha} &= \int_{V_2} \underset{K_1}{\Theta^{\alpha\beta}} \underset{K_1}{n_{2\beta}} \, dV_2 \\ &= \int_{V_2} \underset{K_2 \to K_1}{\Lambda^\alpha{}_\lambda} \underset{K_2}{\Theta^{\lambda\beta}} \underset{K_2}{n_{2\beta}} dV_2 \\ &= \underset{K_2 \to K_1}{\Lambda^\alpha{}_\lambda} \int_{V_2} \underset{K_2}{\Theta^{\lambda\beta}} \underset{K_2}{n_{2\beta}} \, dV_2. \end{aligned} \tag{17}$$

Since, $\underset{K_2}{n_{2\beta}} = (1, 0, 0, 0)$, one obtains

$$\begin{aligned} \underset{K_1}{S^\alpha} &= \underset{K_2 \to K_1}{\Lambda^\alpha{}_\lambda} \int_{V_2} \underset{K_2}{\Theta^{\lambda 0}} dV_2 \\ &= \underset{K_2 \to K_1}{\Lambda^\alpha{}_\lambda} \underset{K_2}{S^\lambda}. \end{aligned} \tag{18}$$

Therefore, $\underset{K_i}{S^\alpha}$ is a 4−vector.

Since in this case Eq. (11) is satisfied, a conservation property can be deduced. Indeed, Eq. (11) can be expressed as

$$\frac{\partial \Theta^{\alpha 0}}{\partial x^0} + \frac{\partial \Theta^{\alpha j}}{\partial x^j} = 0, \tag{19}$$

where $j = 1, 2, 3$, it implies that

$$\begin{aligned} \frac{d\underset{K_i}{S^\alpha}}{dt} &= \int_{V_i} \frac{\partial \underset{K_i}{\Theta^{\alpha 0}}}{\partial t} dV_i = -\int_{V_i} \frac{\partial \underset{K_i}{\Theta^{\alpha j}}}{\partial x^j} dV_i \\ &= -\int_{A_i} \underset{K_i}{\Theta^{\alpha j}} d\overrightarrow{A}_{ij}, \end{aligned} \tag{20}$$

where A_i and $d\overrightarrow{A}_{ij}$ represent the surface of the volume V_i and the $j-$component of its surface vector, respectively. The 3−dimensional Gauss theorem has been used. Since $\underset{K_i}{\Theta^{\alpha\beta}} = 0$ when $|\overrightarrow{x}| \to \infty$ and V_1 represents all the 3−dimensional space at a time t in K_i.

$$\frac{d\underset{K_i}{S^\alpha}}{dt} = 0. \tag{21}$$

The conservation property is that $\underset{K_i}{S^\alpha}$ is a constant.

2.2.2. 4−vector total energy-momentum conservation Let us prove that the energy-momentum tensor satisfies Eq (11). Since, the energy-momentum tensor is well defined in any reference frame, we abandon for the moment the subscript K. Then,

$$\begin{aligned} \frac{\partial T^{\alpha 0}}{\partial x^0} &= \frac{\partial}{\partial t}\left[\sum_{i=1}^{n} p_i^\alpha(t)\delta^3(\overrightarrow{r} - \overrightarrow{r}_i(t))\right] \\ &= \sum_{i=1}^{n} \frac{dp_i^\alpha(t)}{dt}\delta^3(\overrightarrow{r} - \overrightarrow{r}_i(t)) + p_i^\alpha(t)\frac{\partial}{\partial t}\left[\delta^3(\overrightarrow{r} - \overrightarrow{r}_i(t))\right]. \end{aligned} \tag{22}$$

On the other hand, by using the properties of the delta function ($j = 1, 2, 3$)

$$\begin{aligned} \frac{\partial T^{\alpha j}}{\partial x^j} &= \sum_{i=1}^{n} p_i^\alpha(t)\frac{dx_i^j}{dt}\frac{\partial}{\partial x^j}\left[\delta^3(\overrightarrow{r} - \overrightarrow{r}_i(t))\right] \\ &= -\sum_{i=1}^{n} p_i^\alpha(t)\frac{\partial}{\partial t}\left[\delta^3(\overrightarrow{r} - \overrightarrow{r}_i(t))\right]. \end{aligned} \tag{23}$$

From Eqs. (22) and (23), It is clear that

$$\frac{\partial T^{\alpha\beta}}{\partial x^\beta} = \sum_{i=1}^{n} \frac{dp_i^\alpha(t)}{dt}\delta^3(\overrightarrow{r} - \overrightarrow{r}_i(t)) = G^\alpha, \tag{24}$$

where G^α represents the force density. Therefore, if there is no interaction $dp_i^\alpha(t)/dt = 0$, $T^{\alpha\beta}$ satisfies Eq. (11) and consequently $\underset{K}{P^\alpha}$represents a 4−vector and is constant. In fact, as it has been noted above, this is equivalent to the no interaction theorem [36] for classical particles. That is, a 4−vector total energy-momentum can be well defined when there is no interaction and a rest reference frame K_o exists (see Eq. (7)). However, in the general case, the interaction exists and it is needed to add an extra tensor to the energy-momentum tensor in order to obtain a null-divergence of the energy-momentum tensor. Let us consider a system of charged particles where the electromagnetic interaction has to be considered and accordingly G^αdoes not vanish. It is well known that the electromagnetic energy-momentum field tensor $T^{\alpha\beta}_{em}$ can be expressed as

$$T^{\alpha\beta}_{em} = F^\alpha{}_\lambda F^{\beta\lambda} - \frac{1}{4}\eta^{\alpha\beta}F_{\lambda\varepsilon}F^{\lambda\varepsilon}, \tag{25}$$

where $F^{\alpha\beta}$ and $\eta^{\alpha\beta}$ represent the stress electromagnetic tensor and the Minkowskian metric tensor, respectively. We have

$$\frac{\partial T^{\alpha\beta}_{em}}{\partial x^{\beta}} = -F^{\alpha}{}_{\lambda}J^{\lambda}, \tag{26}$$

where

$$J^{\alpha} = \sum_{i=1}^{n} q_i \frac{dx_i^{\alpha}(t)}{dt}\delta^3(\overrightarrow{r} - \overrightarrow{r}_i(t)), \tag{27}$$

being q_i the charge of the $i-$particle. Then,

$$F^{\alpha}{}_{\lambda}J^{\lambda} = \sum_{i=1}^{n} q_i F^{\alpha}{}_{\lambda}\frac{dx_i^{\lambda}(t)}{dt}\delta^3(\overrightarrow{r} - \overrightarrow{r}_i(t)). \tag{28}$$

Lorentz equation permits us to arrive at

$$F^{\alpha}{}_{\lambda}J^{\lambda} = \sum_{i=1}^{n} \frac{dp_i^{\alpha}}{dt}\delta^3(\overrightarrow{r} - \overrightarrow{r}_i(t)) = G^{\alpha}. \tag{29}$$

Then, by adding to the energy-momentum tensor the electromagnetic energy-momentum field tensor, a total energy-momentum tensor can be defined as

$$T^{\alpha\beta}_{T} = T^{\alpha\beta} + T^{\alpha\beta}_{em}, \tag{30}$$

with the interesting property deduced from Eqs. (26) and (29),

$$\frac{\partial T^{\alpha\beta}_{T}}{\partial x^{\beta}} = 0. \tag{31}$$

Thus a 4−vector total energy-momentum can be defined as

$$\underset{K_i}{P^{\alpha}_{T}} = \int \underset{K_i}{T^{\alpha 0}_{T}} dV_i, \tag{32}$$

and a rest reference frame K_o such that

$$\underset{K_o}{P^{\alpha}_{T}} = (p^0, 0, 0, 0). \tag{33}$$

2.2.3. Divergences and reaction force The 4−vector total energy-momentum and the energy-momentum tensor posses some inconveniences. Indeed, Eq. (25) shows that the electromagnetic field must be evaluated at the point particles which leads to divergences in both vector and tensor. Some renormalization technique has been developed in order to avoid such infinities in the deduction of the reaction force [38]; even more in General Relativity, similar techniques were developed to find the expression of the reaction force [39], [40], [41], [42]. Although, there exist different points of view as the perfect absorber universe developed by Wheeler and Feynman [43], [44] and similar theories where action at a distance theories are used, Dirac's result [38] is the most accepted method to deduce the reaction force. However, due to some physical inconsistencies as the runaway´s solutions and the preaccelerations, corrections have been made for the Lorentz-Dirac equation leading principally to two equations, Eliezer equation [45] and Landau-Lifshitz equation [46], which are mathematically different but physically equivalent since their differences cannot be detected within Classical Electrodynamics [47], [48], [49], [50] (the differences are out of Shen's zone [51] where quantum effects are important). Therefore, Eq. (28) can be rewritten as

$$\begin{aligned} F^{\alpha}{}_{\lambda}J^{\lambda} &= \sum_{i=1}^{n} q_i F^{\alpha}{}_{\lambda}\frac{dx_i^{\lambda}(t)}{dt}\delta^3(\overrightarrow{r} - \overrightarrow{r}_i(t)) \\ &= \sum_{k=1}^{n}\sum_{i=1}^{n} q_i F^{\alpha}_{k\;\lambda}\frac{dx_i^{\lambda}(t)}{dt}\delta^3(\overrightarrow{r} - \overrightarrow{r}_i(t)) \\ &= \sum_{k=1}^{n}\sum_{i\neq k}^{n} q_i F^{\alpha}_{k\;\lambda}\frac{dx_i^{\lambda}(t)}{dt}\delta^3(\overrightarrow{r} - \overrightarrow{r}_i(t)) \\ &\quad +\sum_{i=1}^{n} q_i F^{\alpha}{}_{itself\lambda}\frac{dx_i^{\lambda}(t)}{dt}\delta^3(\overrightarrow{r} - \overrightarrow{r}_i(t)). \end{aligned} \tag{34}$$

The term $q_iF^{\alpha}_{k\;\lambda}(dx_i^{\lambda}(t)/d\tau_i)$ represents the force exerted by the $k-$particle on the $i-$particle.. In particular, $q_iF^{\alpha}{}_{iself\lambda}(dx_i^{\lambda}(t)/d\tau_i)$ is the self-force and may be expressed as the radiation force f^{μ}_{i-rad}, such that

$$\begin{aligned} \frac{dp_i^{\alpha}}{d\tau_i} &= \sum_{i\neq k}^{n} q_i F^{\alpha}_{k\;\lambda}\frac{dx_i^{\lambda}(t)}{d\tau_i} + f^{\mu}_{i-rad} \\ &= q_i F'^{\alpha}{}_{\lambda}\frac{dx_i^{\lambda}(t)}{d\tau_i} + f^{\mu}_{i-rad}, \end{aligned} \tag{35}$$

where $F'^{\alpha}{}_{\lambda} = \sum_{i\neq k}^{n} q_i F^{\alpha}_{k\;\lambda}$. Therefore, if we substitute this result on Eq. (35), we obtain

$$\begin{aligned} F^{\alpha}{}_{\lambda}J^{\lambda} &= \sum_{i=1}^{n}\left[q_i F'^{\alpha}{}_{\lambda}\frac{dx_i^{\lambda}(t)}{dt} + f^{\mu}_{i-rad}\frac{d\tau_i}{dt}\right]\delta^3(\overrightarrow{r} - \overrightarrow{r}_i(t)) \\ &= \sum_{i=1}^{n}\frac{dp_i^{\alpha}}{d\tau_i}\delta^3(\overrightarrow{r} - \overrightarrow{r}_i(t)). \end{aligned} \tag{36}$$

Finally, Eq. (36) shows that even if we consider the reaction force (or self-force), Eq. (11) is satisfied. However, the problem is not closed since there are many discussions about the validity of the renormalization process and the expression for the reaction force. Moreover, the total tensor also possesses divergences which complicate the calculation of the the total energy-momentum tensor. The principal problem is due to the divergence of the electromagnetic field in the point particle itself ($T^{\alpha\beta}_{em}$ is not a C^1 type). Frequently, a system of charged particles can be represented with a high degree of approximation by continuous density and current. In such a cases, the divergences disappear and the total energy-momentum is well defined.

3. Discrete Density Representation of the Energy-Momentum Tensor of a system of particles

As we mentioned above, in many cases a continuous description of a system of particles is adequate. However, this does not mean that densities have to be continuous. Indeed, we can describe a system of particles by using a discrete density for a free particle system. First, consider that a rest frame K_o exists as in Eq. (7); then, we can define a discrete energy density ρ_d,

$$\rho_d = \underset{K_o}{T^{00}} = \sum_{i=1}^{n} \underset{K_o}{p_i^0}\delta^3(\overrightarrow{x} - \overrightarrow{x}_i(t)), \tag{37}$$

IOP Conf. Series: Journal of Physics: Conf. Series **845** (2017) 012002 doi:10.1088/1742-6596/845/1/012002

and a discrete pressure p_d,

$$p_d = \frac{1}{3}\sum_{j=1}^{3} \underset{K_o}{T^{jj}} = \frac{1}{3}\sum_{j=1}^{3}\sum_{i=1}^{n} \underset{K_o}{p_i^j} \frac{dx_i^j}{dt}\delta^3(\vec{x} - \vec{x}_i(t)). \tag{38}$$

Being the other components,

$$\begin{aligned} \underset{K_o}{T^{0j}} &= \sum_{i=1}^{n} \underset{K_o}{p_i^0} \frac{dx_i^j}{dt}\delta^3(\vec{x} - \vec{x}_i(t)) \\ &= \sum_{i=1}^{n} \underset{K_o}{p_i^j}\delta^3(\vec{x} - \vec{x}_i(t)) = \underset{K_o}{T^{j0}} \end{aligned} \tag{39}$$

and for $j \neq k$

$$\underset{K_o}{T^{jk}} = \sum_{i=1}^{n} \underset{K_o}{p_i^j} \frac{dx_i^k}{dt}\delta^3(\vec{x} - \vec{x}_i(t)). \tag{40}$$

The energy-momentum is calculated as

$$\begin{aligned} \underset{K_o}{P^0} &= \int \rho_d dV_o \\ &= \int \sum_{i=1}^{n} \underset{K_o}{p_i^0}\delta^3(\vec{x} - \vec{x}_i(t))dV_o \\ &= \sum_{i=1}^{n} \underset{K_o}{p_i^0}. \end{aligned} \tag{41}$$

and

$$\begin{aligned} \underset{K_o}{P^j} &= \int \sum_{i=1}^{n} \underset{K_o}{p_i^j}\delta^3(\vec{x} - \vec{x}_i(t))dV_o \\ &= \sum_{i=1}^{n} \underset{K_o}{p_i^j} = 0, \end{aligned} \tag{42}$$

as expected since K_o is the rest frame. It is important to notice that $\underset{K_o}{T^{0j}} \neq 0$, Eq. (39), but for the purpose of calculating the spatial components of the 4−vector total energy-momentum in the rest frame K_o, it does not contribute, Eq. (42). However, strictly $\underset{K_o}{T^{0j}} \neq 0$ and the other components T^{jk} $(j \neq k)$ do not vanish too. Although in this calculation we have just repeated the results of section 2, the question that arises is whether a system of free particles can actually be described by a 4−vector total energy-momentum and by a energy-momentum tensor with the same structure than the one used in the continuous cases. The answer is contained in the following subsections and sections.

3.1. Three cases of free particle systems

In order to posteriori compare discrete free particle systems with different types of fluids which are continuous, it is necessary to impose some constraints on the energy-momentum tensor.

3.1.1. The imperfect discrete free particle system This is the case of a regular discrete free particle system with the characteristic of not possessing any constraint; that is: $T^{\alpha\beta}$ do not vanish as the general case which is just described above in Eqs. (37), (38), (39) and (40).

3.1.2. The dust discrete free particle system In this case, the particles of the system are at rest in the rest reference frame K_o; that is:

$$\underset{K_o}{p_i^j} = 0 \quad for \quad j = 1, 2, 3. \tag{43}$$

This means that

$$\rho_d = \sum_{i=1}^{n} p_i^0 \delta^3(\overrightarrow{x} - \overrightarrow{x}_i(t)). \tag{44}$$

Consequently,

$$\left\{ \begin{array}{c} \underset{K_o}{T^{00}} = \rho_d \\ \underset{K_o}{T^{\alpha\beta}} = 0 \quad Otherwise \end{array} \right\}. \tag{45}$$

In a subsequent section, it will be shown that this representation of the energy-momentum tensor will generate a 4−vector total energy-momentum similar to the case of the continuous dust fluid in a finite volume. Notice that the pressure vanishes in this case because all the particles are at rest in the rest reference frame K_o.

3.1.3. The perfect discrete free particle system In this case we will impose to a discrete free particle system the following constraints:

$$\left\{ \begin{array}{c} \underset{K_o}{T^{jk}} = p_d \delta_{jk} \quad for \quad j, k = 1, 2, 3 \\ \underset{K_o}{T^{k0}} = 0 \quad for \quad k = 1, 2, 3 \\ \underset{K_o}{T^{00}} = \rho_d \end{array} \right\}, \tag{46}$$

where ρ_d and p_d can be described by Eqs. (37) and (38). However, these constraints impose symmetries on the free particle system. For example, in order to accomplish Eq. (46), in each trajectory of a particle with spatial components of the momentum p_i^j there must exist a set A_l of other particles k ($k \epsilon A_l$, $i \epsilon A_l$) with the same trajectories and with spatial components of the momentum p_k^j such that the sum of all the momenta of the particles in A vanishes; that is:

$$\sum_{k \epsilon A_l} p_k^j = 0. \tag{47}$$

Therefore,

$$\begin{aligned} \sum_{k \epsilon A_l} p_k^j \delta(\overrightarrow{x} - \overrightarrow{x}_k) &= \delta(\overrightarrow{x} - \overrightarrow{x}_i) \sum_{k \epsilon A_l} p_k^j \; with \; i \epsilon A_l \\ &= 0. \end{aligned} \tag{48}$$

The union of all sets A_l being the set of all the particles of the system; that is: $\cup A_l = \{1, 2, 3, ..., n\}$. Therefore,

$$T^{j0} = \sum_{i=1}^{n} \underset{K_o}{p_i^j} \delta^3(\overrightarrow{x} - \overrightarrow{x}_i(t)) = \sum_{\cup A_l} \sum_{k \epsilon A_l} p_k^j \delta(\overrightarrow{x} - \overrightarrow{x}_k) = 0. \tag{49}$$

This cannot be accomplished unless we consider that the differential of the volume dV_o contains a number of particles with this kind of properties; that is: if we consider a continuous fluid. Finally, the energy-momentum tensor of a discrete set of particles cannot satisfy Eq. (46) unless it is considered as a continuous fluid. Moreover, we will see that the representation of the energy-momentum tensor in Eq. (46) does not satisfy Eq. (11).

IOP Conf. Series: Journal of Physics: Conf. Series **845** (2017) 012002 doi:10.1088/1742-6596/845/1/012002

3.2. Relativistic transformations with the discrete energy-momentum tensor

In order to know how useful are the discrete energy-momentum tensors just described above, let us analyze their relativistic transformations and the 4−vector energy-momenta obtained from them.

3.2.1. Transformation of the total dust discrete energy-momentum The discrete dust energy-momentum in a reference frame K is obtained by applying Lorentz transformations to Eq. (45),

$$\underset{K}{T^{\alpha\beta}} = \rho_d u^\alpha u^\beta, \tag{50}$$

where u^α represents the 4−vector which describes the motion between K and K_o and ρ_d is measured in the rest reference frame K_o. We have to remember that in the definition of the dust discrete free particle system, Eq. (45), all the particles are at rest in a particular reference frame which is considered as the rest frame K_o. Therefore, the existence of the rest frame is assured by definition in the case of the dust discrete free particle system. Let us calculate the total 4−vector energy-momentum. Let us see that the total 4−vector energy-momentum is a 4−vector. We have,

$$\begin{aligned} \underset{K}{P^\alpha} &= \int \rho_d u^\alpha u^0 dV = \int \rho_d u^\alpha u^0 \frac{dV_o}{\gamma} \\ &= \int \rho_d u^\alpha \gamma \frac{dV_o}{\gamma} = u^\alpha \int \rho_d dV_o = u^\alpha \underset{K_o}{P^0}, \end{aligned} \tag{51}$$

as expected. But this result must be consistent with the conservation theorem. If a 4−tensor vanishes in a particular reference frame, it vanishes in all the reference frame. Therefore, we can analyze the conservation theorem in the rest frame K_o. Since in such a frame $u^\alpha = (1, 0, 0, 0)$, we need just to analyze the component "0"; that is:

$$\begin{aligned} \frac{\partial \underset{K_o}{T^{0\beta}}}{\partial x^\beta} &= \frac{\partial \rho_d}{\partial t} = \frac{\partial}{\partial t}\left[\sum_i^n p_i^0 \delta^3(\vec{x} - \vec{x}_i)\right] \\ &= \left[\sum_i^n \frac{dp_i^0}{dt}\delta^3(\vec{x} - \vec{x}_i) - p_i^0 \frac{dx_i^j}{dt}\frac{\partial}{\partial x^j}\delta^3(\vec{x} - \vec{x}_i)\right] = 0. \end{aligned} \tag{52}$$

Since we know that in the case of a dust discrete free particle system both $\frac{dp_i^0}{dt}$ and $\frac{dx_i^j}{dt}$ vanish (remember that all the particles are in rest in K_o). Therefore, the conservation theorem is accomplished. We will see that for a dust continuous fluid, the result holds for a finite volume.

3.2.2. Transformation of the total energy-momentum for a perfect discrete free particle system In the case of a perfect discrete free particle system, by using Eq. (46) and the Lorentz transformations, it is easy to see that the energy-momentum tensor may be written in any reference frame K as

$$\underset{K}{T^{\alpha\beta}} = (\rho_d + p_d) u^\alpha u^\beta - p_d n^{\alpha\beta}, \tag{53}$$

where ρ_d and p_d must be measured in the rest frame K_o and $n^{\alpha\beta}$ is the Minkowskian metric. In the case of infinite volume, let us see if this formalism leads to the same result obtained in Eq. (10).

$$\underset{K}{P^\alpha} = \int \left[(\rho_d + p_d) u^\alpha u^0 - p_d n^{\alpha 0}\right] dV. \tag{54}$$

For $\alpha = 0$, considering $u^\alpha = (\gamma, \gamma u)$, we have

$$\begin{aligned} \underset{K}{P^0} &= \int T^{00} dV = \int \left[(\rho_d + p_d) u^0 u^0 - p_d n^{00} \right] dV \\ &= \int \left[(\rho_d + p_d)\gamma\gamma - p_d \right] \frac{dV_o}{\gamma} \\ &= \gamma \underset{K_o}{P^0} + \gamma (\vec{u})^2 \frac{1}{3} \sum_{j=1}^{3} \sum_{i=1}^{n} \underset{K_o}{p_i^j} \frac{dx_i^j}{dt}. \end{aligned} \quad (55)$$

We can conclude that it does not coincide with the transformation described in Eq. (10). For the corresponding spatial coordinates, we have, $j = 1, 2, 3$

$$\begin{aligned} \underset{K}{P^j} &= \int \left[(\rho_d + p_d) u^j u^0 - p_d n^{j0} \right] dV \\ &= \int \left[(\rho_d + p_d) \gamma^2 u_{x_j} \right] dV \\ &= \int \left[(\rho_d + p_d) \gamma^2 u_{x_j} \right] \frac{dV_o}{\gamma} \\ &= \gamma u_{x_j} \left(\underset{K_o}{P^0} + \frac{1}{3} \sum_{k=1}^{3} \sum_{i=1}^{n} \underset{K_o}{p_i^k} \frac{dx_i^k}{dt} \right), \end{aligned} \quad (56)$$

where u_{x_j} represents the $j-$component of $\vec{u}$. We notice that this 4$-$vector energy-momentum does not transform as the 4$-$vector in Eq.(10). For the perfect discrete system of free particles, the 4$-$divergence of the energy-momentum defined in Eq. (53) does not vanish. Indeed, as we noticed above, if a 4$-$tensor vanishes in a reference frame, it will vanish in all other reference frames. Therefore, we just need to show that the energy-momentum tensor in the rest frame, Eq. (46), does not satisfy Eq. (11); that is:

$$\frac{\partial \underset{K_o}{T^{\alpha\beta}}}{\partial x^\beta} \neq 0. \quad (57)$$

Let us proved it. We calculate

$$\frac{\partial \underset{K_o}{T^{\alpha\beta}}}{\partial x^\beta} = \frac{\partial}{\partial x^\beta} \left[(\rho_d + p_d) u^\alpha u^\beta - p_d n^{\alpha\beta} \right]. \quad (58)$$

In this case u^α represents the 4$-$velocity between the rest frame K_o and itself. Consequently, u^α is equal to $u^\alpha = (1, 0)$. Therefore,

$$\frac{\partial \underset{K_o}{T^{\alpha\beta}}}{\partial x^\beta} = \left[\frac{\partial}{\partial x^\beta} (\rho_d + p_d) \right] u^\alpha u^\beta - \left[\frac{\partial}{\partial x^\beta} p_d \right] n^{\alpha\beta}. \quad (59)$$

Let us consider $\alpha = 0$,

$$\begin{aligned} \frac{\partial \underset{K_o}{T^{0\beta}}}{\partial x^\beta} &= \left[\frac{\partial}{\partial x^\beta} (\rho_d + p_d) \right] u^\beta - \left[\frac{\partial}{\partial x^\beta} p_d \right] n^{0\beta} \\ &= \frac{\partial}{\partial t} (\rho_d + p_d) - \frac{\partial}{\partial t} p_d \end{aligned}$$

$$
\begin{aligned}
&= \frac{\partial}{\partial t}\rho_d = \frac{\partial}{\partial t}\left[\sum_{i=1}^{n} \underset{K_o}{p_i^0}\delta^3(x - x_i(t))\right] \\
&= \sum_{i=1}^{n}\left[\frac{d\underset{K_o}{p_i^0}}{dt}\delta^3(x - x_i(t)) - \underset{K_o}{p_i^0}\frac{dx_i^j}{dt}\frac{\partial}{\partial x_i^j}\delta^3(x - x_i)\right] \\
&= \sum_{i=1}^{n}\left[\frac{d\underset{K_o}{p_i^0}}{dt}\delta^3(x - x_i(t)) + \frac{\partial}{\partial x^j}\left[\underset{K_o}{p_i^j}\delta^3(x - x_i)\right]\right]. \qquad (60)
\end{aligned}
$$

Since we are considering that there is no interaction, $\frac{d\underset{K_o}{p_i^0}}{dt} = 0$,

$$
\begin{aligned}
\frac{\partial \underset{K_o}{T^{0\beta}}}{\partial x^\beta} &= \frac{\partial}{\partial x^j}\left[\sum_{i=1}^{n}\left[\underset{K_o}{p_i^j}\delta^3(x - x_i)\right]\right] \\
&= \frac{\partial}{\partial x^j}T^{j0}, \qquad (61)
\end{aligned}
$$

which does not vanish in general. However, in order to have $T^{j0} = 0$, as in Eq. (46), Eq. (49) must be satisfied which will imply that the divergence in Eq.(61) vanishes. As we noted above, this cannot be accomplished unless a continuous fluid is considered instead of a system of free particles. The reason is because in Eq. (23), it is clear that the term $T^{0\beta}$ contributes to nullify the divergence but in the case of this representation the non-diagonal elements are not considered.

Therefore, in this scheme, the conservation theorem is not assured and P^α is not defined as a regular 4−vector. Even more, the existence of the rest frame is not assure. However, the formalism can be used to define a 4−vector but based in defining the vector on the concept of instantaneity. That is, defining the vector on a reference frame where the quantity is measured in certain reference frame K. The existence of the rest frame K_o is assured when the system of particles or the fluid is constrained to a physical volume. Moreover, this will permit using a finite volume which is an important application in Relativistic Thermodynamics.

4. 4−Vector Total Energy-Momentum in a Finite Volume

Until now we have shown that for a system of free particles it is possible to define a 4−vector total energy-momentum by using an infinite volume. A representation of a perfect discrete free particle system does not gives a universal definition of a 4−vector total energy-momentum due to the invalidity of the conservation theorem unless the acceptance of symmetries that implies a continuous description of the system of particles, Eq. (61). When a finite volume (not physical) is considered with a system of particles, the particles will abandon the volume and the energy-momentum of the system contained in the volume will change or if the volume represents a physical volume due to the interaction of the particles with the walls of the volume, the conservation theorem will not apply. However, for finite volume and perfect fluid, it is possible to define a 4−vector total energy-momentum as long as it is related to the instantaneity in a particular reference frame. Similar results have been obtained by Callen and Horwitz [52] by defining a 4-vector enthalpy-momentum. Moreover, as we will see, the result obtained with the perfect fluid in a finite volume corresponds to the one deduced with the discrete free particle system in an infinite volume provided certain terms are substituted as follows [53]:

$$
\underset{K_o}{P^0} = \int \rho_d dV_o \rightarrow E_o \quad and \quad \frac{1}{3}\sum_{j=1}^{3}\sum_{i=1}^{n}\underset{K_o}{p_i^j}\frac{dx_i^j}{dt} \rightarrow pV_o. \qquad (62)
$$

Let us begin by considering a perfect fluid which expression for the energy-momentum tensor similar to Eq. (53), but with the difference that the density ρ and the pressure p correspond to a continuous uniform system which is in thermal equilibrium. ρ and p are constant and they are measured in a reference frame K_o where the fluid is contained in a finite volume at rest in K_o; that is:

$$\underset{K}{T^{\alpha\beta}} = (\rho + p)u^\alpha u^\beta - pn^{\alpha\beta}. \tag{63}$$

Since ρ and p are constant, the conservation theorem is satisfied; that is,

$$\frac{\partial}{\partial x^\beta}\underset{K}{T^{\alpha\beta}} = 0. \tag{64}$$

Remember that in this scheme u^α represents the velocity between the frames K and K_o and it is constant. By using Eq. (64), it can be thought out that a 4−vector total energy-momentum could be defined as in Eq. (18) but first it has to be remembered that we are dealing with a finite volume. Accordingly, the energy-momentum on the finite time-like 3−hypersurface σ (see Eq. (12)) will not vanish and therefore a universal 4−vector energy-momentum cannot be obtained. This can be corrected by forcing ρ to vanish in the finite time-like 3−hypersurface σ which corresponds to having a non constant ρ in the of surface of the volume. This implies that Eq. (64) is not valid. Physically, this represents a fluid contained in a volume and there must exist interaction between the particles of the fluid and the walls of the volume. Therefore, in order to define a 4−quantity $\underset{K,K}{P^\alpha}$ in a reference frame K, it is necessary to resort to the concept of instantaneity defined in the projected volume V_K of V_o (see Figure 4).

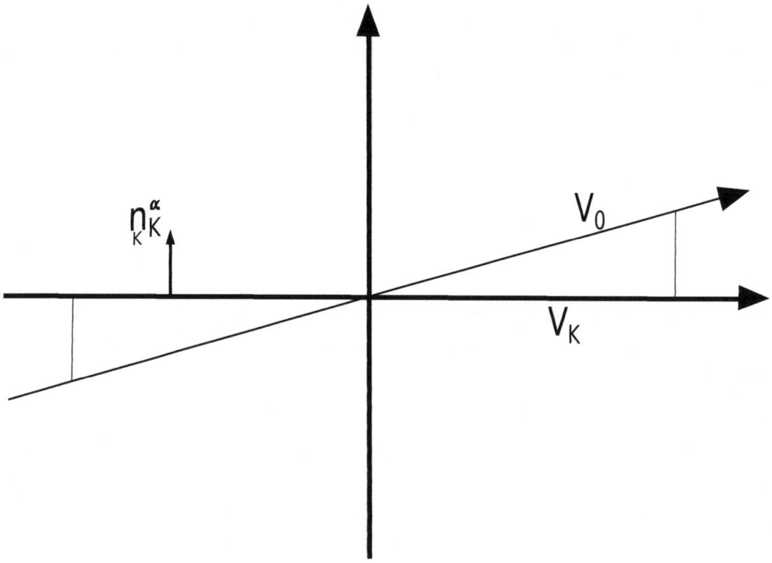

Figure 4. Finite Volume

Hence, the instantaneity along V_K is described by $\underset{K}{n_{K\alpha}} = (1, 0, 0, 0)$ and $\underset{K,K}{P^\alpha}$ by

$$\begin{aligned}\underset{K,K}{P^\alpha} &= \int_{V_K} \left[(\rho + p)u^\alpha u^0 - pn^{\alpha 0}\right] dV_K \\ &= \int_{V_K} \left[(\rho + p)u^\alpha u^\beta - pn^{\alpha\beta}\right] \underset{K}{n_{K\beta}}\, dV_K\end{aligned}$$

$$= \underset{K_o \to K}{\Lambda^\alpha{}_\lambda} \int_{V_o} \left[(\rho + p)u^\lambda u^\beta - pn^{\lambda\beta}\right] \underset{K_o}{n_{K\beta}} \frac{dV_o}{\gamma}, \tag{65}$$

with

$$\underset{K_o}{n_{K\alpha}} = (\gamma, \gamma\vec{u}) \quad and \quad u^\alpha = (1,0,0,0). \tag{66}$$

Since in the parenthesis of Eq. (65), u^α represents the relative motion of K_o with itself. Of course $\underset{K_o}{n_{K\alpha}}$ represents the instantaneity in K measured in K_o. Therefore,

$$\underset{K,K}{P^\alpha} = \Lambda^\alpha{}_\lambda \int_{V_o} \left[(\rho + p)u^\lambda\gamma - p\underset{K_o}{n^\lambda_K}\right] \frac{dV_o}{\gamma} \tag{67}$$

Finally,

$$\begin{aligned} \underset{K,K}{P^0} &= \Lambda^0{}_\lambda \int_{V_o} \left[(\rho + p)u^\lambda\gamma - p\underset{K_o}{n^\lambda_K}\right] \frac{dV_o}{\gamma} \\ &= \Lambda^0{}_0 \int_{V_o} [(\rho + p)\gamma - p\gamma] \frac{dV_o}{\gamma} \\ &\quad + \Lambda^0{}_j \int_{V_o} \left[(\rho + p)0\gamma - p\gamma u_{x_j}\right] \frac{dV_o}{\gamma} \\ &= \gamma\left(E_o + (\vec{u})^2 pV_o\right), \end{aligned} \tag{68}$$

where

$$E_o = \int_{V_o} \rho dV_o. \tag{69}$$

For the spatial component

$$\begin{aligned} \underset{K,K}{P^j} &= \Lambda^j{}_\lambda \int_{V_o} \left[(\rho + p)u^\lambda\gamma - p\underset{K_o}{n^\lambda}\right] \frac{dV_o}{\gamma} \\ &= \Lambda^j{}_0 \int_{V_o} \left[(\rho + p)u^0\gamma - p\underset{K_o}{n^0}\right] \frac{dV_o}{\gamma} \\ &\quad + \Lambda^j{}_k \int_{V_o} \left[(\rho + p)u^k\gamma - p\underset{K_o}{n^k}\right] \frac{dV_o}{\gamma} \\ &= \gamma u_{x_j} \int_{V_o} [(\rho + p)\gamma - p\gamma] \frac{dV_o}{\gamma} \\ &\quad + \gamma \int_{V_o} \left[(-p\gamma u_{x_j}\right] \frac{dV_o}{\gamma} \\ &= \gamma u_{x_j} (E_o + pV_o). \end{aligned} \tag{70}$$

Eqs. (68) and (70) does not correspond to what it was expected. Indeed, the vector in K_o is not $(E_o, 0)$ but $(E_o, p\vec{u}V_o)$. We were expecting to obtain

$$\underset{K,K}{P^\alpha} = \underset{K_o \to K}{\Lambda^\alpha{}_\lambda} (E_o, 0) = \underset{K_o \to K}{\Lambda^\alpha{}_\lambda} \underset{K_o,K_o}{P^\alpha}. \tag{71}$$

The reason why for a finite volume the energy-momentum does not behave as in Eq. (71), consists of noticing that the existence of a volume implies an interaction with the walls of the volume and consequently Eq. (11) is not anymore valid. Therefore,

$$\underset{K,K}{P^\alpha} = \left\{ \begin{array}{c} \gamma\left(\underset{K_o}{P^0} + u^2 pV_o\right) \\ \gamma u_{x_j}\left(\underset{K_o}{P^0} + pV_o\right) \end{array} \right\}, \tag{72}$$

corresponds to the Lorentz transformation of the 4−quantity $(E_o, p\overrightarrow{u}V_o)$. Landau and Lifshitz [54] does not check the 0−component of the vector and simply affirmed that pV_o must be considered as part of the mass of the system. This can be considered for low velocities. However, for relativistic velocities, the case we are concerned, the 0−component possesses a $(\overrightarrow{u})^2$ term in the parenthesis. A similar result happens when the electromagnetic moment is calculated as Gamba [17], [24] has noticed by considering a universal definition of the electromagnetic energy-momentum vector.

Before continuing, let us make two comments:

A- First, it has to be noticed that if we substitute in Eq. (72) the corresponding changes in Eq. (62), we will obtain Eqs. (55) and (56), and vice versa. This shows that both technique are based on defining a 4−vector depending on which reference frame the instantaneity is considered.

B- Second, when continuous systems are considered, each part of the fluid possesses a velocity; that is, u^α represents the velocity of an infinitesimal part of the fluid contained in dV at a point $\overrightarrow{r}$ and a time t with respect a reference frame K. This means that there are commoving frames for each part of the fluid and $u^\alpha = u^\alpha(\overrightarrow{r}, t)$. Consequently, their derivatives do not have to vanish. u^α is a function of the position and time. Therefore, some constraints can be used for forcing the conservation theorem. That is:

$$\begin{aligned} \frac{\partial \underset{K_o}{T^{\alpha\beta}}}{\partial x^\beta} &= 0 = \frac{\partial}{\partial x^\beta}\left[(\rho + p)\, u^\alpha u^\beta - p n^{\alpha\beta}\right] \\ &= \left[\frac{\partial}{\partial x^\beta}(\rho + p)\right] u^\alpha u^\beta - \left[\frac{\partial}{\partial x^\beta} p\right] n^{\alpha\beta} \\ &\quad + (\rho + p)\frac{\partial}{\partial x^\beta}\left(u^\alpha u^\beta\right). \end{aligned} \tag{73}$$

This equation allows to conserve the energy-momentum for a continuous perfect fluid by imposing some constraint to the velocities of each element of the fluid (see Weinberg [35]). Notice that a generalization of the energy-momentum tensor to General Relativity includes the derivation of $g^{\alpha\beta}$ $(n^{\alpha\beta} \to g^{\alpha\beta})$ complicating the equations. Nevertheless, Relativistic Hydrodynamics will imply interaction with the walls of the volume and the conservation theorem will not apply too and it will lead to a non universal definition of the 4− vector momentum.

If we are considering a system of particles which are concentrated in a volume (not physical), these particles eventually will abandon the finite volume and there will be no conservation relation and the result of Sec. 1 will not apply. For a finite time, the volume can be considered enough big and the theory of Sec. 1 will be valid but by using the particle energy-momentum tensor, Eq. (8), and considering that at a moment the particle will abandon the volume.

Once we know that the discrete density model does not permit to universally define a 4−vector total energy-momentum and that the regular energy-momentum tensor, Eq. (18) does not match with the continuous energy-momentum tensor, Eq. (53), we can abandon the discrete representation and put our attention just in the continuous case. The density and pressure of a thermodynamic system are constant and the relative velocity between two inertial reference frame, $\overrightarrow{u}$, is constant. It is clear that Eq. (11) is accomplished for such constant energy-momentum tensor. However, the existence of a physical volume implies some interaction between the particles and the walls of the volume. When we obligate the particles to live in a finite volume, the problem of defining a 4−vector total energy-momentum persists. Indeed, a finite volume implies that the interaction between the walls of the physical volume and the particles exists. Consequently, the conservation equation, Eq. (11) is not anymore valid. Therefore, we need to develop a theory where a 4−vector total energy-momentum can be defined for finite volume even if it depends on a particular reference frame where the instantaneity is considered.

Let us begin by defining a 4−quantity in a reference frame where the instantaneity is taken. The hypersurface is represented by a 4−vector $\underset{K}{n_{K\alpha}} = (1,0,0,0)$ which represents the volume where the physical quantities will be measured. Then

$$\underset{K,K}{P^\alpha} = \int_V \underset{K}{T^{\alpha\beta}}\, \underset{K}{n_{K\beta}}\, dV_K. \tag{74}$$

This represents a 4−vector measured in a reference frame K where the instantaneity is in K. Without loosing generality, just two dimensions are considered in the following. Let us calculate it: $u^\alpha = (\gamma, \gamma u)$ in K and $\underset{K}{n^\alpha_K} = (1,0)$ in K.

$$\begin{aligned}
\underset{K,K}{P^\alpha} &= \int_{V_o}\left[(\rho+p)u^\alpha\gamma - p\begin{pmatrix}1\\0\end{pmatrix}\right]\frac{dV_o}{\gamma} \\
&= \int_{V_o}\left[(\rho+p)\begin{pmatrix}\gamma\\ \gamma u\end{pmatrix}\gamma - p\begin{pmatrix}1\\0\end{pmatrix}\right]\frac{dV_o}{\gamma} \\
&= \begin{pmatrix}\gamma\int_{V_o}\rho dV_o + \frac{\gamma^2-1}{\gamma}\int_{V_o}pdV_o \\ \gamma u\int_{V_o}(\rho+p)dV_o\end{pmatrix} \\
&= \begin{pmatrix}\gamma\int_{V_o}\rho dV_o + \gamma u^2\int_{V_o}pdV_o \\ \gamma u\int_{V_o}(\rho+p)dV_o\end{pmatrix} \\
&= \begin{pmatrix}\gamma\left(E_o + u^2 pV_o\right) \\ \gamma u(E_o + pV_o)\end{pmatrix}
\end{aligned} \tag{75}$$

Eq. (75) represents the Einstein-Planck-Tolman transformation of the 4−vector energy-momentum. It has to be noted that the 4−quantity

$$\begin{pmatrix}E_o \\ upV_o\end{pmatrix} \tag{76}$$

must represent the 4−vector energy-momentum of instantaneity in K measured in K_o. That is, the 4−quantity $\underset{K,K}{P^\alpha}$ must transform into the 4−quantity in Eq. (76) by a Lorentz transformation. Let us confirm this assertion.

$$\begin{aligned}
\underset{K,K_o}{P^\alpha} &= \underset{K\to K_o}{\Lambda^\alpha{}_\lambda}\,\underset{K,K}{P^\lambda} \\
&= \underset{K\to K_o}{\Lambda^\alpha{}_\lambda}\int_{V_K}\underset{K}{T^{\lambda\beta}}\,\underset{K}{n_{K\beta}}\,dV_K \\
&= \int_{V_K}\underset{K\to K_o}{\Lambda^\alpha{}_\lambda}\,\underset{K}{T^{\lambda\beta}}\,\underset{K}{n_{K\beta}}dV_K \\
&= \int_{V_o}\underset{K_o}{T^{\alpha\beta}}\,\underset{K_o}{n_{K\beta}}\,\frac{dV_o}{\gamma},
\end{aligned}$$

where it has to be remembered that even if $\underset{K_o}{n_{K\beta}}$ is measured in K_o and the instantaneity is taken on K; that is:

$$\underset{K_o}{n_{K\beta}} = (\gamma, \gamma\beta), \tag{77}$$

because $\underset{K_o}{n_{K\beta}}$ is the unit time-like 4−vector which represents the reference frame K observed from the reference frame K_o. We have continuously been using the contraction of the volume, $dV_o = \gamma dV_K$ and we can generalized it by introducing the 4−velocity $\underset{KK_o}{u}{}^{\alpha} = (\gamma, \gamma\beta)$, $\underset{K_oK_o}{u}{}^{\alpha} = (1, 0)$. Therefore, in Eq. (77), γ can be substituted by

$$\gamma = \underset{K_oK_o}{u}{}^{\lambda} \underset{K_o}{n_{K\lambda}} . \tag{78}$$

We arrive at:

$$\underset{K,K_o}{P^{\alpha}} = \int_{V_o} \underset{K_o}{T^{\alpha\beta}} \underset{K_o}{n_{K\beta}} \frac{dV_o}{\underset{K_oK_o}{u}{}^{\lambda} \underset{K_o}{n_{K\lambda}}} . \tag{79}$$

Therefore, by using the continuous definition of the energy-momentum tensor for a perfect fluid, we have:

$$\underset{K,K_o}{P^{\alpha}} = \int_{V_o} \left[(\rho + p) \underset{K_oK_o}{u}{}^{\alpha} \underset{K_oK_o}{u}{}^{\beta} - pn^{\alpha\beta} \right] \underset{K_o}{n_{K\beta}} \frac{dV_o}{\underset{K_oK_o}{u}{}^{\lambda} \underset{K_o}{n_{K\lambda}}} . \tag{80}$$

Finally, substituting the values of $\underset{K_oK_o}{u}{}^{\alpha}$ and $\underset{K_o}{n_{K\beta}}$, we obtain

$$\begin{aligned} \underset{K,K_o}{P^{\alpha}} &= \int_{V_o} \left[(\rho + p) \begin{pmatrix} 1 \\ 0 \end{pmatrix} \gamma - p \begin{pmatrix} \gamma \\ \gamma u \end{pmatrix} \right] \frac{dV_o}{\gamma} \\ &= \begin{pmatrix} \int_{V_o} \rho dV_o \\ \int_{V_o} pudV_o \end{pmatrix} = \begin{pmatrix} E_o \\ upV_o \end{pmatrix} \end{aligned} \tag{81}$$

which corresponds to

$$\underset{K,K}{P^{\alpha}} = \underset{K_o \to K}{\Lambda^{\alpha}{}_{\lambda}} \underset{K,K_o}{P^{\alpha}} , \tag{82}$$

or term by term

$$\begin{aligned} \begin{pmatrix} \underset{K,K}{P^0} \\ \underset{K,K}{P^j} \end{pmatrix} &= \begin{pmatrix} \gamma & \gamma u \\ \gamma u & \gamma \end{pmatrix} \begin{pmatrix} E_o \\ upV_o \end{pmatrix} \\ &= \begin{pmatrix} \gamma \left(E_o + u^2 pV_o \right) \\ \gamma u \left(E_o + pV_o \right) \end{pmatrix} , \end{aligned} \tag{83}$$

which shows that the representation of a 4−vector energy-momentum for a finite volume must be done by defining the instantaneity of the measurement.

If the theory is consistent, we must be able to calculate any 4−energy-momentum in a reference frame $K_i$considering the instantaneity in K_j. That is:

$$\underset{K_j,K_i}{P^{\alpha}} = \int_{V_o} \left[(\rho + p) \underset{K_iK_j}{u}{}^{\alpha} \underset{K_iK_j}{u}{}^{\beta} - pn^{\alpha\beta} \right] \underset{K_i}{n_{K_j\beta}} \frac{dV_o}{\underset{K_iK_j}{u}{}^{\lambda} \underset{K_i}{n_{K_j\lambda}}} , \tag{84}$$

where K_j and K_i may be K or K_o. We recall that in the notation of the volume vector

$$\underset{K_i}{n^{\beta}_{K_j}} , \tag{85}$$

where K_j denotes the reference where the instantaneity is considered and K_i represents the reference frame where the volume vector is expressed.
Therefore, for $K_j = K_o$ and $K_i = K_o$, we have

$$\underset{K_o,K_o}{P^\alpha} = \int_{V_o} \left[(\rho + p) \underset{K_oK_o}{u}{}^\alpha \underset{K_oK_o}{u}{}^\beta - pn^{\alpha\beta} \right] \underset{K_o}{n_{K_o\beta}} \frac{dV_o}{\underset{K_oK_o}{u}{}^\lambda \underset{K_o}{n_{K_o\lambda}}}, \tag{86}$$

where $\underset{K_oK_o}{u}{}^\alpha = \binom{1}{0}$ and $\underset{K_o}{n_{K_o\beta}} = (1,0)$. We have

$$\begin{aligned} \underset{K_o,K_o}{P^\alpha} &= \int_{V_o} \left[(\rho + p) \underset{K_oK_o}{u}{}^\alpha - p\underset{K_o}{n^\alpha_{K_o}} \right] dV_o \\ &= \begin{pmatrix} \int_{V_o} \rho dV_o \\ 0 \end{pmatrix} = \begin{pmatrix} E_o \\ 0 \end{pmatrix} \end{aligned} \tag{87}$$

as expected. However, to be consistent we need to prove that

$$\underset{K_o,K}{P^\alpha} = \underset{K_o \to K}{\Lambda^\alpha{}_\lambda} \underset{K_o,K_o}{P^\alpha} \tag{88}$$

using the definition of $\underset{K_o,K}{P^\alpha}$. That is:

$$\underset{K_o,K}{P^\alpha} = \int_{V_o} \left[(\rho + p) \underset{KK_o}{u}{}^\alpha \underset{KK_o}{u}{}^\beta - pn^{\alpha\beta} \right] \underset{K}{n_{K_o\beta}} \frac{dV_o}{\underset{KK_o}{u}{}^\lambda \underset{KK_o}{n_{K_o\lambda}}}, \tag{89}$$

where

$$\underset{KK_o}{u}{}^\alpha = \begin{pmatrix} \gamma \\ \gamma u \end{pmatrix} \quad and \quad \underset{K}{n_{K_o\beta}} = (\gamma, \gamma u). \tag{90}$$

Therefore,

$$\begin{aligned} \underset{K_o,K}{P^\alpha} &= \int_{V_o} \left[(\rho + p) \begin{pmatrix} \gamma \\ \gamma u \end{pmatrix} - p \begin{pmatrix} \gamma \\ \gamma u \end{pmatrix} \right] dV_o \\ &= \begin{pmatrix} \gamma \\ \gamma u \end{pmatrix} \int_{V_o} \rho dV_o = \begin{pmatrix} \gamma \\ \gamma u \end{pmatrix} E_o, \end{aligned} \tag{91}$$

which shows the validity of Eq. (88).

If we choose to measure a thermodynamic system which is at rest in K_o but we are looking at it from reference frame K, and the instantaneity is taken on K, that is, we are measuring $\underset{K,K}{P^\alpha}$. We can compare it with $\underset{K,K_o}{P^\alpha}$ which represents the transformed $\underset{K,K}{P^\alpha}$ to K_o. This is the Planck-Einstein-Tolman point of view [30]. If we take the simultaneity in K_o, we will look for $\underset{K_o,K_o}{P^\alpha}$ and its transformation to reference frame K is $\underset{K_o,K}{P^\alpha}$ which represents one of the options described by Rohrlich [30]. We are now in a position to develop a Relativistic Thermodynamics which includes a relativistic moving system.

5. Relativistic Redefined Thermodynamics

We are interested to reproduce the laws of Thermodynamics in a moving frame with respect the reference frame where the volume of the thermodynamic system is at rest. We have now develop the basis to express such laws in a covariant form. Let us constrain our point of view just to two systems, a lab frame K and the rest frame K_o. Since we are interested in expressing the laws in the lab frame K, the instantaneity is considered in the lab frame itself and the notation can be simplified by just putting

$$\underset{K}{n^{\alpha}_{K}} = w^{\alpha}. \tag{92}$$

It is clear that for calculating the 4−vector total energy-momentum , it is necessary just to use Eq. (84). The transformation laws of thermodynamics are reached by defining the redefined thermodynamic quantities in the following way [30]:

A- In order to reproduce the first law of Thermodynamics, it has to be noticed that for constant ρ and p, the 4−vector energy-momentum may be expressed as:

$$P^{\mu} = (E_o + pV_o)\, u^{\mu} - pV_o \frac{w^{\mu}}{w_{\lambda}u^{\lambda}}, \tag{93}$$

with $E_o = \rho V_o$. It is necessary to define the 4−vector G^{μ} as:

$$\begin{aligned} G^{\mu} &= \left[(\rho+p)\, u^{\mu}u^{\nu} - (\rho+p)n^{\mu\nu}\right] \frac{w_{\nu}}{w_{\lambda}u^{\lambda}} V_o \\ &= (E_o + pV_o)\, u^{\mu} - (E_o + pV_o) \frac{w^{\mu}}{w_{\lambda}u^{\lambda}}. \end{aligned} \tag{94}$$

Then, we define the 4−vector redefined energy-momentum by

$$\begin{aligned} \xi^{\mu} &= P^{\mu} - G^{\mu} \\ &= (E_o + pV_o)\, u^{\mu} - pV_o \frac{w^{\mu}}{w_{\lambda}u^{\lambda}} \\ &\quad - (E_o + pV_o)\, u^{\mu} + (E_o + pV_o) \frac{w^{\mu}}{w_{\lambda}u^{\lambda}} \\ &= \frac{w^{\mu}}{w_{\lambda}u^{\lambda}} E_o. \end{aligned} \tag{95}$$

It is easy to show that with the redefined energy-momentum, we arrive at

$$d\xi^{\mu} = đ\Theta^{\mu} - dW^{\mu}, \tag{96}$$

where

$$đ\Theta^{\mu} = w^{\mu} \frac{đQ_o}{u^{\lambda}w_{\lambda}} \quad and \quad dW^{\mu} = w^{\mu} \frac{pdV_o}{u^{\lambda}w_{\lambda}}. \tag{97}$$

Eq. (96) represents the first law of Thermodynamics but in a covariant form. However, a rapid inspection shows that it represents just the first law multiplied by the factor $w^{\mu}/u^{\lambda}w_{\lambda}$. However, when the entropy is written in this formalism, the theory shows its applicability.

B- To obtain the second law, it is just necessary to define the entropy and show that it represents an invariant. Let us begin by defining the 4−vector temperature and inverse 4−vector temperature as [11]

$$T^{\mu} = w^{\mu} \frac{T_o}{u^{\lambda}w_{\lambda}} \quad and \quad \beta_{\mu} = \frac{u_{\mu}}{kT_0}. \tag{98}$$

The entropy may be written as

$$\begin{aligned} dS &= \beta_\mu \frac{V_o w_\nu}{u^\lambda w_\lambda} dT^{\mu\nu} + \beta_\mu p dV^\mu \\ &= \frac{u_\mu}{kT_0} \frac{V_o w_\nu}{u^\lambda w_\lambda} dT^{\mu\nu} + \frac{u_\mu}{kT_0} p \frac{w^\mu}{w_\lambda u^\lambda} dV_o \\ &= \frac{u_\mu}{kT_0} \frac{V_o w_\nu}{u^\lambda w_\lambda} d\left[(\rho + p)\, u^\mu u^v - p n^{\mu\nu}\right] + \frac{p dV_o}{kT_0} \\ &= \frac{1}{kT_o} dE_o + \frac{p}{kT_o} dV_o. \end{aligned} \tag{99}$$

Consequently, we arrive to an invariant definition of the entropy which is compatible with the formalism used for the first law of Thermodynamics. The third law of thermodynamics follows immediately from Eq. (99). This theory has been used to analyze the black body radiation from a moving frame [27], [28], to solve the Thermodynamics of the mixture of two black bodies [30] including the conservation of the heat in any reference frame, to propose a relativistic thermometer [30] [31], to obtain a Statistical Relativistic Mechanics [32], to deduce the Jüttner distribution [32] and finally to show the invariance of the equations of states [33].

6. Conclusion

The following results have been obtained:

1- An alternative and simple proof of the no interaction theorem.

2- The inclusion of the self-force gives a conservation of the total energy-momentum but the divergences still exist.

3- Using the concept of instantaneity, a 4−vector total energy-momentum can be defined for a thermodynamic system (continuous) or a discrete system.

4- The perfect discrete energy-momentum tensor $T^{\alpha\beta}$ does not satisfy the conservation theorem. Although it is different from the energy-momentum tensor for a perfect fluid $(\rho + p)\, u^\alpha u^\beta - n^{\alpha\beta} p$, a 4−vector total energy-momentum can be defined using an specific instantaneous system for both infinite volume in the case of the discrete free particles system and finite volume for the perfect fluid. The results are similar.

5- A Relativistic Redefined Thermodynamics is derived.

Some applications for Astrophysical Jets and for TOKAMAKS will be showed in future works. Moreover, in the last years, Gill and Zachary have developed the theory of proper time [56] and by using their formalism Ares de Parga et al [57] obtained an equation of motion which includes an special reaction force for each particle of a system of charged particles. The interaction consists of an action at a distance with retarded potentials which gives a constant total momentum. Therefore, a rest frame can be defined and the above theory for a system of interacting charged particles may be used to calculate the 4−vector total energy-momentum for each frame where instantaneity is considered.

Acknowledgments
This work was partially supported by COFAA, EDI SIP - IPN and SNI CONACYT.

7. References

[1] Einstein A 1907 Über das Relativitätsprinzip und die aus demselben gezogenen Folgerungen. *Jahrbuch der Radioaktivität und Elektronik* **4**, 411–62.
[2] Planck M 10908 Zur Dynamik bewegter Systeme. *Annalen der Physik* **331** 1–34.
[3] Liu C 1992 Einstein and relativistic thermodynamics in 1952: A historical and critical study of a strange episode in the history of modern physics. *Br. J. Hist. Sci.* **25**, 185–206.
[4] Ott H 1963 Lorentz-Transformtion der Warme und der Temperatur. *Zeitschrift für Physik A* **175**, 70–104.
[5] Arzeliès H 1965 Transformation Relativiste de la Temperature et de quelques autres grandeurs Thermodynamiques. *Il Nuovo Cimento*, **35**, 792–804.
[6] Landsberg P T 1966 Does a Moving Body Appear Cool? *Nature* **212**, 571–72.
[7] Staruszkiewicz A 1966 Relativistic transformation laws for thermodynamical variables with applications to classical electron theory. *Il Nuovo Cimento* A **45**, 684–88.
[8] Rohrlich F 1966 True and apparent transformations, classical electrons, and relativistic thermodynamics. *Il Nuovo Cimento* B **45**, 76–83.
[9] Noerdlinger P D 1967 A Moving Body must "appear" cool. *Nature* **213**, 1117–18.
[10] Balescu R 1968 Relativistic Statistical Thermodynamics. *Physica* **40**, 309–38.
[11] Van Kampen N G 1968 Relativistic Thermodynamics of Moving Systems. *Phys. Rev.* **173**, 295–301.
[12] Möller C 1972 *The Theory of Relativity; International Series of Monographs on Physics* Clarendon Press, (Oxford, UK) Chapter 7.
[13] Israel W 1976 Nonstationary irreversible thermodynamics: A causal relativistic theory. *Ann. Phys.* **100**, 310–331.
[14] Israel W 1981 Thermodynamics of relativistic systems. *Physica* A **106**, 204–214.
[15] Kibble T W B 1966 Relativistic Transformation Laws for Thermodynamic Variables. *Il Nuovo Cimento* B **41**, 72–78.
[16] Arzeliès H 1966 Comment on Dr. Kibble's Article. *Il Nuovo Cimento* B **41**, 81–82.
[17] Gamba A 1966 Remarks to the Preceding Letter by Kibble. *Il Nuovo Cimento* B **41**, 79–80.
[18] Sieniutycz S 1998 Thermodynamic and relativistic aspects of thermal inertia in fluids. *Phys. Rev.* E **58** , 7027–39.
[19] Landsberg P T, Matsas G E A 1996 Laying the ghost of the relativistic temperature transformation. *Phys. Lett.* A **223**, 401–462.
[20] Landsberg P T, Matsas G E A 2004 The impossibility of a universal relativistic temperature transformation. *Physica* A **340**, 92–94.
[21] Requardt M 2008 Thermodynamics meets Special Relativity–or what is real in Physics? arXiv:0801.2639.
[22] Przanowski M, Tosiek J 2011 Notes on thermodynamics in special relativity. *Physica Scripta* **84**.
[23] Segre E, Fermi E 1962 *Note e Memorie-Collected Papers* University of Chicago Press (Chicago IL, USA) Volume I.
[24] Gamba A 1967 Physical Quantities in Different Reference Systems According to Relativity. *Am. J. Phys.* **35**, 83–89.
[25] Nakamura T K 2006 Relativistic Energy-Momentum of a Body with a Finite Volume. *Space Sci. Rev.* **122**, 271–78.
[26] Nakamura T K Three Views of a Secret in Relativistic Thermodynamics. *Prog. Theor. Phys.* **128**, 463–75.
[27] Ares de Parga G, López-Carrera B, Angulo-Brown F 2005 A proposal for relativistic transformations in thermodynamics. *J. Phys.* A **38**, 2821–34.
[28] López-Carrera B and Ares de Parga G 2008 Relativistic transformation of the canonical distribution function in relativistic statistical mechanics. *Physica* A **387** 1099–1109.
[29] López-Carrera B, Granados V and Ares de Parga G 2008 An alternative deduction of relativistic transformations in thermodynamics. *Rev. Mex. Fís.* 54 **1** 15–19.
[30] Ares de Parga G, López-Carrera B 2009 Redefined relativistic thermodynamics based on the Nakamura formalism. *Physica* A **388**, 4345–56.
[31] López-Carrera B, Rosales M A and Ares de Parga G 2010 The 2.7 K blackbody radiation background reference frame. *Chin. Phys.* B **19** 4.
[32] Ares de Parga G, López-Carrera B 2011 Relativistic Statistical Mechanics vs. Relativistic Thermodynamics. *Entropy* **13**, 1664–93.
[33] Ares de Parga G, Ávalos V A and López-Carrera B 2013 On a Self-Consistency Thermodynamical Criterion for Equations of the State of Gases in Relativistic Frames. *Entropy* **15** 1271-88
[34] Ares de Parga A M, Angulo-Brown F and Ares de Parga G 2015 Deduction of Lorentz Transformations from Classical Thermodynamics. *Entropy* **17**, 197-213.
[35] Weinberg S 1972 *Gravitation and Cosmology, Principles and Applications of the General Theory of Relativity* Wiley (Hoboken NJ USA) p. 48.

[36] Currie D G , Jordan T F and Sudarshan E C G 1963 Relativistic invariance and hamiltonian theories of interacting particles. *Rev. Mod. Phys.* **35** 2 , 350–375.
[37] Pryce M H L 1948 The Mass-centre in the Restricted Theory of Relativity and its Connexion with Quantum Theory of Elementary Particles. *Proc. Roy. Soc. London* A **195**, 62.
[38] Dirac P M 1938 Classical Theory of Radiating Electrons *Proc. Roy. Soc. London* **167** , 148.
[39] DeWitt B S and Brehme R W 1960 Radiation Damping in a Gravitational Field. *Annals of Physics* **9**, 220-259.
[40] Hobbs J M 1968 A Vierbein Formalism of Radiation Damping. *Annals of Physics* **47**, 141-165.
[41] Detweiler S and Whiting B F 2003 Self-force via a Green's function decomposition. *Phys. Rev.* D **67**, 024025.
[42] Poisson E, Pound A and Vega I 2011 The Motion of Point Particles in Curved Space-time. *Living Rev. Relativity* **14** 7.
[43] Wheeler J A and Feynman R P 1945 Interaction with the Absorber as the Mechanism of Radiation *Revs. Modern Phys.* **17** 157.
[44] Wheeler J A and Feynman R P 1949 Classical Electrodynamics in Terms of Direct Interparticle Action *Revs. Modern Phys.* **21** 425.
[45] Eliezer C J 1948 On the Classical Theory of Particles *Proc. Roy. Lond.* A **194** , 1039, 543.
[46] Landau L D and Lifshitz E M 1962 *The Classical Theory of Fields* 2nd ed (Pergamon London) §76.
[47] Di Piazza A, Müller C, K. Z. Hatsagortsyan K Z, Keitel C. H. 2012 Extremely High-intensity Laser Interactions with Fundamental Quantum Systems *Rev. Mod. Phys.* **84** 1177.
[48] Kravets Y, Noble A, Jaroszynski D 2013 Radiation reaction effects on the interaction of an electron with an intense laser pulse *Phys. Rev.* E **88** 011201(R).
[49] Di Piazza A 2008 Radiation Reaction Effects on the Interaction of an Electron with an Intense Laser Pulse *Lett. Math. Phys.* **83** 305.
[50] García-Camacho J, Salinas E, Avalos-Vargas A and Ares de Parga G 2015 Mathematical Differences and Physical Similarities Between Eliezer-Ford-O'Connell Equation and Landau-Lifshitz Equation *Rev. Fís. Mex.* **61** 363–371.
[51] Shen C S 1972 Magnetic Bremsstrahlung in an Intense Magnetic Field *Phys. Rev.* D **6** 10-15, 2736.
[52] Callen H, Horwitz G 1971 Relativistic Thermodynamics *A. J. P.* **39** 938–947.
[53] Tolman R C 1938 *The Principal of Statistical Mechanics* (Oxford University Press)
[54] Landau L D and Lifshitz E M 1962 *The Classical Theory of Fields* 2nd ed. (Pergamon: London) §35
[55] Ares de Parga G, Avalos-Vargas A and López-Carrera B 2013 On a Self-Consistency Thermodynamical Criterion for Equations of the State of Gases in Relativistic Frames *Entropy* **15** 1271.
[56] Gill T L and Zachary W W 2011 Two Mathematically Equivalent Versions of Maxwell's Equations *Found. Phys.* **41** 1 99.
[57] Ares de Parga G , Gill T L and Zachary W W 2013 The Thomas Program and the Canonical Proper-time Theory *Journal of Computational Methods in Sciences and Engineering* **13** 117.

Introducing surface tension to spacetime

H A Perko[1]
[1]Koppa Research, Office 11, 140 E. 4th Street, Loveland, CO, USA 80537

E-mail: hperko@kopparesearch.com

Abstract. Concepts from physical chemistry of surfaces and surface tension are applied to spacetime. More specifically, spacetime is modeled as a spatial fluid continuum bound together by a multi-dimensional membrane of time. A metric tensor that relates empty flat spacetime to energetic curved spacetime is found. Equations of motion for an infinitesimal unit of spacetime are derived. The equation of motion in a time-like direction is a Klein-Gordon type equation. The equations of motion in space-like directions take the form of Schrodinger's equation where Plank's constant is related to membrane elastic modulus. Although much work remains, it is suggested that the spacetime surface tension may serve as a mechanical model for many phenomena in quantum mechanics and atomic particle physics.

1. Introduction

The goal of this study is to start from empty spacetime, and theoretically attempt to construct matter. Empty spacetime itself must be the most basic building material from which all things are constructed. Yet, how does one construct something from nothing? The proposed solution is to introduce concepts from physical chemistry and surface tension to space and time. Spacetime is modeled as a continuum of space held together by a multi-dimensional membrane of time. From this simple mathematical construct, a veritable universe of particles and waves can be manufactured from nothing except spacetime curvature.

Ordinary continuum mechanics describes well the behavior of bulk fluids. However, understanding the behavior of fluid interfaces and small-scale interactions requires concepts from physical chemistry and surface tension. Examples of ordinary small-scale interactions governed by surface tension include the quantized size of water droplets, corpuscular formation, wavelets, adhesion, meniscus formation, and capillarity. By analogy, it is suggested here that including surface tension in the treatment of spacetime may provide new ways of relating general relativity, quantum mechanics, and atomic particle physics.

2. Membrane geometry

Consider an arbitrary, non-relativistic distribution of matter contained within a small volume of spacetime. If time and position of each point mass are plotted relative to a set of nearby clocks and coordinates in space, the resulting mass distribution would align in time and take the form of a fabric or membrane. An example of a matter distribution relative to time and two coordinates of space is shown in Figure 1. The distribution of non-relativistic matter is dispersed over the two-dimensions of space and concentrated in a thin increment of time. The matter distribution forms an interface in time. The same plot can be constructed relative to any two coordinates of space.

$$x_0 = ict \qquad dx_0 = ic\,dt$$

$$t = \tau\, {g_{00}}^{1/2} \qquad dt = d\tau\, {g_{00}}^{1/2}$$

$$u_0 = \frac{\partial x_0}{\partial \tau} = ic\frac{\partial t}{\partial \tau}$$

$$u_i = \frac{\partial x_i}{\partial \tau}$$

$$\begin{bmatrix} g_{00} & 0 & 0 & 0 \\ 0 & g_{ij} & g_{ij} & g_{ij} \\ 0 & g_{ij} & g_{ij} & g_{ij} \\ 0 & g_{ij} & g_{ij} & g_{ij} \end{bmatrix}$$

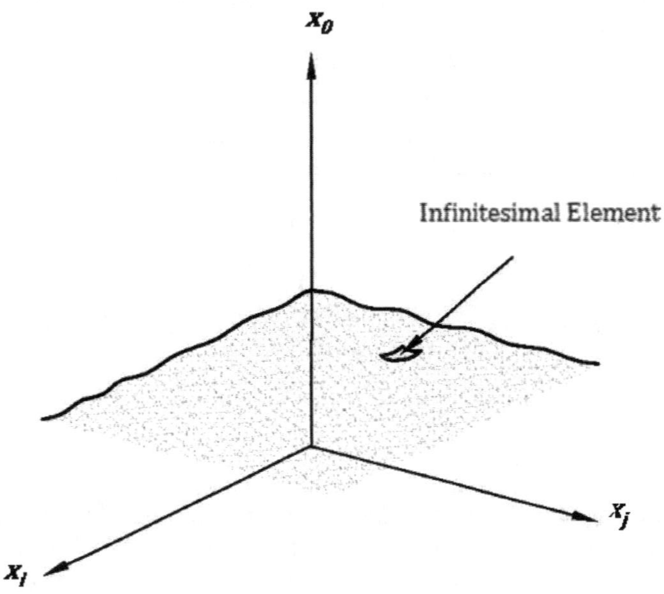

Figure 1. Geometry of an arbitrary matter distribution in a small volume of spacetime displaying membrane-like properties

Consider the same matter distribution from a relativistic perspective. At the start of most physics problems/experiments, clocks are typically set to zero. At the instant clocks are set to zero, the distribution of an arbitrary volume of relativistic matter also would appear as a fabric or membrane. Each point in the fabric would correspond to a particle of mass with space-like, time-like, or light-like trajectories. Constant resetting clocks to zero can be achieved by a reference frame moving along a time-like hypersurface such that all points remain separated by space-like intervals. Such a reference frame is shown by the metric in Figure 1 and is referred to as a *synchronous* frame. The synchronous frame can be found by appropriate transformations for any distribution of matter [3]. In the synchronous frame, the time component of the metric, g_{00}, approaches unity, $g_{00}=1$. Surface tension allows for small perturbations in g_{00}.

The interface of matter in time for an observer can be treated as a boundary layer to which physical chemistry of surfaces and membrane mechanics can be applied. In four dimensions, the distribution of matter is more difficult to visualize. Matter is dispersed in the continuum of space and held together by a multi-dimensional membrane at the interface with time.

3. Surface tension

Each dot along the planar fabric in Figure 1 depicts a point mass. Due to equivalence of matter and energy, each point in Figure 1 also is a concentration of energy. Let the Greek symbol Koppa[1], *Ϙ*, represent surface energy per unit area. A fundamental theorem of physical chemistry of surfaces is that surface energy of an interface is mathematically equivalent to surface tension. A crude

[1] In the study of physical chemistry, the Greek letter gamma, *γ*, is typically used to represent surface tension. To avoid confusion with the Lorentz factor in theoretical physics assignment of a new symbol was necessary. The forgotten Greek letter Koppa, *Ϙ*, was chosen. Before being adopted and then omitted from the modern Greek alphabet, the letter Koppa, *Ϙ*, had its origins in ancient Hebrew. The ancestor Koppa symbol was drawn as the sun on the horizon and it signified light and the edge of time. Thus, Koppa seemed a fitting parameter to describe tension at the boundary of time. In the Roman alphabet, Koppa corresponds to the letter *Q*.

mathematical derivation showing the equivalence of surface energy and surface tension is shown in Table 1 below.

Table 1. Classical and Multi-Dimensional Surface Tension

3D Space	4D Spacetime
$Surface\ Energy = ϙ = \dfrac{Energy}{Area}$	$Surface\ Energy = ϙ = \dfrac{Energy}{Volume}$
$Surface\ Tension = \dfrac{Force}{Width} = \dfrac{Energy\ /L}{Area\ /L} = ϙ$	$Surface\ Tension = \dfrac{Force}{Area} = \dfrac{Energy\ /L}{Volume\ /L} = ϙ$

where L is a space-like increment of length, $ϙ$ is free surface energy per unit area (3D) or per unit volume (4D), and $ϙ$ is also surface tension. Surface energy is equivalent to surface tension. In ordinary three dimensional membrane problems, surface energy has units of J/m^2, which is equivalent to surface tension with units of N/m. For a multi-dimensional membrane in spacetime, surface energy has units of J/m^3, which is equivalent to surface tension with units of N/m^2.

Stable existence of an interface requires that the surface energy be positive; were it negative or zero, fluctuations would lead to continuous expansion of the interface and eventual complete dispersion [1]. Surface tension forms a contractile skin which gives rise to membrane-like behavior. It is important to note that a physical membrane is not required in spacetime, rather matter/energy need only be organized in a surface for an observer. Spacetime surface tension may be the fundamental force of nature, which acts to keep matter and energy together within the construct of general relativity. Analogous to general relativity [2], surface tension tells spacetime how to curve, and spacetime curvature tells surface tension (energy/matter) how to flow.

4. Fluidity

Spacetime is generally assumed to exhibit fluid-like properties. For an ideal fluid, shear stresses are zero.

$$ϙ_{ij} = 0 \qquad for\ i \neq j$$

It follows from traditional mechanics that if shear stress is zero everywhere, then surface tension must be uniform in all directions. The Cauchy stress tensor for spacetime surface tension is given by,

$$ϙ_{ij} = ϙ\delta_{ij} \qquad (1)$$

where δ_{ij} is the Kronecker delta symbol.

5. Equations of motion

The equations of motion are derived here using figures drawn with time and two coordinates of space to aid in visualization. Equations are constructed in manner to include the hidden third dimension of space for full four-dimensional treatment of spacetime.

An infinitesimal element of spacetime membrane is taken from Figure 1 and represented at larger scale in Figure 2. In order to account for potential curvature of spacetime, the element is modeled as a curved manifold such that stresses are not co-linear. Surface tension is assigned to four sides of the element.

IOP Conf. Series: Journal of Physics: Conf. Series **845** (2017) 012003 doi:10.1088/1742-6596/845/1/012003

It should be noted that the infinitesimal element of spacetime can be moving through space and time at some arbitrary four velocity, u_0 and u_i. All lengths and forces in the equations of motion are taken in the material reference frame moving with the *membrane* itself. To understand that which would be experienced for an observer in a spatial/inertial reference frame, Lorentz transformations apply.

Dynamic equations of motion are derived by summing surface tension acting on the infinitesimal element in various directions and setting the result equal to the change of momentum with proper time, τ. Forces acting in a spatial direction are summed as follows,

$$\begin{aligned}\Sigma F_{x_i} &= -\text{♀}\, dx_j\, dx_k \cos\alpha_i + \text{♀}\, dx_j\, dx_k \cos\alpha_i + \frac{\partial \text{♀}}{\partial x_i} dx_i\, dx_j\, dx_k \cos\alpha_i \\ &= \frac{1}{c^2}\frac{\partial}{\partial \tau}(\text{♀}\, u_i)\, dx_i\, dx_j\, dx_k\end{aligned}$$

where α_i is the small angle between planar spacetime and perturbed spacetime, and c is the speed of light. In the foregoing equation, traditional mass has been replaced with surface energy using the relation for energy per volume, $\text{♀}\, dx_i\, dx_j\, dx_k = mc^2$.

Recognizing that for slight curvatures, the cosine of a small angle is approximately equal to one, $\cos\alpha_i \approx 1$, reduces the equation of motion in the spatial direction to,

$$\frac{\partial \text{♀}}{\partial x_i} = \frac{\text{♀}}{c^2}\frac{\partial u_i}{\partial t} + \frac{u_i}{c^2}\frac{\partial \text{♀}}{\partial \tau}$$

The last term in the equation above is the change in surface tension with time. In physical chemistry of surfaces, change in surface tension with time is associated with gradual fluctuations in temperature and is typically assumed to be negligible for short duration and small scale events. The change in surface energy over cosmic scale events is likely of importance and may be the subject of future studies. Omission of the time rate of change of surface energy further reduces the equation of spatial motion of spacetime to,

$$\frac{\partial \text{♀}}{\partial x_i} = \frac{\text{♀}}{c^2}\frac{\partial u_i}{\partial \tau} \qquad (2)$$

The dynamic equation of motion through time has a different appearance. Summing forces in the time direction yields,

$$\begin{aligned}\Sigma F_{x_0} &= 2\text{♀} dx_j dx_k \sin\alpha_i + \frac{\partial \text{♀}}{\partial x_i} dx_i dx_j dx_k \sin\alpha_i + 2\text{♀} dx_i dx_k \sin\alpha_j + \frac{\partial \text{♀}}{\partial x_j} dx_i dx_j dx_k \sin\alpha_j \\ &\quad + 2\text{♀} dx_i dx_j \sin\alpha_k + \frac{\partial \text{♀}}{\partial x_k} dx_i dx_j dx_k \sin\alpha_k - P\, dx_i dx_j dx_k \\ &= \frac{1}{c^2}\frac{\partial}{\partial \tau}(\text{♀}\, u_0)\, dx_i\, dx_j\, dx_k\end{aligned}$$

where P is an applied external volumetric force acting on the fabric of spacetime. If both sides of the equation are divided by differential volume, $dx_i\, dx_j\, dx_k$, and the change in ♀ over an infinitesimal distance and time is assumed small with respect to ♀, then the equation of motion in the temporal direction reduces to,

$$2\text{♀}\frac{1}{dx_i}\sin\alpha_i + 2\text{♀}\frac{1}{dx_j}\sin\alpha_j + 2\text{♀}\frac{1}{dx_k}\sin\alpha_k - P = \frac{\text{♀}}{c^2}\frac{\partial u_0}{\partial \tau}$$

For small angles of curvature, basic trigonometry provides that $\sin\alpha_i \approx dx_i/2r_i$ where r_i is the radius of curvature of the surface in the i'th coordinate. For a thin membrane, the following relation holds,

$$\frac{1}{r_i} = \frac{\partial^2 x_0}{\partial {x_i}^2}$$

By substituting this relation into the dynamic equation of motion in the temporal direction and noting that u_0 is $\partial x_0/\partial\tau$, one obtains,

$$2\text{♀}\frac{dx_i}{2dx_i}\frac{\partial^2 x_0}{\partial {x_i}^2} + 2\text{♀}\frac{dx_j}{2dx_j}\frac{\partial^2 x_0}{\partial {x_j}^2} + 2\text{♀}\frac{dx_k}{2dx_k}\frac{\partial^2 x_0}{\partial {x_k}^2} - P = \frac{\text{♀}}{c^2}\frac{\partial^2 x_0}{\partial \tau^2}$$

which reduces to the following expression,

$$\frac{\partial^2 x_0}{\partial {x_i}^2} + \frac{\partial^2 x_0}{\partial {x_j}^2} + \frac{\partial^2 x_0}{\partial {x_k}^2} - \frac{P}{\text{♀}} = \frac{1}{c^2}\frac{\partial^2 x_0}{\partial \tau^2} \quad (3)$$

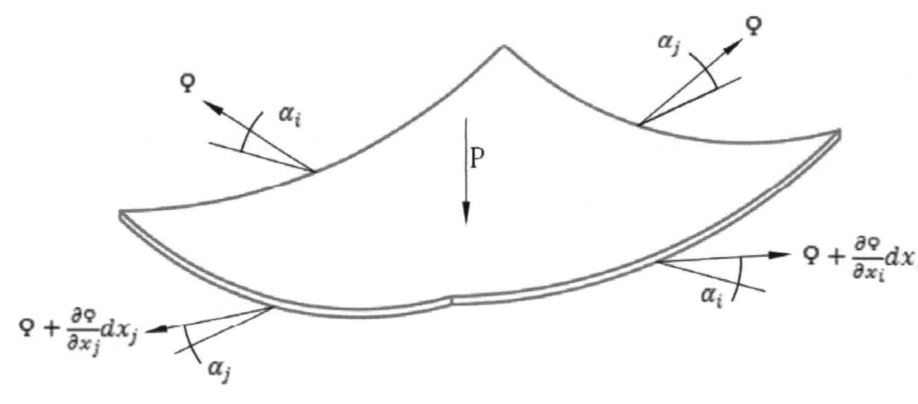

Figure 2. Surface tension about an infinitesimal element of spacetime membrane

6. Strain-displacement

Displacement and strain of an infinitesimal membrane of spacetime is shown in Figure 3. Here again, the infinitesimal element of spacetime can be moving through space and time at some arbitrary four velocity, u_0 and u_i. Displacement in time and space is shown as an upward and right hand translation in the figure. Excellent derivations of strain-displacement equations for membranes can be found in common engineering texts [4, 5, 6, 7]. However, the differential unit of time in these derivations is assumed to be unchanging, set equal to 1, and dropped from equations without mention. Equations for strain-displacement are re-derived here in order to better account for changes in perceived time.

The strain-displacement tensor is the Lie derivative of the metric on space and represents the rate of geometric change of a Riemannian manifold [8]. In this derivation, it represents the dynamical change in the metric as observed from a synchronous frame. There are several definitions of strain-displacement. Here the Green-Lagrange strain, ϵ, is used, as defined by [4],

$$\epsilon\, d\tau = \frac{ds^2 - {ds_o}^2}{2{ds_o}^2} \quad (4)$$

where ds_o is an infinitesimal space-like increment at some time τ and ds is the same space-like increment at time $\tau+d\tau$. The differential increment of proper time $d\tau$ has been added to the definition of Green-Lagrange strain to be rigorous.

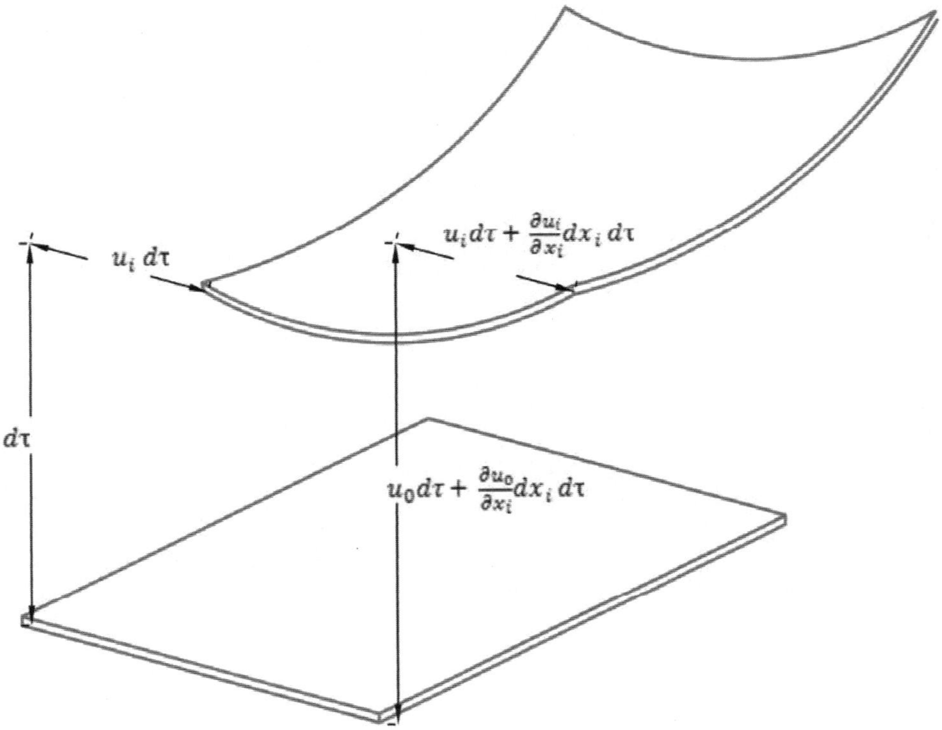

Figure 3. Displacement and strain of an infinitesimal element of spacetime membrane

Referring to Figure 3, the initial length of the infinitesimal fabric of spacetime may be taken as,

$$ds_o{}^2 = dx_i{}^2$$

The length of the element at time $+d\tau$ may be found using the Pythagorean theorem since an imaginary time coordinate is being used in a Gaussian normal (i.e. material) reference frame.

$$ds^2 = \left(dx_i + \frac{\partial u_i}{\partial x_i} dx_i d\tau\right)^2 + \left(\frac{\partial u_0}{\partial x_i} dx_i d\tau\right)^2$$

Expanding the above equation yields,

$$ds^2 = dx_i{}^2 + 2\frac{\partial u_i}{\partial x_i} dx_i{}^2 d\tau + \left(\frac{\partial u_i}{\partial x_i}\right)^2 dx_i{}^2 d\tau^2 + \left(\frac{\partial u_0}{\partial x_i}\right)^2 dx_i{}^2 d\tau^2$$

In membrane mechanics, in-plane strains, $\partial u_i/\partial x_i$ are typically very much smaller than the out-of-plane membrane displacements, so the second order term, $(\partial u_i/\partial x_i)^2$, is omitted leaving,

$$ds^2 = dx_i{}^2 + 2\frac{\partial u_i}{\partial x_i} dx_i{}^2 d\tau + \left(\frac{\partial u_0}{\partial x_i}\right)^2 dx_i{}^2 d\tau^2$$

Substitution of the above equations for ds_o^2 and ds^2 into (4) for the i'th component in strain results in,

$$\epsilon_{ii}\, d\tau = \frac{2\frac{\partial u_i}{\partial x_i} dx_i^{\,2} d\tau + \left(\frac{\partial u_0}{\partial x_i}\right)^2 dx_i^{\,2} d\tau^2}{2\, dx_i^{\,2}}$$

which reduces to,

$$\epsilon_{ii} = \frac{\partial u_i}{\partial x_i} + \frac{1}{2}\left(\frac{\partial u_0}{\partial x_i}\right)^2 d\tau \qquad (5)$$

A similar derivation can be performed to include shear strain-displacements, with the following result obtained [4]. Equation (6) represents a more complete form of the strain tensor for any i and j,

$$\epsilon_{ij} = \frac{1}{2}\left(\frac{\partial u_i}{\partial x_j} + \frac{\partial u_j}{\partial x_i}\right) + \frac{1}{2}\left|\left(\frac{\partial u_0}{\partial x_i}\right)\left(\frac{\partial u_0}{\partial x_j}\right)\right| d\tau \qquad (6)$$

In the principle direction of strain, traction and transverse strains are null.

$$0 = \epsilon_{00} = \epsilon_{i0} = \epsilon_{0j} \qquad (7)$$

As an aside, the metric tensor for space-like separated particles in the synchronous frame for the surface tension model can be found by rearrangement of (4),

$$\left(1 + 2\epsilon_{ij} d\tau\right) ds_o^{\,2} = ds^2$$

Note that the difference in the squares of space-like increment is invariant; the same principle strains can be obtained starting in either an inertial or material reference frame. The strain tensor is covariant, symmetric, and torsion free. The quantity enclosed by parenthesis is somewhat analogous to the Schwarzschild metric except for the positive sign. It is noted that the sign is a convention. Had this derivation been conducted such that surface tension was positive (compression), the sign in the above equation would change.

7. Constitutive relations

If spacetime is homogeneous, isotropic, and linearly elastic, then the constitutive relations between stress and strain are represented by Hooke's laws, given by [5]

$$♀_{ii} = \frac{E\Delta x_0}{1 - \nu^2}\left(\epsilon_{ii} + \nu\epsilon_{jj}\right)$$

where ν is Poisson's Ratio, E is Young's Modulus, and Δx_0 is membrane thickness. In the mechanics of thin shells, Young's Modulus is multiplied by membrane thickness, Δx_0, because surface tension acts over the membrane thickness. According to special relativity, length contraction occurs only in the direction of motion and is not associated with a complimentary dilation in height. Consequently, Poisson's Ratio should be taken as zero, $\nu = 0$, for spacetime. Hence, the constitutive relations take the form,

$$♀_{ii} = E\, ic\Delta t\, \epsilon_{ii} \qquad (8)$$

The constitutive relations for the off-diagonal shear terms with $\nu = 0$ would appear as [5],

$$♀_{ij} = \frac{E\, ic\Delta t}{2} \epsilon_{ij} \qquad (9)$$

8. Analogous Klein-Gordon

In the proposed model of spacetime, perturbations in the coordinate of time, x_0, given by (3) exhibit wave-like properties consistent with the Klein-Gordon equation of quantum mechanics. Setting external pressure, *P*, equal to zero in (3) and evaluating the result in 1-dimension, yields the familiar wave equation given by,

$$\frac{\partial^2 x_0}{\partial {x_i}^2} = \frac{1}{c^2}\frac{\partial^2 x_0}{\partial \tau^2} \qquad (10)$$

In (10), waves in the membrane of spacetime propagate with a constant speed of light, *c*, independent of wavelength and frequency. Since the wave equation was derived for a membrane rather than bulk fluid, the resulting waves travel along hypersurfaces and would exhibit properties of polarization. Electromagnetic radiation behaves in exactly this way.

The equation of motion in the time direction given by (3) has another immediately recognizable solution. The solution is the familiar equation of Young and Laplace which is found when temporal acceleration is zero, $\partial^2 x_0/\partial\tau^2 = 0$, given by,

$$\frac{\partial^2 x_0}{\partial {x_i}^2} + \frac{\partial^2 x_0}{\partial {x_j}^2} + \frac{\partial^2 x_0}{\partial {x_k}^2} = \frac{P}{♀} \qquad (11)$$

where *P* is the applied external volumetric force perhaps best described as encapsulated mass/energy. The Young and Laplace solution is the equation for a corpuscle, which at its lowest energy state is a sphere. The Gauss-Bonnet theorem states that the integral of Gaussian curvature over the surface area of a corpuscle has distinct quantized solutions; this theorem provides for torus and other higher energy corpuscular geometries. Equation (11) suggests three "polarizations" of corpuscular geometry can exist in spacetime; each set of solutions is found by omitting one of the spatial curvature terms.

9. Analogous Schrödinger

Using the constitutive relationship (8), substituting strain (5) for surface tension in the equation of motion (2) yields,

$$E\, ic\Delta t \left(\frac{\partial^2 u_i}{\partial {x_i}^2} + \frac{\partial u_0}{\partial x_i}\left(\frac{\partial^2 u_0}{\partial {x_i}^2}\right)^2 d\tau + \frac{1}{2}\left(\frac{\partial u_0}{\partial x_i}\right)^2 \frac{\partial}{\partial x_i}(d\tau) \right) = \mu \frac{\partial u_i}{\partial \tau}$$

where $♀/c^2$ has been replaced with mass density per unit volume, μ. Here, the thickness of the fabric of spacetime, Δt, is assumed constant. It is proposed that the second higher order term on the left side of the equation can be assumed small compared to the other terms. The third term on the left side of the equation has to do with the curvature of spacetime. For flat spacetime, the third term can be omitted. The above equation can be re-arranged to the familiar dispersion equation of quantum mechanics,

$$E\, ic\Delta t \frac{\partial^2 u_i}{\partial {x_i}^2} = \mu \frac{\partial u_i}{\partial \tau}$$

Young's Modulus and the thickness of the fabric of spacetime are unknown constants. If $Ec\Delta t$ is set equal to $\hbar/2$, it follows,

$$-\frac{\hbar}{2\mu}\frac{\partial^2 u_i}{\partial {x_i}^2} = i\frac{\partial u_i}{\partial \tau} \qquad (12)$$

which is analogous to the non-relativistic, one-dimensional Schrödinger equation. The spatial component of the four velocity replaces the probability wave equation in the Schrödinger equation. Surface tension (matter/energy) within the manifold of spacetime and the corresponding strain (Lie derivative) over space, which is the time rate of change of the manifold to varying tension, move in accordance with a Schrödinger-like equation.

10. Analogous Heisenberg

In the foregoing derivation of the analogous Schrödinger equation, the constants $Ec\Delta t$ were set equal to $\hbar/2$. If this assignment of constants is reinserted into the constitutive relationship (8), the following result is obtained,

$$\text{♀}_{ii} = i\,\frac{\hbar}{2}\,\epsilon_{ii}$$

It is worth noting that the units of surface tension, ♀, are energy per infinitesimal volume and the units of dynamical strain are the inverse of proper imaginary time. This equation holds for a change in surface tension and a change in strain, such that one can write,

$$\Delta Energy\,\Delta\tau = \frac{\hbar}{2} \qquad (13)$$

which has some similarity to the Heisenberg uncertainty principle where the equal sign would be replaced with a greater than or equal sign. Aspects of the uncertainty relation are embedded within Hooke's law for the mechanical medium of spacetime.

11. Analogous Weyl

It follows from the fluidity of spacetime (1), the constitutive relationship (9), and the shear strain (6) that,

$$\frac{\partial u_i}{\partial x_j} + \frac{\partial u_j}{\partial x_i} + \left|\left(\frac{\partial u_0}{\partial x_i}\right)\left(\frac{\partial u_0}{\partial x_j}\right)\right|\, d\tau = 0 \qquad (14)$$

This equation governs the free rotation of geometries manufactured of spacetime which classically represents an eddy and bears some resemblance to the Weyl equation. Rotation itself does not cause stress or strain (change in mass/stored energy) in the membrane of spacetime but does cause a deformation of time thereby altering geodesics. Free rotation may lead to other particle-like geometric solutions in spacetime such as vortices. There are three possible particle-like vortex solutions to (14): unstable, short lived rotational vortices, stable irrotational vortices with field lines ending at a corpuscle, and a stable pair of oppositely rotating irrotational vortices entangled in time.

12. Discussion

The main-stream view of quantum mechanics is based on the Copenhagen interpretation where wave functions have no physical analog, rather quantum mechanical waves are purely mathematical probabilities. The present surface tension model is not the first to suggest a physical interpretation of quantum wave equations. The Copenhagen interpretation edged out a deterministic geometric theory originally proposed by de Broglie and further revised by Bohm, known as pilot wave theory. Pilot wave theory explained that point-like particles moving along hidden fluid-like waves can provide a foundation for quantum mechanics. De Broglie could not explain the fluid-like waves which Einstein called "ghost waves". Still, physicists striving to better understand quantum mechanics continue to look at pilot wave theory even today [9]. The present study proposes the fluid medium suggested by de Broglie may be the metric of spacetime itself under the influence of surface tension.

The present surface tension model is not unique regarding the importance of membrane geometries in attempting a physical explanation for quantum mechanics. In modern string theory, hidden dimensions are conjoined by strings and membranes in a multiverse [10]. The eigenstates of these geometric constructs, specifically including "branes", seem to have the correct constraints to match the spin, mass, and number of particles in the Standard Model. The present work attempts to explain how spacetime might exhibit membrane-like behavior and suggests, as one solution, how string-like entangled micro-vortices are fixed by Dirichlet boundary conditions to the fabric of spacetime. Surface tension may provide a mechanism for how strings, branes, and Dirichlet boundary conditions arise in spacetime. This paper represents only an introduction. Much work remains to compare results with string theory and the Standard Model.

13. Conclusions

From surface tension, the membrane of spacetime can take on many fundamental shapes such as waves, corpuscles, menisci, vortices, agglomerates, and foams. It is suggested that these basic shapes can be arranged in atoms, molecules, and matter. In other words, everything in the universe may be manufactured from the fabric/membrane of spacetime itself. Matter and energy are connected in a virtual membrane at the interface of observed time.

References

[1] Adamson A W 1990 *Physical Chemistry of Surfaces*, 5th Edition (New York: Wiley Interscience)
[2] Wheeler J A 1990 *A Journey Into Gravity and Spacetime* (San Francisco: Scientific American Library)
[3] Landau L D and Lifshitz E M 1988 *Classical Theory of Fields*, 7th Edition, Vol. 2 (Moscow: Nauka)
[4] Wierzbicki T 2007 *Plates and Shells* (Massachusetts: 2.081J/16.230J MIT)
[5] Timoshenko S and Woinowsky-Krieger S 1987 *Theory of Plates and Shells*, 2nd Edition (New York: McGraw-Hill)
[6] Ventsel E and Krauthammer T 2001 *Thin Plates and Shells: Theory, Analysis, and Applications* (New York: Marcel Dekker, Inc.)
[7] Boresi A P and Schmidt R J 2003 *Advanced Mechanics of Materials*, 6th Edition (New York: Wiley)
[8] Marsden J E and Hughes J R 1983 *Mathematical Foundations of Elasticity* (New York: Dover Publications, Inc.)
[9] Chen P and Kleinert H 2016 "Deficiencies of Bohm Trajectories in View of Basic Quantum Principles", Electronic Journal of Theoretical Physics, No. 35
[10] Zwiebach B 2009 *A First Course in String Theory*, 2nd Edition (Cambridge: Cambridge University Press)

IOP Conf. Series: Journal of Physics: Conf. Series **845** (2017) 012004 doi:10.1088/1742-6596/845/1/012004

Recent advancements in conformal gravity

James G. O'Brien[1], Spasen S. Chaykov[2], Jeremy Dentico, Modestas Stulge, Brian Stefanski and Robert J. Moss[3]

[1]Department of Sciences, Wentworth Institute of Technology, Boston, MA 02115, USA
[2]Department of Physics, Brandeis University, Waltham, MA 02451, USA
[3]MIT Lincoln Laboratory, 244 Wood St., Lexington, MA 02421, USA

E-mail: obrienj10@wit.edu[1], chaykovs@brandeis.edu[2], robert.moss@ll.mit.edu[3]

Abstract. In recent years, due to the lack of direct observed evidence of cold dark matter, coupled with the shrinking parameter space to search for new dark matter particles, there has been increased interest in Alternative Gravitational theories. This paper, addresses three recent advances in conformal gravity, a fourth order renormalizable metric theory of gravitation originally formulated by Weyl, and later advanced by Mannheim and Kazanas. The first section of the paper applies conformal gravity to the rotation curves of the LITTLE THINGS survey, extending the total number of rotation curves successfully fit by conformal gravity to well over 200 individual data sets without the need for additional dark matter. Further, in this rotation curve study, we show how MOND and conformal gravity compare for each galaxy in the sample. Second, we look at the original Zwicky problem of applying the virial theorem to the Coma cluster in order to get an estimate for the cluster mass. However, instead of using the standard Newtonian potential, here we use the weak field approximation of conformal gravity. We show that in the conformal case we can get a much smaller mass estimate and thus there is no apparent need to include dark matter. We then show that this calculation is in agreement with the observational data from other well studied clusters. Last, we explore the calculation of the deflection of starlight through conformal gravity, as a first step towards applying conformal gravity to gravitaitonal lensing.

1. Introduction

Conformal gravity is a fourth order renormalizable metric theory of gravity, where the Ricci scalar in the Einstein-Hilbert action is replaced by the square of the conformal Weyl tensor as:

$$I_{\rm W} = -\alpha_g \int d^4x(-g)^{1/2} C_{\lambda\mu\nu\kappa}C^{\lambda\mu\nu\kappa} = -2\alpha_g \int d^4x(-g)^{1/2}\left[R_{\mu\kappa}R^{\mu\kappa} - (1/3)(R^{\alpha}{}_{\alpha})^2\right], \quad (1)$$

where

$$C_{\lambda\mu\nu\kappa} = R_{\lambda\mu\nu\kappa} - \frac{1}{2}\left(g_{\lambda\nu}R_{\mu\kappa} - g_{\lambda\kappa}R_{\mu\nu} - g_{\mu\nu}R_{\lambda\kappa} + g_{\mu\kappa}R_{\lambda\nu}\right) + \frac{1}{6}R^{\alpha}{}_{\alpha}\left(g_{\lambda\nu}g_{\mu\kappa} - g_{\lambda\kappa}g_{\mu\nu}\right) \quad (2)$$

is the conformal Weyl tensor and the gravitational coupling constant α_g is dimensionless. This action, when varied with respect to the metric (as in standard gravity) [1], yields the conformal gravity field equations,

$$4\alpha_g W^{\mu\nu} = 4\alpha_g\left[2C^{\mu\lambda\nu\kappa}{}_{;\lambda;\kappa} - C^{\mu\lambda\nu\kappa}R_{\lambda\kappa}\right] = T^{\mu\nu}. \quad (3)$$

These equations were later solved by Mannheim and Kazanas [2] for a spherical mass, exterior solution, which behaves as a conformal gravity Schwarzchild like solution, and allows for application of the theory to various data sets. In recent years, Mannheim and O'Brien have shown that conformal gravity can be used to accurately describe the rotation curves of a diverse set of galaxies without the need to invoke dark matter [3],[4],[5],[6]. Since all attempts to find any direct evidence of dark matter have been unsuccessful so far, alternative theories of gravity have become increasingly worth considering as an explanation for phenomena ascribed to dark matter. Its success in fitting galactic rotations makes conformal gravity an ideal candidate to explore in an effort to theoretically describe these additional effects. While the rotation curve fitting provides direct data that can be tested against the predictions of an alternative model of gravity, there are other observational motivations for the existence of dark matter, such as the motion of galaxies in clusters and gravitational lensing. The goal of this paper is to first report conformal gravity's fits on the rotation curves of a very recent dwarf galaxy survey called the LITTLE THINGS, to show how conformal gravity can address the cluster motion problem without invoking dark matter, and develop the framework for gravitational lensing formalism in conformal gravity.

2. Rotation Curves of The Little Things Survey

In this paper, we review our recent efforts [7] in fitting the LITTLE THINGS survey [8], a survey of 25 dwarf galaxies offering ultra high resolution of modern rotation curve data. Data from the Local Irregular That Trace Luminosity Extremes in the Hi Nearby Galaxy Survey (LITTLE THINGS) was collected using the NRAO Very Large Array (VLA). The survey is a fairly local compilation, where each of the galaxies featured are within 11 Mpc of the Milky Way. The observations of the galaxies were used to create rotation curves and were combined with Spitzer archival 3.6m and ancillary optical U, B, and V images to construct mass distribution models. The paper calls for more analysis on high-resolution low mass dwarf galaxies to further test the theory of ΛCDM. For our purposes, we use the data collected and make mass model predictions for the rotation curves of the LITTLE THINGS Dwarfs using conformal gravity and MOND. Although MOND is out of the scope of this paper, its inclusion in the fits is a new mechanism for comparison of alternative theories, on a per galaxy, same input parameter basis (see Table 1). We adopt a uniform, non-biased method of adopting the estimated distance from the mean of the NASA Extragalactic Database (NED), and all other input parameters were kept consistent with [8]. In Fig. 1, we present the General Relativity, conformal gravity and MOND rotation curve predictions for each data set. The General Relativity prediction is given by [9],

$$v_{gr}(R) = \sqrt{\frac{N^*\beta^*c^2R^2}{2R_0^3}\left[I_0\left(\frac{R}{2R_0}\right)K_0\left(\frac{R}{2R_0}\right) - I_1\left(\frac{R}{2R_0}\right)K_1\left(\frac{R}{2R_0}\right)\right]}, \tag{4}$$

the conformal gravity prediction is given by [4],

$$v_{cg}(R) = \sqrt{v_{gr}^2 + \frac{N^*\gamma^*c^2R^2}{2R_0}I_1\left(\frac{R}{2R_0}\right)K_1\left(\frac{R}{2R_0}\right) + \frac{\gamma_0c^2R}{2} - \kappa c^2R^2}, \tag{5}$$

and the MOND prediction is given by [10],

$$V_{mond}(R) = \sqrt{V_{gr}^2(R) + \left(V_{gr}^2(R)\frac{\sqrt{1+\frac{4a_0R}{V_{gr}^2}} - 1}{2}\right)}. \tag{6}$$

In each of the above formula, N^* is the estimated number of stars in the galaxy (proportional to the mass) and R_0 is the galactic scale length. In eq. (5), $\gamma^* = 5.42 * 10^{-41}\mathrm{cm}^{-1}$,

IOP Conf. Series: Journal of Physics: Conf. Series **845** (2017) 012004 doi:10.1088/1742-6596/845/1/012004

$\gamma_0 = 3.06 * 10^{-30}\text{cm}^{-1}$, and $\kappa = 9.54 * 10^{-54}$ cm^{-2}. In eq. (6), $a_0 = 1.35 \times 10^{-8}$ $cm\,s^{-2}$ and we refer the reader to [4] and [10] for the derivation and determination of such constants. It should be noted that in each of these models, both the luminous disk and gas were fit, such that for a given galaxy the total velocity contributions are given by

$$v_{tot}(R) = \sqrt{v_{cg}(R)^2 + v_{gas}(R)^2}. \tag{7}$$

The individual galaxies' total gas mass is given in Table 1 (where all quantities with the $\odot$ are in solar quantities). No bulges were included in any of these fits due to lack of observational evidence [8].

It can be easily seen in Fig. 1, that both MOND and conformal gravity capture the essence of the data without the need for dark matter. In particular, conformal gravity fits this sample of dwarf galaxies with high precision without any parameter modification. More specific information about the particular morphology of each galaxy can be found in [8], which helps explain some of the more exotic shapes of the rotation curves, but on the whole, conformal gravity generates mass models that return mass to light ratios consistent with physical observations for dwarf galaxies (see Table 1) . Further, in the last column of Table 1, we see the same trend observed in [4],[3],and [5], where universality can be found in the data, where the number $(\frac{v^2}{c^2 R_l})_{last}$ seems to be of order one, for now over 200 galaxies. The fact that there seems to be this universality in the last data point of each galaxies centripetal acceleration (scaled here to $\frac{1}{c^2}$) provides argument for the recent work of McGaugh et al. [11], where it is stated that such a universality could be explained by new physics outside of GR, instead of by the inclusion of dark matter. Thus, the LITTLE THINGS survey adds a robust, high resolution set of dwarf galaxies to the catalogue of rotation curves successfully fit by conformal gravity without the need for dark matter.

IOP Conf. Series: Journal of Physics: Conf. Series **845** (2017) 012004 doi:10.1088/1742-6596/845/1/012004

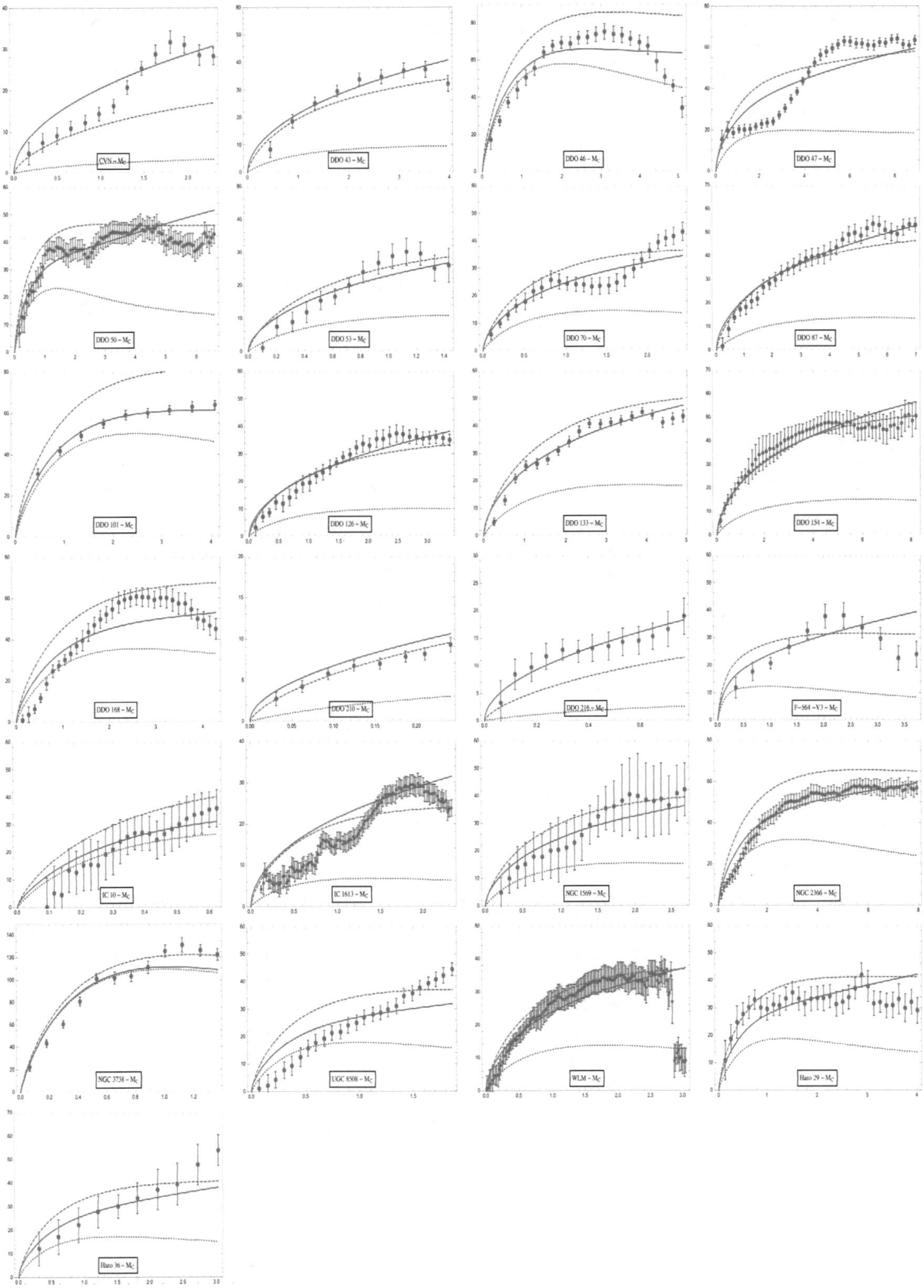

FIG. 1: Fitting to the rotational velocities (km/s) of the LITTLE THINGS sample with quoted errors as a function of radial distance (kpc). Blue dashed curve=Newtonian, Full blue curve= conformal gravity, Full red curve=MOND. No dark matter is assumed.

Table 1. Properties of the LITTLE THINGS Galaxy Sample

Galaxy	i	Dis.	Lum.	R_0	R_l	M_{gas}	M_{disk}	$(M/L)_C$	$(v^2/c^2R_l)_{last}$
	(ø)	(Mpc)	$(10^8L_\odot)$	(kpc)	(kpc)	$(10^7M_\odot)$	$(10^8M_\odot)$	$(M_\odot L_\odot^{-1})$	$(10^{-30}cm^{-1})$
CVN	66.5	3.18	0.11	1.1	2.5	2.26	0.03	0.27	1.16
DDO 43	40.6	7.40	1.54	1.1	4.0	20.94	0.32	0.21	0.94
DDO 46	27.9	8.78	2.45	0.9	5.1	45.77	18.15	7.42	0.83
DDO 47	45.5	6.00	3.28	1.1	8.9	62.31	1.82	0.55	1.62
DDO 50	49.7	2.28	3.06	0.5	6.6	6.08	1.95	0.64	8.92
DDO 53	27.0	3.31	0.44	0.6	1.4	6.34	0.37	0.87	1.69
DDO 70	50.0	1.59	1.01	0.7	2.4	5.67	0.77	0.76	2.75
DDO 87	55.5	7.27	1.40	2.3	7.0	25.96	2.21	1.58	1.44
DDO 101	51.0	11.25	4.85	1.1	4.1	10.82	16.98	3.53	3.60
DDO 126	65.0	4.14	1.01	1.0	3.4	11.73	0.44	0.43	1.30
DDO 133	43.4	4.95	2.59	1.6	4.9	25.72	2.86	1.10	1.37
DDO 154	68.2	3.84	0.80	0.7	8.2	37.99	0.07	0.08	1.01
DDO 168	46.5	4.61	3.41	1.2	4.3	29.75	8.56	2.51	1.70
DDO 210	66.7	0.70	0.22	0.4	0.3	1.05	0.05	2.11	1.28
DDO 216	63.7	0.76	0.21	0.9	0.7	0.20	0.06	0.23	1.66
F-567 V3	56.5	8.73	0.68	0.7	3.7	4.40	0.79	0.11	0.55
IC 10	47.0	0.87	6.85	0.4	0.6	2.20	1.77	0.26	7.45
IC 1613	48.0	0.61	7.96	0.7	0.6	2.20	1.77	0.22	0.75
NGC 1569	69.1	2.87	2.11	0.8	2.7	14.81	0.98	0.05	2.39
NGC 2366	63.0	3.34	7.87	1.4	7.9	6.69	8.50	1.08	1.47
NGC 3738	22.6	3.82	6.54	0.5	1.4	7.65	33.17	5.07	2.95
UGC 8508	83.5	2.60	0.43	0.5	1.9	1.91	0.83	1.92	3.80
WLM	74.0	0.90	0.73	0.8	3.0	6.45	0.79	1.08	0.96
Haro 29	61.2	4.70	0.68	0.4	4.0	5.94	1.20	1.76	7.58
Haro 36	70.0	8.91	3.27	0.6	3.0	10.44	1.00	0.31	3.45

Table 1 Entries, from left to right are: Galaxy name, inclination, NED Distance, Total Blue Luminosity, Disk Scale Length, Distance of the last data point, Mass of the gas (including helium), disk mass, mass to light ratios, scaled centripetal acceleration of the last observed data point.

3. Cluster Motion in Conformal Gravity

Although recent literature on alternative gravitational models have been focused around rotation curves, the original missing mass problem was founded in the study of random motions of galaxies in galactic clusters [10]. In 1933, Zwicky applied the virial theorem to his observations on the Coma cluster and he found that the theoretical mass estimate resulted in a mass to light ratio that was several orders of magnitude larger than theoretical prediction [12]. He reasoned that this result was due to some "missing matter" that could not be seen. Over the following decades, other problems in astrophysics arose that also seemed to point to the existence of this dark matter, such as the rotation curve problem of Rubin [13]. Since the Zwicky missing mass problem was the original calculation that led astronomers to suspect that there was missing matter in the universe, it would be appropriate to revisit the Coma cluster prediction in conformal gravity. This work was originally done by Mannheim [14] but due to the recent work of Mannheim and O'Brien [3], the full fourth order solution (with the inclusion of the global quadratic potential) can be applied to the Coma cluster to give a more representative prediction.

3.1. The Coma Cluster

In order to see whether conformal gravity can solve the problem of dark matter in galactic clusters it would be wise to start with the initial problem as solved by Zwicky. His approach was to use the data for the velocity dispersion of the galaxies within the Coma cluster and apply the virial theorem in order to estimate the mass. The virial theorem for a potential of the form

$V = ar^n$ states that:

$$2\langle T\rangle = n\langle V\rangle, \tag{8}$$

where $\langle T\rangle$ is the average total kinetic energy and $\langle V\rangle$ is the average total potential energy. Zwicky used this fact along with the fact that:

$$\langle v^2\rangle \approx 3\sigma^2, \tag{9}$$

where v^2 is the three dimensional systemic velocity of the cluster and σ is the measured velocity dispersion. The approximation sign is used because if the cluster is rotating or is not completely isotropic this calculation would only be true to an order of magnitude. It should be noted that in a cluster which is virialized this is a decent approximation used commonly in the literature [14]. Since this work is meant to only establish the correct order of magnitudes needed, we will ignore systematic effects such as rotational or tangental motions since these would only account for fine tuning of the numbers within the order of magnitude. For a complete treatment of how these effects will be implemented in the future, see [14]. When we plug eq. (9) along with the Newtonian potential into eq. (8) we arrive at the simple prediction for the mass based on the velocity observations,

$$M \approx \frac{3\sigma^2 R}{G}, \tag{10}$$

where R is the virial radius of the cluster. We can use eq. (10) along with the observations for the Coma Cluster to get an estimate of its mass,

$$M_{Coma} \approx 9.5 * 10^{14}\ M_{\odot}, \tag{11}$$

since for the Coma cluster, the current measured distance and velocity dispersions are $\sigma^2_{Coma} = 3.413 * 10^{12}\ \frac{m^2}{s^2}$ and $R_{Coma} = 1.206\ Mpc$. It should be noted that this is a much larger value than that which Zwicky himself obtained [12], because he used a value for the distance that is much smaller than our current one. However, the larger mass [15] seems to only make his initial claim for missing mass stronger. Since it is estimated that there are around 1000 galaxies in the Coma Cluster and we have an estimate for its total mass, we can say that the average mass of a galaxy in the Coma Cluster is $9.5 * 10^{11}\ M_{\odot}$, which could only be reasonable assuming that the galaxies within the cluster are all large spirals. Also, assuming that an average galaxy has a luminosity of about $5 * 10^9\ L_{\odot}$, we can calculate the mass to light ratio of an average galaxy in the Coma cluster:

$$\gamma = \frac{M}{L} \approx 200\ \gamma_{\odot}, \tag{12}$$

This mass to light ratio is similar to the one Zwicky himself obtained and which caused him to consider that there was some mass inside the cluster that could not be seen. We can now repeat the same calculation using the conformal potential in order to see if this problem could be resolved without invoking a need for non luminous matter.

3.2. The Coma Cluster in Conformal Gravity

To mirror the Zwicky calculation in Conformal Gravity, we first need to establish the single point source potential. This work was originally done by Mannheim and Kazanas [2], who showed that for a spherically symmetric source, the line element for an external solution to the mass:

$$ds^2 = -Bc^2dt^2 + \frac{1}{B}dr^2 + r^2d\theta^2 + r^2\sin^2\theta d\phi^2, \tag{13}$$

can be solved analytically. The exterior fourth order solution yields,

$$B = 1 - \frac{2GM}{c^2 r} + \frac{M\gamma^* r}{M_\odot} + \gamma_0 r - 2kr^2, \tag{14}$$

where $\gamma^* = 5.42 * 10^{-39}\ m^{-1}$, $\gamma_0 = 3.06 * 10^{-28}\ m^{-1}$ and $k = 9.54 * 10^{-50}\ m^{-2}$ are constants of the theory determined through galactic rotation fitting [4]. From this we can immediately get the weak field approximation for the potential:

$$V = -\frac{GM}{r} + \frac{M\gamma^* c^2 r}{2M_\odot} + \frac{\gamma_0 c^2 r}{2} - kc^2 r^2, \tag{15}$$

We can now use this along with eq. (8) to get the conformal estimate for the mass of a cluster:

$$M \approx \frac{\sigma^2 - \frac{\gamma_0 c^2 R}{2} + kc^2 R^2}{\frac{G}{R} + \frac{\gamma^* c^2 R}{2M_\odot}}, \tag{16}$$

which for R on the scale of the virial radius of a cluster can be simplified to

$$M \approx \frac{2k}{\gamma^*} R\ M_\odot. \tag{17}$$

When we plug in the same distance as we did for the original Zwicky problem, we can calculate the conformal estimate for the mass which to yield $M \approx 1.3 * 10^{12}\ M_\odot$ or an average of $1.3 * 10^9\ M_\odot$ which is a much more reasonable average across various morphologies of galaxies (see for example [16]). Finally we can calculate the mass to light ratio which comes out to be:

$$\gamma = \frac{M}{L} \approx 0.26\ \gamma_\odot, \tag{18}$$

As we can see this mass to light ratio in no way leads us to suspect that there might be missing matter.

3.3. Virgo Cluster

A few years after Zwicky, Smith performed a similar calculation for the inner members of the Virgo Cluster [17]. Using eq. (10), Smith found that the mass of an average galaxy from the 500 tracers in the central region of the cluster was $M \approx 2 * 10^{11}\ M_\odot$ which again was about 200 times larger than the Hubble estimate for the mass of an average nebula of $10^9\ M_\odot$. If we use Smith's value for the radius of that cluster of 0.2 Mpc we can plug it into eq. (17) to get the conformal approximation of the mass which comes out to be $2.2 * 10^{11}\ M_\odot$ or $4.4 * 10^8\ M_\odot$ for an average galaxy. Using this and the average luminosity of the virgo cluster [15], we come to a physical mass to light ratio as,

$$\gamma = \frac{M}{L} \approx 1.02\ \gamma_\odot, \tag{19}$$

which again in no way suggests that we need to incorporate dark matter.

3.4. Recent Cluster Observations

We can further this argument since we are now in a position to make predictions about cluster dynamics without having to make a claim about the mass at all. To this end, we note the last quadratic term in the potential is a global term which does not depend on the mass of the

Table 2. Selected SPZ-SZ Survey Clusters

Cluster Name	σ $(\frac{km}{s})$	M_Λ $(10^{14}\ M_\odot)$	R_Λ (Mpc)	M_C $(10^{11}\ M_\odot)$	R_C (Mpc)
SPT-CL J0000-5748	935	4.29	0.707	7.67	0.566
SPT-CL J0014-4952	1004	5.14	0.735	7.97	0.608
SPT-CL J0037-5047	945	3.64	0.587	6.37	0.572
SPT-CL J0040-4407	1171	10.18	1.070	11.60	0.709
SPT-CL J0118-5156	865	3.39	0.653	7.08	0.524
SPT-CL J0205-5829	1101	4.79	0.569	6.17	0.667
SPT-CL J0205-6432	862	3.29	0.638	6.92	0.522
SPT-CL J0233-5819	884	3.71	0.684	7.42	0.535
SPT-CL J0234-5831	1076	7.64	0.951	10.31	0.652
SPT-CL J0240-5946	948	5.29	0.848	9.20	0.574

This table shows selected galaxies taken from the SPZ-SZ survey. The values for the velocity dispersion σ are taken directly from [18]. The mass quoted as M_Λ in the table is the mass derived with dark matter from [18]. The R_C values are the ones derived in this paper from eq.(20). The R_Λ values are derived by using eq.(10) and the values of σ and M_Λ. M_C is the value of the mass when if we used R_Λ in eq. (17).

specific cluster but still becomes the dominant term over large distances. Thus if we take only the last term of eq. (16) and plug it into the virial theorem, we can get the following relation,

$$3\sigma^2 \approx kc^2R^2. \tag{20}$$

Although this is another approximation, it still allows us to make some profound claims on the order of magnitudes of the numbers in question. Moreover, we can use this as a means to make some comparisons to the standard theory. To further assess the validity of the conformal model, we can use the data for the velocity dispersions of clusters from the SPT-SZ survey [18]. In this work, the authors use the measured velocity dispersions, as well as dark matter mass estimates to predict the virial radial distance of the galaxies within the cluster. Since the numbers they obtain using the dark matter paradigm are published and hence trusted as results, we can use eq. (20) to make the same prediction without assuming any dark matter whatsoever, using only the observed measured velocities. It should be noted that the authors here have simply chosen the first 10 clusters studied in [18] and have in no way selected them based on any bias. In Table 1, we show the comparison of the two methods and show that the conformal prediction is right on par with the standard gravity prediction for the virial radius. For completion, we include the expected mass for each cluster within the ΛCDM cosmology paradigm which we reiterate is not used in obtaining the conformal prediction. We also show what the conformal mass estimate from eq. (17) would be if the virial radial distance obtained from the ΛCDM paradigm is taken to be the more precise value. As we can see the mass in the conformal case has to be several orders of magnitude smaller to provide the same prediction. We also note the luminosity estimates for these clusters in [18] are all of order $10^{11}\ L_\odot$, the same order as the conformal mass estimates. Thus we can calculate the mass to light ratio of a cluster for which we have a conformal estimate and a quoted luminosity. SPT-CL J0233-5819 fits both criteria and since its luminosity is $1.21 * 10^{11}\ L_\odot$ and its conformal mass is $7.42 * 10^{11}\ M_\odot$ we get a mass to light ratio of 6.1 $\gamma_\odot$ which again in no way suggests missing matter.

IOP Conf. Series: Journal of Physics: Conf. Series **845** (2017) 012004 doi:10.1088/1742-6596/845/1/012004

4. Gravitational Bending of light in conformal gravity

One of the great observational triumphs of General Relativity was the famous 1919 experiment verifying the gravitational bending of light. Since General Relativity so accurately captured this phenomena, any competing alternative theory of gravity must be able to accommodate this feature. Since conformal gravity is still a metric theory of gravity, it should retain the same qualities of Einstein gravity in the weak field or local level. In order to establish whether conformal gravity can explain the observations from weak lensing without the need for dark matter we must first tackle the simpler problem of establishing what the bending of light by a central source would look like in conformal gravity. We start again by writing the conformal metric (14) for a spherical source. However as an initial proof of concept, we ignore the global terms [19] as we are looking at the local lensing effect. Thus, the metric of eq. (13) becomes:

$$ds^2 = -(1 - \frac{2GM}{c^2 r} + \frac{M\gamma^* r}{M_\odot})c^2dt^2 + \frac{1}{(1 - \frac{2GM}{c^2 r} + \frac{M\gamma^* r}{M_\odot})}dr^2 + r^2d\theta^2 + r^2\sin^2\theta d\phi^2. \quad (21)$$

In order to model the situation of lensing around a spherically symmetric object (such as the sun), we set up the typical coordinate system as pictured in [20]. We look at a two dimensional deflection for simplicity and hence set $\phi = 0$. Following the original deflection calculation in [20], since we are using the sun as our gravitational source, we take the weak approximation of eq. (21) and expand the denominator of the radial metric term, viz:

$$ds^2 = -(1 - \frac{2GM}{c^2 r} + \frac{M\gamma^* r}{M_\odot})c^2dt^2 + (1 + \frac{2GM}{c^2 r} - \frac{M}{M_\odot}\gamma^* r)dr^2 + r^2d\theta^2 \quad (22)$$

The simplest way to approach this problem is to use Huyghens Principle which states that:

$$\frac{d\theta}{dx} = \frac{1}{v}\frac{dv}{dy} \quad (23)$$

where θ is the angle of deflection and v is the coordinate velocity of light in the direction of motion. To exploit eq. (23) , we rewrite the metric of eq. (22) in terms of the x coordinate to yield:

$$ds^2 = -(1 - \frac{2GM}{c^2 r} + \frac{M}{M_\odot}\gamma^* r)c^2dt^2 + (1 + \frac{2GM}{c^2 r} - \frac{M\gamma^* r}{M_\odot})(\frac{dr}{dx})^2dx^2 + r^2(\frac{d\theta}{dx})^2dx^2 \quad (24)$$

We can then plug in the transformation equations between Cartesian and polar coordinates:

$$r = \sqrt{x^2 + y^2}, \quad \cos\theta = \frac{y}{r}, \quad (25)$$

such that the final expression for the line element in this coordinate system becomes:

$$ds^2 = -(1 - \frac{2GM}{c^2 r} + \frac{M}{M_\odot}\gamma^* r)c^2dt^2 + (1 + \frac{2GMx^2}{c^2 r^3} - \frac{M\gamma^* x^2}{M_\odot r})dx^2. \quad (26)$$

Since the light will follow the null geodesics, we set $ds = 0$ and solve for the velocity as,

$$v = \frac{dx}{dt} = [1 + (\frac{M\gamma^* r}{2M_\odot} - \frac{GM}{c^2 r})(1 + \frac{x^2}{r^2})]c. \quad (27)$$

The feature of eq. (27) is that treats the gravitational field as though it was a medium which allows us to essentially calculate an effective gravitational index of refraction. We then use eq.

IOP Conf. Series: Journal of Physics: Conf. Series **845** (2017) 012004 doi:10.1088/1742-6596/845/1/012004

(27) in the polar coordinate transform eq. (25) to calculate the deflection angle in terms of the x coordinate as

$$\frac{d\theta}{dx} = \frac{GM}{c^2}(\frac{3x^2y}{r^5} + \frac{y}{r^3}) + \frac{M\gamma^*}{2M_\odot}(\frac{y}{r} - \frac{yx^2}{r^3}). \tag{28}$$

This expression can be integrated immediately to give the total deflection angle:

$$\theta = \int_{-\infty}^{\infty} \frac{GM}{c^2}(\frac{3x^2y}{r^5} + \frac{y}{r^3})\, dx + \int_{-\infty}^{\infty} \frac{M\gamma^*}{2M_\odot}(\frac{y}{r} - \frac{yx^2}{r^3})\, dx \tag{29}$$

Since we know that the deflection from the straight path would be very small, we can expect that y won't be very different from the impact parameter r_0. Thus we simplify eq. (29) to:

$$\theta \approx \frac{GMr_0}{c^2} \int_{-\infty}^{\infty} (\frac{3x^2}{r^5} + \frac{1}{r^3})\, dx + \frac{M\gamma^* r_0}{2M_\odot} \int_{-\infty}^{\infty} (\frac{1}{r} - \frac{x^2}{r^3})\, dx \tag{30}$$

These integrals are trivial to solve when we make the following substitutions:

$$x = r_0 \tan\theta, \quad r = \frac{r_0}{\cos\theta}, \quad \frac{dx}{d\theta} = \frac{r^2}{r_0} \tag{31}$$

which we can then plug into eq. (30) to get:

$$\begin{aligned} \theta &\approx \frac{2GM}{c^2 r_0} \int_0^{\frac{\pi}{2}} (3\sin^2\theta + 1)\cos\theta\, d\theta + \frac{M\gamma^* r_0}{2M_\odot} \int_0^{\frac{\pi}{2}} \cos\theta\, d\theta \\ \theta &\approx \frac{4GM}{c^2 r_0} + \frac{M\gamma^* r_0}{2M_\odot} \end{aligned} \tag{32}$$

We recognize the first term in eq. (32) as the result for gravitational bending of light in general relativity and the second term is thus the conformal correction. Reassuringly if we take r_0 to be the radius of the sun we see that the conformal correction is negligible and thus we get the same result as general relativity for a beam of light just passing by the sun. We also see that over larger distances we can expect bigger and bigger departures from general relativity which can have a profound effect on the predictions for weak lensing. Over larger distances we would also expect to start noticing the global effects of curvature and thus it would be instructive to derive the full conformal bending of light by following the same procedure described above, but using the metric from eq. (14) and including the global terms. Upon repeating the same procedure outlined above with the added terms, we can calculate the total angular deflection as:

$$\theta \approx \frac{4GM}{c^2 r_0} + \frac{M\gamma^* r_0}{2M_\odot} + \gamma_0 r_0 - \pi k r_0^2. \tag{33}$$

In order to bring this in alignment with the above work on clusters, we can rewrite this equation in terms of the mass:

$$M \approx \frac{\theta - \gamma_0 r_0 + \pi k r_0^2}{\frac{4G}{c^2 r_0} + \frac{\gamma^* r_0}{M_\odot}} \tag{34}$$

which for large impact parameters reduces to:

$$M \approx \left(\frac{\pi k r_0}{\gamma^*}\right) M_\odot \tag{35}$$

This estimate for the mass is consistent with the one seen in eq. (16) using the virial theorem for clusters. It should be noted that authors such as Nesbet have recently been looking for global connections of the conformal gravity constants [21], and here we have shown that we get a consistent framework for the constants which were originally described in terms of rotation curves, with clusters and gravitational bending of light. Further, the work done by Sultana et al. [22] in 2012 has shown that conformal gravity can accurately predict the precession of the perihelion of Mercury. Sultana's result, coupled with the calculation above for the deflection of light around the sun, expand on the applications of conformal gravity to the local solar system. Sulatana's work, similar to Nesbit, has also shown that the global parameter of conformal gravity γ can be bounded by the Mercury observations and is completely consistent with the value originally obtained from rotation curves [22]. These advances in conformal gravity are helping show the universality of the theory as a whole, and the theory is being held consistent at all gravitational scales.

5. Conclusion

In this paper we reviewed three recent important advancements in conformal gravity, which provide the framework for a lot of potential future research. Once again, conformal gravity provides an accurate description to 25 new high resolution rotation curves, increasing the net total of rotation curves fit by the theory to over 200. Further, we have also produced the framework for comparison of conformal gravity to MOND on a one to one galaxy basis. This coupled with the tools described in [23], can be extended to include other models in the future. Second, we show that conformal gravity is equipped to solve the original missing mass problem as found by Zwicky, since it can model cluster dynamics without needing any additional mass. Work on this front is currently being extended to the Abel cluster. In future work, conformal gravity will be put to a similar test in the ultra faint galaxies, studying the radial and tangential motions of individual tracer objects in young galaxies [24]. Lastly, we make the first initial steps towards applying conformal gravity to gravitational lensing, with the reproduction of the gravitational deflection of starlight calculation, which serves as a primer for a more robust lensing discussion.

Acknowledgments

The Authors would like to thank the International Association of Relativistic Dynamics (IARD) for the opportunity to speak at the IARD 2016 Conference in Slovenia. J. G. O'Brien would like to thank Philip Mannheim for his continued guidance in advancing conformal gravity. J. Dentico would like to thank Dr. Joel Brownstein for his insights into MOND phenomenology.

References

[1] Weinberg S 1972 *Gravitation and Cosmology: Principles and Applications of the General Theory of Relativity* (New York,Wiley)

[2] Mannheim P D and Kazanas D 1989 *ApJ* **342** 635–638

[3] Mannheim P D and O'Brien J G 2011 *Phys. Rev. Lett.* **106**(12) 121101

[4] Mannheim P D and O'Brien J G 2012 *Physical Review D* **85** 124020 (*Preprint* `1011.3495`)

[5] OBrien J G and Mannheim P D 2012 *Monthly Notices of the Royal Astronomical Society* **421** 1273–1282 (*Preprint* `http://mnras.oxfordjournals.org/content/421/2/1273.full.pdf+html`) URL `http://mnras.oxfordjournals.org/content/421/2/1273.abstract`

[6] O'Brien J G and Moss R J 2015 *Journal of Physics: Conference Series* **615** 012002 URL `http://stacks.iop.org/1742-6596/615/i=1/a=012002`

[7] O'Brien J G, Dentico J, Stulge M, Stefanski B, Moss R J and Chaykov S 2016 *Monthly Notices of the Royal Astronomical Society* **In Review**

[8] Oh S H, Hunter D A, Brinks E, Elmegreen B G, Schruba A, Walter F, Rupen M P, Young L M, Simpson C E, Johnson M C, Herrmann K A, Ficut-Vicas D, Cigan P, Heesen V, Ashley T and Zhang H X 2015 *The Astronomical Journal* **149** 180 URL `http://stacks.iop.org/1538-3881/149/i=6/a=180`

[9] Mannheim P D 2006 *Progress in Particle and Nuclear Physics* **56** 340–445 (*Preprint* `astro-ph/0505266`)

[10] Famaey B and McGaugh S S 2012 *Living Reviews in Relativity* **15** 10 (*Preprint* `1112.3960`)

[11] McGaugh S, Lelli F and Schombert J 2016 *ArXiv e-prints* (*Preprint* `1609.05917`)

[12] Knill O 2007 *Springer*

[13] Rubin V C, Ford W K J and Thonnard N 1980 *Astrophysical Journal* **238** 471–487

[14] Mannheim P D 1996 *Clusters, Lensing, and the Future of the Universe* (*Astronomical Society of the Pacific Conference Series* vol 88) ed Trimble V and Reisenegger A p 294 (*Preprint* `astro-ph/9508045`)

[15] Biviano 1995 *aap* **297** 610 (*Preprint* `astro-ph/9411025`)

[16] Walter F, Brinks E, de Blok W J G, Bigiel F, Kennicutt R C, Thornley M D and Leroy A 2008 *A. J.* **136** 2563

[17] Smith S 1936 *Contributions from the Mount Wilson Observatory / Carnegie Institution of Washington* **532** 1–8

[18] Ruel J, Bazin G, Bayliss M, Brodwin M, Foley R J, Stalder B, Aird K A, Armstrong R, Ashby M L N, Bautz M, Benson B A, Bleem L E, Bocquet S, Carlstrom J E, Chang C L, Chapman S C, Cho H M, Clocchiatti A, Crawford T M, Crites A T, de Haan T, Desai S, Dobbs M A, Dudley J P, Forman W R, George E M, Gladders M D, Gonzalez A H, Halverson N W, Harrington N L, High F W, Holder G P, Holzapfel W L, Hrubes J D, Jones C, Joy M, Keisler R, Knox L, Lee A T, Leitch E M, Liu J, Lueker M, Luong-Van D, Mantz A, Marrone D P, McDonald M, McMahon J J, Mehl J, Meyer S S, Mocanu L, Mohr J J, Montroy T E, Murray S S, Natoli T, Nurgaliev D, Padin S, Plagge T, Pryke C, Reichardt C L, Rest A, Ruhl J E, Saliwanchik B R, Saro A, Sayre J T, Schaffer K K, Shaw L, Shirokoff E, Song J, uhada R, Spieler H G, Stanford S A, Staniszewski Z, Starsk A A, Story K, Stubbs C W, van Engelen A, Vanderlinde K, Vieira J D, Vikhlinin A, Williamson R, Zahn O and Zenteno A 2014 *The Astrophysical Journal* **792** 45 URL `http://stacks.iop.org/0004-637X/792/i=1/a=45`

[19] Mannheim P D 1995 *Astro - ph* **9504022**

[20] Adler R, Bazin M and Schiffer M 1965 *Introduction to general relativity* International series in pure and applied physics (McGraw-Hill) URL `https://books.google.com/books?id=vHUuAAAAIAAJ`

[21] Nesbet, R K 2015 *EPL* **109** 59001 URL `http://dx.doi.org/10.1209/0295-5075/109/59001`

[22] Sultana J, Kazanas D and Said J L 2012 *Phys. Rev. D* **86**(8) 084008 URL `http://link.aps.org/doi/10.1103/PhysRevD.86.084008`

[23] Moss R J and O'Brien J G 2014 *Journal of Undergrad Research in Physics* **In Review**

[24] Bradford J D, Geha M C and Blanton M R 2015 *APJ* **809** 146 (*Preprint* `1505.04819`)

Self consistent models of deformed neutron stars in the framework of general relativity

Omair Zubairi[1] **and Fridolin Weber**[2,3]

[1] Department of Sciences, Wentworth Institute of Technology, 550 Huntington Ave Boston, MA 02115, USA

[2] Department of Physics, San Diego State University, 5500 Campanile Drive, San Diego, CA 92182, USA

[3] Center for Astrophysics and Space Sciences, University of California, San Diego, 9500 Gilman Drive, La Jolla, CA 92093, USA

E-mail: zubairio@wit.edu[1], fweber@mail.sdsu.edu[2,3]

Abstract. Generally, over the last 75 years, since the publication of the well known papers from from R. C. Tolman (Tolman 1939) and J. R. Oppenheimer and G. M. Volkoff (Oppenheimer and Volkoff 1939), standard models of non-rotating neutron stars are modeled with the assumption that they are perfect spheres. This assumption of perfect spherical symmetry is not correct if the matter inside of neutron stars is described by a non-isotropic equation of state (EoS). Particular classes of neutron stars such as Magnetars and neutron stars that contain color superconducting quark matter cores are expected to be deformed making them oblong spheroids. In this paper, we examine the deformity of these non-spherical neutron stars by deriving the stellar structure equations in the framework of general relativity. Using a non-isotropic model for the equation of state, these stellar structure equations are solved numerically in two dimensions. We then calculate stellar properties such as masses and radii along with pressure and energy-density profiles and investigate any changes from conventional spherical models.

1. Introduction

Non-rotating neutron stars are generally assumed to be perfect spheres, whose stellar properties such as masses and radii are described in the framework of general relativity by the Tolman-Oppenheimer-Volkoff (TOV) equation [1, 2]. Using a given model for the equation of state (EoS), describing the interior composition of a neutron star, this first order differential equation can be solved. However, the assumption of perfect spherical symmetry may not always be correct. It is well known that magnetic fields are present inside of neutron stars. In particular, if the magnetic field is strong (up to around 10^{18} Gauss in the core), such as for magnetars [3, 4, 5], and/or the pressure of the matter in the cores of neutron stars is non-isotropic, as predicted by some models of color superconducting quark matter [6], then deformation of neutron stars can occur [6, 7, 8, 9, 10, 11, 12].

In this work, we investigate the stellar structure of non-rotating deformed neutron stars by deriving TOV like stellar structure equations in the framework of general relativity whose mathematical form is similar to the conventional TOV equation for spherical symmetric stars. We then solve these equations numerically in two dimensions for a given EoS model and produce stellar properties such as masses and radii along with pressure and energy-density profiles and investigate any changes from spherical models of neutron stars.

2. Non-Spherical Symmetry

In order to study deformed neutron stars, illustrated in Fig. 1, we first need to investigate the symmetry of these objects. Our assumption of deformity will lie within the geometry of these

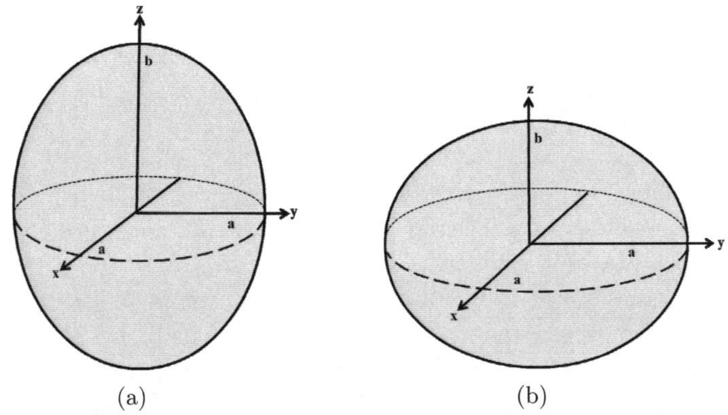

Figure 1. Schematic illustration of the geometry of an prolate (a) and oblate (b) spheroid.

objects. As shown in Fig. 1, the axial symmetry lies in the polar axis (z) being orthogonal with the equatorial axis ($x-y$). Mathematically, this can be described by the well known Weyl metric given by [13, 14],

$$ds^2 = \mathrm{e}^{2\lambda}dt^2 - \mathrm{e}^{-2\lambda}\left[\mathrm{e}^{2\nu}\left(dr^2+dz^2\right)+r^2d\phi^2\right] = g_{\mu\nu}dx^\mu dx^\nu\,, \tag{1}$$

where t is the time component and r, z, and ϕ are the spatial components in cylindrical coordinates. The terms λ and ν are the unknown metric functions that depend on both the radial r and polar z directions such that $\lambda=\lambda(r,z)$ and $\nu=\nu(r,z)$.

Due to the fact that we have distinct polar and radial directions which are both coupled together, as described in Eq. (1), the hydrostatic equilibrium of deformed neutron stars will have to be in two dimensions. Having two dimensional stellar structure equations will allow us to utilize a model for the equation of state of neutron star matter that takes the pressure in both the radial and polar directions into considerations. Our first step is examine each component of Eq. (1) and determine which one makes any contributions to the stellar structure. For this purpose, we will need to make a few assumptions about the stellar configuration. First, we assume that the configuration is non-rotating and not pulsating. Second, we assume that the star is axially symmetric, as shown in Fig. 1. Hence, the only terms that make a contributions to the stellar configuration will be the radial and polar components.

3. A Parameterized Solution

One way of investigating deformity of neutron stars is by applying a parametrization on the polar radius in terms of the equatorial radius along with a deformation constant γ on the metric described by Eq. (1). Such a metric reads as

$$ds^2 = -\mathrm{e}^{2\Phi(r)}dt^2 + \left(1-\frac{2m(r)}{r}\right)^{-\gamma}dr^2 + r^2d\theta^2 + r^2\sin^2(\theta)d\phi^2\,, \tag{2}$$

IOP Conf. Series: Journal of Physics: Conf. Series **845** (2017) 012005 doi:10.1088/1742-6596/845/1/012005

described in [15]. In Eq. (2), γ determines the degree of deformation either in the oblate ($\gamma < 1$) or prolate case ($\gamma > 1$) (see figure 1), where the parametrization is described by $z = \gamma r$. Using Eq. (2), a parameterized TOV equation is then obtained [15]

$$\frac{dP}{dr} = -\frac{(\epsilon + P)\left(\frac{1}{2}r + 4\pi r^3 P - \frac{1}{2}r\left(1 - \frac{2m}{r}\right)^{\gamma}\right)}{r^2\left(1 - \frac{2m}{r}\right)^{\gamma}}\,. \tag{3}$$

In the case when $\gamma = 1$, Eq. (3) simplifies to the well-known Tolman-Oppenheimer-Volkoff equation [1, 2] which describes the stellar structure of perfectly spherically symmetric object. The gravitational mass of a deformed neutron star parameterized with the deformation constant γ is then described by

$$\frac{dm}{dr} = 4\pi r^2 \epsilon \gamma\,, \tag{4}$$

so that the total gravitational mass, M, of a deformed neutron star with an equatorial radius R follows as [14, 15] $M = \gamma\, m(R)$. For a given deformation, γ, this is a very useful and easy-to-use model for the description of deformed compact stars. Using a given isotropic EoS model, we numerically solve these parameterized stellar structure equations and generate stellar properties such as masses and radii along with pressure and energy-density profiles. Extending the work done by [15, 16], we include two more models for the EoS for a total of three isotropic models [17, 18, 19, 20, 21]. The nuclear properties which describe the key differences of each model is summarized in Table 1.

Table 1. Isotropic Equation of state models studied in this work.

Equation of state	Nuclear Composition	Stellar object
MIT Bag Model	u, d, s quarks	Strange Quark Star
Standard Baryonic Model	n, p, e, Σ, Λ, Ξ	Neutron Star
Quark-Hadron Model	n, p, e, Σ, Λ, Ξ, u, d, s quarks	Neutron Star

Using these three different EoS models, we numerically solve Eqs. (3) and (4) and produce mass-radius relations shown in Fig. 2. Using our parameterization of $z = \gamma r$, we can see a trend in all three EoS models–that is we see an increase in mass with increasing oblateness and a decrease in mass with increasing prolateness. When our deformation parameter is equal to one, we obtain the spherically symmetric solution back for original TOV equations.

In order to investigate the actual deformity (shape of the star), we calculate the pressure and energy-density profiles associated with the maximum masses shown in Fig. 2 and examine the convergence of the pressure for both the equatorial and polar directions. The results are illustrated in Fig. 3 for the maximum masses for the quark-hadron EoS model.

From our results shown in Fig. 3, we see the pressure vanishes at the surface for both equatorial and polar directions indicating the distinct shape of the star (oblate or prolate). Using this parameterized model gives us some insight into deformity and shows that it is important and can not be ignored; it thus requires a more detailed description of in two dimensions.

4. Weyl Metric Calculations

In order to investigate the deformation in two dimensions, we need to examine the components of Eq. (1) and see which ones make any contributions to the stellar structure. Due to our assumptions of a non-rotating, axial symmetric stellar configuration, the only components that

IOP Conf. Series: Journal of Physics: Conf. Series **845** (2017) 012005 doi:10.1088/1742-6596/845/1/012005

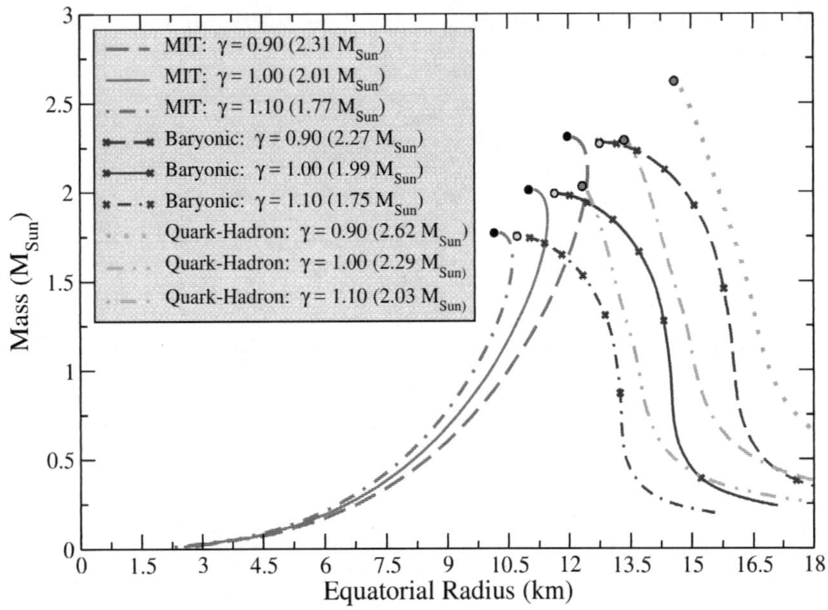

Figure 2. (Color-online) Mass-radius relationships calculated from Eqs. (3) and (4) for various isotropic nuclear equations of state (see also Refs. [22, 23]). The dots on each curve represent the maximum mass for each stellar sequence.

will make any contributions to the stellar structure of our deformed star will be the radial and polar components of Einstein's field equation,

$$G^{\mu}{}_{\nu} \equiv R^{\mu}{}_{\nu} - \frac{1}{2}\delta^{\mu}{}_{\nu} R = -8\pi T^{\mu}{}_{\nu} \,, \tag{5}$$

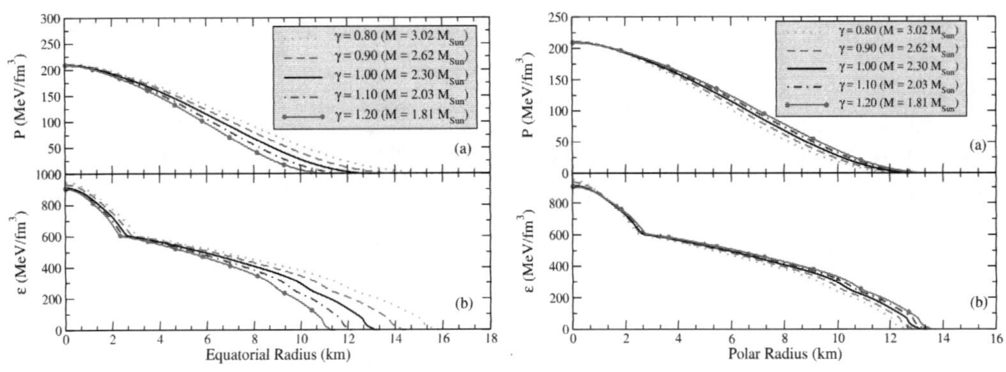

(a) (b)

Figure 3. (Color-online) Pressure (top) and energy-density (bottom) profiles in the equatorial (left) and polar (right) direction for the maximum masses for the quark-hadron EoS model given in Fig. 2 (see also Refs. [22, 23]). As it is shown, a larger equatorial radius for $\gamma < 1$ results in a smaller polar radius thus giving us an oblate star and similarly when $\gamma > 1$, we get a larger polar radius than the equatorial radius resulting in a prolate star.

where $G^{\mu}{}_{\nu}$ denotes the Einstein tensor, $R^{\mu}{}_{\nu}$ is the Ricci tensor, R the scalar curvature, and $T^{\mu}{}_{\nu}$ the energy-momentum tensor. We begin with the Weyl metric of Eq. (1), and calculate the Christoffel symbols to be

$$\Gamma^{r}{}_{tt} = \left[\mathrm{e}^{-2\nu(\mathrm{r,z})+4\lambda(\mathrm{r,z})}\right] \partial_r \lambda(r,z)\,, \tag{6}$$

$$\Gamma^{z}{}_{tt} = \left[\mathrm{e}^{-2\nu(\mathrm{r,z})+4\lambda(\mathrm{r,z})}\right] \partial_z \lambda(r,z)\,, \tag{7}$$

$$\Gamma^{t}{}_{tr} = \partial_r \lambda(r,z),\ \Gamma^{t}{}_{tz} = \partial_z \lambda(r,z)\,, \tag{8}$$

$$\Gamma^{r}{}_{rr} = \partial_r \nu(r,z) - \partial_r \lambda(r,z),\ \Gamma^{z}{}_{rr} = \partial_z \lambda(r,z) - \partial_z \nu(r,z)\,, \tag{9}$$

$$\Gamma^{r}{}_{rz} = \partial_z \nu(r,z) - \partial_z \lambda(r,z),\ \Gamma^{z}{}_{rz} = \partial_r \nu(r,z) - \partial_r \lambda(r,z)\,, \tag{10}$$

$$\Gamma^{z}{}_{rz} = \partial_r \nu(r,z) - \partial_r \lambda(r,z),\ \Gamma^{\phi}{}_{r\phi} = \left(1 - r\partial_r \lambda(r,z)\right)/r\,, \tag{11}$$

$$\Gamma^{r}{}_{zz} = -\Gamma^{r}{}_{rr},\ \Gamma^{z}{}_{zz} = -\Gamma^{z}{}_{rr},\ \Gamma^{\phi}{}_{z\phi} = -\Gamma^{t}{}_{tz}\,, \tag{12}$$

$$\Gamma^{r}{}_{\phi\phi} = \left[\mathrm{e}^{-2\nu(\mathrm{r,z})}\right] r\left(-1 + r\partial_r \lambda(r,z)\right)\,, \tag{13}$$

$$\Gamma^{z}{}_{\phi\phi} = \left[\mathrm{e}^{-2\nu(\mathrm{r,z})}\right] r^2 \partial_z \lambda(r,z)\,, \tag{14}$$

where ∂_r and ∂_z denote partial derivatives with respect to r and z. Using these Christoffel symbols and from Eq. (5) we calculate the components of the Ricci tensor to be

$$R^{t}{}_{t} = \frac{1}{r}\mathrm{e}^{-2\nu+2\lambda}\left(r\partial_r^2\lambda + r\partial_z^2\lambda + \partial_r\lambda\right)\,, \tag{15}$$

$$R^{r}{}_{r} = -\frac{1}{r}\mathrm{e}^{-2\nu+2\lambda}\left(r\partial_r^2\lambda - 2\left(\partial_r\lambda\right)^2 - r\partial_z^2\nu + r\partial_z^2\lambda - r\partial_r^2\nu + \partial_r\lambda + \partial_r\nu\right)\,, \tag{16}$$

$$R^{z}{}_{r} = R^{r}{}_{z} = \frac{1}{r}\mathrm{e}^{-2\nu+2\lambda}\left(2r\left(\partial_r\lambda\right)^2 - \partial_z\nu\right)\,, \tag{17}$$

$$R^{z}{}_{z} = -\frac{1}{r}\mathrm{e}^{-2\nu+2\lambda}\left(r\partial_r^2\nu - 2\left(\partial_z\lambda\right)^2 + r\partial_r^2\lambda - r\partial_z^2\nu + r\partial_z^2\lambda - \partial_r\nu + \partial_r\lambda\right)\,, \tag{18}$$

$$R^{\phi}{}_{\phi} = \frac{1}{r}\mathrm{e}^{2\nu+2\lambda}\left(r\partial_r^2\lambda + r\partial_z^2\lambda + \partial_r\lambda\right)\,, \tag{19}$$

where ∂_r^2 and ∂_z^2 denote second partial derivatives with respect to r and z. The Ricci scalar R is then calculated to be

$$R = -\frac{1}{r}\Big[2\mathrm{e}^{-2\nu+2\lambda}\Big(r\partial_r^2\lambda - r\left(\partial_r\lambda\right)^2 - r\partial_z^2\nu + r\partial_z^2\lambda - r\partial_r^2\nu + \partial_r\lambda - r\left(\partial_z\lambda\right)^2\Big)\Big]\,, \tag{20}$$

Using Eqs. (15) through (20), one obtains for the components of the Einstein tensor the following expressions,

$$G^{t}{}_{t} = \frac{1}{r}\Big[\mathrm{e}^{-2\nu+2\lambda}\Big(2r\partial_r^2\lambda + 2r\partial_z^2\lambda + 2\partial_r\lambda - r\left(\partial_r\lambda\right)^2 - r\partial_z^2\nu - r\partial_r^2\nu - r\left(\partial_z\lambda\right)^2\Big)\Big]\,, \tag{21}$$

$$G^{r}{}_{r} = \frac{1}{r}\mathrm{e}^{-2\nu+2\lambda}\left(r\left(\partial_r\lambda\right)^2 - \partial_r\nu - r\left(\partial_z\lambda\right)^2\right)\,, \tag{22}$$

$$G^z{}_r = G^r{}_z = \frac{1}{r}\mathrm{e}^{-2\nu+2\lambda}\left((2r\partial_r\lambda)\left(\partial_z\lambda\right) - \partial_z\nu\right) , \tag{23}$$

$$G^z{}_z = -\frac{1}{r}\mathrm{e}^{-2\nu+2\lambda}\left(r\left(\partial_r\lambda\right)^2 - \partial_r\nu - r\left(\partial_z\lambda\right)^2\right) , \tag{24}$$

$$G^\phi{}_\phi = -\mathrm{e}^{-2\nu+2\lambda}\left(\left(\partial_r\lambda\right)^2 + \partial_z^2\nu + \partial_r^2\nu + \left(\partial_z\lambda\right)^2\right) . \tag{25}$$

Due to the mathematical form of Eq. (1), there will be off-diagonal terms $T^r{}_z = T^z{}_r \neq 0$ in the energy-momentum tensor,

$$T^\mu{}_\lambda = \begin{pmatrix} \epsilon & 0 & 0 & 0 \\ 0 & P_\| & \tilde{P} & 0 \\ 0 & \tilde{P} & P_\perp & 0 \\ 0 & 0 & 0 & \tilde{P} \end{pmatrix} , \tag{26}$$

where ϵ is the energy-density, $P_\|$ is the pressure in the radial direction, and $P_\perp$ is the pressure in the polar direction. The $\tilde{P}$ terms are the off-diagonal pressure contributions associated with the r–z direction of Eq. (1).

Before evaluating Einstein's field equation (Eq. (5)), we need to define expressions for the total gravitational mass. In cylindrical coordinates, the mass in the equatorial and polar directions will depend on both r and z coordinates such that the gravitational mass has the form of $m = m(r, z)$.

4.1. Mass Formalism

Since our symmetry is axial symmetric, we define two differential masses as shown in Fig. 4, where one of the differential masses corresponds to an infinitesimal cylinder of radius r, height z, and thickness dr, and a second differential mass for an infinitesimal slab of radius r and thickness dz. The differential masses shown in Fig. 4 are then described mathematically by the

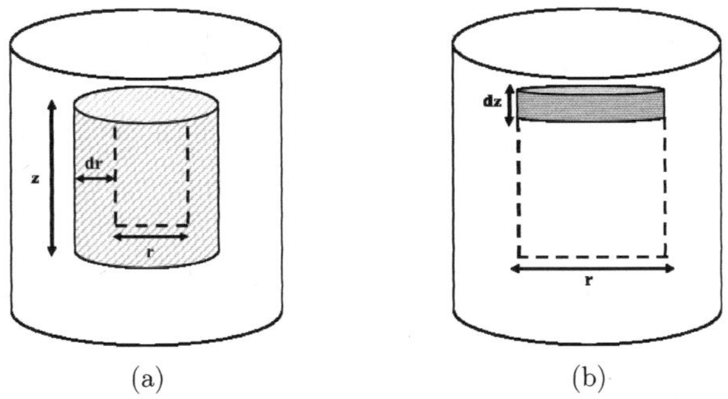

Figure 4. A graphical illustration of the differential masses in cylindrical coordinates for cross-sectional masses in the equatorial direction (a) and in the polar direction (b).

expressions,

$$\partial_r m(r, z) = 2\pi r z \epsilon(r, z) , \tag{27}$$

$$\partial_z m(r, z) = \pi r^2 \epsilon(r, z) . \tag{28}$$

Using Eqs. (27) and (28), we can now utilize Einstein's field equation $G^t{}_t = -8\pi T^t{}_t$, and thus Eq. (27) can be written as

$$\partial_r m(r, z) = \frac{rz}{4} G^t{}_t . \tag{29}$$

Using the other expressions for the Einstein tensor of Eqs. (21) through (25) along with the expression for the energy-momentum tensor given in Eq. (26) leads for Eq. (29) to

$$\partial_r m(r,z) = \partial_r \left(-\mathrm{e}^{-2\nu+2\lambda}\, z\, (\partial_r \lambda)(\partial_r \nu - \partial_r \lambda) \right), \tag{30}$$

Applying an integral on both sides of Eq. (30), we obtain an expression that takes the form

$$\int_0^r (\partial_r m(r,z))\, dr = -\int_0^r \partial_r \left(\mathrm{e}^{-2\nu+2\lambda}\, z\, (\partial_r \lambda)(\partial_r \nu - \partial_r \lambda) \right) dr\,. \tag{31}$$

By examing Eq. (31), is clear that there is a serious inconsistency with the r-component, as the r-term has vanished from Eq. (31). In addition to the r-term vanishing, the unknown metric functions λ'_r and ν'_r are still present in Eq. (31); therefore, we can not solve for the r component $\left(\mathrm{e}^{-2\nu+2\lambda}\right)$ directly. Having no r-term indicates that there is no interior solution of Einstein's field equations that will match the exterior solution for the metric described by Eq. (1) and thus we have lost all the information about the radial coordinate of our deformed compact object.

5. A Self Consistent Model of Deformation

Due to the fact the all the information of the radial coordinate is lost from Eq. (31), we need another way to describe deformity. One way to investigate deformation is to look further into our parameterized one dimensional model.

Using our parametrization of the polar coordinate in terms of the radial coordinate $z = \gamma r$ and assuming our radial and polar step sizes are small (on the order of meters), we can look at the deformation constant γ as a differential ratio of the polar and radial step sizes described by

$$\gamma = \frac{dz}{dr}\,. \tag{32}$$

Hence, we can use Eq. (32), along with the parametrization $z = \gamma r$, and apply a transformation on Eq. (3) to obtain a relationship between pressure and the polar direction. Using Eq. (32), we find that

$$\frac{dP}{dz} = -\frac{(\epsilon + P)\left(\frac{z}{2\gamma} + 4\pi\left(\frac{z}{\gamma}\right)^3 P - \frac{z}{2\gamma}\left(1 - \frac{2m\gamma}{z}\right)^{\gamma}\right)}{\frac{z^2}{\gamma^3}\left(1 - \frac{2m\gamma}{z}\right)^{\gamma}}, \tag{33}$$

where now we can take distinct pressure gradients into account by assuming that the parallel pressure $\left(P_{\|}\right)$ will be associated with dr and the perpendicular pressure $(P_{\perp})$ will be associated with dz. In addition, we will also have to take a look at the total gravitational mass m in Eqs. (3) and (33), since, both Eqs. (3) and (33) need to be coupled together. For the total gravitational mass, we utilize our mass elements described by

$$\partial_r m(r,z) = 2\pi r z \epsilon(r,z)\,, \tag{34}$$

$$\partial_z m(r,z) = \pi r^2 \epsilon(r,z)\,. \tag{35}$$

If we were to integrate then add the expressions given in Eqs. (34) and (35), we would obtain some total gravitational mass described as

$$\begin{aligned} m_{\mathrm{total}}(r,z) &= 2\pi\epsilon(r,z) r^2 z\,, \\ &= 2m(r,z)\,. \end{aligned} \tag{36}$$

IOP Conf. Series: Journal of Physics: Conf. Series **845** (2017) 012005 doi:10.1088/1742-6596/845/1/012005

However the expression for $m_{\text{total}}(r, z)$ does not correctly define the gravitational mass of a oblong spheroid, which is given by

$$M(r, z) = \frac{4}{3}\pi\epsilon(r, z)r^2 z\,. \tag{37}$$

In order to obtain Eq. (37), we must subtract off another cross-sectional term from Eq. (36) that will insure the gravitational confinement of the star, leading to

$$\begin{aligned} M(r, z) &= 2\pi\epsilon(r, z)r^2 z - \frac{2}{3}\pi\epsilon(r, z)r^2 z\,, \\ &= 2m(r, z) - \frac{2}{3}m(r, z)\,, \end{aligned} \tag{38}$$

however, we must break up the top expression in Eq. (38) equally to take the mass in the equatorial and polar directions separately but that are confined and result in a mass described by Eq. (37). Using the expressions from Eqs. (34) and (35) we find the gravitational mass to be

$$\begin{aligned} \mathcal{M}(r, z) &= 2\pi\epsilon(r, z)rz + \pi\epsilon(r, z)r^2 - \frac{1}{3}\pi\epsilon(r, z)r^2 z\,, \\ &= \partial_r m(r, z) + \partial_z m(r, z) - \frac{1}{3}\pi\epsilon(r, z)r^2 z\,, \end{aligned} \tag{39}$$

where we can define

$$\begin{aligned} \partial_r m(r, z) &= m_{\parallel}(r, z)\,, \\ \partial_z m(r, z) &= m_{\perp}(r, z)\,. \end{aligned} \tag{40}$$

The total gravitational mass is then given by

$$\mathcal{M}_{\text{total}}(r, z) = 2\pi\epsilon(r, z)r^2 z - \frac{2}{3}\pi\epsilon(r, z)r^2 z\,. \tag{41}$$

Using the expressions given in Eq. (39), we make substitutions into Eqs. (3) and Eq. (33) along with the substitutions for the parallel and perpendicular pressures and find that the hydrostatic equilibrium equations modify to

$$\frac{\partial P_{\parallel}}{\partial r} = -\frac{(\epsilon + P_{\parallel})\left(\frac{1}{2}r + 4\pi r^3 P_{\parallel} - \frac{1}{2}r\left(1 - \frac{2\mathcal{M}(r,z)}{r}\right)^{\gamma}\right)}{r^2\left(1 - \frac{2\mathcal{M}(r,z)}{r}\right)^{\gamma}}\,, \tag{42}$$

$$\frac{\partial P_{\perp}}{\partial z} = -\frac{(\epsilon + P_{\perp})\left(\frac{z}{2\gamma} + 4\pi\left(\frac{z}{\gamma}\right)^3 P_{\perp} - \frac{z}{2\gamma}\left(1 - \frac{2\mathcal{M}(r,z)\gamma}{z}\right)^{\gamma}\right)}{\frac{z^2}{\gamma^3}\left(1 - \frac{2\mathcal{M}(r,z)\gamma}{z}\right)^{\gamma}}\,, \tag{43}$$

where now the expressions given in Eqs. (42) and (43) are coupled together with the gravitational mass $\mathcal{M}(r, z)$ all governed by distinct parallel $(P_{\parallel})$ and perpendicular $(P_{\perp})$ pressure gradients. It is also important to note that now we do not need to deform the star by γ. For all of our calculations, we fixed γ, so that $\gamma = 1$. Having this fixed constant will allow the anisotropies in the equation of state to dictate the deformation.

Our self consistent two dimensional parameterized stellar model will involve solving the expressions in Eqs. (42) and (43) in conjunction with the expressions given in Eqs. (34) and

(35). The surface of the star will be defined when the pressure gradients both in the parallel and perpendicular directions vanish. That is, we will define the surface of the star by

$$P_{\|}(r = R) = 0\,, \tag{44}$$

$$P_{\perp}(z = Z) = 0\,, \tag{45}$$

where the total gravitational mass of the object is given by

$$M_{\text{Total}}(R, Z) = \mathcal{M}_{\text{total}}\,. \tag{46}$$

In order to solve our coupled system of equations, we now need an equation of state which has distinct pressure gradients in the parallel and perpendicular directions. The next section briefly describes the equation of state used in these two dimensional calculations.

6. Anisotropic Equation of State

In the two dimensional model, we modify the quark-hadron equation of state by changing the pressure by various amounts greater than and less than the spherical case. Modifying the equation of state this way will insure some symmetry in our calculations. We took data presented in [21], and changed the pressure terms by increasing the pressure by 4, 8, and 12 percent. We also decreased the pressure by 4, 8, and 12 percent as illustrated in Fig. 5. Using the data shown

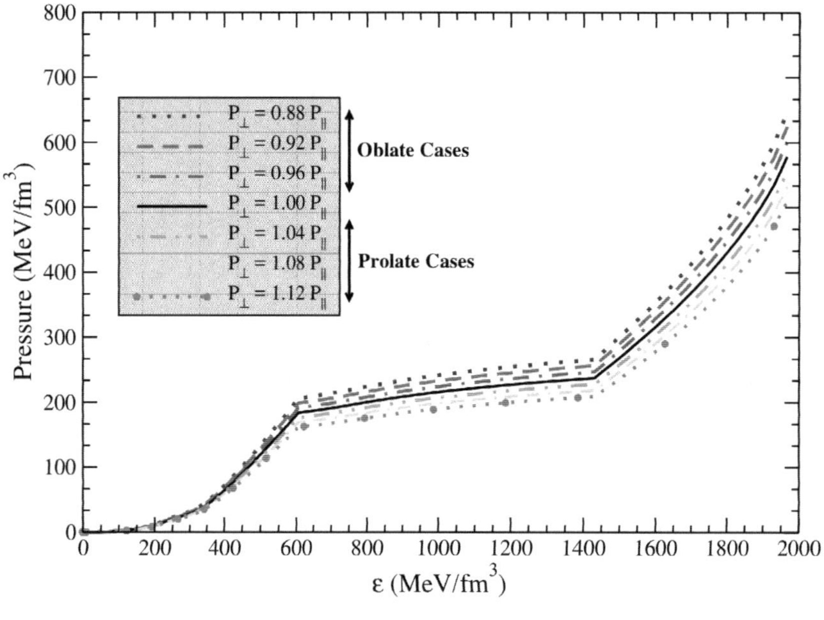

Figure 5. (Color-online) Anisotropic relationship of pressure and energy-density for a quark-hadron equation of state. In this model, the core is comprised of a mixed phase of up, down and strange quarks and hadronic matter. The pressure is varied by 4, 8, and 12 percent to mimic anisotropies in the EoS (for more details, see Refs. [22, 23]), thus giving us distinct pressure gradients.

in Fig. 5, we can now solve our two dimensional system of hydrostatic equilibrium for deformed neutron stars. Hence, with distinct pressure gradients, deformity should result; however, it should not be large. We have only deviated from the isotropic case by a few percent, and therefore the deformation of spheres to oblate and prolate stars should vary slightly.

7. Results and Conclusions

In contrast to traditional spherical models, solving our two dimensional model will require other strategies. In our two dimensional case, we cannot have mass-radius relationships to describe the maximum mass. This will not correctly give us any information of the deformation; however, we can however choose a central density and build our deformed star and calculate the pressure and energy-density profiles associated with that star. From the pressure profiles, we will be able to tell the deformity. Due to the non-isotropic equation of state, we can investigate the change in mass.

We begin with choosing a central density of 933 MeV/fm^3. This central density was chosen from the one-dimensional parameterized model for a spherical object [24]. Using this density, we calculate the pressure and energy-density profiles along with the total mass of the object. We first look at the pressure and energy-density profiles associated with increased pressures of 4, 8, and 12 percent. The results are shown in Figs. 6 and 7 respectively. The pressure profiles for each maximum mass star in Fig. 6 converge to zero at the stellar surfaces for both the equatorial and polar radii.

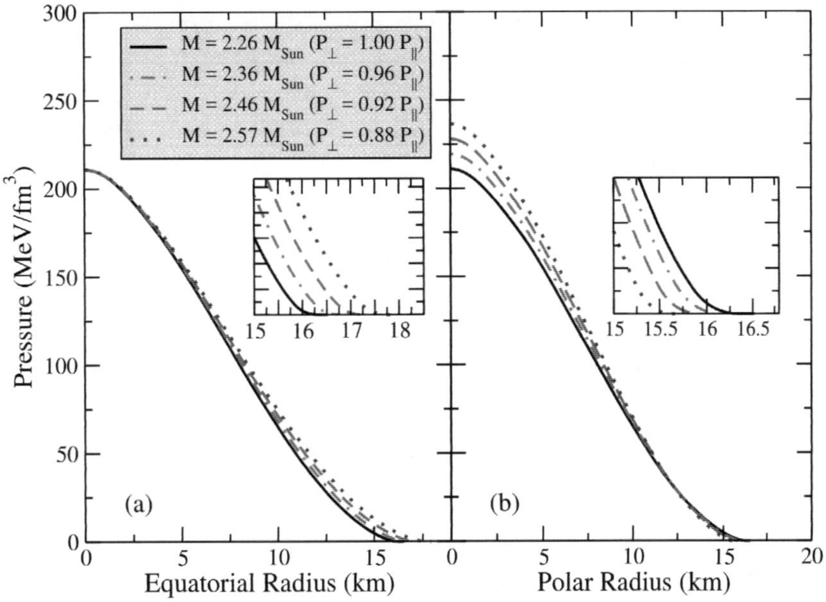

Figure 6. (Color-online) Pressure profiles for both the equatorial (a) and polar (b) directions for pressure gradients in the equatorial directions greater than 4, 8, and 12 percent (see also Refs. [22, 23]). The masses of such stars increase with oblateness.

As shown in Fig. 6, for no change in pressure, the mass is about 2.26 $M_{\odot}$ which is in good agreement with results from [21, 24, 25] for the spherical case. From our calculations, the mass increases as the equatorial radii increases and the polar radius decreases. This increase in mass due to increased oblateness is consistent with our one dimensional parameterized model and validates our results presented using our parametrization. It is also important to note that from the pressure profiles shown in Fig. 6, the pressure decreases monotonically as the equatorial and polar radii increase to the surface and hence the pressure becomes zero as stated by our boundary conditions. The increase in mass is appreciable even though we have small changes in the pressure in our equation of state.

We next investigate the deformation in another way to obtain prolate spheroids and see if the masses decreases as we increase prolateness. We do this by calculating the pressure profiles

IOP Conf. Series: Journal of Physics: Conf. Series **845** (2017) 012005 doi:10.1088/1742-6596/845/1/012005

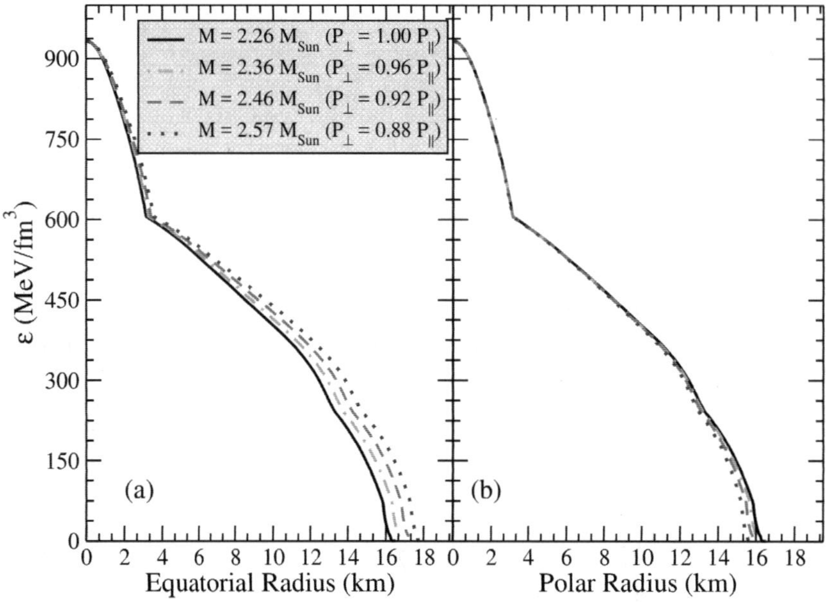

Figure 7. (Color-online) Energy-density profiles for both the equatorial (a) and polar (b) directions for pressure gradients in the equatorial directions (see Refs. [22, 23]) greater than the polar direction by 4, 8, and 12 percent.

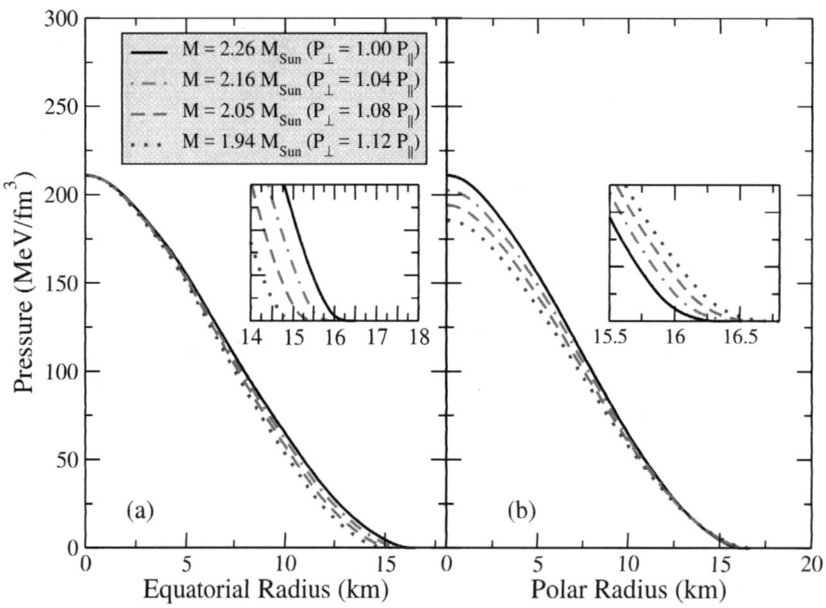

Figure 8. (Color-online) Pressure profiles for both the equatorial (a) and polar (b) directions for pressure gradients in the equatorial directions less than the polar direction by 4, 8, and 12 percent (see also Refs. [22, 23]). The masses of such stars decrease for increasing prolateness.

by decreasing the pressure by 4, 8 and 12 percent from our non-isotropic equation of state. However, we have to be mindful when applying the different pressure gradients. We do not want to rotate the star by any means. Therefore, we decrease the pressure gradients in the equatorial directions, but keep the pressure in the polar direction fixed. This will insure we will obtain prolate spheroid but not an oblate spheroid just rotated by 90 degrees. The results of the pressure and energy-density profiles calculated this way are shown in Figs. 8 and 9 respectively. Just as in the oblate case, we see clearly that the mass decreases as we increase prolateness.

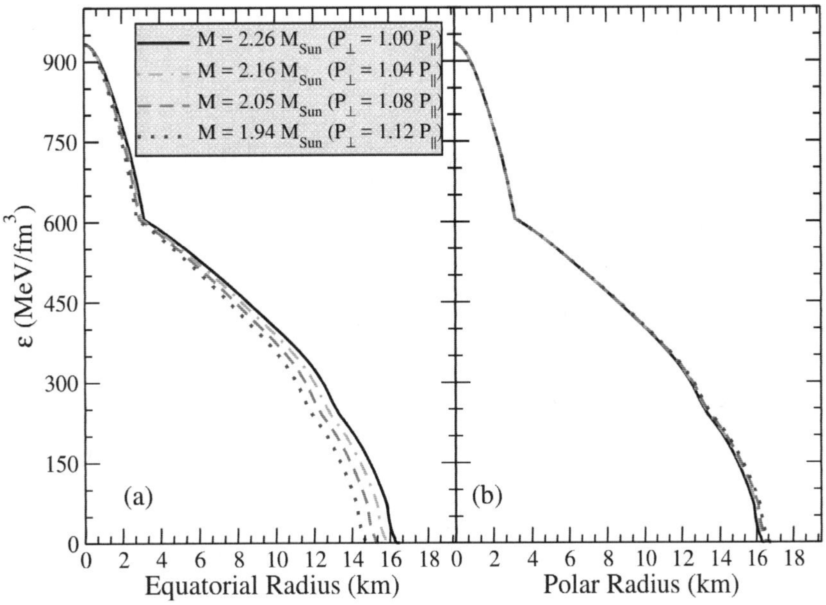

Figure 9. (Color-online) Energy-density profiles for both the equatorial (a) and polar (b) directions (see Refs. [22, 23]) for pressure gradients in the equatorial directions less than the polar direction by 4, 8, and 12 percent.

This is consistent with the results from [15, 24] for the one dimensional parameterized model for prolate stars. We see significant changes in mass due to increasing prolateness. It is important to note that the mass change (either increase or decrease) is roughly the same and it deviates slightly. This implies the two dimensional calculations are more sensitive to the anisotropies in the equation of state. A summary of the stellar properties resulting from our two dimensional model is provided in Table 2.

Table 2. Stellar properties of deformed neutron stars for the EoS given in Fig. 5.

Anisotropic EoS	R (km)	Z (km)	Total Mass ($M_\odot$)	Shape
$P_\perp = 0.88P_\parallel$	17.92	15.77	2.57	Oblate Spheroid
$P_\perp = 0.92P_\parallel$	17.21	15.83	2.46	Oblate Spheroid
$P_\perp = 0.96P_\parallel$	16.73	16.06	2.36	Oblate Spheroid
$P_\perp = 1.00P_\parallel$	16.51	16.51	2.26	Sphere
$P_\perp = 1.04P_\parallel$	16.00	16.67	2.16	Prolate Spheroid
$P_\perp = 1.08P_\parallel$	15.39	16.73	2.05	Prolate Spheroid
$P_\perp = 1.12P_\parallel$	14.80	16.82	1.94	Prolate Spheroid

From our results given in Figs. 6 through 9 along with the values listed in Table 2, we clearly show that deformation plays a vital role in the stellar structure of neutron stars and we can not simply ignore this deformation. From our results, we conclude that anisotropies in the equation of state can deform the star either in the equatorial or polar directions resulting in both oblate or prolate spheroids. Depending on the deformation the mass could either increase or decrease. This change in mass could explain discrepancies in various equation of state models and help us better understand the internal structure of neutron stars.

Acknowledgments
This work was supported through the National Science Foundation under grant PHY-1411708. Additional computing resources were provided by the Computational Science Research Center at San Diego State University and the Department of Sciences at Wentworth Institute of Technology. Travel funding was provided by Wentworth Institute of Technology.

The authors would like to acknowledge Dr. Vivian de la Incera and Dr. Efrain Ferrer for their initial motivation on this work. The authors would also like to thank Dr. Martin Land for the invitation to speak at the biennial 2016 IARD conference.

References
[1] Oppenheimer J R and Volkoff G M 1939 *Phys. Rev.* **55** 374.
[2] Tolman R C 1939 *Phys. Rev.* **55** 364.
[3] Thompson C and Duncan R C 1995 *MNRAS* **275** 255.
[4] Thompson C and Duncan R C 1996 *Astrophys. J.* **473** 322.
[5] Mereghetti S 2008 *Astron. Astrophys. Rev* **15** 225.
[6] Ferrer E, Incera V, Keith J, Portillo I, and Springsteen P 2010 *Phys. Rev. C* **82** 6.
[7] Chandrasekhar S and Fermi E 1953 *Astrophys. J.* **118** 116.
[8] Ferraro V C A 1954 *Astrophys. J.* **119** 407.
[9] Goossens M 1972 *Ap & SS* **16** 286.
[10] Katz J I 1989 *MNRAS* **239** 751.
[11] Payne D J B and Melatos A 2004 *MNRAS* **351** 569.
[12] Haskell B, Samuelsson S, Glampedakis K, and Andersson N 2008 *MNRAS* **385** 531.
[13] Weyl H. 1918 *Ann. Phys.* (Leipzig), **54** 117.
[14] Herrera L, Filipe M, and Santos N O, 1999 *J. Math Phys.* **40** 8.
[15] Zubairi O, Romero A, and Weber F 2015 *J. Phys. Conf.: Ser.* **615** 012003.
[16] Weber F, Contrera G A, Orsaria M, Spinella W, and Zubairi O 2014 *Mod. Phys. Lett. A* **29** No. 23.
[17] Chodos A, Jaffe R L, Johnson K, Thorn C B, and Weisskopf V F 1974a *Phys. Rev. D* **9** 3471.
[18] Chodos A, Jaffe R L, Johnson K, and Thorn C B, 1974b *Phys. Rev. D* **10** 2599.
[19] Farhi E and Jaffe R L 1984 *Phys. Rev. D* **30** 2379.
[20] Weber F 2015 *Prog. Part. Nucl. Phys.* **54** 193.
[21] Contrera G A, Spinella W, Orsaria M, and Weber F 2014 **SLAC eConf/C1309292**.
[22] Zubairi O 2015 *An Investigation of Deformation of the Stellar Structure of Neutron Stars* (Ph.D. Dissertation: Montezuma Publishing, San Diego State University) P 19-68.
[23] Zubairi O and Weber F 2016 *Nonspherical Compact Stellar Objects in Einstein Theory of General Relativity* (Book Chapter: Centennial of General Relativity; A Celebration, World Scientific Publishing) P 77-94.
[24] Zubairi O, Spinella W, Romero A, Mellinger R, Weber F, Orsaria M, and Contrera G A 2015 **arXiv:1504.03006 [astro-ph.SR]**.
[25] Orsaria M, Rodrigues H, Weber F, and Contrera G A 2014 *Phys. Rev. C* **89** 015806.

On a regular modified C-metric

Hristu Culetu
Ovidius University, Department of Physics and Electronics,
Mamaia Avenue 124, 900527 Constanta, Romania

E-mail: hculetu@yahoo.com

Abstract. A particular form of the C-metric is investigated, giving it a non-standard interpretation and removing any singularity at $r = 0$. In the weak field limit of the accelerating black hole, the proper acceleration A of a static observer is constant and the geometry becomes conformally-flat (anti de Sitter). The stress tensor is of Λ-type ($\Lambda = -3a^2/8\pi G$) and its energy density is negative. We propose that Λ is responsible of inertial forces that appear in uniformly accelerated systems (far from the accelerating source m and for $r << 1/a$ the dominant term in the expression of a^r is $-a\ cos\theta$). The components of the stress tensor and all invariants of the conformally-flat Schwarzschild spacetime are regulated by means of the exponential factor $exp(-k/r),\ k > 0$.
Keywords: uniform acceleration; cosmological constant; regular C-metric; inertial forces

1. Introduction

The well-known C-metric belongs to a class of spacetimes with boost-rotation symmetries [1, 2, 3, 4, 5]. Kinnersley and Walker [6] have shown since 1970 that the C-metric describes a pair of uniformly accelerated black holes (BHs) in the Minkowski spacetime. Their acceleration originates from the conical singularities produced by a strut between the two BHs or two semi-infinite strings connecting them to infinity. The pair creation of BHs may be possible in a background with a cosmological constant Λ as this supplies the negative potential energy [7, 8].

To find the physical interpretation of the $\Lambda \neq 0$ case, Podolsky and Griffiths [2] introduced a new coordinate system adapted to the motion of two uniformly accelerating test particles in de Sitter (dS) space. However, the curvature singularity at $r = 0$ is also present, as in the previous papers on the subject. A physical meaning of the C-metric with a negative Λ was given by Podolsky [3]. He showed that this exact solution of Einstein's field equations describes uniformly accelerated BHs in anti de Sitter (AdS) universe, using a convenient coordinate system. Recently Arnfinnsson and Gron [4] found a new source (a singular accelerating mass shell) of the C-metric, using the Israel junction conditions. They took advantage of the C-metric in spherical coordinates, previously used in [9]. The shell consists of a perfect fluid that creates a jump of the extrinsic curvature when the shell is crossed.

Our purpose in this work is to develop the previous studies on the C-metric, giving it a different interpretation and ruling out any singularity at $r = 0$. We get rid of the conical singularity by neglecting $2amG/c^4$ w.r.t. unity (a is the constant acceleration and m is the BH mass). Our basic geometry is conformal to the Schwarzschild (KS) metric and possesses axial symmetry. We found that the const a is the proper acceleration of a static observer and the spacetime is AdS far from the mass m, with a cosmological constant $\Lambda = -3a^2$. To get rid of the curvature singularity at $r = 0$ we make use of the modified KS geometry [10, 11] so that the spacetime becomes AdS both when $r \to 0$ and when $r \to \infty$.

Throughout the paper we use geometrical units $G = c = 1$, unless otherwise specified.

2. Conformal Schwarzschild metric

To begin with, we write down the C-metric in spherical coordinates, following [4, 9]

$$ds^2 = \frac{1}{(1 + arcos\theta)^2}\left(-Qdt^2 + \frac{1}{Q}dr^2 + \frac{r^2}{P}d\theta^2 + Pr^2 sin^2\theta d\phi^2\right), \tag{1}$$

where $Q = (1 - a^2r^2)(1 - 2m/r)$, $P = 1 + 2amcos\theta$, a is a constant and m is the BH (or a point particle) mass. According to the authors of [4] and [9], Eq. (1) can be viewed as a nonlinear combination of the KS and Rindler geometries, representing the metric around an accelerating point particle or a BH. The function P is related to a conical singularity along the symmetry axis where the sources of the acceleration a (a cosmic string or a strut) are supposed to be located.

Let us observe that the term $2amcos\theta$ in the expression of P is very small compared to unity when reasonable values for a and m are used. We have, indeed, $am = amG/c^4 = ma/(c^4/G)$, where $c^4/G \approx 10^{44}N$ is a maximal force. Therefore, we shall work in the approximation $P = 1$. Something similar is valid for the function Q. The term a^2r^2/c^4 contains the factor $1/c^4 \approx 10^{-34}s^4/m^4$ and, therefore, we take $Q = 1 - 2m/r$ from now on, but we will preserve the first order terms ar/c^2 and $2Gm/c^2r$. In other words, our areas of study are $am << 1$ and $a^2r^2 << 1$. The condition $a^2r^2 << 1$ is compatible with the removal of the singularity at $r = 0$, which will be done in Section 4.

The above approximations yield the conformal KS metric

$$ds^2 = \frac{1}{(1 + arcos\theta)^2}\left[-(1 - \frac{2m}{r})dt^2 + \frac{dr^2}{1 - \frac{2m}{r}} + r^2 d\Omega^2\right], \tag{2}$$

where $d\Omega^2$ stands for the metric on the unit 2-sphere. The spacetime (2) has a BH horizon at $r = 2m$. We might consider (2) as a metric by itself and forget that it originates from the C-metric. The Ricci scalar for (2) is given by

$$R^a{}_a = -12a^2 - \frac{12am}{r^2}cos\theta\ (1 - arcos\theta) \tag{3}$$

It is worth noticing that there is a curvature singularity when $r \to 0$ and asymptotically $R^a{}_a = -12a^2$. The singularity will be removed in Section 4, after a slight modification of the geometry (2).

Let us take now a static observer in the geometry (2) with the velocity vector field

$$u^b = \left(\frac{1 + arcos\theta)}{\sqrt{1 - \frac{2m}{r}}}, 0, 0, 0\right), \tag{4}$$

where b labels (t, r, θ, ϕ). The acceleration 4-vector $a^b = u^a\nabla_a u^b$ has the nonzero components

$$a^r = (\frac{m}{r^2} - acos\theta + \frac{3am}{r}cos\theta)(1 + arcos\theta), \quad ra^\theta = asin\theta\ (1 + arcos\theta) \tag{5}$$

with the proper acceleration

$$A \equiv \sqrt{a^b a_b} = \sqrt{\frac{\left[acos\theta(1 - \frac{2m}{r}) - \frac{m}{r^2}(1 + arcos\theta)\right]^2 + (1 - \frac{2m}{r})a^2 sin^2\theta}{1 - \frac{2m}{r}}} \tag{6}$$

IOP Conf. Series: Journal of Physics: Conf. Series **845** (2017) 012006 doi:10.1088/1742-6596/845/1/012006

In the weak field limit $((m = 0), A = a$ and the interpretation of the constant a as the acceleration of a static observer is justified. In addition, the horizon surface gravity is given by

$$\kappa = \sqrt{a^b a_b}\ \sqrt{-g_{tt}}|_{r=2m} = \frac{m}{r^2}|_{r=2m} = \frac{1}{4m}, \tag{7}$$

exactly as for a KS black hole. It does not depend on θ, as it should be.

In the expression of a^r from (5) we observe that the dominant term is $-acos\theta$, when $r >> 2m, ar << 1$ and $ma << 1$. If we consider a test particle of mass m', located on the z-axis ($\theta = 0$), we have in the above conditions $m'ra^\theta = 0$ and $m'a^r \approx -m'a$ (the term mm'/r^2 represents the Newtonian gravitational force from the classical mechanics and $-m'a$ is the inertial force; because of the Newton's constant G, the Newtonian force is obviously much weaker than the inertial force). The standard view asserts that the inertial forces originate from a fictitious field but here the field comes from the nonflat character of the geometry (even when $m = 0$): the nonzero Ricci scalar $R^a_{\ a} = -12a^2$ or the Kretschmann invariant $K = 24a^4$. Therefore, in our opinion, the inertial forces which appear in accelerated systems originates from the negative cosmological constant generated by acceleration (when $m = 0$, the spacetime (2) is AdS, with $\Lambda = -3a^2$). A similar dependence of Λ on the acceleration a has been reached by Cvetic and Griffies [12] in their studies of domain walls with an AdS geometry on one side (see also [13]). Moreover, they showed that asymptotically in the AdS spacetime, geodesic particles moving in the $z = rcos\theta$ direction are Rindler particles whose constant proper acceleration is proportional to $\sqrt{-\Lambda}$. In conclusion, the spacetime in an accelerated coordinate system turns out not to be flat but conformally flat (AdS) and the metric cannot be obtained by a coordinate transformation from the Minkowski space. The curvature is created by the agent which accelerates the test particle. That reminds us the quantum Unruh effect - the heat bath felt by an accelerating observer. We must keep in mind that the thermal bath of particles - which is an intermediary - has energy and, as a consequence, creates curvature (by the agent which put the reference system in accelerated motion). This curvature is, however, neglected (the thermal bath has a negligible temperature) and the metric remains flat.

3. Anisotropic stress-energy tensor

We look now for the sources of the spacetime (2), namely the stress tensor to lie on the r.h.s. of Einsteins' equations $G_{ab} = 8\pi T_{ab}$ in order that (2) to be an exact solution. By means of the software package Maple - GRTensorII, one finds that

$$T^t_{\ t} = -\rho = \frac{3a^2}{8\pi}\left(1 + \frac{2m}{ar^2}cos\theta\right), \quad T^r_{\ r} = p_r = -\rho, T^\theta_{\ \theta} = T^\phi_{\ \phi} = p_\theta = p_\phi = \frac{3a^2}{8\pi}\left(1 - \frac{2m}{r}cos^2\theta\right) \tag{8}$$

Because $T^t_{\ t} = T^r_{\ r}$, and $T^\theta_{\ \theta} = T^\phi_{\ \phi}$, the spacetime has an infinite set of comoving reference frames [14, 15, 16]. One sees also that $a \to 0$ leads to a vanishing T^a_b, even though $m \neq 0$. This is because the metric (2) becomes the KS metric which is an empty space solution of the Einstein equations. Moreover, the weak field limit ($m = 0$) gives us an AdS spacetime, with $\Lambda = -3a^2$, when the anisotropic fluid becomes isotropic, with

$$-\rho = \frac{3a^2}{8\pi} = p_r = p_\theta. \tag{9}$$

In this case T^a_b is Lorentz-invariant ($T^a_b \propto \Lambda\delta^a_b$). The energy density of the gravitational fluid is negative and so the weak energy condition is not observed. However, when $m \neq 0$, for small r and with $cos\theta < 0$ or $a < 0$, one might obtain $\rho > 0$.

With all the fundamental constants introduced in (8), the energy density looks like

$$\rho = -\frac{3a^2}{8\pi G}\left(1 + \frac{2Gm}{ar^2}cos\theta\right) = -\frac{3a^2}{8\pi G}\left(1 + \frac{2g}{a}cos\theta\right) \tag{10}$$

where $g = Gm/r^2$ is the Newtonian acceleration due to the mass m, at the distance r from the point particle or the BH. If we take, for example, $m = 1Kg$, $r = 1m$ and $a = 1m/s^2$, one obtains $g << a$ and $\rho \approx -3a^2/8\pi G = -10^{10} ergs/cm^3$. Consequently, the effect of the acceleration a of the mass m is much greater than that of the gravitational field g due to the mass m. We also note that $\rho \propto a^2$, as in the Newtonian gravitation.

For the tangential pressures we may perform a similar analysis

$$p_\theta = p_\phi = \frac{3a^2}{8\pi G}\left(1 - \frac{2Gm}{rc^2}cos^2\theta\right) = \frac{3a^2}{8\pi G}\left(1 - \frac{r_g}{r}cos^2\theta\right) \tag{11}$$

where r_g is the gravitational radius of the mass m. Outside the source (if the particle is not a BH) we have $r > r_g$ and, therefore, $p_\theta > 0$ for any $\theta \in (0, \pi)$. All the components of the stress tensor are, of course, divergent at $r = 0$ because of the curvature singularity at the origin of coordinates.

4. Regular stress-energy tensor

We try now to modify the metric (2) in order to render all the physical parameters finite, both at the origin and at infinity. For that purpose we apply a previous choice [10, 11] and introduce the factor $exp(-k/r), k > 0$ as a cutoff in the KS metric. Therefore, the geometry (2) becomes

$$ds^2 = \frac{1}{(1 + ar\ cos\theta)^2}\left[-\left(1 - \frac{2m}{r}e^{-\frac{k}{r}}\right)dt^2 + \frac{1}{1 - \frac{2m}{r}e^{-\frac{k}{r}}}dr^2 + r^2 d\Omega^2\right]. \tag{12}$$

A discussion of the geometry inside the square parantheses from (12) in terms of the values of k has been given in [11] and will not be repeated here. The exponential factor, indeed, makes finite all invariants corresponding to the metric (12) and all the components of the stress tensor. The curvature scalar has the form

$$R^a_a = -12a^2 + \frac{2}{r^5}(...)e^{-\frac{k}{r}} \tag{13}$$

where (...) stands for a fourth degree polinomial of r. It is clear from (13) that the Ricci scalar tends to the constant value $-12a^2$ when $r \to 0$ or when $r \to \infty$.

As far as $T^a_{\ b}$ is concerned, we obtain the following expressions for its components

$$T^t_{\ t} = -\rho = \frac{3a^2}{8\pi} - \left[\frac{q^2}{8\pi r^4}(1 + ar\ cos\theta) - \frac{3am}{4\pi r^2}cos\theta\right]e^{-\frac{q^2}{2mr}}, T^r_{\ r} = p_r = -\rho, \tag{14}$$

and

$$T^\theta_{\ \theta} = T^\phi_{\ \phi} = \frac{3a^2}{8\pi} - \left[\frac{3ma^2}{4\pi r}cos^2\theta - \frac{q^2}{8\pi r^4}(1 + ar\ cos\theta)(1 + 3ar\ cos^2\theta) + \frac{q^4}{32\pi m r^5}(1 + ar\ cos\theta)^2\right]e^{-\frac{q^2}{2mr}} \tag{15}$$

where the positive constant k has been replaced with $q^2/2m$ for a particle of charge q (see [11]). We see from Eqs.(14) and (15) that the mass m and the charge q play no role when $r \to 0$ or $r \to \infty$, due to the exponential cutoff. What remains is the first term proportional to the cosmological constant $\Lambda = -3a^2$, i.e. the values of ρ, p_r and p_θ given by (3) when $m = 0$.

A similar effect takes place for the components of the acceleration of a static observer. We have

$$a^r = -acos\theta\ (1 + arcos\theta) + \left[\frac{m}{r^2}(1 + arcos\theta)(1 + 3arcos\theta) - \frac{q^2}{2r^3}(1 + arcos\theta)^2\right] e^{-\frac{q^2}{2mr}} \quad (16)$$

and

$$ra^\theta = asin\theta\ (1 + arcos\theta), \quad a^t = a^\phi = 0. \quad (17)$$

We observe that, at $r = 0$, $a^r = -acos\theta$ and $ra^\theta = asin\theta$, whence $\sqrt{a^b{}_b|_{r=0}} = A_0 = a$. However, for an asymptotic, static observer, both nonzero components are divergent.

Let us remark that, with $P \neq 1, Q \neq 1$ in (1), the Ricci scalar and all nonzero components of T^a_b are not regulated at $r = 0$ even when the exponential factor is used because there are terms in their expressions which do not contain $exp(-k/r)$.

5. Conclusions

Our main assumption in this paper refers to the non-flat character of the metric in an accelerating reference system. Far from a uniformly accelerated BH (or a point particle) of mass m the metric seems to be not flat but conformally-flat, with axial symmetry. The associated energy-momentum tensor is of Λ-type ($\Lambda = -3a^2/8\pi G$), Lorentz invariant, with negative energy density and the Ricci scalar $R^a{}_a = -12a^2$.

We consider that Λ is responsible for the inertial forces that give rise in accelerated reference systems. Therefore, in our view the metric in an accelerating system cannot be obtained from the Minkowski metric through a coordinate transformation.

Acknowledgments

I would like to thank the Organizers of the IARD 2016 Conference, held at Ljubljana, Slovenia, June 6-9, for inviting me to attend it and for financial support.

6. References

[1] Bicak J and Schmidt B 1989 *Phys. Rev.* D **40** 1827
[2] Podolsky J and Griffiths J B 2001 *Phys. Rev.* D **63**, 024006
[3] Podolsky J 2002 *Czech. J. Phys.* **52** 1
[4] Arnfinnsson B and Gron O 2014 A source of the C-metric with perfect translational inertial dragging (*Preprint* gr-qc/1408.4588)
[5] Mann R B 1997 *Class Quantum Grav.* **14** L109
[6] Kinnersley W and Walker M 1970 *Phys. Rev.* D **2** 1359
[7] Plebanski J F and Demianski M 1976 *Ann. Phys. (NY)* **98** 98
[8] Mann R B and Ross S F 1995 *Phys. Rev.* D **52** 2254
[9] Griffiths J B Krtous P and Podolsky J 2006 Interpreting the C-metric *Class. Quantum Grav.* **23** 6745 (*Preprint* gr-qc/0609056)
[10] Culetu H 2013 On a regular modified Schwarzschild spacetime (*Preprint* gr-qc/1305.5964)
[11] Culetu H 2014 On a regular charged black hole with a nonlinear electric source (*Preprint* gr-qc/1408.3334)
[12] Cvetic M and Griffies S 1992 Gravitational Effects in Supersymmetric Domain Wall Backgrounds *Phys. Lett.* B **285** 27 (*Preprint* hep-th/9204031)
[13] Cvetic M Griffies S and Soleng H H 1993 Local and global gravitational aspects of domain wall space-times *Phys. Rev.* D **48** 2613 (*Preprint* gr-qc/9306005)
[14] Dymnikova I 1992 *Gen. Relat. Grav.* **24**, 235
[15] Dymnikova I 2000 The algebraic structure of a cosmological term in spherically symmetric solutions *Phys. Lett.* B **472** 33 (*Preprint* gr-qc/9912116)
[16] Dymnikova I and Korpusik M 2011 *Entropy* **13** 1967

IOP Conf. Series: Journal of Physics: Conf. Series **845** (2017) 012007 doi:10.1088/1742-6596/845/1/012007

Dark matter: a problem in relativistic metrology?

Luca Lusanna
Sezione INFN di Firenze, Via G. Sansone 1, 50019 Sesto Fiorentino (FI), Italy
E-mail: lusanna@fi.infn.it

Abstract.
Besides the tidal degrees of freedom of Einstein general relativity (GR) (namely the two polarizations of gravitational waves after linearization of the theory) there are the inertial gauge ones connected with the freedom in the choice of the 4-coordinates of the space-time, i.e. in the choice of the notions of time and 3-space (the 3+1 splitting of space-time) and in their use to define a non-inertial frame (the inertial ones being forbidden by the equivalence principle) by means of a set of conventions for the relativistic metrology of the space-time (like the GPS ones near the Earth).

The canonical York basis of canonical ADM gravity allows us to identify the Hamiltonian inertial gauge variables in globally hyperbolic asymptotically Minkowskian space-times without super-translations and to define the family of non-harmonic Schwinger time gauges. In these 3+1 splittings of space-time the freedom in the choice of time (the problem of clock synchronization) is described by the inertial gauge variable York time (the trace of the extrinsic curvature of the instantaneous 3-spaces). This inertial gauge freedom and the non-Euclidean nature of the instantaneous 3-spaces required by the equivalence principle need to be incorporated as metrical conventions in a relativistic suitable extension of the existing (essentially Galilean) ICRS celestial reference system.

In this paper I make a short review of the existing possibilities to explain the presence of dark matter (or at least of part of it) as a relativistic inertial effect induced by the non-Euclidean nature of the 3-spaces. After a Hamiltonian Post-Minkowskian (HPM) linearization of canonical ADM tetrad gravity with particles, having equal inertial and gravitational masses, as matter, followed by a Post-Newtonian (PN) expansion, we find that the Newtonian equality of inertial and gravitational masses breaks down and that the inertial gauge York time produces an increment of the inertial masses explaining at least part of what is called dark matter in all its astrophysical signatures.

1. Canonical ADM Tetrad gravity

The results presented in this review derive from the systematic study of the existing Hamiltonian formulations of metric and tetrad ADM gravity [1] in *globally hyperbolic, topologically trivial, asymptotically Minkowskian space-times without super-translations and Killing symmetries* done in the papers [1, 2, 3, 4, 5, 6, 7, 8] (see Refs. [9] for full reviews and Ref.[10] for the inclusion of Killing symmetries). In these papers the matter consists of electrically charged positive-energy scalar particles plus the electro-magnetic field. The canonical theory takes into account all the aspects of Dirac's theory of constraints [11, 12, 13, 14], in particular the use of the

[1] Tetrad gravity is needed for the description of fermion fields.

Shanmugadhasan canonical transformation [6, 13, 15] to find canonical bases adapted to the constraints .

In this class of space-times at spatial infinity there is an asymptotic ADM Poincaré group [16], which becomes the special relativistic Poincaré group of the matter present in the space-time (this is an important condition for the inclusion of particle physics, where elementary particles are irreducible representations of this group) in the limit of vanishing Newton constant ($G = 0$). The canonical Hamiltonian is the weak ADM energy [2], so that there is no frozen picture like in the family of spatially compact without boundary space-times considered in loop quantum gravity, where the Dirac Hamiltonian is a combination of constraints because the canonical Hamiltonian vanishes.

Since the equivalence principle forbids the existence of global inertial frames, we had to define *global non-inertial frames* (with instantaneous non-Euclidean 3-spaces with synchronized clocks) in this class of space-times by adapting to GR the theory of global non-inertial frames in Minkowski space-time developed in Ref. [17] and applied to relativistic atomic physics in Ref.[18].

According to the *3+1 point of view* one gives the world-line of a time-like observer and an admissible 3+1 splitting of the space-time, i.e. a nice foliation whose leaves are instantaneous 3-spaces. *Lorentz-scalar observer-dependent radar 4-coordinates* $\sigma^A = (\tau; \sigma^r)$ are used: τ is an arbitrary increasing function of the observer proper time and σ^r are curvilinear 3-coordinates on the 3-spaces Σ_τ (diffeomorphic to R^3) with the observer as origin. The inverse transformation $\sigma^A \mapsto x^\mu = z^\mu(\tau, \sigma^r)$ defines the embeddings of the 3-spaces Σ_τ into the space-time and the induced 4-metric is $g_{AB}[z(\tau, \sigma^r)] = [z^\mu_A\, z^\nu_B](\tau, \sigma^r)\, g_{\mu\nu}(\tau, \sigma^r)$, where $z^\mu_A = \partial\, z^\mu / \partial\, \sigma^A$ (the signature of the 4-metric is $\epsilon = \pm 1$ according to either the particle physics $\epsilon = 1$ or the general relativity $\epsilon = -1$ convention).

While the 4-vectors $z^\mu_r(\tau, \sigma^u)$ are tangent to Σ_τ, so that the unit normal $l^\mu(\tau, \sigma^u)$ is proportional to $\epsilon^\mu{}_{\alpha\beta\gamma}\, [z^\alpha_1\, z^\beta_2\, z^\gamma_3](\tau, \sigma^u)$, we have $z^\mu_\tau(\tau, \sigma^r) = [N\, l^\mu + N^r\, z^\mu_r](\tau, \sigma^r)$ ($N(\tau, \sigma^r) = \epsilon\, [z^\mu_\tau\, l_\mu](\tau, \sigma^r)$ and $N_r(\tau, \sigma^r) = -\epsilon\, g_{\tau r}(\tau, \sigma^r)$ are the lapse and shift functions).

The foliation is nice and admissible if it satisfies the conditions:
1) $N(\tau, \sigma^r) > 0$ in every point of Σ_τ (the 3-spaces never intersect, avoiding the coordinate singularity of Fermi coordinates);
2) $\epsilon\, {}^4g_{\tau\tau}(\tau, \sigma^r) > 0$, so to avoid the coordinate singularity of the rotating disk, and with the positive-definite 3-metric ${}^3g_{rs}(\tau, \sigma^u) = -\epsilon\, {}^4g_{rs}(\tau, \sigma^u)$ having three positive eigenvalues (as shown in Ref.[17] these are the Møller conditions);
3) all the 3-spaces Σ_τ must tend to the same space-like hyper-plane at spatial infinity. Due to the absence of super-translations [16], [1, 2, 3], the non-Euclidean 3-spaces are orthogonal to the conserved ADM 4-momentum at spatial infinity [6]: this is a *non-inertial rest frame* of the 3-universe (see Ref.[17] for the non-inertial and inertial rest-frames in special relativity). There are asymptotic inertial observers with spatial axes identified by means of the fixed stars of star catalogues.

As shown explicitly in Ref.[8] *the ten quantities* ${}^4g_{AB}(\tau, \sigma^r)$ *are 4-scalars of the space-time due to the use of the 4-scalar radar 4-coordinates.* The same happens for *all the components of "radar tensors" (i.e. tensors expressed in radar 4-coordinates): they are 4-scalars of the space-time.*

As a consequence, the 3-universe (the isolated system "gravitational field plus matter") can be

[2] It is a volume integral over 3-space of a coordinate-dependent energy density. It is weakly equal to the *strong* ADM energy, which is a flux through a 2-surface at spatial infinity.

described as a decoupled non-covariant non-observable external pseudo-particle carrying a pole-dipole structure, whose mass and spin are identified by the ADM weak energy and by the ADM angular momentum. Instead the ADM 3-momentum vanishes due to the rest-frame condition. The vanishing of the ADM Lorentz boosts eliminate the conjugate internal center of mass of the 3-universe. In absence of matter Christodoulou - Klainermann space-times [19] are compatible with this description.

In GR the dynamical variable is the 4-metric, which determines the dynamical chrono-geometrical structure of space-time by means of the line element: it teaches to massless particles which are the allowed trajectories in each point. In tetrad gravity the 4-metric is decomposed in terms of cotetrads, ${}^4g_{AB} = E_A^{(\alpha)}\, {}^4\eta_{(\alpha)(\beta)}\, E_B^{(\beta)}$ [3], and the ADM action, now a functional of the 16 fields $E_A^{(\alpha)}(\tau, \sigma^r)$, is taken as the action for ADM tetrad gravity [2, 3, 6]. This leads to an interpretation of gravity based on a congruence of time-like observers endowed with ortho-normal tetrads: in each point of space-time the time-like axis is the unit 4-velocity of the observer, while the spatial axes are a (gauge) convention for the observer's gyroscopes.

We get the following parametrization of tetrads, cotetrads and 4-metric ($\alpha_{(a)}(\tau, \sigma^r)$ are angles, $\varphi_{(a)}(\tau, \sigma^r)$ are boost parameters, $L^{(\alpha)}{}_{(\beta)}$ are Lorentz matrices, $R_{(a)(b)}$ are rotation matrices, ${}^3e_{(a)r}(\tau, \sigma^r) = \sum_b R_{(a)(b)}(\alpha_{(c)}(\tau, \sigma^r))\, {}^3\bar{e}_{(a)r}(\tau, \sigma^r)$ are cotriads on the 3-space, $N(\tau, \sigma^r) = 1 + n(\tau, \sigma^r)$ and $n_{(a)}(\tau, \sigma^r) = \sum_b R_{(a)(b)}(\alpha_{(c)}(\tau, \sigma^r))\, \bar{n}_{(b)}(\tau, \sigma^r)$ are the lapse and shift functions respectively, ${}^4\overset{\circ}{E}{}^A_{(\beta)}$ and ${}^4\overset{\circ}{\bar{E}}{}^A_{(o)}$ are tetrads adapted to the 3+1 splitting)

$$
\begin{aligned}
{}^4E^A_{(\alpha)} &\overset{def}{=} {}^4\overset{\circ}{E}{}^A_{(\beta)}\, L^{(\beta)}{}_{(\alpha)}(\varphi_{(a)}) \overset{def}{=} {}^4\overset{\circ}{\bar{E}}{}^A_{(o)}\, L^{(o)}{}_{(\alpha)}(\varphi_{(c)}) + \\
&+ \sum_{ab} {}^4\overset{\circ}{\bar{E}}{}^A_{(b)}\, R^T_{(b)(a)}(\alpha_{(c)})\, L^{(a)}{}_{(\alpha)}(\varphi_{(c)}),\\
{}^4g_{AB} &= E_A^{(\alpha)}\, {}^4\eta_{(\alpha)(\beta)}\, E_B^{(\beta)} = \\
&= {}^4\overset{\circ}{E}{}^{(\alpha)}_A\, {}^4\eta_{(\alpha)(\beta)}\, {}^4\overset{\circ}{E}{}^{(\beta)}_B = {}^4\overset{\circ}{\bar{E}}{}^{(\alpha)}_A\, {}^4\eta_{(\alpha)(\beta)}\, {}^4\overset{\circ}{\bar{E}}{}^{(\beta)}_B\,,\\
{}^4\overset{\circ}{\bar{E}}{}^A_{(o)} &= {}^4\overset{\circ}{E}{}^A_{(o)} = \frac{1}{1+n}\,(1; -\sum_a \bar{n}_{(a)}\, {}^3\bar{e}^r_{(a)}) = l^A, \qquad {}^4\overset{\circ}{\bar{E}}{}^A_{(a)} = (0; {}^3\bar{e}^r_{(a)}),\\
{}^4\overset{\circ}{\bar{E}}{}^{(o)}_A &= {}^4\overset{\circ}{E}{}^{(o)}_A = (1+n)\,(1; \vec{0}) = \epsilon\, l_A, \qquad {}^4\overset{\circ}{\bar{E}}{}^{(a)}_A = (\bar{n}_{(a)}; {}^3\bar{e}_{(a)r}),
\end{aligned}
$$

$$
\begin{aligned}
{}^4g_{\tau\tau} &= \epsilon\,[(1+n)^2 - \sum_a \bar{n}^2_{(a)}], \qquad {}^4g_{\tau r} = -\epsilon\, n_r = -\epsilon \sum_a \bar{n}_{(a)}\, {}^3\bar{e}_{(a)r},\\
{}^4g_{rs} &= -\epsilon\, {}^3g_{rs} = -\epsilon \sum_a {}^3\bar{e}_{(a)r}\, {}^3\bar{e}_{(a)s}, \qquad \sqrt{-g} = \sqrt{|{}^4g|} = \frac{\sqrt{{}^3g}}{\sqrt{\epsilon\, {}^4g^{\tau\tau}}} = \sqrt{\gamma}\,(1+n).
\end{aligned}
\tag{1}
$$

As shown in Refs.[2, 3, 6] in canonical ADM tetrad gravity there are 16 fields, 16 conjugate momenta, 14 first-class constraints, generators of Hamiltonian gauge transformations, 14 gauge

[3] (α) are flat indices; the cotetrads $E^{(\alpha)}_A$ are the inverse of the tetrads $E^A_{(\alpha)}$ connected to the world tetrads by $E^\mu_{(\alpha)}(x) = z^\mu_A(\tau, \sigma^r)\, E^A_{(\alpha)}(z(\tau, \sigma^r))$.

variables (the *GR inertial effects*) and 2+2 physical variables, the *tidal effects* (the gravitational waves after linearization). In Ref.[6] a York canonical basis, adapted to ten first-class constraints (not to the super-Hamiltonian and super-momentum ones, whose solution is unknown), was identified by means of a Shanmugadhasan canonical transformations : this allows *for the first time to get the explicit identification of the inertial and tidal variables.* It implements the York map of Ref.[20] and diagonalizes the York-Lichnerowicz approach [21]. Its final form is

$\varphi_{(a)}$	$\alpha_{(a)}$	n	$\bar{n}_{(a)}$
$\pi_{\varphi_{(a)}} \approx 0$	$\pi^{(\alpha)}_{(a)} \approx 0$	$\pi_n \approx 0$	$\pi_{\bar{n}_{(a)}} \approx 0$

θ^r	$\tilde{\phi}$	$R_{\bar{a}}$
$\pi^{(\theta)}_r$	$\pi_{\tilde{\phi}} = \frac{c^3}{12\pi G}\, {}^3K$	$\Pi_{\bar{a}}$

$$\begin{aligned}
{}^3e_{(a)r} &= \sum_b R_{(a)(b)}(\alpha_{(c)})\, V_{rb}(\theta^i)\, \tilde{\phi}^{1/3}\, e^{\sum_{\bar{a}}^{1,2} \gamma_{\bar{a}a} R_{\bar{a}}},\\
{}^4g_{\tau\tau} &= \epsilon\, [(1+n)^2 - \sum_a \bar{n}^2_{(a)}],\\
{}^4g_{\tau r} &= -\epsilon\, \bar{n}_{(a)}\, {}^3\bar{e}_{(a)r},\\
{}^4g_{rs} &= -\epsilon\, {}^3g_{rs} = -\epsilon\, \tilde{\phi}^{2/3} \sum_a V_{ra}(\theta^i)\, V_{sa}(\theta^i)\\
& e^{2\sum_{\bar{a}}^{1,2} \gamma_{\bar{a}a} R_{\bar{a}}},
\end{aligned} \tag{2}$$

In this York canonical basis the *inertial effects* are described by the arbitrary gauge variables $\alpha_{(a)}$, $\varphi_{(a)}$, $1+n$, $\bar{n}_{(a)}$, θ^i, ${}^3K$, while the *tidal effects*, i.e. the physical degrees of freedom of the gravitational field (the two polarizations of gravitational waves (GW) in the linearized theory), by the two canonical pairs $R_{\bar{a}}$, $\Pi_{\bar{a}}$, $\bar{a} = 1, 2$. The momenta $\pi^{(\theta)}_r$ and the 3-volume element $\tilde{\phi} = \sqrt{det\, {}^3g_{rs}}$ have to be found as solutions of the super-momentum [4] and super-Hamiltonian (i.e. the Lichnerowicz equation) constraints, respectively.

Instead the Dirac observables (DO) (gauge invariant under the Hamiltonian gauge transformations generated by all the first class constraints; see Ref.[8]) of the gravitational field are not known: they would be the two pairs of 4-scalar tidal variables in a Shanmugadhasan canonical basis adapted to all the 14 first class constraints.

The gauge variables $\alpha_{(a)}$, $\varphi_{(a)}$ parametrize the extra O(3,1) gauge freedom of the tetrads (the gauge freedom for each observer to choose three gyroscopes as spatial axes and to choose the law for their transport along the world-line). In the Schwinger time gauges where one imposes the gauge fixings $\varphi_{(a)}(\tau, \sigma^r) \approx 0$, $\alpha_{(a)}(\tau, \sigma^r) \approx 0$ so that the tetrads become adapted to the 3+1 splitting (the time-like tetrad coincides with the unit normal to the 3-space).

The gauge angles θ^i (i.e. the director cosines of the tangents to the three coordinate lines in each point of Σ_τ) describe the freedom in the choice of the axes for the 3-coordinates σ^r on each

[4] The name super-momentum is due to the fact that these three constraints determine three components of the momenta conjugated to the metric.

IOP Conf. Series: Journal of Physics: Conf. Series **845** (2017) 012007 doi:10.1088/1742-6596/845/1/012007

3-space: their fixation implies the determination of the shift gauge variables $\bar{n}_{(a)}$, namely the appearances of gravito-magnetism in the chosen 3-coordinate system.

Only one momentum is a gauge variable (a reflection of the Lorentz signature): the *York time*, i.e. the trace ${}^3K(\tau, \sigma^r)$ of the *extrinsic curvature* of the non-Euclidean 3-spaces as 3-sub-manifolds of space-time. This inertial effect (absent in Newtonian gravity with its absolute Euclidean 3-space) describes the GR remnant of the special-relativistic gauge freedom in clock synchronization [18]. Its fixation determines the lapse function.

In the York canonical basis the Hamilton equations generated by the Dirac Hamiltonian $H_D = \hat{E}_{ADM} + (constraints)$ are divided in four groups: A) four contracted Bianchi identities, namely the evolution equations for $\tilde{\phi}$ and $\pi_i^{(\theta)}$ (they say that given a solution of the constraints on a Cauchy surface, it remains a solution also at later times); B) four evolution equation for the four basic gauge variables θ^i and ${}^3K$: these equations determine the lapse and the shift functions once four gauge fixings for the basic gauge variables are added; C) four evolution equations for the tidal variables $R_{\bar{a}}$, $\Pi_{\bar{a}}$; D) the Hamilton equations for matter, when present.

Once a gauge is completely fixed, the Hamilton equations become deterministic. Given a solution of the super-momentum and super-Hamiltonian constraints and the Cauchy data for the tidal variables on an initial 3-space, one can find a solution of Einstein's equations in radar 4-coordinates adapted to a time-like observer. To it there is associated a special 3+1 splitting of space-time with dynamically selected instantaneous 3-spaces in accord with Ref.[5]. Then one can pass to adapted world 4-coordinates ($x^\mu = z^\mu(\tau, \sigma^r) = x^\mu_o + \epsilon^\mu_A\, \sigma^A$) and we can describe the solution in every 4-coordinate system by means of 4-diffeomorphisms.

In Ref.[7] the coupling of N charged scalar particles plus the electro-magnetic field to ADM tetrad gravity is studied in this class of asymptotically Minkowskian space-times without super-translations. To regularize the self-energies both the electric charge and the sign of the energy of the particles are Grassmann-valued. The introduction of the non-covariant radiation gauge allows to reformulate the theory in terms of transverse electro-magnetic fields and to extract the generalization of the Coulomb interaction among the particles in the Riemannian instantaneous 3-spaces of global non-inertial frames.

From the Hamilton equations in the York canonical basis [7], followed by a Hamiltonian Post-Minkowskian (HPM) linearization (disregarding terms of order $O(G^2)$ in the Newton constant and using an ultra-violet cutoff for matter) with the asymptotic flat Minkowski 4-metric at spatial infinity as background, it has been possible to develop a theory of GW's with asymptotic background propagating in the non-Euclidean 3-spaces Σ_τ of a family of *non-harmonic 3-orthogonal* Schwinger time gauges $\alpha_{(a)}(\tau, \sigma^r) \approx 0$, $\varphi_{(a)}(\tau, \sigma^r) \approx 0$, $\theta^i(\tau, \sigma^r) \approx 0$, ${}^3K(\tau, \sigma^r) \approx F(\tau, \sigma^r)$ parametrized by the numerical values $F(\tau, \sigma^r)$ of the York time ${}^3K(\tau, \sigma^r)$ (the left gauge freedom in the shape of the 3-spaces Σ_τ) and having the 3-metric in the 3-spaces diagonal. The lapse and shift functions for these gauges are determined by the Hamilton equations for θ^i and ${}^3K$, which are elliptic equations depending only on the 3-space differently from what happens in the harmonic gauges.

2. Relativistic Metrology and Dark Matter

An extremely important (till now unnoticed) point is that the fixation of the gauge freedom of GR (and of every generally covariant theory of gravity), i.e. the choice of the non-inertial frame and of the axes for the 4-coordinates in each point, is nothing else than *the establishment of conventions for relativistic metrology* [22] (see also Ref.[5] for theoretical considerations concerning the nature of space and time), an operation performed from atomic physicists, NASA engineers and astronomers [23, 24, 25, 26, 27, 28].

The choice of the 4-coordinates is *solved at the experimental level inside the Solar system by the choice of a convention for the description of matter*: a) for satellites near the Earth (like the GPS ones) one uses NASA 4-coordinates compatible with the terrestrial ITFR2003 and geocentric GCRS IAU2000 frames; b) for planets in the Solar System one uses the barycentric BCRS-IAU2000 frame. These frames are compatible with "quasi-inertial frames" in Minkowski space-time. These are metrological choices like the choice of a certain atomic clock as standard of time.

In astronomy the positions of stars and galaxies are determined from the data (luminosity, light spectrum, angles) on the sky as living in a 4-dimensional nearly-Galilei space-time with the celestial ICRS [23, 26, 27, 28] frame considered as a "quasi-inertial frame" (all galactic dynamics is Newtonian gravity), in accord with the standard Friedmann-Robertson-Walker (FRW) ΛCDM cosmological model when the constant intrinsic 3-curvature of 3-spaces is zero (as implied by the CMB data) [29]. To reconcile all the data with this 4-dimensional reconstruction one must postulate the existence of dark matter and dark energy as the dominant components of the classical universe after the recombination 3-surface or of modifications of GR like MOND [30]! Since the celestial reference frame ICRS has diagonal 3-metric, our 3-orthogonal Schwinger time gauges are a good choice.

The linearized PM Hamilton equations for the particles (whose world-lines $x_i^\mu(\tau) = z^\mu(\tau, \eta_i^r(\tau))$ identify radar 3-coordinates $\eta_i^r(\tau)$ due to the 3+1 splitting) and for the electro-magnetic field have been written explicitly in Refs.[7]: among the forces there are both the inertial potentials and the GW's. In the third paper of Ref.[7] we disregarded electro-magnetism and we studied in more detail the HPM equations of motion of the particles. Then we studied the Post-Newtonian (PN) expansion of these regularized HPM equations of motion for the particles and we found that the particle 3-coordinates $\eta_i^r(\tau = ct) = \tilde{\eta}_i^r(t)$ (coinciding with the Newtonian coordinates of the world-lines at this level of approximation) satisfy the equation of motion

$$\frac{d}{dt}\Big[m_i \Big(1 + \frac{1}{c}\frac{d}{dt}\, {}^3\tilde{\mathcal{K}}_{(1)}(t, \vec{\tilde{\eta}}_i(t))\Big) \frac{d\,\tilde{\eta}_i^r(t)}{dt}\Big] \stackrel{\circ}{=} -G\, \frac{\partial}{\partial\, \tilde{\eta}_i^r} \sum_{j\neq i} \eta_j\, \frac{m_i\, m_j}{|\vec{\tilde{\eta}}_i(t) - \vec{\tilde{\eta}}_j(t)|} + \mathcal{O}(G^2). \tag{3}$$

where at the lowest order we find the standard Newton gravitational force $\vec{F}_{i(Newton)}(t) = -m_i\, G\, \frac{\partial}{\partial\, \tilde{\eta}_i^r} \sum_{j\neq i} \frac{m_j}{|\vec{\tilde{\eta}}_i(t) - \vec{\tilde{\eta}}_j(t)|} = -m_i\, \frac{\partial\, \Phi(t, \vec{\tilde{\eta}}_i(t))}{\partial\, \tilde{\eta}_i^r}$.

Since we have $\epsilon\, {}^4g_{\tau\tau}(\tau = ct, \sigma^r) - 1 = 2\, n(\tau = ct, \sigma^r) + O(G^2) = 2\, \frac{\Phi(t, \sigma^r)}{c^2} - \frac{2}{c}\frac{\partial}{\partial t}\, {}^3\tilde{\mathcal{K}}_{(1)}(\tau = ct, \sigma^r) + O(G^2)$, there is a 0.5 PN *inertial effect* (hidden in the lapse function) not existing in the Newton theory where the Euclidean 3-space is an absolute notion like the Newtonian time. It does not depend on the York time ${}^3K_{(1)}$ but on the *non-local York time* (△ is the Laplacian associated to the asymptotic Minkowski 4-metric) ${}^3\tilde{\mathcal{K}}_{(1)}(\tau, \sigma^r) = \Big(\frac{1}{\triangle}\, {}^3K_{(1)}\Big)(\tau, \sigma^r)$. If we put ${}^3K_{(1)} = 0$, the standard results about binaries are reproduced.

We see that the term in the non-local York time can be *interpreted* as the introduction of an *effective (time-, velocity- and position-dependent) inertial mass term* for the kinetic energy of each particle: $m_i \mapsto m_i \left(1 + \frac{1}{c}\frac{d}{dt}\,{}^3\tilde{\mathcal{K}}_{(1)}(t, \vec{\eta}_i(t))\right)$ in each instantaneous 3-space. Instead in the Newton potential there are the gravitational masses of the particles, equal to the inertial ones in the 4-dimensional space-time due to the equivalence principle. Therefore the effect is due to a modification of the effective inertial mass in each non-Euclidean 3-space depending on its shape as a 3-sub-manifold of space-time: *it is the equality of the inertial and gravitational masses of Newtonian gravity to be violated*!

In the 2-body case we get that for Keplerian circular orbits of radius r the modulus of the relative 3-velocity can be written in the form $\sqrt{\frac{G\,(m+\triangle\, m(r))}{r}}$ with $\triangle\, m(r)$ function only of ${}^3\mathcal{K}_{(1)}$. Now the *rotation curves of spiral galaxies* (see Refs.[30, 31, 32]) imply that the relative 3-velocity goes to constant for large r (instead of vanishing like in Kepler theory). This result can be simulated by fitting $\triangle\, m(r)$ (i.e. the non-local York time) to the experimental data: as a consequence $\triangle\, m(r)$ is interpreted as a *dark matter halo* around the galaxy. With our approach this dark matter would be a *relativistic inertial effect* consequence of the a non-trivial shape of the non-Euclidean 3-space as a 3-sub-manifold of space-time. A similar interpretation [7] can be given for the other two main signatures of the existence of dark matter in the observed masses of galaxies and clusters of galaxies, namely the virial theorem and weak gravitational lensing. Therefore there is the possibility of describing part (or maybe all) dark matter as a *relativistic inertial effect*. As we have seen the three main experimental signatures of dark matter can be explained in terms of the non-local York time ${}^3\mathcal{K}_{(1)}(\tau, \vec{\sigma})$, the inertial gauge variable describing the general relativistic remnant of the gauge freedom in clock synchronization.

Therefore there is the concrete possibility to explain the rotation curves of galaxies [31, 32, 33] as a *relativistic inertial effect inside Einstein GR* (choice of a York time compatible with observations) without modifications: a) of Newton gravity like in MOND [30]; b) of GR like in $f(R)$ theories [34, 35]; c) of particle physics with the introduction of WIMPS [36].

Therefore one can propose to define a *Post-Minkowskian ICRS* with non-Euclidean 3-spaces, whose intrinsic 3-curvature (due essentially to gravitational waves) is small, in such a way that the York time be (at least partially) fitted to the observational data implying the presence of dark matter. Then automatically BCRS would be its quasi-Minkowskian approximation for the Solar System. Let us remark that the 3-spaces can be quasi-Euclidean (i.e. with a small 3-curvature tensor), as required by CMB data in the astrophysical context, even when their shape as 3-sub-manifolds of space-time is not trivial and is described by a not-small York time. This would be the way out from the gauge problem in general relativity: the observational conventions for matter would select a reference system of 4-coordinates for PM space-times in the associated 3-orthogonal gauge. A Post-Minkowskian definition of ICRS will be also useful for the ESA-GAIA mission [37] (cartography of the Milky Way) and for the possible anomalies (different from the already explained Pioneer one) inside the Solar System [38].

The open problem is the determination of the non-local York time from the data. From what is known from the Solar System and from inside the Milky Way near the galactic plane, it seems that it is negligible near the stars inside a galaxy. On the other hand, it is non zero near galaxies and clusters of galaxies of big mass. However only a mean value in time of time- and space-derivatives of the non-local York time can be extracted from the data. At this stage it seems that the non-local York time is relevant around the galaxies and the clusters of galaxies where there are big concentrations of mass and the dark matter haloes and that it becomes negligible inside the galaxies where there is a lower concentration of mass. Instead there is no indication on its value in the voids existing among the clusters of galaxies. However if we do not *know the non-local York time on all the 3-universe* at a given τ we cannot get an experimental determination of the York time ${}^3K_{(1)}(\tau, \vec{\sigma}) = \triangle\, {}^3\tilde{\mathcal{K}}_{(1)}(\tau, \vec{\sigma})$. Therefore some phenomenological parametrizations of ${}^3\tilde{\mathcal{K}}_{(1)}(\tau, \vec{\sigma})$ will have to be devised to see the implications for ${}^3K_{(1)}(\tau, \vec{\sigma})$.

3. Conclusion

In Ref.[39] there is a first attempt to fit some data of dark matter by using a Yukawa-like ansatz on York time suggested by the $f(R)$ theories [40]: in this way the good fits of the rotation curves of galaxies obtainable with $f(R)$ theories can be reproduced inside Einstein's GR as an inertial gauge effect. Therefore no conclusion can yet be reached about this proposal for explaining dark matter as a relativistic inertial effect.

The next big challenge after dark matter is *dark energy* in cosmology [41, 42]. Let me remark that in the FRW cosmological solution the Killing symmetries connected with homogeneity and isotropy imply (τ is the cosmic time, $a(\tau)$ the scale factor) ${}^3K(\tau) = -\frac{\dot{a}(\tau)}{a(\tau)} = -H$, namely the York time is no more a gauge variable but coincides with the Hubble constant. However at the first order in cosmological perturbations we have ${}^3K = -H + {}^3K_{(1)}$ with ${}^3K_{(1)}$ being again an inertial gauge variable. Instead in inhomogeneous space-times without Killing symmetries like the Szekeres ones [43, 44, 45] the York time remains an inertial gauge variable.

Therefore the York time has a central position also in the main quantities on which relies the interpretation of dark energy in the standard ΛCDM cosmological model (Hubble constant, the old Hubble redshift-distance relation replaced in FRW cosmology with the velocity distance relation or Hubble law). As a consequence it looks reasonable to investigate on a possible gauge origin also of dark energy by studying the dependence on the York time of quantities like redshift, luminosity distance, gravitational lensing.

References

[1] L.Lusanna, *The Rest-Frame Instant Form of Metric Gravity*, Gen.Rel.Grav. **33**, 1579 (2001)(arXiv gr-qc/0101048).
[2] L.Lusanna and S.Russo, *A New Parametrization for Tetrad Gravity*, Gen.Rel.Grav. **34**, 189 (2002)(arXiv gr-qc/0102074).
[3] R.DePietri, L.Lusanna, L.Martucci and S.Russo, *Dirac's Observables for the Rest-Frame Instant Form of Tetrad Gravity in a Completely Fixed 3-Orthogonal Gauge*, Gen.Rel.Grav. **34**, 877 (2002) (arXiv gr-qc/0105084).
[4] J.Agresti, R.De Pietri, L.Lusanna and L.Martucci, *Hamiltonian Linearization of the Rest-Frame Instant Form of Tetrad Gravity in a Completely Fixed 3-Orthogonal Gauge: a Radiation Gauge for Background-Independent Gravitational Waves in a Post-Minkowskian Einstein Spacetime*, Gen.Rel.Grav. **36**, 1055 (2004) (arXiv gr-qc/0302084).
[5] L.Lusanna and M.Pauri, *Explaining Leibniz equivalence as difference of non-inertial Appearances: Dis-solution of the Hole Argument and physical individuation of point-events*, History and Philosophy of Modern Physics **37**, 692 (2006) (arXiv gr-qc/0604087); *The Physical Role of Gravitational and Gauge Degrees of Freedom in General Relativity. I: Dynamical Synchronization and Generalized Inertial Effects; II: Dirac versus Bergmann Observables and the Objectivity of Space-Time*, Gen.Rel.Grav. **38**, 187 and 229 (2006) (arXiv gr-qc/0403081 and 0407007); *Dynamical Emergence of Instantaneous 3-Spaces in a Class of Models of General Relativity*, in *Relativity and the Dimensionality of the World*, ed. V.Petkov (Springer Series Fundamental Theories of Physics, Berlin, 2007) (arXiv gr-qc/0611045).
[6] D.Alba and L.Lusanna, *The York Map as a Shanmugadhasan Canonical Transformationn in Tetrad Gravity and the Role of Non-Inertial Frames in the Geometrical View of the Gravitational Field*, Gen.Rel.Grav. **39**, 2149 (2007) (arXiv gr-qc/0604086, v2).

[7] D.Alba and L.Lusanna, *The Einstein-Maxwell-Particle System in the York Canonical Basis of ADM Tetrad Gravity: I) The Equations of Motion in Arbitrary Schwinger Time Gauges.*, Canad.J.Phys. **90**, 1017 (2012) (arXiv 0907.4087); *II) The Weak Field Approximation in the 3-Orthogonal Gauges and Hamiltonian Post-Minkowskian Gravity: the N-Body Problem and Gravitational Waves with Asymptotic Background.*, Canad.J.Phys. **90**, 1077 (2012) (arXiv 1003.5143); *III) The Post-Minkowskian N-Body Problem, its Post-Newtonian Limit in Non-Harmonic 3-Orthogonal Gauges and Dark Matter as an Inertial Effect.* , Canad.J.Phys. **90**, 1131 (2012) (arXiv 1009.1794).

[8] L.Lusanna and M.Villani, *Hamiltonian Expression of Curvature Tensors in the York Canonical Basis: I) Riemann Tensor and Ricci Scalars*, Int.J.Geom.Meth.Mod.Phys. **11**, 1450052 (2014) (arXiv 1401.1370); *II) Weyl Tensor, Weyl Scalars, Weyl Eigenvalues and the Problem of the Observables of the Gravitational Field*, Int.J.Geom.Meth.Mod.Phys. **11**, 1450053 (2014) (arXiv 1401.1375).

[9] L.Lusanna, *From Clock Synchronization to Dark Matter as a Relativistic Inertial Effect*, Lecture at the *Black Objects in Supergravity* School BOSS2011, Frascati, 9-13 May 2011 (arXiv 1205.2481), Springer Proc.Phys. **144**, pp.267-343 (Spinger, Berlin, 2013); *Non-Inertial Frames in Special and General Relativity*, to appear in the Proceedings of the School *Gravity: where do we stand?*, 11-15 May 2009, Villa Olmo, Como, Italy, ed.R.Peron(arXiv:1310.4465); *Canonical ADM Tetrad Gravity: from Metrological Inertial Gauge Variables to Dynamical Tidal Dirac Observables*, Int.J.Geom.Meth.Mod.Phys. **12**, 1530001 (2015) (arXiv 1401.1375).

[10] L.Lusanna, *Killing Symmetries as Hamiltonian Constraints* , Int.J.Geom.Meth.Mod.Phys. *13*, 1650044 (2016) (arXiv 1511.05829).

[11] P.A.M.Dirac, *Lectures on Quantum Mechanics*, Belfer Graduate School of Science, Monographs Series (Yeshiva University, New York, N.Y., 1964); *Gauge Invariant Formulation of Quantum Electrodynamics*, Can.J.Phys. **33**, 650 (1955).

[12] M.Henneaux and C.Teitelboim, *Quantization of Gauge Systems* (Princeton University Press, Princeton, 1992).

[13] L.Lusanna, *The Shanmugadhasan Canonical Transformation, Function Groups and the Second Noether Theorem*, Int.J.Mod.Phys. **A8**, 4193 (1993); *The Relevance of Canonical Transformations in Gauge Theories and General Relativity*, Lecture Notes of "Seminario Interdisciplinare di Matematica" (Basilicata Univ.) **5**, 125 (2006).

[14] M.Chaichian, D.Louis Martinez and L.Lusanna, *Dirac's Constrained Systems: the Classification of Second-Class Constraints*, Ann.Phys. (N.Y.) **232**, 40 (1994).

[15] S.Shanmugadhasan, *Canonical Formalism for Degenerate Lagrangians*, J.Math.Phys. **14**, 677 (1973).

[16] R.Beig and Ó Murchadha, *The Poincaré Group as the Symmetry Group of Canonical General Relativity*, Ann.Phys.(N.Y.) **174**, 463 (1987).

[17] D.Alba and L.Lusanna, *Charged Particles and the Electro-Magnetic Field in Non-Inertial Frames: I. Admissible 3+1 Splittings of Minkowski Spacetime and the Non-Inertial Rest Frames*, Int.J.Geom.Methods in Physics **7**, 33 (2010) (arXiv 0908.0213) and *II. Applications: Rotating Frames, Sagnac Effect, Faraday Rotation, Wrap-up Effect*, Int.J.Geom.Methods in Physics, **7**, 185 (2010) (arXiv 0908.0215); *Generalized Radar 4-Coordinates and Equal-Time Cauchy Surfaces for Arbitrary Accelerated Observers* (2005), Int.J.Mod.Phys. **D16**, 1149 (2007) (arXiv gr-qc/0501090).

[18] D.Alba, H.W.Crater and L.Lusanna, *Towards Relativistic Atom Physics. I. The Rest-Frame Instant Form of Dynamics and a Canonical Transformation for a system of Charged Particles plus the Electro-Magnetic Field*, Canad.J.Phys. **88**, 379 (2010) (arXiv 0806.2383) and *II. Collective and Relative Relativistic Variables for a System of Charged Particles plus the Electro-Magnetic Field*, Canad.J.Phys. **88**, 425 (2010) (arXiv 0811.0715).

[19] D.Christodoulou and S.Klainerman, *The Global Nonlinear Stability of the Minkowski Space* (Princeton, Princeton, 1993).

[20] J.Isenberg and J.E.Marsden, *The York Map is a Canonical Transformation*, J.Geom.Phys. **1**, 85 (1984).

[21] I.Ciufolini and J.A.Wheeler, *Gravitation and Inertia* (Princeton Univ.Press, Princeton, 1995).

[22] L.Lusanna, *Relativistic Metrology: from Earth to Astrophysics*, In-Tech E-Book *Modern Metrology Concerns*, 2012 (ISBN 978-953-51-0584-8).

[23] G.H.Kaplan, *The IAU Resolutions on Astronomical Reference Systems, Time Scales and Earth Rotation Models*, U.S.Naval Observatory circular No. 179 (2005) (astro-ph/0602086).

[24] T.D.Moyer, *Formulation for Observed and Computed Values of Deep Space Network Data Types for Navigation* (John Wiley, New York, 2003).

[25] *IERS Conventions (2003)*, eds. D.D.McCarthy and G.Petit, IERS TN 32 (2004), Verlag des BKG.

[26] M.Soffel, S.A.Klioner, G.Petit, P.Wolf, S.M.Kopeikin, P.Bretagnon, V.A.Brumberg, N.Capitaine, T.Damour, T.Fukushima, B.Guinot, T.Huang, L.Lindegren, C.Ma, K.Nordtvedt, J.Ries, P.K.Seidelmann, D.Vokroulicky', C.Will and Ch.Xu, *The IAU 2000 Resolutions for Astrometry, Celestial Mechanics and Metrology in the Relativistic Framework: Explanatory Supplement* Astron.J., **126**, pp.2687-2706, (2003)

(astro-ph/0303376).
[27] J.Kovalevski, I.I.Mueller and B.Kolaczek, *Reference Frames in Astronomy and Geophysics* (Kluwer, Dordrecht, 1989).
[28] O.J.Sovers and J.L.Fanselow, *Astrometry and Geodesy with Radio Interferometry: Experiments, Models, Results*, Rev.Mod.Phs. **70**, 1393 (1998).
[29] M.Bartelmann, *The Dark Universe*, Rev.Mod.Phys. **82**, 331 (2010)(arXiv 0906.5036).
[30] M.Milgrom, *New Physics at Low Accelerations (MOND): An Alternative to Dark Matter*, (aRxiv 0912.2678).
[31] E.Battaner and E.Florido, *The Rotation Curve of Spiral Galaxies and its Cosmological Implications*, Fund.Cosmic Phys. **21**, 1 (2000).
[32] D.G.Banhatti, *Disk Galaxy Rotation Curves and Dark Matter Distribution*, Current Science **94**, 986 (2008).
[33] W.J.G. de Blok and A.Bosma, *High Resolution Rotation Curves of Low Surface Brightness Galaxies*, Astron.Astrophys. **385**, 816 (2002) (astro-ph/0201276).
[34] S.Capozziello, V.F.Cardone and A.Troisi, *Low Surface Brightness Galaxy Rotation Curves in the Low Energy Limit of R^n Gravity: No Need for Dark Matter?*, Mon.Not.R.Astron.Soc. **375**, 1423 (2007) (arXiv astro-ph/0603522).
[35] S.Capozziello, E.De Filippis and V.Salzano, *Modelling Clusters of Galaxies by $f(R)$ Gravity*, Mon.Not.R.Astron.Soc. **394**, 947 (2009) (arXiv 0809.1882).
[36] J.L.Feng, *Dark Matter Candidates from Particle Physics and Methods of Detection* , (arXiv 1003.0904).
[37] S.Jordan, *The GAIA Project: Technique, Performance and Status*, Astron.Nachr. **329**, 875 (2008) (DOI 10.1002/asna.200811065).
[38] S.G.Turyshev and V.T.Toth, *The Pioneer Anomaly* Living Rev.Rel. **13**, n.4 (2010) (arXiv 1001.3686).
[39] M.Villani, *Constraints on ADM tetrad gravity parameter space from S2 star in the center of the Galaxy and from the Solar System*, 2015 (arXiv:1502.06801).
[40] S.Capozziello and M.de Laurentis, *The Dark Matter Problem from f(R) Gravity Viewpoint* , Annalen der Physik **524**, 545 (2012)
[41] R.Durrer, *What do we really Know about Dark Energy?*, Phil.Trans.R.Soc. **A369**, 5102 (2011) (arXiv 1103.5331).
[42] C.Bonvin and R.Durrer, *What Galaxy Surveys Really Measure*, Phys.Rev. D84, 063505 (2011) (arXiv 1105.5280).
[43] W.B.Bonnor, A.H.Sulaiman and N.Tanimura, *Szekeres's Space-Times have no Killing Vectors*, Gen.Rel.Grav. **8**, 549 (1977).
[44] B.K.Berger, D.M.Eardley and D.W.Olson, *Notes on the Spacetimes of Szekeres*, Phys.Rev. **D16**, 3086 (1977).
[45] A.Nwankwo, M.Ishak and J.Thomson, *Luminosity Distance and Redshift in the Szekeres Inhomogeneous Cosmological Models*, JCAP **1105**, 028 (2011) (arXiv 1005.2989).

IOP Conf. Series: Journal of Physics: Conf. Series **845** (2017) 012008 doi:10.1088/1742-6596/845/1/012008

A complete characterization of relativistic uniform acceleration

Tzvi Scarr and Yaakov Friedman

Jerusalem College of Technology
Departments of Mathematics and Physics
P.O.B. 16031 Jerusalem 91160, Israel

E-mail: scarr@g.jct.ac.il, friedman@jct.ac.il

Abstract. We use the Frenet frame to define and completely characterize "uniform acceleration" in flat spacetime. We extend the definition to arbitrary curved spacetime and provide an example in Schwarzschild spacetime.

1. Introduction

Einstein's intuitive definition of *uniform acceleration* is "constant acceleration in the instantaneously co-moving inertial frame." This definition is found widely in the literature, as early as [1] and [2], again in [3], and as recently as [4] and [5]. This definition is natural, since the acceleration in the comoving frame is "precisely the push we feel when sitting in an accelerating rocket" or automobile. Similarly, "by the equivalence principle, the gravitational field in our terrestrial lab is the negative of our proper acceleration, our instantaneous rest frame being an imagined Einstein cabin falling with acceleration g" ([6], page 71).

Historically, however, there have been many difficulties in capturing this definition *mathematically*. Einstein considered the equation

$$F = \frac{dp}{d\tau}, \tag{1}$$

where F is the four-force, p is the four-momentum, and τ is proper time (see [7]). Unfortunately, when F is a constant, as in a homogeneous gravitational field, equation (1) has no solution! This follows from the fact that the four-velocity and the four-acceleration are orthogonal. This was noticed by Planck, who wrote to Einstein about it. This, in turn, prompted Einstein to submit a "correction" [8] to [9]. In the correction, he states that the "concept 'uniformly accelerated' needs further clarification."

In [10, 11], working in flat spacetime, we considered motions which are solutions to

$$c\frac{d\lambda^{\mu}_{(\kappa)}}{d\tau} = \Omega^{\mu}{}_{\nu}\lambda^{\nu}_{(\kappa)}, \tag{2}$$

where $\{K_\tau\}$ is a one-parameter family, indexed by proper time τ, of instantaneously comoving inertial frames, with orthonormal bases $\Lambda(\tau) = \{\lambda_{(\kappa)}(\tau) : \kappa = 0,1,2,3\}$, $u(\tau) = \lambda_{(0)}(\tau)$ is the four-velocity of the motion, and $\Lambda^{-1}\Omega\Lambda$ is constant along the worldline. Equation (2) is fully Lorentz covariant and admits four Lorentz-invariant classes of solutions. The class of *translational acceleration* is the smallest Lorentz invariant class which includes *hyperbolic motion.* In the *null acceleration* class, the worldline of the motion is *cubic* in the time. *Rotational acceleration* covariantly extends pure rotational motion. *General acceleration* is obtained when the translational component of the acceleration is parallel to the axis of rotation.

In [10, 11], we showed that all four classes of solutions to (2) have constant acceleration in the comoving frame. At that time, however, it was not clear whether equation (2) captures *all* uniformly accelerated motions. In other words, do there exist motions which have constant acceleration in the instantaneously co-moving inertial frame but do *not* satisfy equation (2)? In section 5, we settle this question by showing that, in flat spacetime, a motion has constant acceleration in the co-moving frame if and only if it satisfies equation (2).

Having completely characterized uniform acceleration in flat spacetime, we show how to extend the definition of uniform acceleration to *curved spacetime.* We do this with the help of the so-called *Frenet frame.* While the authors of [10, 11, 12, 13] define uniform acceleration *with respect to an inertial frame*, the Frenet frame is more suited for working on a *manifold* and resembles the approach of Mashhoon [14]. It was already shown in [10] that in flat spacetime, the inertial frame and the Frenet frame approaches are, in some sense, equivalent.

The Frenet frame or *Frenet basis* at each point along a smooth future-pointing timelike curve is an orthonormal basis of four-vectors belonging to the tangent space at the point in question. In section 3, working in arbitrary curved spacetime, we provide the explicit construction of the Frenet frame and the derivation of the *Frenet equations.* We show in section 4 that the Frenet equations extend the *geodesic equation.* Despite the fact that the Frenet equations are coupled, we show how to *parallel transport* the individual vectors of the Frenet basis. A detailed example in Schwarzschild spacetime is provided in section 6.

Basic definitions and notation are established in section 2. A discussion appears in section 7. Sections 3 and 4 follow closely the development in [13]. Most of section 6 appeared in [15].

2. Notation

Consider a time-orientable four-dimensional differential manifold $\mathcal{M}$ endowed with a metric $g_{\mu\nu}$ of Lorentzian signature $(+,-,-,-)$. A tangent vector v at a given point of M is *timelike* if $g(v,v)>0$, *spacelike* if $g(v,v)<0$, and *null* if $g(v,v)=0$. Let $\gamma: I\to\mathcal{M}$, I an open interval of $\mathbb{R}$, be a smooth future-pointing timelike curve, parameterized by the arclength $ds=\sqrt{g_{\mu\nu}dx^\mu dx^\nu}$. Note that $ds=cd\tau$. Moreover, under a change of coordinates $x^\mu\to x^{\mu'}$, we have $ds=ds'$. In a local system of coordinates x^μ, the curve $\gamma(s)$ is described by a set of four functions $x^\mu(s)$. We assume that the metric satisfies the metricity condition $dg_{\mu\nu}/ds=0$.

At every point of γ, the *four-velocity* u^μ is defined by

$$u^\mu=\frac{dx^\mu}{ds} \tag{3}$$

and has unit length:

$$u^2=g_{\mu\nu}u^\mu u^\nu=1. \tag{4}$$

The *covariant derivative* $\frac{D}{ds}$ of a four-vector w^μ along $\gamma(s)$ is defined [16] by

$$\frac{Dw^\mu}{ds}=\frac{dw^\mu}{ds}+\Gamma^\mu{}_{\alpha\nu}w^\alpha u^\nu, \tag{5}$$

where the Christoffel symbols $\Gamma^\mu{}_{\alpha\nu}$ are defined by

$$\Gamma^\mu{}_{\alpha\nu}=\frac{1}{2}g^{\rho\mu}(g_{\rho\alpha,\nu}+g_{\rho\nu,\alpha}-g_{\alpha\nu,\rho}). \tag{6}$$

For ease of notation, we often write $\dot{w}$ instead of $\frac{Dw}{ds}$.

The *four-acceleration* a^μ is the covariant derivative of the four-velocity:

$$a = c^2\dot{u}. \tag{7}$$

Differentiating $u^2 = 1$, we have

$$\dot{u} \cdot u = 0, \tag{8}$$

meaning that the four-acceleration is orthogonal to the four-velocity. Since the four-velocity is timelike, the four-acceleration is spacelike.

The *plane of simultaneity* or the *restspace* at the point $\gamma(s)$ is the linear subspace of four-vectors w such that $w \cdot u(s) = 0$.

3. The Frenet Frame

We now construct the Frenet basis $\{\lambda_{(0)}, \lambda_{(1)}, \lambda_{(2)}, \lambda_{(3)}\}$, $\lambda_{(\alpha)} = \lambda_{(\alpha)}(s)$, of the tangent space at the point $\gamma(s)$. The construction uses covariant differentiation and the Gram-Schmidt orthonormalization procedure. In the event that the Gram-Schmidt process breaks down, we are free to complete the orthonormal basis arbitrarily. The orthonormality condition means that

$$\lambda_{(\alpha)} \cdot \lambda_{(\beta)} = \eta_{\alpha\beta}, \tag{9}$$

where $\eta_{\alpha\beta}$ is the Minkowski metric diag$(1, -1, -1, -1)$. Differentiating (9), we obtain

$$\dot{\lambda}_{(\alpha)} \cdot \lambda_{(\beta)} = -\lambda_{(\alpha)} \cdot \dot{\lambda}_{(\beta)}. \tag{10}$$

In particular,

$$\lambda_{(\alpha)} \cdot \dot{\lambda}_{(\alpha)} = 0. \tag{11}$$

First, let $\lambda_{(0)}(s) = u(s)$. The four-acceleration $a = \dot{u}$ is spacelike and orthogonal to the four-velocity u. We assume that $\dot{u} \neq 0$ for all s. Set $\kappa = \sqrt{-a^2}$ and define

$$\lambda_{(1)}(s) = \frac{a(s)}{\kappa}. \tag{12}$$

The unit vector $\lambda_{(1)}(s)$ gives the direction of the four-acceleration. The scalar $\kappa(s)$ is the magnitude of the four-acceleration and is also called the curve's *curvature*. From (12), we trivially get the first Frenet equation

$$c^2\dot{\lambda}_{(0)} = \kappa\lambda_{(1)}. \tag{13}$$

Using the Gram-Schmidt procedure, (9) and (10), we construct a vector $v_{(2)}$ which is orthogonal to both $\lambda_{(0)}$ and $\lambda_{(1)}$:

$$v_{(2)} = c^2\dot{\lambda}_{(1)} - (c^2\dot{\lambda}_{(1)} \cdot \lambda_{(0)})\lambda_{(0)} = c^2\dot{\lambda}_{(1)} - \kappa\lambda_{(0)}. \tag{14}$$

Let $\tau_1 = \sqrt{-(v_{(2)})^2} > 0$, and define

$$\lambda_{(2)} = \frac{v_{(2)}}{\tau_1}. \tag{15}$$

Then, from (14), we have

$$c^2\dot{\lambda}_{(1)} = \kappa\lambda_{(0)} + \tau_1\lambda_{(2)}. \tag{16}$$

Similarly, we construct a vector $v_{(3)}$ orthogonal to $\lambda_{(0)}, \lambda_{(1)}$ and $\lambda_{(2)}$:

$$v_{(3)} = c^2\dot{\lambda}_{(2)} - (c^2\dot{\lambda}_{(2)} \cdot \lambda_{(0)})\lambda_{(0)} + (c^2\dot{\lambda}_{(2)} \cdot \lambda_{(1)})\lambda_{(1)}. \tag{17}$$

Now (10), (13) and (9) imply that $\dot{\lambda}_{(2)}\cdot\lambda_{(0)} = 0$, while (10), (16) and (9) imply that $\dot{\lambda}_{(2)}\cdot\lambda_{(1)} = \tau_1$. Let $\tau_2 = \sqrt{-(v_{(3)})^2} > 0$, and define

$$\lambda_{(3)} = \frac{v_{(3)}}{\tau_2}. \tag{18}$$

From (17) we now obtain

$$c^2\dot{\lambda}_{(2)} = -\tau_1\lambda_{(1)} + \tau_2\lambda_{(3)}. \tag{19}$$

Finally, using (13), (16), (10) and (11), we get that $\dot{\lambda}_{(3)}$is parallel to $\lambda_{(2)}$. Then, using (10), (19) and (9), we get $\lambda_{(2)} \cdot \dot{\lambda}_{(3)} = \tau_2$. Hence,

$$c^2\dot{\lambda}_{(3)} = -\tau_2\lambda_{(2)}. \tag{20}$$

The $\lambda_{(\alpha)}$'s constitute the *Frenet basis* of the trajectory, and the equations (13), (16), (19) and (20) are called the *Frenet equations.* They are a system of coupled ordinary differential equations and can be written compactly as

$$c^2\dot{\lambda}_{(\alpha)}(s) = \lambda_{(\beta)}(s)A^{(\beta)}_{\ (\alpha)}(s), \tag{21}$$

where

$$A^{(\beta)}_{\ (\alpha)}(s) = \begin{pmatrix} 0 & \kappa(s) & 0 & 0 \\ \kappa(s) & 0 & -\tau_1(s) & 0 \\ 0 & \tau_1(s) & 0 & -\tau_2(s) \\ 0 & 0 & \tau_2(s) & 0 \end{pmatrix}. \tag{22}$$

Note that $A(s)$ is a matrix of *scalars.* Under a change of coordinates $x^\mu \to x^{\mu'}$, we have $ds' = ds$, and therefore, $\kappa(s') = \kappa(s)$ and likewise for $\tau_1(s)$ and $\tau_2(s)$. Thus, its two indices are coordinate-free, so we place them in parentheses. We call $A(s)$ the *acceleration matrix.* Note that the lower index on the λ's is a *label* and not a tensorial index.

The physical meaning of $\kappa(s)$ and $\tau_1(s), \tau_2(s)$ is as follows. An observer on $\gamma(s)$ experiences *linear acceleration* of magnitude $\kappa(s)$ in the direction of $\lambda_{(1)}$. The magnitude and the direction of this acceleration can be measured by an accelerometer carried by an observer moving along $\gamma(s)$. The torsion is defined by a 3D vector $\boldsymbol{\omega} = -\tau_2\lambda_{(1)} - \tau_1\lambda_{(3)}$, which can be measured by the precession of gyroscopes carried by the observer. The vector $\boldsymbol{\omega}$ is the axis of the observer's rotational acceleration, with magnitude $\sqrt{\tau_1^2 + \tau_2^2}$. τ_2 is the component of the torsion parallel to the direction of linear acceleration, while τ_1 is the component in the orthogonal direction. The Frenet basis and κ, τ_1 and τ_2 are definable *locally* and are completely determined by the trajectory.

If κ is constant and $\tau_1 = 0$, then equations (13) and (16) imply that

$$\ddot{\lambda}_{(0)} = \frac{\kappa^2}{c^4}\lambda_{(0)}, \tag{23}$$

which is equivalent to [17], equation (6). Equation (23) describes hyperbolic motion. However, to obtain a fully Lorentz covariant theory, we cannot assume that $\tau_1 = 0$. This is because hyperbolic motion itself is not covariant, as shown in [11, 12]. Note that equation (23) is third order in the coordinates x^μ. For a unique solution, one need only specify the initial position $\gamma(0)$, the initial four-velocity $u(0)$, and the initial linear acceleration $a(0)$.

IOP Conf. Series: Journal of Physics: Conf. Series **845** (2017) 012008 doi:10.1088/1742-6596/845/1/012008

4. Analysis of the Frenet equations

Our first goal is to show that the Frenet equations extend the *geodesic equation*

$$\frac{du^\mu(s)}{ds} + \Gamma^\mu{}_{\sigma\rho} u^\sigma(s) u^\rho(s) = 0. \tag{24}$$

First, use (5), (21) and $\lambda_{(0)}(s) = u(s)$ to write the Frenet equations as

$$c^2 \frac{d\lambda^\mu_{(\alpha)}(s)}{ds} + c^2 \Gamma^\mu{}_{\sigma\rho} \lambda^\sigma_{(\alpha)}(s) \lambda^\rho_{(0)}(s) = \lambda^\mu_{(\beta)}(s) A^{(\beta)}{}_{(\alpha)}. \tag{25}$$

Now we show that equation (25) extends the geodesic equation. Along a geodesic, there is zero acceleration, so $A \equiv 0$. Thus, (25) becomes

$$\frac{d\lambda^\mu_{(\alpha)}(s)}{ds} + \Gamma^\mu{}_{\sigma\rho} \lambda^\sigma_{(\alpha)}(s) \lambda^\rho_{(0)}(s) = 0. \tag{26}$$

Setting $\alpha = 0$ and using $\lambda_{(0)}(s) = u(s)$, we obtain the geodesic equation (24).

We now show how to parallel transport the Frenet basis vectors $\lambda_{(\kappa)}$. Recall that a four-vector w^μ is said to be *parallel transported* along $\gamma(s)$ by the Levi-Civita connection if

$$\frac{Dw^\mu}{ds} = 0. \tag{27}$$

If the four-acceleration $a(s) = c^2 \frac{Du(s)}{ds}$ is nonzero, then the four-velocity u is *not* parallel transported along $\gamma(s)$ by the Levi-Civita connection. This means that the Frenet basis is not parallel transported. The covariant derivative along the curve does not preserve the restspaces of an accelerating particle.

A more appropriate transport is defined using the *generalized Fermi-Walker derivative* (*GFW*) $\frac{\widehat{D}w^\mu}{ds}$ of a four vector w^μ (see [16], [17]):

$$\frac{\widehat{D}w^\mu}{ds} = c^2 \frac{Dw^\mu}{ds} - \Omega^\mu{}_\nu w^\nu, \tag{28}$$

where $\Omega_{\mu\nu}$ is a rank 2 tensor, defined along $\gamma(s)$. Hehl [16] shows that the metric compatibility condition, which follows from the transport of the orthonormal basis, implies that $\Omega_{\mu\nu}$ is antisymmetric.

Let $\Lambda(s)$ be the 4×4 matrix whose ith column consists of the components of the vector $\lambda_{(i)}(s)$ in the local basis. Then the Frenet equations (21) can be written as

$$c^2 \frac{D\Lambda(s)}{ds} = \Lambda(s) A(s). \tag{29}$$

In order to ensure parallel transport of the Frenet basis vectors $\lambda_{(\kappa)}$, we choose

$$\Omega(s) = \Lambda(s) A(s) \Lambda^{-1}(s). \tag{30}$$

$\Omega(s)$ is an antisymmetric tensor with the same Lorentz invariants as the matrix $A(s)$. In fact, $\Omega(s)$ is the acceleration matrix $A(s)$ computed in the initial comoving frame. Hence, we refer to $\Omega(s)$ as the *pullback* of the acceleration matrix $A(s)$ along the worldline. Multiplying the right side of (29) by $\Lambda^{-1}(s)\Lambda(s)$, we obtain

$$c^2 \frac{D\Lambda(s)}{ds} = \Omega(s) \Lambda(s). \tag{31}$$

Clearly, the Frenet basis is *GFW parallel transported*. Using (31) and (28), we have

$$\frac{\widehat{D}\lambda_{(\kappa)}(s)}{ds} = c^2\frac{D\lambda_{(\kappa)}(s)}{ds} - \Omega(s)\lambda_{(\kappa)}(s) = 0. \tag{32}$$

In equation (29), the matrix Λ is multiplied by A on the *right*. This means that the time evolution of each basis vector depends on *all* of the basis vectors. However, to have parallel transport, each basis vector must be transported without referring to the other basis vectors. This explains why, in equation (29), Λ is multiplied by Ω on the *left*.

5. Uniform Acceleration

Einstein's intuitive definition of *uniform acceleration* is "constant acceleration in the instantaneously co-moving inertial frame." Historically, however, there have been many difficulties in capturing this definition *mathematically*. For more details, see [12].

In [10, 11], working in flat spacetime, we considered motions which are solutions to

$$c\frac{d\lambda^\mu_{(\kappa)}}{d\tau} = \Omega^\mu{}_\nu\lambda^\nu_{(\kappa)}, \tag{33}$$

where $\{K_\tau\}$ is a one-parameter family, indexed by proper time τ, of instantaneously comoving inertial frames, with orthonormal bases $\Lambda(\tau) = \{\lambda_{(\kappa)}(\tau) : \kappa = 0,1,2,3\}$, $u(\tau) = \lambda_{(0)}(\tau)$ is the four-velocity of the motion, and $\Lambda^{-1}\Omega\Lambda$ is constant along the worldline.

In the $1+3$ decomposition of flat spacetime, the tensor Ω of equation (33) has the form

$$\Omega^\mu{}_\nu = \begin{pmatrix} 0 & \mathbf{g}^T \\ \mathbf{g} & -c\mathbf{B} \end{pmatrix}. \tag{34}$$

where $\mathbf{g}$ is a 3D vector with physical dimension of acceleration, $\boldsymbol{\omega}$ is a 3D vector with physical dimension 1/time, the superscript T denotes matrix transposition, and, for any 3D vector $\boldsymbol{\omega} = (\omega^1, \omega^2, \omega^3)$,

$$\mathbf{B} = \varepsilon_{ijk}\omega^k,$$

where ε_{ijk} is the Levi-Civita tensor. The factor c in Ω provides the necessary physical dimension of acceleration. The vector $\mathbf{g}$ represents *linear* acceleration. If $\boldsymbol{\omega} = 0$, we obtain constant linear acceleration in a fixed direction, otherwise known as *hyperbolic motion*. The vector $\boldsymbol{\omega}$ is the angular velocity of the motion. If $\mathbf{g} = 0$, we obtain *pure rotational motion* with constant angular velocity.

In [10, 11], we showed that the solutions to (33) have constant acceleration in the comoving frame. At that time, however, it was not clear whether equation (33) captures *all* uniformly accelerated motions. In other words, do there exist motions which have constant acceleration in the instantaneously co-moving inertial frame but do *not* satisfy equation (33)? With the help of the Frenet frame, we settle this issue.

Theorem 1. *In flat spacetime with proper time τ, the following are equivalent:*

(i) The future-pointing, timelike worldline $\gamma(\tau)$ has constant acceleration in the instantaneously co-moving inertial frame,

(ii) There exist an antisymmetric tensor $\Omega_{\mu\nu}(\tau)$ and a one-parameter family $\{K_\tau\}$ of instantaneously comoving inertial frames, with orthonormal bases $\Lambda(\tau) = \{\lambda_{(\kappa)}(\tau)\}$ such that

(a) $\Lambda^{-1}\Omega\Lambda$ is constant with respect to τ,

(b) $c\frac{d}{d\tau}\lambda^{\mu}_{(\kappa)}(\tau) = \Omega^{\mu}{}_{\nu}\lambda^{\nu}_{(\kappa)}(\tau)$,
(c) $u(\tau) = \lambda_{(0)}(\tau)$ *is the four-velocity of* $\gamma(\tau)$.

The implication $(ii) \Rightarrow (i)$ was proven in [10, 11].

To prove $(i) \Rightarrow (ii)$, let $\gamma(\tau)$ be the worldline of a motion with constant acceleration in the instantaneously co-moving inertial frame. Let $\Lambda(\tau)$ and $A(\tau)$ be as in the construction of section 3. Since the acceleration is constant in the co-moving frame, we have $A(s) \equiv A$. Using (29) and $ds = cd\tau$, and noting that covariant derivatives becomes normal derivatives, since we are working in flat spacetime, we have

$$c\frac{d\Lambda(\tau)}{d\tau} = \Lambda(\tau)A. \tag{35}$$

Then (a), (b) and (c) of (ii) are satisfied with

$$\lambda^{\mu}_{(\kappa)} = \Lambda^{\mu}_{(\kappa)} \quad \text{and} \quad \Omega^{\mu}{}_{\nu} = \Lambda^{\mu}_{(\alpha)}A^{(\alpha)}_{(\beta)}(\Lambda^{-1})^{(\beta)}_{\nu}. \tag{36}$$

This proves the theorem.

Theorem 1, which is valid in flat spacetime, motivates the following extension of the definition of uniform acceleration to curved spacetime.

Definition 1. *Let* $\gamma(s)$ *be a timelike curve, with acceleration matrix* $A(s)$. *We say that* γ *represents* uniformly accelerated motion *if*

$$\frac{dA(s)}{ds} = 0. \tag{37}$$

We turn now to an example of uniform acceleration in the Schwarzschild metric.

6. Uniform Acceleration in Schwarzschild spacetime

In this section, we provide an example of uniform acceleration in *Schwarzschild spacetime*, with the metric

$$ds^2 = \left(1 - \frac{r_s}{r}\right)c^2dt^2 - \left(1 - \frac{r_s}{r}\right)^{-1}dr^2 - r^2d\Omega^2, \tag{38}$$

where r_s is the Schwarzschild radius and $d\Omega^2 = d\theta^2 + sin^2\theta d\varphi^2$. Here θ is the colatitude (= the angle from north) and φ is longitude.

The known (see [18, 19]) nonzero Christoffel symbols are

$$\Gamma^{0}{}_{01} = \frac{r_s}{2r(r-r_s)} \quad , \quad \Gamma^{1}{}_{00} = \frac{r_s}{2r^2}\left(1 - \frac{r_s}{r}\right) \quad , \quad \Gamma^{1}{}_{11} = \frac{-r_s}{2r(r-r_s)}$$

$$\Gamma^{1}{}_{22} = r_s - r \quad , \quad \Gamma^{1}{}_{33} = (r_s - r)\sin^2\theta \quad , \quad \Gamma^{2}{}_{12} = \frac{1}{r}$$

$$\Gamma^{2}{}_{33} = -\sin\theta\cos\theta \quad , \quad \Gamma^{3}{}_{13} = \frac{1}{r} \quad , \quad \Gamma^{3}{}_{23} = \frac{\cos\theta}{\sin\theta}. \tag{39}$$

For our example, we consider motion in the (t, r) plane and set $\theta = \frac{\pi}{2}, \varphi = 0$. Then

$$\Gamma^{1}{}_{33} = r_s - r \quad , \quad \Gamma^{2}{}_{33} = \Gamma^{3}{}_{23} = 0.$$

Let the acceleration matrix $A = \begin{pmatrix} 0 & \kappa & 0 & 0 \\ \kappa & 0 & 0 & 0 \\ 0 & 0 & 0 & 0 \\ 0 & 0 & 0 & 0 \end{pmatrix}$, representing hyperbolic motion in the radial direction. Equation (25), with $\alpha = 0, \mu = 2$, is

$$c^2 \frac{d\lambda^2_{(0)}}{ds} + \frac{2c^2}{r} \lambda^1_{(0)} \lambda^2_{(0)} = \kappa \lambda^2_{(1)}.$$

Since $\theta = \frac{\pi}{2}$, we have $\lambda^2_{(0)} = \frac{d\theta}{ds} = 0$. Thus, $\lambda^2_{(1)} = 0$. Similarly, since $\varphi = 0$, we have $\lambda^3_{(0)} = \lambda^3_{(1)} = 0$.

Since the $\lambda_{(\alpha)}$'s form an orthonormal basis, we have

$$\left(1 - \frac{r_s}{r}\right) \lambda^0_{(0)} \lambda^0_{(1)} - \left(1 - \frac{r_s}{r}\right)^{-1} \lambda^1_{(0)} \lambda^1_{(1)} = 0, \tag{40}$$

$$\left(1 - \frac{r_s}{r}\right) \left(\lambda^0_{(0)}\right)^2 - \left(1 - \frac{r_s}{r}\right)^{-1} \left(\lambda^1_{(0)}\right)^2 = 1, \tag{41}$$

$$\left(1 - \frac{r_s}{r}\right) \left(\lambda^0_{(1)}\right)^2 - \left(1 - \frac{r_s}{r}\right)^{-1} \left(\lambda^1_{(1)}\right)^2 = -1. \tag{42}$$

Next, we write $\lambda^0_{(0)}, \lambda^0_{(1)}, \lambda^1_{(1)}$ in terms of $\lambda^1_{(0)}$. From (41), we get

$$\lambda^0_{(0)} = \sqrt{\left(\frac{r}{r - r_s}\right) \left(1 + \frac{r}{r - r_s} \left(\lambda^1_{(0)}\right)^2\right)}. \tag{43}$$

From (42), we get

$$\lambda^0_{(1)} = \sqrt{\left(\frac{r}{r - r_s}\right) \left(-1 + \frac{r}{r - r_s} \left(\lambda^1_{(1)}\right)^2\right)}. \tag{44}$$

Substituting (43) and (44) into (40), we get

$$\lambda^1_{(1)} = \sqrt{1 - \frac{r_s}{r} + \left(\lambda^1_{(0)}\right)^2}. \tag{45}$$

Substituting (45) into (44) yields

$$\lambda^0_{(1)} = \frac{r}{r - r_s} \lambda^1_{(0)}. \tag{46}$$

Equation (25), with $\alpha = 0, \mu = 1$, is, after dividing by c^2,

$$\frac{d\lambda^1_{(0)}}{ds} + \frac{r_s}{2r^2} \left(1 - \frac{r_s}{r}\right) \left(\lambda^0_{(0)}\right)^2 - \frac{r_s}{2r(r - r_s)} \left(\lambda^1_{(0)}\right)^2 = \frac{\kappa}{c^2} \lambda^1_{(1)}. \tag{47}$$

Substituting (43) and (45) into this equation, we obtain

$$\left\{ \begin{array}{c} \frac{d\lambda^1_{(0)}}{ds} + \frac{r_s}{2r^2} - \frac{\kappa}{c^2} \sqrt{1 - \frac{r_s}{r} + \left(\lambda^1_{(0)}\right)^2} = 0 \\ \frac{dr}{ds} = \lambda^1_{(0)} \end{array} \right\}, \tag{48}$$

or, equivalently,

$$\frac{d^2 r}{ds^2} + \frac{r_s}{2r^2} - \frac{\kappa}{c^2} \sqrt{1 - \frac{r_s}{r} + \left(\frac{dr}{ds}\right)^2} = 0. \tag{49}$$

IOP Conf. Series: Journal of Physics: Conf. Series **845** (2017) 012008 doi:10.1088/1742-6596/845/1/012008

In the particular case $\kappa = 0$, then, writing $\dot{r}$ for $\frac{dr}{ds}$ and $\ddot{r}$ for $\frac{d^2r}{ds^2}$, equation (49) becomes

$$\ddot{r} + \frac{r_s}{2r^2} = 0. \tag{50}$$

Multiplying by $2\dot{r}$, we obtain

$$2\dot{r}\ddot{r} + \frac{\dot{r}r_s}{r^2} = 0. \tag{51}$$

Integrating, we obtain

$$\dot{r}^2 - \frac{r_s}{r} = \text{constant}. \tag{52}$$

The quantity $E = \dot{r}^2 - \frac{r_s}{r}$ is the *total dimensionless energy*. It is the total energy divided by the maximal kinetic energy $\frac{mc^2}{2}$. Hence, the total energy is conserved, as expected along a geodesic.

We consider now the general case of equation (49) ($\kappa \neq 0$). Differentiating E by t, we have

$$\dot{E} = 2\dot{r}\ddot{r} + \frac{\dot{r}r_s}{r^2}. \tag{53}$$

Dividing by $2\dot{r}$ and using (49), we obtain

$$\frac{\dot{E}}{2\dot{r}} = \frac{\kappa}{c^2}\sqrt{1+E}. \tag{54}$$

Separating variables and integrating, we have

$$\sqrt{1+E} = \frac{\kappa r}{c^2} + C, \tag{55}$$

where C is a constant of integration. Squaring and using the definition of E, we obtain

$$\dot{r}^2 = \frac{r\left(\frac{\kappa r}{c^2} + C\right)^2 + r_s - r}{r}. \tag{56}$$

We now show that there are no *bounded orbits*. Define

$$f(r) = r\left(\frac{\kappa r}{c^2} + C\right)^2 + r_s - r. \tag{57}$$

To have a bounded orbit, say between r_1 and r_2, with $0 < r_1 < r_2$, we must have $f(r_1) = f(r_2) = 0$ and $f(r) > 0$ for $r_1 < r < r_2$. However, $f(r)$ is a cubic polynomial, $f(0) > 0$ and $\lim_{r\to\infty} f(r) = +\infty$. This implies that f has at most two zeroes for $r > 0$ and between these two zeroes, $f(r) < 0$. Hence, there are no bounded orbits.

Figure 1 compares solutions for $r(s)$ in flat spacetime and Schwarzschild spacetime. Here, the Schwarzschild radius $r_s = 3000\ km$. In order to see the difference between the trajectories in flat spacetime and Schwarzschild spacetime, we must use relatively small values of r, since, for large r, the Schwarzschild metric is approximately flat. Thus, we take $r(0) = 2r_s$. Similarly, we use a high acceleration of $\kappa = 8 \times 10^{16}\ ms^{-2}$. Otherwise, the trajectories will be indistinguishable. The need for high acceleration can be seen from the first equation of (48), in which κ is divided by c^2. We take the initial velocity to be $\dot{r}(0) = 0$.

For small values of s, the curves are indistinguishable. For larger s, the curves start to separate, reflecting the difference between flat and curved spacetime. Comparing curves (b) and (d) (respectively, (c) and (e)), we see that an object in Schwarzschild spacetime moves away from (respectively, toward) the attracting mass more slowly than in flat spacetime. Nevertheless, the trajectories are asymptotically parallel. To check this, first use (56) to compute

$$\lim_{r\to\infty}\frac{\dot{r}}{r}=\sqrt{\frac{r\left(\frac{\kappa r}{c^2}+C\right)^2+r_s-r}{r^3}}=\frac{\kappa}{c^2}. \tag{58}$$

For the flat spacetime solution, we have

$$r(s)=\frac{c^2}{\kappa}\left(\cosh\left(\frac{\kappa s}{c^2}\right)-1\right)+2r_s. \tag{59}$$

Hence, $\dot{r}(s)=\sinh\left(\frac{\kappa s}{c^2}\right)$. Since $\lim_{x\to\infty}\tanh(x)=1$, we have

$$\lim_{s\to\infty}\frac{\dot{r}}{r}=\frac{\kappa}{c^2}. \tag{60}$$

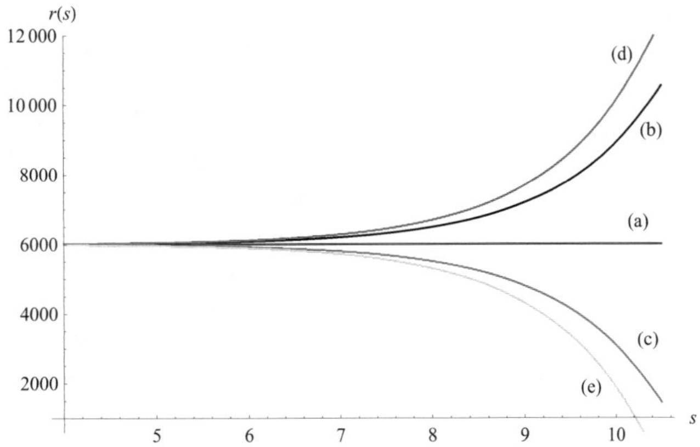

Figure 1. Solutions for $r(s)$, with $r_s=3000$m, $r(0)=2r_s$, $\dot{r}(0)=0$: (a) $\kappa=0$ (a geodesic) (b) $\kappa=8\times10^{16}ms^{-2}$, compared to flat spacetime solution (d), (c) $\kappa=-8\times10^{16}ms^{-2}$, compared to flat spacetime solution (e)

7. Discussion

Uniformly accelerated systems are important because only in *these* systems can all of the rest clocks in the system be synchronized to each other. It was shown in [13] that the rate of a rest clock in a uniformly accelerated system is constant in time but varies with position within the system. The time dilation between rest clocks at different positions can be explained by the difference in the potential energy between the two positions. This is similar to the known "gravitational time shift" for inertial systems with gravity, which, by the Equivalence Principle, are equivalent to accelerating systems. In a uniformly accelerated system, we may synchronize two rest clocks by sending two light signals back and forth between the two clocks. The first light signal is used to synchronize the common $t=0$, and the second light signal is used to adjust for the different *rates* of the clocks. However, once the *rates* have been synchronized, they will remain synchronized, because the rate of each clock is constant. If the system is *not* uniformly accelerated, then, in general, the rate of each rest clock *changes* with time. Therefore, it will not be enough to synchronize the rates of two clocks only once since their relative time dilation changes with time.

We plan to extend the example in Schwarzschild spacetime to include planetary motion. One may consider the gravitational pull of, say, Jupiter, on the Earth to be uniform acceleration. A solution of our equations would then predict the perturbation on the Earth's orbit caused by Jupiter. Alternatively, one could predict the perturbation of the Moon's orbit around the Earth caused by the Sun.

Our theory of uniform acceleration may prove useful in both *Relativistic Newtonian Dynamics (RND)* [20, 21, 22] and *Extended Relativity (ER)* [23, 24]. The dynamics of Special Relativity describes the influence of velocity, or *kinetic energy*, on spacetime. RND, on the other hand, modifies Newtonian dynamics by considering the effect on spacetime due to *potential energy*. This theory has successfully and accurately predicted the precession of Mercury *without* General Relativity [20]. RND also handles the trajectories of binary stars and accurately predicts the Hulse-Taylor pulsar's periastron advance [22]. Currently, however, we have only the three-dimensional version of RND. We plan to apply the ideas of [12] to derive a fully Lorentz covariant 4D version.

ER extends Special Relativity by examining the influence of high accelerations on relativistic dynamics. ER has been successfully applied to both the hydrogen atom [23] and the harmonic oscillator [24]. Thus far, however, only one-dimensional acceleration has been treated. The next step is to extend ER to full Lorentz covariance by treating first the case of *rotations*.

Rotations, even at constant angular velocity, are not trivial. The search for the spacetime transformations between a disk, rotating with constant angular velocity, and a (non-rotating) inertial lab frame has a long and rich history, dating back to Einstein and continuing to the present day. The recent book [25] makes it clear that *there is still no universally accepted theory.* More specifically, as pointed out in [26], a good and complete theory must have a *global metric* and a *unique time at each event.* There is no such theory known today which also agrees with recent experiments [27, 28]. Moreover, many of the current approaches make arbitrary assumptions about the form of the transformation of the *radial* coordinate. There is no theory today which derives the transformations from first principles.

We are currently applying our theory of uniform acceleration to the case of rotations. We hope to obtain explicit spacetime transformations using only the basic tenets of Special Relativity, the inherent symmetries of the problem, and the results of [10, 11, 13]. We conjecture that we can avoid both the *time gap* and the *horizon problem.*

In [10, 11, 13], working in flat spacetime, we derived spacetime transformations, velocity transformations, and acceleration transformations from a uniformly accelerated system to an inertial frame. We plan to extend these transformations to *curved* spacetime. We also want to determine whether the spacetime transformations between uniformly accelerated systems form a *group.* If yes, we want to characterize this group, which will be an extension of the Lorentz group.

References

[1] Minkowski H 1915 *Ann. Phys. Lpz.* **47** p. 927
[2] Born M 1909 *Ann. Phys. Lpz.* **30** p. 1
[3] Landau L and Lifshitz E 1971 *The Classical Theory of Fields* Course of Theoretical Physics vol. 2 (Pergamon: Addison-Wesley)
[4] Rohrlich F 2007 *Classical Charged Particles, 3rd edition* (London: World Scientific)
[5] Lyle S 2008 *Uniformly Accelerating Charged Particles* (Berlin: Springer-Verlag)
[6] Rindler W 2001 *Relativity: Special, General, and Cosmological* (New York: Oxford University Press)
[7] Goldstein H, Poole C and Safko J 2002 *Classical Mechanics, 3rd edition* (New Jersey: Pearson Educational International) p. 299
[8] Einstein A 1908 *Jahrb. Rad. Elektr.* **5** p. 98, translated in 1989 *The Collected Papers of Albert Einstein* (Princeton University Press) vol 2 p. 316

[9] Einstein A 1907 Über das relativitäts prinzip und die aus demselben gezogenen folgerungen *Jahrbuch der Radioaktivität und Elektronik* **4** p. 411 - 62
[10] Friedman Y and Scarr T 2013 Spacetime Transformations from a Uniformly Accelerated Frame *Phys. Scr.* **87** 055004
[11] Friedman Y and Scarr T 2013 Covariant Uniform Acceleration *Journal of Physics: Conference Series* **437** 012009
[12] Friedman Y and Scarr T 2012 Making the relativistic dynamics equation covariant: explicit solutions for motion under a constant force *Phys. Scr.* **86** 065008
[13] Friedman Y and Scarr T 2015 Uniform Acceleration in General Relativity *Gen. Relativ. and Gravit.* **47** 121
[14] Mashhoon B and Muench U 2002 Length measurement in accelerated systems *Ann. Phys.* **11** p. 532-547
[15] Scarr T and Friedman Y 2016 Solutions for Uniform Acceleration in General Relativity *Gen. Relativ. and Gravit.* **48** 65
[16] Hehl F W, Lemke J and Mielke E W 1991 Two lectures on fermions and gravity *Geometry and Theoretical Physics* Debrus J and Hirshfeld A C (eds.) (Berlin:Springer) p. 56140
[17] de la Fuente D and Romero A 2015 Uniformly accelerated motion in General Relativity: completeness of inextensible trajectories *Gen. Relativ. Gravit.* 47:33
[18] Misner C W, Thorne K S and Wheeler J A 1973 *Gravitation* (San Francisco: W.H. Freeman and Company)
[19] Grøn Ø and Hervik, S 2007 *Einstein's General Theory of Relativity with Modern Applications in Cosmology* (New York: Springer-Verlag)
[20] Friedman Y and Steiner, J 2016 Predicting Mercurys precession using simple relativistic Newtonian dynamics *Eur. Phys. Lett.* **113** 39001
[21] Friedman Y 2016 Relativistic Newtonian Dynamics under a central force *Eur. Phys. Lett.* **116** 19001
[22] Friedman Y and Steiner J 2016 Predicting the Hulse-Taylor pulsar's periastron advance without curving spacetime, to appear
[23] Friedman Y and Resin E 2012 Dynamics of hydrogen-like atom bounded by maximal acceleration *Phys. Scr.* **86** 015002
[24] Friedman Y 2013 Extending the Relativity of Time *J. Phys. Conference Series* **437** 012017
[25] Rizzi G and Ruggiero M, eds. 2004 *Relativity in Rotating Frames* (Dordrecht: Kluwer Academic)
[26] Klauber R 2007 Relativistic Rotation *Found. Phys.* **37** 198
[27] Antonini P, Okhapkin M, Göklü E and Schiller S 2005 Test of constancy of speed of light with rotating cyrogenic optical resonators *Phys. Rev. A* **71** 050101
[28] Herrmann S et al 2005 Test of the isotropy of the speed of light using a continuously rotating optical resonator *Phys. Rev. Lett.* **95** 150401

Preliminary stability analysis of a Friedman-Lemaitre-Robertson-Walker universe

Asher Yahalom
Ariel University, Ariel 40700, Israel
E-mail: asya@ariel.ac.il

Abstract. It is stated in many text books that the any metric appearing in general relativity should be locally Lorentzian i.e. of the type $g_{\mu\nu} = \text{diag}\ (1, -1, -1, -1)$ this is usually presented as an independent axiom of the theory, which cannot be deduced from other assumptions. The meaning of this assertion is that a specific coordinate (the temporal coordinate) is given a unique significance with respect to the other spatial coordinates. It was shown that the above assertion is a consequence of requirement that the metric of empty space should be linearly stable and need not be assumed. In this work we remove the empty space assumption and investigate the consequences of spatially uniform matter on the stability of a locally Lorentzian space-time that is the Friedman-Lemaitre-Robertson-Walker space-time. It is shown that a partial stability analysis restricts the type of allowable solutions to the Friedman-Lemaitre-Robertson-Walker space-time. In particular it is shown that an open section universe is stable while an Euclidean and a closed section universes are not in accordance with observation. It will be suggested that in the presence of matter an upper limit scale to the size of a locally Lorentzian universe exists which incidentally is about the size of the observable universe.

1. Introduction

It is well known that our daily space-time is approximately of Lorentz (Minkowski) type that is, it possesses the metric $\eta_{\mu\nu} = \text{diag}\ (1, -1, -1, -1)$. The above statement is taken as one of the central assumptions of the theory of special relativity and has been supported by numerous experiments. But why should it be so?

Many textbooks [3, 4] state that in the general theory of relativity any space-time is locally of the type $\eta_{\mu\nu} = \text{diag}\ (1, -1, -1, -1)$, although it can not be presented so globally due to the effect of matter. This is a part of the demands dictated by the well known equivalence principle. The above principle is taken to be one of the assumptions of general relativity. Other assumption such as diffeomorphism invariance, and the requirement that theory reduce to Newtonian gravity in the proper regime lead to the Einstein equations[1]:

$$G_{\mu\nu} = -\frac{8\pi G}{c^4} T_{\mu\nu} \tag{1}$$

in which $G_{\mu\nu}$ is the Einstein tensor, $T_{\mu\nu}$ is the stress-energy tensor, G is the gravitational constant and c is the velocity of light.

[1] This specific form of Einstein equations is taken from Narlikar [3].

Generic solution of Einstein equations for empty (and non-empty) spaces usually show different signatures as was shown by Eddington [5, page 25]. Among those possible solutions it was shown that for empty space-times [1, 2] only the Minkowskian solutions are stable.

Thus one need not assume that space-time is locally Minkowskian based on an empirical (unexplained) fact, rather one can derive this property from the field equations themselves based on the stability of the Minkowskian solutions. The number of assumptions needed to obtain the celebrated results of general relativity is reduced; by making the theory more compact we enhance its predictive strength. Moreover, it is shown that the existence of a temporal coordinate is a necessary consequence of the geometrical structure of the four dimensional space and not a separate ad-hoc assumption.

A different approach is due to Mukohyama & Uzan [6]. Those authors have derived a Lorentzian metric assuming a that at the microscopic level the metric is Riemannian, i.e., locally Euclidean, and that the Lorentzian structure, that we usually consider as fundamental, is in fact an effective property that emerges in some regions of a four dimensional space with a positive definite metric. In such a model, there is no dynamics nor signature flip across some hypersurface; instead, all the fields develop a Lorentzian dynamics in these regions because they propagate in an effective metric. They have shown that one can construct a decent classical field theory for scalars, vectors, and (Dirac) spinors in flat spacetime. And that gravity can be included but that the theory for the effective Lorentzian metric is not general relativity but of the covariant Galilean type. The constraints arising from stability, the equivalence principle, and the constancy of fundamental constants are detailed and a phenomenological picture of the emergence of the Lorentzian metric is also given. This construction, while restricted to classical fields, offers a new view on the notion of time. Unfortunately this approach is not consistent with Occam's razor as one is forced to introduce several scalar fields (one of them a "clock field") with fine tuned coupling in order to obtain the effective sign change.

What is deficient in the above approach (and similar approaches) is that additional theoretical structures and assumptions are needed in order to justify what appears to be a fundamental property of space-time. In previous works [1, 2, 7] it was claimed otherwise. It was shown that general relativistic equations and some "old fashioned" linear stability analysis will lead to a unique choice of the Lorentzian metric being the only one which is linearly stable in empty space-time. Other metrics are allowed but are unstable and thus can exist in only a limited region of empty space-time. The implications of this are discussed elsewhere [8, 9, 10]. It should be mentioned that the choice of coordinates in the Fisher approach to physics [11] can be also be justified using the stability approach [7]. The nonlinear stability question of the Lorentzian metric for empty space-times was settled by D. Christodoulou and S. Klainerman [13].

This paper assumes that space-time must have four dimensions, it does not explain why this is so. For a possible explanation derived from string theory one can consult a paper by S. K. Rama [14]. The analysis discussed in this work is classical. For quantum aspects of space time difference one can consult the book by S. Hawking and R. Penrose [15]. Quantum field considerations made by A. White, S. Weinfurtner and M. Visser [16] have shown that regardless of the underlying classical theory, there are severe problems associated with any quantum field theory residing on a signature-changing background which does not respect the space-time difference (Such as the production of what is naively an infinite number of particles, with an infinite energy density). Those authors raise the question as to whether signature change transitions could be fully understood and dynamically generated within (modified) classical general relativity, or whether they require the knowledge of a theory of quantum gravity.

The plan of this paper is as follows: in the first section we outline the stability analysis of empty space-times, the next is dedicated to the modifications of those stability equations in the presence of a spatially uniform matter; this is essentially the stability analysis of a Friedman-Lemaitre-Robertson-Walker Universe. Finally we suggest an upper scale for a stable locally

IOP Conf. Series: Journal of Physics: Conf. Series **845** (2017) 012009 doi:10.1088/1742-6596/845/1/012009

Minkowskian space-time and compare this scale with the size of the observable universe.

2. Stability analysis for empty space-times

In what follows I give an outline of the proof of the linear stability of Lorentzian space-time and the linear instability of other possible flat space-times. Unfortunately the details cannot be given here for the lack of space, but can be found elsewhere [1].

Consider a flat space-time, in this case one can find a set of coordinates such that the metric will be constant everywhere. A constant metric is nothing but a 4 by 4 symmetric matrix, therefore by a suitable choice of coordinates it can be made diagonal with eigenvalues all real. A final step would be to choose the scaling of the coordinates such that the metric will have the form:

$$\eta = \text{diag } (\pm 1, \pm 1, \pm 1, \pm 1). \tag{2}$$

Notice that no zero eigen values are allowed by virtue of the four dimensionality of space-time.

Once the family of possible canonical flat metrics is established we make a small perturbation $h_{\mu\nu}$[2] of the metric and consider a new metric $g_{\mu\nu} = \eta_{\mu\nu} + h_{\mu\nu}$ in which $h_{\mu\nu} << 1$. Inserting $g_{\mu\nu}$ into equation (1) and keeping only first order terms in $h_{\mu\nu}$ while taking $T_{\mu\nu} = 0$ we arrive at homogenous linear equations for $h_{\mu\nu}$. Solving those equations with **local** but otherwise arbitrary boundary conditions we see that only in the case of a Lorentzian metric the perturbation does not exponentially grow while in any other case it does. This explains why space-time is Lorentzian and why time is unique among the coordinates of space-time [1].

3. Stability analysis for a locally Lorentzian non empty space-time

Let us now assume that space-time is locally Lorentzian and contains a spatial uniform distribution of matter; this is the case of a Friedman-Lemaitre-Robertson-Walker metric. We choose following Narlikar [3], the non-perturbed metric to be:

$$g^{(0)}_{\mu\nu} = \text{diag } (1, -\frac{S^2(t)}{1-kr^2}, -S^2(t)r^2, -S^2(t)r^2 \sin^2\theta) \tag{3}$$

for the coordinates $x^\nu = (ct, r, \theta, \phi)$. Hence for radial distances such that $r < \frac{1}{\sqrt{k}}$ (in a closed universe) and anywhere in a Euclidean or open universes. the metric is locally (up to scaling) $\eta_{\mu\nu} =$ diag $(+1, -1, -1, -1)$ that is Lorentzian. We assume a fluid energy momentum tensor of the form [3]:

$$T_{\mu\nu} = (p + \rho c^2) u_\mu u_\nu - p\ g_{\mu\nu} \tag{4}$$

In the above p is the pressure, ρ is the density and $u_\mu = \frac{dx_\mu}{ds}$ in which the interval ds is defined as:

$$ds^2 = g_{\mu\nu} dx^\mu dx^\nu \tag{5}$$

We notice that $T_{\mu\nu}$ depend on metric perturbations both directly through the term $-pg_{\mu\nu}$ and through the u_μ term. The metric $g^{(0)}_{\mu\nu}$ is a solution of the Einstein equation (1):

$$G^{(0)}_{\mu\nu} = -\frac{8\pi G}{c^4} T^{(0)}_{\mu\nu} \tag{6}$$

in which $G^{(0)}_{\mu\nu}, T^{(0)}_{\mu\nu}$ are the Einstein and energy-momentum tensors calculated using the metric $g^{(0)}_{\mu\nu}$. The solution is valid provided that p and ρ are spatially (but not temporally) uniform. We

[2] In this paper Greek indices have the values $(0, 1, 2, 3)$ while Latin indices are spatial and have the values $(1, 2, 3)$.

IOP Conf. Series: Journal of Physics: Conf. Series **845** (2017) 012009 doi:10.1088/1742-6596/845/1/012009

now look at a perturbed metric $g_{\mu\nu} = g^{(0)}_{\mu\nu} + g^{(1)}_{\mu\nu}$. Inserting this metric into equation (1) and keeping only first order terms in $g^{(1)}_{\mu\nu}$ we arrive at:

$$G^{(0)}_{\mu\nu} + G^{(1)}_{\mu\nu} = -\frac{8\pi G}{c^4}(T^{(0)}_{\mu\nu} + T^{(1)}_{\mu\nu}) \tag{7}$$

in which $G^{(1)}_{\mu\nu}, T^{(1)}_{\mu\nu}$ are the first order corrections to the Einstein and energy-momentum tensors due to the metric perturbation $g^{(1)}_{\mu\nu}$. Combining equation (7) with equation (6) we arrive at the equation:

$$G^{(1)}_{\mu\nu} = -\frac{8\pi G}{c^4}T^{(1)}_{\mu\nu} \tag{8}$$

Since the Einstein tensor can be calculated from the Ricci tensor $R_{\mu\nu}$ and the curvature scalar R as follows [3]:

$$G_{\mu\nu} = R_{\mu\nu} - \frac{1}{2}g_{\mu\nu}R \tag{9}$$

Thus the first order correction to $G_{\mu\nu}$ will be:

$$G^{(1)}_{\mu\nu} = R^{(1)}_{\mu\nu} - \frac{1}{2}g^{(1)}_{\mu\nu}R^{(0)} - \frac{1}{2}g^{(0)}_{\mu\nu}R^{(1)} \tag{10}$$

However, it follows from equation (1) and equation (4) that:

$$R = -g^{\mu\nu}G_{\mu\nu} = \frac{8\pi G}{c^4}(\rho c^2 - 3p) \tag{11}$$

Hence assuming that the density and pressure are not perturbed it follows that the scalar curvature is metric independent, that is:

$$R^{(0)} = R = \frac{8\pi G}{c^4}(\rho c^2 - 3p), \qquad R^{(1)} = 0 \tag{12}$$

and thus:

$$G^{(1)}_{\mu\nu} = R^{(1)}_{\mu\nu} - g^{(1)}_{\mu\nu}\frac{4\pi G}{c^4}(\rho c^2 - 3p) \tag{13}$$

The Ricci tensor can be calculated in terms of the affine connection

$$R_{\mu\nu} = \partial_\nu \Gamma^\alpha_{\mu\alpha} - \partial_\alpha \Gamma^\alpha_{\mu\nu} + \Gamma^\alpha_{\mu\beta}\Gamma^\beta_{\alpha\nu} - \Gamma^\beta_{\mu\nu}\Gamma^\alpha_{\alpha\beta} \tag{14}$$

in terms of which the first order correction is:

$$R^{(1)}_{\mu\nu} = \partial_\nu \Gamma^{(1)\alpha}_{\mu\alpha} - \partial_\alpha \Gamma^{(1)\alpha}_{\mu\nu} + \Gamma^{(1)\alpha}_{\mu\beta}\Gamma^{(0)\beta}_{\alpha\nu} + \Gamma^{(0)\alpha}_{\mu\beta}\Gamma^{(1)\beta}_{\alpha\nu} - \Gamma^{(1)\beta}_{\mu\nu}\Gamma^{(0)\alpha}_{\alpha\beta} - \Gamma^{(0)\beta}_{\mu\nu}\Gamma^{(1)\alpha}_{\alpha\beta} \tag{15}$$

where $\partial_\nu = \frac{\partial}{\partial x^\nu}$. The affine connection can be calculated as:

$$\Gamma^\alpha_{\mu\nu} = \frac{1}{2}g^{\alpha\beta}(\partial_\mu g_{\beta\nu} + \partial_\nu g_{\mu\beta} - \partial_\beta g_{\mu\nu}) \tag{16}$$

Hence to first order we have:

$$\Gamma^{(1)\alpha}_{\mu\nu} = \frac{1}{2}g^{(0)\alpha\beta}(\partial_\mu g^{(1)}_{\beta\nu} + \partial_\nu g^{(1)}_{\mu\beta} - \partial_\beta g^{(1)}_{\mu\nu}) + \frac{1}{2}g^{(1)\alpha\beta}(\partial_\mu g^{(0)}_{\beta\nu} + \partial_\nu g^{(0)}_{\mu\beta} - \partial_\beta g^{(0)}_{\mu\nu}) \tag{17}$$

which can also be written as:

$$\Gamma^{(1)\alpha}_{\mu\nu} = \frac{1}{2}g^{(0)\alpha\beta}(\partial_\mu g^{(1)}_{\beta\nu} + \partial_\nu g^{(1)}_{\mu\beta} - \partial_\beta g^{(1)}_{\mu\nu}) + g^{(1)\alpha\beta}g^{(0)}_{\beta\gamma}\Gamma^{(0)\gamma}_{\mu\nu} \tag{18}$$

IOP Conf. Series: Journal of Physics: Conf. Series **845** (2017) 012009 doi:10.1088/1742-6596/845/1/012009

It can be easily shown that up to first order:

$$g^{(1)\mu\nu} = -g^{(0)\alpha\nu} g^{(1)}_{\alpha\beta} g^{(0)\beta\mu} \tag{19}$$

Inserting equation (19) into equation (18) we obtain:

$$\Gamma^{(1)\alpha}_{\mu\nu} = \frac{1}{2} g^{(0)\alpha\beta} (\partial_\mu g^{(1)}_{\beta\nu} + \partial_\nu g^{(1)}_{\mu\beta} - \partial_\beta g^{(1)}_{\mu\nu}) - g^{(1)}_{\gamma\beta} g^{(0)\alpha\beta} \Gamma^{(0)\gamma}_{\mu\nu} \tag{20}$$

We will now consider a fluid which is at rest in a comoving frame hence $u^\mu = 0$ for $\mu \neq 0$. Hence according to equation (5): $g_{00}(u^0)^2 = 1$ and we choose $u^0 = \frac{1}{\sqrt{g_{00}}}$:

$$u_\mu = g_{\mu\nu} u^\nu = g_{\mu 0} u^0 = \frac{g_{\mu 0}}{\sqrt{g_{00}}} \tag{21}$$

By inserting $g_{\mu\nu} = g^{(0)}_{\mu\nu} + g^{(1)}_{\mu\nu}$ we obtain:

$$u^{(0)}_\mu = \frac{g^{(0)}_{\mu 0}}{\sqrt{g^{(0)}_{00}}}, \qquad u^{(1)}_\mu = \frac{g^{(1)}_{\mu 0} g^{(0)}_{00} - \frac{1}{2} g^{(1)}_{00} g^{(0)}_{\mu 0}}{\sqrt{g^{(0)}_{00}}^3} \tag{22}$$

For a diagonal metric $g^{(0)}$[3]: $u^{(0)}_a = 0$, $u^{(0)}_0 = \sqrt{g^{(0)}_{00}}$ and $u^{(1)}_a = \frac{g^{(1)}_{a0}}{\sqrt{g^{(0)}_{00}}}$, $u^{(1)}_0 = \frac{1}{2} \frac{g^{(1)}_{00}}{\sqrt{g^{(0)}_{00}}}$. Thus zeroth and first order contributions to the energy-momentum tensor are thus:

$$T^{(0)}_{\mu\nu} = (p + \rho c^2) u^{(0)}_\mu u^{(0)}_\nu - p \, g^{(0)}_{\mu\nu} \tag{23}$$

$$T^{(1)}_{\mu\nu} = (p + \rho c^2)(u^{(0)}_\mu u^{(1)}_\nu + u^{(1)}_\mu u^{(0)}_\nu) - p \, g^{(1)}_{\mu\nu} \tag{24}$$

From this one can easily calculate:

$$T^{(1)}_{ab} = -p \, g^{(1)}_{ab}, \qquad T^{(1)}_{\mu 0} = \rho c^2 \, g^{(1)}_{\mu 0} \tag{25}$$

Hence combining the above equations with equation (13) and equation (8) we have

$$R^{(1)}_{ab} - g^{(1)}_{ab} \frac{4\pi G}{c^4} (\rho c^2 - p) = 0, \qquad R^{(1)}_{\mu 0} + g^{(1)}_{\mu 0} \frac{4\pi G}{c^4} (\rho c^2 + 3p) = 0 \tag{26}$$

We will now assume a perturbation of the type $h \equiv g^{(1)}_{11}(t, r)$ and will look only at the $(1,1)$ component of equation (26). This will take the form:

$$\frac{1}{2} \partial_0^2 h + \frac{(1 - kr^2)}{r S(t)^2} \partial_1 h + \frac{(S'(t)^2 - 2c^2 k)}{c^2 S(t)^2} h - \frac{4\pi G}{c^2} \rho h = 0 \tag{27}$$

in case pressure is negligible with respect to ρc^2 as is the case in the present epoch. This can also be written as:

$$\partial_t^2 h + \frac{2c^2 (1 - kr^2)}{r S(t)^2} \partial_1 h + 2 \left(H^2(t) - 2c^2 \frac{k}{S(t)^2} - 4\pi G \rho \right) h = 0 \tag{28}$$

[3] We remind the reader that Latin indices are spatial.

IOP Conf. Series: Journal of Physics: Conf. Series **845** (2017) 012009 doi:10.1088/1742-6596/845/1/012009

in which $H(t) = \frac{S'(t)}{S(t)}$ is the Hubble function. Assuming that the spatial derivative of h are negligible we arrive at the following equation.

$$\partial_t^2 h + 2\left(H^2(t) - 2c^2 \frac{k}{S(t)^2} - 4\pi G\rho\right) h = 0 \tag{29}$$

The solution of which will be stable if:

$$H^2(t) - 2c^2 \frac{k}{S(t)^2} - 4\pi G\rho > 0 \tag{30}$$

At the present epoch this takes the form:

$$\frac{H_0^2}{4\pi G} - c^2 \frac{k}{2\pi G S_0^2} - \rho_0 > 0 \tag{31}$$

Since the critical density is $\rho_c = \frac{3H_0^2}{8\pi G}$, we can write:

$$\frac{2}{3}\rho_c - c^2 \frac{k}{2\pi G S_0^2} - \rho_0 > 0 \tag{32}$$

This is indicates that an Euclidean section universe with $k = 0$ is unstable since it requires $\rho_0 = \rho_c$, the same can be said about a universe with a closed section in which $k = 1$. This universe requires $\rho_0 > \rho_c$, this clearly cannot be satisfied at the same time with equation (32). The open section universe with $k = -1$ is stable to this type of perturbation provided that its density satisfies:

$$\frac{2}{3}\rho_c + \frac{c^2}{2\pi G S_0^2} > \rho_0 \tag{33}$$

Since $\rho_c = 1.83 \cdot 10^{-26} \frac{Kg}{M^3}$ while ρ_0 is estimated to be [17] $4.5 \cdot 10^{-28} \frac{Kg}{M^3}$, the stability condition for this type of perturbation is satisfied. Thus the current stability analysis is consistent with observations of an expanding universe.

4. The scale of the observable universe

Although the current analysis is not comprehensive, we would like to point out a general feature which will negatively effect the stability of any spatial perturbation, that is the existence of a critical wave number in equation (26) [20]:

$$k_{crit} = \frac{4}{c}\sqrt{2\pi G\rho} \tag{34}$$

In terms of wavelengths we see that an upper scale for stable perturbations is:

$$\lambda < \lambda_{max} = \frac{2\pi}{k_{crit}} = \frac{c}{2}\frac{\sqrt{\pi}}{\sqrt{2G\rho}} \tag{35}$$

The effect of an upper stability wavelength is well known in astrophysical systems [21, 22, 23]. For example Jeans [21] has studied the stability of a self gravitating uniform medium and has determined that if the system is smaller that a certain size instability will not set in to change the system configuration. But if it is bigger than a certain size (Jeans length) the system will become unstable and move to different configuation thus structure will spontaneously appear in uniform medium such structures are galaxies and stars.

The same happens in a cosmological scale. If the universe is smaller than a certain scale the instability will not set in but if it is bigger than this scale the instability will set in and push the universe to a different stable equilibrium which may not be uniform and may effect the metric locally in certain space-time events.

Inserting the density of the universe as estimated by [17]: $\rho \simeq 4.5 \cdot 10^{-28} \frac{Kg}{M^3}$ which leads to a $\lambda_{max} \simeq 1.08 \cdot 10^{27} Meter$. This is slightly larger than the radius of the observable universe: [18] $14 \cdot 10^9 Parsec \simeq 4.32 \cdot 10^{26} Meter$. Hence in "surprising" coincidence the diameter of the observable universe is about the size of the largest scale stable perturbation. At this time the size of the universe is not known [19] but it is suspected [9, 10] that above the stability scale the metric of space-time and hence physics may be quite different.

5. Conclusion

Mathematically speaking one of the main differences between time and space is encapsulated in the flat metric of our space-time which is **locally** of the Lorentzian type $\eta_{\mu\nu} =$ diag (1,$-1, -1, -1$). But this is an empirical fact or a mathematical postulate thus unexplained. One can imagine also other flat metrics such as the Euclidian metric: diag $(1, 1, 1, 1)$. In Euclidian metrics there is no restriction on the speed of any moving body as the speed of light restricts the speed of propagation only in the presence of a Lorentzian metric. Why is our space-time Lorentzian and not Euclidean? The answer is that only the Lorentzian metric is stable [1] for an (almost) empty space. But space-time is not empty and thus the notion of time always progressing forward with the increase of entropy is probably just a consequence of the scales of reality that we are exposed to. In huge cosmological scale the Friedman-Lemaitre-Robertson-Walker universe may lose its Lorentzian character [9, 10]. The horizon problem related to the homogeneity of cosmic microwave background can be solved if one takes into account the superluminal motion of particles for $r > r_c$ and the same particles moving into the $r < r_c$ Lorentzian domain. What is shown here is that a **locally** Lorentzian type universe although stable for all scales in an empty universe has a size limitation for a non empty universe and hence the suggested solution to the horizon problem is not only plausible but in fact strongly suggested by the analysis presented here.

It should be noted that the stability analysis given here is rather restricted as we only allowed h_{11} perturbations. A full stability analysis is expected to involve a set of partial differential equations for the ten free linear metric components which probably can only be solved numerically. However, we can also learn something about the structure of the equations and there solution by looking at specific perturbations as was done here.

References

[1] Yahalom, A. (2008). The Geometrical Meaning of Time, Foundations of Physics, Volume 38, Number 6, Pages 489-497.

[2] Yahalom, A. (2009). The Gravitational Origin of the Distinction between Space and Time, International Journal of Modern Physics D, Vol. 18, Issue: 14, pp. 2155-2158. Gravity Research Foundation - 2009 Honorable Mention.

[3] Narlikar, J. V. (1993). Introduction to Cosmology, Second Edition. Cambridge University Press.

[4] Weinberg, S. (1995). The Quantum Theory of Fields, Cambridge University Press.

[5] Eddington, A. S. (1923). The mathematical theory of relativity, Cambridge University Press.

[6] Mukohyama, S. & Uzan, J. P. (2013). From configuration to dynamics: Emergence of Lorentz signature in classical field theory. Physical Review D, 87(6), 065020.

[7] Yahalom, A. (2011). Advances in Classical Field Theory, Chapter 6, Bentham eBooks eISBN: 978-1-60805-195-3.

[8] Yahalom A. (2013). Gravity and Faster than Light Particles. Journal of Modern Physics (JMP), Vol. 4 No. 10 PP. 1412-1416. DOI: 10.4236/jmp.2013.410169.

[9] Yahalom, A. (2014). On the Difference between Time and Space. Cosmology Vol. 18. 466-483. Cosmology.com.

[10] Yahalom, A. (2015). The Geometrical Meaning of Time - Some Cosmological Implications. Proceedings of 3rd International Conference on Mathematical Modelling in Physical Sciences (IC-MSQUARE 2014), 28-31 August 2014, Madrid, Spain. Journal of Physics: Conference Series (IOP Publishing), Volume 574, 012061. doi:10.1088/1742-6596/574/1/012061.
[11] Frieden, B. R. (2004). Science from Fisher Information - a Unification, Cambridge University Press.
[12] Yahalom, A. (2010). Gravity and the Complexity of Coordinates in Fisher Information, International Journal of Modern Physics D, Vol. 19, No. 14 1-5, World Scientific Publishing Company.
[13] Christodoulou, D. & Klainerman, S. (1989-1990). The global nonlinear stability of the Minkowski space, Seminaire Equations aux derivees partielles (dit "Goulaouic-Schwartz"), Exp. No. 13, p. 29.
[14] Kalyana Rama, S. (2006). Los Alamos Archive hep-th/0610071 v2.
[15] Hawking, S. & Penrose, R. (2010). The Nature of Space and Time, Princeton University Press.
[16] White, A. Weinfurtner, S. and Visser, M. (2010). Signature change events: A challenge for quantum gravity?. Classical and Quantum Gravity 27.4 : 045007.
[17] "Universe 101: What is the Universe Made Of?". NASA: WMAP's Universe. Jan 24, 2014. Retrieved 2015-02-17.
[18] Bars, I. & Terning, J. (2009). Extra Dimensions in Space and Time. Springer. pp. 27. ISBN 978-0-387-77637-8. Retrieved 2011-05-01.
[19] Greene, B. (2011). The Hidden Reality. Alfred A. Knopf.
[20] Yahalom, A. "The Geometrical Meaning of Time in the Presence of Matter" proceedings of the 4th International Conference on Mathematical Modelling in Physical Sciences, IC-MSQUARE 2015, June 5-8, 2015, Mykonos, Greece.
[21] Jeans, J. H. (1929). Astronomy and Cosmogony 2nd ed., Cambridge University Press.
[22] Toomre, A. (1964). Astrophysical Journal **139** 1217.
[23] Yahalom, A. (2011) Stability in the Weak Variational Principle of Barotropic Flows and Implications for Self-Gravitating Discs. [Los-Alamos Archives - astro-ph/9501080] Monthly Notices of the Royal Astronomical Society 418, 401426 . doi:10.1111/j.1365-2966.2011.19492.x.

Back to epicycles – relativistic Coulomb systems in velocity space[1]

Uri Ben-Ya'acov

School of Engineering, Kinneret Academic College on the Sea of Galilee,
D.N. Emek Ha'Yarden 15132, Israel

E-mail: uriby@kinneret.ac.il

Abstract. The study of relativistic Coulomb systems in velocity space is prompted by the fact that the study of Newtonian Kepler/Coulomb systems in velocity space, although less familiar than the analytic solutions in ordinary space, provides a much simpler (also more elegant) method. The simplicity and elegance of the velocity-space method derives from the linearity of the velocity equation, which is the unique feature of $1/r$ interactions for Newtonian and relativistic systems alike. The various types of possible trajectories are presented, their properties deduced from the orbits in velocity space, accompanied with illustrations. In particular, it is found that the orbits traversed in the relativistic velocity space (which is hyperbolic (H^3) rather than Euclidean) are epicyclic – circles whose centres also rotate – thus the title.

PACS numbers : 03.30.+p
Keywords : hodograph, relativistic Coulomb system, relativistic velocity space, Hamilton's vector, rapidity

1. Introduction

The motivation for the present work stems from the success of studying Newtonian Kepler/Coulomb (KC) 2-body systems in velocity space. This method was originally presented by Hamilton in 1847 [1], and elaborated years later mainly by Maxwell [2] and Feynmann [3]. It discusses the dynamics of the system by following the orbit traced by the tip of the velocity vector (termed *hodograph* by Hamilton). Although hardly known, the virtue of the hodograph method is that its application to classical systems with $1/r$ potential provides, in a very simple and elegant way, the full solution – all the necessary information regarding the dynamics of the system, including spatial trajectories – just from the discussion in velocity space. Its merits have been discusses on several occasions in the last decades [4, 5, 6, 7, 8, 9, 10].

The success of the hodograph method with the Newtonian KC systems triggers an interest in its possible application to relativistic systems. The simplest extension of Newtonian KC systems to relativity are relativistic Coulomb systems – the limit of EM 2-body systems when one of the charges is much heavier than the other. Such systems were studied in the literature ([11, 12] and references therein), generally following standard methods in configuration space. The present work complements these studies, with a thorough discussion in relativistic velocity space.

[1] Dedicated to the memory of J. D. Bekenstein – physicist, teacher and human

The essentials of the hodograph method are presented in the following. It is worthwhile to emphasize that the hodograph method, Newtonian and relativistic alike, is not a study in Hamiltonian phase space $\{(x,p)\}$, but an analysis confined only to the velocities, making use of the rotational symmetry of the system and the associated angular-momentum conservation to transform spatial dependencies to velocity dependencies. The use of rotational symmetry in this way is certainly an essential part of the method. Also it is to be noted that the relativistic velocity space is a 3-D hyperboloid H^3 embedded in a 1+3 pseudo-Euclidean space [13] (unlike the 3-D Euclidean velocity space for Newtonian systems), and the results depend heavily on this property.

The talk started with reviewing the classical KC hodograph method (see section 2), followed with an account of the relativistic velocity space, the relativistic hodograph equations and general properties of the hodographs in section 3. The main body of the talk then consists of an illustrative account of the various particular cases.

A fuller account of the results is in preparation to be published elsewhere.

Notation. The convention $c = 1$ is used throughout, unless specified otherwise. Events in Minkowski space-time are $x^\mu = (x^0, x^1, x^2, x^3)$, with metric tensor $g_{\mu\nu} = \mathrm{diag}\,(-1,1,1,1)$, $\mu,\nu = 0,1,2,3$. For any 4-vectors $a^\mu = (a^0, \vec{a})$ and $b^\mu = (b^0, \vec{b})$, their inner product is then $a \cdot b = -a^0 b^0 + \vec{a}\cdot\vec{b}$.

2. The hodograph in classical Kepler/Coulomb systems

Classical KC systems are determined by the Hamiltonian

$$H(\vec{r},\vec{p}) = \frac{\vec{p}^{\,2}}{2m} + \frac{\kappa}{r} \tag{1}$$

with the velocity equation of motion and the conserved angular momentum

$$m\frac{d\vec{v}}{dt} = \frac{\kappa}{r^3}\vec{r} \quad , \quad \ell = mr^2\frac{d\theta}{dt} \,. \tag{2}$$

Coordinates x, y are assumed the for the plane of motion, with the conserved angular momentum vector $\vec{\ell} = \vec{r}\times\vec{p} = \ell\hat{z}$ perpendicular to it. Polar coordinates (r,θ) are also used, with the polar-planar unit vectors

$$\hat{r} = \cos\theta\hat{x} + \sin\theta\hat{y} \quad , \quad \hat{\theta} = -\sin\theta\hat{x} + \cos\theta\hat{y} \tag{3}$$

satisfying

$$\hat{r} = -\hat{z}\times\hat{\theta} = -\frac{d\hat{\theta}}{d\theta} \quad , \quad \hat{\theta} = \hat{z}\times\hat{r} = \frac{d\hat{r}}{d\theta} \,. \tag{4}$$

Eliminating the time variable from both equations in (2) yields the velocity-angular equation

$$\frac{d\vec{v}}{d\theta} = \frac{\kappa}{\ell}\hat{r} = -\frac{\kappa}{\ell}\frac{d\hat{\theta}}{d\theta} \,, \tag{5}$$

with the general hodograph solution

$$\vec{v}(\theta) = \vec{B}_o - \frac{\kappa}{\ell}\hat{\theta} \,. \tag{6}$$

The constant of integration $\vec{B}_o$ is known as *Hamilton's vector.* The solution (6) describes a circle (or at least a circular arc for unbound systems) in velocity space, centred around $\vec{B}_o$ with radius $|\kappa|/\ell$.

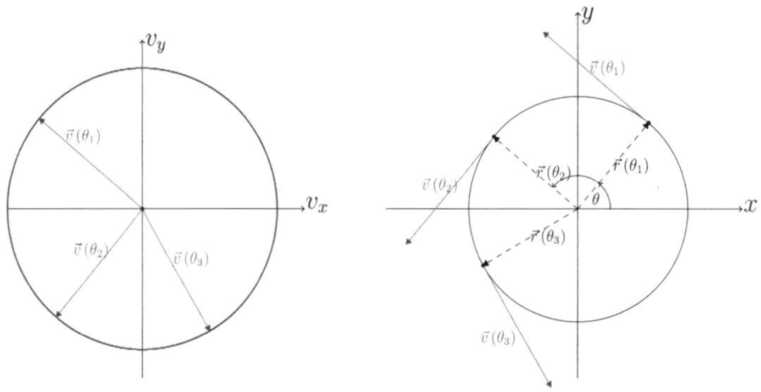

Figure 1. Newtonian hodograph method : For spatial circular motion (right) the hodograph (left) is simply a circle centred at the origin of the velocity space.

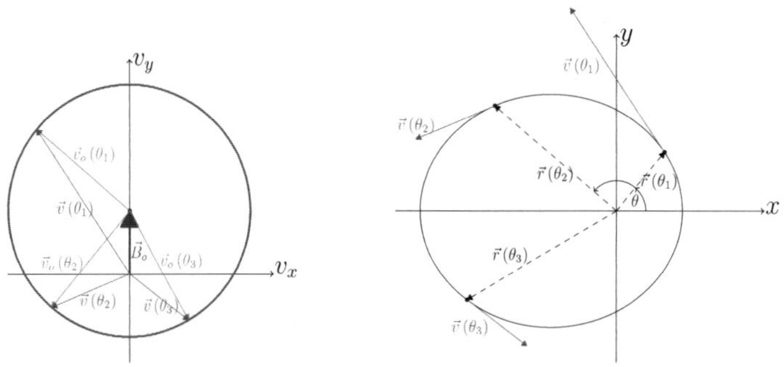

Figure 2. Newtonian hodograph method : For spatial bound non-circular motion (elliptic – right) the hodograph (left) is still a circle but shifted from the origin of the velocity space.

The spatial trajectory $\vec{r}(\theta) = r(\theta)\,\hat{r}(\theta)$ derives from (6) using the relation

$$v_\theta = \vec{v} \cdot \hat{\theta} = \frac{\ell}{mr}\,, \tag{7}$$

from which also follows the energy equation, completely confined to velocity space variables,

$$\frac{m\vec{v}^{\,2}}{2} + \frac{\kappa}{r} = \frac{m\vec{v}^{\,2}}{2} + \frac{m\kappa}{\ell} v_\theta = E'\,. \tag{8}$$

The nature of the solution depends on Hamilton's vector. Substituting (6) in (8), the relation

$${B_o}^2 = \frac{2E'}{m} + \frac{\kappa^2}{\ell^2}\,, \tag{9}$$

follows. The minimal energy hodograph, with $\vec{B}_o = 0$, is a canonical circle $C_o(\ell)$ drawn by the tips of the vectors $\vec{v}_o(\theta) = -(\kappa/\ell)\,\hat{\theta}$ centred at the origin of velocity space (Figure 1), corresponding to circular spatial motion. With increasing energy $\vec{B}_o$ becomes non-zero, the corresponding hodographs (6) remain circles with the same radius $|\kappa|/\ell$ but shifted from the origin of velocity space by $\vec{B}_o$ (Figure 2 & 3), and the spatial orbits become conic sections. In particular, for unbound systems the hodographs reduce the infinite spatial trajectory into a finite circular arc (Figure 3). The endpoints correspond to $v_\theta = 0 \Leftrightarrow r \to \infty$, where $\vec{v}$ is tangent to the hodograph.

IOP Conf. Series: Journal of Physics: Conf. Series **845** (2017) 012010 doi:10.1088/1742-6596/845/1/012010

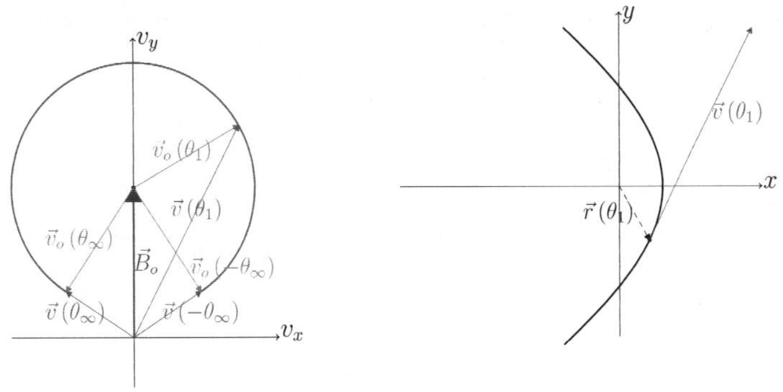

(a) Attraction.

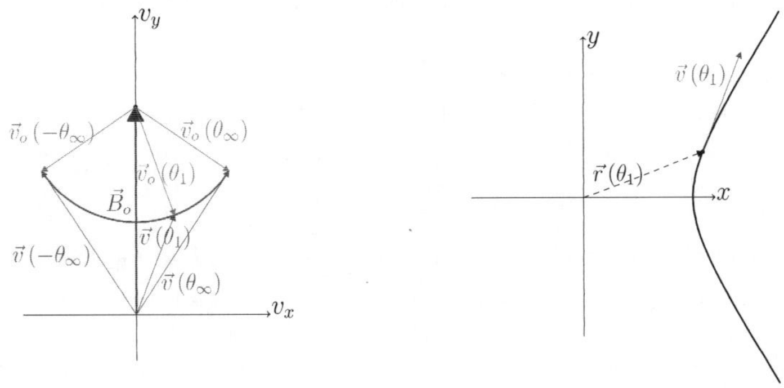

(b) Repulsion.

Figure 3. Newtonian hodograph method : For spatially unbound motion (right) the hodographs (left) are finite circular arcs with the centre shifted from the origin of the velocity space. The velocity vectors are tangent to the arcs at both ends.

The interested reader may find more details regarding the hodograph method in classical systems in several publications [4, 5, 6, 7, 8, 9, 10].

3. The hodograph method in relativistic Coulomb systems

We use for the variables of the relativistic velocity space unit velocity (time-like future-directed) 4-vectors $u^\mu = \gamma(v)\,(1, \vec{v})$, all satisfying $u \cdot u = -1$. The relativistic velocity space is then the space of all unit velocity 4-vectors, namely the 3-D unit hyperboloid (H^3) [13]

$$\mathcal{V}_{\text{rel}} \equiv \left\{ u^\mu = \left(u^0, \vec{u}\right) | u^0 = \sqrt{1 + \vec{u}^{\,2}} \right\} \tag{10}$$

embedded in a 4-D pseudo-Euclidian space

$$E^{(1,3)} = \left\{ w^\mu = \left(w^0, \vec{w}\right) \in \mathbb{R}^4 | g_{\mu\nu} = \text{diag}\,(-1, 1, 1, 1) \right\} . \tag{11}$$

Relativistic Coulomb systems are determined by the Hamiltonian

$$H\left(\vec{r}, \vec{p}\right) = \sqrt{\vec{p}^{\,2} + m^2} + \frac{\kappa}{r} \tag{12}$$

The hodograph equations that follow from Hamilton's equations may be formulated for all $w^\mu \in E^{(1,3)}$:

$$\frac{d\vec{w}}{d\theta} = \frac{\kappa}{\ell} w^0 \hat{r} \quad , \quad \frac{dw^0}{d\theta} = \frac{\kappa}{\ell} \vec{w} \cdot \hat{r} \tag{13}$$

and may be combined Lorentz-covariantly as

$$\frac{dw^\mu}{d\theta} = \frac{\kappa}{\ell}\Omega^\mu{}_\nu w^\nu \tag{14}$$

with the 4-D rotation matrix

$$\Omega^\mu{}_\nu = \left(\begin{array}{c|c} 0 & \hat{r} \\ \hline \hat{r} & 0 \end{array}\right) = \left(\begin{array}{c|ccc} 0 & \cos\theta & \sin\theta & 0 \\ \hline \cos\theta & 0 & 0 & 0 \\ \sin\theta & 0 & 0 & 0 \\ 0 & 0 & 0 & 0 \end{array}\right) \tag{15}$$

Since $\Omega_{\mu\nu} = -\Omega_{\nu\mu}$, it follows that for any two solutions $w_1^\mu(\theta), w_2^\mu(\theta)$ of (14) in $E^{(1,3)}$ the inner product $w_1 \cdot w_2$ is constant. In particular, any solution of the hodograph equations in $E^{(1,3)}$ is of constant magnitude.

The hodograph equations induce pseudo-rotation in $E^{(1,3)}$, relative to an axis that itself rotates with θ. The axis of rotation is identified as being along the vector

$$v_o^\mu = \left(1, -\frac{\kappa}{\ell}\hat{\theta}\right) \tag{16}$$

It is easily verified that the vector (16) is preserved by (14). The fact that the classical basic circle $\vec{v}_o = -(\kappa/\ell)\hat{\theta}$ is the spatial component of v_o^μ is very significant and will be discussed in the following.

The hodographs of relativistic Coulomb systems are the solutions of (14) that are confined to $\mathcal{V}_{\mathrm{rel}}$. The confinement is insured by the constant norm of the solution. Such general solutions, depending on the total energy E and angular momentum ℓ as parameters, are denoted in the following $u^\mu(\theta|E,\ell)$. The principal significance of the axis vector (16) is then in determining the energy integral, and thus the energy dependence of the hodograph :

$$-u \cdot v_o = u^0 + \frac{\kappa}{\ell}u_\theta = \frac{E}{m} \tag{17}$$

($u_\theta = \vec{u} \cdot \hat{\theta}$), so the energy is the (constant) projection of the kinematical momentum mu^μ on the rotation axis.

It is fairly straight-forward to obtain the general energy dependence of the hodographs. Consider the infinitesimal variation between two hodographs whose energy differ by δE,

$$\delta u^\mu = u^\mu(\theta|E+\delta E,\ell) - u^\mu(\theta|E,\ell) = \frac{\partial u^\mu}{\partial E}(\theta|E,\ell)\,\delta E \tag{18}$$

The constraints $u \cdot v_o = -E/m$ and $u \cdot u = -1$ imply that

$$\frac{\partial u^\mu}{\partial E} = \frac{Eu^\mu - mv_o^\mu}{\Lambda^2(E,\ell)} \tag{19}$$

with

$$\Lambda^2(E,\ell) = (Eu - mv_o)^2 = E^2 + m^2\left(\frac{\kappa^2}{\ell^2} - 1\right). \tag{20}$$

The vector $\partial u^\mu/\partial E$ is a space-like 4-vector, being orthogonal to u^μ and tangent to $\mathcal{V}_{\mathrm{rel}}$ at u^μ, therefore $\Lambda^2(E,\ell) = (\partial u/\partial E)^{-2} > 0$. Integrating (19) as a 1st order ODE in E then yields the generic form of the hodographs,

$$u^\mu(\theta|E,\ell) = w_o^\mu(\theta|E,\ell) + \Lambda(E,\ell)\,n_o^\mu(\theta|\ell) \tag{21}$$

where

$$w_o^\mu(\theta|E,\ell) = \begin{cases} \dfrac{E}{\left(1-\kappa^2/\ell^2\right)m} v_o^\mu(\theta|\ell) & \text{for } \ell \neq |\kappa| \\ \dfrac{m}{2E} v_o^\mu(\theta|\ell) & \text{for } \ell = |\kappa| \end{cases}, \tag{22}$$

and $n_o^\mu(\theta|\ell)$ is an energy-independent solution of the hodograph equation (14) (whose explicit form, depending on κ/ℓ, will be given in the following), and also linearly independent of v_o^μ. Equation (21) demonstrates $u^\mu(\theta|E,\ell)$ as the linear combination of an on-axis component w_o^μ and an off-axis component

$$B^\mu(\theta|E,\ell) \equiv \Lambda(E,\ell)\, n_o^\mu(\theta|\ell) \,. \tag{23}$$

Both have constant magnitudes but changing directions. Being both solutions of the hodograph equation (14), they also maintain constant relative angle. The combined motion is therefore epicyclic, or precession – a rotation (B^μ) imposed on another rotation (w_o^μ). $B^\mu(\theta|E,\ell)$ will be recognized in the following as the *relativistic Hamilton vector.*

Once the hodograph solution $u^\mu(\theta|E,\ell)$ is available, it is also straight-forward to obtain the corresponding spatial trajectory. In simile to (7), from the definition of the conserved angular momentum $\ell = mru_\theta$ then follows

$$\vec{r}(\theta) = \frac{\ell}{mu_\theta(\theta)}\hat{r}(\theta)\,. \tag{24}$$

In return, since $r > 0$, the requirement that u_θ must be non-negative, $u_\theta(\theta) \geq 0$, with equality possible only for unbound configurations at $r \to \infty$, determines the possible θ-range.

The ratio $|\kappa|/\ell$ determines the nature of the rotation axis, and consequently of the hodograph itself. From the magnitude of the axis vector (16) we get

$$v_o{}^2 = \frac{\kappa^2}{\ell^2} - 1 \quad \begin{array}{lcl} \nearrow & \ell > |\kappa| & \Leftrightarrow \; v_o^\mu \text{ time-like} \\ \rightarrow & \ell = |\kappa| & \Leftrightarrow \; v_o^\mu \text{ light-like} \\ \searrow & \ell < |\kappa| & \Leftrightarrow \; v_o^\mu \text{ space-like} \end{array} \tag{25}$$

$|\kappa|/\ell$ is the magnitude of the spatial velocity on the basis circle – minimal energy circle – formed by v_o^μ, $C_o(\ell) = \{v_o^\mu(\theta)\}$ (see Figure 4). The relativistic requirement that particles' velocities cannot reach the velocity of light necessarily implies that the relativistic solution can have non-relativistic limit only for time-like v_o^μ. The following results split accordingly.

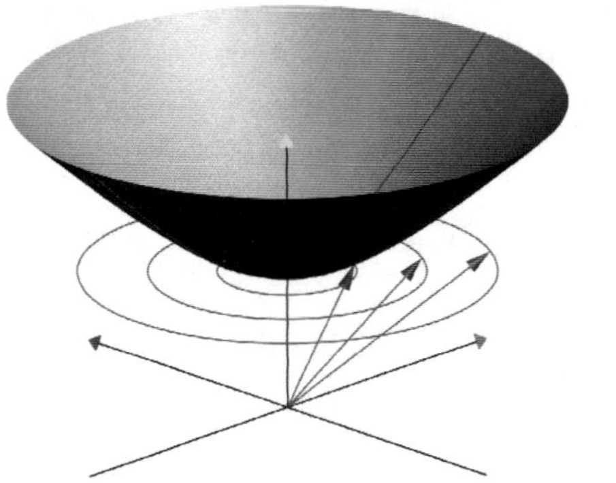

Figure 4. Relativistic hodograph method : The axis vector v_o^μ and the basis circle $C_o(\ell)$ that it traverses, relative to the velocity space $\mathcal{V}_{\rm rel}$ hyperboloid, for the 3 possible cases : (i) left arrow – v_o^μ time-like (for $\ell > |\kappa|$), if continued the vector punches the hyperboloid into its interior; (ii) middle arrow – v_o^μ light-like (for $\ell = |\kappa|$), if continued the vector approaches the hyperboloid asymptotically; (iii) right arrow – v_o^μ space-like (for $\ell < |\kappa|$), if continued the vector recedes from the hyperboloid.

A note regarding the hodograph illustrations in the following : Since the relativistic hodographs are 3-D curves, each of the cases discussed is illustrated with 3 projected views of a segment of the hodograph – from above, from the side (horizontal) and an oblique-diagonal view

IOP Conf. Series: Journal of Physics: Conf. Series **845** (2017) 012010 doi:10.1088/1742-6596/845/1/012010

– accompanied by an illustration of the corresponding spatial orbit. All these illustrations share a common colour code : The axes with red, blue and green arrowhead correspond, respectively, to u_x, u_y and u^0 (the u_z axis suppressed); the thick red line is the hodograph itself; the blue circle is the basis circle $C_o(\ell) = \{v_o^\mu(\theta)\}$; brown is the colour for the vector w_o^μ and the circle $\{w_o^\mu(\theta)\}$ it draws; the orange line is a cross-section of the velocity space hyperboloid $\mathcal{V}_{\text{rel}}$ in the $(u_x = u_y)$-u^0 plane; and the black line is the Hamilton vector in each case. These colours cannot be seen in the printed version, and the reader is advised to use the on-line or PDF versions. The scales are different according to case.

4. Hodographs with Newtonian limit ($\ell > |\kappa|$)

For $\ell > |\kappa|$ the solution of the hodograph equation (14) (with an arbitrary shift angle put to zero) is

$$\begin{aligned} \vec{u} &= B_o \sin(\beta\theta)\,\hat{r} + \left[-\frac{\kappa E}{m\ell\beta^2} + \frac{B_o}{\beta}\cos(\beta\theta)\right]\hat{\theta}\,, \\ u^0 &= \frac{E}{m} - \frac{\kappa}{\ell}u_\theta = \frac{E}{m\beta^2} - \frac{\kappa B_o}{\beta\ell}\cos(\beta\theta) \end{aligned} \tag{26}$$

with the notations

$$\beta = \sqrt{1 - \frac{\kappa^2}{\ell^2}} \quad , \quad B_o = \sqrt{\frac{E^2}{m^2\beta^2} - 1} = \frac{\Lambda(E,\ell)}{m\beta}\,. \tag{27}$$

It may be cast into the generic hodograph form (21) as

$$u^\mu(\theta|E,\ell) = \frac{E}{m\beta^2}v_o^\mu + B_o n_1^\mu \tag{28}$$

with the vector

$$n_1^\mu(\theta|\ell) \equiv \left(-\frac{\kappa}{\beta\ell}\cos(\beta\theta)\,, \sin(\beta\theta)\,\hat{r} + \frac{1}{\beta}\cos(\beta\theta)\,\hat{\theta}\right) \tag{29}$$

which is a space-like unit 4-vector, tangent to $\mathcal{V}_{\text{rel}}$ at u_o^μ, satisfying $n_1 \cdot n_1 = 1\,,\; u_o \cdot n_1 = 0$.

As is evident from Figure 4, the axis vector v_o^μ, if continued, punches the velocity space hyperboloid. The punching point is at $u_o^\mu(\theta|\ell) = \beta^{-1}v_o^\mu(\theta|\ell)$, which is the hodograph that corresponds, for any given value of $\ell > |\kappa|$, to the state of minimal energy with $E_{\text{min}} = \beta m$ and $B_o = 0$, describing a uniform circular orbit in $\mathcal{V}_{\text{rel}}$ and corresponding to spatial circular motion.

For $E > \beta m$, the vector $w_o^\mu = \left(E/m\beta^2\right)v_o^\mu$, being time-like, punching through the velocity hyperboloid $\mathcal{V}_{\text{rel}}$ at u_o^μ, describes a circle which lies horizontally within $\mathcal{V}_{\text{rel}}$, with its centre right above the origin of the embedding space $E^{(1,3)}$. Then, adding $B_o n_1^\mu$ at the tip of w_o^μ (and recalling that both are orthogonal as 4-vectors), the superposed vector reaches $\mathcal{V}_{\text{rel}}$ at u^μ.

The representation (28) strongly resembles the Newtonian solution (6) : Not only is $B_o n_1^\mu$ superposed upon the uniformly rotating vector w_o^μ, but in the nonrelativistic limit $w_o^\mu \to (1, \vec{v}_o)$ and $B_o n_1^\mu \to \left(0, \vec{B}_o\right)$, the latter thus reducing to the classical Hamilton vector. It seems therefore appropriate to regard $B^\mu \equiv B_o n_1^\mu$ as the *relativistic Hamilton vector.*

The vector B^μ is not constant, but rotates with constant magnitude. Its derivative tangent to $\mathcal{V}_{\text{rel}}$ at u_o^μ follows from

$$\frac{Dn_1^\mu}{d\theta} = \frac{dn_1^\mu}{d\theta} - \left(n_1 \cdot \frac{du_o}{d\theta}\right)u_o^\mu = -\frac{\kappa^2}{\beta\ell^2}n_2^\mu\,. \tag{30}$$

where

$$n_2^\mu(\theta|\ell) = \left(\frac{\kappa}{\beta\ell}\sin(\beta\theta)\,, \cos(\beta\theta)\,\hat{r} - \frac{1}{\beta}\sin(\beta\theta)\,\hat{\theta}\right) \tag{31}$$

IOP Conf. Series: Journal of Physics: Conf. Series **845** (2017) 012010 doi:10.1088/1742-6596/845/1/012010

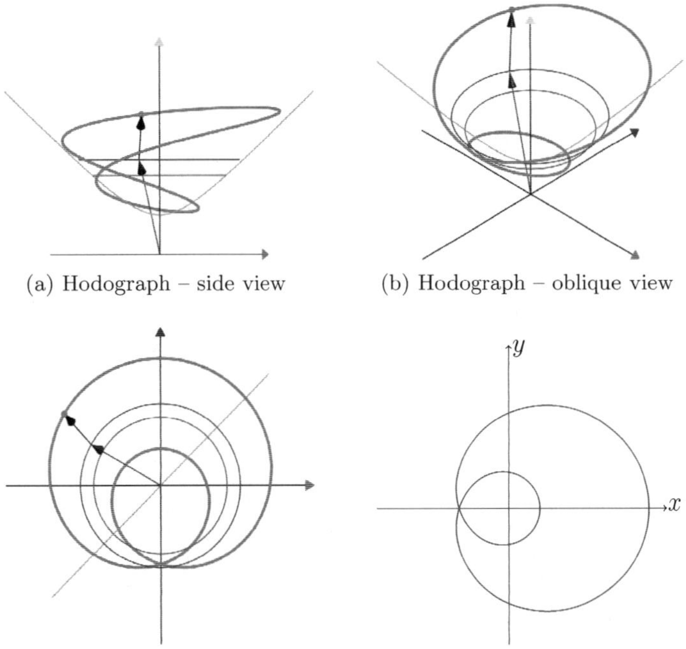

(a) Hodograph – side view (b) Hodograph – oblique view

(c) Hodograph – above view (d) Spatial orbit

Figure 5. Hodograph and spatial orbit for the bound state for v_o^μ time-like ($E = 0.6m, \kappa/\ell = -\sqrt{3}/2$). The trajectory is closed (periodical) because in this particular case $\beta = 1/2$.

is another unit 4-vector tangent to $\mathcal{V}_{\rm rel}$ at u_o^μ, orthogonal to $n_1^\mu(\theta|\ell)$ and satisfying a similar equation,

$$\frac{Dn_2^\mu}{d\theta} = \frac{dn_2^\mu}{d\theta} - \left(n_2 \cdot \frac{du_o}{d\theta}\right) u_o^\mu = \frac{\kappa^2}{\beta\ell^2} n_1^\mu \,. \tag{32}$$

It follows that B^μ rotates uniformly around the axis of rotation.

4.1. Bound states

Bound states are obtained only for $\kappa < 0$ (attraction) and $\beta m \leq E < m$. Extrema are obtained when $u_r = 0$, namely $\beta\theta = n\pi$, $n \in \mathbb{Z}$, with

$$\frac{|\kappa|\, E}{m\ell\beta^2} - \frac{B_o}{\beta} \leq u_\theta \leq \frac{|\kappa|\, E}{m\ell\beta^2} + \frac{B_o}{\beta} \tag{33}$$

u_θ cannot vanish, thus determining, via the relation $u_\theta = \ell/mr$, corresponding lower and upper bounds for r :

$$\frac{|\kappa|\, E - \ell\Lambda}{m^2 - E^2} \leq r \leq \frac{|\kappa|\, E + \ell\Lambda}{m^2 - E^2} \,. \tag{34}$$

Since $\beta \neq 1$ the hodograph is not periodic 2π in θ, therefore not circular and the spatial trajectories are rotating ellipses. Periodicity is obtained only when β is a rational number.

An exemplary hodograph, using the values $E = 0.6m$, $\kappa/\ell = -\sqrt{3}/2$ ($\beta = 0.5$) together with the corresponding spatial orbit obtained via (24) is demonstrated in Figure 5.

IOP Conf. Series: Journal of Physics: Conf. Series **845** (2017) 012010 doi:10.1088/1742-6596/845/1/012010

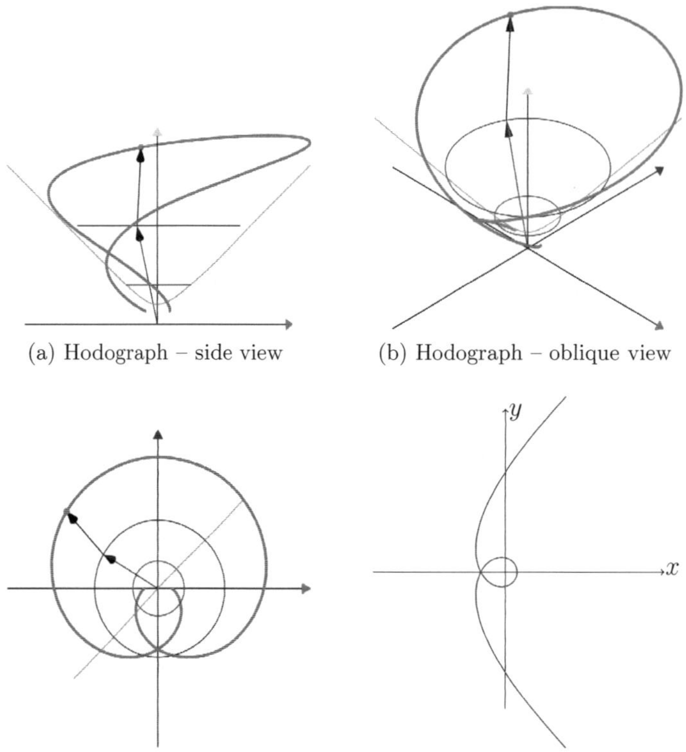

(a) Hodograph – side view (b) Hodograph – oblique view

(c) Hodograph – above view (d) Spatial orbit

Figure 6. Hodograph and spatial orbit for the attraction unbound state for v_o^μ time-like $(E = 1.25m, \kappa/\ell = -\sqrt{3}/2)$.

4.2. Unbound states

For unbound systems, with $E \geq m$, the infinite limit $r \to \infty$ corresponds, according to (24), to $u_\theta = 0$. The infinite spatial trajectory turns into a finite-size orbit in velocity space, another merit of the hodograph method. The finite angular range of the spatial trajectory is $-\theta_\infty < \theta < \theta_\infty$, with the endpoints (at which $u_\theta = 0$) at

$$\theta_\infty = \begin{cases} \dfrac{\psi_\infty}{\beta} & \kappa > 0 \\ \dfrac{\pi - \psi_\infty}{\beta} & \kappa < 0 \end{cases} \quad , \quad \psi_\infty = \sin^{-1}\left(\frac{\sqrt{E^2 - m^2}}{mB_o}\right) \tag{35}$$

The distance of closest approach is also found from $\max(u_\theta)$ via (24),

$$r_{\min} = \frac{\ell\Lambda - \kappa E}{E^2 - m^2} \tag{36}$$

With the numerical values $E = 1.25m$, $\kappa/\ell = \pm\sqrt{3}/2$ $(\beta = 0.5)$, the angular endpoints are at $\theta_\infty = 1.79\pi$ [rad] for attraction $(\kappa < 0)$ and at $\theta_\infty = 0.212\pi$ [rad] for repulsion $(\kappa > 0)$. Exemplary hodographs with these values, together with the corresponding spatial trajectories, are given in Figure 6 for attraction and Figure 7 for repulsion.

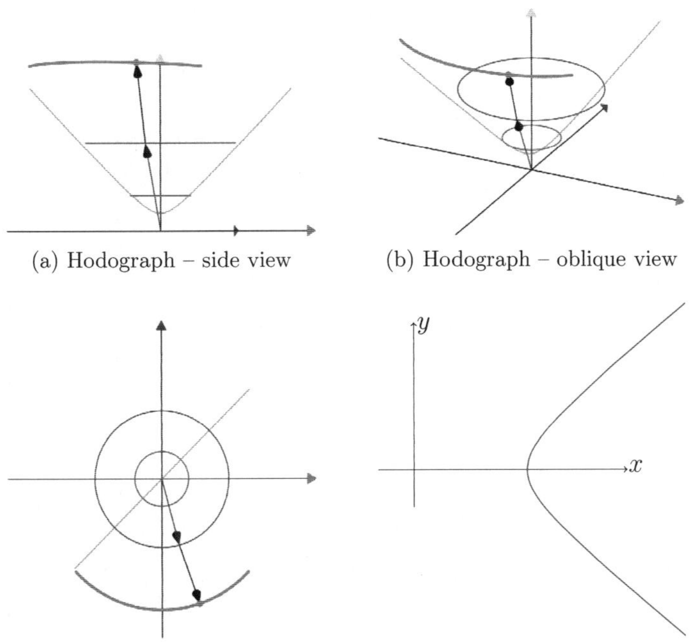

(a) Hodograph – side view (b) Hodograph – oblique view

(c) Hodograph – above view (d) Spatial orbit

Figure 7. Hodograph and spatial orbit for the repulsion unbound state for v_o^μ time-like ($E = 1.25m, \kappa/\ell = \sqrt{3}/2$).

5. Hodographs for exclusively relativistic cases

The cases with $\ell \leq |\kappa|$ don't have a nonrelativistic counterpart, because that would imply, as already pointed out, luminal or superluminal velocities in the Newtonian limit. The solutions for $\ell = |\kappa|$ and $\ell < |\kappa|$ are presented separately because of the different mathematical expressions, although the general features of the hodographs and the particle's spatial trajectories are similar in both cases.

5.1. Hodographs for $\ell = |\kappa|$

The hodograph solution in the light-like case is (with an arbitrary shift angle put to zero)

$$\begin{aligned} \vec{u} &= \frac{\epsilon E}{m}\theta\hat{r} - \epsilon\left(\frac{E}{2m}\theta^2 - \frac{E^2 - m^2}{2mE}\right)\hat{\theta} \\ u^0 &= \frac{E^2 + m^2}{2mE} + \frac{E}{2m}\theta^2 \end{aligned} \tag{37}$$

with $\epsilon = \text{sign}(\kappa)$. It may be cast in the form of the general hodograph solution (21) as

$$u^\mu(\theta|E,\ell) = \frac{m}{2E}v_o^\mu(\theta|E,\ell) + \frac{E}{2m}n^\mu(\theta|E,\ell) \tag{38}$$

with the vector

$$n^\mu = \left(1 + \theta^2, 2\epsilon\theta\hat{r} + \epsilon\left(1 - \theta^2\right)\hat{\theta}\right). \tag{39}$$

Here both v_o^μ and n^μ are light-like, $v_o \cdot v_o = n \cdot n = 0$, and $n \cdot v_o = -2$.

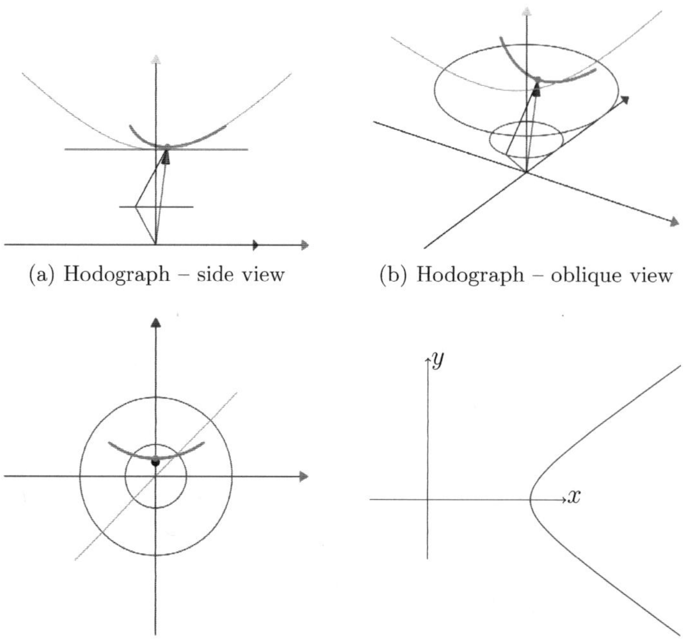

(a) Hodograph – side view (b) Hodograph – oblique view

(c) Hodograph – above view (d) Spatial orbit

Figure 8. Hodograph and spatial orbit for repulsion for v_o^μ light-like ($E = 1.25m, \ell = \kappa$).

For the angular ranges of the trajectory, determined by $u_\theta > 0$, we distinguish between the case of repulsion ($\kappa = \ell$ and $E > m$) for which

$$-\theta_\infty < \theta < \theta_\infty \tag{40}$$

and two possible cases of attraction ($\kappa = -\ell$)

$$\begin{cases} -\infty < \theta < \infty & \text{when } E < m \\ \theta > \theta_\infty \quad \text{or} \quad \theta < -\theta_\infty & \text{when } E > m \end{cases} \tag{41}$$

where (in both (40) and (41))

$$\theta_\infty = \frac{\sqrt{E^2 - m^2}}{E}. \tag{42}$$

In the case of repulsion ($\kappa = \ell$) the particle comes from infinity ($r \to \infty$, $u_\theta = 0$) to a minimal distance at $\theta = 0$

$$r_{\min} = \frac{\ell}{m u_{\theta\max}} = \frac{2\ell E}{E^2 - m^2} \tag{43}$$

and then back again to infinity. The hodograph is a finite segment between $u^\mu_{-\infty}$ and u^μ_∞, where

$$u^\mu_{\pm\infty} = u^\mu(\pm\theta_\infty) = \left(\frac{E}{m}, \pm\frac{\sqrt{E^2 - m^2}}{m}\hat{r}(\pm\theta_\infty)\right) \tag{44}$$

For the particular choice $E = 1.25m$ the endpoints are at $\theta_\infty = 0.6$ [rad]. The hodograph for these values, together with the corresponding spatial trajectory, is given in Figure 8.

IOP Conf. Series: Journal of Physics: Conf. Series **845** (2017) 012010 doi:10.1088/1742-6596/845/1/012010

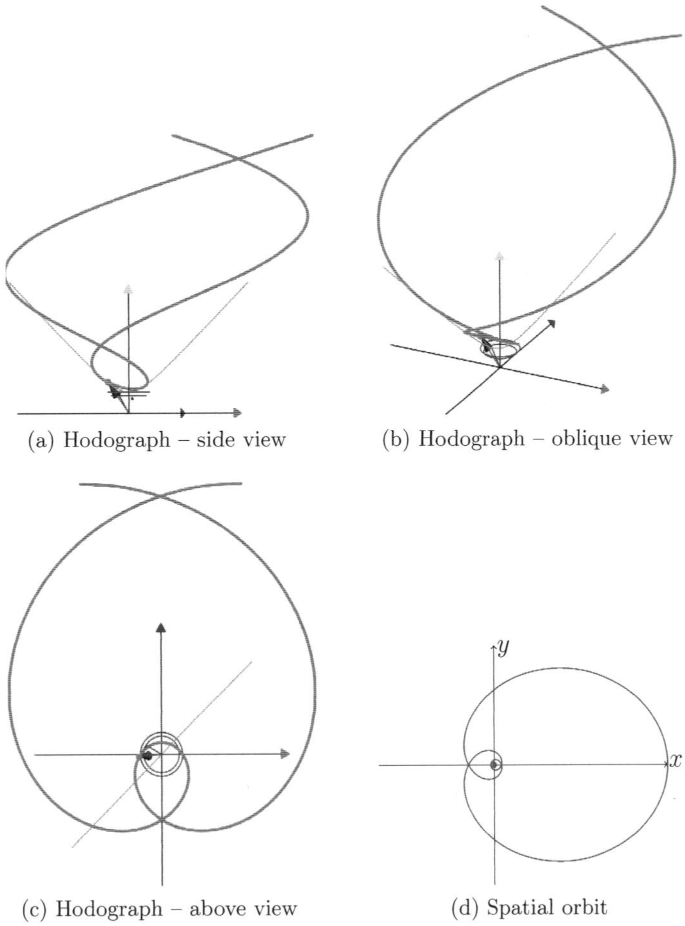

Figure 9. Hodograph and spatial orbit for 1st attraction (unstable bound-like) case for v_o^μ light-like ($E = 0.6m, \ell = -\kappa$).

The first attraction case ($\kappa = -\ell$, $E < m$) resembles an unstable bound state : θ may get any value without bounds ($-\infty < \theta < \infty$), but the particle is spatially bound within the maximum distance from the centre

$$r_{\text{max}} = \frac{\ell}{m u_{\theta\,\text{min}}} = \frac{2E\ell}{m^2 - E^2} \tag{45}$$

The particle may start spatially at some distance from the centre of force, then move in a spiral trajectory towards the centre. At the same time the hodograph starts close to the centre of velocity space and spirals away to infinity (velocity space infinity). In the limit $\theta \to \infty$ then also $\vec{u} \to \infty$ ($|\vec{v}| \to c$), and the particle collapses into the centre of force. The hodograph with the choice $E = 0.6m$, together with the corresponding spatial trajectory, is given in Figure 9.

In the second attraction case ($\kappa = -\ell$, $E > m$) a situation similar to gravitational singularity is encountered : Either (i) $\theta_\infty < \theta < \infty$, then the hodograph ia a semi-infinite segment, the particle arrives from spatial infinity ($u_\theta = 0$) and collapses into the centre of force for $\theta \to \infty$ ($u_\theta \to \infty$, $r \to 0$), or (ii) $-\infty < \theta < -\theta_\infty$, again a semi-infinite hodograph, and the particle bursts out of the centre of force corresponding to $\theta \to -\infty$ ($u_\theta \to \infty$, $r \to 0$) and escapes to infinity ($u_\theta \to 0$) (although the latter possibility is bizarre and unlikely, it is mentioned

IOP Conf. Series: Journal of Physics: Conf. Series **845** (2017) 012010 doi:10.1088/1742-6596/845/1/012010

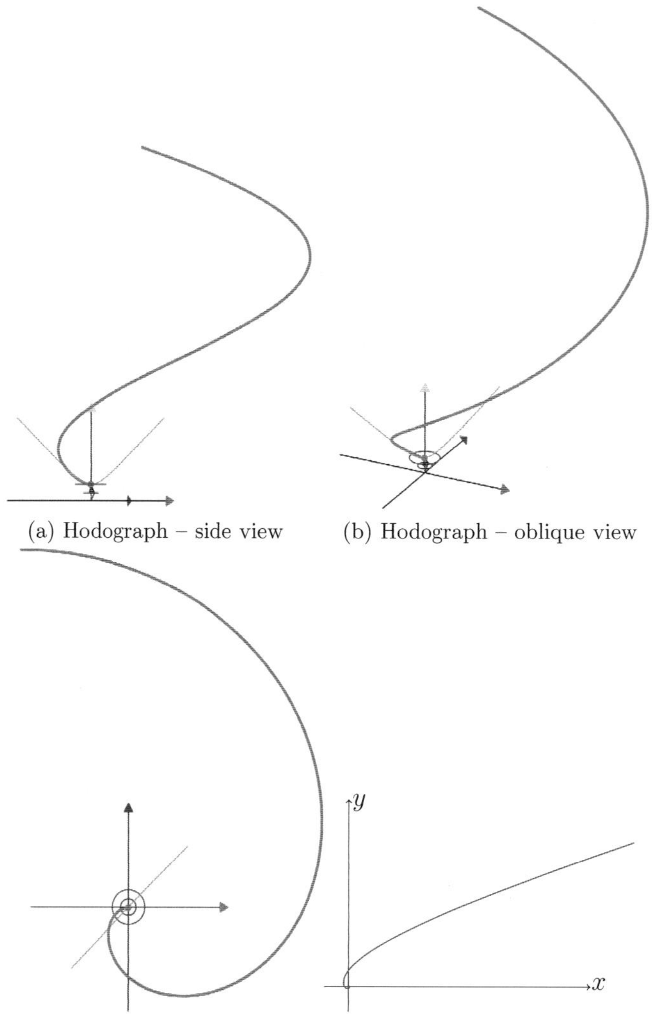

(a) Hodograph – side view (b) Hodograph – oblique view

(c) Hodograph – above view (d) Spatial orbit

Figure 10. Hodograph and spatial orbit for 2nd attraction (collapse) case for v_o^μ light-like ($E = 1.05m, \ell = -\kappa$).

because of its theoretical possibility). The hodograph with the choice $E = 1.05m$, for which $\theta_\infty = 0.01$ [rad], together with the corresponding spatial trajectory, is given in Figure 10.

5.2. Hodographs for $\ell < |\kappa|$

The hodograph solution in the space-like case (with an arbitrary shift angle put to zero) is

$$\begin{aligned} \vec{u} &= -\epsilon A_o \sinh\left(\bar{\beta}\theta\right)\hat{r} + \left[\frac{\kappa E}{m\ell\bar{\beta}^2} - \frac{\epsilon A_o}{\bar{\beta}}\cosh\left(\bar{\beta}\theta\right)\right]\hat{\theta} \\ u^0 &= -\frac{E}{m\bar{\beta}^2} + \frac{|\kappa|\,A_o}{\bar{\beta}\ell}\cosh\left(\bar{\beta}\theta\right) \end{aligned} \tag{46}$$

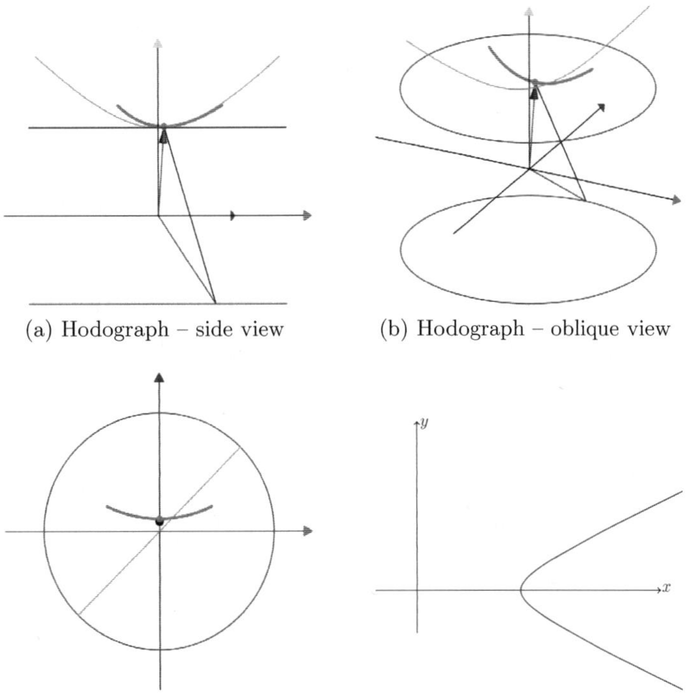

(a) Hodograph – side view (b) Hodograph – oblique view

(c) Hodograph – above view (d) Spatial orbit

Figure 11. Hodograph and spatial orbit for repulsion for v_o^μ space-like ($E = 1.25m, \kappa/\ell = 1.5$).

where $\epsilon = \text{sign}(\kappa)$ and with the notations

$$\bar{\beta} = \sqrt{\frac{\kappa^2}{\ell^2} - 1} \quad , \quad A_o = \sqrt{\frac{E^2}{m^2\bar{\beta}^2} + 1} = \frac{\Lambda(E, \ell)}{m\bar{\beta}} . \tag{47}$$

As with the previous cases, it may be cast into the general hodograph form (21) as

$$u^\mu(\theta|E, \ell) = -\frac{E}{m\bar{\beta}^2} v_o^\mu + A_o n^\mu \tag{48}$$

with

$$n^\mu(\theta|\ell) = \left(\frac{|\kappa|}{\bar{\beta}\ell} \cosh\left(\bar{\beta}\theta\right), \epsilon \sinh\left(\bar{\beta}\theta\right) \hat{r} - \frac{\epsilon}{\bar{\beta}} \cosh\left(\bar{\beta}\theta\right) \hat{\theta}\right) \tag{49}$$

The rôles of v_o^μ and n^μ are now interchanged, since v_o^μ is space-like in the present case and n^μ is a time-like unit 4-vector. The negative coefficient of v_o^μ turns it up-side down, pointing downwards rather than upwards. This behaviour is clear in Figures 11, 12 & 13. Although this case doesn't have a Newtonian counter-part, the composition in (48) suggests, in analogy with (28), to regard the vector $A_o n^\mu$ as the Hamilton vector in the present cases.

The angular ranges of the possible trajectories are, in principle, the same as in the former case, given by (40) for repulsion and (41) for attraction, this time with

$$\theta_\infty = \frac{1}{\bar{\beta}} \cosh^{-1}\left(\frac{\kappa E}{\ell\Lambda}\right) . \tag{50}$$

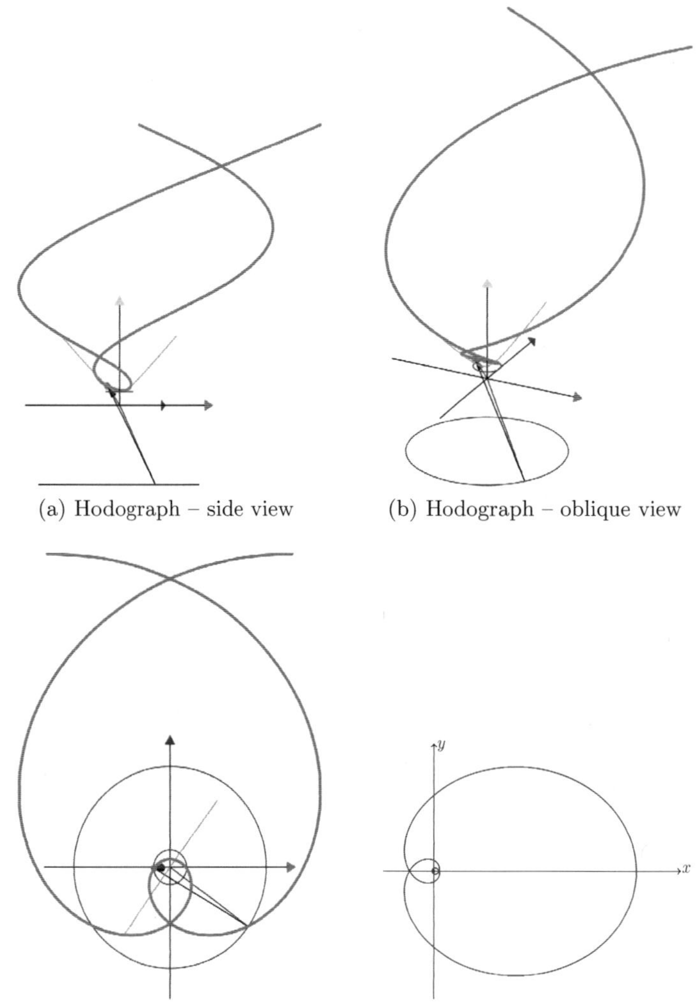

(a) Hodograph – side view (b) Hodograph – oblique view

(c) Hodograph – above view (d) Spatial orbit

Figure 12. Hodograph and spatial orbit for 1st attraction (unstable bound-like) case for v_o^μ space-like ($E = 0.6m, \kappa/\ell = -1.05$).

In the case of repulsion ($\kappa > \ell$ and $E > m$), with the angular range $-\theta_\infty < \theta < \theta_\infty$, the particle comes from spatial infinity to closest approach at $\theta = 0$, with the minimal distance given in this case by

$$r_{\min} = \frac{\ell}{mu_{\theta\,\max}} = \frac{\ell\Lambda + \kappa E}{E^2 - m^2} \tag{51}$$

and then back again to infinity. The hodograph is a finite segment between $u^\mu_{-\infty}$ and u^μ_∞, with $u^\mu_{\pm\infty}$ given again by (44). For the particular choice $E = 1.25m$ the endpoints are, again, at $\theta_\infty = 0.6\,[\text{rad}]$. An hodograph for these values and $\kappa = 1.5\ell$, together with the corresponding spatial trajectory, is given in Figure 11.

The first case of attraction in (41) ($\kappa < -\ell$, $E < m$), with the angular range $-\infty < \theta < \infty$, again resembles an unstable bound state with the maximum distance from the centre of force

$$r_{\max} = \frac{\ell}{mu_{\theta\,\min}} = \frac{\ell\Lambda + |\kappa|\,E}{m^2 - E^2} \tag{52}$$

IOP Conf. Series: Journal of Physics: Conf. Series **845** (2017) 012010 doi:10.1088/1742-6596/845/1/012010

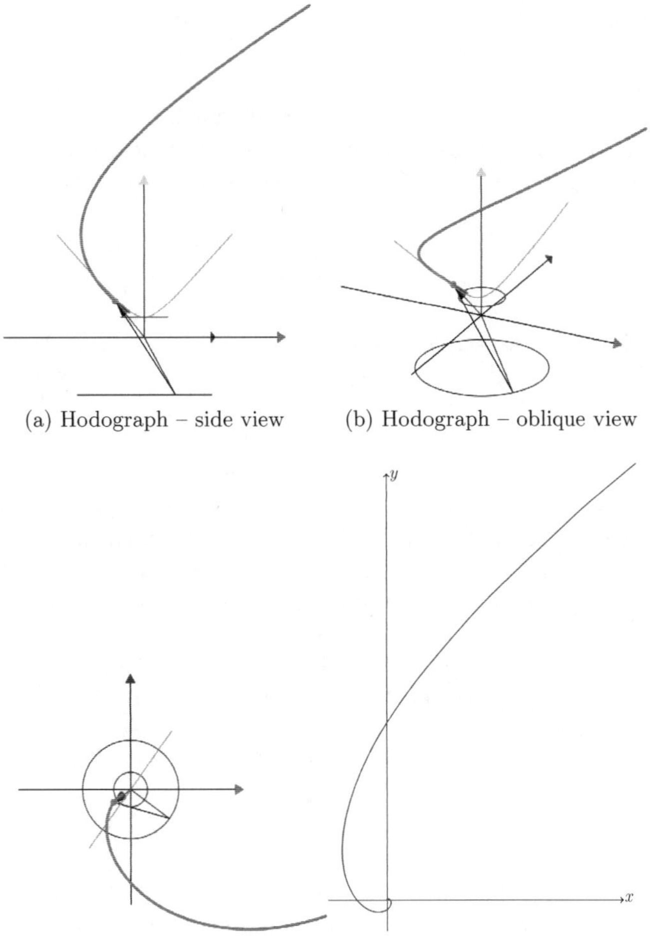

Figure 13. Hodograph and spatial orbit for 2nd attraction (collapse) case for v_o^μ space-like ($E = 1.25m, \kappa/\ell = -1.2$).

The particle starts spatially at some distance from the centre, then moves in a spiral trajectory towards the centre, eventually collapsing into it. The hodograph starts close to the centre of velocity space and spirals away to (velocity space) infinity. These hodographs are illustrated, with the choice $E = 0.6m$ and $\kappa = -1.05\ell$, together with the corresponding spatial trajectory, in Figure 12.

In the second attraction case ($\kappa < -\ell$, $E > m$) then, again, as with the light-like case : Either (i) $\theta_\infty < \theta < \infty$, then the hodograph is a semi-infinite segment, the particle arrives from spatial infinity ($u_\theta = 0$) and collapses into the centre of force for $\theta \to \infty$ ($u_\theta \to \infty$, $r \to 0$), or (ii) $-\infty < \theta < -\theta_\infty$, again a semi-infinite hodograph where the particle bursts out of the centre of force corresponding to $\theta \to -\infty$ ($u_\theta \to \infty$, $r \to 0$) and escapes to infinity ($u_\theta \to 0$). These hodographs are illustrated with the choice $E = 1.25m$, for which $\theta_\infty = 0.6$ [rad], and $\kappa = -1.2\ell$, together with the corresponding spatial trajectory, in Figure 13.

6. Concluding remarks

The purpose of the talk was mainly to present the application of the generally non-familiar *hodograph method* – studying the dynamics of a system in velocity space – as it applies to a charged relativistic particle in Coulomb field. The hodograph method has the merit that when applied to systems with $1/r$ potentials, Newtonian and relativistic alike, the velocity equations are simply linear, providing a very straight-forward and elegant means to analyze the dynamics of the system.

The Newtonian Kepler/Coulomb (KC) hodograph is always a circle, in general displaced from the origin of velocity space, with a constant displacement that depends on the energy state and determines the shape of the (conic section) spatial trajectory. For relativistic Coulomb systems this structure is preserved, in principle – a basic circle associated with the minimum energy state, uniformly displaced depending on the actual energy state. The unique relativistic feature is that the displacement is itself rotating, so the whole phenomenon is of rotation superimposed on rotation – precession on the velocity hyperboloid – thus the title with "back to epicycles".

Results of the analysis were presented in the form of an excursion in the garden of the many possible manifestations of the system, mostly in an illustrative way. Explicit distinction was made between the cases which have a non-relativistic limit and others which are exclusively relativistic, demonstrating the particular characteristics of each.

This analysis, whose details will be published elsewhere, is sought to assist attempts to advance the solution of the relativistic EM 2-body problem : Unlike the Newtonian case, the relativistic (non-quantum) EM 2-body problem doesn't have, despite all the years, a satisfactory solution. Some simple particular solutions have been found, mainly in the 60's and the 70's [15], but since then no real advancement has been marked and a general solution is missing.

However, as is well known, the $1/r$ interaction allows Newtonian KC systems to enjoy a very special and simple form of the Laplace-Runge-Lenz (LRL) symmetry, which is also directly associated with the spatial trajectories being conic sections [16, 17, 18]. The combination of rotational symmetry with the extra LRL symmetry provides the full solution for these systems in configuration space. The application of the hodograph method to Newtonian KC systems, providing the full solution in velocity space, may be demonstrated to be equivalent with the application of the LRL symmetry in ordinary space. In particular, the vector displacing the basic hodograph circle in velocity space, the Hamilton vector, is a constant vector directly related to the constant LRL vector in the same system [9].

The hodograph method therefore displays what may be termed *Hamilton's symmetry*, complementary to the LRL symmetry. These symmetries are mathematically equivalent, but physically they behave differently : the LRL symmetry transforms between states of same energy but different internal angular momentum, while the Hamilton symmetry transforms between states of same angular momentum while changing the energy.

The success of using the extra symmetries with Newtonian KC systems gives rise to the hope that by studying the corresponding extra symmetries in relativistic EM systems a general solution may be advanced. It is in this context that the present work has been performed.

Newtonian KC 2-body systems may be reduced to 1-body systems by transition to the centre-of-mass (CM) reference frame. Such a simple and direct procedure is impossible, in general, for relativistic 2-body systems because the interaction is not instantaneous. It is therefore convenient to start with Coulomb systems, which may be regarded as the limit of EM 2-body systems when one of the charges is much heavier than the other. The LRL symmetry in relativistic Coulomb systems was already studied to some extent in recent years [12, 19, 20]. Here we considered the other face of these internal symmetries – the relativistic Hamilton symmetry – demonstrating its usefulness for Coulomb systems.

In futuristic view, an answer to the question *How may non-instantaneous interactions be handled on velocity space, especially in view of the fact that time doesn't appear in the structure of the velocity space ?* may provide a key to the whole issue of the 2-body EM problem. It is hoped to come back to such questions soon.

References

[1] Hamilton W R 1847 The hodograph, or a new method of expressing in symbolical language the Newtonian law of attraction *Proc. R. Irish Acad.* **3** 344-53
[2] Maxwell J C 1925 *Matter and Motion* (London: Sheldon Press) p107-9
[3] Goodstein D L and Goodstein J R 1996 *Feynman's Lost Lecture: The Motion of Planets Around The Sun* (New York: Norton)
[4] Milnor J 1983 On the geometry of the Kepler orbits *Am. Math. Monthly* **90** 353-65
[5] González-Villanueva A, Guillaumín-España E, Núñez-Yépez H N and Salas-Brito A L 1996 In velocity space the Kepler orbits are circular *Eur. J. Phys.* **17** 168-71
González-Villanueva A, Guillaumín-España E, Martínez-Romero R P, Núñez-Yépez H N and Salas-Brito A L 1998 From circular paths to elliptic orbits: a geometric approach to Kepler's motion *Eur. J. Phys.* **19** 431-8
[6] Butikov E I 2000 The velocity hodograph for an arbitrary Keplerian motion *Eur. J. Phys.* **21** 1-6
[7] Derbes D 2001 Reinventing the wheel: hodographic solutions of the Kepler problems *Am. J. Phys.* **69(4)** 481-9
[8] Kowen M and Mathur H 2003 On Feynman's analysis of the geometry of Keplerian orbits *Am. J. Phys.* **71(4)** 397-401
[9] Muñoz G 2003 Vector constants of the motion and orbits in the Coulomb/Kepler problem *Am. J. Phys.* **71(12)** 1292-3
[10] Cariñena J F, Rañada M F and Santander M 2016 A new look at the Feynman 'hodograph' approach to the Kepler first law *Eur. J. Phys.* **37** 025004
[11] Boyer T H 2004 Unfamiliar trajectories for a relativistic particle in a Kepler or Coulomb potential *Am. J. Phys.* **72(8)** 992-7
[12] Stahlhofen A A 2005 Relativistic trajectories and the Runge-Lenz vector *Am. J. Phys.* **73(7)** 581
[13] Rhodes J A and Semon M D 2004 Relativistic velocity space, Wigner rotation and Thomas precession *Am. J. Phys.* **72(7)** 943-60
[14] Ben-Ya'acov U Relativistic Coulomb systems in velocity space, in preparation
[15] Stephas P 1978 Relativistic two-body electrodynamnics *Am. J. Phys.* **46(4)** 360-5
[16] Goldstein H 1975 On the prehistory of the "Runge-Lenz" vector *Am. J. Phys.* **43(8)** 737-8.
Goldstein H 1976 More on the prehistory of the Laplace or Runge-Lenz vector *Am. J. Phys.* **44(11)** 1123-4.
[17] Goldstein H, Poole C and Safko J 2000 *Classical Mechanics* (New York: Addison-Wesley)
[18] Khachidze T T and Khelashvili A A 2008 *Dynamical Symmetry of the Kepler-Coulomb Problem in Classical and Quantum Mechanics* (New-York : Nova Science)
[19] Yoshida T 1988 Rotating Laplace-Runge-Lenz vector leading to two relativistic Kepler's equations *Phys. Rev. A* **38(1)** 19-25
[20] Ben-Ya'acov U 2010 Laplace-Runge-Lenz symmetry in general rotationally symmetric systems *J. Math. Phys.* **51** 122902

IOP Conf. Series: Journal of Physics: Conf. Series **845** (2017) 012011 doi:10.1088/1742-6596/845/1/012011

Spin 1/2 properties of massless particles of any spin

Alexander Gersten and Amnon Moalem

Department of Physics, Ben Gurion University of the Negev, Beer-Sheva, Israel

Abstract. In previous papers we have derived wave equations for massless particles of any spin. In this paper we found that all these equations, are equations of the helicity operator and are similar in form to the spin one-half equation. In these equations the helicity operators are reducible representations of the helicity operator of spin one-half. The helicity of massless particles of any spin can have only projections in the forward and backward direction of the momentum. In order to have equations with this property, subsidiary conditions have to be imposed. We have developed a first quantized equations for massless particles of any spin for which the subsidiary conditions were implicitly included in the main equation. The helicity operators of these equations are reducible representations of the spin one half helicity operator. A new feature of this paper is to use this reducibility to present higher spin equation with its lower spin equivalent. As an application we recover the bi-quaternionic form of Maxwell's equations (spin 1 equation in a form of spin one-half equation).

Key words: wave equations, massless particles, any spin, helicity equations, bi-quaternions.

1. Introduction

In our previous papers [5], [6], [7] we have derived wave equations for massless particles of any spin. In this paper we found that all these equations, are equations of the helicity operator and are similar in form to the spin one-half equation. In these equations the helicity operators are reducible representations of the helicity operator of spin one-half. Moreover, for any spin, there are only two helicity equations similar to the spin one-half equations.

The wavefunction of particle with spin s should describe the $2s+1$ components of the spin. For free particles the helicity is the projection of the spin in the direction of the momentum. For massless particles only the helicities in the direction and opposite to the direction of motion are different from zero, all other components vanish [1], [2],[3] Therefore for spins $s > \frac{1}{2}$ in addition to the $2s+1$ equations for the spin components, one has to add $2s-1$ conditions to eliminate the vanished helicities, these are the subsidiary conditions. Our helicity equations include implicitly the subsidiary conditions.

In the following we will work in the flat Minkowski space of special relativity with coordinates $x_\mu, \quad \mu = 0, 1, 2, 3$ and use the short hand notation for functions,

$$\psi \equiv \psi\left(x_\mu\right).$$

In Section 2 we present the Maxwell's equations as wave equations of the helicity operator. In Section 3 we generalize this result for any spin. In Section 4 we give the basic properties of quaternions and bi-quaternions and we show how bi-quaternions can be used in the description of the helicity operators.

IOP Conf. Series: Journal of Physics: Conf. Series **845** (2017) 012011 doi:10.1088/1742-6596/845/1/012011

In Section 5 We derive the wave bi-quaternionic Maxwell's equations. Here the Helicity operator is the same as the spin one-half helicity operator, the difference is the form of the wave function which is a bi-quaternion.

In Section 6 we summarize our results.

2. Maxwell's equations as wave equations of the helicity operator

Maxwell's equations in the covariant form [4] in Gaussian system units are,

$$\partial_\mu F^{\mu\nu} = \frac{4\pi}{c} j^\nu, \quad \partial_\mu \tilde{F}^{\mu\nu} = 0, \quad \nu = 0,1,2,3, \tag{1}$$

where $\partial_0 = \frac{1}{c}\frac{\partial}{\partial t}$ and the antisymmetric tensor $F^{\mu\nu}$ and its dual $\tilde{F}^{\mu\nu}$ are defined via the electric and magnetic fields $\mathbf{E}$ and $\mathbf{B}$ respectively as,

$$(F^{\mu\nu}) = \begin{pmatrix} 0 & -E_x & -E_y & -E_z \\ E_x & 0 & -B_z & B_y \\ E_y & B_z & 0 & -B_x \\ E_z & -B_y & B_x & 0 \end{pmatrix}, \tag{2}$$

$$\tilde{F}^{\mu\nu} = \frac{1}{2}\epsilon^{\mu\nu\alpha\beta} F_{\alpha\beta} = \begin{pmatrix} 0 & -B_x & -B_y & -B_z \\ B_x & 0 & E_z & -E_y \\ B_y & -E_z & 0 & E_x \\ B_z & E_y & -E_x & 0 \end{pmatrix}, \tag{3}$$

where $\epsilon^{\mu\nu\alpha\beta}$ is the totally antisymmetric tensor ($\epsilon^{0123} = 1, \quad \epsilon^{\mu\nu\alpha\beta} = -\epsilon_{\mu\nu\alpha\beta}$). The sum of the $F^{\mu\nu}$ and $i\tilde{F}^{\mu\nu}$ is the self-dual antisymmetric tensor, it depends only on the combination,

$$\mathbf{\Psi} = (\mathbf{E}+i\mathbf{B}), \tag{4}$$

$$F^{\mu\nu} + i\tilde{F}^{\mu\nu} = \begin{pmatrix} 0 & -E_x - iB_x & -E_y - iB_y & -E_z - iB_z \\ E_x + iB_x & 0 & iE_z - B_z & B_y - iE_y \\ E_y + iB_y & B_z - iE_z & 0 & iE_x - B_x \\ E_z + iB_z & iE_y - B_y & B_x - iE_x & 0 \end{pmatrix}, \tag{5}$$

$$= \begin{pmatrix} 0 & -\Psi_x & -\Psi_y & -\Psi_z \\ \Psi_x & 0 & i\Psi_z & -i\Psi_y \\ \Psi_y & -i\Psi_z & 0 & i\Psi_x \\ \Psi_z & i\Psi_y & -i\Psi_x & 0 \end{pmatrix} \tag{6}$$

The self-dual Maxwell equations take the form,

$$\partial_\mu \left(F^{\mu\nu} + i\tilde{F}^{\mu\nu}\right) = -\left(\alpha_\mu\right)^{\nu k} \partial_\mu \Psi_k = \frac{4\pi}{c} j^\nu, \quad \mu = 0, x, y, z, = 0,1,2,3 \quad k = x, y, z, = 1,2,3 \tag{7}$$

where,

$$\alpha_0 = \begin{pmatrix} 1 & 0 & 0 & 0 \\ 0 & 1 & 0 & 0 \\ 0 & 0 & 1 & 0 \\ 0 & 0 & 0 & 1 \end{pmatrix}, \quad \alpha_1 = \begin{pmatrix} 0 & -1 & 0 & 0 \\ -1 & 0 & 0 & 0 \\ 0 & 0 & 0 & -i \\ 0 & 0 & i & 0 \end{pmatrix},$$

,

IOP Conf. Series: Journal of Physics: Conf. Series **845** (2017) 012011 doi:10.1088/1742-6596/845/1/012011

$$\alpha_2 = \begin{pmatrix} 0 & 0 & -1 & 0 \\ 0 & 0 & 0 & i \\ -1 & 0 & 0 & 0 \\ 0 & -i & 0 & 0 \end{pmatrix}, \quad \alpha_3 = \begin{pmatrix} 0 & 0 & 0 & -1 \\ 0 & 0 & -i & 0 \\ 0 & i & 0 & 0 \\ -1 & 0 & 0 & 0 \end{pmatrix} \tag{8}$$

and the Ψ_k is one of the the three components of the wave function,

$$\Phi = \begin{pmatrix} 0 \\ \Psi_x \\ \Psi_y \\ \Psi_z \end{pmatrix} = \begin{pmatrix} 0 \\ E_x + iB_x \\ E_y + iB_y \\ E_z + iB_{zz} \end{pmatrix}. \tag{9}$$

Multiplying Eq. (7) by $i\hbar c$, where $\hbar$ is the reduced Planck constant, we obtain,

$$(\alpha_\mu)^{\nu k} \left(i\hbar c\partial_\mu\right) \Psi_k = -i\hbar c\frac{4\pi}{c} j^\nu. \tag{10}$$

Replacing $i\hbar\partial_\mu$ by their wave energy-momentum operators, Eq. (10) takes the form,

$$(\alpha_0 E - c\boldsymbol{\alpha} \cdot \boldsymbol{p})\, \Phi = -i\hbar 4\pi J, \tag{11}$$

where J is the column of the four vector j^ν. Maxwell's equations without sources (the photon equation) are,

$$(\alpha_0 E - c\boldsymbol{\alpha} \cdot \boldsymbol{p})\, \Phi = 0. \tag{12}$$

The structure of the α matrices is,

$$\alpha_0 = \begin{pmatrix} 1 & 0\ 0\ 0 \\ 0 & \\ 0 & (I_0) \\ 0 & \end{pmatrix}, \quad \alpha_1 = \begin{pmatrix} 0 & -1\ 0\ 0 \\ -1 & \\ 0 & (S_x) \\ 0 & \end{pmatrix}, \tag{13}$$

$$\alpha_2 = \begin{pmatrix} 0 & 0\ -1\ 0 \\ 0 & \\ -1 & (S_y) \\ 0 & \end{pmatrix}, \quad \alpha_3 = \begin{pmatrix} 0 & 0\ 0\ -1 \\ 0 & \\ 0 & (S_z) \\ -1 & \end{pmatrix} \tag{14}$$

$$S_x = \begin{pmatrix} 0 & 0 & 0 \\ 0 & 0 & -i \\ 0 & i & 0 \end{pmatrix}, \quad S_y - \begin{pmatrix} 0 & 0 & i \\ 0 & 0 & 0 \\ -i & 0 & 0 \end{pmatrix}, \quad S_z = \begin{pmatrix} 0 & -i & 0 \\ i & 0 & 0 \\ 0 & 0 & 0 \end{pmatrix}, \quad I_0 = \begin{pmatrix} 1 & 0 & 0 \\ 0 & 1 & 0 \\ 0 & 0 & 1 \end{pmatrix} \tag{15}$$

where S_i are the spin 1 matrices with the properties,

$$[S_x, S_y] = iS_z, \quad [S_z, S_x] = iS_y, \quad [S_y, S_z] = iS_x, \quad \mathbf{S}^2 = 2I_0. \tag{16}$$

Substituting Eqs. (13-15) in Eq. (12) we obtain,

$$-c\boldsymbol{p} \cdot \boldsymbol{\Psi} = \mathbf{0}, \tag{17}$$

$$(I_0 E - c\boldsymbol{S} \cdot \boldsymbol{p})\, \boldsymbol{\Psi} = \mathbf{0}. \tag{18}$$

Above

$$\mathbf{\Psi} = \begin{pmatrix} \Psi_x \\ \Psi_y \\ \Psi_z \end{pmatrix} = \begin{pmatrix} E_x + iB_x \\ E_y + iB_y \\ E_z + iB_z \end{pmatrix}. \tag{19}$$

Eq. (18) is the unconstrained helicity equation and Eq. (17) is the subsidiary condition (the perpendicularity condition) which eliminates the zero helicity component.

The α matrices satisfy the same relations as the Pauli matrices, namely

$$(\alpha_1)^2 = (\alpha_2)^2 = (\alpha_3)^2 = (\alpha_0)^2 = \alpha_0, \tag{20}$$

$$\alpha_1\alpha_2 = i\alpha_3 \quad \text{(and cyclic permutation)}, \tag{21}$$

$$\alpha_k\alpha_l = -\alpha_l\alpha_k, \quad \text{for } \ k \neq l, \quad k,l = 1,2,3, \tag{22}$$

therefore the α matrices form a reducible representation of the Pauli matrices. Furthermore

$$(\alpha_0 E - c\boldsymbol{\alpha}\cdot\boldsymbol{p})(\alpha_0 E + c\boldsymbol{\alpha}\cdot\boldsymbol{p}) = \left(E^2 - c^2\mathbf{p}^2\right)\alpha_0, \tag{23}$$

As a result we obtain for the photon the two helicity equations:

$$[E\alpha_0 - c\mathbf{p}\cdot\alpha]\,\psi_R = 0, \quad \text{Helicity}/s = 1. \tag{24}$$

$$[E\alpha_0 + c\mathbf{p}\cdot\alpha]\,\psi_L = 0, \quad \text{Helicity}/s = -1. \tag{25}$$

One should note that both equations are equivalent to the free Maxwell's equations.

3. Generalization to any spin

In previous papers we have generalized the helicity equations to any spin [5],[6],[7]. We have used the angular momentum basis of the $D^{(s-1/2,1/2)}$ representation of the Lorentz group. This basis is the sum of the bases of spins s and spin $s-1$ with

$$(2s+1) + (2s-1) = 4s \text{ components}.$$

In this basis the wavefunction is,

$$\Phi^{(4s)} = \begin{pmatrix} \psi_{s-1}^{(2s-1)} \\ \vdots \\ \psi_{-s+1}^{(2s-1)} \\ \psi_s^{(2s+1)} \\ \vdots \\ \psi_{-s}^{(2s+1)} \end{pmatrix} \Longrightarrow \begin{pmatrix} 0 \\ \vdots \\ 0 \\ \psi_s^{(2s+1)} \\ \vdots \\ \psi_{-s}^{(2s+1)} \end{pmatrix}. \tag{26}$$

The advantage of this formalism, as we have shown in Refs. [6],[7] is that by equating to zero the spin $(s-1)$ components of the wavefunction $\psi_\mu^{(2s-1)}$, the $2s-1$ subsidiary conditions (needed to eliminate the non-forward and non-backward helicities) are automatically satisfied. The equations for any spin have the form

$$\left[\hat{E}\Gamma_0^{(4s)} - c\mathbf{\Gamma}^{(4s)}\cdot\hat{\boldsymbol{p}}\right]\Phi_R^{(4s)} = 0, \tag{27}$$

$$\left[\hat{E}\Gamma_0^{(4s)} + c\boldsymbol{\Gamma}^{(4s)} \cdot \hat{\boldsymbol{p}}\right] \Phi_L^{(4s)} = 0, \tag{28}$$

where the $(4s \times 4s)$ gamma matrices have the same multiplication table as the Pauli matrices, therefore they are reducible representation of the Pauli matrices. The equations have built in subsidiary conditions, which are needed to eliminate the non-forward and non-backward helicities. By equating to zero the spin s-1 components of the wave function, the 2s-1 subsidiary conditions (needed to eliminate the non-forward and non-backward helicities), are automatically satisfied. For spins $\frac{1}{2}$ and spin 1 they are the σ and the α matrices respectively. For spin 2 we have found,

$$\boldsymbol{\Gamma}_x^{(8)} = \frac{1}{2}\begin{pmatrix} 0 & \frac{1}{\sqrt{2}} & 0 & \sqrt{3} & 0 & -\frac{1}{\sqrt{2}} & 0 & 0 \\ \frac{1}{\sqrt{2}} & 0 & \frac{1}{\sqrt{2}} & 0 & \sqrt{\frac{3}{2}} & 0 & -\sqrt{\frac{3}{2}} & 0 \\ 0 & \frac{1}{\sqrt{2}} & 0 & 0 & 0 & \frac{1}{\sqrt{2}} & 0 & -\sqrt{3} \\ \sqrt{3} & 0 & 0 & 0 & -1 & 0 & 0 & 0 \\ 0 & \sqrt{\frac{3}{2}} & 0 & -1 & 0 & -\sqrt{\frac{3}{2}} & 0 & 0 \\ -\frac{1}{\sqrt{2}} & 0 & \frac{1}{\sqrt{2}} & 0 & -\sqrt{\frac{3}{2}} & 0 & -\sqrt{\frac{3}{2}} & 0 \\ 0 & -\sqrt{\frac{3}{2}} & 0 & 0 & 0 & -\sqrt{\frac{3}{2}} & 0 & -1 \\ 0 & 0 & -\sqrt{3} & 0 & 0 & 0 & -1 & 0 \end{pmatrix}, \tag{29}$$

$$\boldsymbol{\Gamma}_y^{(8)} = \frac{i}{2}\begin{pmatrix} 0 & \frac{1}{\sqrt{2}} & 0 & -\sqrt{3} & 0 & -\frac{1}{\sqrt{2}} & 0 & 0 \\ -\frac{1}{\sqrt{2}} & 0 & \frac{1}{\sqrt{2}} & 0 & -\sqrt{\frac{3}{2}} & 0 & -\sqrt{\frac{3}{2}} & 0 \\ 0 & -\frac{1}{\sqrt{2}} & 0 & 0 & 0 & -\frac{1}{\sqrt{2}} & 0 & -\sqrt{3} \\ \sqrt{3} & 0 & 0 & 0 & -1 & 0 & 0 & 0 \\ 0 & \sqrt{\frac{3}{2}} & 0 & 1 & 0 & -\sqrt{\frac{3}{2}} & 0 & 0 \\ \frac{1}{\sqrt{2}} & 0 & \frac{1}{\sqrt{2}} & 0 & \sqrt{\frac{3}{2}} & 0 & -\sqrt{\frac{3}{2}} & 0 \\ 0 & \sqrt{\frac{3}{2}} & 0 & 0 & 0 & \sqrt{\frac{3}{2}} & 0 & -1 \\ 0 & 0 & \sqrt{3} & 0 & 0 & 0 & 1 & 0 \end{pmatrix}, \tag{30}$$

$$\boldsymbol{\Gamma}_z^{(8)} = \frac{1}{2}\begin{pmatrix} 1 & 0 & 0 & 0 & -\sqrt{3} & 0 & 0 & 0 \\ 0 & 0 & 0 & 0 & 0 & -2 & 0 & 0 \\ 0 & 0 & -1 & 0 & 0 & 0 & -\sqrt{3} & 0 \\ 0 & 0 & 0 & -2 & 0 & 0 & 0 & 0 \\ -\sqrt{3} & 0 & 0 & 0 & -1 & 0 & 0 & 0 \\ 0 & -2 & 0 & 0 & 0 & 0 & 0 & 0 \\ 0 & 0 & -\sqrt{3} & 0 & 0 & 0 & 1 & 0 \\ 0 & 0 & 0 & 0 & 0 & 0 & 0 & 2 \end{pmatrix}, \tag{31}$$

and $\boldsymbol{\Gamma}_0^{(8)}$ is the (8×8) unit matrix.

4. Quaternions, bi-quaternions and Paulions

The quaternions usually are written in terms of three complex numbers as,

$$\mathbf{Q} = c_0\mathbf{1} + c_1\mathbf{i} + c_2\mathbf{j} + c_3\mathbf{k}, \tag{32}$$

$$\mathbf{i}^2 = \mathbf{j}^2 = \mathbf{k}^2 = -\mathbf{1}, \quad \mathbf{ij} = -\mathbf{ji}, \quad \mathbf{ik} = -\mathbf{ki}, \quad \mathbf{kj} = -\mathbf{jk}, \quad \mathbf{ijk} = -\mathbf{1}, \tag{33}$$

where the c_μ are real coefficients. Let us rewrite Eq. (32) as,

$$\mathbf{Q} = c_0\mathbf{q}_0 + c_1\mathbf{q}_1 + c_2\mathbf{q}_2 + c_3\mathbf{q}_3, \tag{34}$$

$$\mathbf{q}_0 \equiv \mathbf{1}, \quad \mathbf{q}_1 \equiv \mathbf{i}, \quad \mathbf{q}_2 \equiv \mathbf{j}, \quad \mathbf{q}_3 \equiv \mathbf{k}. \tag{35}$$

The conjugate quaternion is defined as,

$$\mathbf{Q}^C = c_0\mathbf{q}_0 - c_1\mathbf{q}_1 - c_2\mathbf{q}_2 - c_3\mathbf{q}_3. \tag{36}$$

Using Eqs. (32-36) we obtain,

$$\mathbf{Q}^C\mathbf{Q} = \mathbf{q}_0\left(c_0^2 + c_1^2 + c_2^2 + c_3^2\right). \tag{37}$$

Therefore quaternions are suitable for the metric $[1, 1, 1, 1]$. In Minkowski's space of special relativity we will use the metric $[1, -1, -1, -1]$. Wave equations for massless particles can be derived by factorizing (the Klein-Gordon operator),

$$p_0^2 - p_1^2 - p_2^2 - p_3^2 = 0, \quad p_0 = E/c \tag{38}$$

The quaternions are not suitable for factorizing this equation. The factorization can be done with bi-quaternions, for which the coefficients c_μ in Eq. (32) or Eq.(35) can be complex. The bi-quaternions suitable for the Minkowski metric, which can factorize Eq.(38), are

$$\mathbf{R} = c_0\mathbf{r}_0 + c_1\mathbf{r}_1 + c_2\mathbf{r}_2 + c_3\mathbf{r}_3, \tag{39}$$

where the c_μ are real coefficients and their relation to quaternions is,

$$\mathbf{r}_0 = \mathbf{q}_0 = \mathbf{1}, \quad \mathbf{r}_1 = i\mathbf{q}_1, \quad \mathbf{r}_2 = i\mathbf{q}_2, \quad \mathbf{r}_3 = i\mathbf{q}_3, \tag{40}$$

The conjugate is,

$$\mathbf{R}^C = c_0\mathbf{r}_0 - c_1\mathbf{r}_1 - c_2\mathbf{r}_2 - c_3\mathbf{r}_3, \tag{41}$$

and

$$\mathbf{R}^C\mathbf{R} = \mathbf{R}\mathbf{R}^C = \mathbf{r}_0\left(c_0^2 - c_1^2 - c_2^2 - c_3^2\right), \tag{42}$$

which allows factorization of Eq.(38). We will call these special bi-quaternions as Paulions as they have a matrix representation in terms of Pauli matrices,

$$\sigma_0 = \begin{pmatrix} 1 & 0 \\ 0 & 1 \end{pmatrix}, \quad \sigma_x = \begin{pmatrix} 0 & 1 \\ 1 & 0 \end{pmatrix}, \quad \sigma_y = \begin{pmatrix} 0 & -i \\ i & 0 \end{pmatrix}, \quad \sigma_z = \begin{pmatrix} 1 & 0 \\ 0 & -1 \end{pmatrix}. \tag{43}$$

$$\mathbf{R}_\sigma = c_0\sigma_0 + c_1\sigma_x + c_2\sigma_y + c_3\sigma_z, \quad \mathbf{R}_\sigma^C = c_0\sigma_0 - c_1\sigma_x - c_2\sigma_y - c_3\sigma_z, \tag{44}$$

and

$$\mathbf{R}_\sigma\mathbf{R}_\sigma^C = \mathbf{R}_\sigma^C\mathbf{R}_\sigma = \left(c_0^2 - c_1^2 - c_2^2 - c_3^2\right)\sigma_0.$$

The two component massless neutrino equations can be derived from the decomposition,

$$\left(E^2 - c^2\mathbf{p}^2\right)\sigma_0\psi = \left[E\sigma_0 - c\mathbf{p}\cdot\sigma\right]\left[E\sigma_0 + c\mathbf{p}\cdot\sigma\right]\psi = 0, \tag{45}$$

where E is the energy and $\boldsymbol{p}$ the momentum operators.

From Eq. (45) two helicity equations (for positive energies) are obtained,

$$[E\sigma_0 - c\mathbf{p}\cdot\sigma]\,\psi_R = 0, \quad \text{Helicity}/s = 1. \tag{46}$$

$$[E\sigma_0 + c\mathbf{p}\cdot\sigma]\,\psi_L = 0, \quad \text{Helicity}/s = -1. \tag{47}$$

5. The bi-quaternionic Maxwell's equations

In Section 2 the free Maxwell's equations were converted to Eq.(12),

$$(\alpha_0 E - c\boldsymbol{\alpha}\cdot\boldsymbol{p})\,\Phi = 0. \quad \Phi = \begin{pmatrix} 0 \\ \Psi_x \\ \Psi_y \\ \Psi_z \end{pmatrix} = \begin{pmatrix} 0 \\ E_x + iB_x \\ E_y + iB_y \\ E_z + iB_{zz} \end{pmatrix},$$

whdere the α matrices are given by Eq.(8).

One can check that the equation,

$$(\alpha_0 E - c\boldsymbol{\alpha}\cdot\boldsymbol{p})\,(-\alpha_1\Psi_x - \alpha_2\Psi_y - \alpha_3\Psi_z) = 0, \quad \boldsymbol{\Psi} = \begin{pmatrix} \Psi_x \\ \Psi_y \\ \Psi_z \end{pmatrix} = \begin{pmatrix} E_x + iB_x \\ E_y + iB_y \\ E_z + iB_z \end{pmatrix}, \tag{48}$$

is also equivalent to Maxwell's equations. The difference between Eq.(12) and Eq.(48) is the way the wave function is presented. In Eq.(12) the wave function is a vector with complex coefficients, while in Eq.(48) the wave function has a property of a bi-quaternion, namely as the α_i matrices form a reducible representation of of the Pauli σ_i matrices, Eq. (48) is a reducible representation of the following bi-quaternionic Maxwell's equations,

$$(\sigma_0\dot{E} - c\boldsymbol{\sigma}\cdot\boldsymbol{p})\,(-\boldsymbol{\sigma}\cdot\boldsymbol{\Psi}) = 0. \tag{49}$$

Eq. (49) is similar to the spin one-half massless neutrino equation Eq. (46), namely the helicity operator is the same as for the spin one-half equation, but the wave function is different, it is a bi-quaternion.

It is worthwhile to note that Maxwell has derived his early 20 equations using quaternions [8]. In modern presentations a different form of Maxwell's original equations is used [4] . Our work is related to modern developments.

6. Summary and conclusions

First quantized equations for massless particles of any spin s were developed. The helicities of massless particles of any spin can have only projections in the forward and backward direction of the momentum. In order to have equations with this property, subsidiary conditions have to be imposed. A formalism has been developed in which the subsidiary conditions were implicitly included in the main equation. The resulting equations for the two helicities are,

$$\left[\hat{E}\Gamma_0^{(4s)} - c\boldsymbol{\Gamma}^{(4s)}\cdot\hat{\boldsymbol{p}}\right]\Phi_R^{(4s)} = 0, \quad \left[\hat{E}\Gamma_0^{(4s)} + c\boldsymbol{\Gamma}^{(4s)}\cdot\hat{\boldsymbol{p}}\right]\Phi_L^{(4s)} = 0, \tag{50}$$

and they are similar to the spin one-half massless (neutrino) equations,

$$[E\sigma_0 - c\mathbf{p}\cdot\sigma]\,\psi_R = 0, \quad [E\sigma_0 + c\mathbf{p}\cdot\sigma]\,\psi_L = 0, \tag{51}$$

IOP Conf. Series: Journal of Physics: Conf. Series **845** (2017) 012011 doi:10.1088/1742-6596/845/1/012011

The helicity operator in Eq. (50) for any spin, has (multiple) eigenvalues for forward and backward momentum projections only, and is accordingly a reducible representation of the spin one-half helicity operator.

As an application we have recovered the bi-quaternionic form of Maxwell's equations Eq. (49), similar to the spin one-half massless neutrino equations Eq. (51),

$$(\sigma_0 E - c\boldsymbol{\sigma} \cdot \boldsymbol{p})\,(-\boldsymbol{\sigma} \cdot \boldsymbol{\Psi}) = 0,$$

namely the helicity operator is the same as for the spin one-half equation, but the wave function is a bi-quaternion.

Additional work is needed to understand the mechanism of presenting the spin equations with a lower spin equivalent equations.

References

[1] Wigner E.P., On Unitary Representations of the Inhomogeneous Lorentz Group. *Annals of Mathematics*, **40**, 1939, pp.149-204
[2] Weinberg S. *The Quantum Theory of Fields. Vol. I.* Cambridge University Press, New York (1995), pp. 55-91
[3] LIU ChangLi and GE FengJun, arXiv:1403.2698v4 [physics.gen-ph] 28 Apr. 2014
[4] Jackson J.D., *Classical Electrodynamics, Third Edition,* John Wiley & sons, New York (1999), p.781
[5] Gersten A. and Moalem A., *Journal of Physics*: Conference Series **330** (2011) 012010.
[6] Gersten A. and Moalem A., Journal of Physics: Conference Series **437** (2013), 012019.
[7] Gersten A. and Moalem A., Journal of Physics: Conference Series **615** (2015) 012011.
[8] Maxwell, James Clerk, *A treatise on electricity and magnetism* , Oxford, Clarendon Press (1873).

IOP Conf. Series: Journal of Physics: Conf. Series **845** (2017) 012012 doi:10.1088/1742-6596/845/1/012012

Skewon modified electrodynamics

Yakov Itin

Inst. Mathematics, The Hebrew Univ. of Jerusalem and
Jerusalem College of Technology, Jerusalem, Israel

E-mail: itin@math.huji.ac.il

Abstract. Premetric electrodynamics is a representation of classical electrodynamics based on topological conservation laws. This model predicts a covariant extension of classical theory by dilaton, axion, and skewon as copartners of photon. In this paper, we report on some recent results on skewon modification of classical electrodynamics. We present the skewon modified dispersion relation for electromagnetic wave propagation. It yields several kinds of the birefringence effect of topologically different types. The superluminal character of wave propagation and the Higgs-type effect for the symmetric skewon are indicated. We present the skewon modified photon propagator and discuss the corresponding modification of Coulomb's law.

1. Introduction

Interior and exterior symmetries play an important role in classical and quantum field theory. Violations of some of these symmetries may be considered as a straightforward way of extension of field theories. Moreover, the dynamical violation of the interior symmetry (Higgs effect) is well-known as a necessary ingredient of the Standard Model.

Many modern field models in loop quantum gravity and string theory predict a modification of the light cone structure expressed by an anisotropic dispersion relation, a birefringent vacuum, and a violation of local Lorentz and CPT invariance. Such violations of the exterior symmetry, if they really exist, must yield crucial modifications of the very basis of physics. Since these models are still very far from their complete form, it is important to have a phenomenological model that predicts the indicated optics phenomena. Moreover, it is possible that even when we will already have a good quantum gravity, we will need a phenomenological model in order to give the observational predictions. It is due to a huge gap between the energies that are available in experiments and the energies for which quantum gravity effects come to be relevant.

Modification of the electrodynamics action by an 1-parametric axion was applied by Wei Tou Ni [1] already in 1977 for studying possible violation of the equivalent principle. In Carroll-Field-Jackiw electrodynamics [2], modified energy-momentum tensor and birefringent dispersion relation for axion model was derived and the astronomical data were applied for deducing the upper magnitude of the axion.

In Kostelecky's model [3], an extension of the axion model was proposed in the form of a modified Lagrangian

$$L = -\frac{1}{4}\eta^{ik}\eta^{jl}F_{ij}F_{kl} - \frac{1}{4}\kappa^{ijkl}F_{ij}F_{kl}\,. \tag{1}$$

Here, the fixed set of the coefficients κ^{ijkl} represents the small perturbations of the standard electrodynamics model. The symmetries of κ^{ijkl} essentially restrict the number of independent components. Moreover, the leading term is usually absorbed by a field redefinition, thus 20 independent parameters of κ^{ijkl} are left over. An extension of the construction (1) to the standard model, called Standard Model Extension (SME), was worked out. Theoretical work in this model has spanned a vast number of areas that include string theory, gravity theory, quantum field theory and cosmology. Several exceptionally accurate Lorentz tests have recently been conducted, and other experiments are currently under way.

In this paper, we report on an alternative construction termed premetric electrodynamics. Premetric construction can be considered as a model inspired by the ideas from solid state physics and thus it can serve as a bridge between macroscopic and microscopic physics. In this Lorentz violation model, the individual terms can be expressed via the well-known physics notations such as dielectric permittivity and magnetic permeability. Since premetric electrodynamics is based on topological conservation laws instead of a Lagrangian, it predicts an extra set of parameters that are not accounted for in (1). This set forms a pseudotensor of 15 independent components called skewon. In this paper, we discuss some recent results on the skewon modification of classical electrodynamics.

The organization of the paper is as follows: In Section 2, we present a brief account of premetric electrodynamics and introduce our basic notations. In Section 3, we discuss the wave propagation in the skewon modified electrodynamics. We show that the modified dispersion relation predicts superluminal wave motion for all types of the skewon. For the antisymmetric skewon, it describes the birefringence effect similar to one known from optics. For the symmetric skewon, the wavefront is of a topologically different type and the wave propagation is forbidden for small skewon parameters (Higgs-type behavior). In Section 4, we present the skewon modified photon propagator and describe the contribution of a skewon to the Coulomb law. Some additional information on the observational status of the skewon modified electrodynamics and its possible applications to electromagnetism technology is presented in Conclusion section.

2. Premetric electrodynamics

In this section, we give a brief account of premetric electrodynamics. For a comprehensive introduction to this subject , see the book of Hehl and Obukhov [5] and the references given therein. An account of the recent developments in this area is presented in [6].

2.1. Field equations

The basic variables of the premetric electrodynamics are the even 2-form of the *field strength* F and the odd 3-form of the *electric current* J. Both of them are defined on a 4-dimensional differential manifold without metric and connection.

For an electric current localized on a closed hypersurface C_3, the *electric charge conservation law* is expressed in the integral and differential forms, respectively

$$\int_{C_3} J = 0\,, \qquad dJ = 0\,. \tag{2}$$

For topologically simple spacetime regions, this equation yields existence of an odd 2-form of the *excitation* $\mathcal{H}$

$$d\mathcal{H} = J\,. \tag{3}$$

This relation is a differential form representation of the inhomogeneous Maxwell equation.

The homogeneous Maxwell equation is treated as a result of the *magnetic flux conservation law*. For a closed 2-dimensional surface C, one postulates the relations

$$\int_{C} F = 0\,, \qquad dF = 0\,. \tag{4}$$

With the definitions above the tensorial form of Maxwell's equations is given by

$$\epsilon^{ijkl}F_{jk,l} = 0\,, \qquad \mathcal{H}^{ij}{}_{,i} = \mathcal{J}^{j}\,. \tag{5}$$

The system (5) is explicitly general relativistic covariant and does not require a metric or a connection for its formulation.

2.2. Constitutive pseudotensor

Field equations (5) describe only the formal topological structure of electromagnetism. They are endowed with a physical content only when a constitutive relation is postulated. We consider *linear, local constitutive relation:*

$$\mathcal{H}^{ij} = \frac{1}{2}\chi^{ijkl}F_{kl}\,. \tag{6}$$

Due to its definition, the *constitutive pseudotensor* χ^{ijkl} possess the symmetries

$$\chi^{ijkl} = -\chi^{jikl} = -\chi^{ijlk}\,. \tag{7}$$

Hence, in a 4-dimensional space, χ^{ijkl} has 36 independent components. A multi-component tensor, such as χ^{ijkl}, can be decomposed into the sum of simpler independent sub-tensors with a fewer number of components. In group theory, such a decomposition is derived by applying the *Young diagram* technique. For the constitutive pseudotensor with the symmetries (7), only 3 different diagrams are relevant

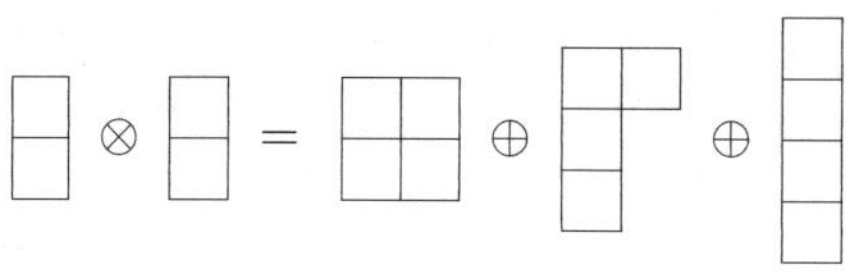
(8)

It is a decomposition of the tensor space into the direct sum of invariant subspaces with the dimensions 36 = 20 + 15 + 1, respectively. This canonical decomposition of the constitutive pseudotensor is *unique and irreducible.* The decomposition depicted in (8) reads

$$\chi^{ijkl} = {}^{(1)}\chi^{ijkl} + {}^{(2)}\chi^{ijkl} + {}^{(3)}\chi^{ijkl}\,. \tag{9}$$

The *principal part* of 20 independent components corresponds to the first diagram

$${}^{(1)}\chi^{ijkl} = \frac{1}{6}\left(2\chi^{ijkl} + 2\chi^{klij} - \chi^{iklj} - \chi^{ljik} - \chi^{iljk} - \chi^{jkil}\right). \tag{10}$$

It is a generalization of the standard Maxwell-Lorentz constitutive tensor. Two other parts do not emerge in Maxwell's theory and present its covariant extension.

The *skewon part* ${}^{(2)}\chi^{ijkl}$ of 15 independent components corresponds to the middle diagram of (8). It reads

$${}^{(2)}\chi^{ijkl} = \frac{1}{2}\left(\chi^{ijkl} - \chi^{klij}\right). \tag{11}$$

The *axion part* ${}^{(3)}\chi^{ijkl}$ corresponds to the third diagram of (8). It has only 1 component

$${}^{(3)}\chi^{ijkl} = \chi^{[ijkl]} = \alpha\varepsilon^{ijkl}\,. \tag{12}$$

2.3. Lorentz force, enegy-momentum current, and action

In premetric electrodynamics, the Lorentz force density is presented by a twisted covector valued 4-form

$$f_i = (e_i \rfloor F) \wedge J\,. \tag{13}$$

This expression can be considered as an operational definition of the electromagnetic field strength F.

Energy-momentum current in premetric electrodynamics can be treated as a pre-potential of the Lorentz force density. It is presented by a covector-valued 3-form

$$\Sigma_i = \frac{1}{2}\left[F \wedge (e_i \rfloor H) - H \wedge (e_i \rfloor H)\right]\,. \tag{14}$$

When the constitutive relation (6) with the decomposition (9) is used here the axion part is canceled. The principle and the skewon part contribute to the energy-momentum current. As it was demonstrated in [8], the skewon field contributes non-trivially to the electromagnetic energy. It induces an asymmetric electromagnetic energy-momentum tensor, which can cause specific source term in the Einstein-Cartan-Maxwell system.

The electromagnetic Lagrangian 4-form of the model can be expressed in the premetric form

$$\Lambda = -\frac{1}{2} F \wedge H\,. \tag{15}$$

Using $F = dA$ we can derive the field equation $dH = J$ and the current (14) from the Lagrangian (15). Observe that (15) is equivalent to the Kostelecky's model (1). The skewon part does not give a contribution to this expression, thus it can be treated as a non-Lagrangian extension of electromagnetism. Moreover, the skewon contribution prevents the conservation of the energy-momentum tensor. Thus skewon part of the energy-momentum tensor can be treated as a dissipative term, that in general cannot be described within the Lagrangian framework, [5], [7], [8].

3. Wave propagation in skewon-modified electrodynamics

In this section, we present the results related to the wave propagation in the skewon modified electrodynamics. For the details, see [9], [11], [12], [13], [14].

3.1. Skewon field

In order to extract the pure skewon contribution to electrodynamics, we forbid the axion part ${}^{(3)}\chi^{ijkl}$ of the constitutive pseudo-tensor and assume its principal part to be presented in the simplest metric form

$${}^{(1)}\chi^{ijkl} = g^{ik}g^{jl} - g^{il}g^{gk}\,. \tag{16}$$

Moreover, we assume the flat Minkowski metric $g^{ij} = \mathrm{diag}(1, +1, +1, +1)$. The skewon part ${}^{(2)}\chi^{ijkl}$, however, will be considered in the most general form. Since the skewon has 15 independent components, it can be expressed as a traceless tensor of the second order. We use the definition as in [5]

$$S_i{}^j = \frac{1}{4}\epsilon_{iklm}{}^{(2)}\chi^{klmj}\,. \tag{17}$$

The inverse relation reads

$${}^{(2)}\chi^{ijkl} = \epsilon^{ijm[k}S_m{}^{l]} - \epsilon^{klm[i}S_m{}^{j]}\,. \tag{18}$$

These definitions are metric-free. On a space endowed with a metric tensor g_{ij}, the mixed tensor $S_i{}^j$ can be replaced by the covariant and contravariant tensors

$$S^{ij} = g_{ik}S_k{}^j\,, \quad \text{and} \quad S_{ij} = g_{jk}S_i{}^k\,. \tag{19}$$

These tensors are decomposed into a sum of symmetric and antisymmetric parts. Thus, we can define a special case of a symmetric skewon that satisfies $S_{ij} = S_{ji}$ and of an antisymmetric skewon with $S_{ij} = -S_{ji}$.

3.2. Electromagnetic waves

In order to study the electromagnetic waves in skewon modified electrodynamics, we start with the generic premetric jump conditions, [10]. These conditions are resulted from Maxwell equations and hold on an arbitrary hypersurface $\varphi(x^i) = 0$ in spacetime. In differential forms, they take the form of the *wavefront conditions*

$$F \wedge d\varphi = 0\,, \qquad \mathcal{H} \wedge d\varphi = 0\,. \tag{20}$$

The hypersurface $\varphi(x^i) = 0$ is referred to as a *wavefront*. In the tensorial form, this characteristic system with the *wave covector* $q_i = \varphi_{,i}$ reads

$$\epsilon^{ijkl} f_{kl} q_j = 0; \qquad \chi^{ijkl} f_{kl} q_j = 0 \tag{21}$$

The most general solution of the first equation is written by the use of an arbitrary covector a_k

$$f_{kl} = \frac{1}{2}\left(a_k q_l - a_l q_k\right)\,. \tag{22}$$

The covector a_k is an analog of the electromagnetic potential. Substituting (22) in the second equation of (22) we remain with the characteristic equation

$$M^{ik} a_k = 0 \qquad \text{where} \qquad M^{ik} = \chi^{ijkl} q_l q_j. \tag{23}$$

Due to the gauge invariance, this equation has a non-trivial solution if and only if its adjoint matrix is equal to zero,

$$\mathrm{Adj}(M^{ij}) = \lambda(q) q_i q_j = 0\,. \tag{24}$$

Thus the dispersion relation is represented as

$$\lambda(q) = 0\,. \tag{25}$$

With the use of Eq.(9), the characteristic matrix M^{ik} is irreducibly decomposed as

$$M^{ik} = P^{ik} + Q^{ik} \tag{26}$$

with

$$P^{ik} = {}^{(1)}\chi^{ijkl} q_j q_l\,, \qquad Q^{ik} = {}^{(2)}\chi^{ijkl} q_j q_l\,. \tag{27}$$

Moreover, P^{ik} is the symmetric part and Q^{ik} is the antisymmetric part of the characteristic matrix M^{ik}

$$P^{ik} = M^{(ik)}\,, \qquad Q^{ik} = M^{[ik]}\,. \tag{28}$$

We recall that the axion part ${}^{(3)}\chi^{ijkl}$ does not contribute to M^{ik}.

The relation

$$Q^{ij} q_j = 0\,. \tag{29}$$

results in the expression

$$Q^{ij} = -\frac{1}{2}\epsilon^{ijrs}(Y_r q_s - Y_s q_r) = \epsilon^{ijkl} q_k Y_l \tag{30}$$

with an arbitrary covector Y_i. Consequently, the skewon contribution to the dispersion relation is completely determined by the *skewon optic covector* Y_i. Calculating the adjoint matrix $\mathrm{Adj}(M)$ of the optic tensor $M^{ij} = P^{ij} + Q^{ij}$ we have for the most generic linear constitutive pseudo-tensor χ^{ijkl}, the dispersion relation of the form

$$\Lambda(P) + P^{ij} Y_i Y_j = 0\,, \tag{31}$$

where the scalar expression $\Lambda(P)$ is defined via the relation $\mathrm{Adj}(P^{ij}) = \Lambda(P) q_i q_j$.

3.3. Superluminal wave motion

Using the metric principle part (16) we obtain from Eq.(31) the dispersion relation for the most generic skewon media in the (pseudo) Riemannian vacuum

$$q^4 - Y^2 q^2 + (Y, q)^2 = 0\,, \tag{32}$$

where the scalar products are constructed with the use of the metric tensor g^{ij}. An analysis of this dispersion relation yields the following remarkable fact [11]:

In the Lorentz signature metric space, the solution of the skewon modified dispersion relation (32) are spacelike or null.

This means that the skewon generates superluminal wave motion and thus yields strong violation of Lorentz invariance. This consideration remains completely unaltered in the isotropic medium with two standard parameters ε and μ. In this case, however, the wave velocity for the skewon modified medium only exceeds the value $\sqrt{\varepsilon\mu}$ instead of the universal speed of light. Such behavior is certainly can not be excluded by any fundamental principle. Thus the skewon part could exist in a material medium as a phenomenological quantity. For instance, it can be applicable for a description of dissipative media. Moreover, the observation of the electromagnetic waves with velocity higher than $\sqrt{\varepsilon\mu}$ could point towards the presence of the skewon.

3.4. Symmetric skewon — Higgs-type effect

On a space endowed with a metric tensor, a special symmetric traceless skewon can be extracted. In 4-dimensional spacetime, it has 9 independent components. For a generic skewon, the dispersion relation (32) can be written in the form

$$q^2 = \frac{1}{2}\left(Y^2 \pm \sqrt{Y^4 - (q, Y)^2}\right). \tag{33}$$

In order to have a real Lorentz square q^2, the expression under the square root must be non-negative. For an antisymmetric skewon, the relation $(q, Y) = 0$ holds identically, thus there are no additional restrictions on the wave propagation. For a symmetric skewon, both terms under the square root are non-zero and the inequality

$$Y^4 - (q, Y)^2 \geq 0 \tag{34}$$

must be imposed. Recall that the covector Y is linear in the skewon matrix S_{ij}. If we rescale this tensor, $S_{ij} \to CS_{ij}$, Eq.(34) takes the form

$$C^2 Y^4 - (q, Y)^2 \geq 0\,. \tag{35}$$

For small values of C, the first term approaches zero and the inequality is broken. Consequently, we came to the following result [13]:

Let the skewon have a non-zero symmetric part. For sufficient small magnitudes of the skewon parameters, there are no solutions of the dispersion relation. Thus there is a gap for values of the skewon parameters near zero, where the wave propagation is forbidden.

This way we constructed a model of an electromagnetic media with Higgs-type spontaneously broken transparency. This result is in a correspondence with an alternative description presented in [9], where the similar effect is treated as holes in wavefront surface.

4. Photon propagator in skewon electrodynamics

4.1. Propagaor for linear response media

In geometrical optic approximation, Eq.(5) could be rewritten as a general system of four linear equations for four variables

$$M^{im}a_m = j^i\,. \tag{36}$$

The matrix M is singular and thus for given functions $M^{im}(q)$ and $j^i(q)$, the solution of Eq.(2.8) is not unique. A formal solution of the linear system (36) can be written as

$$a_i = D_{ik}j^k\,. \tag{37}$$

The tensor D_{ik} is called the *propagator tensor.* In standard quantum electrodynamics based on pure Minkowski constitutive tensor, it suffices to consider symmetric propagators, $D_{ik} = D_{ki}$. Such symmetry, however, does not represent any physical requirement. Indeed, in the case of a generic linear constitutive relation, the propagator tensor is asymmetric. Using the second adjoint technique [14], we derive the photon propagator in Feynman gauge

$$D_{ij} = \frac{1}{\lambda q^2}B^m{}_{ijm}\,. \tag{38}$$

Here B_{ijkl} is the second adjoint of the matrix M^{ij}. It is obtained by removing two rows and two columns from the initial matrix. The function $\lambda(q)$ represents the left-hand side of the dispersion relation (25).

The tensor $B^m{}_{ijm}$ is quadratic in the entries of the matrix M^{ij} , hence it is a homogeneous 4-th order polynomial in the components of q_i. Also $\lambda(q)$ is a homogeneous 4-th order polynomial in q_i. Consequently the propagator $D_{ij} \sim q^{-2}$ as the ordinary non-modified photon propagator.

4.2. Propagator for skewon modified electrodynamics

We calculate the second adjoint with the metric principle part (16) and use the representation (30) of the skewon part of the characteristic matrix M^{ij} via the covector Y_i. The result is

$$D_{ij} = -\frac{g_{ij}q^2 + \epsilon_{ikmj}q^kY^m + Y_iY_j}{q^4 - q^2Y^2 + (q,Y)^2}\,. \tag{39}$$

Let us observe some basic properties of this propagator:

(1) For a vanishing skewon field, the standard vacuum form of Feynman propagator is reinstated

$$D_{ij} = -\frac{g_{ij}}{q^2}\,. \tag{40}$$

(2) Skewon propagator has nontrivial symmetric and antisymmetric parts:

$$D_{(ij)} = -\frac{g_{ij}q^2 + Y_iY_j}{q^4 - q^2Y^2 + (q,Y)^2}\,, \qquad \text{and} \qquad D_{[ij]} = -\frac{\epsilon_{ikmj}q^kY^m}{q^4 - q^2Y^2 + (q,Y)^2}\,, \tag{41}$$

respectively.

(3) In general, the propagator is singular for four roots of the denominator. Thus, there are at most two light cones at every spacetime point – birefringence.

(4) In the first order approximation for a small skewon field, Eq. (41) reads

$$D_{(ij)} = -\frac{g_{ij}}{q^2}\,, \qquad \text{and} \qquad D_{[ij]} = -\frac{\epsilon_{ikmj}q^kY^m}{q^4}\,. \tag{42}$$

Thus, the small skewon field contributes only to the antisymmetric part of the propagator.

4.3. Skewon and modified Coulomb law

Using the propagator expression with the skewon modifications, we are able to calculate the corresponding modification of the Coulomb law. For this, we need an expression of propagator in Coulomb gauge. Moreover, in the static case, it is enough to have only the temporal component of the propagator. It takes the form

$$^{C}D_{00} = \frac{q^2(k^2 - Y_0^2)}{k^4(q^2 - Y^2)}\,, \tag{43}$$

where the 3-dimensional notations $q = (\omega, k_\alpha)$ and $Y = (Y_0, Y_\alpha)$ are used.

We consider now an explicit model of the antisymmetric skewon field

$$S_{0\mu} = -S_{\mu 0} = \alpha_\mu\,. \tag{44}$$

Correspondingly, we are left with

$$^{C}D_{00} = \frac{1}{k^2}\frac{k^2 - (\alpha, k)^2}{k^2 + (\alpha, k)^2}\,. \tag{45}$$

The electromagnetic potential for a point-wise charge Q is calculated by the inverse Fourier transform of (45) and results in

$$\varphi(r) = \frac{Q}{4\pi r}\left(\left(1 - \alpha^2\right) + \frac{(\alpha, r)^2}{r^2}\right)\,. \tag{46}$$

The corresponding equipotential surfaces are 2-dimensional ellipsoids. They are similar to the wave-front surfaces of the skewon model [13].

5. Results and discussion

In this paper, we discussed some theoretical results related to the skewon modification of the classical electrodynamics.

The vacuum birefringence effect presents in all skewon models and can be of topologically different types. The antisymmetric skewon produces two wavefronts similar to those known from anisotropic optics. Alternatively, the symmetric skewon provides very strong modification of the wavefront structure with rotation of the time axis and some forbidden directions of wave propagation – "topological holes".

Another remarkable property is the superluminal wave propagation for all types of the skewon. This fact, probably, indicates that the skewon must be forbidden in vacuum (but not in a medium). The skewon modified photon propagator shows a proper behavior, and can be used for modeling the anisotropic modifications of the Coulomb law.

An important issue in skewon electrodynamics is the estimation of the magnitude of the skewon and its comparison to the observation data. In the recent publications of Ni [16], [17], [18], the first order approximation of the skewon contributions to electrodynamics was worked out. From the dispersion relation, it is shown that the requirement of no dissipation/no amplification in propagation implies that the skewon field must antisymmetric (Type II, in the nomenclature of Ni). For the symmetric (Type I) skewon field, the dissipation/amplification in propagation comes to be proportional to the frequency and the CMB spectrum would deviate from Planck spectrum. This result is in a correspondence with our restrictions coming from the topology of the modified wavefront.

Another interesting issue is the discussion of possible applications of the skewon in engineering electromagnetism. In the publications of Lindell and Sihvola [19],[20],[21],[22], the skewon (and axion) additions to electrodynamics was used for realization the DB and SHDB boundary conditions. This approach provides a deeper understanding of the nature of metamaterials.

Acknowledgments
I thank F.W. Hehl (Cologne), Y. Obukhov (Moscow) and both referees for their useful comments and suggestions.

References

[1] W. -T. Ni, "Equivalence Principles and Electromagnetism," Phys. Rev. Lett. **38** (1977) 301.
[2] S. M. Carroll, G. B. Field and R. Jackiw, "Limits on a Lorentz and Parity Violating Modification of Electrodynamics," Phys. Rev. D **41**, 1231 (1990).
[3] V. A. Kostelecky and M. Mewes, "Signals for Lorentz violation in electrodynamics," Phys. Rev. D **66**, 056005 (2002).
[4] E. J. Post, *Formal Structure of Electromagnetics – General Covariance and Electromagnetics* (North Holland: Amsterdam, 1962).
[5] F. W. Hehl and Yu. N. Obukhov, *Foundations of Classical Electrodynamics: Charge, Flux, and Metric* (Birkhäuser: Boston, MA, 2003).
[6] F. W. Hehl, Y. Itin and Y. N. Obukhov, "On Kottler's path: origin and evolution of the premetric program in gravity and in electrodynamics," Int. J. Mod. Phys. D **25**, no. 11, 1640016 (2016) doi:10.1142/S0218271816400162 [arXiv:1607.06159 [gr-qc]].
[7] F. W. Hehl, Y. N. Obukhov and G. F. Rubilar, "On a possible new type of a T odd skewon field linked to electromagnetism," gr-qc/0203096. 1st Mexican Meeting on Mathematical and Experimental Physics 10-14 Sep 2001. Mexico City, Mexico CNUM: C01-09-10.11 URL: http://uxmal.uam.mx/colegio/
[8] F. W. Hehl, Y. N. Obukhov, G. F. Rubilar and M. Blagojevic, "On the theory of the skewon field: From electrodynamics to gravity," Phys. Lett. A **347**, 14 (2005) doi:10.1016/j.physleta.2005.06.033 [gr-qc/0506042].
[9] Y. N. Obukhov and F. W. Hehl, "On possible skewon effects on light propagation," Phys. Rev. D **70**, 125015 (2004) doi:10.1103/PhysRevD.70.125015 [physics/0409155 [physics.class-ph]].
[10] Y. Itin, "Covariant jump conditions in electromagnetism," Annals Phys. **327**, no. 2, 359 (2012) doi:10.1016/j.aop.2011.09.005 [arXiv:1403.7700 [gr-qc]].
[11] Y. Itin, "Skewon no-go theorem," Phys. Rev. D **88**, no. 10, 107502 (2013) doi:10.1103/PhysRevD.88.107502
[12] Y. Itin, "Skewon modification of the light cone structure," Phys. Rev. D **91**, no. 8, 085002 (2015) doi:10.1103/PhysRevD.91.085002 [arXiv:1407.6722 [hep-th]].
[13] Y. Itin, "Electromagnetic media with Higgs-type spontaneously broken transparency," Phys. Lett. A **379**, 619 (2014) doi:10.1016/j.physleta.2014.12.024 [arXiv:1406.3442 [hep-th]].
[14] Y. Itin, "Photon propagator in skewon electrodynamics," Phys. Rev. D **93**, no. 2, 025023 (2016) doi:10.1103/PhysRevD.93.025023 [arXiv:1511.01840 [hep-th]].
[15] A. Favaro and F. W. Hehl, "Light propagation in local and linear media: Fresnel-Kummer wave surfaces with 16 singular points," Phys. Rev. A **93**, no. 1, 013844 (2016) doi:10.1103/PhysRevA.93.013844 [arXiv:1510.05566 [physics.optics]].
[16] W. T. Ni, "Skewon field and cosmic wave propagation," Phys. Lett. A **378**, 1217 (2014) doi:10.1016/j.physleta.2014.02.039 [arXiv:1312.3056 [gr-qc]].
[17] W. T. Ni, "Dilaton field and cosmic wave propagation," Phys. Lett. A **378**, 3413 (2014) doi:10.1016/j.physleta.2014.09.049 [arXiv:1410.0126 [gr-qc]].
[18] W. T. Ni, "Spacetime structure and asymmetric metric from the premetric formulation of electromagnetism," Phys. Lett. A **379**, 1297 (2015) doi:10.1016/j.physleta.2015.03.004 [arXiv:1411.0460 [gr-qc]].
[19] Lindell, Ismo V., "Inverse for the skewon medium dyadic," Progress In Electromagnetics Research **63**, 21-32 (2006).
[20] Lindell, Ismo V., and Ari Sihvola, "Soft-and-hard/DB boundary conditions defined by a skewon-axion medium." IEEE Transactions on Antennas and Propagation **61**, 768-774 (2013).
[21] Lindell, Ismo Veikko, and Ari Henrik Sihvola, "Simple skewon medium realization of DB boundary conditions." Progress In Electromagnetics Research Letters **30**, 29-39 (2012).
[22] Lindell, I. V., and A. Sihvola. "Skewon-axion medium and SHDB boundary conditions." Electromagnetics in Advanced Applications (ICEAA), 2012 International Conference on. IEEE (2012).

The electromagnetic field equations for moving media

T Ivezić

Ruđer Bošković Institute, P.O.B. 180, 10002 Zagreb, Croatia

E-mail: ivezic@irb.hr

Abstract.

In this paper a formulation of the field equation for moving media is developed by the generalization of an axiomatic geometric formulation of the electromagnetism in vacuum (Ivezić T 2005 *Found. Phys. Lett.* **18** 401). First, the field equations with bivectors $F(x)$ and $\mathcal{M}(x)$ are presented and then these equations are written with the 4D vectors $E(x)$, $B(x)$, $P(x)$ and $M(x)$. The latter contain both the 4D velocity vector u of a moving medium and the 4D velocity vector v of the observers who measure E and B fields. They do not appear in previous literature. All these equations are also written in the standard basis and compared with Maxwell's equations with 3D vectors. In this approach the Ampère-Maxwell law and Gauss's law are inseparably connected in one law and the same happens with Faraday's law and the law that expresses the absence of magnetic charge. It is shown that Maxwell's equations with 3D vectors and the field equations with 4D geometric quantities are not equivalent in 4D spacetime

PACS numbers: 03.30.+p, 03.50.De

1. Introduction

The field equations for moving media in a relativistically covariant formulation were first presented by Minkowski [1]. An axiomatic geometric formulation of electromagnetism in vacuum is presented in [2]. That formulation is with only one axiom: the field equation for the bivector field F. In this paper the formulation from [2] is generalized to moving media. The geometric approach to special relativity (SR) that is used in [2], and in this paper as well, exclusively deals either with the abstract, coordinate-free, four-dimensional (4D) geometric quantities, e.g., vectors (4-vectors in the usual notation) $E(x)$, $B(x)$, .. (x is the position vector), or with their representations in some basis, the 4D coordinate-based geometric quantities (CBGQs) comprising *both* components and *a basis*, e.g., $E = E^{\nu}\gamma_{\nu}$. In that approach, which is called "invariant special relativity" (ISR), an independent physical reality is attributed to 4D geometric quantities and not, as usual, to 3D quantities. Every 4D CBGQ is *invariant* under the passive Lorentz transformations (LT); the components transform by the LT and the basis by the inverse LT leaving the whole CBGQ unchanged. The invariance of a 4D CBGQ under the passive LT reflects the fact that such 4D geometric quantity represents *the same physical quantity* for relatively moving inertial observers. This invariance is the reason for the name ISR. The principle of relativity is naturally satisfied, if the physical laws are written with the abstract 4D geometric quantities or with 4D CBGQs. There is no need to postulate it outside the mathematical formulation of the theory as in Einstein's formulation of SR [3]. The paper is organized as follows.

In Sec. 2 the basic field equation for moving media (Eq. (7)) is expressed in terms of the bivector $F = F(x)$ that represents the electromagnetic field and the generalized magnetization-polarization bivector $\mathcal{M} = \mathcal{M}(x)$. Then that equation with abstract quantities (AQs) is written with CBGQs in the standard basis. The resulting equation is separated into two equations, the equation with sources and the equation without sources. The field equation with sources is also written in the "source representation" both with AQs and with CBGQs in the $\{\gamma_\mu\}$ basis, according to which the sources of F field are the true currents $j^{(C)}$ and the magnetization-polarization current density $\partial \cdot \mathcal{M}$.

In Sec. 3 the decomposition of F is presented. F is decomposed into the electric field vector E, the magnetic field vector B and the velocity vector v of the observers who measure E and B fields (in the usual notation E, B, v, ... are called 4-vectors). Similarly, $\mathcal{M}$ is decomposed into the polarization vector $P(x)$, the magnetization vector $M(x)$ and the bulk velocity vector u of the medium. Inserting these decompositions into the basic field equation with F and $\mathcal{M}$ (Eq. (7)) we find the general form of the field equations for a magnetized and polarized moving medium expressed in terms of $E(x)$, $B(x)$, $P(x)$ and $M(x)$, which are named the field equations in the Ampèrian form. In equation with the geometric product (Eq. (26)), i.e., in its vector part (Eq. (27)), there are two different velocities u and v and, as I am aware, these field equations do not appear in previous literature. They are important results that are obtained in this paper. All AQs E, B, P and M are represented in the standard basis $\{\gamma_\mu\}$ in order to compare these basic field equations with usual formulations that deal with 3D vectors. The equation (27) is also written in the "source representation" both with AQs and with CBGQs. It is visible from that representation that the sources of E and B fields are the true current density $j^{(C)}$ and the P and M vectors. *The field equations with AQs* (Eq. (7)) *and* (Eq. (26)) *or the corresponding equations with CBGQs comprise and generalize all usual Maxwell's equations (with 3D vectors) for moving media.*

In Sec. 4 we first present a brief review of the existence of the fundamental difference between the usual transformations (UT) of the electric and magnetic fields as 3D vectors and the mathematically correct LT of 4D geometric quantities that represent the electric and magnetic fields in 4D spacetime. Then, the basic field equations with CBGQs are compared with usual Maxwell's equations with 3D vectors for the case when the observers are at rest in a stationary medium. It is shown that these two formulations are not equivalent, because the UT are not the LT.

In Sec. 5 the similar comparison is presented for the case when the observers are at rest in the laboratory frame, but material medium is moving. Again the same result is obtained as in Sec. 4 that these two formulations are not equivalent since 3D quantities do not properly transform under the LT.

In Sec. 6 the comparison with Galilean Electromagnetism is presented. It is shown that both Galilean limits are ill-defined in 4D spacetime and that they are not a quasi-static approximation of the relativistically correct field equations.

In Secs. 7, 7.1 and 7.2 the motional emf ε is calculated using 3D quantities and their UT, Sec. 7.1, and 4D geometric quantities and their mathematically correct LT, Sec. 7.2. In Sec. 7.1 it is shown that if ε is defined by the usual definition with 3D quantities then ε is different in relatively moving inertial frames, $\varepsilon = UBl$ in S and $\varepsilon' = \gamma UBl$ in S', which means that the principle of relativity is not satisfied in the usual formulation of electromagnetism with 3D quantities and their UT of **E** and **B** (and the UT of **P** and **M**). In Sec. 7.2 it is shown that if ε is defined as an invariant 4D quantity, the Lorentz scalar, then always the same value for ε is obtained, $\varepsilon = \gamma UBl$. This result unambiguously shows that the principle of relativity is naturally satisfied in the approach with 4D geometric quantities and their LT.

In Sec. 8 the discussion of the results and the conclusions are presented.

2. The basic field equation for moving media in terms of F and $\mathcal{M}$

In this paper, the geometric algebra formalism [4] will be used. In order to compare Eq. (26), i.e., Eqs. (27) and (28), with the usual formulations of electromagnetism with 3D vectors we shall represent all abstract quantities in (26), i.e., (27) and (28), with 4D CBGQs in the standard basis $\{\gamma_\mu\}$. Thus, for the reader's convenience, all equations will be written not only with the abstract multivectors but also with CBGQs in the standard basis. Therefore, *the knowledge of the geometric algebra is not required for the understanding of this presentation.*

However, a very brief summary of the geometric algebra will be provided here. The geometric (Clifford) product is written by simply juxtaposing multivectors AB. The geometric product of a grade-r multivector A_r with a grade-s multivector B_s decomposes into $A_rB_s = \langle AB\rangle_{\ r+s} + \langle AB\rangle_{\ r+s-2} \cdots + \langle AB\rangle_{\ |r-s|}$. The inner and outer (or exterior) products are the lowest-grade and the highest-grade terms respectively of the above series; $A_r \cdot B_s \equiv \langle AB\rangle_{\ |r-s|}$ and $A_r \wedge B_s \equiv \langle AB\rangle_{\ r+s}$. For vectors a and b we have: $ab = a\cdot b + a\wedge b$, where $a\cdot b \equiv (1/2)(ab+ba)$, $a\wedge b \equiv (1/2)(ab-ba)$. Usually the above mentioned standard basis is introduced. The generators of the spacetime algebra (the Clifford algebra generated by Minkowski spacetime) are taken to be four basis vectors $\{\gamma_\mu\}$, $\mu = 0, ..., 3$, satisfying $\gamma_\mu\cdot\gamma_\nu = \eta_{\mu\nu} = diag(+---)$. The basis vectors γ_μ generate by multiplication a complete basis for the spacetime algebra: 1, γ_μ, $\gamma_\mu\wedge\gamma_\nu$, $\gamma_\mu\gamma_5$, γ_5 ($2^4 = 16$ independent elements). γ_5 is the right-handed unit pseudoscalar, $\gamma_5 = \gamma_0\wedge\gamma_1\wedge\gamma_2\wedge\gamma_3$. Any multivector can be expressed as a linear combination of these 16 basis elements of the spacetime algebra.

This basis, the standard basis $\{\gamma_\mu\}$, is a right-handed orthonormal frame of vectors in the Minkowski spacetime M^4 with γ_0 in the forward light cone, $\gamma_0^2 = 1$ and $\gamma_k^2 = -1$ $(k = 1, 2, 3)$. The $\{\gamma_\mu\}$ basis corresponds to Einstein's system of coordinates in which the Einstein synchronization of distant clocks [3] and Cartesian space coordinates x^i are used in the chosen inertial frame of reference.

The field equation in vacuum in the geometric algebra formalism is:

$$\partial F = j/\varepsilon_0 c, \quad \partial\cdot F + \partial\wedge F = j/\varepsilon_0 c, \tag{1}$$

i.e., with the CBGQs in the $\{\gamma_\mu\}$ basis, that equation becomes

$$\partial_\alpha F^{\alpha\beta}\gamma_\beta - \partial_\alpha\ {}^*F^{\alpha\beta}\gamma_5\gamma_\beta = (1/\varepsilon_0 c)j^\beta\gamma_\beta, \tag{2}$$

where ${}^*F^{\alpha\beta} = (1/2)\varepsilon^{\alpha\beta\gamma\delta}F_{\gamma\delta}$ is the usual dual tensor, $\varepsilon^{\alpha\beta\gamma\delta}$ is the totally skew-symmetric Levi-Civita pseudotensor. The usual covariant form of Eq. (2), i.e., *only the basis components in the* $\{\gamma_\mu\}$ *basis,* are two equations, the equation with sources $\partial_\alpha F^{\alpha\beta} = j^\beta/\varepsilon_0 c$, and that one without sources $\partial_\alpha\ {}^*F^{\alpha\beta} = 0$. It is shown in [2] that the bivector $F = F(x)$, which represent the electromagnetic field, can be taken as the primary quantity for electromagnetism and the field equation for F, Eq. (1), is the basic equation. As shown in [2], the bivector field F yields

IOP Conf. Series: Journal of Physics: Conf. Series **845** (2017) 012013 doi:10.1088/1742-6596/845/1/012013

a complete description of the electromagnetic field and, in fact, there is no need to introduce either the field vectors or the potentials. For the given sources the Clifford algebra formalism enables one to find in a simple way the electromagnetic field F, see Eqs. (7) and (8) in [2].

If j is the total current density then (1), i.e., (2), *holds unchanged in moving medium as well.* The equation (1) can be separated into the field equation with sources and that one without sources as

$$\partial \cdot F = j/\varepsilon_0 c, \quad \partial \wedge F = 0, \tag{3}$$

i.e., with CBGQs in the $\{\gamma_\mu\}$ basis they are

$$\partial_\alpha F^{\alpha\beta}\gamma_\beta = (1/\varepsilon_0 c) j^\beta \gamma_\beta, \quad \partial_\alpha \,{}^*F^{\alpha\beta}\gamma_5\gamma_\beta = 0. \tag{4}$$

Since j is a vector the trivector part is identically zero (in the absence of a magnetic charge).

The total current density vector j can be decomposed as

$$j = j^{(C)} + j^{(\mathcal{M})}, \tag{5}$$

where $j^{(C)}$ is the conduction current density of the *free* charges and $j^{(M)}$ is the magnetization-polarization current density of the *bound* charges

$$j^{(\mathcal{M})} = -c\partial\mathcal{M} = -c\partial \cdot \mathcal{M} \tag{6}$$

($\partial \wedge \mathcal{M} = 0$, since $j^{(\mathcal{M})}$ is a vector). $\mathcal{M}$ is the generalized magnetization-polarization bivector $\mathcal{M} = \mathcal{M}(x)$.

Then (1) (i.e., (3)) can be written as

$$\partial(\varepsilon_0 F + \mathcal{M}) = j^{(C)}/c; \quad \partial \cdot (\varepsilon_0 F + \mathcal{M}) = j^{(C)}/c, \ \partial \wedge F = 0. \tag{7}$$

The trivector part, i.e., the field equation without sources, remained unchanged, because it is not affected by the separation of the current density vector j into free and bound parts; that part does not contain j . The equations (7) are the primary equations for electromagnetism in moving media. In most materials $\mathcal{M}$ is a function of the field F and this dependence is determined by the constitutive relations. Recently, they are discussed in detail in [5]. In that case (7) are well-defined equations for F. The constitutive relations from [5] with 4D geometric quantities, which correctly transform under the LT, are compared with Minkowski's constitutive relations with 3D vectors and several essential differences are pointed out. They are caused by the fact that, as in the case with the field equations that are investigated here, the UT of 3D vectors **E**, **B**, **P**, **M**, etc. are not the LT. Furthermore, in [5], the physical explanation is presented for the existence of the magnetoelectric effect in moving media that essentially differs from the traditional explanation.

If (7) is written with CBGQs in the standard basis it becomes

$$\partial_\alpha(\varepsilon_0 F^{\alpha\beta} + \mathcal{M}^{\alpha\beta})\gamma_\beta - \partial_\alpha(\varepsilon_0 \,{}^*F^{\alpha\beta})\gamma_5\gamma_\beta = c^{-1} j^{(C)\beta}\gamma_\beta, \tag{8}$$

which can be separated into two equations, the equation with sources

$$\partial_\alpha(\varepsilon_0 F^{\alpha\beta} + \mathcal{M}^{\alpha\beta})\gamma_\beta = c^{-1} j^{(C)\beta}\gamma_\beta \tag{9}$$

IOP Conf. Series: Journal of Physics: Conf. Series **845** (2017) 012013 doi:10.1088/1742-6596/845/1/012013

and the equation without sources, which is the same as in the vacuum

$$\partial_\alpha\ {}^*F^{\alpha\beta}\gamma_5\gamma_\beta = 0. \tag{10}$$

Instead of dealing with the axiomatic formulation of electromagnetism for moving media that uses only the *local* form of the field equation (7) one can construct the equivalent integral form simply replacing F by $F + \mathcal{M}/\varepsilon_0$ in Eqs. (18), (21), (22) and also j by $j^{(C)}$ in (21), (22) in [2]. However, the integral form will not be investigated here.

Proceeding in the same way as in [2] one can derive from (7) the stress-energy vector $T(n)$ for a moving medium simply replacing F by $F + \mathcal{M}/\varepsilon_0$ in Eqs. (26), (37-47) in [2]. For example, Eq. (26) in [2] becomes

$$T(n) = T(n(x), x) = -(\varepsilon_0/2)\left\langle (F + \mathcal{M}/\varepsilon_0)n(F + \mathcal{M}/\varepsilon_0)\right\rangle_1 . \tag{11}$$

$T(n)$ is a vector-valued linear function on the tangent space at each spacetime point x describing the flow of energy-momentum through a hypersurface with unit normal vector $n = n(x)$. The expression for $T(n)$, $T(n) = Un + (1/c)S$, Eq. (41) in [2], will remain unchanged, but the energy density U and the Poynting vector S will change according to the described replacement. All this with $T(n)$ will not be discussed in this paper, but in a separate paper in which the Abraham-Minkowski controversy will also be examined in a new way.

Another form of the field equation with sources (9) is the "source representation"

$$\partial \cdot \varepsilon_0 F = j^{(C)}/c - \partial \cdot \mathcal{M}, \tag{12}$$

i.e., with the CBGQs in the $\{\gamma_\mu\}$ basis

$$\partial_\alpha(\varepsilon_0 F^{\alpha\beta})\gamma_\beta = (c^{-1}j^{(C)\beta} - \partial_\alpha \mathcal{M}^{\alpha\beta})\gamma_\beta \tag{13}$$

according to which the sources of the fundamental electromagnetic field F are the true currents $j^{(C)}$ and the magnetization-polarization current density $\partial \cdot \mathcal{M}$, i.e., the space-time changes of the generalized magnetization-polarization bivector $\mathcal{M}$.

In previous formulations of electromagnetism in media (at rest, or moving), starting with Minkowski (his f_{hk}), [1], the electromagnetic excitation tensor is introduced, see, e.g., a modern textbook on classical electromagnetism, [6], or the papers [7-9] and references therein, in the recent - Annalen der Physik, Special Topic Issue 9-10/2008: The Minkowski spacetime of special relativity - 100 years after its discovery. Here, in (7), $\mathcal{H}$ can be introduced as

$$\mathcal{H} = \varepsilon_0 F + \mathcal{M}. \tag{14}$$

However, it is worth noting that (14) is in some sense unsatisfactory, since physically different kind of entities are mixed in it; an electromagnetic field F and a matter field $\mathcal{M}$, i.e., the magnetization-polarization bivector. Moreover, as will be seen in the next section, in general, two different velocity vectors, v - the velocity of the observers and u - the velocity of the moving medium, enter into the decompositions of F and $\mathcal{M}$, Eqs. (17) and (21), respectively. This fact causes that *the usual decomposition of $\mathcal{H}$ into the electric and magnetic excitations, Eq. (25), is not possible in the general case but only in the case if* $u = v$, or if both decompositions (17) and (21) are made with the same velocity vector, either u or v. In that case $\mathcal{H}$ can be introduced in (7) and the usual form of the field equations in moving media is obtained

$$\partial \cdot \mathcal{H} = j^{(C)}/c, \quad \partial \wedge F = 0, \tag{15}$$

i.e., with CBGQs in the $\{\gamma_\mu\}$ basis,

$$\partial_\alpha \mathcal{H}^{\alpha\beta}\gamma_\beta = c^{-1}j^{(C)\beta}\gamma_\beta, \quad \partial_\alpha\ {}^*F^{\alpha\beta}\gamma_5\gamma_\beta = 0. \tag{16}$$

3. The basic field equation for moving media in terms of E, B and P, M

In this paper instead of using (14), (15) and (16) we deal with (7), i.e., with (8), as the basic field equations. In that equation bivectors F and $\mathcal{M}$ can be decomposed. First, the decomposition of F is considered. There is *a mathematical theorem according to which any antisymmetric tensor of the second rank can be decomposed into two space-like vectors and the unit time-like vector.* For the proof of that theorem in geometric terms see, e.g., [10]. When applied to the bivector F, e.g., Eq. (13) in [2], this yields

$$F = E \wedge v/c + (IcB) \cdot v/c, \tag{17}$$

where the electric and magnetic fields are represented by vectors $E(x)$ and $B(x)$. The unit pseudoscalar I is defined algebraically without introducing any reference frame as in Sec. 1.2 in [4] (Hestenes D and Sobczyk G). We choose I in such a way that when I is represented in the $\{\gamma_\mu\}$ basis it becomes $I = \gamma_0 \wedge \gamma_1 \wedge \gamma_2 \wedge \gamma_3 = \gamma_5$. With such choice for I, $\{\gamma_1, \gamma_2, \gamma_3\}$ form a right-handed orthonormal set, as usual for a 3D Cartesian frame. The LT (boosts) do not change the orientation for spacetime. Here, in the whole paper, under the name LT we shall only consider - boosts.

If (17) is written with CBGQs in the $\{\gamma_\mu\}$ basis it becomes

$$F = (1/2)F^{\mu\nu}\gamma_\mu \wedge \gamma_\nu,\ F^{\mu\nu} = (1/c)(E^\mu v^\nu - E^\nu v^\mu) + \varepsilon^{\mu\nu\alpha\beta} v_\alpha B_\beta, \tag{18}$$

where $\gamma_\mu \wedge \gamma_\nu$ is the bivector basis. In the same way as for any other CBGQ it holds that bivector F *is the same 4D quantity* for relatively moving inertial observers and for all bases chosen by them, $F = (1/2)F^{\mu\nu}\gamma_\mu \wedge \gamma_\nu = (1/2)F'^{\mu\nu}\gamma'_\mu \wedge \gamma'_\nu = \ldots$.

Minkowski, Sec. 11.6, [1], see also [11], was the first who introduced vectors (4-vectors in the usual notation) of the electric and magnetic fields and the velocity vector, Φ, Ψ and w, respectively, in his notation, and presented the decomposition of F, his equation (55), that corresponds to (17). Note that he considered that w, Φ and Ψ are 1×4 matrices and F is a 4×4 matrix. Thus he worked with components of the geometric quantities taken in the standard basis $\{\gamma_\mu\}$.

The vector v in the decomposition (17) is interpreted as the velocity vector of the observers who measure E and B fields. Then $E(x)$ and $B(x)$ are defined with respect to v, i.e., with respect to the observer, as

$$E = F \cdot v/c, \quad B = -(1/c)I(F \wedge v/c). \tag{19}$$

It also holds that $E \cdot v = B \cdot v = 0$; both E and B are space-like vectors. It is visible from (19) that *E and B depend not only on F but on v as well.*

If (19) is written with CBGQs in the $\{\gamma_\mu\}$ basis it becomes

$$E = E^\mu\gamma_\mu = (1/c)F^{\mu\nu}v_\nu\gamma_\mu, \quad B = B^\mu\gamma_\mu = (1/2c^2)\varepsilon^{\mu\nu\alpha\beta}F_{\nu\alpha}v_\beta\gamma_\mu. \tag{20}$$

As F is antisymmetric it holds that $E^\mu v_\mu = B^\mu v_\mu = 0$. Only three components of E and B in any basis are independent. However, as E and B depend not only on F but on v as well this result does not mean that three spatial components of E, or B, are necessarily independent components. The form of v in a given inertial frame will determine which three components are independent. The relations (17) - (20) are mathematically correct definitions.

Similarly, using the same theorem, the bivector $\mathcal{M}(x)$ can be decomposed into two space-like vectors, the polarization vector $P(x)$ and the magnetization vector $M(x)$ and the unit time-like vector u/c

$$\mathcal{M} = P \wedge u/c + (MI) \cdot u/c^2. \tag{21}$$

There is the rest frame for a medium, i.e., for $\mathcal{M}$, or P and M , and therefore the vector u in the decomposition (21) may be identified with bulk velocity vector of the medium in spacetime. Integral curves of u define the averaged world-lines of identifiable constituents of the medium. If (21) is written with CBGQs in the $\{\gamma_\mu\}$ basis it becomes

$$\mathcal{M} = (1/2)\mathcal{M}^{\mu\nu}\gamma_\mu \wedge \gamma_\nu,\ \mathcal{M}^{\mu\nu} = (1/c)(P^\mu u^\nu - P^\nu u^\mu) + (1/c^2)\varepsilon^{\mu\nu\alpha\beta}M_\alpha u_\beta. \tag{22}$$

The vectors $P(x)$ and $M(x)$ are determined by $\mathcal{M}(x)$ and the unit time-like vector u/c as

$$P = \mathcal{M} \cdot u/c, \quad M = cI(\mathcal{M} \wedge u/c) \tag{23}$$

and it holds that $P \cdot u = M \cdot u = 0$. As in the case with F, it can be seen from (23) that P and M depend not only on $\mathcal{M}$ but on u as well. $P(x)$ and $M(x)$ from (23) can be written as CBGQs in the $\{\gamma_\mu\}$ basis

$$P = (1/c)\mathcal{M}^{\mu\nu}u_\nu\gamma_\mu, \quad M = (1/2)\varepsilon^{\mu\nu\alpha\beta}\mathcal{M}_{\alpha\nu}u_\beta\gamma_\mu, \tag{24}$$

with $P^\mu u_\mu = M^\mu u_\mu = 0$.

Usually, only the velocity vector u of the moving medium is taken into account, or the case $u = v$ is considered, i.e., it is supposed that the observer frame is comoving with medium, or both decompositions (17), i.e., (18), and (21), i.e., (22), are made with the same velocity vector, either u or v.

Such assumptions enable the introduction of the electromagnetic excitation bivector $\mathcal{H}$, Eq. (14), and, by using (17) and (21), one finds the decomposition of $\mathcal{H}$ into the electric and magnetic excitations (other names of which are "electric displacement" and "magnetic field intensity")

$$\mathcal{H} = D \wedge u/c + (IH) \cdot u/c^2, \tag{25}$$

where, as usual, the electric displacement vector $D = \varepsilon_0 E + P$ and the magnetic field intensity vector $H = (1/\mu_0)B - M$ are introduced. The bivector $\mathcal{H}$ in (25) can be written as a CBGQ in the $\{\gamma_\mu\}$ basis

$$\mathcal{H} = (1/2)\mathcal{H}^{\mu\nu}\gamma_\mu \wedge \gamma_\nu,\ \mathcal{H}^{\mu\nu} = (1/c)(D^\mu u^\nu - D^\nu u^\mu) + (1/c^2)\varepsilon^{\mu\nu\alpha\beta}u_\alpha H_\beta.$$

The decomposition (25) was first introduced by Minkowski, Eq. (56) Sec. 11.6, [1]. Notice that Minkowski dealt only with bulk velocity vector of the medium u; in both his Eqs. (55) (our Eq. (17) but with $v = u$) and (56) (our Eq. (25)) the vector w (our u) appears. The same treatment with the decomposition of $\mathcal{H}$ and consequently with only one velocity, the velocity u, is used in several textbooks, e.g., [12], and papers, e.g., [13, 14]. However, in general, $u \neq v$, e.g., the observers are at rest in the laboratory frame ($v = c\gamma_0$) in which the considered medium is moving with velocity u ($u \neq c\gamma_0$). Therefore, we continue with an alternative approach which deals with two different velocity vectors v and u, i.e., with Eqs. (17) and (21).

Inserting (17) and (21) into the field equation (7) one gets the general form of the field equation for a magnetized and polarized moving medium expressed in terms of $E(x)$, $B(x)$, $P(x)$ and $M(x)$

$$\partial\{\varepsilon_0[E \wedge v/c + (IB) \cdot v] + [P \wedge u/c + (1/c^2)(MI) \cdot u]\} = j^{(C)}/c. \tag{26}$$

IOP Conf. Series: Journal of Physics: Conf. Series **845** (2017) 012013 doi:10.1088/1742-6596/845/1/012013

In the same way as in (7), Eq. (26) with the geometric product can be divided into the vector part (with sources)

$$\partial \cdot \{\varepsilon_0[E \wedge v/c + (IB) \cdot v] + [P \wedge u/c + (1/c^2)(MI) \cdot u]\} = j^{(C)}/c \tag{27}$$

and the trivector part (without sources)

$$\partial \wedge [E \wedge v/c + (IB) \cdot v] = 0. \tag{28}$$

The field equation without sources (28) remains unchanged relative to the corresponding equation for vacuum, because the same assertion holds for the trivector part of (7). We call (26), i.e., (27) and (28), the field equations in the Ampèrian form, in analogy with Maxwell's equations when they are written in terms of the 3D vectors **E**, **B**, **P** and **M**; for the latter ones and the name see, e.g., Eqs. (4.5) in [15]. Observe that in (26), i.e., (27), there are two different velocities u and v. The equation (26) is a fundamental result, which is not previously reported in the physics literature, as I am aware.

If the geometric product is used then there is *only one equation for electromagnetism in moving media, Eq.* (7), i.e., in the Ampèrian form Eq. (26). They are written with abstract 4D geometric quantities and *they comprise and generalize all usual Maxwell's equations (with 3D vectors) for moving media.*

If Eq. (27) is written in the $\{\gamma_\mu\}$ basis, it becomes

$$\partial_\alpha\{\varepsilon_0[\delta^{\alpha\beta}{}_{\mu\nu}E^\mu v^\nu + c\varepsilon^{\alpha\beta\mu\nu}v_\mu B_\nu] + [\delta^{\alpha\beta}{}_{\mu\nu}P^\mu u^\nu + (1/c)\varepsilon^{\alpha\beta\mu\nu}M_\mu u_\nu]\}\gamma_\beta = j^{(C)\beta}\gamma_\beta, \tag{29}$$

where $\delta^{\alpha\beta}{}_{\mu\nu} = \delta^\alpha_\mu\delta^\beta_\nu - \delta^\alpha_\nu\delta^\beta_\mu$. Similarly, in the $\{\gamma_\mu\}$ basis, (28) becomes

$$\partial_\alpha(c\delta^{\alpha\beta}{}_{\mu\nu}B^\mu v^\nu + \varepsilon^{\alpha\beta\mu\nu}E_\mu v_\nu)\gamma_5\gamma_\beta = 0. \tag{30}$$

Again, as for (28), Eq. (30) is the same as in vacuum. In (29), as in (26), i.e., (27), there are two different velocities u and v. The equation (29) does not appear in the entire previous literature. The equation (26), i.e., (27) and (28) and also (29) and (30) are the fundamental results that are obtained in this paper and they enable an alternative, but viable, treatment of electromagnetism of moving media.

The equation (27) can be written in another form, i.e., in the "source representation" as with F and $\mathcal{M}$, Eq. (12),

$$\partial \cdot \{\varepsilon_0[E \wedge v/c + (IB) \cdot v]\} = j^{(C)}/c - \partial \cdot [P \wedge u/c + (1/c^2)(MI) \cdot u], \tag{31}$$

according to which the sources of E and B fields are the true current density $j^{(C)}$ and the P and M vectors.

If the abstract quantities, e.g., $j^{(C)}$, ∂, E , etc. in (31) are replaced by their representations, i.e., CBGQs in the standard basis, $j^{(C)} = j^{(C)\beta}\gamma_\beta$, $\partial = \gamma^\beta\partial_\beta$, $E = E^\mu\gamma_\mu$, ..., then (31) becomes

$$\begin{aligned} &\partial_\alpha\{\varepsilon_0[\delta^{\alpha\beta}{}_{\mu\nu}E^\mu v^\nu + c\varepsilon^{\alpha\beta\mu\nu}v_\mu B_\nu]\}\gamma_\beta \\ = \;&\{j^{(C)\beta} - \partial_\alpha[\delta^{\alpha\beta}{}_{\mu\nu}P^\mu u^\nu + (1/c)\varepsilon^{\alpha\beta\mu\nu}M_\mu u_\nu]\}\gamma_\beta. \end{aligned} \tag{32}$$

From the equations (27) or (31), i.e., with CBGQs, (29) or (32), it is clear that *in 4D spacetime it is not possible to separate the field equation with sources for the E field from that one for the*

B field. Thus, *the usual Ampère-Maxwell law and Gauss's law are inseparably connected in one law - Eq.* (27) or *Eq.* (31), i.e., (29) or (32). Similarly, *in* (28), i.e., Eq. (30), *Faraday's law and the law that expresses the absence of magnetic charge are also inseparably connected in one law.* This is an essential difference relative to Maxwell's equations with 3D vectors **E**, **B**, **P** and **M**. Of course, the same statement holds for the original Eq. (3), i.e., for the vacuum as well.

The mathematical reason for such an inseparability is that, e.g., the gradient operator ∂ is a vector field defined on 4D spacetime. If represented in some basis then its vector character remains unchanged only when *all its components together with associated basis vectors* are taken into account in the considered equation. The same holds for other vectors E, B, j, P, etc. and multivectors like F, $\mathcal{M}$, ... For example, in general, in 4D spacetime, the current density vector j is a well-defined physical quantity, but it is not the case with the usual charge density ρ and the usual current density $\mathbf{j}$ as a 3-vector. Similarly, in general, the gradient operator ∂ cannot be divided into the usual time derivation and the spatial derivations. In 4D spacetime, an independent physical reality is attributed to the position vector x, the gradient operator ∂, the current density vector j , the vectors of the electric and magnetic fields E and B, respectively, etc., but not to the 3-vector $\mathbf{r}$ and the time t, to 3D vectors $\mathbf{j}$, $\mathbf{E}$, $\mathbf{B}$, etc. Therefore, in 4D spacetime, it is not possible to speak about the static case in electromagnetism, i.e., about the electrostatics and magnetostatics.

An important consequence stems from the above mentioned inseparability of 4D spacetime into the 3D space and the time and therefore from the inseparability of Eq. (27), i.e. (31), into two laws, and similarly for Eq. (28). It can be seen from Maxwell's equations with 3D vectors, e.g., Eqs. (43) and (44), and also (46) and (47), which all are given below, that *in the static case* the electric and magnetic fields, **E** and **B**, respectively, are completely decoupled. However, as already stated, in 4D spacetime there is no static case. The equations (27), i.e., (31), and (28) reveal that *the vectors of the electric and magnetic fields E and B*, respectively, *are never decoupled.* This statement holds for the vacuum as well. Thus if, for example, we have a magnetization M (a permanent magnet) but with a negligible permanent polarization P and without $j^{(C)}$, then, as can be seen from (31), M will induce *both B and E*. Such a result is completely understandable because E and B are derived from *one* fundamental quantity, the electromagnetic field bivector F, by the decomposition of F (17) and by (19), and similarly P and M are derived from *one* quantity, the generalized magnetization-polarization bivector $\mathcal{M}$, by the decomposition of $\mathcal{M}$ (21) and by (23). The equations (7), i.e., their "source representation" (12), are the basic field equations with the bivectors F and $\mathcal{M}$; F unites E and B and $\mathcal{M}$ unites P and M. Besides, F is independent on v and $\mathcal{M}$ is independent on u. The formulation of electromagnetism of moving media could be done exclusively in terms of F and $\mathcal{M}$ in the same way as in [2] for vacuum.

It is worth mentioning that in the integral form the equation that corresponds to the local equation (27) can be obtained from Eq. (21) in [2] replacing F by $F + \mathcal{M}/\varepsilon_0$, j by $j^{(C)}$ and inserting into it the decompositions (17) and (21), and similarly for (28) and Eq. (18) in [2].

4. Observers are at rest in a stationary medium. Comparison with the usual formulation with 3D vectors

Recently, [16 - 21], [11], it is proved that, contrary to the general belief, the UT of the 3D $\mathbf{E}(\mathbf{r},t)$ and $\mathbf{B}(\mathbf{r},t)$, see, e.g., Eqs. (11.148) and (11.149) in [6], or Eq. (6) in [5], i.e., Eq. (33) here, *differ* from the LT (boosts) of the 4D E and B vector fields. Note that, in previous literature, starting with Einstein [3], the UT of **E** and **B** are always considered to be the relativistically correct LT. The same holds for the UT of **P** and **M**, Eq. (35) below, and the LT of vectors P and M. For a recent more detailed review see Secs. 3.1 and 3.2 in [22]. The essential point is that in the UT *the transformed* $\mathbf{E}'$ *is expressed by the mixture of the 3D vectors* **E** *and* **B**, Eq. (11.149) in [6], which is

$$\mathbf{E}' = \gamma(\mathbf{E} + \beta \times c\mathbf{B}) - (\gamma^2/(1+\gamma))\beta(\beta \cdot \mathbf{E}), \tag{33}$$

and similarly for $\mathbf{B}'$. For the components implicitly taken in the standard basis it holds that

$$E_1' = E_1, \quad E_2' = \gamma(E_2 - \beta c B_3), \quad E_3' = \gamma(E_3 + \beta c B_2), \tag{34}$$

and similarly for the components of $\mathbf{B}'$, Eq. (11.148) in [6].

The same holds for the couple of the 3D vectors **P** and **M** and their UT

$$\begin{aligned} \mathbf{P} &= \gamma(\mathbf{P}' + \beta \times \mathbf{M}'/c) - (\gamma^2/(1+\gamma))\beta(\beta \cdot \mathbf{P}'), \\ \mathbf{M} &= \gamma(\mathbf{M}' - \beta \times c\mathbf{P}') - (\gamma^2/(1+\gamma))\beta(\beta \cdot \mathbf{M}'). \end{aligned} \tag{35}$$

These UT of **P** and **M**, Eq. (35), are given, e.g., by Eq. (4.2) in [15].

On the other hand, in ISR, as shown in [16 - 21], [11], *the correct LT always transform the 4D algebraic object representing, e.g., the electric field only to the electric field; there is no mixing with the magnetic field.* These correct LT are given by, e.g., Eq. (8) in [5], i.e., Eq. (13) in [11]. The same happens with P and M. The LT of the components E^μ (in the $\{\gamma_\mu\}$ basis) of $E = E^\mu \gamma_\mu$ are given as

$$E'^0 = \gamma(E^0 - \beta E^1), \ E'^1 = \gamma(E^1 - \beta E^0), \ E'^{2,3} = E^{2,3}, \tag{36}$$

for a boost along the x^1 axis. As already mentioned, any CBGQ is unchanged under the LT, i.e., it holds that $E = E^\nu \gamma_\nu = E'^\nu \gamma'_\nu = E_r^\nu r_\nu = E_r'^\nu r'_\nu$, where the primed quantities in both bases $\{\gamma_\mu\}$ and $\{r_\mu\}$ are the Lorentz transforms of the unprimed ones. For the $\{r_\mu\}$ basis, with the "radio," "r"synchronization, and the LT in that basis, see [23] and [24]. The same LT hold for any other vector, e.g., x, B, P, M, etc.

A short derivation of the above LT is presented in [21]. Let us introduce the γ_0-frame in which the standard basis is chosen and $v = c\gamma_0$. Therefore, in the γ_0-frame, E from (19) becomes $E = F \cdot \gamma_0$. If, in the γ_0-frame, that E is written as a CBGQ, then the components are given as $E^0 = B^0 = 0$ and only the spatial components remain, $E^i = F^{i0}$, $B^i = (1/2c)\varepsilon^{ijk0} F_{jk}$; the same components as in, e.g., Eq. (11.137) in [6]. It is proved by Minkowski [1], and reinvented and generalized in [16 - 21], [11], see also Sec. 5 in [22], that in the mathematically correct procedure for the derivation of the LT of E *both* F *and* the velocity vector v have to be transformed by the LT, e.g., for the LT from the γ_0-frame;

$$E = E^\mu \gamma_\mu = [(1/c)F^{i0} v_0]\gamma_i = [(1/c)F'^{\mu\nu} v'_\nu]\gamma'_\mu = E'^\mu \gamma'_\mu. \tag{37}$$

The velocity vector v transforms as any other vector and it holds that $v = v^0 \gamma_0 = c\gamma_0 = v'^\mu \gamma'_\mu$. Hence, *the components* E^μ *transform by the LT again to the components* E'^μ *of the same electric field vector*, i.e., the above quoted LT (36) of the components E'^μ are obtained. *The main point is that the transformed components* E'^μ *are not determined only by* $F'^{\mu\nu}$, *as in all usual approaches*, e.g., Eqs. (11.147) and (11.148) in [6], *but also by* v'^μ.

As already stated, Minkowski, section 11.6 in [1], was the first who derived these mathematically correct LT. He assumed that v, E and B are 1×4 matrices and F is a 4×4 matrix; their components are implicitly determined in the standard basis. He described how v and F separately transform under the LT A (the matrix of the LT is denoted as A)

$$v \longrightarrow v' = vA, \quad F \longrightarrow F' = A^{-1}FA. \tag{38}$$

Then, as shown by Minkowski, the mathematically correct LT of $E = vF$ is

$$E = vF \longrightarrow E' = (vA)(A^{-1}FA) = (vF)A = EA. \tag{39}$$

Thus, under the LT *both* quantities, the field-strength tensor F (4×4 matrix) *and the 4-velocity* v (1×4 matrix) are transformed and their product transforms as any 1×4 matrix transforms. It is already mentioned that this mathematically correct procedure is reinvented and generalized using 4D geometric quantities both in the tensor formalism and in the geometric algebra formalism in [16 - 21], [11].

The comparison with experiments in electromagnetism, the motional emf [17] and Secs. 7 - 7.2 here, the Faraday disk [18], and the Trouton-Noble experiment [2, 25], shows that the approach with 4D geometric quantities and their LT, i.e., the ISR, always agrees with the principle of relativity and it is in a true agreement (independent of the chosen inertial reference frame and of the chosen system of coordinates in it) with all experiments in electromagnetism. Also, it is shown in the mentioned papers that such a true agreement does not exist in the usual formulations of SR, e.g., [6, 15, 26], in which the electric and magnetic fields are represented by the 3D vectors $\mathbf{E}(\mathbf{r},t)$ and $\mathbf{B}(\mathbf{r},t)$ that transform according to the UT. The same conclusion about the true agreement between the approach with 4D geometric quantities, the ISR, and the well-known experiments that test special relativity is obtained in [27]. There, in [27], and particularly in [23], it is explicitly shown that the relativity of simultaneity, the Lorentz contraction and the time dilation are not well-defined in 4D spacetime. They are not the intrinsic relativistic effects, because they depend on the chosen synchronization. But, *every synchronization is only a convention and physics must not depend on conventions. This true agreement of ISR with experiments directly proves the physical reality of 4D geometric quantities.*

Having briefly discussed the LT and the UT we go back to the discussion of Eqs. (29) (i.e., (32)) and (30). In all relatively moving inertial frames of reference and for any system of coordinates in them every term in the considered equations is always the same, because all CBGQs are the Lorentz invariant quantities, e.g., $j^{(C)\beta}\gamma_\beta = j'^{(C)\beta}\gamma'_\beta = j_r^{(C)\beta}\gamma_{r,\beta} = \ldots$. Observe that, in 4D spacetime, *only if all components, together with the associated basis vectors,* are taken into account in every term then all terms are invariant under the passive LT and thus Eqs. (29) ((32)) and (30) remain unchanged for different relatively moving inertial frames and for different systems of coordinates in them. Only in that case the physical quantities and the equations with them are correctly defined in 4D spacetime and the principle of relativity is naturally satisfied. This means that, in general, it is not allowed to consider separately some parts of 4D geometric quantities, or some parts of the equations with them, e.g., to take the part with γ_0 separately from those ones with γ_i in Eqs. (29) ((32)) and (30). Thus, for example, in 4D spacetime, only the whole current density $j^{(C)}$, the abstract vector from (27) and (31), or some of its representations, e.g., that one in the standard basis, the CBGQ, $j = j^{(C)\beta}\gamma_\beta$, is well-defined physical quantity, but not the charge density, $j^{(C)0}$ component, or the spatial components $j^{(C)i}$ taken alone. From the viewpoint of this 4D geometric approach the physical meaning of the charge density ρ is not well-defined. It is the temporal component j^0/c for one observer, but it transforms by the LT to the temporal component *and* the spatial component as well for the relatively moving observer. The same holds for the gradient operator ∂ and its CBGQ in the standard basis $\partial = \gamma^\beta \partial_\beta$ and for all other 4D geometric quantities.

This is particularly visible going to some nonstandard basis, like the $\{r_\mu\}$ basis, i.e. with the "radio," "r," synchronization, see, e.g., [23], Preprints in [27] and [24]. The "r" synchronization is commonly used in everyday life. If the observers who are at different distances from the studio clock set their clocks by the announcement from the studio then they have synchronized their clocks with the studio clock according to the "r" synchronization. The unit vectors in the $\{\gamma_\mu\}$ basis and the $\{r_\mu\}$ basis are connected as $r_0 = \gamma_0$, $r_i = \gamma_0 + \gamma_i$. The components of any vector are connected in the same way as the components of the position vector x are connected, $x_r^0 = x^0 - x^1 - x^2 - x^3$, $x_r^i = x^i$, e.g. for the components of vector E it also holds that $E_r^0 = E^0 - E^1 - E^2 - E^3$, $E_r^i = E^i$. The inverse relations are $\gamma_0 = r_0$, $\gamma_i = r_i - r_0$ and, e.g., for the components of the current density vector j, $j^0 = j_r^0 + j_r^1 + j_r^2 + j_r^3$, $j^i = j_r^i$. Thus, even in the same frame, the charge density in the $\{\gamma_\mu\}$ basis ($j^0 = c\rho$) loses its usual meaning; it is expressed by the sum of all components in the $\{r_\mu\}$ basis. However, observe that, as already stated, $j = j^\mu \gamma_\mu = j_r^\mu r_\mu$ and the same holds for $E = E^\mu \gamma_\mu = E_r^\mu r_\mu$, for B, for x, etc. This reveals that in the $\{r_\mu\}$ basis the space and time cannot be separated. Hence, in 4D spacetime the usual interpretations of the physical quantities, e.g., the charge density ρ and the current density as a 3-vector $\mathbf{j}$, are inappropriate.

An independent physical reality can be attributed either to the abstract geometric quantities, e.g., *vectors x, E, B, P, M , j, .. bivectors F, $\mathcal{M}$, ..., or to their representations in different bases, the CBGQs, like $j^\mu \gamma_\mu$, $E_r^\mu r_\mu$, $M'^\beta \gamma'_\beta$,* etc.

Now, let us consider Eq. (29) in the case when $u = v$, i.e., the observer frame is comoving with medium. In that case it can be taken that $v = u = c\gamma_0$ ($u^\mu = v^\mu = (c, 0, 0, 0)$), i.e., that the observers who measure fields are at rest in a stationary medium. Then, Eq. (29) becomes

$$\begin{aligned} \partial_\alpha \{\varepsilon_0 [\delta^{\alpha\beta}{}_{\mu\nu} E^\mu (\gamma_0)^\nu + c\varepsilon^{\alpha\beta\mu\nu}(\gamma_0)_\mu B_\nu] \quad &+ \\ [\delta^{\alpha\beta}{}_{\mu\nu} P^\mu (\gamma_0)^\nu + (1/c)\varepsilon^{\alpha\beta\mu\nu} M_\mu (\gamma_0)_\nu]\}\gamma_\beta \quad &= \quad c^{-1} j^{(C)\beta}\gamma_\beta, \end{aligned} \tag{40}$$

The equation (40) can be also written in the "source representation" as

$$\begin{aligned} &\partial_\alpha \{\varepsilon_0 [\delta^{\alpha\beta}{}_{\mu\nu} E^\mu (\gamma_0)^\nu + c\varepsilon^{\alpha\beta\mu\nu}(\gamma_0)_\mu B_\nu]\}\gamma_\beta \\ &= \quad \{c^{-1} j^{(C)\beta} - \partial_\alpha [\delta^{\alpha\beta}{}_{\mu\nu} P^\mu (\gamma_0)^\nu + (1/c)\varepsilon^{\alpha\beta\mu\nu} M_\mu (\gamma_0)_\nu]\}\gamma_\beta. \end{aligned} \tag{41}$$

As already mentioned, the sources of both fields together, E and B, are the true current density $j^{(C)}$ and the polarization and magnetization vectors, P and M respectively.

The observer frame is the γ_0-frame, $v = c\gamma_0$, which, with (19), yields that $E^0 = B^0 = 0$ and $E^i = F^{i0}$, $B^i = (1/2c)\varepsilon^{ijk0} F_{jk}$. Furthermore, in the considered case, the γ_0-frame coincides with the rest frame of the medium. Hence, in that frame and with (23), it also holds that $P^0 = M^0 = 0$, $P^i = \mathcal{M}^{i0}$, $M^i = (c/2)\varepsilon^{0ijk}\mathcal{M}_{jk}$. Then, Eq. (40) becomes

$$\begin{aligned} [\partial_k (E^k + P^k/\varepsilon_0) - j^{(C)0}/c\varepsilon_0]\gamma_0 + \quad &\{c\varepsilon^{ijk0}\partial_j [(B_k - \mu_0 M_k] \\ -j^{(C)i}/c\varepsilon_0 - \partial_0 (E^i + P^i/\varepsilon_0)\}\gamma_i \quad &= \quad 0 \end{aligned} \tag{42}$$

In vacuum, Eq. (42) coincides with the first two terms, i.e., the terms with γ_0 and γ_i, in Eq. (8) in [18]. In the approach with 4D geometric quantities, i.e., in the ISR, it is not possible to make any further simplification. *In 4D spacetime, only the whole Eq.* (42) *is physically meaningful and there is no physical sense in some parts of it, for example, to take the part with γ_0 separately* *from those ones with* γ_i. Note that in the approach with 3D vectors there is not any Maxwell's equation that corresponds to Eq. (42).

Let us write Eq. (42) as $a^\alpha\gamma_\alpha = 0$, in which, as can be easily recognized, the coefficients a^α correspond to the usual Maxwell's equations in the component form. There are *two Maxwell equations* in the *component form*; the coefficient a^0 corresponds to the *component form* of the Gauss law for the electric field and the coefficients a^i correspond to the Ampère-Maxwell law in the *component form.* In [18], for the first time, a fundamental discovery is achieved that the usual Maxwell's equations with 3D vectors (for vacuum) are not covariant under the LT. In Sec. 2.3 in [18], the active LT (Eq. (16) in [18]) are applied to Eq. (8) in [18] (in vacuum, as already stated, our Eq. (42) corresponds to the first two terms of Eq. (8) from [18]). There, in that section, it is obtained that the coefficient a^0, which corresponds to the *component form* of the Gauss law for the electric field, does not transform by the LT again to the Gauss law but to a'^0, $a'^0 = \gamma a^0 - \beta\gamma a^1$, which is a combination of the Gauss law and a part of the Ampère-Maxwell law (a^1). (In our case, for the material medium, $a^0 = \partial_k(E^k + P^k/\varepsilon_0) - j^{(C)0}/c\varepsilon_0$.) If the Lorentz transformed Eq. (42), similarly as in Eqs. (21)-(24) in [18], is expressed in terms of Lorentz transformed derivatives and Lorentz transformed vectors E, B, P, M and j, then we find the same equation for a'^0 as it is Eq. (24) in [18], $a'^0 = \{[\gamma(\partial'_k E'^k) - j'^0/c\varepsilon_0] + \beta\gamma[\partial'_1 E'^0 + c(\partial'_2 B'_3 - \partial'_3 B'_2)]\}$, but E'^α has to be replaced by $E'^\alpha + P'^\alpha/\varepsilon_0$, B'_k by $B'_k - \mu_0 M'_k$, and j'^0 by $j'^{(C)0}$. The same discussion holds here as it is presented after Eq. (24) in [18]. From that discussion, and the above mentioned replacements, one concludes that the LT do not transform the Gauss law into the "primed" Gauss law but into quite different law; a'^0 contains the time component $E'^0 + P'^0/\varepsilon_0$, whereas $E^0 = P^0 = 0$, and also the new "Gauss law" includes the derivatives of the magnetic field. The same situation happens with other Lorentz transformed terms, which again explicitly shows that *the Lorentz transformed Maxwell's equations are not of the same form as the original ones.* Hence, contrary to all previous considerations, and contrary to the general opinion, *the usual Maxwell's equations are not Lorentz covariant equations either in vacuum or in a material medium.* This result proves in another way that *in 4D spacetime* only the whole Eq. (42) is physically meaningful and not its separate parts. Remember that Eq. (42) is derived from Eq. (40), i.e. from (29), for which it also holds that $a^\beta\gamma_\beta = a'^\beta\gamma'_\beta = a_r^\beta\gamma_{r,\beta} = \ldots$. Here, as before, the primed quantities are the Lorentz transforms of the unprimed ones in the $\{\gamma_\mu\}$ basis, whereas the last expression refers to the $\{r_\mu\}$ basis with the "r" synchronization.

Let us see how Eq. (42) could be compared with the usual form of Maxwell's equations for stationary media, which deals with 3D vectors, e.g., [15, 26, 28, 29], etc. Obviously, the comparison will be possible *only* if the term with γ_0 is considered separately from those ones with γ_i and if in Eq. (42) *only the components are taken into account.* But, as explained above, *from the viewpoint of the geometric approach,* i.e., *the ISR, such a procedure is not correct in 4D spacetime.*

If in Eq. (42), i.e., in $a^0\gamma_0 + a^i\gamma_i = 0$, one takes that $a^0 = 0$, and multiply the spatial components of E, P and $j^{(C)}$ from a^0 by *the unit 3D vectors* **i**, **j**, **k**, then the term with γ_0 will become the equation

$$\nabla \cdot \varepsilon_0 \mathbf{E}(\mathbf{r}, t) = \rho^{(C)}(\mathbf{r}, t) - \nabla \cdot \mathbf{P}(\mathbf{r}, t), \tag{43}$$

what is Eq. (4.5) (3) in [15], or Eq. (9-6)(1) in [28], or Eq. (4.139) in [29], etc. In the same way it will be obtained that the terms with γ_i will become the equation

$$\nabla \times \mathbf{B}(\mathbf{r}, t) = \mu_0[\mathbf{j}^{(C)}(\mathbf{r}, t) + \partial(\varepsilon_0\mathbf{E}(\mathbf{r}, t) + \mathbf{P}(\mathbf{r}, t))/\partial t + \nabla \times \mathbf{M}(\mathbf{r}, t)], \tag{44}$$

what is Eq. (4.5) (2) in [15], or Eq. (9-6)(4) in [28], or Eq. (4.142) in [29], etc. The way in

IOP Conf. Series: Journal of Physics: Conf. Series **845** (2017) 012013 doi:10.1088/1742-6596/845/1/012013

which the equations with 3D vectors (43) and (44) are constructed clearly shows that Eq. (42) is essentially different than Eqs. (43) and (44).

Similarly, we find that in the γ_0-frame Eq. (30) becomes

$$(c^2\partial_k B^k)\gamma_5\gamma_0 - (c\partial_0 B^i + \varepsilon^{ijk0}\partial_j E_k)\gamma_5\gamma_i = 0. \tag{45}$$

The equation (45) coincides, without any changes, with the last two terms, i.e., the terms with $\gamma_5\gamma_0$ and $\gamma_5\gamma_i$, in the equation for vacuum, Eq. (8) in [18]. As already stated, in (45), Faraday's law and the law that expresses the absence of magnetic charge are *inseparably* connected in one law. But, as in the discussion of Eq. (42), we can make the comparison of Eq. (45) with the usual form of Maxwell's equations for stationary media, which deals with 3D vectors. The equation with 3D vectors that expresses the absence of magnetic charge

$$\nabla \cdot \mathbf{B}(\mathbf{r}, t) = 0, \tag{46}$$

can be constructed from the term with $\gamma_5\gamma_0$ in (45) in the same way as Eqs. (43) and (44) are constructed from Eq. (42). The obtained Eq. (46) is Eq. (4.5) (4) in [15], or Eq. (9-6)(2) in [28], or Eq. (4.140) in [29], or Eq. (7.55)(ii) in [26], etc. Similarly, the terms with $\gamma_5\gamma_i$ will give the equation with 3D vectors, Faraday's law,

$$\nabla \times \mathbf{E}(\mathbf{r}, t) = -\partial \mathbf{B}(\mathbf{r}, t)/\partial t, \tag{47}$$

what is Eq. (4.5) (1) in [15], or Eq. (9-6)(3) in [28], or Eq. (4.141) in [29], or Eq. (7.55)(iii) in [26], etc. (The components of vectors E, B, P, M with superscripts (E^i, B^i, P^i, M^i) from (42) are identified with the components of the usual 3D vectors and $\varepsilon^{0123} = 1$.) As already mentioned, Maxwell's equations in terms of **E**, **B**, **P** and **M**, Eqs. (4.5) in [15], i.e., (43), (44), (46) and (47) here, are said to be in the Ampèrian form.

If, as usual, the electric displacement 3D vector $\mathbf{D} = \varepsilon_0\mathbf{E} + \mathbf{P}$ has been introduced together with the magnetic field intensity 3D vector $\mathbf{H} = (1/\mu_0)\mathbf{B} - \mathbf{M}$ then Eqs. (43) and (44) become

$$\begin{aligned} \nabla \cdot \mathbf{D}(\mathbf{r}, t) &= \rho^{(C)}(\mathbf{r}, t), \\ \nabla \times \mathbf{H}(\mathbf{r}, t) &= \mathbf{j}^{(C)}(\mathbf{r}, t) + \partial \mathbf{D}/\partial t. \end{aligned} \tag{48}$$

In (48), the first equation is Eq. (9-7)(1) in [28], or Eq. (4.139a) in [29], or Eq. (7.55)(i) in [26], etc., whereas the second one is Eq. (9-7)(4) in [28], or Eq. (4.142a) in [29], or Eq. (7.55)(iv) in [26], etc.

According to this discussion there is an essential difference between Maxwell's equations (43), (44), or (48), with the 3D vectors and Eq. (42), i.e., the equations from which (42) is derived, (40), (29) and (27). In the 4D geometric approach, i.e., in the ISR, there is *one law*, Eq. (42), i.e., (40), or (29), or (27), whereas in the approach with 3D vectors there are *two laws*, Eqs. (43) and (44), or (48). In order to obtain two laws (43), (44), or (48), from Eq. (42) we had to make several steps. First, the term with γ_0 is taken separately from those ones with γ_i, then only the components in these terms are taken into account and finally the components are multiplied by the unit 3D vectors **i**, **j**, **k**. But, in 4D spacetime, as explained above, these steps are not mathematically correct. This consideration clearly shows that Eq. (42), i.e., Eqs. (40), or (29), or (27), from which (42) is derived, is not equivalent to Eqs. (43), (44), or (48). The equation (42) is more general and, strictly speaking, it is not possible to obtain Eqs. (43), (44), or (48) from (42) by a mathematically correct procedure in 4D spacetime. The same consideration holds in the same measure for the relation between (45) and the equations with 3D vectors (46) and (47).

Furthermore, in the usual approach with 3D vectors one can speak about the static case. Then, Eqs. (43) and (44), or the first and the second equation in (48), are completely decoupled, i.e., in the static case the electric and magnetic fields as 3D vectors are decoupled. In the 4D geometric approach such a decoupling is never possible, because there is only one law in which there are both together E and B as vectors. The same consideration holds for Maxwell's equations (46) and (47) with 3D vectors and Eq. (45), i.e., the equations from which (45) is derived, (30) and (28).

The most important difference is the following. The quantities entering into (42) and (45) are representations in the standard basis of the abstract 4D quantities from Eq. (26), i.e., Eqs. (27) and (28). All these quantities are correctly defined in 4D spacetime and they correctly transform under the LT, e.g., for the components (36), or for the vector E, Eq. (8) in [5], i.e., Eq. (13) in [11], whereas it is not the case with quantities appearing in (43),(44), (46), (47) and (48), which transform according to the UT, e.g., (33), of the 3D vectors **E**, **B**, **D**, **H**.

5. Observers are at rest in the laboratory frame, but material medium is moving. Comparison with the usual formulation with 3-vectors

Let us examine Eqs. (29) and (30) in the case of a moving material medium, but the observers are at rest in the laboratory frame, which will be denoted as the S frame. Then, in S, $v = c\gamma_0$, $v^\mu = (c, 0, 0, 0))$. Thus the laboratory frame is the γ_0 -frame in which it holds that $E^0 = B^0 = 0$ and $E^i = F^{i0}$, $B^i = (1/2c)\varepsilon^{ijk0}F_{jk}$. Obviously, Eq. (30) becomes the same as Eq. (45), i.e., the same as in vacuum and the whole discussion about the comparison of (45) with Maxwell's equations with 3D vectors remains unchanged. But, it is not so for Eq. (29).

In the γ_0-frame the considered medium is moving with velocity u , $u \neq c\gamma_0$, i.e., some of u^i are $\neq 0$. The rest frame of the medium will be denoted as the S' frame. For the sake of comparison with the usual formulation we present the considered equation in an expanded form in which the term with γ_0 and the terms with γ_i are explicitly written. Then, in the laboratory frame, which is the γ_0-frame, Eq. (29) becomes

$$\begin{aligned}\{\partial_k[\varepsilon_0 E^k + c^{-1}(P^k u^0 - P^0 u^k) + c^{-2}\varepsilon^{kij0}M_i u_j] - c^{-1}j^{(C)0}\}\gamma_0 &\quad + \\ \{-c^{-1}j^{(C)i} + (c\mu_0)^{-1}\varepsilon^{ijk0}\partial_j B_k - \varepsilon_0\partial_0 E^i &\quad + \\ c^{-1}\partial_\mu(P^\mu u^i - P^i u^\mu) - c^{-2}\varepsilon^{i\mu\alpha\beta}\partial_\mu M_\alpha u_\beta]\}\gamma_i &= 0 \end{aligned} \qquad (49)$$

Again, as in the discussion of Eq. (42), it can be argued that *in 4D spacetime, only the whole Eq.* (49) *is physically meaningful and there is no physical sense in some parts of it, for example, to take the part with* γ_0 *separately from those ones with* γ_i. Observe that in (49) there are terms with P^0 and M^0, which cannot exist in the usual formulation with 3D vectors.

What will be obtained from (49) for the case of low velocities of the medium, i.e., for $\beta_u \ll 1$, $\gamma_u = (1-\beta_u^2)^{-1/2} \simeq 1$, where, in S and in the $\{\gamma_\mu\}$ basis, $u = u^\nu\gamma_\nu$, $u^\nu = (\gamma_u c, \gamma_u U^1, \gamma_u U^2, \gamma_u U^3)$, U^k are the same as the components of the 3-velocity $\mathbf{U}$ and $\beta_u = |\mathbf{U}|/c$. To determine and compare P^0 and P^k in S *we use the LT of* P'^μ *from* S', the rest frame of the medium, and, for simplicity, it is taken that the medium, the S' frame, is moving along the common $+x^1$, x'^1 axes, i.e., $u^\nu = (\gamma_u c, \gamma_u U^1, 0, 0)$. In S' , $P'^\mu = (0, P'^1, P'^2, P'^3)$. Then, using the LT, the same as in Eq. (36), $P^\mu = (\beta_u\gamma_u P'^1, \gamma_u P'^1, P'^2, P'^3)$. Since $\beta_u \ll 1$ and $\gamma_u \simeq 1$ it follows that $P^0 \ll P^1$

IOP Conf. Series: Journal of Physics: Conf. Series **845** (2017) 012013 doi:10.1088/1742-6596/845/1/012013

and $P^k u^0 - P^0 u^k$ in (49) becomes $\simeq cP^k$, i.e., in that approximation $P^0 u^k$ can be neglected relative to $P^k u^0$. In the same way it can be concluded that $M^0 u^k$ can be neglected relative to $M^k u^0$. Therefore, with these approximations, Eq. (49) can be written as

$$\begin{aligned}\partial_k \varepsilon_0 E^k \gamma_0 + (c\mu_0)^{-1}\varepsilon^{ijk0}\partial_j B_k \gamma_i \simeq [c^{-1}j^{(C)0} - \partial_k P^k - \\ c^{-2}\varepsilon^{kij0}\partial_k M_i U_j]\gamma_0 + [c^{-1}j^{(C)i} + \partial_0(\varepsilon_0 E^i + P^i) \quad + \\ c^{-1}((U^k\partial_k)P^i - U^i(\partial_k P^k)) + c^{-2}\varepsilon^{ijk0}(\partial_0 M_j U_k + c\partial_j M_k]\gamma_i,\end{aligned} \tag{50}$$

Notice that (50) *is obtained from* (49) *using the LT of the vectors* $P^\mu \gamma_\mu$ and $M^\mu \gamma_\mu$ *and not the UT of the 3D vectors* **P** and **M**, Eq. (35). We see that $D^k = \varepsilon_0 E^k + P^k$ *and* $H^k = B^k/\mu_0 - M^k$ *can be introduced into Eq.* (50), *whereas such replacement is not possible for Eq.* (49), *due to the existence of the terms with* P^0 *and* M^0 *in* (49). Then the part with γ_0 in (50) becomes

$$\partial_k D^k \gamma_0 = [c^{-1}j^{(C)0} - c^{-2}\varepsilon^{kij0}\partial_k M_i U_j]\gamma_0, \tag{51}$$

whereas the part with γ_i is

$$\varepsilon^{ijk0}\partial_j H_k \gamma_i = [j^{(C)i} + c\partial_0 D^i + ((U^k\partial_k)P^i - U^i(\partial_k P^k)) + c^{-1}\varepsilon^{ijk0}\partial_0 M_j U_k]\gamma_i. \tag{52}$$

The equation (50), i.e., Eqs. (51) and (52), can be compared with the usual form of Maxwell's equations for moving media, which deal with 3D vectors, e.g., [15, 28, 29], etc. Again, as in Sec. 4, the comparison will be possible *only* if the term with γ_0 is considered separately from those ones with γ_i, as in (42) and (45) and if in Eq. (50) *only the components are taken into account.* As before, we argue that *from the viewpoint of the geometric approach,* i.e., *the ISR, such a procedure is not correct in 4D spacetime.* If, as in Sec. 4, in Eq. (50), i.e., in $a^0\gamma_0 + a^i\gamma_i = 0$, one takes that $a^0 = 0$, and multiply the spatial components of E, P, M and $j^{(C)}$ from a^0 by *the unit 3D vectors* **i**, **j**, **k** then the term with γ_0 will become the equation

$$\nabla \cdot \varepsilon_0 \mathbf{E}(\mathbf{r},t) = \rho^{(C)}(\mathbf{r},t) - \nabla \cdot [\mathbf{P}(\mathbf{r},t) - c^{-2}(\mathbf{M}(\mathbf{r},t) \times \mathbf{U})]. \tag{53}$$

Usually, as, e.g., in [28] (the derivation of Eqs. (9-18) (1-4)) the case of: "a non-magnetized medium moving with a velocity **u** which is small compared with velocity of light," is considered. This means that in (49) one has to take not only $\beta_u \ll 1$, which leads to (50), but also $M_i = 0$. Then, instead of the part with γ_0 from (50), i.e., Eq. (51), one gets the equation $\partial_k D^k \gamma_0 = \rho^{(C)}\gamma_0$. In the formulation with 3D vectors that equation corresponds to, e.g., Eq. (9-18)(1) in [28], or Eq. (7.55)(i) in [26], $\nabla \cdot \mathbf{D} = \rho^{(C)}$.

In the same way, the terms with γ_i from (50), i.e., Eq. (52), can be compared with the usual form of Maxwell's equations for moving media, which deals with 3D vectors, e.g., Eq. (9-18) (4) in [28], or the equations in Problem 6.8 in [29]. Then, the terms with γ_i from (50) correspond to the following equation with 3D vectors

$$\begin{aligned}\nabla \times \mathbf{B}(\mathbf{r},t) &= \mu_0[\mathbf{j}^{(C)}(\mathbf{r},t) + \partial(\varepsilon_0\mathbf{E}(\mathbf{r},t) + \mathbf{P}(\mathbf{r},t))/\partial t + \nabla \times (\mathbf{P}(\mathbf{r},t) \times \mathbf{U}) \\ &+(1/c^2)\partial(\mathbf{U} \times \mathbf{M}(\mathbf{r},t))/\partial t + \nabla \times \mathbf{M}(\mathbf{r},t)],\end{aligned} \tag{54}$$

or, from (52), the equation (54) can be written in the following form

$$\nabla \times \mathbf{H}(\mathbf{r},t) = \mathbf{j}^{(C)}(\mathbf{r},t) + \partial\mathbf{D}(\mathbf{r},t)/\partial t + (1/c^2)\partial(\mathbf{U} \times \mathbf{M}(\mathbf{r},t))/\partial t + \nabla \times (\mathbf{P}(\mathbf{r},t) \times \mathbf{U}), \tag{55}$$

where the 3D vectors $\mathbf{D} = \varepsilon_0\mathbf{E} + \mathbf{P}$ and $\mathbf{H} = (1/\mu_0)\mathbf{B} - \mathbf{M}$ have been introduced. Taking in Eq. (50) that not only $\beta_u \ll 1$ but that $M_i = 0$ as well, i.e., that a non-magnetized medium is considered, then instead of Eq. (55) we find

$$\nabla \times \mathbf{B}(\mathbf{r},t) = \mu_0[\mathbf{j}^{(C)}(\mathbf{r},t) + \partial\mathbf{D}(\mathbf{r},t)/\partial t + \nabla \times (\mathbf{P}(\mathbf{r},t) \times \mathbf{U})]. \tag{56}$$

This equation is the fourth equation in Problem 6.8 in [29]. It differs from Eq. (9-18) (4) in [28], which contains an additional term $\mu_0\rho^{(C)}\mathbf{U}$. As seen from (56) the appearance of that additional term is not justified.

The whole consideration on the difference between Maxwell's equations with 3D vectors and the equations with 4D quantities that is presented at the end of Sec. 4 holds in the same measure here. However, the difference between these two approaches (3D vectors versus 4D geometric quantities) is even bigger for the case examined in this section. Namely, due to the existence of the terms with P^0 and M^0 in (49) that equation *cannot* be compared with Maxwell's equations with 3D vectors. The comparison can be made only for low velocities of the medium when Eq. (49) reduces to Eq. (50).

There is also an additional difference between Maxwell's equations with 3D vectors, e.g., Eqs. (9-18) in [28], and our Eqs. (49) and (50). It is stated in [28] (under Eqs. (9-18)): "Note that Maxwell's equations for moving (nonmagnetic) media in the form given by Eq. (9-18) (4) are "mixed," i.e., the sources $\mathbf{j}_{true}$, $\mathbf{P}$, ρ_{true}, are measured in the moving medium, while the fields are given in the stationary frame." On the other hand, as already stated above, *all quantities* in (49) and (50) are determined in the laboratory frame, which is the γ_0-frame. Moreover, as already explained, all quantities in (49) and (50) are correctly defined in 4D spacetime and they correctly transform under the LT, which is not the case with Maxwell's equations with 3D vectors.

6. Comparison with Galilean Electromagnetism

At this place it is worth mentioning that in a recent review [30] under the title "Forty years of Galilean Electromagnetism (1973 - 2013)" and in the references therein, e.g., [31, 32], it is argued that, [30], "Galilean Electromagnetism is precisely the low-velocity limit of Special Relativity when applied to Classical Electromagnetism." There, it is also stated that in a Galilean limit the usual Maxwell-Minkowski equations with the 3D vectors **E**, **B**, **D** and **H**, Eq. (1) in [30], have to be replaced by two Galilean limits, the magnetic and electric limits, i.e., with two sets of low-velocity formulae, Eqs. (16) and (18) in [30], respectively. Furthermore, in [30], it is considered that the UT, Eqs. (33) and (35) here and the similar ones for **D** and **H**, Eq. (3) in [30] or in [31], are the relativistically correct LT, but that in Galilean approximation they have to be replaced by two sets of low-velocity formulae, the magnetic and electric limits, which are presented, e.g., in Sec. 3 in [30]. The same happens with Minkowski's constitutive equations, Eq. (9), i.e., Eq. (10) in [30] or in [31], which are replaced by the Galilean magnetic constitutive equations, Eq. (13) in [30], and the Galilean electric constitutive equations, Eq. (14) in [30]. In addition, it is considered in [30], as in almost all other usual approaches, that the contraction of lengths and the dilation of time are "the phenomena inherent to Special Relativity."

However, in 4D spacetime the physical quantities are represented in a mathematically correct way by 4D geometric quantities that properly transform under the LT, e.g., (36), and not by 3D quantities that transform by the UT, (33) and (35) and the similar UT for **D** and **H**, Eq. (3) in [30] or in [31]. As discussed in Sec. 4, according the UT, (33) and (35), i.e., Eq. (3) in [30] or in [31], e.g., the transformed $\mathbf{E}'$ is expressed by the mixture of the 3D vectors **E** and

B. Using that essential feature of the UT of the fields as 3D vectors a Galilean limit of the field transformations is derived in two steps in, e.g., [32]. First, the quasi-static approximation, $\beta \ll 1$, is taken and then the assumption on the relative magnitude of $|\mathbf{E}|$ and $c|\mathbf{B}|$ is taken into account. If the magnetic field is dominant then the magnetic limit, Eq. (7) in [32] is obtained, whereas in the opposite case the electric limit, Eq. (8) in [32], is obtained. On the other hand, as discussed in Sec. 2, the essential feature of the mathematically correct LT of 4D fields, like (36), is that the LT transform, e.g., the electric field vector only to the electric field vector; there is no mixing with the magnetic field vector. This means that both Galilean limits, the magnetic limit and the electric limit of the field transformations, Eqs. (7) and (8) in [32] are meaningless in 4D spacetime in which the fields are represented by 4D geometric quantities that correctly transform under the LT.

In [5] the constitutive relations are formulated in terms of coordinate-free quantities that correctly transform under the LT. First, in Sec. 3 they are formulated as the relations between $\mathcal{M}$ and F, Eqs. (11) and (12) in [5]. Then, using the decompositions of F (17) and $\mathcal{M}$ (21) the basic constitutive relations for $P(x)$ and $M(x)$, Eqs. (13) and (14) in [5], are obtained. They show how $P(x)$ and $M(x)$ depend on E, B and two different velocity vectors, u and v. These constitutive relations differ from all previous expressions and they are not reported in any previous approach. In Secs. 5 - 5.2 in [5] it is explained that Minkowski's constitutive relations, Eqs. (23) and (24), or (25) in [5], i.e., Eq. (11) in [30], or Eq. (10) in [31], are not the *relativistic* constitutive equations. They are the relations with 3D vectors that transform according to the UT, like (33) and (35), and these transformations, as already stated several times, are not the LT. Therefore, the constitutive relations, Eqs. (13) and Eq. (14) in [30] are not any kind of a quasi-static approximation of the relativistically correct constitutive relations. As stated in [5], there is only one mathematically correct quasi-static approximation of the constitutive relations for $P(x)$ and $M(x)$, which is given by Eqs. (19) and (21) in [5] in which for the low velocities of the medium it is only taken that $\beta_u \ll 1$, i.e., $\gamma_u \simeq 1$, see Secs. 4, 5 - 5.2 in [5].

It can be concluded from the preceding discussion that both Galilean limits, the Galilean magnetic Maxwell - Minkowski equations and the Galilean electric Maxwell - Minkowski equations, e.g., Eqs. (16) and (18) in [30], respectively, are ill-defined in 4D spacetime; they are not a quasi-static approximation of the relativistically correct field equations. As shown in Sec. 4 and in this section the usual Maxwell-Minkowski equations with the 3D vectors $\mathbf{E}$, $\mathbf{B}$, $\mathbf{D}$ and $\mathbf{H}$, Eq. (1) in [30], *are not* the relativistic form of Maxwell's equations, because in 4D spacetime there is no room for 3D vectors and their UT. The mathematically correct field equation is, e.g., Eq. (49) and the equations from which (49) is derived and there is only one its quasi-static approximation, i.e., the approximation for the case of low velocities of the medium, it is Eq. (50).

Furthermore, as discussed in Sec. 4, it is exactly proved in [27] and in [23] that the relativity of simultaneity, the Lorentz contraction and the time dilation *are ill-defined in 4D spacetime*, because they depend on the chosen synchronization. Hence, they are not, as argued in [30] "the phenomena inherent to Special Relativity." This dependence on chosen synchronization holds in the same measure for the usual Maxwell-Minkowski equations, Eq. (1) in [30], for the UT of the fields as 3D vectors, Eq. (3) in [30] or in [31], for Minkowski's constitutive relations, Eq. (11) in [30] and for the whole Galilean Electromagnetism including the comparison with experiments that is presented in Secs. 5 - 8.2 in [30].

IOP Conf. Series: Journal of Physics: Conf. Series **845** (2017) 012013 doi:10.1088/1742-6596/845/1/012013

7. Motional electromotive force in the approaches with 3D quantities and with 4D geometric quantities

On the other hand, as already mentioned in Sec. 4, the approach with 4D geometric quantities and their LT, always agrees with the principle of relativity and it is in a true agreement with all experiments in electromagnetism and all experiments that test SR, see [17, 18, 2, 25, 27]. In Secs. 7.1 and 7.2, instead of to discuss the experiments which are presented in Secs. 5 - 8.2 in [30], we shall briefly present the discussion of the motional electromotive force (emf) that is exposed in Secs. 5 - 5.2 in [17]. However, in Sec. 7.2, some changes relative to Sec. 5.2 in [17] will be introduced (F instead of E and B). It is a nice example that illustrates the fundamental difference between the LT, e.g., (36), and the UT (33) and (35), i.e., between the approach with 4D geometric quantities and the usual approach with 3D vectors.

7.1. Motional emf with 3D quantities

The motional emf is produced in an electrical circuit when a circuit or part of a circuit moves in a magnetic field. In Sec. 5.1 in [17] the emf ε is calculated using 3D quantities, the 3D Lorentz force $\mathbf{F}_L = q\mathbf{E} + q\mathbf{U} \times \mathbf{B}$, the 3D $\mathbf{E}$ and $\mathbf{B}$ and their UT (33) as in Secs. 6.4 in [29], 7.2 in [33], 9-5 in [28], 5.6 in [15] and in all other calculations in the usual approaches. The emf ε of a complete circuit is defined by means of the Lorentz force $\mathbf{F}_L$ that acts on a charge q, which is at rest relative to the section $\mathbf{dl}$ of the circuit,

$$\varepsilon = \oint (\mathbf{F}_L/q) \cdot \mathbf{dl}, \tag{57}$$

Eq. (26) in [17]. The important remark is that it is implicitly assumed in these equations for ε that the integral is taken over the whole circuit at the same moment of time in S, say $t = 0$. Let us take that in the laboratory frame S a conducting bar is moving in a steady uniform magnetic field (3-vector) $\mathbf{B} = -B\mathbf{k}$ with velocity 3-vector $\mathbf{U}$ parallel to the x axis. The length of the bar is l and it moves parallel to the y axis. There is no external applied electric field in S, $\mathbf{E}$ =0. Since in S $\mathbf{E}$ =0 and the components of $\mathbf{B}$ are $(0, 0, -B)$ the emf ε is determined by the contribution of the magnetic part of $\mathbf{F}_L$, i.e., $q\mathbf{U} \times \mathbf{B}$, as $\varepsilon = \int_o^l UBdy = UBl$, which is Eq. (27) in [17].

In S' the conducting bar is at rest. Then, *according to the UT* (33) of the 3D $\mathbf{E}$ and $\mathbf{B}$ the observer in the S' frame "sees" $E'_y = \gamma UB$ and $B'_z = -\gamma B$. Hence, *in* S', there is not only the magnetic field but *an electric field* $\mathbf{E}'$ as well. The contribution of the magnetic part (due to B'_z) of the Lorentz force $\mathbf{F}'_L$ to the emf ε' is zero and only the contribution of the electric part (due to E'_y) of the Lorentz force remains, which is $\varepsilon' = \int_o^l \gamma UBdy = \gamma UBl$, Eq. (29) in [17]. Observe that the integral is taken at the same moment of time t' in S', which can be arbitrarily chosen, say $t' = 0$, or $t' = 10s$, ... The moments of time t in S and t' in S' are not connected in any way. The LT cannot transform the moment of time t in S again, exclusively, to some t' in S'. According to the LT, to one t in S will correspond many t' in S' depending on the spatial position in S'; $t = \gamma(t' + Ux'/c^2)$. This remark clearly shows that the usual definition of ε, Eq. (57), is not relativistically correct definition. Obviously, *the emf* ε' *in* S' *is not equal to the emf* ε, *determined in* S;

$$\varepsilon = UBl, \; \varepsilon' = \gamma UBl, \quad \varepsilon' \neq \varepsilon. \tag{58}$$

Consequently, *the principle of relativity is not satisfied;* the emf obtained by the application of the UT (33) is different for relatively moving 4D observers. This explicitly shows that the conventional calculation of ε and the UT (33) of the 3D $\mathbf{E}$ and $\mathbf{B}$ are not relativistically correct, i.e., the UT (33) are not the LT. The fact that ε and ε' do not significantly differ for low velocities, $U \ll c$, is completely irrelevant; the principle of relativity is not satisfied in the usual approach. Of course, the same holds for the Galilean Electromagnetism.

7.2. Motional emf with 4D geometric quantities

In Sec. 5.2 in [17] the emf ε is calculated using 4D geometric quantities, the vectors E and B, the vector of the Lorentz force K, etc. There, it is found that in the approach with 4D geometric quantities and their LT, e.g., (36), the same value for ε is always obtained, $\varepsilon = \gamma UBl$, which means that the principle of relativity is naturally satisfied. Here, we shall obtain the same result as in Sec. 5.2 in [17] but dealing with F and not with its decompositions (17) and (18). The Lorentz force K is defined as an abstract vector and as a CBGQ in the standard basis by the following relations

$$K = (q/c)F \cdot u, \quad K = K^{\mu}\gamma_{\mu} = (q/c)(F^{\mu\nu}u_{\nu})\gamma_{\mu}. \tag{59}$$

where u is the velocity vector of the considered charge. *The emf ε is defined as an invariant 4D quantity, the Lorentz scalar,*

$$\varepsilon = \int_{\Gamma}(K/q) \cdot dl, \quad \varepsilon = \int_{\Gamma}(K^{\mu}/q)dl_{\mu} = (1/c)\int_{\Gamma}F^{\mu\nu}u_{\nu}dl_{\mu}, \tag{60}$$

where *vector dl* is the infinitesimal spacetime length and Γ is the spacetime curve.

In the laboratory frame S with the standard basis in it the components of u and dl are $u^{\mu} = (\gamma c, \gamma U, 0, 0)$, $dl^{\mu} = (0, 0, dl^2 = dy, 0)$. It can be seen from Eq. (20) that in the considered case all components $F^{\mu\nu}$ are zero except F^{21} (F^{12}), which is $F^{21} = cB^3$. This result and Eq. (59) yield that the components of the Lorentz force K are $K^0 = K^1 = K^3 = 0$, but $K^2 = (q/c)F^{21}(-\gamma U)$. Hence, the emf ε is

$$\varepsilon = (1/c)\int_0^l F^{21}(-\gamma U)dy = (1/c)F^{21}(-\gamma U)l. \tag{61}$$

This result can be compared with that one from Sec. 5.2 in [17], $K^2 = \gamma qUB$ and $\varepsilon = \gamma UBl$, using the relations $F^{21} = cB^3$, and $B^3 = B_z = -B$. Then

$$\varepsilon = (1/c)\int_0^l F^{21}(-\gamma U)dy = \gamma UBl, \tag{62}$$

which is the same as in [17]. In contrast to the usual approaches with 3D quantities the expression for ε (60) is independent of the chosen reference frame and of the chosen basis in it; ε is the same in S and in the relatively moving S' frame,

$$\varepsilon = \int_{\Gamma}(K^{\mu}/q)dl_{\mu} = \int_{\Gamma}(K'^{\mu}/q)dl'_{\mu} = \gamma UBl, \tag{63}$$

which is Eq. (37) in [17]. This result for ε can be checked directly performing the LT of all vectors from S to S'. Observe that in S' the 3-velocity $\mathbf{U}$ is zero, but the velocity vector u *is not*, $u = c\gamma_0$, i.e., $u'^{\mu} = (c, 0, 0, 0)$. The same value for ε will be obtained if another basis, e.g., the $\{r_{\mu}\}$ basis, will be used in both frames. Obviously, *in the 4D geometric approach, the principle of relativity is naturally satisfied.*

The result that the conventional theory with the 3D **E** and **B** and their UT (33) yields different values for ε for relatively moving inertial observers, $\varepsilon = UBl$ in S and $\varepsilon' = \gamma UBl$ in S', whereas ISR, i.e., the approach with 4D geometric quantities and their LT, e.g., (36), yields always the same value for ε, $\varepsilon = \gamma UBl$, Eq. (63), is very strong evidence that the usual approach is not relativistically correct. It is for the experimentalists to find the way to measure the emf ε with a great precision and to see that in the laboratory frame $\varepsilon = \gamma UBl$ and not simply $\varepsilon = UBl$.

Observe that in this calculation with 4D geometric quantities all quantities are invariant under the passive LT, e.g., $u = u^\nu \gamma_\nu = u'^\nu \gamma'_\nu = u^\nu_r r_\nu = u'^\nu_r r'_\nu$, $K = K^\nu \gamma_\nu = K'^\nu \gamma'_\nu = K^\nu_r r_\nu = K'^\nu_r r'_\nu$, etc., which means that the observers in S and S' are "looking" at the same quantity, for example, the Lorentz force K. This consideration shows that ISR, i.e., the approach with 4D geometric quantities and their LT (36) is substantially different than the usual approaches with the 3D quantities and their UT (33). The former is completely suited to the symmetry of 4D spacetime, which is not the case with the latter.

There are many similar examples in the literature; it will be always found that there is a fundamental difference between the UT of the 3D **E** and **B** (33) and the LT of 4D geometric quantities representing the electric and magnetic fields, e.g., (36), and that the 4D geometric approach, i.e., ISR, correctly describes electromagnetic phenomena in all relatively moving 4D inertial frames of reference. One important experiment, the Faraday disk, which leads to the same conclusions, is considered in detail in [18].

8. Discussion and conclusions

There are several important differences between the field equations reported here and all others in previous literature including the modern textbook on classical electrodynamics [34], which uses the calculus of exterior forms.

First, instead of dealing with the electromagnetic excitation $\mathcal{H}$ (14) and the field equation with it (15) we exclusively deal with the equations (7) for the electromagnetic field F and a matter field $\mathcal{M}$ as the primary equations for electromagnetism in moving media. As discussed in Secs. 2 and 3, the expression for $\mathcal{H}$ (14) in terms of F and $\mathcal{M}$ is in some sense unsatisfactory, since F and $\mathcal{M}$ are physically different kind of entities. Furthermore, what is particularly important, in general, two different velocity vectors, v - the velocity of the observers and u - the velocity of the moving medium, enter into the decompositions of F and $\mathcal{M}$, Eqs. (17) and (21), respectively. For this reason we also do not deal with the decomposition of $\mathcal{H}$ (25) into the electric and magnetic excitations D and H, respectively, where $D = \varepsilon_0 E + P$ and $H = (1/\mu_0)B - M$. As stated in Sec. 3, such a decomposition as (25) is possible if only one velocity, the velocity of the medium u, is taken into account, or the case $u = v$ is considered, or both decompositions (17) and (21) are made with the same velocity vector, either u or v. Recently, the last case is considered in [35], but with F and $\mathcal{H}$.

The second important difference refers to the interpretation of the field equations. The basic field equation (26) contains two different velocities u and v. It is also written as Eqs. (27) and (28), i.e., with CBGQs, (29) and (30), respectively. From these equations it is visible that *in 4D spacetime*, in contrast to the formulation of electromagnetism in terms of Maxwell's equations with the 3D vectors **E**, **B**, **P** and **M**, *there are no two laws, the Ampère-Maxwell law and Gauss's law, but only one law, that is expressed by Eq.* (27) ((29)), i.e., *Eq.* (31) ((32)), *and the same for Eq.* (28) ((30)) and *Faraday's law and the law that expresses the absence of magnetic charge.*

Furthermore, the interesting results are obtained in Secs. 4 and 5. There, the field equations, written in the standard basis (29), i.e., (32), and (30), are compared with the usual form (with 3D vectors) of Maxwell's equations for moving media. In Sec. 4, it is shown that the comparison is possible *only* if the term with γ_0 is considered *separately* from those ones with γ_i and if in Eqs. (42) and (45) *only the components* are taken into account. In order to get the usual equations with 3D vectors, (43),(44), (46), (47) and (48), these components have to be multiplied by *the unit 3D vectors* **i**, **j**, **k**. Moreover, as shown in Sec. 5, such a procedure is not applicable to Eq. (49), but only to Eq. (50), which is derived from (49) for the case of low velocities of the medium. As explained in Sec. 4, *the above mentioned steps in the comparison are not mathematically correct in 4D spacetime.* Hence, in 4D spacetime, the equations (26), (27), (28), (29), (30), ... , with the 4D geometric quantities E, B, P and M that correctly transform under the LT, e.g., for the components (36), or for the vector E, Eq. (8) in [5], i.e., Eq. (13) in [11], *are not equivalent* to the usual Maxwell equations (43), (44), (46), (47), (48), (53), (54), ... , with the 3D vectors **E**, **B**, **P** and **M** that transform according to UT, e.g., Eqs. (33) and (35).

It will be important for physics to examine the theoretical and experimental consequences of the results that are obtained in this paper. An interesting consequence that can be experimentally examined is already mentioned in connection with Eq. (31). There, it is stated that if we have a magnetization M, a permanent magnet, moving *or stationary*, but with a negligible permanent polarization P and without $j^{(C)}$, then, as can be seen from (31), M will induce *both B and E*. In Sec. 8 in [22] the existence of the electric field from a stationary permanent magnet is investigated in detail and that field is used in the consideration of the "charge-magnet paradox" in [22] and [36].

At this place it is worth mentioning a recent paper, [37], under the title "Nature of Electric and Magnetic Fields; How the Fields Transform." In that paper, in 3.1 and 3.3 the mathematically correct proofs are given that the electric and magnetic fields, $E(x)$ and $B(x)$, respectively are properly defined vectors on 4D spacetime and not the usual 3D vectors **E** and **B**. Furthermore, it is proved in [37] that the correct LT of the electric field are given by (36) and not by the UT of the 3D vectors Eqs. (11.148) and (11.149) in [1]. The proof 3.3 from [37] is based on the mathematical theorem presented in Sec. 3 here, i.e., it is given by the relations (17) - (20) here. The proof 3.1 is a simple but very strong mathematical argument, which is stated by Oziewicz, e.g., in [38]: *What is essential for the number of components of a vector field is the number of variables on which that vector field depends, i.e., the dimension of its domain.* In general, the dimension of a vector field that is defined on a n-dimensional space is equal - n. *The electric and magnetic fields are defined on a 4D space, i.e., the spacetime. They are always functions of the position vector x. This means that they are not the usual 3D fields, but they are properly defined vectors on 4D spacetime,* $E(x)$ and $B(x)$. The same holds for the polarization vector $P(x)$ and the magnetization vector $M(x)$. Since E, B, P and M are 4D vectors they must transform *under the LT as any other vector transforms*, e.g., *the electric field vector must transform again to the electric field vector* like in (36). In [39] the same result is obtained for the electric field as a bivector and for the magnetic field as well.

Acknowledgments

It is a pleasure to acknowledge to Larry Horwitz and Martin Land for inviting me to the IARD conferences and for their continuous support of my work. I am also very grateful to Zbigniew Oziewicz for numerous and very useful discussions during years and to him and to Alex Gersten for the continuous support of my work.

References

[1] Minkowski H 1908 *Nachr. Ges. Wiss. Göttingen* 53; Reprinted in: 1910 *Math. Ann.* **68** 472; English translation in: M N Saha and S N Bose 1920 *The Principle of Relativity: Original Papers by A. Einstein and H. Minkowski* (Calcutta: Calcutta University Press)

[2] Ivezić T 2005 *Found. Phys. Lett.* **18** 401

[3] Einstein A 1905 *Annalen der Physik* **17** 891; English translation in: Perrett W and Jeffery G B 1952 *The Principle of Relativity* (New York: Dover)

[4] Hestenes D 1966 *Space-Time Algebra* (New York: Gordon & Breach); Hestenes D and Sobczyk G 1984 *Clifford Algebra to Geometric Calculus* (Dordrecht: Reidel); Doran C and Lasenby A 2003 *Geometric algebra for physicists* (Cambridge: Cambridge University Press)

[5] Ivezić T 2012 *Int. J. Mod. Phys. B* **26** 1250040

[6] Jackson J D 1998 *Classical Electrodynamics* 3rd edn (New York: Wiley)

[7] Hehl F W 2008 *Annalen der Physik* **17** 691

[8] Obukhov Y N 2008 *Annalen der Physik* **17** 830

[9] Itin Y and Friedman Y 2008 *Annalen der Physik* **17** 769

[10] Ludvigsen M 1999 *General Relativity, A Geometric Approach* (Cambridge: Cambridge University Press); Sonego S and Abramowicz M A J 1998 *J. Math. Phys.* **39** 3158

[11] Ivezić T 2010 *Phys. Scr.* **82** 055007

[12] Møller C 1972 *The Theory of Relativity* 2nd edn (Oxford: Clarendon Press)

[13] Hillion P 1993 *Phys. Rev. E* **48** 3060;

[14] Dereli T, Gratus J and Tucker R W 2007 *Phys. Lett. A* **361** 190

[15] Bladel J Van 1984 *Relativity and Engineering* (Berlin: Springer-Verlag)

[16] Ivezić T 2003 *Found. Phys.* **33** 1339

[17] Ivezić T 2005 *Found. Phys.* Lett. **18** 301

[18] Ivezić T 2005 *Found. Phys.* **35** 1585

[19] Ivezić T 2008 *Fizika A* **17** 1

[20] Ivezić T 2008 *arXiv:* 0809.5277

[21] Ivezić T 2007 *Phys. Rev. Lett.* **98** 108901

[22] Ivezić T 2013 *J. Phys.: Conf. Ser.* **437** 012014

[23] Ivezić T 2001 *Found. Phys.* **31** 1139

[24] Ivezić T 2010 *Phys. Scr.* **81** 025001

[25] Ivezić T 2007 *Found. Phys.* **37** 747

[26] Griffiths D J 2013 *Introduction to Electrodynamics* 4th edn (Boston: Pearson)

[27] Ivezić T 2002 *Found. Phys. Lett.* **15** 27; Ivezić T 2001 *Preprint* physics/0103026; Ivezić T 2001 *Preprint* physics/0101091

[28] Panofsky W K H and Phillips M 1962 *Classical electricity and magnetism* 2nd edn (Reading: Addison-Wesley)

[29] Rosser W G W 1968 *Classical Electromagnetism via Relativity* (New York: Plenum)

[30] Rousseaux G 2013 *Eur. Phys. J. Plus* **128** 81

[31] Rousseaux G 2008 *Europhys. Lett.* **84** 20002

[32] de Montigny M and Rousseaux G 2006 *Eur. Phys. J. Plus* **27** 755

[33] Purcell E M 1985 *Electricity and Magnetism* 2nd edn (New York: McGraw-Hill)

[34] Hehl F W and Obukhov Yu N 2003 *Foundations of Classical Electrodynamics: Charge, flux, and metric* (Boston: Birkhäuser)

[35] Goto Shin-itiro, Tucker R W and Walton T J 2011 *Proc. Roy. Soc. A* **467** 59

[36] Ivezić T 2012 *Preprint* 1212.4684

[37] Ivezić T 2015 *Preprint* 1508.04802

[38] Oziewicz Z 2011 *J. Phys.: Conf. Ser.* **330** 012012

[39] Ivezić T 2008 *Fizika A* **17** 1

On continuum dynamics and the electromagnetic field in the scalar ether theory of gravitation

Mayeul Arminjon

Laboratory "Soils, Solids, Structures, Risks" (CNRS & Université Grenoble-Alpes), Grenoble, France

E-mail: Mayeul.Arminjon@3sr-grenoble.fr

Abstract. Our study of aspects of a scalar ether theory of gravitation is formulated in a preferred reference frame in a four-dimensional spacetime endowed with a curved "physical" metric. The dynamics of a test particle is defined by an extension of the special-relativistic form of Newton's second law. This determines the dynamical equation verified by the energy-momentum tensor **T** of a "dust" continuum, also in the presence of a non-gravitational external force. The dynamical equation for **T** thus obtained is assumed valid for a general continuous medium or a system of fields. When the non-gravitational force is the Lorentz force, this equation in turn determines the Maxwell equations in a gravitational field for the present theory. They are consistent with the dynamics of photons i.e. with the geometrical optics of the theory. Except for a constant gravitational field, they seem to imply some local production or destruction of electric charge. The possible amounts are yet to be assessed.

1. Introduction

A theory based on an alternative view of gravity has been proposed and investigated. The idea of the theory lies in a semi-heuristic interpretation of gravitation as a pressure force in an imagined perfect fluid or "ether" and in the introduction of a finite velocity of propagation through a compressibility of the ether [1]. [1] The construction of a theory based on this idea and accounting for special relativity, leads to a scalar theory written in a preferred reference frame in a four-dimensional spacetime endowed with a flat "background" metric and a curved "physical" metric [6, 7, 8]. Thus, although in its concept this is a very different theory from general relativity (GR) and from most alternative theories, nevertheless this theory also considers a curved Lorentzian spacetime as the main mathematical tool. (The curvature scalar and tensors are not needed, however.) Also like GR, this theory coincides with special relativity if the gravitational field vanishes.

[1] Our approach differs in spirit from numerous works (some of them rather recent), in which it is aimed at deriving classical equations of electromagnetism and/or gravitation from some mechanical model of ether. For instance, Zareski [2] considered an ether being a special type of elastic solid and found that the equations governing this medium are identical in form with the classical Maxwell equations. Later on, he investigated a generalization of this model in which Einstein gravity is assumed to be present [3]. An elastic-solid ether and its relation with electromagnetism and gravitation were also studied in some detail by Dmitriyev [4, 5]. In our approach, the interpretation of gravity as a pressure force in a compressible fluid leads us to state a self-consistent set of equations which builds a new theory of gravitation. The Maxwell equations discussed in the present paper are derived, not from some assumed mechanical constitutive relation in the ether, but from continuum dynamics applied to a material medium having a charge density and being subjected to the Lorentz force.

The aim of this paper is to summarize the development [9] of the theory to describe continuum dynamics and the electromagnetic (e.m.) field in the presence of gravitation. The equations of the e.m. field are indeed obtained as an application of the equation for continuum dynamics in the presence of gravitation and a non-gravitational external field.

In contrast with [9], the present paper does not discuss the motivations for the theory nor its experimental check, and it does not present in detail the calculations related either to Newton's second law in a general continuum or to the e.m. field. However, it summarizes the results while exposing with enough detail the basic points: e.g., the general notion of a reference fluid, the associated physical space, and the local time along a trajectory when such a reference fluid is given; Newton's second law for a test particle in a general reference fluid in a curved spacetime; and the transition from Newton's second law for a test particle to the dynamical equation satisfied by a dust continuum.

2. Assumptions used in this work

We note first that the equation for the scalar field (which is the flat-spacetime wave equation [8]) is *not* used in this work. In fact, the very existence of the flat background spacetime metric is not needed, except for the comparison with the geometrical optics i.e. the transition to the dynamics of a free photon (Subsect. 4.3).

2.1. Assumptions common with general relativity

Two assumptions of this work are common with GR:

- *a*) Our space and time measurements may be arranged so as to be described by a Lorentzian metric γ, i.e., a metric with $(+---)$ signature on a 4-dimensional manifold V (the spacetime).

- *b*) A continuous medium or a continuous field is defined by the expression of its energy-momentum tensor $\mathbf{T}$, which is a spacetime tensor depending on some state variables. The dynamical laws can be universally expressed in terms of $\mathbf{T}$, independently of the specific medium/field. To introduce this tensor, we follow Fock [10]: there is no need for a Lagrangian. (This point is explained in detail in Ref. [9]: Assumption (*b*) in Sect. 2 there.)

Moreover, several notions used in this theory can also be defined in GR:

- i) The notion of a reference fluid. This is a 3-D network $\mathcal{F}$ of *reference points,* each of them being defined by its world line. Each reference world line has to be time-like for an admissible reference fluid [11]. An admissible reference fluid $\mathcal{F}$ may thus be defined by the associated unit tangent 4-vector field on V, $\boldsymbol{U} = \boldsymbol{U}_\mathcal{F}$ (e.g. Ref. [12]): the reference world lines are the integral curves of $\boldsymbol{U}_\mathcal{F}$. For a given reference fluid $\mathcal{F}$, a system of *adapted coordinates* on V, or an adapted chart of V, is a system of spacetime coordinates x^μ $(\mu = 0, ..., 3)$ for which the vector of the spatial coordinates, $\mathbf{x} \equiv (x^i) \in \mathbb{R}^3$, is constant on any reference world line (e.g. Refs. [13, 14]). The notion of a reference fluid was thus first defined for GR, but it is not commonly used explicitly there, although coordinate systems are of course routinely used, and the data of one coordinate system does define a reference fluid (in the domain of definition of the coordinate system) — namely, the one whose reference world lines are defined to be the lines $\mathbf{x} \equiv (x^i) = \text{Constant}$.

- ii) Synchronized local time. Given an admissible reference fluid $\mathcal{F}$, consider an open curve in spacetime, defined as a one-to-one smooth mapping $C : \xi \mapsto C(\xi) = X \in \mathrm{V}$ from a real interval I into the spacetime. Up to a constant shift, we can define uniquely the *synchronized local time* on this trajectory. This is the proper time measured at the successive reference points of $\mathcal{F}$ met along the trajectory, synchronized along that trajectory according to the (Poincaré-Einstein) standard synchronization procedure in $\mathcal{F}$ [11, 15, 16]. Because in an adapted chart a reference point of $\mathcal{F}$ can be defined by the vector $\mathbf{x} \in \mathbb{R}^3$, we denote the synchronized local time by $t_{\mathbf{x}}(\xi)$ — even though in general the different values of ξ correspond with different reference points $\mathbf{x} = \mathbf{x}(\xi)$, unless the trajectory in question is a reference world line. Explicitly [11, 16]:

$$\frac{\mathrm{d}t_{\mathbf{x}}}{\mathrm{d}\xi} = \frac{\sqrt{\gamma_{00}}}{c}\left(\frac{\mathrm{d}x^0}{\mathrm{d}\xi} + \frac{\gamma_{0i}}{\gamma_{00}}\frac{\mathrm{d}x^i}{\mathrm{d}\xi}\right). \tag{1}$$

(*We are considering adapted coordinates henceforth.*) The sense in which $t_{\mathbf{x}}(\xi)$ is indeed the standardly synchronized proper time along the trajectory is explained by Landau & Lifshitz [15], after Eq. (88.9). (The line of reasoning there does not depend on the assumption of stationarity.) Note that (1) is invariant under any regular coordinate change for which the spatial coordinate change does not depend on the time coordinate [11], hence it is well defined for a given reference fluid $\mathcal{F}$ and a given curve C.

- iii) Space manifold associated with a reference fluid. This notion too can be defined in GR also but, to our knowledge, it has not been used yet. The "physical space" associated with a reference fluid $\mathcal{F}$ can be defined to be the set of the reference world lines of $\mathcal{F}$ [16]. It thus depends on the reference fluid which is considered. Any non-vanishing vector field $\boldsymbol{U}$ on V (whether $\boldsymbol{U}$ is time-like or not) defines a unique reference fluid: the one whose reference world lines are the maximal integral curves of $\boldsymbol{U}$ [17]. The set of these curves is denoted by $\mathrm{N}_{\boldsymbol{U}}$. (These curves, hence that set, are not changed if $\boldsymbol{U}$ is replaced by $\boldsymbol{U}' \equiv \lambda \boldsymbol{U}$ with λ any smooth, non-vanishing real function defined on V [18].) If $\boldsymbol{U}$ is "normal" in a precise sense, then adapted charts do exist for that reference fluid $\mathcal{F}_{\boldsymbol{U}}$ [17]. They allow one to endow the space $\mathrm{N}_{\boldsymbol{U}}$ with a canonical structure of *3-D differentiable manifold* [17]. So, given a reference fluid, we can define *spatial fields*, i.e., fields over the space manifold $\mathrm{N}_{\boldsymbol{U}}$ — either scalar, vector, or tensor fields. In particular, the spatial metric defined from $\boldsymbol{\gamma}$ by Landau & Lifshitz [15] or Møller [13] is a (0 2) spatial tensor field $\boldsymbol{g} = \boldsymbol{g}_{\boldsymbol{U}}$ (it depends on the reference fluid). It makes $\mathrm{N}_{\boldsymbol{U}}$ a Riemannian manifold — but the spatial metric $\boldsymbol{g}_{\boldsymbol{U}}$ depends on time.

- iv) Globally-synchronized reference fluid. We shall consider a special admissible reference fluid $\mathcal{E}$, with four-velocity vector field $\boldsymbol{U}_{\mathcal{E}}$, and we shall assume that it is globally synchronized. By this, we mean that there is a *global* space-time coordinate system — i.e., the mapping $X \mapsto (x^\mu)$ is defined for any event X — which is adapted to $\mathcal{E}$, and in which the components of the spacetime metric $\boldsymbol{\gamma}$ verify

$$\gamma_{0i} = 0 \quad (i = 1, 2, 3). \tag{2}$$

We note that the existence of a *local* chart (defined in a neighborhood of an arbitrary given event), in which (2) is true, is warranted in a generic spacetime (e.g. Ref. [15]). Condition (2) alone does not specify a unique reference fluid, even less a unique coordinate system. The preferred character of the reference fluid $\mathcal{E}$ assumed in the theory appears with the dynamics.

2.2. Dynamics of a test particle based on Newton's second law

- i) Newton's second law in a curved spacetime. The original dynamics assumed in this theory is a central point of the latter. We assume an extension to a general Lorentzian spacetime [16] of the special-relativistic form of Newton's second law. It may be defined for any admissible reference fluid, and it too depends on the reference fluid. It has the following form:

$$\mathbf{F} + (E/c^2)\mathbf{g} = D\mathbf{P}/Dt_{\mathbf{x}}. \tag{3}$$

In this equation, $\mathbf{F}$ is the non-gravitational force; $\mathbf{g}$ is the gravity acceleration; E is the energy of the test particle: $E = m(v)c^2$ for a mass point, with

$$m(v) \equiv m_0\gamma_v, \qquad \gamma_v \equiv 1/\sqrt{1-(v^2/c^2)}. \tag{4}$$

This is the rest-mass plus kinetic energy and it may be called the "purely material" energy of the particle, for it does not take into account the potential energy of the particle in either the gravitational field or the electromagnetic field. Further, $\mathbf{v}$ is the 3-velocity (relative to the admissible reference fluid $\mathcal{F}$ which is considered, the latter being characterized by the four-velocity field $\boldsymbol{U}$): it is measured with the synchronized local time $t_{\mathbf{x}}$, and its modulus v is defined with the space metric $\boldsymbol{g} = \boldsymbol{g}_{\boldsymbol{U}}$:

$$v^i \equiv \frac{\mathrm{d}x^i}{\mathrm{d}t_{\mathbf{x}}}, \quad v \equiv [\boldsymbol{g}(\mathbf{v},\mathbf{v})]^{1/2} = (g_{ij}v^i v^j)^{1/2}. \tag{5}$$

In the case of a synchronized reference fluid, thus if Condition (2) is satisfied, the synchronized local time (1) has a simple relation to the coordinate time $t \equiv x^0/c$ [16]:

$$\mathrm{d}t_{\mathbf{x}}/\mathrm{d}t = \sqrt{\gamma_{00}} \equiv \beta(t,\mathbf{x}). \tag{6}$$

In that case, it suggests itself to parameterize the trajectory of the particle with the coordinate time t and we have

$$v^i(t) = \frac{1}{\beta(t,\mathbf{x}(t))}\frac{\mathrm{d}x^i}{\mathrm{d}t}. \tag{7}$$

Still, $\mathbf{P} \equiv (E/c^2)\mathbf{v}$ is the momentum, and $D/Dt_{\mathbf{x}}$ is the relevant time-derivative of a vector on a curve in the space manifold $\mathrm{N}_{\boldsymbol{U}}$ endowed with the time-dependent spatial metric $\boldsymbol{g} = \boldsymbol{g}_{\boldsymbol{U}}$ [16]. It is given by

$$\frac{D\mathbf{P}}{Dt_{\mathbf{x}}} \equiv \frac{\mathrm{d}t}{\mathrm{d}t_{\mathbf{x}}}\left(\frac{D_0\mathbf{P}}{Dt} + \frac{1}{2}\mathbf{t}.\mathbf{P}\right), \quad \mathbf{t} \equiv \boldsymbol{g}^{-1}.\frac{\partial\,\boldsymbol{g}}{\partial\,t}, \tag{8}$$

where D_0/Dt is the absolute derivative relative to the space metric of the time t at which the derivative is to be calculated; explicitly:

$$\left(\frac{D_0\mathbf{P}}{Dt}\right)^i \equiv \frac{\mathrm{d}P^i}{\mathrm{d}t} + \Gamma^i_{jk}P^j\frac{\mathrm{d}x^k}{\mathrm{d}t}, \tag{9}$$

where the Γ^i_{jk}'s are the Christoffel symbols of the metric $\boldsymbol{g}$ of the time t. [The definition (8) does not need that a synchronized reference fluid is considered.]

- ii) The gravity acceleration. Newton's second law (3) is compatible with the geodesic motion of GR, provided a peculiar (velocity-dependent) form of the gravity acceleration $\mathbf{g}$ is assumed [16]. However, we assume a different, simpler form [19]:

$$\mathbf{g} = -c^2\frac{\mathrm{grad}_g\beta}{\beta}, \quad \left(\mathrm{grad}_g\beta\right)^i \equiv g^{ij}\beta_{,j} \quad \left(g^{ij}\right) \equiv \left(g_{ij}\right)^{-1}. \tag{10}$$

IOP Conf. Series: Journal of Physics: Conf. Series **845** (2017) 012014 doi:10.1088/1742-6596/845/1/012014

(Recall $\beta \equiv \sqrt{\gamma_{00}}$.) This form came out from a mechanism/interpretation of gravity as a pressure force [1, 6]. It can also be *derived* by demanding that (i) the metric field $\boldsymbol{\gamma}$ should be a spatial potential for a space vector $\mathbf{g}$, and (ii) the law of motion (3) should imply geodesic motion for "free" test particles in a *static* metric [16]. The form (10) is valid only in the preferred reference fluid $\mathcal{E}$. It is stable under the coordinate transformations (either global or local) that are internal to $\mathcal{E}$ and preserve the synchronization condition (2). The general form of these coordinate changes is [16]:

$$x'^0 = \varphi(x^0), \quad x'^i = \psi^i(x^1, x^2, x^3). \tag{11}$$

- iii) Four-acceleration of a mass test particle. This is defined to be the absolute derivative of the four-velocity, relative to the spacetime metric (e.g. [15]). Using Newton's second law (3) with the gravity acceleration (10) and the explicit definition (8) of the derivative $D\mathbf{P}/Dt_{\mathbf{x}}$, we can compute [9]

$$A^0 = \frac{1}{2\beta^2}\, g_{jk,0}\, U^j\, U^k + \frac{\gamma_v}{\beta}\frac{\mathbf{F.v}}{m_0\, c^3}, \qquad A^i = \frac{1}{2}\, g^{ij} g_{jk,0}\, U^0\, U^k + \gamma_v \frac{F^i}{m_0\, c^2}. \tag{12}$$

Here $\mathbf{U}$, with components U^μ, is the 4-velocity of the mass point. (This 4-velocity $\mathbf{U}$ should not be confused with $\boldsymbol{U}$, that is the 4-velocity field of the preferred reference fluid.)

3. Continuum dynamics

3.1. Four-acceleration and dynamical equation for a dust continuum

"Dust" is defined to be a continuum made of coherently moving, non-interacting particles, each of which conserves its rest mass. Thus, Newton's second law (3) applies to each individual particle which constitutes that continuum. The energy-momentum tensor for the dust continuum is [10]:

$$T^{\mu\nu} = \rho^*\, c^2\, U^\mu U^\nu, \tag{13}$$

where ρ^* is the rest-mass density field in the rest frame of the continuum. (The U^μ's are now the components of the 4-velocity field of the dust continuum.) Mass conservation writes: $(\rho^* U^\nu)_{;\nu} = 0$ [10]. With (13), we get thus:

$$T^{\mu\nu}_{\ \ ;\nu} = \rho^* c^2\, U^\mu_{\ ;\nu}\, U^\nu = \rho^* c^2\, A^\mu. \tag{14}$$

The four-acceleration (12) deduced from Newton's second law (3) is valid, as is the latter, for the constitutive particles of the dust. To write this for the dust when it is seen as a continuum, we substitute into (12) $\mathbf{F} = \mathbf{f}\delta V$ and $m_0 = \rho_0 \delta V$, with $\mathbf{f}$ and ρ_0 the volume density fields of the non-gravitational external force and the rest-mass, respectively. ($\delta V \equiv \sqrt{g}\, \mathrm{d}x^1 \mathrm{d}x^2 \mathrm{d}x^3$, with $g \equiv \det(g_{ij})$, is the volume element of the continuum, as measured from the reference fluid $\mathcal{E}$.) Inserting this modified version of (12) into (14), and accounting for the relation $\rho_0 = \gamma_v \rho^*$ [15, 19], we get:

$$T^{0\nu}_{\ \ ;\nu} = b^0(\mathbf{T}) + \frac{\mathbf{f.v}}{c\beta}, \qquad T^{i\nu}_{\ \ ;\nu} = b^i(\mathbf{T}) + f^i, \tag{15}$$

where

$$b^0(\mathbf{T}) \equiv \frac{1}{2\beta^2}\, g_{ij,0}\, T^{ij}, \quad b^i(\mathbf{T}) \equiv \frac{1}{2}\, g^{ij} g_{jk,0}\, T^{0k}. \tag{16}$$

3.2. Induction from a dust to a general continuum

Dynamical equation for a general continuum: We assume that the same dynamical equation for the **T** tensor should apply to any kind of continuum. This expresses the mass-energy equivalence and the universality of the gravitation force in the framework of Assumption (*b*). Thus, we assume that the dynamical equation derived for a dust in the presence of a field of external force density **f**, Eq. (15) with (16), is true for a general continuum.

3.3. Newton's second law for a general continuum

Instead of fully inducing it from a dust, the dynamical equation for a general continuum may also be obtained by writing Newton's second law for a volume element of the general continuum. The induction from a dust is then used only to define the "purely material" energy content of a volume element: for a dust, this is unambiguously defined as $\delta E \equiv \delta m_0 \gamma_v c^2 = \rho_0 \gamma_v \, c^2 \delta V$, and one finds easily that this is equal to $T^0{}_0 \delta V$. The latter expression is thus defined to be the relevant energy to write Newton's second law for a volume element of a general continuum. When writing this, the volume density $\mathbf{f}'$ of the non-gravitational force must now include the volume density of the *internal* forces in the continuum, added to the volume density **f** of the external forces. (Only external forces act on a dust, by definition.) The equation of motion thus obtained coincides with $(15)_2$ for a dust, i.e., when $\mathbf{f}' = \mathbf{f}$ [9]. (This latter fact is proved by using the bimetric nature of the theory.)

4. Maxwell equations in a gravitational field

4.1. First group, Lorentz force

- i) Field tensor and first group of Maxwell equations. As usual, we assume that the field tensor $\boldsymbol{F}$ derives from a 4-potential **A**:

$$F_{\mu\nu} \equiv A_{\nu,\mu} - A_{\mu,\nu} = A_{\nu;\mu} - A_{\mu;\nu}. \tag{17}$$

This is equivalent to assuming:

  - that $\boldsymbol{F}$ is antisymmetric ($F_{\mu\nu} = -F_{\nu\mu}$);
  - and that the first group of the Maxwell equations is satisfied:

$$F_{\lambda\mu\,,\nu} + F_{\mu\nu,\lambda} + F_{\nu\lambda,\mu} = F_{\lambda\mu\,;\nu} + F_{\mu\nu;\lambda} + F_{\nu\lambda;\mu} = 0. \tag{18}$$

- ii) Lorentz force in a gravitational field. The expression of the Lorentz force should:
  - be a space vector, invariant by the transformations $x'^0 = \varphi(x^0), \quad x'^i = \psi^i(x^1, x^2, x^3)$;
  - reduce to that valid in SR, when the gravitational field vanishes.

This leads straightforwardly [9] to

$$\mathbf{F} = q\left(\mathbf{E} + \mathbf{v} \wedge \frac{\mathbf{B}}{c}\right), \qquad (\mathbf{a} \wedge \mathbf{b})^i \equiv e^i{}_{jk}\, a^j\, b^k, \tag{19}$$

where e_{ijk} is the usual antisymmetric spatial tensor, whose indices are raised or lowered using the spatial metric $\boldsymbol{g}$ in the preferred reference fluid $\mathcal{E}$. We have $e_{ijk} = \sqrt{g}\, \varepsilon_{ijk}$ in spatial coordinate systems whose natural basis is direct, ε_{ijk} being the signature of the permutation $(i\,j\,k)$. In Eq. (19), **E** and **B** are the electric and magnetic spatial vector fields in the reference fluid $\mathcal{E}$, with components

$$E^i \equiv \frac{F^i{}_0}{\beta}, \quad B^k \equiv -\frac{1}{2} e^{ijk} F_{ij}. \tag{20}$$

IOP Conf. Series: Journal of Physics: Conf. Series **845** (2017) 012014 doi:10.1088/1742-6596/845/1/012014

4.2. Second group of Maxwell equations in a gravitational field

- i) Equation for the energy-momentum tensor of the electromagnetic field. Let $\mathbf{T}_{\text{field}}$ be the energy-momentum tensor of the e.m. field [10]:

$$T^{\mu\nu}_{\text{field}} \equiv \left(-F^{\mu}{}_{\lambda} F^{\nu\lambda} + \frac{1}{4}\gamma^{\mu\nu} F_{\lambda\rho} F^{\lambda\rho} \right) / 4\pi. \tag{21}$$

The *total* energy-momentum is $\mathbf{T} = \mathbf{T}_{\text{charged medium}} + \mathbf{T}_{\text{field}}$. Since we assume that Eq. (15) is universal, $\mathbf{T}$ obeys Eq. (15) for continuum dynamics, without any non-gravitational external force. The charged medium obeys Eq. (15), with the non-gravitational force being the Lorentz force (19) — more precisely, with the Lorentz force volume density $\mathbf{f} \equiv \delta\mathbf{F}/\delta V$, got by substituting $\delta q = \rho_{\text{el}} \delta V$ for q into (19), where ρ_{el} is the volume density of electric charge in the preferred reference fluid. Combining the two using their linearity, we get:

$$T^{0\nu}_{\text{field}\ ;\nu} = b^0(\mathbf{T}_{\text{field}}) - \frac{\mathbf{f.v}}{c\beta}, \qquad T^{i\nu}_{\text{field}\ ;\nu} = b^i(\mathbf{T}_{\text{field}}) - f^i. \tag{22}$$

That is, **the e.m. field behaves as a material continuum subjected to the gravitation and to the opposite of the Lorentz force**.

- ii) Gravitationally-modified second group. Equation (22), after a short algebra, can be shown [9] to be equivalent to the following *gravitationally-modified second group:*

$$F^{\mu}{}_{\lambda} F^{\lambda\nu}{}_{;\nu} = 4\pi b^{\mu} \left(\mathbf{T}_{\text{field}}\right) - 4\pi F^{\mu}{}_{\lambda} \frac{J^{\lambda}}{c}. \tag{23}$$

Our main comments about this new set of equations are as follows:

  - If the gravitational field is time-independent in the preferred reference fluid, we have $b^{\mu} = 0$ from (16). If moreover the matrix $(F^{\mu}{}_{\nu})$ is invertible, then (23) hence reduces to the gravitationally-modified Maxwell equations in metric theories of gravity:

$$F^{\mu\nu}{}_{;\nu} = -4\pi \frac{J^{\mu}}{c}. \tag{24}$$

  - When is the matrix $(F^{\mu}{}_{\nu})$ invertible? It is easy to check that $\det(F^{\mu}{}_{\nu}) \neq 0$ is equivalent to $\mathbf{E.B} \equiv \boldsymbol{g}(\mathbf{E},\mathbf{B}) \neq 0$, i.e., one invariant of the field does not vanish. This is true at a generic point for a generic e.m. field, but it is not true for the simplest examples of such fields: purely electric fields, purely magnetic fields, and "simple e.m. waves".

  - In a variable gravitational field such that $b^{\mu} \neq 0$, and if the matrix $(F^{\mu}{}_{\nu})$ is invertible, Eq. (23) may lead to **macroscopic charge production/destruction**:

$$\hat{\rho} \equiv (J^{\mu})_{;\mu} = c\left((\boldsymbol{F}^{-1})^{\mu}{}_{\nu}\, b^{\nu}(\mathbf{T}_{\text{field}})\right)_{;\mu}. \tag{25}$$

It had previously been found [20, 8] that, in a variable gravitational field, this theory predicts macroscopic matter production/destruction in general (i.e. not especially for a charged medium: e.g. for a perfect fluid). It had also been easily shown that the amounts are extremely tenuous in usual situations, hence are compatible with the experimental evidence on mass conservation. In contrast with this, it is difficult to assess the amounts of charge production/destruction that are predicted in realistic situations. We hope to be able to give estimates soon. By integrating (25) in a spatial domain one finds that the time variation of the charge in that domain happens to be expressible as a surface integral on the boundary of that domain, and that, if the charged system is isolated, then the total charge is conserved. Note that this macroscopic theory does not say which elementary processes would act at the particles' scale. If this phenomenon turns out to be real, it should have some implications for the magnetic fields of the astronomical objects, which currently are not fully understood.

4.3. Consistency with trajectories of photons
The photon trajectories are defined by Newton's second law (3), with $E = h\nu \quad (\nu \equiv \mathrm{d}n/\mathrm{d}t_{\mathbf{x}}$ is the frequency of the e.m. wave as measured with the local time of an observer of the reference fluid, n being the number of periods), with zero non-gravitational force $\mathbf{F}$. Therefore, to make the link with the equations for the e.m. field, we have to consider a "dust of photons". I.e., the energy-momentum tensor of the e.m. field should have the following form:

$$T^{\mu\nu}_{\text{field}} = V^\mu V^\nu. \tag{26}$$

This happens to mean exactly that $\boldsymbol{F}$ is a "null" field, i.e., the two classical invariants of $\boldsymbol{F}$ are zero. For such a field, one shows (using the bimetric nature of the theory) that Maxwell's second group (23) says exactly [9]:

- that the trajectories of the e.m. energy flux, with velocity $\mathbf{u}$ defined from $\mathbf{T}_{\text{field}}$ by $T^{00}\, u^i = cT^{i0}$, are photon trajectories;
- that one has the continuous form for dust of the energy equation.

5. Conclusion
In the investigated theory of gravitation, particle dynamics is defined by an extension to curved spacetime of the relativistic form of Newton's second law, with the gravity acceleration $\mathbf{g}$ being derived from a spatial potential: Eq. (3) with (10). The equation for continuum dynamics is got by applying (3) to a dust and by postulating that the equation thus obtained, Eq. (15) with (16), is valid for a general continuum or system of fields, i.e., for a general form of the energy-momentum tensor $\mathbf{T}$.

The Maxwell equations in a gravitational field are obtained by applying the equation for continuum dynamics to the energy-momentum tensor of the e.m. field. They are consistent with photon dynamics as defined by Newton's second law. The theory predicts macroscopic charge production/destruction in a variable gravitational field. This is a dangerous point for the theory, but there may be a relation with the magnetic fields of the astronomical objects.

References
[1] Arminjon M 1993 A theory of gravity as a pressure force. I. Newtonian space and time *Rev. Roum. Sci. Techn. - Méc. Appl.* **38** 3-24
[2] Zareski D 2001 The elastic interpretation of electrodynamics *Found. Phys. Lett.* **14** 447–469
[3] Zareski D 2014 Unification of physics theories by the ether elasticity theory *Glob. J. Sci. Front. Res. A* **14** Issue 6 Version 1.0
[4] Dmitriyev V P 1992 The elastic model of physical vacuum *Mech. Solids* (N.Y.) **26** 60–71 [1992 *Izv. Akad. Nauk SSSR – Mekhanika Tverdogo Tela* **26**, 66].
[5] Dmitriyev V P 2004 Elasticity and electromagnetism *Meccanica* **39** 511-520
[6] Arminjon M 1993 A theory of gravity as a pressure force. II. Lorentz contraction and 'relativistic' effects *Rev. Roum. Sci. Techn. - Méc. Appl.* **38** 107-128
[7] Arminjon M 2004 Gravity as Archimedes' thrust and a bifurcation in that theory *Found. Phys.* **34** 1703-1724

IOP Conf. Series: Journal of Physics: Conf. Series **845** (2017) 012014 doi:10.1088/1742-6596/845/1/012014

[8] Arminjon M 2006 Space isotropy and weak equivalence principle in a scalar theory of gravity *Braz. J. Phys.* **36** 177-189

[9] Arminjon M 2016 Continuum dynamics and the electromagnetic field in the scalar ether theory of gravitation *Open Physics* **14** 395-409

[10] Fock V 1964 *The theory of space, time and gravitation* (2nd English edition), Pergamon, Oxford

[11] Cattaneo C 1958 General relativity: relative standard mass, momentum, energy and gravitational field in a general system of reference *Nuovo Cim.* **10** 318-337

[12] Rodrigues W A Jr, Capelas de Oliveira E 2007 *The many faces of Maxwell, Dirac and Einstein equations. A Clifford bundle approach* Lecture Notes in Physics No. 722, Springer International Publishing, p. 174

[13] Møller C 1952 *The theory of relativity*, Clarendon Press, Oxford

[14] Cattaneo C 1969 Problèmes d'interprétation en relativité générale, in *Fluides et Champ Gravitationnel en Relativité Générale* (Lichnerowicz A, Tonnelat M-A, eds.), pp. 227-235, Ed. du CNRS, Paris

[15] Landau L D and Lifshitz E M 1971 *The classical theory of fields* (3rd English edition), Pergamon, Oxford

[16] Arminjon M 1996 On the extension of Newton's second law to theories of gravitation in curved space-time *Arch. Mech.* **48** 551-576

[17] Arminjon M 2016 Defining the space in a general spacetime *Int. J. Geom. Meth. Mod. Phys.* **13** 1650031

[18] Calcut J S and Gompf R E 2013 Orbit spaces of gradient vector fields *Ergodic Theory and Dynam. Syst.* **33** 1732–1747

[19] Arminjon M 1996 Energy and equations of motion in a tentative theory of gravity with a privileged reference frame *Arch. Mech.* **48** 25-52

[20] Arminjon M 1998 On the possibility of matter creation/destruction in a variable gravitational field *Analele Universit. Bucureşti – Fizică* **47** 3-21

IOP Conf. Series: Journal of Physics: Conf. Series **845** (2017) 012015 doi:10.1088/1742-6596/845/1/012015

Experimental result on the propagation of Coulomb fields

R. de Sangro, G. Finocchiaro, P. Patteri, M. Piccolo, G. Pizzella

Istituto Nazionale di Fisica Nucleare, Laboratori Nazionali di Frascati

E-mail: guido.pizzella@lnf.infn.it

Abstract.

In planetary systems the problem arises whether gravity attracting the planets towards the central star has an instantaneous action or propagates with finite velocity. Laplace noticed that, if gravity propagated with finite velocity, planets motion would become unstable due to a torque originating from time lag of the gravitational interactions. Given that actions describing gravitational interaction are formally the same as that describing electrostatic interactions, we have performed an experiment meant to measure the time/space evolution of the electric field generated by an uniformly moving set of electrons. The results we obtain seem compatible with an electric field rigidly carried by the beam itself.

1. Introduction

The idea of performing an experiment to measure Coulomb fields propagation speed bears its rationale from the puzzling phenomenon of the gravitational interaction for planetary systems. In the solar system, for instance, the force attracting Earth towards the Sun is not directed where we see the Sun but where the Sun really is, that is where we shall see it about eight minutes later. In other words it is as the gravitational force propagated with infinite velocity, or, at least, with a speed much greater than the speed of light; The reason being that a time lag between actual position and position where *we see* the Sun would generate a torque which would not allow a stable orbit. This fact is well known since the Newton's time [1]. Laplace [2] calculated that the propagation velocity of the Newtonian force should have been more than a million times the velocity of light.

Such an intriguing behavior occurs in electromagnetism when the field of an electric charge moving with constant velocity is computed. One finds that in such a case the electric field at a given point $\mathbf{P}(\mathbf{x}, \mathbf{y}, \mathbf{z}, \mathbf{t})$ evaluated with the Liénard-Wiechert (L.W.) potentials is identical to that calculated by assuming that the Coulomb field travels with infinite velocity. Feynman [3] points out that a straight uniform motion continues indefinitely and that the uniform speed can be used to determine times and positions.

To verify if the Feynman interpretation of the L.W. potentials holds in case of a charge moving with constant velocity for a finite time, we have performed an experiment to measure the time evolution of the electric field produced by an electron beam in our laboratory, covering a wide range of transverse distances with respect to the beam line (up to 55 cm). The result have been published in [4] and might be interpreted as compatible with an instantaneous propagation of the Coulomb field.

IOP Conf. Series: Journal of Physics: Conf. Series **845** (2017) 012015 doi:10.1088/1742-6596/845/1/012015

In this paper in want to discuss in more detail the experimental results and interpretation.

2. Theoretical Considerations

The electric field at $\mathbf{r}(x, y, z)$ from a charge e traveling with constant velocity $\mathbf{v}$, at a time t can be written, using the Liénard-Wiechert retarded potentials as [5, 6, 7]:

$$\mathbf{E}(\mathbf{r}, t) = \frac{e}{4\pi\epsilon_o} \frac{1 - v^2/c^2}{\left(R(t') - \frac{\mathbf{R}(t')\cdot\mathbf{v}}{c}\right)^3} \left(\mathbf{R}(t') - \mathbf{v}\frac{R(t')}{c}\right), \tag{1}$$

where

$$\mathbf{R}(t') = \mathbf{r} - \mathbf{v}t' \tag{2}$$

is the distance between the moving charge and the space point where one measures the field at time t, and

$$t' = t - \frac{R(t')}{c}. \tag{3}$$

The field from a steadily moving charge can also be written (as easily deducible from eqn 1 in case of constant velocity) [3, 5, 6, 7] as

$$\mathbf{E}(t) = \frac{e}{4\pi\epsilon_o} \frac{\mathbf{R}(t)}{R(t)^3} \frac{1 - v^2/c^2}{(1 - \frac{v^2}{c^2} sin^2(\theta(t))^{\frac{3}{2}}} \tag{4}$$

where $\mathbf{R(t)}$ is the vector joining the charge position and the point at which we evaluate the e.m. field at time t (eqns. 38.8 and 38.9 of [5]) and $\theta(t)$ is the angle between $\mathbf{v}$ and $\mathbf{R(t)}$

A pictorial view of the various quantities mentioned in eqns. 1 and 4 can be seen in Fig.1.

If we indicate with y the generic transverse coordinate, using eqn.1 we can compute the maximum transverse electric field w.r.t. the direction of motion, given by ($\gamma \equiv 1/\sqrt{1 - v^2/c^2}$):

$$E_{max} = \frac{e}{4\pi\epsilon_o} \frac{\gamma}{y^2}, \tag{5}$$

a value obtained when the charge is at a distance γy at a time

$$t' = t - \frac{\gamma y}{c} \tag{6}$$

.

Using the L.W. potentials as given by eqn.1 we have calculated the field, normalized to E_{max}, generated by a relativistic electron ($E = 500\,\mathrm{MeV}$) moving along the z axis, at a transverse distance $y = 30\,\mathrm{cm}$, as shown in Fig. 2.

The sensor is located at the zero abscissa and measures the field launched by the charge when it is at the z location. We observe that the maximum value of the field, in this particular case, is obtained if the field is launched by the charge at a distance of z=300 meters away from the sensor. By the time it takes to the field to reach the sensor, traveling with velocity c, the electric charge, traveling with its own velocity smaller than c, reaches a point just under the sensor located at a transverse coordinate $y = 30$ cm. Thus, from the point of view of the observer, he does not know whether the signal is due to the field launched by the charge at a distance of 300 m or to an instantaneous field due to the charge when has reached the closest distance, as calculated with eqn.4.

Should one evaluate the field, assuming that just the last 10 m of the beam path would be *active* in launching the field itself, one then would be in the situation depicted in the lower graph of Fig.2: the response of the sensor would be smaller by a factor $\approx 10^{-4}$.

IOP Conf. Series: Journal of Physics: Conf. Series **845** (2017) 012015 doi:10.1088/1742-6596/845/1/012015

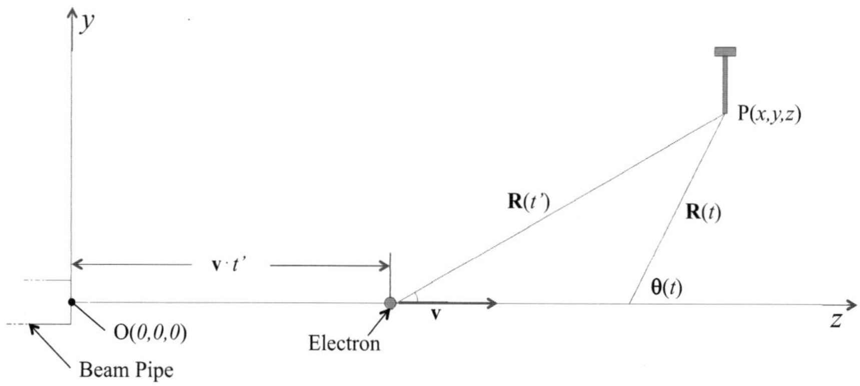

Figure 1. A pictorial view of various quantities mentioned in eqns 1 and 4

3. The Experiment

In our experiment we measure the electric field generated by the electron beam produced at the DAΦNE Beam Test Facility (BTF) [8], a beam line built and operated at the Frascati National Laboratory to produce a well-defined number of electrons (or positrons) with energies between 50 and 800 MeV. At maximum intensity the facility yields, at a 50 Hz repetition rate, 10 nsec long beams with a total charge up to several hundreds pCoulomb. The electron beam is delivered to the 7 m long experimental hall in a beam pipe of about 10 cm diameter, closed by a 40 μm Kapton window. Tests were carried out shielding the exit window with a thin copper layer, but we did not observe any change in the experimental situation. At the end of the hall a lead beam dump absorbs the beam particles. In our measurements we used 500 MeV beams of $0.5 \div 5.0 \times 10^8$ electrons/pulse.($\gamma \simeq 10^3$).

A schematic view of the experimental setup is shown in Fig. 3. At the beam pipe exit flange, electrons go through a fast toroidal transformer measuring total charge and providing redundancy on our LINAC-RF based trigger.

To measure the electric field we used as sensors 14.5 cm long, 0.5 cm diameter Copper round bars, connected to our Data Acquisition System by means of fast, terminated coax cables. Signals out of the sensors were stored on a Switched Capacitor Array(SCA) working at 5 GHz samplig frequency.

The Coulomb field acts on the sensor quasi-free electrons, generating a current. An example of the recorded signals is shown in Fig.4. The pulse shape depends on the inductance, capacitance and resistance (L, C, R) of the detectors.

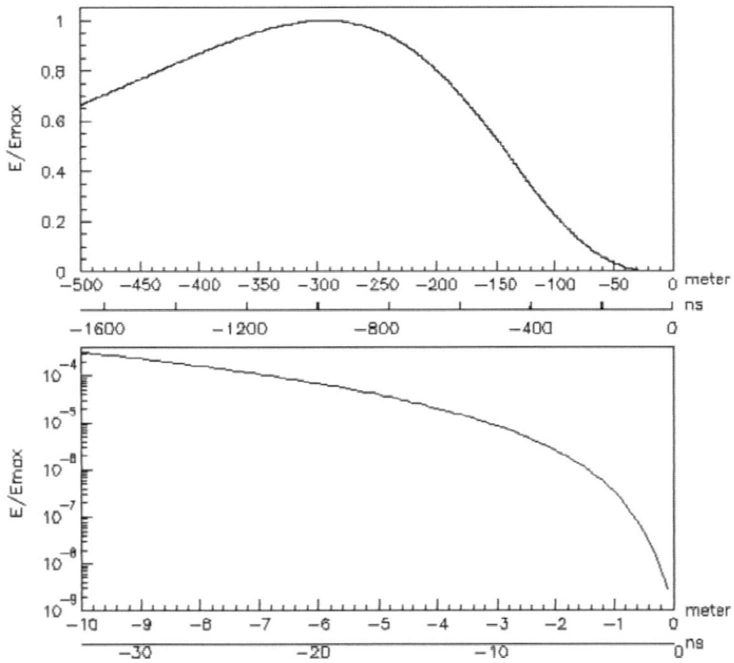

Figure 2. The electric field from eqn. 1 normalized to its maximum value, $E_y(R)/E_{max}$, generated by 500 MeV electrons as a function of z' (or t', lower abscissa scales), expected at ($z = 0\,\mathrm{cm}, y = 30\,\mathrm{cm}$). z' and t' are defined in eqns. 2 and 3. The horizontal scale of the upper graph (a) is such to include the point where $E_y(R) = E_{max}$; the lower graph (b) is a close-up of the region $z \in [-10, 0]$ m typical of our experiment (note the different vertical scales).

The sensor response $V(t)$ for a step excitation V_0 can be written as:

$$V(t) = V_o \cdot e^{-\frac{R}{2L}t} sin(\omega t). \tag{7}$$

The natural frequency of the detectors is $\approx$ 250 MHz. The voltage difference between the bar ends for the maximum value of the Coulomb field, obtained suitably modifying eqn. 5 for a finite longitudinal extent of the charge distribution, (in our case, the electron beam is $\approx$ 3 m. long) is:

$$V^t_{max} = \eta \frac{\lambda}{2\pi\epsilon_o} ln\left(\frac{y + 14.5\ cm}{y}\right), \tag{8}$$

where λ is the charge per unit length of the incoming beam and η is the sensor calibration constant. In the electric field calculations, the image charges appearing on the flange as the beam exits the pipe have also been included. However, as their effect decreases rapidly with the distance from the flange, it is completely negligible in our experiment (minimum distance $\approx$ 1 m). The sensor calibration has been carried out using a known field generated by a parallel plate capacitor. We find experimentally $\eta = 7.5 \times 10^{-2} \pm 3\%$, however due to various systematic effects we believe our calibration to be good to $\approx$ 20%, in absolute terms.

Assuming that the L.W. formula, (eqn 8), holds (which should apply only if the uniform charge motion would last indefinitely and the charges generating the field would not be shielded by conductors) we expect, in our typical beam operating conditions, pulse heights of the order of 10 mV out of our sensors.

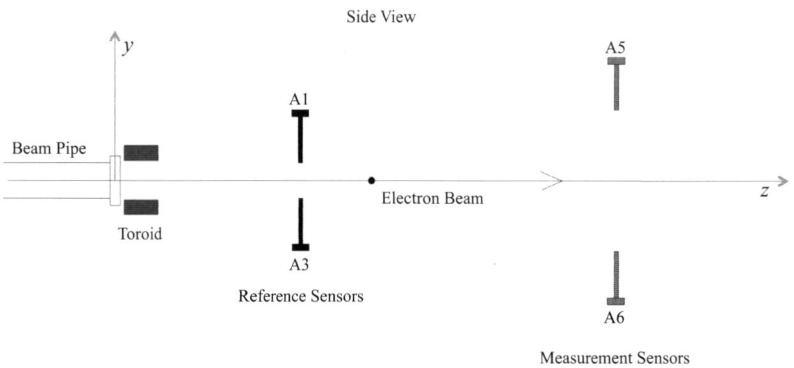

Figure 3. Schematic side of our experimental apparatus. Sensors A2, A4, orthogonal to the figure plane are not shown.

In the more realistic hypothesis that because of the beam pipe shielding and the finite lifetime for the charges uniform motion, as it is in our experiment, the expected amplitude due to the field launched from short distances, cfr. fig.2, would be of the order of few nanoVolt and hence unmeasurable.

We used six sensors: four of them, A1,A2,A3 and A4 in the following, were located at a (longitudinal) distance of 92 cm from the beam exit flange (cfr. fig 5), in a cross configuration, each at a transverse distance of 5 cm from the beam line. The main purpose of these four sensors is to provide reference for the other two detectors A5 and A6 located, through out the measurements, at various longitudinal and transverse coordinates along the beam trajectory.

4. Measurements and data base

Electron beams were delivered by BTF operators at a rate of few Hertz; data were collected in different runs, identified by given longitudinal and transverse position of the movable detectors (A5 and A6).

We collected a total of eighteen runs, spanning six transverse positions and three longitudinal positions of A5 and A6 for a total of about 15,000 triggers. Through out the data taking, the references sensors (A1,A2,A3,A4) were left at the same location (92 cm. from the beam exit flange) in order to extract a timing and amplitude reference. As mentioned before, we collected data with the movable sensor at 172 cm, 329.5 cm and 552.5 cm longitudinal distance from the beam exit flange. For each of the longitudinal positions we collected data on six transverse positions: 3, 5, 10, 20, 40 and 55 cm from the nominal beam line. For each run, A5 and A6 were positioned symmetrically with respect to the nominal beam line; spatial precision in the

IOP Conf. Series: Journal of Physics: Conf. Series **845** (2017) 012015 doi:10.1088/1742-6596/845/1/012015

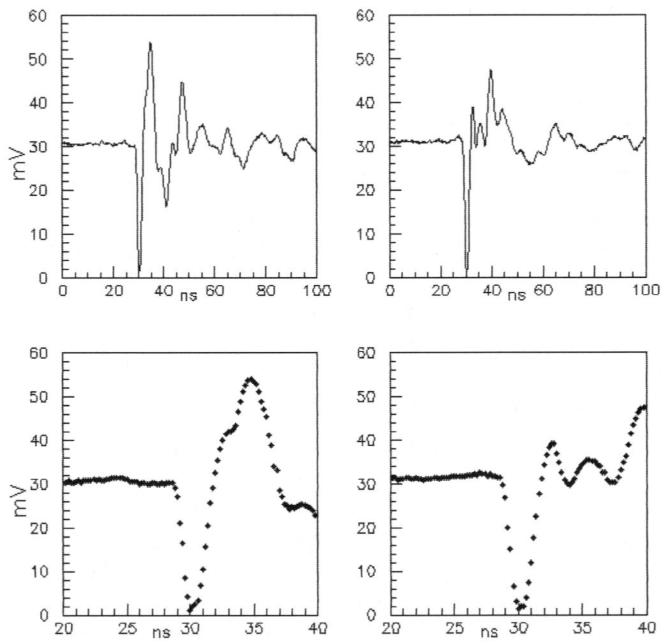

Figure 4. Typical A5 (left) and A6 (right) sensor responses. The lower plots show in detail the granularity of our time measurements.(200 psec./bin)

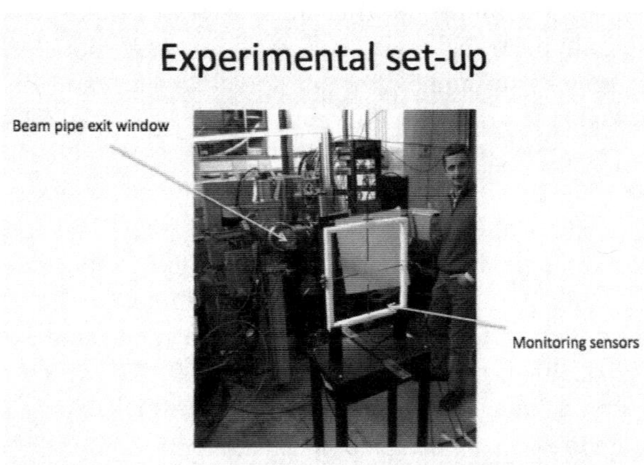

Figure 5. A photograph of the beam pipe exit window and of the four reference sensors in the cross configuration.

sensor positioning was of the order few mm in the longitudinal coordinate and about 1 mm. in the transverse one.

We define:

$$\mathbf{Sn} = \frac{V_{max} \times 10^8}{N_{elec,}}, \tag{9}$$

where V_{max} is the peak signal recorded by the SCA and $N_{elec.}$ is the total number of electrons in the beam, as measured by the fast toroid. The factor 10^8 in eqn.9 takes into account the typical beam charge. The details of the analysis have been described at length in our paper [4], so we will just show our final results.

We have made several measurements, keeping the reference sensors A1...A4 at the same location and the moving sensors A5 and A6 at different y and z locations in order to verify the response of the reference sensors in different environmental situations,

The measured values of Sn for the reference sensors 1 and 3 are Sn(1)=(8.98±0.54)mV and Sn(3)=(10.25±0.59)mV and do not depend on the different environmental situations, as expected. These figures are consistent with the potential calculated with the L.W. formula of eqn.8 on the same reference sensors, located at y=5cm: V^t_{max} = 9.78 mV, the calculated value for a charge indefinitely moving with constant speed. It is important to remark that no normalization is applied to equalize the absolute scales of the measured and calculated values. We find the agreement between measurements and prediction very remarkable.

We stress again that the amplitude we measure is many orders of magnitude higher than the one would expect from the *unshielded beam charge.* Were we sensitive only to fields generated by the electron beam once they exited the beam pipe, our pulse height would have been, as mentioned in the previous section, in the few nanoVolt range, much less than the noise of our apparatus, and then undetectable. Instead we do detect large signals in the milliVolt region, so they cannot be due to the unshielded beam charges.

In figg. 6, 7 and 8 we show the amplitude ratios between sensors A5 and A1 (A6 and A3) as a function of transverse distance from the beam line. Also in this case, data are completely consistent with the logarithmic behavior of eqn 8.

We remark that the amplitudes measured at all transverse positions of the detectors do not depend on the different longitudinal distances, showing, once more, that the field does not decreases moving far away from the beam exit pipe.

4.1. Timing measurements

Our 200 psec/chn SCA provides timing information for detector outputs, so that it is possible to measure both longitudinal and transverse position-time correlations. As a reminder we stress again that, in the hypothesis of stationary constant speed motion, no time difference is expected as a function of transverse distance, while different longitudinal positions should exhibit delays consistent with particles traveling at $\gamma \approx 1000$.

The timing performance of our detectors are shown in fig.10 where the time distribution of each sensor with respect to the beam trigger is depicted. Also for the time measurements we will have to rescale the errors yielded by standard procedures extracting central values from quasi Gaussian distributions; in this case we impose that, by symmetry, the time difference between A5 and A6 be independent of the transverse distance between detector and nominal beam line.

The upper graph of fig 10 shows the time difference relative to 172 cm longitudinall distance data. The amount of rescaling, in this case, is about a factor of 10 and the overall resolution on time difference measurements is of the order of 50 psec.

The data show no time dependence of the sensor signal on transverse distance: the reduced χ^2 for the hypothesis of a constant delay as function of y is always below 2 at each longitudinal positions. Furthermore, adding a linear term depending on transverse distance for the sensor

IOP Conf. Series: Journal of Physics: Conf. Series **845** (2017) 012015 doi:10.1088/1742-6596/845/1/012015

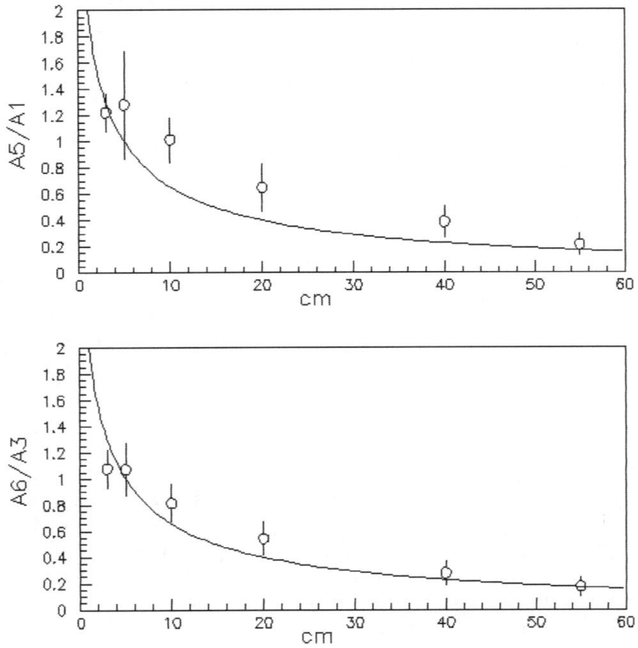

Figure 6. Upper graph: the points show the ratio $\frac{V_{max}(A5)}{V_{max}(A1)}$ at $z_{A5,A6}$=172.0 cm versus the transverse distance. Lower graph: $\frac{V_{max}(A6)}{V_{max}(A3)}$. The continuous lines represent eqn. 8 for the depicted ratios. The two reduced χ^2 are respectively 1.82 and 1.06. No fit has been performed on the data: the reduced χ^2 has been evaluated from eqn 8 and the experimental data.

time delay, the *inverse velocity* we obtain has a value smaller than $3 \times 10^{-9}\ \frac{sec}{m}$ at 95% confidence level.

We summarize the time distance correlations in Table 1, where the data obtained at the three different longitudinal positions are shown.

Table 1. Timing measurements. The expected differences are calculated for 500 MeV electrons. In the last column the average timing value between A5 and A6 (see text).

longitudinal distances between two sensors [cm]	expected [ns]	experimental $\overline{A5, A6}$ [ns]
(552.5-329.5) 223.0±1.5	7.43 ± 0.05	7.40 ± 0.06
(552.5-172.0) 380.5±1.5	12.68 ± 0.05	12.73 ± 0.09
(329.5-172.0) 157.5 ±15.	5.19 ± 0.05	5.19 ± 0.07

The time measurement reported on our original paper were criticized by A. Shabad [12]: actually one single measurement was discussed at length. The author speculated that it might infringe the speed of light boundary. The effect amounts to a bit less than three standard deviations, usually considered in the realm of fluctuations; however, if one analyzes ALL the measurements reported in table 1 of ref [4] one finds that measurements concerning sensor 5 come *early* , while the measurements concerning sensor 6 tend to be *late*. There is probably a

IOP Conf. Series: Journal of Physics: Conf. Series **845** (2017) 012015 doi:10.1088/1742-6596/845/1/012015

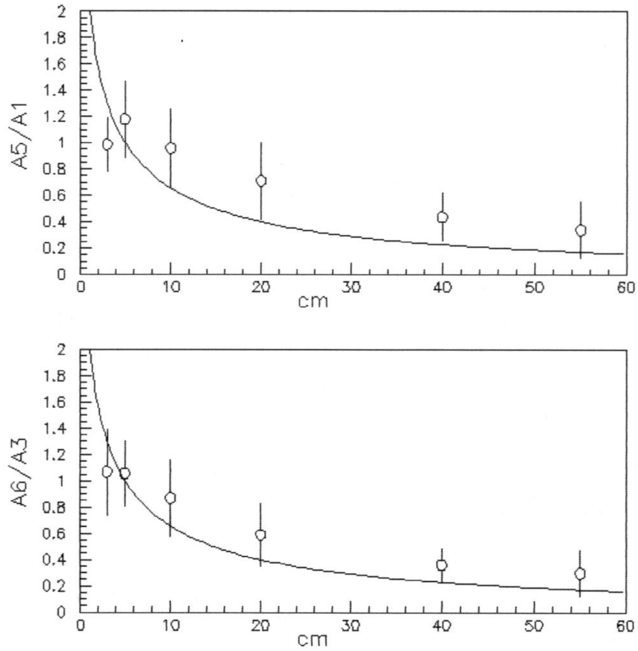

Figure 7. The same plot as in fig.6 at $z_{A5,A6}$=329.5 cm. The two reduced χ^2 are respectively 1.36 and 0.66.

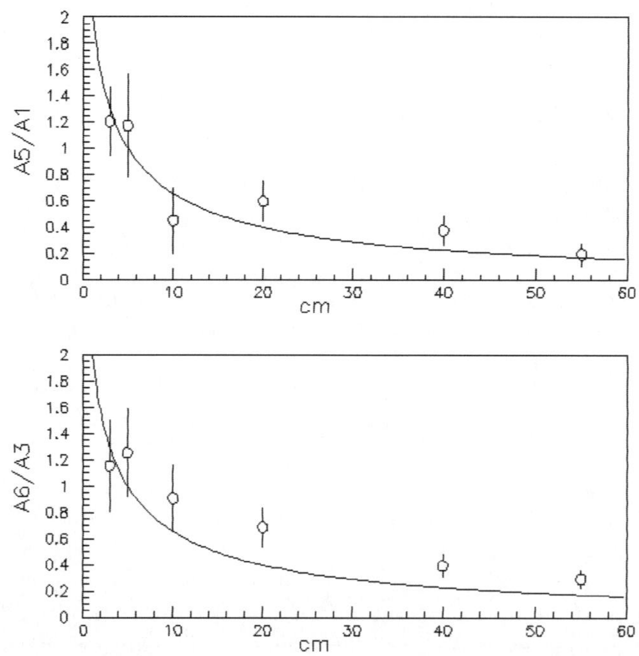

Figure 8. The same plot as in fig.6 at $z_{A5,A6}$=552.5 cm The two reduced χ^2 are respectively 0.91 and 2.48.

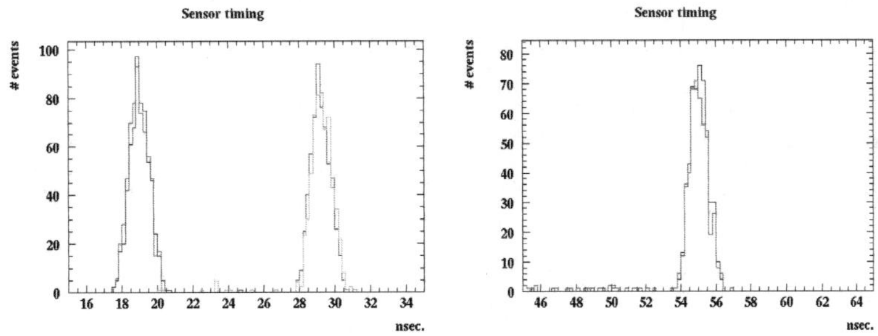

Figure 9. Time measurements accuracy. On the left hand side the four stationary sensors are shown.The time difference for the different pair is due to different cabling: a 10 nsec. delay was added to the radial sensors for calibration purposes. On the right hand side the movable sensors are shown. The widths of the distributions are remarkably similar with a σ of $\approx .5$ nsec. With the available statistics the accuracy with which we can determine the centroid of the distribution is ≈ 50 psec.

systematic effect having to do with a less than perfect flatness of the experimental hall floor wich translates in a non vertical support for the sensors: ≈ 35 mrad angle would cause the reported difference[1].

4.2. Do our sensor see E.M. radiation ?

We performed different tests in order to ascertain that E.M. radiation coming from the interaction of the electron beam with its environment was not the original cause of our sensors' response.

With the beam steering system, we changed the launch angle in the experimental hall; varying the current of the beam line magnet(s) one can predict the amplitude ratio of two detectors located right and left of the beam line, according to the calculated beam position at the sensors' longitudinal coordinates. Special runs were taken to this purpose and the results are completely consistent with the expected horizontal beam displacement w.r.t. the nominal position.

Other E.M. phenomena are related to boundary crossings: as the beam travels between different media (*e.g.* the beam exit flange) E.M. radiation can be generated which, in turn, might mimic pulses we assume due to the interaction of the beam itself with our sensors. The experimental situation can be schematized as a Tamm [9] problem: a beam of particles traveling inside the LINAC vacuum pipe, suddenly appears out of the end flange of the accelerator, moves with uniform velocity through out the experimental hall (≈ 7 m.) and disappears in the concrete wall of the hall. A calculation of the expected effect, using the formulae reported in [10] lent us confidence that this background was not relevant; however, in order to demonstrate that such a phenomenon does not contribute (or contributes very little) to our sensors' signal, we had a dedicated run during April 2014.

We collected data in two different modes:

(i) *Calibration* runs in order to match the data collected during the 2012 campaign to the latest (2014) runs.

(ii) *Beam dump* runs in which the electron beam was stopped in a 40 X_0 lead dump before reaching the vertical detectors A5 and A6.

[1] A detailed confutation of the Shabad criticism is reported in ref.[13].

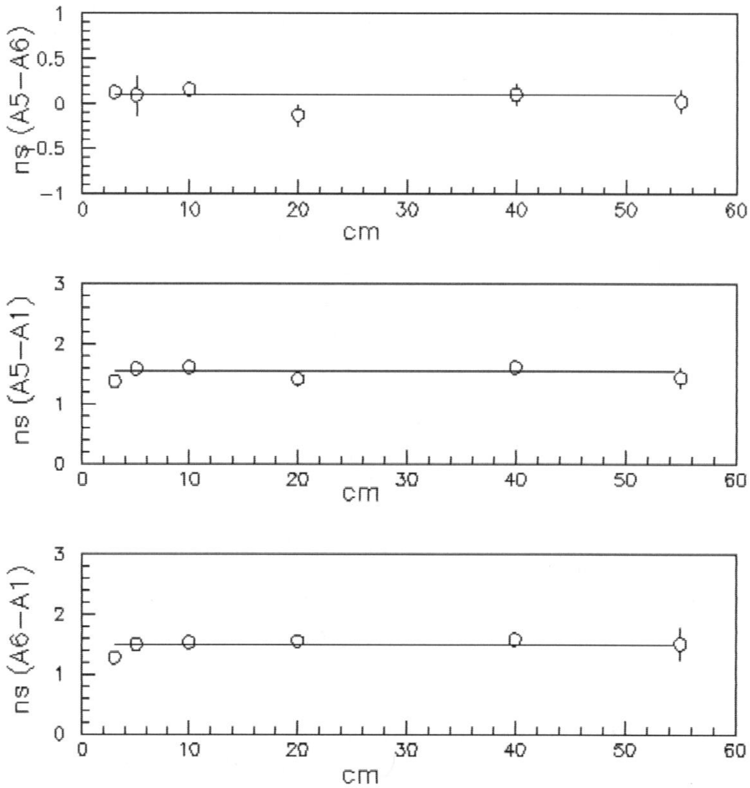

Figure 10. Top graph: time difference between A5 and A6 versus the transverse distance. Errors have been rescaled according to the procedure described in the text. Middle and bottom plots: time difference between the movable sensors A5 and A6 (respectively) and one of the fixed sensors, A1. $z_{A5,A6} = 172$ cm., $z_{A1} = 92$ cm. The line at 1.549±0.036 nsec.(middle) and 1.500±0.036 nsec (lower) indicate the weighted average of our measurements, once the reduced χ^2 is rescaled according to the procedure described in the text (par. 4.2). The values obtained for the A5-A1, A6-A1 absolute delays have to be corrected for the cables of different lengths (1.1 nsec.).

The underlying idea was that runs taken with the beam dump, would yield the response of the A5,A6 detectors in a no-beam situation thus allowing us to map the pulse height of our detectors when just backgrounds were present in the experimental hall. From fig.11 one can infer that the main features of the previous (2012) measurements are retained in the (2014) latest run; small difference in the absolute values for the given ratios can be attributed to a less than perfect alignment of the two sets of detector on the beam line.

Fig 12 shows the measurements when the 40 X_0 lead absorber is inserted between the A1...A4 and the A5,A6 sensors. The dot shows the average measurement of the two vertical sensors in the cross configuration before the beam stopper. The triangles show the average measurements made with the movable detectors A5 and A6 for different vertical positions, after the beam has been stopped. The vertical sensors responses are, with the beam dump in place, reduced by a factor ≈ 10, at 5 cm (transverse) distance, with practically no dependence on (transverse) distance from the beam line. Such behavior lends itself to the interpretation that the overall amount

IOP Conf. Series: Journal of Physics: Conf. Series **845** (2017) 012015 doi:10.1088/1742-6596/845/1/012015

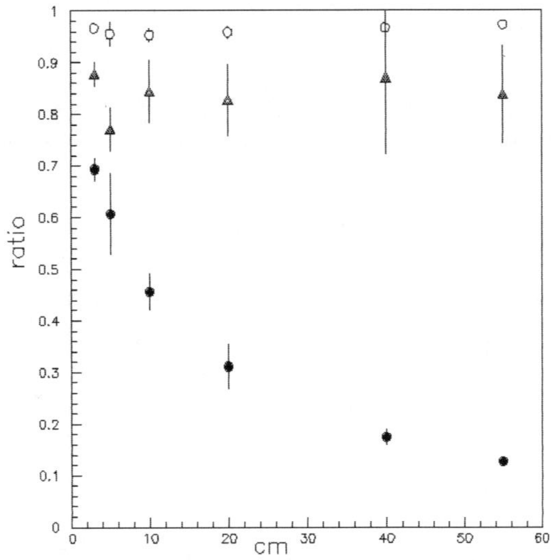

Figure 11. Amplitude ratios vs transverse distance of sensors A5,A6 for *calibration runs*: open circles: A1 A3 ratio, triangles: A5 A6 ratio, full dots: A5 A1 ratio.

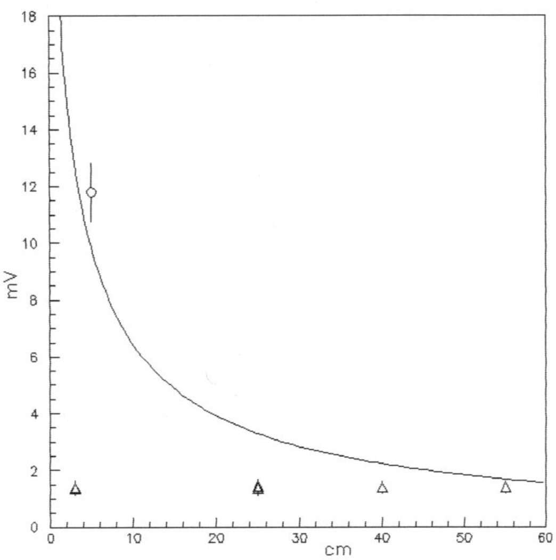

Figure 12. Measurements with the beam stopper. The open circle shows the average measurement of the two vertical sensors in the cross configuration before the beam stopper. The triangles show the average measurements made with the movable detectors A5 and A6 for different vertical positions, after the beam has been stopped.

IOP Conf. Series: Journal of Physics: Conf. Series **845** (2017) 012015 doi:10.1088/1742-6596/845/1/012015

E.M. background originating either at the transition flange or at the beam dump entrance is indeed small w.r.t. the response obtained when beams unimpeded go through the experimental hall.

Speculations have been made [12] that signals out of our sensors might come from electron beams radiation in the last bend of the experimental hall. On this count we have to point out that:

(i) no dependence has been seen of the sensor signals on the longitudinal distance from the last bend.

(ii) A complete calculation has been carried out of the total energy electron beams radiate in the last bend and, in the frequency range our sensors cover, it has been found to be completely negligible with respect to the response our detectors provide as the electron beams go through the experimental hall. A detailed report in in preparation soon to be published soon.[13].

5. Discussion

With reference to table 1, we notice that the longitudinal time differences are completely consistent with the hypothesis of a beam traveling along the z axis with a Lorentz factor $\gamma \approx 1000$.

Such a situation agrees with the Liénard-Weichert model. Retarded potentials, however, predict that most of the *virtual photons*[11] responsible for the field detected at coordinates z and y are emitted several hundred meters before the sensor positions and at different times according to the detectors transverse distances. Conversely, assuming that such virtual photons are emitted in a physically meaningful region (between the beam exit window and our detectors), the amplitude response of the sensors should be several order of magnitude smaller than what is being measured (cfr. Fig.2). Our result, obtained with a well definite set of boundary conditions (longitudinal and transverse distance between beam line and sensors, details of the beam delivery to the experimental hall etc.) matches precisely (within the experimental uncertainties) the expected value of the maximum field calculated according to L.W. theory, that is also the value calculated with Eq.4 when the beam is at the minimum distance from the sensor.

We again point out that the consistency of our measurement with eqn 8 has been obtained without any adjustment of normalization.

6. Conclusions

Assuming that the electric field of the electron beams we used would act on our sensor only after the beam itself has exited the beam pipe and given that Cerenkov and/or transition radiation effects are negligible, the L.W. model would predict sensors responses orders of magnitudes smaller than the ones we measure. The Feynman interpretation of the Liénard-Weichert formula for uniformly moving charges does not show consistency with our experimental data. Even if the steady state charge motion in our experiment lasted few tens of nanoseconds, our measurements indicate that everything behaves as if this state lasted indefinitely.

Apart from complete consistency of the sensors's amplitude with L.W. calculation for an *unshielded* beam, it is extremely difficult, in our opinion, to reconcile any ballistic theory with the results of the beam dump measurement (cfr. section 4.2). To summarize our finding in few words, one might say that our measurements are consistent with the hypothesis that the Coulomb field is carried *rigidly* by the electron beam.

We welcome new experimental work on the subject as well as new theoretical interpretations of the reported measurements.

Acknowledgments

We gratefully acknowledge Angelo Loinger for stimulating discussions, Ugo Amaldi for his interest in this research line and his advice and Francesco Ronga pointing out and discussing Cerenkov and transition radiation effects.

We stress as well the relevant contribution of our colleagues from the Frascati National Laboratory Accelerator Division and in particular of Bruno Buonomo and Giovanni Mazzitelli. Technical support of Giuseppe Mazzenga and Giuseppe Pileggi was also very valuable in the preparation and running of the experiment. A special acknowledgment is in order for Paolo Valente, who in different stages of the experiment, has provided, with his advice and ingenuity, solutions for various experimental challenges we faced.

References

[1] A.Eddington *Space, Time and Gravitation,* Harper Torchbooks, pag 94 (1959)

[2] Laplace, P., *Mechanique Celeste* , volumes published from 1799-1825, English translation reprinted by Chelsea Publ., New York (1966)

[3] R. Feynman, R.B. Leighton, M.L. Sands, *The Feynman Lectures on Physics,* Addison-Wesley, Redwood City, vol. II, Chapters 21 and 26.2 (1989)

[4] R. de Sangro, G. Finocchiaro, P. Patteri, M. Piccolo, G. Pizzella *Euro Physics Journal* **C75** 137 (2015)

[5] L.D.Landau and E.M.Lifshitz *The classical theory of fields,* pag.162, Pergamon Press, Oxford (1971)

[6] R.Becker *Teoria della elettricità,* pagg. 73-77 Sansoni Ed. Scientifiche (1950)

[7] J.D.Jackson *Classical Electrodynamics,* pagg. 654-658 John Wiley & Sons Inc. (1962,1975)

[8] A. Ghigo, G. Mazzitelli, F. Sannibale, P.Valente, G. Vignola *N.I.M.* A**515** 524-542 (2003)

[9] I.E. Tamm Jour. Phys. USSR **1** 439 1939

[10] D. Garcia-Fernandez, J. Alvarez-Muniz, W. R. Carvalho, A. Romero-Wolf and E. Zas, "Calculations of electric fields for radio detection of Ultra-High Energy particles," Phys. Rev. D **87**, 023003 (2013) [arXiv:1210.1052 [astro-ph.HE]]

[11] W. K. H.. Panofsky and M. Philips,, *Classical Electricity and Magnetism* pagg. 350-351 Addison Wesley Publishing Co.(1955, 1962)

[12] A. Shabad arXiv:1606.00710v 29 May 2016

[13] R. de Sangro, G. Finocchiaro, P. Patteri, M. Piccolo, G. Pizzella *Why the interpretation of "Measuring propagation speed of Coulomb fields" stands* arXiv:1611.06935 (2016)

IOP Conf. Series: Journal of Physics: Conf. Series **845** (2017) 012016 doi:10.1088/1742-6596/845/1/012016

Does Pizzella's experiment violate causality?

Eugene V Stefanovich
1763 Braddock Court, San Jose, CA 95125, USA
E-mail: eugene_stefanovich@usa.net

Abstract. Recent experiments, performed by Prof. Pizzella's team with relativistic electron bunches, indicate that Coulomb field is rigidly attached to the charge's instantaneous position. Despite a widespread opinion, this fact does not violate causality in moving reference frames. To see that, one should apply the Wigner – Dirac theory of relativistic dynamics and take into account that the Lorentz boost generator depends on interaction. Then one can show that interaction remains instantaneous in all reference frames.

1. Experiment at Frascati

In 2012 the team of Prof. Pizzella at the Frascati National Laboratory performed a remarkable experiment [1] that observed rigid connection between electric charge and its Coulomb field. In order to see how surprising this result is from the point of view of the traditional Maxwell–Liénard–Wiechert electrodynamics, let us turn to Fig. 1.

In three panels 1 (a)→(b)→(c) we show the sequence of events expected in the traditional theory. The snapshot 1 (a) is taken just before the electron bunch left the accelerator's pipe. In this case the bunch's field is shielded by the pipe's metal, so there is no electric field in the surrounding space.

In two frames 1 (b) - (c) the bunch has left the pipe. The emerged force field has two components. First, there is an expanding spherical electromagnetic wave (radiation field) centered at the pipe's exit. Second, there is a non-radiating bound Coulomb field, which is squeezed (or Lorentz-contracted [2]) in the direction of the bunch's motion, thus forming a narrow disk.

According to special relativity, electric field lines of the Coulomb disk cannot "leak" outside the "light cone" bounded by the expanding sphere of the electromagnetic wave. Therefore, the transverse dimension d of the disk grows with time according to formula

$$d \approx 2ct/\gamma \approx 2L/\gamma \tag{1}$$

where L is the distance traveled by the bunch from the pipe's exit and $\gamma \equiv (1-v^2/c^2)^{-1/2}$ is the usual relativistic factor. In the Frascati experiment, the energies of electrons in the beam were 500 MeV, which corresponded to the Lorentz contraction factor of $\gamma \approx 1000$. At the longest experimental travel distance of 5.5 m the expected size of the Coulomb disk was not greater than 11 mm — too small to have any effect on the field sensors placed around the beam.

In the parallel sequence of frames 1 (d)→(e)→(f) we present actual results of the Frascati experiment. In contrast to theoretical expectations, the size of the Coulomb field's disk did not

IOP Conf. Series: Journal of Physics: Conf. Series **845** (2017) 012016 doi:10.1088/1742-6596/845/1/012016

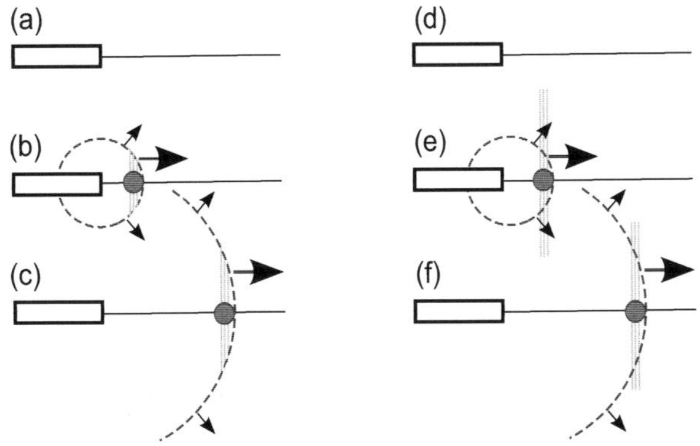

Figure 1. Time evolution of the electric field of an electron bunch (small circle) leaving the accelerator pipe (rectangle): (a)→(b)→(c) Maxwell–Liénard–Wiechert theory; (d)→(e)→(f) Frascati experiment. Broken line circle fragments show the expanding spherical electromagnetic wave. Groups of lines perpendicular to the beam's axis represent the Coulomb field disk viewed from its side.

grow in a linear fashion (1). Instead, measurements suggested that the field's disk emerged from the accelerator fully formed in the entire space (even beyond the light cone) and did not change with time, apart from the uniform movement together with the bunch.

This fact can be interpreted as an indication of an instantaneous force between charges in the bunch and in the field sensors. The idea about electromagnetic interactions being composed of both instantaneous (bound) and retarded (radiation) parts is not new. It was repeatedly expressed theoretically [3, 4, 5], and electromagnetic superluminal effects were seen in experiments as well [6, 7, 8]. However, these ideas and experiments are usually met with scepticism, because they violate the special relativistic ban on faster-than-light propagation of signals.

2. Ban of superluminal signals

To understand why special relativity does not tolerate such superluminal effects, consider the time evolution of the instantaneous Coulomb field shown in three consecutive snapshots in Fig. 2 (a)→(b)→(c).[1] Frame 2 (a) is taken just before the bunch has left the accelerator's pipe, frame 2 (b) shows the exact instant of the exit, and 2 (c) is taken a moment after this event. This time, for clarity, we chose not to show the spherical electromagnetic waves, because according to the experiment [1], their signals are weaker by an order of magnitude. We also placed two electric field sensors next to the pipe's exit.

From the point of view of the observer at rest O, the field's disk forms instantaneously (panel 2 (b)), so the three events A (the bunch's exit from the pipe), B and C (clicks of the sensors) occur simultaneously, despite the fact that B and C are caused by the event A and separated from it by a considerable distance.[2]

In three frames 2 (d)→(e)→(f) we show how this situation looks from the point of view of an inertial observer O' moving with high velocity in the vertical direction from the bottom of the page to its top.[3] To draw these panels we employed Lorentz transformations of special relativity

$$t' = t\cosh\theta - (y/c)\sinh\theta \quad (2)$$
$$x' = x \quad (3)$$
$$y' = y\cosh\theta - ct\sinh\theta \quad (4)$$
$$z' = z \quad (5)$$

[1] they are basically the same as in Fig. 1(d)→(e)→(f)
[2] from 3 cm to 55 cm in the Frascati setup
[3] To be consistent with [1], we call it y-direction.

IOP Conf. Series: Journal of Physics: Conf. Series **845** (2017) 012016 doi:10.1088/1742-6596/845/1/012016

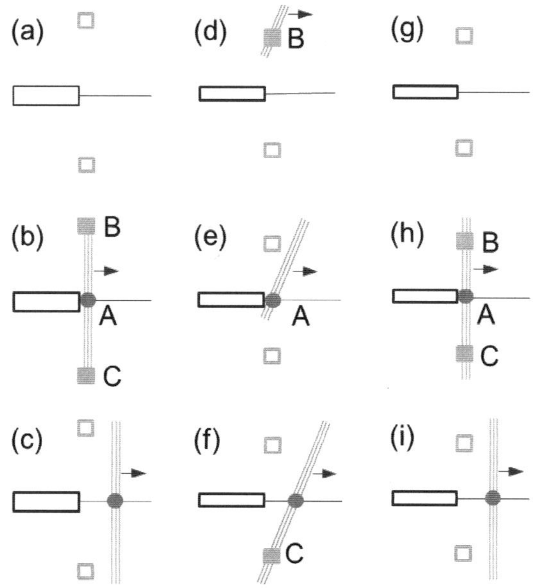

Figure 2. Time evolution of the bunch's instantaneous Coulomb field: (a)→(b)→(c) in the rest frame; (d)→(e)→(f) in the moving frame, as predicted by Lorentz transformations (2) – (5); (g)→(h)→(i) in the moving frame, as predicted by the Wigner–Dirac theory. Squares mark positions of electric field sensors. Filled squares indicate "clicking" sensors.

where parameter θ was related to the observer's velocity by formula $v = c \tanh \theta$.

The absurdity of these drawings is clear already from the panel 2 (d), where a portion of the Coulomb field's disk has emerged even before the bunch has left the accelerator's pipe. This means that the "effect" B occurred earlier than the "cause" A in a clear violation of the principle of causality. Two other panels show the further sequence of events: the bunch leaves the accelerator 2 (e), the lower sensor clicks 2 (f).

To avoid such blatant violations of causality, special relativity forbids faster-than-light formation of the field's disk. However, in a clear disrespect of this ban, the Frascati experiment seems to show exactly such a behavior in the laboratory. We are not in a position to judge the validity of conclusions drawn by the Frascati team. Right now the most important next step would be to double check their results by repeating these remarkable measurements, preferably in a different laboratory.[4] Our goal here is more modest: We are going to show that there is no inherent contradiction between instantaneously propagating interactions and the principle of causality.

3. Relativistic interactions between particles

Our claim is that charges in the bunch and in field sensors may, indeed, interact via an instantaneous action-at-a-distance potential. The usual attitude is that such potentials are impossible because (i) they cannot be relativistically invariant and (ii) they violate causality. We will challenge the latter statement in section 4. Here we would like to mention that the former statement is not correct as well.

First explicit construction of a relativistically invariant multiparticle model with instantaneous forces was undertaken by Bakamjian and Thomas [11]. They used the theory of unitary representations of the Poincaré group developed earlier by Wigner [12] and Dirac [13]. According to this theory [14], interactions must modify generators of the Poincaré group representation, so that the commutators of generators remain intact. The full Hamiltonian (=

[4] Hopefully, this will answer all outstanding questions regarding the interpretation of the Frascati experiment [9, 10].

the generator of time translations) is obtained by adding the "potential energy" operator V to the free Hamiltonian: $\mathrm{H} = \mathrm{H}_0 + \mathrm{V}$. In Dirac's instant form of relativistic dynamics, theory is consistent with adding an interacting "potential boost" $\mathbf{Z}$ to the noninteracting boost generator: $\mathbf{K} = \mathbf{K}_0 + \mathbf{Z}$. Generators of space translations and rotations remain interaction-free: $\mathbf{P} = \mathbf{P}_0$, $\mathbf{J} = \mathbf{J}_0$.

If this construction is performed in a two-particle system,[5] then the Hamiltonian H is a function of positions $\mathbf{r}_j$ and momenta $\mathbf{p}_j$ of both particles. Time dependencies of these observables are obtained by standard quantum mechanical formulas

$$\begin{aligned} \mathbf{r}_j(t) &= e^{\frac{i}{\hbar}\mathrm{H}t}\mathbf{r}_j e^{-\frac{i}{\hbar}\mathrm{H}t} \\ \mathbf{p}_j(t) &= e^{\frac{i}{\hbar}\mathrm{H}t}\mathbf{p}_j e^{-\frac{i}{\hbar}\mathrm{H}t} \end{aligned} \tag{6}$$

where $j = 1, 2$ is particle label. These expressions can be viewed either as quantum mechanical time-dependent operators in the Heisenberg picture or as particle trajectories in the $\hbar \to 0$ classical limit. In the latter case one should replace quantum commutators with classical Poisson brackets.

By its usual (non-covariant) definition, the force acting on the particle 2 is given by the time derivative of this particle's momentum

$$\mathbf{f}_2(t) \equiv \frac{d}{dt}\mathbf{p}_2(t) = -\frac{i}{\hbar}[\mathbf{p}_2(t), \mathrm{H}]$$

This is a function that depends on positions and momenta of both particles at the same time instant t

$$\mathbf{f}_2(t) = \mathbf{f}_2(\mathbf{r}_1(t), \mathbf{p}_1(t); \mathbf{r}_2(t), \mathbf{p}_2(t)) \tag{7}$$

This demonstrates that interaction propagates instantaneously in the reference frame O at rest.[6]

4. Causality

Now consider the above two-particle system from the point of view of the moving reference frame O'. Particle trajectories in this frame are[7]

$$\begin{aligned} \mathbf{r}_j(\theta, t') &= e^{-\frac{ic}{\hbar}\mathrm{K}_y\theta} e^{\frac{i}{\hbar}\mathrm{H}t'}\mathbf{r}_j e^{-\frac{i}{\hbar}\mathrm{H}t'} e^{\frac{ic}{\hbar}\mathrm{K}_y\theta} \\ \mathbf{p}_j(\theta, t') &= e^{-\frac{ic}{\hbar}\mathrm{K}_y\theta} e^{\frac{i}{\hbar}\mathrm{H}t'}\mathbf{p}_j e^{-\frac{i}{\hbar}\mathrm{H}t'} e^{\frac{ic}{\hbar}\mathrm{K}_y\theta} \end{aligned} \tag{8}$$

The Hamiltonian in the reference frame O' is $\mathrm{H}(\theta) = \exp\left(-\frac{ic}{\hbar}\mathrm{K}_y\theta\right)\mathrm{H}\exp\left(\frac{ic}{\hbar}\mathrm{K}_y\theta\right)$. Therefore, the force acting on the particle 2

[5] For example, we can assume that charge 1 represents an electron in the bunch and charge 2 is a part of the electric field sensor. Of course, replacing Pizzella's experimental setup by a couple of charges requires a significant stretch of imagination, but we believe that this model captures essential features of the involved interaction.

[6] Indeed, in our pure 2-particle system there is no place for fields or virtual particles that could serve as "interaction carriers" and store momentum and energy travelling from one charge to the other. Thus momentum-energy lost by one charge should be immediately picked up by the other, so there can be no retardation in our model.

[7] Here t' is time measured by the clock of the observer O'.

$$\begin{aligned}
\mathbf{f}_2'(t') &\equiv \frac{d}{dt'}\mathbf{p}_2(\theta,t') = -\frac{i}{\hbar}\left[\mathbf{p}_2(\theta,t'),\mathrm{H}(\theta)\right] \\
&= -\frac{i}{\hbar}\left[e^{-\frac{ic}{\hbar}\mathrm{K_y}\theta}e^{\frac{i}{\hbar}\mathrm{Ht}'}\mathbf{p}_2 e^{-\frac{i}{\hbar}\mathrm{Ht}'}e^{\frac{ic}{\hbar}\mathrm{K_y}\theta}, e^{-\frac{ic}{\hbar}\mathrm{K_y}c\theta}\mathrm{H}e^{\frac{ic}{\hbar}\mathrm{K_y}\theta}\right] \\
&= -\frac{i}{\hbar}e^{-\frac{ic}{\hbar}\mathrm{K_y}\theta}\left[e^{\frac{i}{\hbar}\mathrm{Ht}'}\mathbf{p}_2 e^{-\frac{i}{\hbar}\mathrm{Ht}'},\mathrm{H}\right]e^{\frac{ic}{\hbar}\mathrm{K_y}\theta} \\
&= -\frac{i}{\hbar}e^{-\frac{ic}{\hbar}\mathrm{K_y}\theta}\left[\mathbf{p}_2(0,t'),\mathrm{H}\right]e^{\frac{ic}{\hbar}\mathrm{K_y}\theta} \\
&= e^{-\frac{ic}{\hbar}\mathrm{K_y}\theta}\mathbf{f}_2(0,t')e^{\frac{ic}{\hbar}\mathrm{K_y}\theta} \\
&= e^{-\frac{ic}{\hbar}\mathrm{K_y}\theta}\mathbf{f}_2\left(\mathbf{r}_1(0,t'),\mathbf{p}_1(0,t');\mathbf{r}_2(0,t'),\mathbf{p}_2(0,t')\right)e^{\frac{ic}{\hbar}\mathrm{K_y}\theta} \\
&= \mathbf{f}_2(\mathbf{r}_1(\theta,t'),\mathbf{p}_1(\theta,t');\mathbf{r}_2(\theta,t'),\mathbf{p}_2(\theta,t')) \qquad (9)
\end{aligned}$$

is a function of positions and momenta of both particles at the same time instant t'. Thus, from the point of view of the moving observer O', the interaction propagates instantaneously, exactly as for the observer at rest O. Moreover, in agreement with the principle of relativity, the force function $\mathbf{f}_2$ in (9) has the same form as in the rest frame (7).

In regard to the Frascati experiment, the above derivation means that clicks of the sensors (events B and C) in the frame O' must occur exactly at the time of the bunch's exit from the pipe (event A). This prediction of the Wigner – Dirac theory is shown in the sequence of snapshots 2 (g)→(h)→(i). In other words, in this theory the three events A, B, C remain simultaneous in all frames, the "effects" B and C never occur before the "cause" A, and instantaneous potentials do not violate the principle of causality.

5. Discussion

The above discussion implies that in the presence of interaction ($V \neq 0$, $\mathbf{Z} \neq 0$), particle trajectories in the moving frame (8) cannot be obtained from rest-frame trajectories (6) by applying Lorentz formulas (2) - (5). One can say that interacting Wigner – Dirac theory does not have "invariant worldlines" [15, 16]. This idea is highly controversial, and many theoreticians believe that the invariant worldline condition must be respected in all dynamical theories. Based on this belief, a multitude of approaches were formulated [17, 18, 19, 20, 21, 22, 23, 24, 25], which deviated from the relativistic Hamiltonian dynamics of Wigner and Dirac. So far, the predictive power of these approaches remains rather limited.

Another idea is that particle equations of motion (including boost transformations) should be extracted from our most successful theory – quantum electrodynamics (QED). At the first sight, it seems that Maxwell's equations, as well as Liénard–Wiechert retarded potentials, must follow from the QED field Lagrangian in the classical limit. However, this conclusion is not so obvious, because time evolution is an ill-defined concept in the theory formulated in terms of fictitious bare particles. Moreover, the Hamiltonian of QED contains divergent renormalization counterterms. These interpretational difficulties are conducive to opinions that observables (positions and momenta) of interacting particles have no meaning in QED [26] or even that *"there are no particles, there are only fields"* [27]. For more works on the same theme see [28, 29, 30, 31].

A somewhat different and more pragmatic attitude is taken in research programs that try to replace the field-based language of quantum field theory with ideas of physical particles interacting through effective relativistic potentials. One way to obtain such potentials is to fit them to the known and experimentally verified S-matrix of the renormalized QED. This can be done not only in the lowest second perturbation order, resulting in the classic Coulomb–Darwin–Breit potential [26, 32], but also taking into account radiative corrections [33, 34, 35, 36]. Another way forward is to apply a unitary dressing transformation [37, 38, 39] to the QED Hamiltonian. This method has the added advantage that it explicitly preserves the relativistic invariance of the theory [40].

The resulting effective electromagnetic interactions between dressed charges separate into radiation and bound types [39]. As expected, the radiation force field (like radio signals) is transmitted by real photons and propagates with the speed of light [41, 42, 43]. However, this is not true for the bound (Coulomb and magnetic) potentials. Even in high perturbation orders they are expressed by functions that depend on instantaneous positions and momenta of the charges. These potentials are well suited to describe most experimental observations in classical electrodynamics [39], including the electron beam's field dynamics observed at Frascati. Moreover, these potentials are in full agreement with the principle of causality.

Acknowledgments
The author is thankful to Rainer Grobe, Guido Pizzella, and Charles Su for illuminating discussions.

References
[1] de Sangro R, Finocchiaro G, Patteri P, Piccolo M and Pizzella G 2015 Measuring propagation speed of Coulomb fields *Eur. Phys. J.* C **75** 137
[2] Jackson J D 1999 *Classical Electrodynamics* (New York: Wiley)
[3] Chubykalo A E and Vlaev S J Necessity of simultaneous co-existence of instantaneous and retarded interactions in classical electrodynamics 1999 *Int. J. Mod. Phys.* A **14** 3789
[4] Kholmetskii A L 2006 Momentum-energy of the non-radiating electromagnetic field: open problems? *Phys. Scr.* **73** 620
[5] Field J H 2006 Classical electromagnetism as a consequence of Coulomb's law, special relativity and Hamilton's principle and its relationship to quantum electrodynamics *Phys. Scr.* **74** 702
[6] Giakos G C and Ishii T K 1991 Anomalous microwave propagation in open space *Microw. Opt. Techn. Let.* **4** 79
[7] Ranfagni A, Fabeni P, Pazzi G P and Mugnai D 1993 Anomalous pulse delay in microwave propagation: A plausible connection to the tunneling time *Phys. Rev.* E **48** 1453
[8] Kholmetskii A L , Missevitch O V, Smirnov-Rueda R, Tzonchev R I, Chubykalo A E and Moreno I 2007 Experimental test on the applicability of the standard retardation condition to bound magnetic fields *J. Appl. Phys.* **101** 023532
[9] Shabad A 2016 A different interpretation of "Measuring propagation speed of Coulomb fields" by R. de Sangro, G. Finocchiaro, P. Patteri, M. Piccolo, G. Pizzella *Preprint* arXiv:1606.00710v1
[10] de Sangro R, Finocchiaro G, Patteri P, Piccolo M and Pizzella G 2016 Why the interpretation of "Measuring propagation speed of Coulomb fields" stands *Preprint* arXiv:1611.06935v1
[11] Bakamjian B and Thomas L H 1953 Relativistic particle dynamics. II *Phys. Rev.* **92** 1300
[12] Wigner E P 1939 On unitary representations of the inhomogeneous Lorentz group *Ann. Math.* **40** 149
[13] Dirac P A M 1949 Forms of relativistic dynamics *Rev. Mod. Phys.* **21** 392
[14] Weinberg S 1995 *The Quantum Theory of Fields* vol. 1 (Cambridge: Cambridge University Press)
[15] Currie D G, Jordan T F and Sudarshan E C G 1963 Relativistic invariance and Hamiltonian theories of interacting particles *Rev. Mod. Phys.* **35** 350
[16] Stefanovich E V 2002 Is Minkowski space-time compatible with quantum mechanics? *Found. Phys.* **32** 673
[17] Van Dam H and Wigner E P 1965 Classical relativistic mechanics of interacting point particles *Phys. Rev.* B **138** 1576
[18] Van Dam H and Wigner E P 1966 Instantaneous and asymptotic conservation laws for classical relativistic mechanics of interacting point particles *Phys. Rev.* **142** 838
[19] Currie G 1966 Poincaré-invariant equations of motion for classical particles *Phys. Rev.* **142** 817
[20] Hill R N 1967 Instantaneous action-at-a-distance in classical relativistic mechanics *J. Math. Phys.* **8** 201
[21] Komar A 1978 Interacting relativistic particles *Phys. Rev.* D **18** 1887
[22] Crater H W and Alstine P V 1983 Two-body Dirac equations *Ann. Phys.* **148** 57
[23] Sudarshan E C G and Mukunda N 1983 Forms of relativistic dynamics and world line condition and separability *Found. Phys.* **13** 385
[24] Polyzou W N 1985 Manifestly covariant, Poincaré invariant quantum theories of directly interacting particles *Phys. Rev.* D **32** 995

IOP Conf. Series: Journal of Physics: Conf. Series **845** (2017) 012016 doi:10.1088/1742-6596/845/1/012016

[25] Sokolov S N 1997 Mechanics with retarded interactions *Preprint IHEP Protvino* IHEP 97-84 http://web.ihep.su/library/pubs/prep1997/ps/97-84.pdf
[26] Berestetskii V B, Livshitz E M and Pitaevskii L P 1982 *Quantum Electrodynamics* (Oxford: Butterworth-Heinemann)
[27] Hobson A 2013 There are no particles, there are only fields *Am. J. Phys.* **81** 211
[28] Malament D B 1996 In defence of dogma: Why there cannot be a relativistic quantum mechanics of (localizable) particles *Perspectives on Quantum Reality* ed R Clifton (Dordrecht: Springer) p 1
[29] Wilczek F 1999 Quantum field theory *Rev. Mod. Phys.* **71** S58
[30] Halvorson H and Clifton R 2002 No place for particles in relativistic quantum theories? *Phil. Sci.* **69** 1
[31] Strocchi F 2004 Relativistic quantum mechanics and field theory *Found. Phys.* **34** 501
[32] Itoh T 1965 Derivation of nonrelativistic Hamiltonian for electrons from quantum electrodynamics *Rev. Mod. Phys.* **37** 159
[33] Gupta S N and Radford S F 1980 Quantum field-theoretical electromagnetic and gravitational two-particle potentials *Phys. Rev.* D **21** 2213
[34] Feinberg G and Sucher J 1988 Two-photon-exchange force between charged systems: Spinless particles *Phys. Rev.* D **38** 3763
[35] Pineda A and Soto J 1998 The Lamb shift in dimensional regularization *Phys. Lett.* B **420** 391
[36] Holstein B R 2004 Effective interactions and the hydrogen atom *Am. J. Phys.* **72** 333
[37] Shebeko A V and Shirokov M I 2001 Unitary transformations in quantum field theory and bound states *Phys. Part. Nucl.* **32** 15
[38] Stefanovich E V 2001 Quantum field theory without infinities *Ann. Phys.* **292** 139
[39] Stefanovich E 2015 *Relativistic Quantum Theory of Particles* vol. 2 (Saarbrücken: Lambert Academic)
[40] Shebeko A V and Shirokov M I 1998 Relativistic quantum field theory (RQFT) treatment of few-body systems *Nucl. Phys.* A **631** 564c
[41] Shirokov M I 1978 Signal velocity in quantum electrodynamics *Sov. Phys. Usp.* **21** 345
[42] Kaup D J and Rupasov V I 1996 Exactly solvable one-dimensional model of resonance energy transfer *J. Phys. A: Math. Gen.* **29** 2149
[43] Kaup D J and Rupasov V I 1996 Exactly solvable 3D model of resonance energy transfer *J. Phys. A: Math. Gen.* **29** 6911

IOP Conf. Series: Journal of Physics: Conf. Series **845** (2017) 012017 doi:10.1088/1742-6596/845/1/012017

Spin-charge-family theory is offering next step in understanding elementary particles and fields and correspondingly universe

Norma Susana Mankoč Borštnik

FMP, University of Ljubljana, Jadranska 19, 1000 Ljubljana,Slovenia

E-mail: norma.mankoc@fmf.uni-lj.si

Abstract. More than 40 years ago the *standard model* made a successful new step in understanding properties of fermion and boson fields. Now the next step is needed, which would explain what the *standard model* and the cosmological models just assume: **a.** The origin of quantum numbers of massless one family members. **b.** The origin of families. **c.** The origin of the vector gauge fields. **d.** The origin of the Higgses and Yukawa couplings. **e.** The origin of the dark matter. **f.** The origin of the matter-antimatter asymmetry. **g.** The origin of the dark energy. **h.** And several other open problems. The *spin-charge-family* theory, a kind of the Kaluza-Klein theories in $(d = (2n-1)+1)$-space-time, with $d = (13+1)$ and the two kinds of the spin connection fields, which are the gauge fields of the two kinds of the Clifford algebra objects anti-commuting with one another, may provide this much needed next step. The talk presents: **i.** A short presentation of this theory. **ii.** The review over the achievements of this theory so far, with some not published yet achievements included. **iii.** Predictions for future experiments.

1. Introduction

The *standard model* made a great step in understanding properties of fermion and boson fields by: **i.** Starting with massless fields. **ii.** Assuming quantum numbers of one family of massless quarks and leptons and relating handedness with charges. **iii.** Postulating the existence of several families. **iv.** Postulating the existence of the vector gauge fields of the charges of quarks and leptons. **v.** Postulating a simple action for fermions and vector bosons under the requirement of the gauge invariance. **vi.** Postulating the existence of the scalar field, which breaks the weak and the hyper charges of the vacuum making fermions and heavy bosons massive. **viii.** Postulating the Yukawa couplings.

Properties of fermions and bosons in the *standard model* are presented and commented on in Appendix B.

Although the assumptions from **i.- v.** are elegant, in particular the assumption that all the elementary fields are massless gaining masses through the interactions only, as well as the choice of simple actions for massless fermion and vector boson fields, yet these assumptions need the explanation, why has nature "decided" to make this particular choice of fermions and vector gauge fields and what steps to take in the evolution.

The assumption (**vi.**) that there is the massive scalar field, carrying the charges in the fundamental representations of the groups, while all the other bosons (vector bosons) carry

IOP Conf. Series: Journal of Physics: Conf. Series **845** (2017) 012017 doi:10.1088/1742-6596/845/1/012017

charges in the adjoint representations of the groups, and the assumption (**viii.**) that the Yukawa couplings take care of the fermion properties, without explaining the origin of these couplings, do not seem either elegant or simple.

The experiments have confirmed so far the existence of three families of fermions with properties required by the *standard model*, of the vector fields, which are the gauge fields of the charges $SU(3), SU(2)$ and $U(1)$, and of the Higgs, all in accordance with the *standard model* assumptions.

To be able to predict the outcome of future experiments the next step beyond the *standard model* is needed.

Just adding several new fields by repeating ideas of the *standard model*, without explaining the assumptions of this so far so successful model, has to my understanding a little chance to be the right step.

The *spin-charge-family* theory [1, 2, 3, 4, 5, 6, 7, 8, 9, 10, 11, 12, 13, 14, 15] does explain all the assumptions of the *standard model*: **i.** The charges of the left and of the right handed quarks and leptons of one family - the right handed neutrinos are in this theory regular members of each family - and of their antiquarks and antileptons. **ii.** The appearance and properties of families. **iii.** The appearance and properties of the vector gauge fields of the family members charges. **iv.** The appearance and properties of scalars fields, explaining the properties of the Higgses carrying charges in the fundamental representations of the groups and the Yukawa couplings.

The *spin-charge-family* theory is offering also the explanation for the existence of the phenomena not explained by the *standard model*: **a.** For the dark matter [13]. **b.** For the (ordinary) matter-antimatter asymmetry [2].

This theory predicts: **a.** At the low energy regime two decoupled groups of four families; **a.i.** The fourth [1, 5, 4, 9, 12] to the already observed three families of quarks and leptons will be measured at the LHC [14]; **a.ii.** The lowest of the upper four families might be considered as constituting the dark matter [13]. **b.** New scalar fields with the weak and the hyper charges of the Higgs [1, 4], some of them will be measured at the LHC. **c.** New $SU(2)$ vector gauge fields, explaining the appearance of the hyper charge, and its $U(1)$ gauge field. **d.** New scalar fields, which are in the fundamental representations of the colour charge (triplets), explaining the ordinary matter-antimatter asymmetry and causing the proton decay.

Within this theory many consequences of the *standard model*, like the "miraculous" cancellation of the triangle anomalies, can straightforwardly be explained (Subsect. 4.3).

2. Short presentation of the *spin-charge-family* theory

This section follows a lot the similar one from Ref. [2]. It briefly presents the *spin-charge-family* theory ([1, 2, 4, 5], and the references therein). The details of the theory will follow in Sect. (3).

Let me start with the *assumptions* on which the theory is built. Comments, following the assumptions, will explain the meaning of each of the assumptions.

A i. Fermions (ψ) carry in $d = (13+1)$ as the *internal degrees of freedom only two kinds of spins*, no charges, determined by the *two kinds of the Clifford objects* [1], (γ^a and $\tilde{\gamma}^a$), and *interact correspondingly with the two kinds of the spin connection fields* - $\omega_{ab\alpha}$ and $\tilde{\omega}_{ab\alpha}$, (the *gauge fields* of $S^{ab} = \frac{i}{4}\,(\gamma^a\gamma^b - \gamma^b\gamma^a)$, the generators of $SO(13,1)$ and $\tilde{S}^{ab} = \frac{i}{4}\,(\tilde{\gamma}^a\tilde{\gamma}^b - \tilde{\gamma}^b\tilde{\gamma}^a)$,

[1] There exist only two kinds of the Clifford algebra objects, connected with the left and right multiplication ([1], Sect. IV. Eq.(28)).

the generators of $\widetilde{SO}(13,1)$), and the *vielbeins* $f^{\alpha}{}_{a}$.

$$\begin{aligned}
\mathcal{A} &= \int d^d x \, E \, \mathcal{L}_f + \\
& \int d^d x \, E \, (\alpha \, R + \tilde{\alpha} \, \tilde{R}) \,, \\
\mathcal{L}_f &= \frac{1}{2} (\bar{\psi} \, \gamma^a p_{0a} \psi) + h.c. \,, \\
p_{0a} &= f^{\alpha}{}_a p_{0\alpha} + \frac{1}{2E} \{ p_\alpha, E f^{\alpha}{}_a \}_- \,, \\
p_{0\alpha} &= p_\alpha - \frac{1}{2} S^{ab} \omega_{ab\alpha} - \frac{1}{2} \tilde{S}^{ab} \tilde{\omega}_{ab\alpha} \,, \\
R &= \frac{1}{2} \{ f^{\alpha [a} f^{\beta b]} \, (\omega_{ab\alpha,\beta} - \omega_{ca\alpha} \, \omega^{c}{}_{b\beta}) \} + h.c. \,, \\
\tilde{R} &= \frac{1}{2} \{ f^{\alpha [a} f^{\beta b]} \, (\tilde{\omega}_{ab\alpha,\beta} - \tilde{\omega}_{ca\alpha} \, \tilde{\omega}^{c}{}_{b\beta}) \} + h.c. \,. \qquad (1)
\end{aligned}$$

Here [2] $f^{\alpha[a} f^{\beta b]} = f^{\alpha a} f^{\beta b} - f^{\alpha b} f^{\beta a}$. R and $\tilde{R}$ are the two scalars (R is a curvature).

A ii. The manifold $M^{(13+1)}$ breaks first into $M^{(7+1)} \times M^{(6)}$ (which manifests as $SO(7,1)$ $\times SU(3)$ $\times U(1)$), affecting both internal degrees of freedom - the one represented by γ^a and the one represented by $\tilde{\gamma}^a$. Since the left handed (with respect to $M^{(7+1)}$) spinors couple differently to scalar (with respect to $M^{(7+1)}$) fields than the right handed ones, the break can leave massless and mass protected $2^{((7+1)/2-1)}$ families [34]. The rest of families get heavy masses [3].

A iii. There is additional breaking of symmetry: The manifold $M^{(7+1)}$ breaks further into $M^{(3+1)} \times$ $M^{(4)}$.

A iv. There is a scalar condensate (Table 1) of two right handed neutrinos with the family quantum numbers of the upper four families, bringing masses of the scale $\propto 10^{16}$ GeV or higher to all the vector and scalar gauge fields, which interact with the condensate [2].

A v. There are the scalar fields with the space index $(7,8)$ carrying the weak (τ^{1i}) and the hyper charges ($Y = \tau^{23} + \tau^4$, τ^{1i} and τ^{2i} are generators of the subgroups of $SO(4)$, τ^4 and τ^{3i} are the generators of $U(1)_{II}$ and $SU(3)$, respectively, which are subgroups of $SO(6)$), which with their nonzero vacuum expectation values change the properties of the vacuum and break the weak charge and the hyper charge. Interacting with fermions and with the weak and hyper bosons, they bring masses to heavy bosons and to twice four groups of families. Carrying no electromagnetic ($Q = \tau^{13} + Y$) and colour (τ^{3i}) charges and no $SO(3,1)$ spin, the scalar fields leave the electromagnetic, colour and gravity fields in $d = (3+1)$ massless.

Comments (**C**) *on the assumptions* (**A**):

[2] $f^{\alpha}{}_{a}$ are inverted vielbeins to $e^{a}{}_{\alpha}$ with the properties $e^{a}{}_{\alpha} f^{\alpha}{}_{b} = \delta^{a}{}_{b}$, $e^{a}{}_{\alpha} f^{\beta}{}_{a} = \delta^{\beta}_{\alpha}$, $E = \det(e^{a}{}_{\alpha})$. Latin indices $a, b, .., m, n, .., s, t, ..$ denote a tangent space (a flat index), while Greek indices $\alpha, \beta, .., \mu, \nu, ..\sigma, \tau, ..$ denote an Einstein index (a curved index). Letters from the beginning of both the alphabets indicate a general index ($a, b, c, ..$ and $\alpha, \beta, \gamma, ..$), from the middle of both the alphabets the observed dimensions $0, 1, 2, 3$ ($m, n, ..$ and $\mu, \nu, ..$), indices from the bottom of the alphabets indicate the compactified dimensions ($s, t, ..$ and $\sigma, \tau, ..$). We assume the signature $\eta^{ab} = diag\{1, -1, -1, \cdots, -1\}$.

[3] A toy model [34, 35] was studied in $d = (5+1)$ with the same action as in Eq. (1). The break from $d = (5+1)$ to $d = (3+1) \times$ an almost S^2 was studied. For a particular choice of vielbeins and for a class of spin connection fields the manifold $M^{(5+1)}$ breaks into $M^{(3+1)}$ times an almost S^2, while $2^{((3+1)/2-1)}$ families remain massless and mass protected. Equivalent assumption, although not yet proved how does it really work, is made in the $d = (13+1)$ case. This study is in progress.

IOP Conf. Series: Journal of Physics: Conf. Series **845** (2017) 012017 doi:10.1088/1742-6596/845/1/012017

C i. The simple starting action, Eq.(1), of the *spin-charge-family* theory leads in the low energy regime - after the breaks of the starting symmetry - to the effective action, which is the *standard model* action with the right handed neutrinos included, what offers the explanation for all the *standard model* assumptions, as well as for the appearance of the families, of the Higgs and of the Yukawa couplings:

C i.a. One Weyl (massless) representation of $SO(13,1)$ contains [5, 3, 4, 9, 1], if analyzed with respect to the subgroups $SO(3,1)$, $SU(2)_I$, $SU(2)_{II}$, $SU(3)$ and $U(1)$ (Eqs. (A1) - (A6) of Ref. [1]), all the family members and anti-members assumed by the *standard model*, with the right handed neutrinos as the regular members of each family in addition: It contains the left handed weak ($SU(2)_I$) charged and $SU(2)_{II}$ chargeless colour triplet quarks and colourless leptons and right handed weak chargeless and $SU(2)_{II}$ charged coloured quarks and colourless leptons, as well as the right handed weak charged and $SU(2)_{II}$ chargeless anti-coloured (anti-triplet) antiquarks and (anti)colourless antileptons, and left handed weak chargeless and $SU(2)_{II}$ charged antiquarks and antileptons. (The anti-fermion states are reachable from the fermion states by the application of the discrete symmetry operator $\mathbb{C}_{\mathcal{N}}\,\mathcal{P}_{\mathcal{N}}$, presented in Ref. [36].)

C i.b. Before the electroweak break all observable gauge fields are massless: the gravity, the colour octet vector gauge fields (of the group $SU(3)$ from $SO(6)$), the weak triplet vector gauge fields (of the group $SU(2)_I$ from $SO(4)$), and the hyper singlet vector gauge field (a superposition of $U(1)$ from $SO(6)$ and the third component of $SU(2)_{II}$ triplet gauge fields). All are the superposition of the $f^{\alpha}{}_{c}\,\omega_{ab\alpha}$ spin connection gauge fields.

C i.c. Before the electroweak break there are two decoupled massless groups of four families of quarks and leptons, in the fundamental representations of $\widetilde{SU}(2)_{R,\widetilde{SO}(3,1)}\times\widetilde{SU}(2)_{II,\widetilde{SO}(4)}$ and $\widetilde{SU}(2)_{L,\widetilde{SO}(3,1)}\times\widetilde{SU}(2)_{I,\widetilde{SO}(4)}$ groups, respectively - the subgroups of $\widetilde{SO}(3,1)$ and $\widetilde{SO}(4)$ (Table 5). These eight families remain massless up to the electroweak break due to the "mass protection mechanism": Right handed members have no left handed partners with the same charges.

C i.d. There are scalar fields [2, 1] with the space index $(7,8)$ and with respect to the space index with the weak charge and the hyper charge of the Higgs's scalar (Eq. (21)). They have additional quantum numbers, belonging either to one of the two groups of two triplets or to three singlets.

One group of the two triplets belong to the family groups $\widetilde{SU}(2)_{R\,\widetilde{SO}(3,1)}$ and $\widetilde{SU}(2)_{II\,\widetilde{SO}(4)}$ and couple to the upper four families. The second group of the two triplets belong to the family groups $\widetilde{SU}(2)_{L\,\widetilde{SO}(3,1)}$ and $\widetilde{SU}(2)_{I\,\widetilde{SO}(4)}$ and couple to the lower four families. All these scalars are the superposition of $f^{\sigma}{}_{s}\,\tilde{\omega}_{ab\sigma}$ [4]. Scalars, belonging to three singlets, are the gauge fields of the charges (Q, Q', Y') [5] and couple to the family members of both groups of families. They are the superposition of $f^{\sigma}{}_{s}\,\omega_{ab\sigma}$ [6]. Both kinds of scalar fields determine the fermion masses (Eq. (31)), offering the explanation for the Yukawa couplings and the heavy bosons masses.

C i.e. The starting action contains also additional $SU(2)_{II}$ (from $SO(4)$, Eq. (A.21)) vector gauge fields (one of the components contributes to the hyper charge gauge fields as explained above), as well as the scalar fields with the space index $s \in (5,6)$ and $t \in (9,10,\dots,14)$. All these fields gain masses of the scale of the condensate (Table 1) with which they interact. They

[4] $\vec{\tilde{A}}^{\tilde{1}}_{s} = (\tilde{\omega}_{\tilde{5}\tilde{8}s} - \tilde{\omega}_{\tilde{6}\tilde{7}s},\, \tilde{\omega}_{\tilde{5}\tilde{7}s} + \tilde{\omega}_{\tilde{6}\tilde{8}s},\, \tilde{\omega}_{\tilde{5}\tilde{6}s} - \tilde{\omega}_{\tilde{7}\tilde{8}s})$, $\vec{\tilde{A}}^{\tilde{N}_{\tilde{L}}}_{s} = (\tilde{\omega}_{\tilde{2}\tilde{3}s} + i\,\tilde{\omega}_{\tilde{0}\tilde{1}s},\, \tilde{\omega}_{\tilde{3}\tilde{1}s} + i\,\tilde{\omega}_{\tilde{0}\tilde{2}s},\, \tilde{\omega}_{\tilde{1}\tilde{2}s} + i\,\tilde{\omega}_{\tilde{0}\tilde{3}s})$, $\vec{\tilde{A}}^{\tilde{2}}_{s} = (\tilde{\omega}_{\tilde{5}\tilde{8}s} + \tilde{\omega}_{\tilde{6}\tilde{7}s},\, \tilde{\omega}_{\tilde{5}\tilde{7}s} - \tilde{\omega}_{\tilde{6}\tilde{8}s},\, \tilde{\omega}_{\tilde{5}\tilde{6}s} + \tilde{\omega}_{\tilde{7}\tilde{8}s})$ and $\vec{\tilde{A}}^{\tilde{N}_{\tilde{R}}}_{s} = (\tilde{\omega}_{\tilde{2}\tilde{3}s} - i\,\tilde{\omega}_{\tilde{0}\tilde{1}s},\, \tilde{\omega}_{\tilde{3}\tilde{1}s} - i\,\tilde{\omega}_{\tilde{0}\tilde{2}s},\, \tilde{\omega}_{\tilde{1}\tilde{2}s} - i\,\tilde{\omega}_{\tilde{0}\tilde{3}s})$, where $(s \in (7,8))$ (Ref. [1], Eq. (A8)).

[5] $Q := \tau^{13} + Y$, $Q' := -Y\tan^2\vartheta_1 + \tau^{13}$, $Q' := -Y\tan^2\vartheta_1 + \tau^{13}$, $Y := \tau^4 + \tau^{23}$, $Y' := -\tau^4\tan^2\vartheta_2 + \tau^{23}$, $Q := \tau^{13} + Y$ (Ref. [1], Eq. (A7).

[6] $A^4_s = g^Q\,Q A^Q_s + g^{Q'}\,Q' A^{Q'}_s + g^{Y'}\,Y' A^{Y'}_s$, $A^4_s = -(\omega_{9\,10\,s} + \omega_{11\,12\,s} + \omega_{13\,14\,s})$, $A^{13}_s = (\omega_{56s} - \omega_{78s})$, $A^{23}_s = (\omega_{56s} + \omega_{78s})$, $A^Q_s = \sin\vartheta_1\,A^{13}_s + \cos\vartheta_1\,A^Y_s$, $A^{Q'}_s = \cos\vartheta_1\,A^{13}_s - \sin\vartheta_1\,A^Y_s$, $A^{Y'}_s = \cos\vartheta_2\,A^{23}_s - \sin\vartheta_2\,A^4_s$, with $(s \in (7,8))$ (Re. [1], Eq. (A9)).

all are expressible as superposition of $f^{\mu}{}_{m}\,\tilde{\omega}_{ab\mu}$. In the case of free fields (if no spinor source, carrying their quantum numbers, is present) both $f^{\mu}{}_{m}\,\omega_{ab\mu}$ and $f^{\mu}{}_{m}\,\tilde{\omega}_{ab\mu}$ are expressible with vielbeins, Eq. (C9) in Ref. [1], correspondingly only one kind of the three gauge fields are the propagating fields.

C ii., C iii.: There are many ways of breaking symmetries from $d = (13+1)$ to $d = (3+1)$. The assumed breaks explain why the weak and the hyper charges are connected with the handedness of spinors, manifesting correspondingly the properties of the family members - the quarks and the leptons, left and right handed (Table 2) - and of the vector gauge fields.
Since the left handed members are weak charged while the right handed ones are weak chargeless [7], the family members remain massless and mass protected up to the electroweak break, when the nonzero vacuum expectation values of the scalar fields with the space index $(7,8)$ break the weak and the hyper charge symmetry.
Antiparticles are accessible from particles by the application of the operator $\mathbb{C}_{\mathcal{N}} \cdot \mathcal{P}_{\mathcal{N}}$, as explained in Refs. [36, 37]. This discrete symmetry operator does not contain $\tilde{\gamma}^{a}$'s degrees of freedom: To each family member there corresponds the anti-member, with the same family quantum number.

C iv.: It is the condensate of two right handed neutrinos with the quantum numbers of the upper four families (Table 1), which makes massive all the scalar gauge fields (with the index $(5,6,7,8)$, as well as those with the index $(9,\dots,14)$) and the vector gauge fields, manifesting nonzero τ^{4}, τ^{23}, $\tilde{\tau}^{4}$, $\tilde{\tau}^{23}$, $\tilde{N}^{3}_{R}$ [8]. Only the vector gauge fields of Y, $SU(3)$ and $SU(2)_{I}$ remain massless, since they do not interact with the condensate, their corresponding quantum numbers are zero ($Y = 0$, $\tau^{3i} = 0$ and $\tau^{1i} = 0$).

C v.: At the electroweak break the scalar fields with the space index $s = (7,8)$ - originating in $\tilde{\omega}_{abs}$ [9], as well as some superposition of $\omega_{s's"s}$ with the quantum numbers (Q, Q', Y') (footnotes in this paper and Ref. [1], Eq. (22)), conserving the electromagnetic charge - change their mutual interaction, and gaining nonzero vacuum expectation values change correspondingly also their masses. They contribute to the mass matrices of twice four families, as well as to the masses of the heavy vector bosons (to the two members of the weak triplet and the superposition of the third member of the weak triplet with the hyper vector field (Ref. ([1], Eqs. (17-20)).

All the rest scalar fields keep masses of the scale of the condensate and are correspondingly unobservable in the low energy regime.

The fourth family to the observed three ones is *predicted to be observed* at the LHC. Its properties are under consideration [14, 15]. The *baryons of the stable family of the upper four families offer the explanation for the dark matter* [13]. The *triplet and anti-triplet scalar fields* contribute together with the condensate to the *matter/anti-matter asymmetry* [2].

[7] The left handed anti-members are weak chargeless while the right handed anti-members are weak charged.

[8] $\vec{\tau}^{1} := \frac{1}{2}(S^{58} - S^{67}, S^{57} + S^{68}, S^{56} - S^{78})$, $\vec{\tau}^{2} := \frac{1}{2}(S^{58} + S^{67}, S^{57} - S^{68}, S^{56} + S^{78})$, $\vec{N}_{\pm}(= \vec{N}_{(L,R)}) : \frac{1}{2}(S^{23} \pm i S^{01}, S^{31} \pm i S^{02}, S^{12} \pm i S^{03})$, $\vec{\tau}^{3} := \frac{1}{2}\{S^{9\,12} - S^{10\,11}, S^{9\,11} + S^{10\,12}, S^{9\,10} - S^{11\,12}, S^{9\,14} - S^{10\,13}, S^{9\,13} + S^{10\,14}, S^{11\,14} - S^{12\,13}, S^{11\,13} + S^{12\,14}, \frac{1}{\sqrt{3}}(S^{9\,10} + S^{11\,12} - 2S^{13\,14})\}$, $\tau^{4} := -\frac{1}{3}(S^{9\,10} + S^{11\,12} + S^{13\,14})$, $\vec{\tilde{\tau}}^{1} := \frac{1}{2}(\tilde{S}^{58} - \tilde{S}^{67}, \tilde{S}^{57} + \tilde{S}^{68}, \tilde{S}^{56} - \tilde{S}^{78})$, $\vec{\tilde{\tau}}^{2} := \frac{1}{2}(\tilde{S}^{58} + \tilde{S}^{67}, \tilde{S}^{57} - \tilde{S}^{68}, \tilde{S}^{56} + \tilde{S}^{78})$, $\vec{\tilde{N}}_{L,R} := \frac{1}{2}(\tilde{S}^{23} \pm i\tilde{S}^{01}, \tilde{S}^{31} \pm i\tilde{S}^{02}, \tilde{S}^{12} \pm i\tilde{S}^{03})$, $\tilde{\tau}^{4} := -\frac{1}{3}(\tilde{S}^{9\,10} + \tilde{S}^{11\,12} + \tilde{S}^{13\,14})$ Ref. ([1], Eqs. (A1-A6)).

[9] $\tilde{A}^{N_L \boxed{\pm}}_{\substack{78\\(\pm)}} = \tilde{A}^{N_L 1}_{\substack{78\\(\pm)}} \boxed{\pm}\, i\, \tilde{A}^{N_L 2}_{\substack{78\\(\pm)}}$, $\tilde{A}^{\tilde{N}_L \boxed{\pm}}_{\substack{78\\(\pm)}} = \{(\tilde{\omega}_{\tilde{2}\tilde{3}\substack{78\\(\pm)}} + i\,\tilde{\omega}_{\tilde{0}\tilde{1}\substack{78\\(\pm)}}) \boxed{\pm}\, i\,(\tilde{\omega}_{\tilde{3}\tilde{1}\substack{78\\(\pm)}} + i\,\tilde{\omega}_{\tilde{0}\tilde{2}\substack{78\\(\pm)}})\}$, $\tilde{A}^{\tilde{N}_L 3}_{\substack{78\\(\pm)}} = (\tilde{\omega}_{\tilde{1}\tilde{2}\substack{78\\(\pm)}} + i\,\tilde{\omega}_{\tilde{0}\tilde{3}\substack{78\\(\pm)}})$, $A^{Q}_{\substack{78\\(\pm)}} = \omega_{56\substack{78\\(\pm)}} - (\omega_{9\,10\substack{78\\(\pm)}} + \omega_{11\,12\substack{78\\(\pm)}} + \omega_{13\,14\substack{78\\(\pm)}})$, with $N^{3}_{L}\, \tilde{A}^{\tilde{N}_L \boxed{\pm}}_{\substack{78\\(\pm)}} = \boxed{\pm}\, \tilde{A}^{\tilde{N}_L \boxed{\pm}}_{\substack{78\\(\pm)}}$, $N^{3}_{L}\, \tilde{A}^{\tilde{N}_L 3}_{\substack{78\\(\pm)}} = 0$ and $Q\, A^{Q}_{\substack{78\\(\pm)}} = 0$ (Ref. [1], Eq. (22)).

state	S^{03}	S^{12}	τ^{13}	τ^{23}	τ^{4}	Y	Q	$\tilde{\tau}^{13}$	$\tilde{\tau}^{23}$	$\tilde{\tau}^{4}$	$\tilde{Y}$	$\tilde{Q}$	$\tilde{N}^3_L$	$\tilde{N}^3_R$
$(\vert\nu^{\mathbf{VIII}}_{\mathbf{1R}}>_{\mathbf{1}}\ \vert\nu^{\mathbf{VIII}}_{\mathbf{2R}}>_{\mathbf{2}})$	0	0	0	1	−1	0	0	0	1	−1	0	0	0	1
$(\vert\nu^{VIII}_{1R}>_1\ \vert e^{VIII}_{2R}>_2)$	0	0	0	0	−1	−1	−1	0	1	−1	0	0	0	1
$(\vert e^{VIII}_{1R}>_1\ \vert e^{VIII}_{2R}>_2)$	0	0	0	−1	−1	−2	−2	0	1	−1	0	0	0	1

Table 1. This table is taken from [2]. The condensate of the two right handed neutrinos ν_R, with the VIIIth family quantum numbers, coupled to spin zero and belonging to a triplet with respect to the generators τ^{2i}, is presented together with its two partners. The right handed neutrino has $Q = 0 = Y$. The triplet carries $\tau^4 = -1$, $\tilde{\tau}^{23} = 1$, $\tilde{\tau}^4 = -1$, $\tilde{N}^3_R = 1$, $\tilde{N}^3_L = 0$, $\tilde{Y} = 0$, $\tilde{Q} = 0$. The family quantum numbers are presented in Table 5.

3. Quarks, leptons and vector and scalar gauge fields in the *spin-charge-family* theory

I shall formally rewrite the part of the action in Eq.(1), which determines the spinor degrees of freedom, in the way that we can clearly see that the action does manifest in the low energy regime by the *standard model* required degrees of freedom of the fermions (Table B1), of the vector gauge fields (Table B2) and of the scalar gauge fields (Table B3) [5, 3, 4, 10, 1, 9, 6, 7, 8, 11, 12, 13, 14].

$$\begin{aligned}
\mathcal{L}_f &= \bar{\psi}\gamma^m(p_m - \sum_{A,i} g^{Ai}\tau^{Ai}A^{Ai}_m)\psi + \\
&\quad \{\sum_{s=7,8} \bar{\psi}\gamma^s p_{0s}\,\psi\} + \\
&\quad \{\sum_{t=5,6,9,\dots,14} \bar{\psi}\gamma^t p_{0t}\,\psi\}\,,
\end{aligned} \tag{2}$$

where $p_{0s} = p_s - \frac{1}{2}S^{s's"}\omega_{s's"s} - \frac{1}{2}\tilde{S}^{ab}\tilde{\omega}_{abs}$, $p_{0t} = p_t - \frac{1}{2}S^{t't"}\omega_{t't"t} - \frac{1}{2}\tilde{S}^{ab}\tilde{\omega}_{abt}$, with $m \in (0,1,2,3)$, $s \in (7,8)$, $(s',s") \in (5,6,7,8)$, (a,b) (appearing in $\tilde{S}^{ab}$) run within either $(0,1,2,3)$ or $(5,6,7,8)$, t runs $\in (5,\dots,14)$, $(t',t")$ run either $\in (5,6,7,8)$ or $\in (9,10,\dots,14)$. The spinor function ψ represents all family members of all the $2^{\frac{7+1}{2}-1} = 8$ families.

The first line of Eq. (2) determines (in $d = (3+1)$) the kinematics and dynamics of spinor (fermion) fields, coupled to the vector gauge fields. The generators τ^{Ai} of the charge groups are expressible in terms of S^{ab} through the complex coefficients $c^{Ai}{}_{ab}$ (definition of τ^{Ai} can be found in the footnote [8], the rest in [10],

$$\tau^{Ai} = \sum_{a,b} c^{Ai}{}_{ab}\, S^{ab}\,, \tag{3}$$

fulfilling the commutation relations

$$\{\tau^{Ai}, \tau^{Bj}\}_- = i\delta^{AB} f^{Aijk}\tau^{Ak}\,. \tag{4}$$

They represent the colour (τ^{3i}), the weak (τ^{1i}) and the hyper (Y) charges, as well as the $SU(2)_{II}$ (τ^{2i}) and $U(1)_{II}$ (τ^4) charges, the gauge fields of these last two groups gain masses interacting with the condensate, Table 1. The condensate leaves massless, besides the colour and gravity

[10] After the electroweak break the nonconserved charges manifest $Y := \tau^4 + \tau^{23}$, $Y' := -\tau^4 \tan^2\vartheta_2 + \tau^{23}$, $Q' := -Y\tan^2\vartheta_1 + \tau^{13}$, θ_1 is the electroweak angle, breaking the weak $SU(2)_I$ and the hyper charge, θ_2 is the angle of the break of $SU(2)_{II}$ from $SU(2)_I \times SU(2)_{II}$. $Q := \tau^{13} + Y$ remains the conserved charge.

gauge fields, the weak and the hyper charge vector gauge fields. The corresponding vector gauge fields A^{Ai}_m are expressible with the spin connection fields ω_{stm}.

$$A^{Ai}_m = \sum_{s,t} c^{Ai}{}_{st}\, \omega^{st}{}_m\,, \tag{5}$$

with (s,t) either in $(5,6,7,8)$ or in $(9,\ldots,14)$, in agreement with the assumptions **A ii.** and **A iii.**. I demonstrate [1, 18] in Subsect. 3.2 the equivalence between the usual Kaluza-Klein procedure leading to the vector gauge fields through the vielbeins and the procedure with the spin connections used by the *spin-charge-family* theory.

All the vector gauge fields, appearing in the first line of Eq. (2), except $A^{2\pm}_m$ and $A^{Y'}_m$ ($= \cos\vartheta_2\, A^{23}_m - \sin\vartheta_2\, A^4_m$, Y' and τ^4 are defined in the footnote [11]), are massless before the electroweak break. $\vec{A}^3_m$ carries the colour charge $SU(3)$ (originating in $SO(6)$), $\vec{A}^1_m$ carries the weak charge $SU(2)_I$ ($SU(2)_I$ and $SU(2)_{II}$ are the subgroups of $SO(4)$) and A^Y_m ($= \sin\vartheta_2\, A^{23}_m + \cos\vartheta_2\, A^4_m$) carries the corresponding $U(1)$ charge ($Y = \tau^{23} + \tau^4$, τ^4 originates in $SO(6)$ and τ^{23} is the third component of the second $SU(2)_{II}$ group, A^4_m and $\vec{A}^2_m$ are the corresponding vector gauge fields). The fields $A^{2\pm}_m$ and $A^{Y'}_m$ get masses of the order of the condensate scale through the interaction with the condensate of the two right handed neutrinos with the quantum numbers of one of the group of four families (the assumption **A iv.**, Table 1). (See Ref. [1].)

3.1. Quarks and leptons in the spin-charge-family theory

To offer the explanation for the origin of quantum numbers of one (anyone) family of massless quarks and leptons, assumed by the *standard model* (Table B1), the *spin-charge-family* theory must answer the question, where do the *standard model* charges originate and why are the weak and the hyper charge connecting with the spin of quarks and leptons. The theory must answer also the question: Where do families of quarks and leptons originate?

This section demonstrates that spinors, which carry in $d = (13+1)$ nothing but spins of two kinds, determined by the two kinds of the Clifford algebra objects (Sect. IV. and App. B in [1]), explain the origin of spins and charges and of families: One kind of spins manifests in $d = (3+1)$ at low energies the spin and all the charges, connecting the spin (the handedness) and the charges [6, 7, 8, 11, 31, 32, 33]. The second kind explains the origin of families.

To explain the appearance of the electroweak break, which causes that all the families become massive, the properties of the scalar fields must be explained. This is done in Subsects. 3.3, where also the appearance of the Yukawa couplings is explained, while masses of the two groups of four families, predicted by the *spin-charge-family* theory, will be discussed in Subsect. 4.1.

In Table 2 one Weyl representation of spinors in $d = (13 + 1)$ is presented. The *technique* [6, 7, 31, 32], Appendix A, is used, which makes that the states themselves demonstrate properties of spinors. Besides the states also the quantum numbers of the members are presented with respect to the groups $SO(3,1)$, $SU(2)_I$, $SU(2)_{II}$, $U(1)_{II}$ and $SU(3)$, which are the subgroups of the group $SO(13,1)$. One easily sees that the states of one Weyl representation include all the quarks and the leptons, and the antiquarks and the antileptons of one family of quarks and leptons, with just the quantum as assumed by the *standard model*, Table B1.

Table 2 demonstrates that left handed quarks and leptons carry the weak charge ($SU(2)_I$ with τ^{1i} as generators) and the hyper charge ($Y = \tau^{23} + \tau^4$, τ^{2i} are generators of $SU(2)_{II}$, which is a subgroup of $SO(4)$, τ^4 are generators of $U(1)_{II}$, which is a subgroup of $SO(6)$) just as required by the *standard model*, Table B1, while the right handed quarks and leptons are weak

[11] $Y' := -\tau^4 \tan^2\vartheta_2 + \tau^{23}$, $\tau^4 = -\frac{1}{3}(S^{9\,10} + S^{11\,12} + S^{13\,14})$.

Table 2. The left handed ($\Gamma^{(13,1)} = -1$) ($= \Gamma^{(7,1)} \times \Gamma^{(6)}$) multiplet of spinors - the members of the $SO(13,1)$ group representation, manifesting the subgroup $SO(7,1)$ - of the colour charged quarks and antiquarks and the colourless leptons and antileptons, is presented in the massless basis using the technique presented in Appendix A. It contains the left handed ($\Gamma^{(3,1)} = -1$) weak charged ($\tau^{13} = \pm\frac{1}{2}$) and $SU(2)_{II}$ chargeless ($\tau^{23} = 0$) quarks and the right handed weak chargeless and $SU(2)_{II}$ charged ($\tau^{23} = \pm\frac{1}{2}$) quarks of three colours ($c^i = (\tau^{33}, \tau^{38})$) with the "spinor" charge ($\tau^4 = \frac{1}{6}$) and the colourless left handed weak charged and right handed weak chargeless leptons with the "spinor" charge ($\tau^4 = -\frac{1}{2}$). S^{12} defines the ordinary spin $\pm\frac{1}{2}$. Table contains also the corresponding anti-states with opposite charges, reachable from the particle states by the application of the discrete symmetry operator $\mathbb{C}_{\mathcal{N}}\, \mathcal{P}_{\mathcal{N}}$, presented in Refs. [36, 37]. The vacuum state, on which the nilpotents and projectors operate, is not shown. The reader can find this Weyl representation also in Refs. [2, 10, 1]. Table is separated into three parts.

i		$\vert^a\psi_i>$	$\Gamma^{(3,1)}$	S^{12}	$\Gamma^{(4)}$	τ^{13}	τ^{23}	τ^{33}	τ^{38}	τ^4	Y	Q
		(Anti)octet, $\Gamma^{(1,7)} = (-1)\,1\,,\ \Gamma^{(6)} = (1)\,-1$ of (anti)quarks and (anti)leptons										
1	u_R^{c1}	$\overset{03}{(+i)}\overset{12}{(+)}\vert\overset{56}{(+)}\overset{78}{(+)}\Vert\overset{9\,10}{(+)}\;\overset{11\,12}{(-)}\;\overset{13\,14}{(-)}$	1	$\frac{1}{2}$	1	0	$\frac{1}{2}$	$\frac{1}{2}$	$\frac{1}{2\sqrt{3}}$	$\frac{1}{6}$	$\frac{2}{3}$	$\frac{2}{3}$
2	u_R^{c1}	$\overset{03}{[-i]}\overset{12}{[-]}\vert\overset{56}{(+)}\overset{78}{(+)}\Vert\overset{9\,10}{(+)}\;\overset{11\,12}{(-)}\;\overset{13\,14}{(-)}$	1	$-\frac{1}{2}$	1	0	$\frac{1}{2}$	$\frac{1}{2}$	$\frac{1}{2\sqrt{3}}$	$\frac{1}{6}$	$\frac{2}{3}$	$\frac{2}{3}$
3	d_R^{c1}	$\overset{03}{(+i)}\overset{12}{(+)}\vert\overset{56}{[-]}\overset{78}{[-]}\Vert\overset{9\,10}{(+)}\;\overset{11\,12}{(-)}\;\overset{13\,14}{(-)}$	1	$\frac{1}{2}$	1	0	$-\frac{1}{2}$	$\frac{1}{2}$	$\frac{1}{2\sqrt{3}}$	$\frac{1}{6}$	$-\frac{1}{3}$	$-\frac{1}{3}$
4	d_R^{c1}	$\overset{03}{[-i]}\overset{12}{[-]}\vert\overset{56}{[-]}\overset{78}{[-]}\Vert\overset{9\,10}{(+)}\;\overset{11\,12}{(-)}\;\overset{13\,14}{(-)}$	1	$-\frac{1}{2}$	1	0	$-\frac{1}{2}$	$\frac{1}{2}$	$\frac{1}{2\sqrt{3}}$	$\frac{1}{6}$	$-\frac{1}{3}$	$-\frac{1}{3}$
5	d_L^{c1}	$\overset{03}{[-i]}\overset{12}{(+)}\vert\overset{56}{[-]}\overset{78}{(+)}\Vert\overset{9\,10}{(+)}\;\overset{11\,12}{(-)}\;\overset{13\,14}{(-)}$	-1	$\frac{1}{2}$	-1	$-\frac{1}{2}$	0	$\frac{1}{2}$	$\frac{1}{2\sqrt{3}}$	$\frac{1}{6}$	$\frac{1}{6}$	$-\frac{1}{3}$
6	d_L^{c1}	$\overset{03}{(+i)}\overset{12}{[-]}\vert\overset{56}{[-]}\overset{78}{(+)}\Vert\overset{9\,10}{(+)}\;\overset{11\,12}{(-)}\;\overset{13\,14}{(-)}$	-1	$-\frac{1}{2}$	-1	$-\frac{1}{2}$	0	$\frac{1}{2}$	$\frac{1}{2\sqrt{3}}$	$\frac{1}{6}$	$\frac{1}{6}$	$-\frac{1}{3}$
7	u_L^{c1}	$\overset{03}{[-i]}\overset{12}{(+)}\vert\overset{56}{(+)}\overset{78}{[-]}\Vert\overset{9\,10}{(+)}\;\overset{11\,12}{(-)}\;\overset{13\,14}{(-)}$	-1	$\frac{1}{2}$	-1	$\frac{1}{2}$	0	$\frac{1}{2}$	$\frac{1}{2\sqrt{3}}$	$\frac{1}{6}$	$\frac{1}{6}$	$\frac{2}{3}$
8	u_L^{c1}	$\overset{03}{(+i)}\overset{12}{[-]}\vert\overset{56}{(+)}\overset{78}{[-]}\Vert\overset{9\,10}{(+)}\;\overset{11\,12}{(-)}\;\overset{13\,14}{(-)}$	-1	$-\frac{1}{2}$	-1	$\frac{1}{2}$	0	$\frac{1}{2}$	$\frac{1}{2\sqrt{3}}$	$\frac{1}{6}$	$\frac{1}{6}$	$\frac{2}{3}$
9	u_R^{c2}	$\overset{03}{(+i)}\overset{12}{(+)}\vert\overset{56}{(+)}\overset{78}{(+)}\Vert\overset{9\,10}{[-]}\;\overset{11\,12}{[+]}\;\overset{13\,14}{(-)}$	1	$\frac{1}{2}$	1	0	$\frac{1}{2}$	$-\frac{1}{2}$	$\frac{1}{2\sqrt{3}}$	$\frac{1}{6}$	$\frac{2}{3}$	$\frac{2}{3}$
10	u_R^{c2}	$\overset{03}{[-i]}\overset{12}{[-]}\vert\overset{56}{(+)}\overset{78}{(+)}\Vert\overset{9\,10}{[-]}\;\overset{11\,12}{[+]}\;\overset{13\,14}{(-)}$	1	$-\frac{1}{2}$	1	0	$\frac{1}{2}$	$-\frac{1}{2}$	$\frac{1}{2\sqrt{3}}$	$\frac{1}{6}$	$\frac{2}{3}$	$\frac{2}{3}$
...												

chargeless, carrying the $SU(2)_{II}$ charge with τ^{2i} as generators, determining the hyper charges of quarks and leptons, again in agreement with the *standard model*.

Quarks carry the colour charge in the fundamental representation and the "fermion charge" $\tau^4 = \frac{1}{6}$, leptons are colourless carrying the "fermion charge" $\tau^4 = -\frac{1}{2}$. Correspondingly is the hyper charge of either left handed or right handed quarks and leptons in agreement with the *standard model* assumptions.

Table 2 demonstrates that left handed antiquarks and antileptons are weak chargeless and right handed antiquarks and antileptons are weak charged. Antiquarks carry antitriplet charges, while leptons are anticolourless. Correspondingly fermions and anti-fermions carry opposite colour, hyper, electromagnetic and "fermion" charges than fermions.

Handedness is in the one Weyl representation of $SO(13,1)$ strongly related to the weak charge and the hyper charge.

Left and right handed neutrinos (carrying nonzero $Y' = -\tau^4 \tan^2\vartheta_2 + \tau^{23}$ quantum number) are the regular members of each family, and so are the antineutrinos.

Let me call attention to the reader that the term $\gamma^0 \overset{78}{(-)}\, \tau^{Ai}\, A^{Ai}_{\underset{(-)}{78}}$, where τ^{Ai} represent the superposition of either S^{ab} or $\tilde{S}^{ab}$ and $A^{Ai}_{\underset{(-)}{78}}$ represent correspondingly the superposition of either the scalar fields ω_{abs} or the scalar fields $\tilde{\omega}_{abs}$, $s \in (7,8)$, as presented in Subsect. 3.3, Eq. (20), transforms the right handed u_R^{c1} quark from the first line of Tables 2–4 into the left handed

Table 3. Continuation of Table2

i		$\vert^a\psi_i>$	$\Gamma^{(3,1)}$	S^{12}	$\Gamma^{(4)}$	τ^{13}	τ^{23}	τ^{33}	τ^{38}	τ^4	Y	Q
		(Anti)octet, $\Gamma^{(1,7)} = (-1)\,1\,,\ \Gamma^{(6)} = (1)\,-1$ of (anti)quarks and (anti)leptons										
17	u_R^{c3}	$\stackrel{03}{(+i)}\stackrel{12}{(+)}\vert\stackrel{56}{(+)}\stackrel{78}{(+)}\Vert\stackrel{9\,10}{[-]}\stackrel{11\,12}{(-)}\stackrel{13\,14}{[+]}$	1	$\frac{1}{2}$	1	0	$\frac{1}{2}$	0	$-\frac{1}{\sqrt{3}}$	$\frac{1}{6}$	$\frac{2}{3}$	$\frac{2}{3}$
18	u_R^{c3}	$\stackrel{03}{[-i]}\stackrel{12}{[-]}\vert\stackrel{56}{(+)}\stackrel{78}{(+)}\Vert\stackrel{9\,10}{[-]}\stackrel{11\,12}{(-)}\stackrel{13\,14}{[+]}$	1	$-\frac{1}{2}$	1	0	$\frac{1}{2}$	0	$-\frac{1}{\sqrt{3}}$	$\frac{1}{6}$	$\frac{2}{3}$	$\frac{2}{3}$
...												
25	ν_R	$\stackrel{03}{(+i)}\stackrel{12}{(+)}\vert\stackrel{56}{(+)}\stackrel{78}{(+)}\Vert\stackrel{9\,10}{(+)}\stackrel{11\,12}{[+]}\stackrel{13\,14}{[+]}$	1	$\frac{1}{2}$	1	0	$\frac{1}{2}$	0	0	$-\frac{1}{2}$	0	0
26	ν_R	$\stackrel{03}{[-i]}\stackrel{12}{[-]}\vert\stackrel{56}{(+)}\stackrel{78}{(+)}\Vert\stackrel{9\,10}{(+)}\stackrel{11\,12}{[+]}\stackrel{13\,14}{[+]}$	1	$-\frac{1}{2}$	1	0	$\frac{1}{2}$	0	0	$-\frac{1}{2}$	0	0
27	e_R	$\stackrel{03}{(+i)}\stackrel{12}{(+)}\vert\stackrel{56}{[-]}\stackrel{78}{[-]}\Vert\stackrel{9\,10}{(+)}\stackrel{11\,12}{[+]}\stackrel{13\,14}{[+]}$	1	$\frac{1}{2}$	1	0	$-\frac{1}{2}$	0	0	$-\frac{1}{2}$	-1	-1
28	e_R	$\stackrel{03}{[-i]}\stackrel{12}{[-]}\vert\stackrel{56}{[-]}\stackrel{78}{[-]}\Vert\stackrel{9\,10}{(+)}\stackrel{11\,12}{[+]}\stackrel{13\,14}{[+]}$	1	$-\frac{1}{2}$	1	0	$-\frac{1}{2}$	0	0	$-\frac{1}{2}$	-1	-1
29	e_L	$\stackrel{03}{[-i]}\stackrel{12}{(+)}\vert\stackrel{56}{[-]}\stackrel{78}{(+)}\Vert\stackrel{9\,10}{(+)}\stackrel{11\,12}{[+]}\stackrel{13\,14}{[+]}$	-1	$\frac{1}{2}$	-1	$-\frac{1}{2}$	0	0	0	$-\frac{1}{2}$	$-\frac{1}{2}$	-1
30	e_L	$\stackrel{03}{(+i)}\stackrel{12}{[-]}\vert\stackrel{56}{[-]}\stackrel{78}{(+)}\Vert\stackrel{9\,10}{(+)}\stackrel{11\,12}{[+]}\stackrel{13\,14}{[+]}$	-1	$-\frac{1}{2}$	-1	$-\frac{1}{2}$	0	0	0	$-\frac{1}{2}$	$-\frac{1}{2}$	-1
31	ν_L	$\stackrel{03}{[-i]}\stackrel{12}{(+)}\vert\stackrel{56}{(+)}\stackrel{78}{[-]}\Vert\stackrel{9\,10}{(+)}\stackrel{11\,12}{[+]}\stackrel{13\,14}{[+]}$	-1	$\frac{1}{2}$	-1	$\frac{1}{2}$	0	0	0	$-\frac{1}{2}$	$-\frac{1}{2}$	0
32	ν_L	$\stackrel{03}{(+i)}\stackrel{12}{[-]}\vert\stackrel{56}{(+)}\stackrel{78}{[-]}\Vert\stackrel{9\,10}{(+)}\stackrel{11\,12}{[+]}\stackrel{13\,14}{[+]}$	-1	$-\frac{1}{2}$	-1	$\frac{1}{2}$	0	0	0	$-\frac{1}{2}$	$-\frac{1}{2}$	0

Table 4. Continuation of Table2

i		$\vert^a\psi_i>$	$\Gamma^{(3,1)}$	S^{12}	$\Gamma^{(4)}$	τ^{13}	τ^{23}	τ^{33}	τ^{38}	τ^4	Y	Q
		(Anti)octet, $\Gamma^{(1,7)} = (-1)\,1\,,\ \Gamma^{(6)} = (1)\,-1$ of (anti)quarks and (anti)leptons										
33	$\bar{d}_L^{\bar{c}1}$	$\stackrel{03}{[-i]}\stackrel{12}{(+)}\vert\stackrel{56}{(+)}\stackrel{78}{(+)}\Vert\stackrel{9\,10}{[-]}\stackrel{11\,12}{[+]}\stackrel{13\,14}{[+]}$	-1	$\frac{1}{2}$	1	0	$\frac{1}{2}$	$-\frac{1}{2}$	$-\frac{1}{2\sqrt{3}}$	$-\frac{1}{6}$	$\frac{1}{3}$	$\frac{1}{3}$
34	$\bar{d}_L^{\bar{c}1}$	$\stackrel{03}{(+i)}\stackrel{12}{[-]}\vert\stackrel{56}{(+)}\stackrel{78}{(+)}\Vert\stackrel{9\,10}{[-]}\stackrel{11\,12}{[+]}\stackrel{13\,14}{[+]}$	-1	$-\frac{1}{2}$	1	0	$\frac{1}{2}$	$-\frac{1}{2}$	$-\frac{1}{2\sqrt{3}}$	$-\frac{1}{6}$	$\frac{1}{3}$	$\frac{1}{3}$
35	$\bar{u}_L^{\bar{c}1}$	$\stackrel{03}{[-i]}\stackrel{12}{(+)}\vert\stackrel{56}{[-]}\stackrel{78}{[-]}\Vert\stackrel{9\,10}{[-]}\stackrel{11\,12}{[+]}\stackrel{13\,14}{[+]}$	-1	$\frac{1}{2}$	1	0	$-\frac{1}{2}$	$-\frac{1}{2}$	$-\frac{1}{2\sqrt{3}}$	$-\frac{1}{6}$	$-\frac{2}{3}$	$-\frac{2}{3}$
36	$\bar{u}_L^{\bar{c}1}$	$\stackrel{03}{(+i)}\stackrel{12}{[-]}\vert\stackrel{56}{[-]}\stackrel{78}{[-]}\Vert\stackrel{9\,10}{[-]}\stackrel{11\,12}{[+]}\stackrel{13\,14}{[+]}$	-1	$-\frac{1}{2}$	1	0	$-\frac{1}{2}$	$-\frac{1}{2}$	$-\frac{1}{2\sqrt{3}}$	$-\frac{1}{6}$	$-\frac{2}{3}$	$-\frac{2}{3}$
37	$\bar{d}_R^{\bar{c}1}$	$\stackrel{03}{(+i)}\stackrel{12}{(+)}\vert\stackrel{56}{(+)}\stackrel{78}{[-]}\Vert\stackrel{9\,10}{[-]}\stackrel{11\,12}{[+]}\stackrel{13\,14}{[+]}$	1	$\frac{1}{2}$	-1	$\frac{1}{2}$	0	$-\frac{1}{2}$	$-\frac{1}{2\sqrt{3}}$	$-\frac{1}{6}$	$-\frac{1}{6}$	$\frac{1}{3}$
38	$\bar{d}_R^{\bar{c}1}$	$\stackrel{03}{[-i]}\stackrel{12}{[-]}\vert\stackrel{56}{(+)}\stackrel{78}{[-]}\Vert\stackrel{9\,10}{[-]}\stackrel{11\,12}{[+]}\stackrel{13\,14}{[+]}$	1	$-\frac{1}{2}$	-1	$\frac{1}{2}$	0	$-\frac{1}{2}$	$-\frac{1}{2\sqrt{3}}$	$-\frac{1}{6}$	$-\frac{1}{6}$	$\frac{1}{3}$
39	$\bar{u}_R^{\bar{c}1}$	$\stackrel{03}{(+i)}\stackrel{12}{(+)}\vert\stackrel{56}{[-]}\stackrel{78}{(+)}\Vert\stackrel{9\,10}{[-]}\stackrel{11\,12}{[+]}\stackrel{13\,14}{[+]}$	1	$\frac{1}{2}$	-1	$-\frac{1}{2}$	0	$-\frac{1}{2}$	$-\frac{1}{2\sqrt{3}}$	$-\frac{1}{6}$	$-\frac{1}{6}$	$-\frac{2}{3}$
40	$\bar{u}_R^{\bar{c}1}$	$\stackrel{03}{[-i]}\stackrel{12}{[-]}\vert\stackrel{56}{[-]}\stackrel{78}{(+)}\Vert\stackrel{9\,10}{[-]}\stackrel{11\,12}{[+]}\stackrel{13\,14}{[+]}$	1	$-\frac{1}{2}$	-1	$-\frac{1}{2}$	0	$-\frac{1}{2}$	$-\frac{1}{2\sqrt{3}}$	$-\frac{1}{6}$	$-\frac{1}{6}$	$-\frac{2}{3}$
41	$\bar{d}_L^{\bar{c}2}$	$\stackrel{03}{[-i]}\stackrel{12}{(+)}\vert\stackrel{56}{(+)}\stackrel{78}{(+)}\Vert\stackrel{9\,10}{(+)}\stackrel{11\,12}{(-)}\stackrel{13\,14}{[+]}$	-1	$\frac{1}{2}$	1	0	$\frac{1}{2}$	$\frac{1}{2}$	$-\frac{1}{2\sqrt{3}}$	$-\frac{1}{6}$	$\frac{1}{3}$	$\frac{1}{3}$
...												
49	$\bar{d}_L^{\bar{c}3}$	$\stackrel{03}{[-i]}\stackrel{12}{(+)}\vert\stackrel{56}{(+)}\stackrel{78}{(+)}\Vert\stackrel{9\,10}{(+)}\stackrel{11\,12}{[+]}\stackrel{13\,14}{(-)}$	-1	$\frac{1}{2}$	1	0	$\frac{1}{2}$	0	$\frac{1}{\sqrt{3}}$	$-\frac{1}{6}$	$\frac{1}{3}$	$\frac{1}{3}$
...												
57	$\bar{e}_L$	$\stackrel{03}{[-i]}\stackrel{12}{(+)}\vert\stackrel{56}{(+)}\stackrel{78}{(+)}\Vert\stackrel{9\,10}{[-]}\stackrel{11\,12}{(-)}\stackrel{13\,14}{(-)}$	-1	$\frac{1}{2}$	1	0	$\frac{1}{2}$	0	0	$\frac{1}{2}$	1	1
58	$\bar{e}_L$	$\stackrel{03}{(+i)}\stackrel{12}{[-]}\vert\stackrel{56}{(+)}\stackrel{78}{(+)}\Vert\stackrel{9\,10}{[-]}\stackrel{11\,12}{(-)}\stackrel{13\,14}{(-)}$	-1	$-\frac{1}{2}$	1	0	$\frac{1}{2}$	0	0	$\frac{1}{2}$	1	1
59	$\bar{\nu}_L$	$\stackrel{03}{[-i]}\stackrel{12}{(+)}\vert\stackrel{56}{[-]}\stackrel{78}{[-]}\Vert\stackrel{9\,10}{[-]}\stackrel{11\,12}{(-)}\stackrel{13\,14}{(-)}$	-1	$\frac{1}{2}$	1	0	$-\frac{1}{2}$	0	0	$\frac{1}{2}$	0	0
60	$\bar{\nu}_L$	$\stackrel{03}{(+i)}\stackrel{12}{[-]}\vert\stackrel{56}{[-]}\stackrel{78}{[-]}\Vert\stackrel{9\,10}{[-]}\stackrel{11\,12}{(-)}\stackrel{13\,14}{(-)}$	-1	$-\frac{1}{2}$	1	0	$-\frac{1}{2}$	0	0	$\frac{1}{2}$	0	0
61	$\bar{\nu}_R$	$\stackrel{03}{(+i)}\stackrel{12}{(+)}\vert\stackrel{56}{[-]}\stackrel{78}{(+)}\Vert\stackrel{9\,10}{[-]}\stackrel{11\,12}{(-)}\stackrel{13\,14}{(-)}$	1	$\frac{1}{2}$	-1	$-\frac{1}{2}$	0	0	0	$\frac{1}{2}$	$\frac{1}{2}$	0
62	$\bar{\nu}_R$	$\stackrel{03}{[-i]}\stackrel{12}{[-]}\vert\stackrel{56}{[-]}\stackrel{78}{(+)}\Vert\stackrel{9\,10}{[-]}\stackrel{11\,12}{(-)}\stackrel{13\,14}{(-)}$	1	$-\frac{1}{2}$	-1	$-\frac{1}{2}$	0	0	0	$\frac{1}{2}$	$\frac{1}{2}$	0
63	$\bar{e}_R$	$\stackrel{03}{(+i)}\stackrel{12}{(+)}\vert\stackrel{56}{(+)}\stackrel{78}{[-]}\Vert\stackrel{9\,10}{[-]}\stackrel{11\,12}{(-)}\stackrel{13\,14}{(-)}$	1	$\frac{1}{2}$	-1	$\frac{1}{2}$	0	0	0	$\frac{1}{2}$	$\frac{1}{2}$	1
64	$\bar{e}_R$	$\stackrel{03}{[-i]}\stackrel{12}{[-]}\vert\stackrel{56}{(+)}\stackrel{78}{[-]}\Vert\stackrel{9\,10}{[-]}\stackrel{11\,12}{(-)}\stackrel{13\,14}{(-)}$	1	$-\frac{1}{2}$	-1	$\frac{1}{2}$	0	0	0	$\frac{1}{2}$	$\frac{1}{2}$	1

u_L^{c1} quark from the seventh line of the same table [12], which can, due to the properties of the scalar fields (Eq. (21)), be interpreted also in the *standard model* way, namely, that $A^{Ai}_{\stackrel{78}{(-)}}$ "dress" u_R^{c1} giving it the weak and the hyper charge of the left handed u_L^{c1} quark, while γ^0 changes handedness. Equivalently happens to ν_R from the 25th line, which transforms under the action of $\gamma^0\ \stackrel{78}{(-)}\ \tau^{Ai}\ A^{Ai}_{\stackrel{78}{(-)}}$, into ν_L from the 31th line.

[12] This transformation of the right handed family members into the corresponding left handed partners can easily be calculated by using Eqs. (A.12, A.10, A.26).

Table 5. Eight families of the right handed u^{c1}_R (2–4) quark with the spin $\frac{1}{2}$, colour charge ($\tau^{33} = 1/2$, $\tau^{38} = 1/(2\sqrt{3})$, and of the colourless right handed neutrino ν_R of spin $\frac{1}{2}$ are presented in the left and in the right column, respectively. They belong to two groups of four families, one (I) is a doublet with respect to ($\vec{\tilde{N}}_L$ and $\vec{\tilde{\tau}}^{(1)}$) and a singlet with respect to ($\vec{\tilde{N}}_R$ and $\vec{\tilde{\tau}}^{(2)}$), the other (II) is a singlet with respect to ($\vec{\tilde{N}}_L$ and $\vec{\tilde{\tau}}^{(1)}$) and a doublet with with respect to ($\vec{\tilde{N}}_R$ and $\vec{\tilde{\tau}}^{(2)}$). All the families of each of the two groups follow from the starting one by the application of one of the two operators ($\tilde{N}^{\pm}_{R,L}$, $\tilde{\tau}^{(2,1)\pm}$), Eq. (A.26), respectively. The generators ($N^{\pm}_{R,L}$, $\tau^{(2,1)\pm}$) (Eq. (A.26)) transform u_{1R} of one family to all the members of the same family of the same colour. The same generators transform equivalently the right handed neutrino ν_{1R} to all the colourless members of the same family. The table is taken from Ref. [2]

					$\tilde{\tau}^{13}$	$\tilde{\tau}^{23}$	$\tilde{N}^3_L$	$\tilde{N}^3_R$	$\tilde{\tau}^4$
I	$u^{c1}_{R\,1}$	$\stackrel{03}{(+i)}\stackrel{12}{[+]}\mid\stackrel{56}{[+]}\stackrel{78}{(+)}\mid\mid\stackrel{9\,10}{(+)}\;\stackrel{11\,12}{(-)}\;\stackrel{13\,14}{(-)}$	$\nu_{R\,2}$	$\stackrel{03}{(+i)}\stackrel{12}{[+]}\mid\stackrel{56}{[+]}\stackrel{78}{(+)}\mid\mid\stackrel{9\,10}{(+)}\;\stackrel{11\,12}{[+]}\;\stackrel{13\,14}{[+]}$	$-\frac{1}{2}$	0	$-\frac{1}{2}$	0	$-\frac{1}{2}$
I	$u^{c1}_{R\,2}$	$\stackrel{03}{[+i]}\stackrel{12}{(+)}\mid\stackrel{56}{[+]}\stackrel{78}{(+)}\mid\mid\stackrel{9\,10}{(+)}\;\stackrel{11\,12}{(-)}\;\stackrel{13\,14}{(-)}$	$\nu_{R\,2}$	$\stackrel{03}{[+i]}\stackrel{12}{(+)}\mid\stackrel{56}{[+]}\stackrel{78}{(+)}\mid\mid\stackrel{9\,10}{(+)}\;\stackrel{11\,12}{[+]}\;\stackrel{13\,14}{[+]}$	$-\frac{1}{2}$	0	$\frac{1}{2}$	0	$-\frac{1}{2}$
I	$u^{c1}_{R\,3}$	$\stackrel{03}{(+i)}\stackrel{12}{[+]}\mid\stackrel{56}{(+)}\stackrel{78}{[+]}\mid\mid\stackrel{9\,10}{(+)}\;\stackrel{11\,12}{(-)}\;\stackrel{13\,14}{(-)}$	$\nu_{R\,3}$	$\stackrel{03}{(+i)}\stackrel{12}{[+]}\mid\stackrel{56}{(+)}\stackrel{78}{[+]}\mid\mid\stackrel{9\,10}{(+)}\;\stackrel{11\,12}{[+]}\;\stackrel{13\,14}{[+]}$	$\frac{1}{2}$	0	$-\frac{1}{2}$	0	$-\frac{1}{2}$
I	$u^{c1}_{R\,4}$	$\stackrel{03}{[+i]}\stackrel{12}{(+)}\mid\stackrel{56}{(+)}\stackrel{78}{[+]}\mid\mid\stackrel{9\,10}{(+)}\;\stackrel{11\,12}{(-)}\;\stackrel{13\,14}{(-)}$	$\nu_{R\,4}$	$\stackrel{03}{[+i]}\stackrel{12}{(+)}\mid\stackrel{56}{(+)}\stackrel{78}{[+]}\mid\mid\stackrel{9\,10}{(+)}\;\stackrel{11\,12}{[+]}\;\stackrel{13\,14}{[+]}$	$\frac{1}{2}$	0	$\frac{1}{2}$	0	$-\frac{1}{2}$
II	$u^{c1}_{R\,5}$	$\stackrel{03}{[+i]}\stackrel{12}{[+]}\mid\stackrel{56}{[+]}\stackrel{78}{[+]}\mid\mid\stackrel{9\,10}{(+)}\;\stackrel{11\,12}{(-)}\;\stackrel{13\,14}{(-)}$	$\nu_{R\,5}$	$\stackrel{03}{[+i]}\stackrel{12}{[+]}\mid\stackrel{56}{[+]}\stackrel{78}{[+]}\mid\mid\stackrel{9\,10}{(+)}\;\stackrel{11\,12}{[+]}\;\stackrel{13\,14}{[+]}$	0	$-\frac{1}{2}$	0	$-\frac{1}{2}$	$-\frac{1}{2}$
II	$u^{c1}_{R\,6}$	$\stackrel{03}{(+i)}\stackrel{12}{(+)}\mid\stackrel{56}{[+]}\stackrel{78}{[+]}\mid\mid\stackrel{9\,10}{(+)}\;\stackrel{11\,12}{(-)}\;\stackrel{13\,14}{(-)}$	$\nu_{R\,6}$	$\stackrel{03}{(+i)}\stackrel{12}{(+)}\mid\stackrel{56}{[+]}\stackrel{78}{[+]}\mid\mid\stackrel{9\,10}{(+)}\;\stackrel{11\,12}{[+]}\;\stackrel{13\,14}{[+]}$	0	$-\frac{1}{2}$	0	$\frac{1}{2}$	$-\frac{1}{2}$
II	$u^{c1}_{R\,7}$	$\stackrel{03}{[+i]}\stackrel{12}{[+]}\mid\stackrel{56}{(+)}\stackrel{78}{(+)}\mid\mid\stackrel{9\,10}{(+)}\;\stackrel{11\,12}{(-)}\;\stackrel{13\,14}{(-)}$	$\nu_{R\,7}$	$\stackrel{03}{[+i]}\stackrel{12}{[+]}\mid\stackrel{56}{(+)}\stackrel{78}{(+)}\mid\mid\stackrel{9\,10}{(+)}\;\stackrel{11\,12}{[+]}\;\stackrel{13\,14}{[+]}$	0	$\frac{1}{2}$	0	$-\frac{1}{2}$	$-\frac{1}{2}$
II	$u^{c1}_{R\,8}$	$\stackrel{03}{(+i)}\stackrel{12}{(+)}\mid\stackrel{56}{(+)}\stackrel{78}{(+)}\mid\mid\stackrel{9\,10}{(+)}\;\stackrel{11\,12}{(-)}\;\stackrel{13\,14}{(-)}$	$\nu_{R\,8}$	$\stackrel{03}{(+i)}\stackrel{12}{(+)}\mid\stackrel{56}{(+)}\stackrel{78}{(+)}\mid\mid\stackrel{9\,10}{(+)}\;\stackrel{11\,12}{[+]}\;\stackrel{13\,14}{[+]}$	0	$\frac{1}{2}$	0	$\frac{1}{2}$	$-\frac{1}{3}$

The operator $\gamma^0 \stackrel{78}{(+)} \tau^{Ai} A^{Ai}_{\stackrel{78}{(+)}}$ transforms d^{c1}_R from the third line of Tables 2–4 into d^{c1}_L from the fifth line of this table, or e_R from the 27th line into e_L from the 29th line, where $A^{Ai}_{\stackrel{78}{(+)}}$ belong to the scalar fields from Eq. (19).

The operator τ^{Ai}, if representing the first three operators in Eq. (19), (only) multiplies the right handed family member with its eigenvalue.

The term $\gamma^0 \stackrel{78}{(\mp)} \tau^{Ai} A^{Ai}_{\stackrel{78}{(\mp)}}$ of the action (Eqs. (1, 2)) determines the Yukawa couplings (3.3, 4.1).

Since spinors (fermions) carry besides the family members quantum numbers also the family quantum numbers, determined by $\tilde{S}^{ab} = \frac{i}{4}(\tilde{\gamma}^a\tilde{\gamma}^b - \tilde{\gamma}^b\tilde{\gamma}^a)$, there are correspondingly $2^{(7+1)/2-1} = 8$ families [1], which split into two groups of families, each manifesting the $(\widetilde{SU}(2)_{\widetilde{SO}(3,1)} \times \widetilde{SU}(2)_{\widetilde{SO}(4)} \times U(1))$ symmetry.

The eight families of the right handed u_{1R} quark (the first member of the eight-plet of quarks from Tables 2 – 4) and of the right handed ν_{1R} leptons (the first member of the eight-plet of leptons from Tables 2 – 4) are presented as an example in Table 5 [4].

All the other members of any of the eight families of quarks or leptons follow from any member of a particular family by the application of the operators $N^{\pm}_{R,L}$ and $\tau^{(2,1)\pm}$ on this particular member.

The eight families separate into two groups of four families: One group contains doublets with respect to $\vec{\tilde{N}}_R$ and $\vec{\tilde{\tau}}^2$, these families are singlets with respect to $\vec{\tilde{N}}_L$ and $\vec{\tilde{\tau}}^1$.Another group of four families contains doublets with respect to $\vec{\tilde{N}}_L$ and $\vec{\tilde{\tau}}^1$, these families are singlets with respect to $\vec{\tilde{N}}_R$ and $\vec{\tilde{\tau}}^2$.

If τ^{Ai} represents the last four operators of Eq. (19) in Subsect. 3.3, the operators $\gamma^0 \stackrel{78}{(\mp)} \tau^{Ai} A^{Ai}_{\stackrel{78}{(\mp)}}$ (($\mp$) for (u_R, ν_R) and (d_R, e_R), respectively) transform the right handed family member of one family into the left handed partner of another family within the same group of four families, since these four operators manifest the symmetry twice $(\widetilde{SU}(2)_{\widetilde{SO}(3,1)} \times \widetilde{SU}(2)_{\widetilde{SO}(4)})$. One group of four families carries the family quantum numbers $(\vec{\tilde{\tau}}^1, \vec{\tilde{N}}_L)$, the other group of four families carries the family quantum numbers $(\vec{\tilde{\tau}}^2, \vec{\tilde{N}}_R)$.

The contribution of the scalar fields to masses of fermions and to the Yukawa couplings will be discussed in Subsects. (4.1, 3.3), respectively.

At each break of symmetry fermions can gain masses [38] of the order of the scale of the condensate. In the Refs. [34, 35] we discuss possible conditions under which fermion remain massless at the break. This discussion concerns in our case the break of $d = (2(2n+1)-1, 1)$, for $n = 9$ or larger, down to $d = (13+1)$ and from $d = (13+1)$ to $d = (7+1)$ before the symmetry between spinors and antispinors is broken. After that the massless fermions are mass protected, since the left handed and right handed members differ in the weak and hyper charges, until the weak and the hyper charges are no longer conserved quantum numbers.

Let me point out here that there are scalar fields, the gauge scalars of $\vec{\tilde{N}}_R$ and $\vec{\tilde{\tau}}^2$, which couple only to the four families which are doublets with respect to these two groups, while the scalar fields, which are the gauge scalars of $\vec{\tilde{N}}_L$ and $\vec{\tilde{\tau}}^1$, couple only to the four families which are doublets with respect to these last two groups. Each of the two kinds of scalar contribute after the electroweak transition to their own group of four families, while the scalar gauge fields of (Q, Q', Y') couple to family members of all eight families.

3.2. Vector gauge fields in the spin-charge-family theory

In the starting action of Eq. (1) of the *spin-charge-family* theory all the gauge fields are the gravitational ones: the vielbeins and the spin connections of two kinds. Both kinds of the spin connection fields are uniquely determined by the vielbeins (Ref. [1], Eqs. ((30)-(32), (C9))), if there are no spinor sources present. Spinors (fermions) interact with the vielbeins and the two kinds of the spin connection fields. After the break of the starting symmetry $SO(13,1)$ the starting action manifests at low energies the effective action. Eq. (2) represents the effective action for fermions in $d = (3+1)$ interacting with the vector gauge fields, which are the superposition of the spin connection gauge fields with the vector index m $(m = (0,1,2,3))$ - $A^{Ai}_m = \sum_{s,t} c^{Aist}\,\omega_{stm}$ - and the scalar gauge fields, which are the superposition of the spin connection gauge fields of both kinds ω_{abs}'s and $\tilde{\omega}_{abs}$'s with the scalar index s $(s \geq 5)$ - $A^{(Q,Q',Y')}_s = \sum_{s,t} c^{(Q,Q",Y')st}\,\omega_{stm}$ (for (Q, Q', Y'), respectively), $\tilde{A}^{\tilde{A}i)}_s = \sum_{s,t} \tilde{c}^{\tilde{A}ist}\,\tilde{\omega}_{stm}$.

I comment in this section that the vector gauge fields in the *spin-charge-family* theory appear equivalently either from vielbeins $f^{\sigma}{}_m$ - like it is usually proceeded in the Kaluza-Klein-like theories [40, 39] (with which the *spin-charge-family* theory has many things in common) - or from the spin connection fields $\sum_{s,t} c^{Ai}{}_{st}\,\omega^{st}{}_m$. This is indeed known for a long time [39, 40, 41].

This section reviews Refs. [1, 18, 42], where this equivalence is demonstrated when the spaces of $d \geq 5$ have the metric tensor $g_{\sigma\tau} = \eta_{\sigma\tau} f^{-2}$, where (x^{σ}, x^{τ}) determine the coordinates of the (almost [34]) compactified space, $\eta_{\sigma\tau}$ is the diagonal matrix in this space and f is any scalar function of these coordinates.

Let the space with $s \geq 5$ have the symmetry allowing the infinitesimal transformations of the kind

$$x'^{\mu} = x^{\mu}, \quad x'^{\sigma} = x^{\sigma} - i \sum_{A,i,s,t} \varepsilon^{Ai}(x^{\mu})\, c_{Ai}{}^{st} M_{st}\, x^{\sigma}, \tag{6}$$

where $M^{st} = S^{st} + L^{st}$, $L^{st} = x^s p^t - x^t p^s$, S^{st} concern internal degrees of freedom of boson and fermion fields, $\{M^{st}, M^{s't'}\}_- = i(\eta^{st'} M^{ts'} + \eta^{ts'} M^{st'} - \eta^{ss'} M^{tt'} - \eta^{tt'} M^{ss'})$. From Eq. (6) it

IOP Conf. Series: Journal of Physics: Conf. Series **845** (2017) 012017 doi:10.1088/1742-6596/845/1/012017

follows

$$
\begin{aligned}
-i \sum_{s,t} c_{Ai}{}^{st} M_{st}\, x^{\sigma} &= E^{\sigma}_{Ai} = \sum_{s,t} c_{Ai}{}^{st} \, (x_s \, f^{\sigma}{}_t - x_t \, f^{\sigma}{}_s)\,, \\
\sum_{s,t} c_{Ai}{}^{st} \, M_{st}{}^{\sigma} : &= i E^{\sigma}_{Ai}\,,
\end{aligned} \tag{7}
$$

and correspondingly: $\tau_{Ai} = E^{\sigma}_{Ai} p_{\sigma}$, where $\tau^{Ai} = \sum_{s,t} c_{Ai}{}^{st} \, M_{st}$ with the commutation relations $\{\tau_{Ai}, \tau_{Bj}\}_- = i\delta^{AB} \, f^{Aijk} \, \tau_{Ak}$, f^{Aijk} are the structure constants of the symmetry group A. One derives, when taking into account Eq. (7) and the commutation relations among generators of the infinitesimal transformations τ_{Ai} the equation for the Killing vectors E^{σ}_{Ai} [39]

$$
E^{\sigma}_{Ai} p_{\sigma} E^{\tau}_{Bj} p_{\tau} - E^{\sigma}_{Bj} p_{\sigma} E^{\tau}_{Ai} p_{\tau} \;=\; i\delta^{AB} \, f^{Aijk} \, E^{\tau}_{Ak} p_{\tau}\,, \tag{8}
$$

and the Killing equation

$$
\begin{aligned}
&D_{\sigma} E_{\tau Ai} + D_{\tau} E_{\sigma Bj} = 0\,, \\
&D_{\sigma} E_{\tau Ai} = \partial_{\sigma} E_{\tau Ai} - \Gamma^{\tau'}_{\tau\sigma} E_{\tau' Ai}\,.
\end{aligned} \tag{9}
$$

Let the corresponding background field ($g_{\alpha\beta} = e^{a}{}_{\alpha} \, e^{a}{}_{\beta}$) be

$$
e^{a}{}_{\alpha} = \begin{pmatrix} \delta^{m}{}_{\mu} & e^{m}{}_{\sigma} = 0 \\ e^{s}{}_{\mu} & e^{s}{}_{\sigma} \end{pmatrix}, \quad f^{\alpha}{}_{a} = \begin{pmatrix} \delta^{\mu}{}_{m} & f^{\sigma}{}_{m} \\ 0 = f^{\mu}{}_{s} & f^{\sigma}{}_{s\,,} \end{pmatrix}, \tag{10}
$$

so that the background field in $d = (3+1)$ is flat. From $e^{a}{}_{\mu} f^{\sigma}{}_{a} = \delta^{\sigma}_{\mu} = 0$ it follows

$$
e^{s}{}_{\mu} \;=\; -\delta^{m}_{\mu} e^{s}{}_{\sigma} f^{\sigma}{}_{m}\,. \tag{11}
$$

This leads to

$$
g_{\alpha\beta} = \begin{pmatrix} \eta_{mn} + f^{\sigma}{}_{m} f^{\tau}{}_{n} e^{s}{}_{\sigma} e_{s\tau} & -f^{\tau}{}_{m} e^{s}{}_{\tau} e_{s\sigma} \\ -f^{\tau}{}_{n} e^{s}{}_{\tau} e_{s\sigma} & e^{s}{}_{\sigma} e_{s\tau} \end{pmatrix}, \tag{12}
$$

and

$$
g^{\alpha\beta} = \begin{pmatrix} \eta^{mn} & f^{\sigma}{}_{m} \\ f^{\sigma}{}_{m} & f^{\sigma}{}_{s} f^{\tau s} + f^{\sigma}{}_{m} f^{\tau m} \end{pmatrix}. \tag{13}
$$

We have: $\Gamma^{\tau'}_{\tau\sigma} = \frac{1}{2} g^{\tau'\sigma'} (g_{\sigma\sigma'},_{\tau} + g_{\tau\sigma'},_{\sigma} - g_{\sigma\tau},_{\sigma'})$.

Let us make a choice for the vielbein

$$
f^{\sigma}{}_{m} \;=\; \sum_{A} \vec{\tau}^{A\sigma} \, \vec{\mathcal{A}}^{A}_{m}\,, \tag{14}
$$

where we expect $\vec{\mathcal{A}}^{A}_{m}$ that they manifest in $d = (3+1)$ as the gauge fields of the charges τ^{Ai}. To prove this we must compare the gauge fields $A^{A}_{m} = c_{Ai}{}^{st} \omega_{stm}$, appearing in Eq. (2), with the gauge fields $\vec{\mathcal{A}}^{A}_{m}$.

If there are no fermions present then the vector gauge fields of the family members charges and of the family charges - ω_{abm} and $\tilde{\omega}_{abm}$, respectively - are uniquely expressible with the vielbeins [2, 1, 42]. We are interested in the vector gauge fields in $d = (3+1)$, for which we find

$$
\begin{aligned}
\omega_{stm} \;=\; & \frac{1}{2E} \{ f^{\sigma}{}_{m} \, [e_{t\sigma} \partial_{\tau} (E f^{\tau}{}_{s}) - e_{s\sigma} \, \partial_{\tau} (E f^{\tau}{}_{t})] \\
& + e_{s\sigma} \partial_{\tau} [E (f^{\sigma}{}_{m} f^{\tau}{}_{t} - f^{\tau}{}_{m} f^{\sigma}{}_{t})] - e_{t\sigma} \partial_{\tau} [E (f^{\sigma}{}_{m} f^{\tau}{}_{s} - f^{\tau}{}_{m} f^{\sigma}{}_{s})] \}\,.
\end{aligned} \tag{15}
$$

We must show that if we calculate $c_{Ai}{}^{st}\,\omega_{stm}$ by taking ω_{stm} from Eq. (15), and if we put on the right hand side of this equation the vielbeins $f^{\sigma}{}_{m} = \sum_{A} \vec{\tau}^{A\sigma}\, \vec{\mathcal{A}}^{A}_{m}$ from Eq. (14), we must end up with the equality relation

$$A^{Ai}_{m} \;=\; \mathcal{A}^{Ai}_{m}\,. \qquad (16)$$

It is not difficult to check that Eq. (16) follows, if we take into account that $e^{s}{}_{\mu} = -\delta^{m}_{\mu}$ and make a choice of the symmetry of space $(d-4)$: $f^{\sigma}{}_{s} = f\,\delta^{\sigma}_{s}$ (Ref. [1], Sect. II.).

Calculating from Eqs. (12, 13) the Riemann curvature $R^{(d)}$ in d-dimensional space by taking into account that $(d=(3+1))$ space is flat, one obtains

$$\begin{aligned} R^{(d)} &= R^{(d-4)} - \frac{1}{4} g_{\sigma\tau} E^{\sigma}{}_{Ai} E^{\tau}{}_{A'i'}\, F^{Ai}{}_{mn} F_{A'i'}{}^{mn}\,,\\ F^{Ai}{}_{mn} &= \partial_m A^{Ai}_{n} - \partial_n A^{Ai}_{m} - i f^{Aijk}\, A^{Aj}_{m}\, A^{Ak}_{n}\,,\\ A^{Ai}_{m} &= \sum_{st} c^{Ai}{}_{st}\, \omega^{st}{}_{m}\,,\\ \tau^{Ai} &= \sum_{st} c^{Aist}\, M_{st}\,. \end{aligned} \qquad (17)$$

The integration of the action $\int E\, d^4x d^{(d-4)}x\, R^{(d)}$ over $(d-4)$ space (in which it turns out that only even functions of coordinates x^{σ} give nonzero contributions) leads to the well known effective action for the vector gauge fields in $d=(3+1)$ space: $\int d^4x\,\{-\frac{1}{4}F^{Ai}{}_{\mu\nu}\,F^{Ai\mu\nu}\}$.

The quadratic form for the vector gauge fields, Eq. (17), in $d=(3+1)$, obtained from the curvature $R^{(d)}$ can be found in many text book [39]. In Ref. ([21], Sect. 5.3) the Lagrange function for the gauge vector fields is derived by using the Clifford algebra space. The author allows besides the curvature R also its quadratic form R^2 (Eq. (240)).

3.3. Scalar fields in the spin-charge-family theory explain the origin of the Higgs and Yukawa couplings

In the *spin-charge-family* theory the spin connection fields of both kinds, ω_{abs} and $\tilde{\omega}_{abs}$, carrying the space index $s=(7,8)$, explain the Higgs and the Yukawa couplings of the *standard model.* They all belong to the weak charge doublets ([1, 2] and references therein), as will be demonstrated in this section.

After gaining nonzero vacuum expectation values these scalar fields break the weak and the hyper charges of the vacuum (the assumption **A v.** and comments **C v.**) making all the fermions, massive due to the interaction with the vacuum. Also the heavy bosons gain masses, while interacting with the vacuum.

The gauge scalar fields with the space index $s>8$ contribute to the matter-antimatter asymmetry in the universe [2].

This section follows mainly the equivalent sections in Refs. [2, 1].

It turns out [2] that all scalars (the gauge fields with the space index $s \geq 5$) of the action (Eq. (1)) carry charges in fundamental representations due to the space index: They are either doublets (Table 6), $s=(5,6,7,8)$, or triplets (Table 9, and Sect. II, Table I in Ref. [2]), $s=(9,10,..,13,14)$. Scalars with the space indices $s\in(7,8)$ and $s\in(5,6)$ are the $SU(2)$ *doublets* (Table 6) with respect to this degree of freedom.

All scalars carry additional quantum numbers: Besides the quantum numbers determined by the space index s they carry also the quantum numbers Ai, Eq. (3), the states of which belong to the adjoint representations. They originate in S^{ab} or $\tilde{S}^{ab}$, Eq. (A.3), $S^{ab}=\frac{i}{4}(\gamma^a\gamma^b-\gamma^b\gamma^a)$, $\tilde{S}^{ab}=\frac{i}{4}(\tilde{\gamma}^a\tilde{\gamma}^b-\tilde{\gamma}^b\tilde{\gamma}^a)$, the gauge fields of which are ω_{abs} and $\tilde{\omega}_{abs}$, respectively. S^{ab} determine family members spin and charges, $\tilde{S}^{ab}$ determine family charges.

The infinitesimal generators $\mathcal{S}^{ab}$, which apply on the spin connections ω_{bde} $(= f^{\alpha}{}_{e}\, \omega_{bd\alpha})$ and $\tilde{\omega}_{\tilde{b}\tilde{d}e}$ $(= f^{\alpha}{}_{e}\, \tilde{\omega}_{\tilde{b}\tilde{d}\alpha})$, on either the space index e or any of the indices $(b, d, \tilde{b}, \tilde{d})$, operates as follows

$$\mathcal{S}^{ab}\, A^{d\ldots e\ldots g} \;=\; i\,(\eta^{ae}\, A^{d\ldots b\ldots g} - \eta^{be}\, A^{d\ldots a\ldots g})\,, \tag{18}$$

(see Section IV. and Appendix B in Ref. [1]).

The expressions for the infinitesimal operators of the subgroups of the starting group (presented in Eq. (3) and the footnote [8] and determined by the coefficients $c^{Ai}{}_{ab}$ in Eq. (3)) are the same for all three kinds of degrees of freedom, S^{ab}, $\tilde{S}^{ab}$ or $\mathcal{S}^{ab}$. Correspondingly the commutation relations are also the same.

At the electroweak break all the scalar fields with the space index $(7, 8)$, those which belong to one of twice two triplets carrying the family quantum numbers $(\tilde{\tau}^{\tilde{A}i})$ and those which belong to one of the three singlets carrying the family members quantum numbers (Q, Q', Y'), Eq. (19), start to self interact, gaining nonzero vacuum expectation values and breaking the weak charge, the hyper charge and the family charges.

Let me introduce a common notation A^{Ai}_{s} for all the scalar fields with $s = (7, 8)$, independently of whether they originate in ω_{abs} - in this case $Ai = (Q, Q', Y')$ - or in $\tilde{\omega}_{\tilde{a}\tilde{b}s}$ - in this case all the family quantum numbers of all eight families contribute.

$$\begin{aligned} A^{Ai}_{s} &\quad \text{represents} \quad (\, A^{Q}_{s}\,, A^{Q'}_{s}\,, A^{Y'}_{s}\,, \vec{\tilde{A}}^{\tilde{1}}_{s}\,, \vec{\tilde{A}}^{\tilde{N}_{\tilde{L}}}_{s}\,, \vec{\tilde{A}}^{\tilde{2}}_{s}\,, \vec{\tilde{A}}^{\tilde{N}_{\tilde{R}}}_{s}\,)\,, \\ \tau^{Ai} &\quad \text{represents} \quad (Q,\, Q',\, Y',\, \vec{\tilde{\tau}}^{1},\, \vec{\tilde{N}}_{L},\, \vec{\tilde{\tau}}^{2},\, \vec{\tilde{N}}_{R})\,. \end{aligned} \tag{19}$$

Here τ^{Ai} represent all the operators, which apply on the spinor states. These scalars, the gauge scalar fields of the generators τ^{Ai} and $\tilde{\tau}^{Ai}$, are expressible in terms of the spin connection fields (Ref. [1], Eqs. (10, 22, A8, A9)).

Let me demonstrate [1] that all the scalar fields with the space index $(7, 8)$ carry with respect to this space index the weak and the hyper charge $(\mp\frac{1}{2}, \pm\frac{1}{2})$, respectively. This means that all these scalars have properties as required for the Higgs in the *standard model.*

Let me make a choice of the superposition of the scalar fields so that they are eigenstates of $\tau^{13} = \frac{1}{2}(\mathcal{S}^{56} - \mathcal{S}^{78})$ (Eq (3) and footnotes at one page before). I rewrite for this purpose the second line of Eq. (2) as follows (the momentum p_s is left out [13])

$$\begin{aligned} &\sum_{s=(7,8),A,i} \bar{\psi}\, \gamma^{s}\, (-\tau^{Ai}\, A^{Ai}_{s}\,)\, \psi = \\ &-\bar{\psi}\,\{\, \overset{78}{(+)}\; \tau^{Ai}\,(A^{Ai}_{7} - i\, A^{Ai}_{8}) + \overset{78}{(-)}\; (\tau^{Ai}\,(A^{Ai}_{7} + i\, A^{Ai}_{8})\,\}\, \psi\,, \\ &\overset{78}{(\pm)} = \frac{1}{2}\,(\gamma^{7} \pm i\,\gamma^{8})\,, \quad A^{Ai}_{\overset{78}{(\pm)}} := (A^{Ai}_{7} \mp i\, A^{Ai}_{8})\,, \end{aligned} \tag{20}$$

with the summation over A and i performed, since A^{Ai}_{s} represent the scalar fields (A^{Q}_{s}, $A^{Q'}_{s}$, $A^{Y'}_{s}$, $\tilde{A}^{\tilde{4}}_{s}$, $\vec{\tilde{A}}^{\tilde{1}}_{s}$, $\vec{\tilde{A}}^{\tilde{2}}_{s}$, $\vec{\tilde{A}}^{\tilde{N}_{R}}_{s}$ and $\vec{\tilde{A}}^{\tilde{N}_{L}}_{s}$).

The application of the operators Y $(Y = \tau^{23} + \tau^{4}$, $\tau^{23} = \frac{1}{2}(\mathcal{S}^{56} + \mathcal{S}^{78})$, $\tau^{4} = -\frac{1}{3}(\mathcal{S}^{9\,10} + \mathcal{S}^{11\,12} + \mathcal{S}^{13\,14}))$, Q $(Q = \tau^{13} + Y$ and τ^{13} $(\tau^{13} = \frac{1}{2}(\mathcal{S}^{56} - \mathcal{S}^{78}))$ on the fields $(A^{Ai}_{7} \mp i\, A^{Ai}_{8})$ gives ($\mathcal{S}^{ab}$ is defined in Eq. (18))

[13] It is expected that solutions with nonzero momenta lead to higher masses of fermion fields in $d = (3+1)$ [34, 35]. We pay correspondingly no attention to the momentum p_s, $s \in (5, \ldots, 8)$, when having in mind the lowest energy solutions, manifesting at low energies.

Table 6. The two scalar weak doublets, one with $\tau^{23} = -\frac{1}{2}$ and the other with $\tau^{23} = +\frac{1}{2}$, both with the "spinor" quantum number $\tau^4 = 0$, are presented. In this table all the scalar fields carry besides the quantum numbers determined by the space index also the quantum numbers A and i from Eq. (19).

	state	τ^{13}	τ^{23}	spin	τ^4	Q
$A^{Ai}_{\substack{78\\(-)}}$	$A^{Ai}_7 + iA^{Ai}_8$	$+\frac{1}{2}$	$-\frac{1}{2}$	0	0	0
$A^{Ai}_{\substack{56\\(-)}}$	$A^{Ai}_5 + iA^{Ai}_6$	$-\frac{1}{2}$	$-\frac{1}{2}$	0	0	-1
$A^{Ai}_{\substack{78\\(+)}}$	$A^{Ai}_7 - iA^{Ai}_8$	$-\frac{1}{2}$	$+\frac{1}{2}$	0	0	0
$A^{Ai}_{\substack{56\\(+)}}$	$A^{Ai}_5 - iA^{Ai}_6$	$+\frac{1}{2}$	$+\frac{1}{2}$	0	0	+1

$$
\begin{aligned}
\tau^{13}\,(A^{Ai}_7 \mp i\,A^{Ai}_8) &= \pm\frac{1}{2}\,(A^{Ai}_7 \mp i\,A^{Ai}_8)\,,\\
Y\,(A^{Ai}_7 \mp i\,A^{Ai}_8) &= \mp\frac{1}{2}\,(A^{Ai}_7 \mp i\,A^{Ai}_8)\,,\\
Q\,(A^{Ai}_7 \mp i\,A^{Ai}_8) &= 0\,.
\end{aligned}
\qquad (21)
$$

Since τ^4, Y, τ^{13} and τ^{1+}, τ^{1-} give zero if applied on $(A^Q_s$, $A^{Q'}_s$ and $A^{Y'}_s)$ with respect to the quantum numbers (Q, Q', Y'), and since Y and τ^{13} commute with the family quantum numbers, one sees that the scalar fields A^{Ai}_s ($=(A^Q_s, A^Y_s, A^{Y'}_s, \tilde{A}^{\tilde{4}}_s, \tilde{A}^{\tilde{Q}}_s, \vec{\tilde{A}}^{\tilde{1}}_s, \vec{\tilde{A}}^{\tilde{2}}_s, \vec{\tilde{A}}^{\vec{\tilde{N}}_R}_s, \vec{\tilde{A}}^{\vec{\tilde{N}}_L}_s)$), rewritten as $A^{Ai}_{\substack{78\\(\pm)}} = (A^{Ai}_7 \mp i\,A^{Ai}_8)$, are eigenstates of τ^{13} and Y, having the quantum numbers of the *standard model* Higgs' scalar.

These superposition of $A^{Ai}_{\substack{78\\(\pm)}}$ are presented in Table 6 as two doublets with respect to the weak charge τ^{13}, with the eigenvalue of τ^{23} (the second $SU(2)_{II}$ charge) equal to either $-\frac{1}{2}$ or $+\frac{1}{2}$, respectively.

The operators $\tau^{1\substack{\boxplus\\\boxminus}} = \tau^{11} \pm i\tau^{12}$

$$
\tau^{1\substack{\boxplus\\\boxminus}} = \frac{1}{2}[(\mathcal{S}^{58} - \mathcal{S}^{67}) \substack{\boxplus\\\boxminus}\, i\,(\mathcal{S}^{57} + \mathcal{S}^{68})]\,, \qquad (22)
$$

transform one member of a doublet from Table 6 into another member of the same doublet, keeping τ^{23} $(= \frac{1}{2}\,(\mathcal{S}^{56} + \mathcal{S}^{78}))$ unchanged, clarifying the above statement.

It is shown in Ref. ([1], Eq. (22)) that the scalar fields $A^{Ai}_{\substack{78\\(\pm)}}$ are either *triplets* as the gauge fields of the *family quantum numbers* $(\vec{\tilde{N}}_R, \vec{\tilde{N}}_L, \vec{\tilde{\tau}}^2, \vec{\tilde{\tau}}^1)$ or they are singlets as the gauge fields of $Q = \tau^{13} + Y$, $Q' = -\tan^2\vartheta_1 Y + \tau^{13}$ and $Y' = -\tan^2\vartheta_2 \tau^4 + \tau^{23}$.

One finds

$$
\begin{aligned}
\tilde{N}^3_L\,\tilde{A}^{\tilde{N}_L\substack{\boxplus\\\boxminus}}_{\substack{78\\(\pm)}} &= \substack{\boxplus\\\boxminus}\,\tilde{A}^{\tilde{N}_L\substack{\boxplus\\\boxminus}}_{\substack{78\\(\pm)}}\,, \quad \tilde{N}^3_L\,\tilde{A}^{\tilde{N}_L 3}_{\substack{78\\(\pm)}} = 0\,,\\
Q\,A^{Q}_{\substack{78\\(\pm)}} &= 0\,.
\end{aligned}
\qquad (23)
$$

with $Q = \mathcal{S}^{56} + \tau^4 = \mathcal{S}^{56} - \frac{1}{3}(\mathcal{S}^{9\,10} + \mathcal{S}^{11\,12} + \mathcal{S}^{13\,14})$, and with τ^4 defined in the footnote [8] (if one replaces S^{ab} by $\mathcal{S}^{ab}$ from Eq. (18)).

Similarly one finds properties with respect to the Ai quantum numbers for all the scalar fields $A^{Ai}_{\stackrel{78}{(\pm)}}$.

After the appearance of the condensate (Table 1), which breaks the $SU(2)_{II}$ symmetry (bringing masses to all the scalar fields), the weak charge $\vec{\tau}^1$ and the hyper charge Y remain the conserved charges [14].

The nonzero vacuum expectation values of the scalar fields of Eq. (19) break the mass protection mechanism of quarks and leptons and determine correspondingly the mass matrices (Eq. (31)) of the two groups of quarks and leptons.

Obviously the scalar fields in the *spin-charge-family* theory have all the properties of the Higgs. I show below and in Subsect. 4.1 that these scalar fields explain also the Yukawa couplings of the *standard model.*

All other scalar fields: A^{Ai}_s, $s \in (5,6)$ and A^{Ai}_t, $t \in (9,\dots,14)$ have masses of the order of the condensate scale and contribute to matter-antimatter asymmetry [2].

Effective action for scalar fields with the space index $(7,8)$

It would be possible, at least in principle, to derive the low energy effective action for scalars from the starting action (Eq. 1) by guessing the boundary conditions, under which the universe evolved, since all the scalar fields, as well as their Lagrange density, are included in the starting action. This is an extremely demanding project.

In what follows the effective Lagrange density for the scalar fields is assumed to be changed from the Lagrange density before the electroweak break $\mathcal{L}_s = E\,\{(p_m\, A^{Ai}_s)^\dagger\,(p^m\, A^{Ai}_s) - (m'_{Ai})^2\, A^{Ai\dagger}_s\, A^{Ai}_s\}$ to

$$\begin{aligned}\mathcal{L}_{sg} &= E\sum_{A,i}\{(p_m\, A^{Ai}_s)^\dagger\,(p^m\, A^{Ai}_s) - (-\lambda^{Ai} + (m'_{Ai})^2))\, A^{Ai\dagger}_s A^{Ai}_s \\ &+ \sum_{B,j} \Lambda^{AiBj}\, A^{Ai\dagger}_s A^{Ai}_s\, A^{Bj\dagger}_s A^{Bj}_s\}\,, \end{aligned} \tag{24}$$

where $-\lambda^{Ai} + m'^2_{Ai} = m^2_{Ai}$ and m_{Ai} manifests as the mass of the A^{Ai}_s scalar.

The Lagrange density leads to the coupled equations of motion for many scalar fields with, in this assumption, harmonic interactions. It requires a lot of effort to extract the dependence of the eigen modes on the parameters of the Lagrange density to see the influence of the parameters on the properties of fermions. This work has not yet been done. First attempts are in progress.

Yukawa couplings in the spin-charge-family theory

Let $\psi^\alpha_{(L,R)}$ denote massless and $\Psi^\alpha_{(L,R)}$ massive four vectors for each family member $\alpha = (u_{L,R}, d_{L,R}, \nu_{L,R}, e_{L,R})$, let say for the group of four families among which there are the observed three families, after taking into account loop corrections in all orders.

$$\psi^\alpha_{(L,R)} = V^\alpha_{(L,R)}\, \Psi^\alpha_{(L,R)}\,, \tag{25}$$

and let $(\psi^{\alpha\, k}_{(L,R)}\,,\ \Psi^{\alpha\, k}_{(L,R)})$ be any component of the four vectors, massless and massive, respectively. On the tree level we have $\psi^\alpha_{(L,R)} = V^\alpha_{(o)}\, \Psi^{\alpha\,(o)}_{(L,R)}$ and

$$<\psi^\alpha_L|\gamma^0\, \mathcal{M}^\alpha_{(o)}\, |\psi^\alpha_R> = <\Psi^{\alpha\,(o)}_L|\gamma^0\, V^{\alpha\,\dagger}_{(o)}\, \mathcal{M}^\alpha_{(o)}\, V^\alpha_{(o)}\, |\Psi^\alpha_{R\,(o)}>, \tag{26}$$

[14] It is τ^{23} which determines the hyper charge Y ($Y = S^{23} + \tau^4$) of these scalar fields, since τ^4, if applies on the scalar index of these scalar fields, gives zero, according to equations in the footnote [8].

with $\mathcal{M}^{\alpha}_{(o)k\,k'} = \sum_{A,i}(-g^{Ai}\,v_{Ai\mp})\,C^{\alpha}_{k\,k'}$. Here $g^{Ai}\,v_{Ai\mp}$ represent the nonzero vacuum expectation values of the scalar fields. In this case the coefficients $C^{\alpha}_{k\,k'}$ are determined by the mass matrices, Eq.(31), on the tree level. It then follows

$$\begin{aligned}\overline{\Psi}^{\alpha}\,V^{\alpha\dagger}_{(o)}\,\mathcal{M}^{\alpha}_{(o)}\,V^{\alpha}_{(o)}\,\Psi^{\alpha} &= \overline{\Psi}^{\alpha}\,\mathrm{diag}(m^{\alpha}_{(o)1},\cdots,m^{\alpha}_{(o)4})\,\Psi^{\alpha}\,,\\ V^{\alpha\dagger}_{(o)}\,\mathcal{M}^{\alpha}_{(o)}\,V^{\alpha}_{(o)} &= \Phi^{\alpha}_{\Psi(o)}\,.\end{aligned} \tag{27}$$

On the tree level the coupling constants $m^{\alpha}_{(o)k}$ (in some units) of the dynamical scalar fields $\Phi^{\alpha}_{\Psi(o)\,k}$ - the superposition of A^{Ai}_{s} - to the family member $\Psi^{\alpha\,k}$ belonging to the kth family are equal to

$$(\Phi^{\alpha}_{\Psi(o)})_{k\,k'}\,\Psi^{\alpha\,k'} = \delta_{k\,k'}\,m^{\alpha}_{(o)k}\,\Psi^{\alpha\,k}\,. \tag{28}$$

The superposition of scalar fields $(\Phi^{\alpha}_{\Psi(o)})$, which couple to fermions and depend on the quantum numbers α and k, are in general different from the superposition, which are their mass eigenstates. Each family member α of each massive family k couples in general to different superposition of scalar fields.

It turns out that mass matrices of both - quarks and leptons - behave in a very similar way. No additional neutrinos, offering a "sea-saw" mechanism, are needed. All this is already included in the starting action.

3.4. The condensate in the spin-charge-family theory

The appearance of the condensate of two right handed neutrinos with properties presented in Table 1 is in this paper assumed so that in the low energy regime the *spin-charge-family* theory leads to the effective action, explaining the assumptions of the *standard model* and consequently the observed phenomena. The condensate should appear during the expansion of the universe, due to particular boundary conditions and the conditions in the universe in the time of the appearance of the condensate. This study has not yet been done.

The condensate, presented on Table 1, does not influence the colour, the weak and the hyper charges ($\vec{\tau}^{3}$, $\vec{\tau}^{1}, Y$, respectively) of the corresponding gauge fields. Since these vector gauge fields don't interact with the condensate, the colour, the weak and the hyper charges remain the conserved quantities up to the electroweak phase transition.

The condensate changes the properties of the scalar fields, which are before the appearance of the condensate massless scalar gauge fields. Interaction with the condensate makes all the scalar fields massive.

After the electroweak break, when the scalar fields with the space index $s = (7,8)$ - those with the family quantum numbers $(\tilde{N}^{i}_{(L,R)}, \tilde{\tau}^{(1,2)i})$ and those with the family members quantum numbers (Q, Q', Y') - start to strongly self interact (Eq. (24)), gaining nonzero vacuum expectation values (Eq. (24)) and correspondingly changing their own masses as well as the properties of the vacuum, so that the weak charge and the hyper charge are no longer conserved quantities. The only conserved charges are then the colour and the electromagnetic charges.

4. Summary of the *spin-charge-family* theory achievements so far

To understand better the history of our universe the explanation of the *standard model* assumptions is certainly needed. It is also needed to know the number of families in the low energy regime and to understand the appearance of phenomena like the existence of the dark matter, the matter-antimatter asymmetry and the dark energy.

IOP Conf. Series: Journal of Physics: Conf. Series **845** (2017) 012017 doi:10.1088/1742-6596/845/1/012017

I have demonstrated so far, that the *spin-charge-family* theory, starting with the simple action in $d = (13+1)$ for fermions (carrying only two kinds of spins, no charges) and for the gauge fields to which fermions are coupled (vielbeins and two kinds of spin connection fields) offers the explanation for all the assumptions of the *standard model*:
a. The theory explains all the properties of the family members - quarks and leptons, left and right handed, relating handedness and charges, and their right and left handed antiquarks and antileptons.
b. It explains the appearance and properties of the families of family members.
c. It explains the existence of the gauge vector fields of the family members charges.
d. It explains the appearance and properties of the scalar field (the Higgs) and the Yukawa couplings.

The *spin-charge-family* theory predicts that there are at the low energy regime two decoupled groups of four families of quarks and leptons, what means that besides the observed three there is the fourth not yet observed family of quarks and leptons.

The existence of two decoupled groups of four families also means that the stable of the upper four families must also be observed. In Subsect. 4.4 [13] the possibility is discussed that the stable of the upper four families constitutes the dark matter.

In this section I overview the *spin-charge-family* theory achievements, explaining:
i. The properties of the lower four families, the three of which have already been observed, as they follow from the properties of the scalar fields of this theory, Subsect. 4.1,
i. a. presenting the results of the calculations (not yet published),
i. b. discussing whether or not present experiments speak or not against the existence of the fourth family (not yet published), in particular I shall comment the contribution of the fourth family to the production of the Higgs in the quark-fusion process, Subsect. 4.2, the topics which seem to speak the most against the existence of the fourth family.
ii. The fact that this theory easily explains the "miraculous" cancellation of the triangle anomalies in the *standard model*, Subsect. 4.3 (also this topics is not yet published).
iii. The existence of the dark matter, Subsect. 4.4.
iv. The explanation for the matter-antimatter asymmetry, Subsect. 4.5.

4.1. Masses of the lower four families of quarks and leptons in the spin-charge-family theory [12, 14, 15]

This subsection is a short report of the not yet published results of Ref. [15] (published in the Proceedings).

There are two groups of four families. The mass matrix of each family member of each of the group of four families demonstrates in the massless basis the $U(1) \times \widetilde{SU}(2) \times \widetilde{SU}(2)$ symmetry (each of the two $\widetilde{SU}(2)$ is a subgroup, one of $\widetilde{SO}(3,1)$ and the other of $\widetilde{SO}(4)$).

The scalars with the family quantum numbers split the eight families into twice four families. To the masses of the lower four families the scalar fields, which are the gauge fields of $\vec{\tilde{N}}_L$ and $\vec{\tilde{\tau}}^1$ contribute. To the masses of the upper four families the gauge fields of $\vec{\tilde{N}}_R$ and $\vec{\tilde{\tau}}^2$ contribute. The scalars with the family members quantum numbers (Q, Q', Y') contribute to the masses of the lower and upper four families.

I discuss here properties of quarks and leptons of the lower four families, Eq. (31).

Let ψ_i, $i = (1,2,3,4)$, denote the massless basis for a particular family member α. And let us denote the two kinds of the operators, which transform the basis vectors into one another as

$$\tilde{N}^i_L \,,\; i = (1,2,3) \,, \qquad \tilde{\tau}^i_L \,,\; i = (1,2,3) \,. \tag{29}$$

One finds

$$\begin{aligned}
\tilde{N}^3_L\,(\psi_1,\psi_2,\psi_3,\psi_4) &= \frac{1}{2}(-\psi_1,\ \psi_2, -\psi_3,\ \psi_4)\,,\\
\tilde{N}^+_L\,(\psi_1,\psi_2,\psi_3,\psi_4) &= \ (\psi_2,\ 0,\ \psi_4,\ 0)\,,\\
\tilde{N}^-_L\,(\psi_1,\psi_2,\psi_3,\psi_4) &= \ (0\ ,\ \psi_1,\ 0,\ \psi_3)\,,\\
\tilde{\tau}^{13}\ (\psi_1,\psi_2,\psi_3,\psi_4) &= \frac{1}{2}(-\psi_1, -\psi_2,\ \psi_3,\ \psi_4)\,,\\
\tilde{\tau}^{1+}\ (\psi_1,\psi_2,\psi_3,\psi_4) &= \ (\psi_3,\ \psi_4,\ 0,\ 0)\,,\\
\tilde{\tau}^{1-}\ (\psi_1,\psi_2,\psi_3,\psi_4) &= \ (\ 0,\ 0,\ \psi_1,\ \psi_2)\,.
\end{aligned} \tag{30}$$

This is indeed what the two $SU(2)$ operators in the *spin-charge-family* theory do on the lower four families. The gauge scalar fields $A^{Ai}_{\substack{78\\(\pm)}}$, $(A,i) = [\tilde{N}^i_L, i = (\pm,3), \tilde{\tau}^{1i}, i = (\pm,3)]$, Eqs. (20, 19), of these operators determine the off diagonal and diagonal matrix elements after the electroweak phase transition in which scalar fields gain nonzero vacuum expectation values.

In addition to these two kinds of $SU(2)$ scalars there are three $U(1)$ scalars, which distinguish among the family members, contributing on the tree level the same diagonal matrix elements for all the families.

In loop corrections in all orders the symmetry of mass matrices remains unchanged, while the three $U(1)$ scalars manifest in off diagonal elements as well.

All the scalars, the two triplets and the three singlets, are doublets with respect to the weak charge, contributing to the weak and the hyper charge of the fermions so that they transform the right handed members into the left handed onces with the phases presented in Table 2.

$$\mathcal{M}^\alpha = \begin{pmatrix} -a_1 - a & e & d & b \\ e & -a_2 - a & b & d \\ d & b & a_2 - a & e \\ b & d & e & a_1 - a \end{pmatrix}^\alpha . \tag{31}$$

Although any accurate 3×3 submatrix of the 4×4 unitary matrix determines the 4×4 matrix uniquely, neither the quark nor (in particular) the lepton 3×3 mixing matrix are measured accurately enough that it would be possible to determine three complex phases of the 4×4 mixing matrix as well as the mixing matrix elements of the fourth family members to the lower three.

We therefore assumed in our calculations [12, 14, 15] that the mass matrices are symmetric and real. Correspondingly the mixing matrices are orthogonal. We fitted the 6 free parameters of each quark mass matrix, Eq. (31), to twice three, that is 6, measured quark masses, and to the 6 (from the experimental data extracted) parameters of the corresponding 4×4 mixing matrix.

While the experimental accuracy of the quark masses of the lower three families does not influence the calculated mass matrices considerably, it turned out that the experimental accuracy of the 3×3 quark mixing matrix is not good enough to trustworthy determine the mass intervals for the fourth family quarks.

Taking into account our calculations, in which we fit parameters of Eq. (31) to the experimental data for masses and mixing matrices for quarks and the meson decays evaluations in the literature, as well as our own evaluations, we estimated that the fourth family quarks masses might be above 1 TeV. Choosing the masses of the fourth family quarks we were able not only to calculate the fourth family matrix elements to the lower three families, but also predict towards which values will the matrix elements of the 3×3 submatrix move in more accurate experiments [15].

IOP Conf. Series: Journal of Physics: Conf. Series **845** (2017) 012017 doi:10.1088/1742-6596/845/1/012017

The two fitted mass matrices, Ref. ([15], Eqs. (23, 27)) lead to masses of Eq. (32) for the choice of $M_{u_4}/\text{MeV/c}^2 = 700\,000 = M_{d_4}/\text{MeV/c}^2$

$$\begin{aligned}\mathbf{M}^u/\text{MeV/c}^2 &= (1.3, 620.0, 172\,000., 700\,000.)\,,\\ \mathbf{M}^d/\text{MeV/c}^2 &= (2.88508, 55.024, 2\,899.99, 700\,000.)\,,\end{aligned} \tag{32}$$

and to masses of Eq. (33) for the choice of $M_{u_4}/\text{MeV/c}^2 = 1\,200\,000 = M_{d_4}/\text{MeV/c}^2$

$$\begin{aligned}\mathbf{M}^u/\text{MeV/c}^2 &= (1.3, 620.0, 172\,000., 1\,200\,000.)\,,\\ \mathbf{M}^d/\text{MeV/c}^2 &= (2.88508, 55.024, 2\,899.99, 1\,200\,000.)\,.\end{aligned} \tag{33}$$

They lead to the 4×4 mixing matrix in which we fit two kinds of the experimental - the old data (exp_o) and the new data (exp_n) - each used in calculations for the choice $m_{u_4} = m_{d_4} = 700$ GeV (old_1, new_1) and $m_{u_4} = m_{d_4} = 1\,200$ GeV (old_2, new_2), Eq. (34)

$$|V_{(ud)}| = \begin{pmatrix} exp_o & 0.97425\pm0.00022 & 0.2252\pm0.0009 & 0.00415\pm0.00049\\ exp_n & 0.97425\pm0.00022 & 0.2253\pm0.0008 & 0.00413\pm0.00049\\ \hline old_1 & 0.97423 & 0.22531 & 0.003\\ old_2 & 0.97425 & 0.22536 & 0.00301\\ new_1 & 0.97423 & 0.22531 & 0.00299\\ new_2 & 0.97423 & 0.22538 & 0.00299\\ \hline exp_o & 0.230\pm0.011 & 1.006\pm0.023 & 0.0409\pm0.0011\\ exp_n & 0.225\pm0.008 & 0.986\pm0.016 & 0.0411\pm0.0013\\ \hline old_1 & 0.22526 & 0.97338 & 0.042\\ old_2 & 0.22534 & 0.97336 & 0.04239\\ new_1 & 0.22534 & 0.97335 & 0.04245\\ new_2 & 0.22531 & 0.97336 & 0.04248\\ \hline exp_o & 0.0084\pm0.0006 & 0.0429\pm0.0026 & 0.89\pm0.07\\ exp_n & 0.0084\pm0.0006 & 0.0400\pm0.0027 & 1.021\pm0.032\\ \hline old_1 & 0.00663 & 0.04197 & 0.9991\\ old_2 & 0.00663 & 0.04198 & 0.9991\\ new_1 & 0.00667 & 0.04203 & 0.99909\\ new_2 & 0.00667 & 0.04206 & 0.99909 \end{pmatrix}. \tag{34}$$

It was noticed [15] that the mass matrix elements of the u and d quarks matrices change by a factor of ≈ 1.5 in the average, when the masses of the fourth family members grow from 700 GeV to 1 200 GeV. The two mass matrices become more "democratic" (the matrix elements become closer to each other). Although the mixing matrix elements of the 3×3 submatrix of the 4×4 matrix do not change a lot with the masses of the fourth family quarks, they do change so that they agree better with the newer [19] than with the older [20] experimental values.

From the above results it follows:

i. The prediction of the calculated mixing matrix elements, obtained by fitting the symmetry of the mass matrices (Eq. (31)) to the experimental data [20], was confirmed by more accurate experimental data [19]. In all cases are the calculated 3×3 matrix elements closer to the new experimental values than to the old experimental values.

ii. The fourth family masses change the mass matrices considerably, while their influence on the 3×3 submatrix of the 4×4 mixing matrix is much weaker.

iii. We expect that more accurate experiments will bring a slightly smaller values for ($V_{u_1 d_1}$, $V_{u_1 d_3}$, $V_{u_3 d_3}$), smaller ($V_{u_2 d_2}$, $V_{u_3 d_1}$), ($V_{u_1 d_2}$, $V_{u_2 d_1}$) will slightly grow and ($V_{u_2 d_3}$) $V_{u_3 d_2}$ will grow.

iv. The matrix elements $V_{u_i d_4}$ and $V_{u_4 d_i}$ change considerably with the mass of the fourth family members, and they differ quite a lot also when using new instead of the old experimental data

IOP Conf. Series: Journal of Physics: Conf. Series **845** (2017) 012017 doi:10.1088/1742-6596/845/1/012017

for the mixing matrix.
v. Fitting (twice 6) free parameters of the mass matrices to the new experimental data [19] gives smaller uncertainty in fitting procedure than when fitting to the old experimental data [20], while the masses of the fourth family members do not influence the uncertainty of the calculations considerably. Only very accurate mixing matrix elements would allow to determine fourth family quarks masses more accurately.
vi. It is difficult to predict the interval for the masses of the fourth family members since the choice of the fourth family quark masses does not appreciably influence either the fitting procedure or the obtained 3×3 mixing submatrix, and also not the accuracy of the masses of the three lower families. Other experimental data, like decays of mesons, speaks for masses of the fourth family quarks close or above 1 TeV.
vii. For the masses of the fourth family members above 1 TeV the mass matrices are close to the democratic matrix: The matrix elements are closer to one another the higher is the mass of the fourth family member. In such a case are the fourth family masses mostly determined by the scalars carrying the family quantum numbers. Correspondingly the masses of the u_4-quarks are closer to the masses of the d_4-quarks.

The complex mass matrices would lead to unitary and not to orthogonal mixing matrices. The more accurate experimental data for quarks mixing matrix would allow us to extract also the phases of the unitary mixing matrix, allowing us to predict the fourth family masses.

4.2. Is the existence of the fourth family in agreement with the present experiments?

This part is following equivalent part in my contribution to the Proceedings to the Conference on New Physics at the Large Hadron Collider, 29 February – 4 March, **2016**, Nanyang Executive Centre, NTU, Singapore. The paper, a more elaborated version of this notice, is in preparation.

The *spin-charge-family* theory predicts the existence of the fourth family to the observed three, while there has been no direct observation of the fourth family quarks with the masses below 1 TeV. The fourth family quarks with masses above 1 TeV contribute according to the *standard model* (the *standard model* Yukawa couplings of quarks to the scalar Higgs are proportional to $\frac{m_4^\alpha}{v}$, where m_4^α is the fourth family member ($\alpha = u, d$) mass and v the vacuum expectation value of the scalar) to either the quark-gluon fusion production of the scalar field (the Higgs) or to the scalar field decay into two photons $\approx$ 10 times too much in comparison with the observations. Correspondingly the high energy physicists do not expect the existence of the fourth family members at all [26].

I am stressing [30] in this subsection that the u_i-quarks and d_i-quarks of an ith family, if they couple with the opposite sign (with respect to the "$\pm$" degree of freedom) to the scalar fields carrying the family $(\tilde{A}, i)$ quantum numbers [15] and have the same masses, do not contribute to either the quark-gluon fusion production of the scalar fields with the family quantum numbers or to the decay of these scalars into two photons. Since the u_4-quarks and d_4-quarks might have similar masses (Subsect. 4.1) and since their masses are for $m_{u_4} > 1$ TeV and $m_{d_4} > 1$ TeV mostly determined by the scalars with the family quantum numbers, the observations so far are consequently not in contradiction with the *spin-charge-family* theory prediction that there exists the fourth family coupled to the observed three.

The couplings of u_i and d_i to the scalars carrying the family members quantum numbers are determined besides by the corresponding couplings also by the eigenvalues of the operators (Q, Q', Y') on the quarks states (which do distinguish between u_i and d_i).

The strong influence of the scalar fields carrying the family members quantum numbers to the masses of the lower (observed) three families manifests in the huge differences in the masses

[15] The contribution of the scalar fields $\tilde{A}_{\pm}^{\tilde{A}i}$, $(\tilde{A}, i) = (\tilde{\tau}^{1i}, \tilde{N}_L^i)$, Eq. 19), are the same for all the family members.

Table 7. The weak, hyper and electromagnetic charges for quarks in their massless basis are presented. The colour charge is not shown. These and other properties of quarks and leptons can be read from Tables 2–4.

state	τ^{13}	Y	Q
u_{Ri}	0	$\frac{2}{3}$	$\frac{2}{3}$
u_{Li}	$\frac{1}{2}$	$\frac{1}{6}$	$\frac{2}{3}$
d_{Ri}	0	$-\frac{1}{3}$	$-\frac{1}{3}$
d_{Li}	$-\frac{1}{2}$	$\frac{1}{6}$	$-\frac{1}{3}$

of u_i and d_i, $i = (1, 2, 3)$, among families (i) and family members (u, d). For the fourth family quarks, which are more and more decoupled from the observed three families the higher are their masses [15, 14], the influence of the scalar fields carrying the family members quantum numbers on their masses is expected to be much weaker. Correspondingly the u_4 and d_4 masses become closer to each other the higher are their masses and the weaker are their couplings (the mixing matrix elements) to the lower three families.

If the masses of the fourth family quarks are close to each other, then u_4 and d_4 contribute in the quark-gluon fusion very little to the production of the scalar field - the Higgs - which is mostly superposition of the scalar fields with the family members quantum numbers, what is in agreement with the observation: In the quark-gluon fusion production of the Higgs mostly the top (u_3) contributes.

In Tables 2–4 the phases of all the states are chosen to be 1. Here I use different phases, those which enable the usual presentation of fermions under the change of spin and under $\mathbb{C}_{\mathcal{N}} \cdot \mathcal{P}_{\mathcal{N}}$.

In Table 7 the properties of u and d quarks (and their antiquarks) needed in Fig. 1 are presented. Fig. 1 presents the properties of u and d quarks, contributing in the quark-gluon fusion to the production of the scalar fields $\Phi^{Ai}_{\substack{78\\(\pm)}}$, Eq.(19), ($A^{Ai}_{\substack{78\\(\pm)}} = \Phi^{Ai}_{\substack{78\\(\pm)}} + v^{Ai}_{\substack{78\\(\pm)}}$). One notices the opposite signs of the couplings of u_i with respect to d_i for ether Φ_- or for Φ_+. Correspondingly the fourth family quarks of almost the same mass contribute very little to the production of any scalar, with the scalar Higgs included, which is in agreement with the observation. Also to the decay of the Higgs can the fourth family quarks contribute very little: It is, as observed, the u_3- quark (the top), which contributes the most to either the production or to the decay of the Higgs.

The fourth family quarks can still contribute to the production and to the decay of the scalars of the masses of a few TeV, to which they couple stronger than to the Higgs.

The figures are valid for any Ai and correspondingly also for any superposition of $\Phi^{Ai}_{\pm}$.

4.3. Anomaly cancellation in the spin-charge-family theory [16]

In the *standard model* the triangle anomalies "miraculously" disappear due to the fact that the sum of all possible traces $Tr[\tau^{Ai}\tau^{Bj}\tau^{Ck}]$, where $(\tau^{Ai}, \tau^{Bi}, \tau^{Ck})$ are the generators of one, of two or of three of the groups of $SU(3), SU(2)$ and $U(1)$) over the representations of one family of the left handed fermions and their anti-fermions (and separately of the right handed fermions and their anti-fermions), contributing to the triangle currents, are equal to zero [22, 23, 24, 25].

Let me demonstrate that this cancellation of the *standard model* triangle anomaly follows straightforwardly, if the $SO(3,1), SU(2), U(1)$ and $SU(3)$ are considered as subgroups of the

[16] A more elaborated version of this report is in preparation.

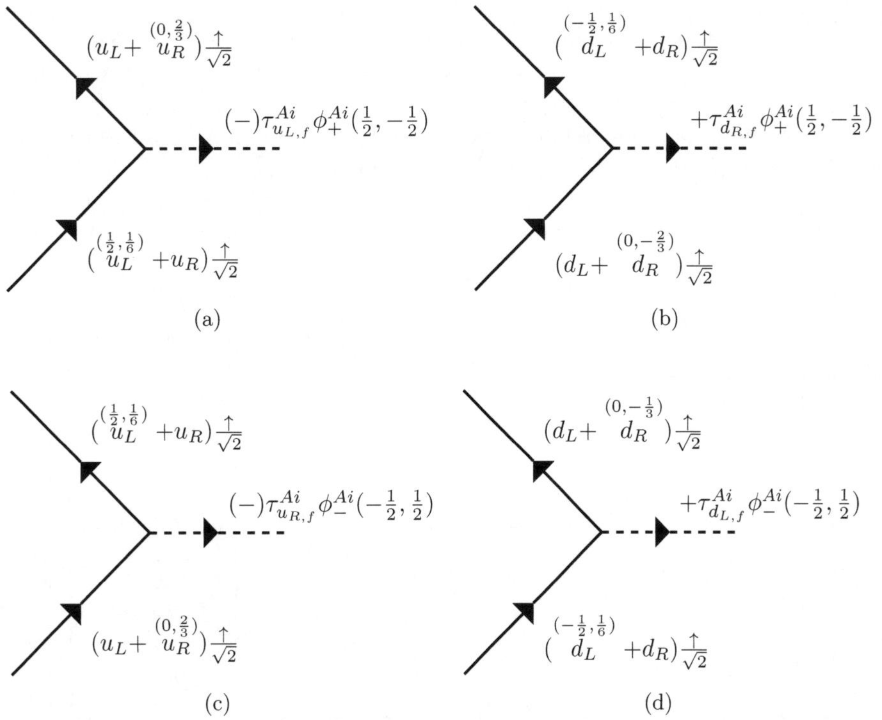

Figure 1. The contributions of u and d quarks to the production of the scalar fields Φ^{Ai}_{-} and Φ^{Ai}_{+}, when τ^{Ai} represent the family quantum numbers (which are for the lower four families $\tilde{\tau}^{1i}$ and $\tilde{N}^{i}_{L}$) or the family members quantum numbers (Q, Q', Y'), are presented: (a) the u-quark contribution to the scalar fields Φ^{Ai}_{+}, (b) the d-quark contribution to Φ^{Ai}_{+}, (c) the u-quark contribution to Φ^{Ai}_{-}, (d) the d-quark contribution to Φ^{Ai}_{-}. $\tau^{Ai}_{u_{(L,R)},f}$ and $\tau^{Ai}_{d_{(L,R)},f}$ denote the application values of the operators $\tilde{\tau}^{1i}$, $\tilde{N}^{i}_{L}$ and of (Q, Q', Y') on the states. While $\tilde{\tau}^{1i}$ and $\tilde{N}^{i}_{L}$ do not distinguish among family members u and d so that in this case the contribution of u and d have opposite signs, (Q, Q', Y') do, influencing the signs and the values.

orthogonal group $SO(13, 1)$.

To the triangle anomaly the right-handed spinors (fermions) and antispinors contribute with the opposite sign than the left handed spinors and their antispinors. Their common contribution to anomalies is proportional to [24]

$$\Big(\sum_{(A,i,B,j,C,k)_{L\,\bar{L}}} Tr[\tau^{Ai}\,\tau^{Bj}\,\tau^{Ck}] - \sum_{(A,i,B,j,C,k)_{R\,\bar{R}}} Tr[\tau^{Ai}\,\tau^{Bj}\,\tau^{Ck}]\Big)\,, \tag{35}$$

where τ^{Ai} are in the *standard model* the generators of the infinitesimal transformation of the groups $SU(3), SU(2)$ and $U(1)$, while in the *spin-charge-family* theory τ^{Ai} are irreducible subgroups of the starting orthogonal group $SO(2(2n+1)-1, 1)$, $n = 3$. The indexes $_{L\,\bar{L}}$ $(_{R\,\bar{R}})$ denote the left (right) handed spinors and their antispinors (right (left)), respectively.

In the first seven columns (up to $||$) of Table 8 the properties of one family members assumed by the *standard model*, running in the triangle, are presented. The last two columns - taken from Table 2 - describe additional properties which quarks and leptons (and antiquarks and antileptons) would have, if the *standard model* groups $SO(3, 1), SU(2), SU(3)$ and $U(1)$ are embedded into the $SO(13, 1)$ group. To demonstrate that the "miraculous" cancellation of the

Table 8. Properties of the left handed quarks and leptons and antiqurks and antileptons, and of the right handed quarks and leptons and antiquarks and antileptons, as assumed by the *standard model*, are presented in the first seven columns. In the last two columns the two quantum numbers are added, which the fermions and anti-fermions would have if the *standard model* groups $SO(3,1), SU(2), SU(3)$ and $U(1)$ are embedded into the $SO(13,1)$ group. The whole quark part appears, due to the colour charges, three times. One can check that the hyper charge is the sum of $\tau^{4}_{i_{L,R}} + \tau^{23}_{i_{L,R}}$ (Table 2). The quantum numbers are the same for all the families.

i_L	name	hand-edness $\Gamma^{(3,1)}$	weak charge τ^{13}	hyper charge Y	colour τ^{33}	charge τ^{38}	elm charge Q	$SU(2)_{II}$ charge τ^{23}	$U(1)_{II}$ charge τ^{4}
1_L	u_L	-1	$\frac{1}{2}$	$\frac{1}{6}$	$\frac{1}{2}$	$\frac{1}{2\sqrt{3}}$	$\frac{2}{3}$	0	$\frac{1}{6}$
2_L	d_L	-1	$-\frac{1}{2}$	$\frac{1}{6}$	$\frac{1}{2}$	$\frac{1}{2\sqrt{3}}$	$-\frac{1}{3}$	0	$\frac{1}{6}$
3_L	u_L	-1	$\frac{1}{2}$	$\frac{1}{6}$	$-\frac{1}{2}$	$\frac{1}{2\sqrt{3}}$	$\frac{2}{3}$	0	$\frac{1}{6}$
4_L	d_L	-1	$-\frac{1}{2}$	$\frac{1}{6}$	$-\frac{1}{2}$	$\frac{1}{2\sqrt{3}}$	$-\frac{1}{3}$	0	$\frac{1}{6}$
5_L	u_L	-1	$\frac{1}{2}$	$\frac{1}{6}$	0	$-\frac{1}{\sqrt{3}}$	$\frac{2}{3}$	0	$\frac{1}{6}$
6_L	d_L	-1	$-\frac{1}{2}$	$\frac{1}{6}$	0	$-\frac{1}{\sqrt{3}}$	$-\frac{1}{3}$	0	$\frac{1}{6}$
7_L	ν_L	-1	$\frac{1}{2}$	$-\frac{1}{2}$	0	0	0	0	$-\frac{1}{2}$
8_L	e^L	-1	$-\frac{1}{2}$	$-\frac{1}{2}$	0	0	-1	0	$-\frac{1}{2}$
9_L	$\bar{u}L$	-1	0	$-\frac{2}{3}$	$-\frac{1}{2}$	$-\frac{1}{2\sqrt{3}}$	$-\frac{2}{3}$	$-\frac{1}{2}$	$-\frac{1}{6}$
10_L	$\bar{d}L$	-1	0	$\frac{1}{3}$	$-\frac{1}{2}$	$-\frac{1}{2\sqrt{3}}$	$\frac{1}{3}$	$\frac{1}{2}$	$-\frac{1}{6}$
11_L	$\bar{u}L$	-1	0	$-\frac{2}{3}$	$\frac{1}{2}$	$-\frac{1}{2\sqrt{3}}$	$-\frac{2}{3}$	$-\frac{1}{2}$	$-\frac{1}{6}$
12_L	$\bar{d}L$	-1	0	$\frac{1}{3}$	$\frac{1}{2}$	$-\frac{1}{2\sqrt{3}}$	$\frac{1}{3}$	$\frac{1}{2}$	$-\frac{1}{6}$
13_L	$\bar{u}L$	-1	0	$-\frac{2}{3}$	0	$\frac{1}{\sqrt{3}}$	$-\frac{2}{3}$	$-\frac{1}{2}$	$-\frac{1}{6}$
14_L	$\bar{d}L$	-1	0	$\frac{1}{3}$	0	$\frac{1}{\sqrt{3}}$	$\frac{1}{3}$	$\frac{1}{2}$	$-\frac{1}{6}$
15_L	$\bar{\nu}_L$	-1	0	0	0	0	0	$-\frac{1}{2}$	$\frac{1}{2}$
16_L	$\bar{e}_L$	-1	0	1	0	0	1	$\frac{1}{2}$	$\frac{1}{2}$
1_R	u_R	1	0	$\frac{2}{3}$	$\frac{1}{2}$	$\frac{1}{2\sqrt{3}}$	$\frac{2}{3}$	$\frac{1}{2}$	$\frac{1}{6}$
2_R	d_R	1	0	$-\frac{1}{3}$	$\frac{1}{2}$	$\frac{1}{2\sqrt{3}}$	$-\frac{1}{3}$	$-\frac{1}{2}$	$\frac{1}{6}$
3_R	u_R	1	0	$\frac{2}{3}$	$-\frac{1}{2}$	$\frac{1}{2\sqrt{3}}$	$\frac{2}{3}$	$\frac{1}{2}$	$\frac{1}{6}$
4_R	d_R	1	0	$-\frac{1}{3}$	$-\frac{1}{2}$	$\frac{1}{2\sqrt{3}}$	$-\frac{1}{3}$	$-\frac{1}{2}$	$\frac{1}{6}$
5_R	u_R	1	0	$\frac{2}{3}$	0	$-\frac{1}{\sqrt{3}}$	$\frac{2}{3}$	$\frac{1}{2}$	$\frac{1}{6}$
6_R	d_R	1	0	$-\frac{1}{3}$	0	$-\frac{1}{\sqrt{3}}$	$-\frac{1}{3}$	$-\frac{1}{2}$	$\frac{1}{6}$
7_R	ν_R	1	0	0	0	0	0	$\frac{1}{2}$	$-\frac{1}{2}$
8_R	e_R	1	0	-1	0	0	-1	$-\frac{1}{2}$	$-\frac{1}{2}$
9_R	$\bar{u}_R$	1	$-\frac{1}{2}$	$-\frac{1}{6}$	$-\frac{1}{2}$	$-\frac{1}{2\sqrt{3}}$	$-\frac{2}{3}$	0	$-\frac{1}{6}$
10_R	$\bar{d}_R$	1	$\frac{1}{2}$	$-\frac{1}{6}$	$-\frac{1}{2}$	$-\frac{1}{2\sqrt{3}}$	$\frac{1}{3}$	0	$-\frac{1}{6}$
11_R	$\bar{u}_R$	1	$-\frac{1}{2}$	$-\frac{1}{6}$	$\frac{1}{2}$	$-\frac{1}{2\sqrt{3}}$	$-\frac{2}{3}$	0	$-\frac{1}{6}$
12_R	$\bar{d}_R$	1	$\frac{1}{2}$	$-\frac{1}{6}$	$\frac{1}{2}$	$-\frac{1}{2\sqrt{3}}$	$\frac{1}{3}$	0	$-\frac{1}{6}$
13_R	$\bar{u}_R$	1	$-\frac{1}{2}$	$-\frac{1}{6}$	0	$\frac{1}{\sqrt{3}}$	$-\frac{2}{3}$	0	$-\frac{1}{6}$
14_R	$\bar{d}_R$	1	$\frac{1}{2}$	$-\frac{1}{6}$	0	$\frac{1}{\sqrt{3}}$	$\frac{1}{3}$	0	$-\frac{1}{6}$
15_R	$\bar{\nu}_R$	1	$-\frac{1}{2}$	$\frac{1}{2}$	0	0	0	0	$\frac{1}{2}$
16_R	$\bar{e}_R$	1	$\frac{1}{2}$	$\frac{1}{2}$	0	0	1	0	$\frac{1}{2}$

triangle anomalies is "trivial", one takes into account that the *standard model* groups can easily be interpreted (unified) by making the next step beyond the *standard model*. In the *standard model* the triangle anomaly for the Feynman triangle diagrams, in which the gauge vector fields of the charges

$$\begin{aligned}
&U(1) \times U(1) \times U(1)\,, \\
&SU(2) \times SU(2) \times U(1)\,, \\
&SU(3) \times SU(3) \times SU(3)\,, \\
&SU(3) \times SU(3) \times U(1)\,, \\
&U(1) \times \text{gravitational}
\end{aligned} \tag{36}$$

contribute to the triangle anomaly, occurs if the traces in Eq.(35) are not zero for either the left handed quarks and leptons and their left handed antiparticles or the right handed quarks and leptons and their right handed antiparticles.

To see that embedding the *standard model* groups into the orthogonal group $SO(13,1)$ makes the cancellation of the triangle anomalies self evident, let us recognize: The subgroups of the $SO(13,1)$ group are $SO(7,1)$ and $SO(6)$. The subgroups of $SO(6)$ are the colour group $SU(3)$ with the generators denoted by τ^{3i}, $i=(1,\ldots,8)$ and the $U(1)$ (we shall call it $U(1)_{II}$) group with the generator τ^4. One sees that all the quarks have $\tau^4=\frac{1}{6}$, all the antiquarks have $\tau^4=-\frac{1}{6}$, while the leptons have $\tau^4=-\frac{1}{2}$ and the antileptons have $\tau^4=\frac{1}{2}$. Correspondingly the trace of τ^4 over all the family members is equal to zero.

The subgroups of the $SO(7,1)$, as seen in Table 2, have as subgroups $SO(3,1)$, $SU(2)_I$ and $SU2_{II}$, with the generators τ^{1i} (representing the weak group operators) and τ^{2i} (representing the generators of the additional $SU(2)$ group), respectively. The left handed spinors are $SU(2)_I$ (weak) doublets and $SU(2)_{II}$ singlets, while the right handed spinors are the $SU(2)_I$ (weak) singlets and $SU(2)_{II}$ doublets. Correspondingly are the left handed antispinors the $SU(2)_I$ (weak) singlets and $SU(2)_{II}$ doublets, while the right handed antispinors are the $SU(2)_I$ (weak) doublets and the $SU(2)_{II}$ singlets.

The hypercharge of the *standard model* corresponds to the sum of τ^4 and τ^{23}

$$Y = \tau^4+\tau^{23}\,. \tag{37}$$

To the triangle Feynman diagram, to which three hyper $U(1)$ boson fields contribute, the sum $\sum_i Tr(Y_i)^3$ runs over all the members (i) of the left handed spinors and antispinors, and of the right handed spinors and antispinors separately. Embedding the *standard model* groups into $SO(13,1)$ it follows

$$\begin{aligned}\sum_{i_{L,R}}(Y_{i_{L,R}})^3 &= \sum_{i_{L,R}}(\tau^4_{i_{L,R}}+\tau^{23}_{i_{L,R}})^3\\ &= \sum_{i_{L,R}}(\tau^4_{i_{L,R}})^3+\sum_{i_{L,R}}(\tau^{23}_{i_{L,R}})^3\\ &+ \sum_{i_{L,R}}3\cdot(\tau^4_{i_{L,R}})^2\cdot\tau^{23}_{i_{L,R}}+\sum_{i_{L,R}}3\cdot\tau^4_{i_{L,R}}\cdot(\tau^{23}_{i_{L,R}})^2\,,\end{aligned} \tag{38}$$

for either the left, i_L, or the right, i_R, handed members. Table 8 demonstrates clearly (last column) that $\sum_{i_L}(\tau^4_{i_L})^3=0$, when the contribution of left (right) handed spinors and antispinors are taken into account.

Table 8 also demonstrates (the last but one column) even more trivially that $\sum_{i_L}3\cdot(\tau^4_{i_L})^2\cdot\tau^{23}_{i_L}=0$, since the contribution of either spinors or antispinors - left or right handed - separately are equal to zero.

The easiest is to evaluate $\sum_{i_{L,R}}(\tau^{23}_{i_L})^3=0$ and $\sum_{i_L}3.(\tau^4_{i_L})^2\cdot\tau^{23}_{i_L}=0$ since, as seen from Table 8, the summation separately within the quarks and lepton representations give zero.

Since all the members belong to one spinor representation, it is straightforwardly that all the triangle traces are zero, if the *standard model* groups are the subgroups of the orthogonal group $SO(13,1)$.

From only the *standard model* assumptions point of view the cancellation of the triangle anomalies does look miraculously. For our $\sum_{i_{L,R}}(Y_{i_{L,R}})^3$ one obtains for the left handed members: $[3\cdot2\cdot(\frac{1}{6})^3+2\cdot(-\frac{1}{2})^3+3\cdot((-\frac{2}{3})^3+(\frac{1}{3})^3)+1^3)\ ]$, and for the right handed members: $[3\cdot((\frac{2}{3})^3+(-\frac{1}{3})^3)+(-1)^3)+3\cdot2\cdot(-\frac{1}{6})^3+2\cdot(\frac{1}{2})^3\ ]$.

4.4. Dark matter in the spin-charge-family theory

As discussed in Sect. 2 the *spin-charge-family* theory [5, 3, 6, 7, 8, 9, 10, 4, 11, 12, 13, 14, 15, 16, 17, 2, 1] predicts in the low energy region two decoupled groups of four families. In Ref. [13] the

IOP Conf. Series: Journal of Physics: Conf. Series **845** (2017) 012017 doi:10.1088/1742-6596/845/1/012017

possibility that the dark matter consists of clusters of the fifth family - the stable heavy family of quarks and leptons (with (amost) zero Yukawa couplings to the lower group of four families) - is discussed.

I review here briefly the estimation done in Ref. [13].

We used in Ref. [13] the simple hydrogen-like model to evaluate properties of the fifth family heavy baryons, taking into account that for masses of the order of a few TeV or larger the one gluon exchange determines the force among the constituents of the fifth family baryons ([13], Sect. II). We estimated the fifth family neutron as the most stable nucleon.

Due to very large masses of the fifth family baryons "the nuclear interaction" among these baryons has very interesting properties.

We also estimated the behaviour of the neutral clusters when scattering among themselves and with the ordinary matter. We studied possible limitations on the family properties due to the cosmological evidences, the direct experimental evidences ([13], Sect. IV) and all others known properties of the dark matter.

We followed the behaviour of the fifth family quarks and antiquarks in the plasma of the expanding universe, through the freezing out procedure, solving the Boltzmann equations, through the colour phase transition, while forming neutrons, up to the present dark matter ([13], Sect. III).

The cosmological evolution suggested the limits for the masses of the fifth family quarks

$$10\ \mathrm{TeV} < m_{q_5}\, c^2 < \mathrm{a\,few} \cdot 10^2\ \mathrm{TeV} \tag{39}$$

and for the scattering cross sections

$$10^{-8}\,\mathrm{fm}^2 < \sigma_{c_5} < 10^{-6}\,\mathrm{fm}^2\,, \tag{40}$$

while the measured density of the dark matter does not put much limitation on the properties of heavy enough clusters.

The direct measurements limited the fifth family quark mass to ([13], Sect. IV.)

$$\mathrm{several}\,10\,\mathrm{TeV} < m_{q_5} c^2 < 10^5\,\mathrm{TeV}\,. \tag{41}$$

We also find that our fifth family baryons of the mass of a few 10 TeV$/c^2$ have for a factor more than 100 times too small scattering amplitude with the ordinary matter to cause a measurable heat flux on the Earth's surface.

4.5. Matter-antimatter in the spin-charge-family theory

I shortly overview in this section the properties, quantum numbers, and discrete symmetries of those scalar and vector gauge fields appearing in the starting action (Eqs. (1, 2) which cause transitions of antileptons into quarks and back, and antiquarks into quarks and back. The appearance of the condensate breaks this symmetry making possible under non thermal conditions the ordinary (mostly made of the first family members) matter-antimatter asymmetry. The reader can find details in Ref. [2].

Scalar gauge fields, contributing to matter-antimatter asymmetry and causing also the proton decay, carry the triplet or antitriplet colour charges (see Table 9) and the fractional hyper and electromagnetic charge.

The Lagrange densities from Eqs. (1, 2) manifest $\mathbb{C}_{\mathcal{N}} \cdot \mathcal{P}_{\mathcal{N}}$ invariance [36]. All the vector and the spinor gauge fields are massless before the appearance of the condensate (Subsect. 3.4) and reactions creating particles from antiparticles and back go in both directions equivalently. Correspondingly there is no matter-antimatter asymmetry. It is the condensate, which breaks this symmetry.

Let me analyze the Lagrange density of Eq. (2) before the appearance of the condensate. The term $\gamma^t \frac{1}{2} S^{s's"} \omega_{s's"t}$ can be rewritten as follows

$$
\begin{aligned}
\gamma^t \frac{1}{2} S^{s's"} \omega_{s's"t} &= \sum_{+,-} \sum_{(t\,t')} \overset{tt'}{(\oplus)} \frac{1}{2} S^{s's"} \omega_{s"s"\overset{tt'}{(\oplus)}} ,\\
\omega_{s"s"\overset{tt'}{(\oplus)}} : &= \omega_{s"s"\overset{tt'}{(\pm)}} = (\omega_{s's"t} \mp i\, \omega_{s's"t'}) ,\\
\overset{tt'}{(\oplus)} : &= \overset{tt'}{(\pm)} = \frac{1}{2}(\gamma^t \pm \gamma^{t'}) ,\\
(t\,t') &\in ((9\,10), (11\,12), (13\,14)) .
\end{aligned} \tag{42}
$$

I introduced the notations $\overset{tt'}{(\oplus)}$ and $\omega_{s"s"\overset{tt'}{(\oplus)}}$ to distinguish among different superposition of states in equations below.

The expression $\overset{tt'}{(\oplus)} \frac{1}{2} S^{s's"} \omega_{s"s"\overset{tt'}{(\oplus)}}$ can be further rewritten as follows

$$
\begin{aligned}
&\overset{tt'}{(\oplus)} \frac{1}{2} S^{s's"} \omega_{s"s"\overset{tt'}{(\oplus)}} = \\
&\overset{tt'}{(\oplus)} \{ \tau^{2+} A^{2+}_{\overset{tt'}{(\oplus)}} + \tau^{2-} A^{2-}_{\overset{tt'}{(\oplus)}} + \tau^{23} A^{23}_{\overset{tt'}{(\oplus)}} + \tau^{1+} A^{1+}_{\overset{tt'}{(\oplus)}} + \tau^{1-} A^{1-}_{\overset{tt'}{(\oplus)}} + \tau^{13} A^{13}_{\overset{tt'}{(\oplus)}} \} ,\\
A^{2\pm}_{\overset{tt'}{(\oplus)}} &= (\omega_{58\overset{tt'}{(\oplus)}} + \omega_{67\overset{tt'}{(\oplus)}}) \pm i (\omega_{57\overset{tt'}{(\oplus)}} - \omega_{68\overset{tt'}{(\oplus)}}) , \quad A^{23}_{\overset{tt'}{(\oplus)}} = (\omega_{56\overset{tt'}{(\oplus)}} + \omega_{78\overset{tt'}{(\oplus)}}) ,\\
A^{1\pm}_{\overset{tt'}{(\oplus)}} &= (\omega_{58\overset{tt'}{(\oplus)}} - \omega_{67\overset{tt'}{(\oplus)}}) \pm i (\omega_{57\overset{tt'}{(\oplus)}} + \omega_{68\overset{tt'}{(\oplus)}}) , \quad A^{13}_{\overset{tt'}{(\oplus)}} = (\omega_{56\overset{tt'}{(\oplus)}} - \omega_{78\overset{tt'}{(\oplus)}}) .
\end{aligned} \tag{43}
$$

Equivalently one expresses the term $\gamma^t \frac{1}{2} \tilde{S}^{ab} \tilde{\omega}_{abt}$ in Eq. (2) with $\tilde{S}^{ab}$ as the infinitesimal generators of either $\widetilde{SO}(3,1)$ or $\widetilde{SO}(4)$ and $\tilde{\omega}_{abt}$ belonging to the corresponding gauge fields with $t = (9, \ldots, 14)$, by using Eqs. (A.23 - A.26), as

$$
\begin{aligned}
&\gamma^t \frac{1}{2} \tilde{S}^{ab} \tilde{\omega}_{abt} = \overset{tt'}{(\oplus)} \frac{1}{2} \tilde{S}^{ab} \tilde{\omega}_{ab\overset{tt'}{(\oplus)}} = \\
&\overset{tt'}{(\oplus)} \{ \tilde{\tau}^{2+} \tilde{A}^{2+}_{\overset{tt'}{(\oplus)}} + \tilde{\tau}^{2-} \tilde{A}^{2-}_{\overset{tt'}{(\oplus)}} + \tilde{\tau}^{23} \tilde{A}^{23}_{\overset{tt'}{(\oplus)}} + \\
&\tilde{\tau}^{1+} \tilde{A}^{1+}_{\overset{tt'}{(\oplus)}} + \tilde{\tau}^{1-} \tilde{A}^{1-}_{\overset{tt'}{(\oplus)}} + \tilde{\tau}^{13} \tilde{A}^{13}_{\overset{tt'}{(\oplus)}} + \\
&\tilde{N}^{+}_{R} \tilde{A}^{N_R +}_{\overset{tt'}{(\oplus)}} + \tilde{N}^{-}_{R} \tilde{A}^{N_R -}_{\overset{tt'}{(\oplus)}} + \tilde{N}^{3}_{R} \tilde{A}^{N_R 3}_{\overset{tt'}{(\oplus)}} + \\
&\tilde{N}^{+}_{L} \tilde{A}^{N_L +}_{\overset{tt'}{(\oplus)}} + \tilde{N}^{-}_{L} \tilde{A}^{N_L -}_{\overset{tt'}{(\oplus)}} + \tilde{N}^{3}_{L} \tilde{A}^{N_L 3}_{\overset{tt'}{(\oplus)}} \} ,\\
\tilde{A}^{N_R \pm}_{\overset{tt'}{(\oplus)}} &= (\tilde{\omega}_{23\overset{tt'}{(\oplus)}} - i\, \tilde{\omega}_{01\overset{tt'}{(\oplus)}}) \pm i (\tilde{\omega}_{31\overset{tt'}{(\oplus)}} - i\, \tilde{\omega}_{02\overset{tt'}{(\oplus)}}) , \quad \tilde{A}^{N_R 3}_{\overset{tt'}{(\oplus)}} = (\tilde{\omega}_{12\overset{tt'}{(\oplus)}} - i\, \tilde{\omega}_{03\overset{tt'}{(\oplus)}}) ,\\
\tilde{A}^{N_L \pm}_{\overset{tt'}{(\oplus)}} &= (\tilde{\omega}_{23\overset{tt'}{(\oplus)}} + i\, \tilde{\omega}_{01\overset{tt'}{(\oplus)}}) \pm i (\tilde{\omega}_{31\overset{tt'}{(\oplus)}} + i\, \tilde{\omega}_{02\overset{tt'}{(\oplus)}}) , \quad \tilde{A}^{N_R 3}_{\overset{tt'}{(\oplus)}} = (\tilde{\omega}_{12\overset{tt'}{(\oplus)}} + i\, \tilde{\omega}_{03\overset{tt'}{(\oplus)}}) .
\end{aligned} \tag{44}
$$

The expressions for $\tilde{A}^{2\pm}_{\overset{tt'}{(\oplus)}}$, $\tilde{A}^{23}_{\overset{tt'}{(\oplus)}}$, $\tilde{A}^{1\pm}_{\overset{tt'}{(\oplus)}}$ and $\tilde{A}^{13}_{\overset{tt'}{(\oplus)}}$ can easily be obtained from Eq.(43) by replacing in expressions for $A^{2\boxplus}_{\substack{tt'\\(\oplus)}}$, $A^{23}_{\substack{tt'\\(\oplus)}}$, $A^{1\boxplus}_{\substack{tt'\\(\oplus)}}$ and $A^{13}_{\substack{tt'\\(\oplus)}}$, respectively, $\omega_{s's"\substack{tt'\\(\oplus)}}$ by $\tilde{\omega}_{s's"\substack{tt'\\(\oplus)}}$.

The term $\gamma^t \, \frac{1}{2} \, S^{t't"} \, \omega_{t't"t}$ in Eq. (2) can be rewritten with respect to the generators $S^{t't"}$ and the corresponding gauge fields $\omega_{s's"t}$ as one colour octet scalar field and one $U(1)_{II}$ singlet scalar field (Eq. A.22)

$$\begin{aligned}
\gamma^t \, \frac{1}{2} \, S^{t"t'"} \, \omega_{t"t'"t} &= \sum_{+,-} \sum_{(t\,t')} \overset{tt'}{(\oplus)} \, \{ \vec{\tau}^3 \cdot \vec{A}^3_{\substack{tt'\\(\oplus)}} + \tau^4 \cdot A^4_{\substack{tt'\\(\oplus)}} \} \,, \\
(t\,t') &\in ((9\,10), 11\,12), 13\,14)) \,.
\end{aligned} \tag{45}$$

Considering all the above equations (42 - 45), and leaving out $p_{\substack{tt'\\(\oplus)}}$ since in the low energy limit the momentum does not play any role, the $\mathcal{L}_{f"}$ follows

$$\begin{aligned}
\mathcal{L}_{f"} &= \psi^\dagger \gamma^0 (-) \{ \sum_{+,-} \sum_{(t\,t')} \overset{tt'}{(\oplus)} \cdot \\
& \quad [\, \tau^{2+} \, A^{2+}_{\substack{tt'\\(\oplus)}} + \tau^{2-} \, A^{2-}_{\substack{tt'\\(\oplus)}} + \tau^{23} \, A^{23}_{\substack{tt'\\(\oplus)}} \\
&+ \quad \tau^{1+} \, A^{1+}_{\substack{tt'\\(\oplus)}} + \tau^{1-} \, A^{1-}_{\substack{tt'\\(\oplus)}} + \tau^{13} \, A^{13}_{\substack{tt'\\(\oplus)}} \\
&+ \quad \tilde{\tau}^{2+} \, \tilde{A}^{2+}_{\substack{tt'\\(\oplus)}} + \tilde{\tau}^{2-} \, \tilde{A}^{2-}_{\substack{tt'\\(\oplus)}} + \tilde{\tau}^{23} \, \tilde{A}^{23}_{\substack{tt'\\(\oplus)}} \\
&+ \quad \tilde{\tau}^{1+} \, \tilde{A}^{1+}_{\substack{tt'\\(\oplus)}} + \tilde{\tau}^{1-} \, \tilde{A}^{1-}_{\substack{tt'\\(\oplus)}} + \tilde{\tau}^{13} \, \tilde{A}^{13}_{\substack{tt'\\(\oplus)}} \\
&+ \quad \tilde{N}^{+}_{R} \, \tilde{A}^{N_R +}_{\substack{tt'\\(\oplus)}} + \tilde{N}^{-}_{R} \, \tilde{A}^{N_R -}_{\substack{tt'\\(\oplus)}} + \tilde{N}^{3}_{R} \, \tilde{A}^{N_R 3}_{\substack{tt'\\(\oplus)}} \\
&+ \quad \tilde{N}^{+}_{L} \, \tilde{A}^{N_L +}_{\substack{tt'\\(\oplus)}} + \tilde{N}^{-}_{L} \, \tilde{A}^{N_L -}_{\substack{tt'\\(\oplus)}} + \tilde{N}^{3}_{L} \, \tilde{A}^{N_L 3}_{\substack{tt'\\(\oplus)}} \\
&+ \quad \sum_{i} \tau^{3i} \, A^{3i}_{\substack{tt'\\(\oplus)}} + \tau^4 \, A^4_{\substack{tt'\\(\oplus)}} \\
&+ \quad \sum_{i} \tilde{\tau}^{3i} \, \tilde{A}^{3i}_{\substack{tt'\\(\oplus)}} + \tilde{\tau}^4 \, \tilde{A}^4_{\substack{tt'\\(\oplus)}} \,] \} \, \psi \,,
\end{aligned} \tag{46}$$

where (t, t') run in pairs over $[(9, 10), \dots (13, 14)]$ and the summation must go over $+$ and $-$ of $\overset{tt'}{(\oplus)}$.

On Table 9, taken from Ref. [2], the quantum numbers of the scalar and vector gauge fields, appearing in Eq. (2), are presented, where is taken into account that the spin of gauge fields is determined according to Eq. (47),

$$(S^{ab})^c{}_e \, A^{d \dots e \dots g} = i \, (\eta^{ac} \delta^b_e - \eta^{bc} \delta^a_e) \, A^{d \dots e \dots g} \,, \tag{47}$$

for each index ($\in (d \dots g)$) of a bosonic field $A^{d \cdots g}$ separately. We must take into account also the relation among S^{ab} and the charges (the relations are, of course, the same for bosons and fermions), presented in Eqs. (A.20, A.21, A.22)).

The scalar fields with the scalar index $s = (9, 10, \cdots, 14)$, presented in Table 9, carry one of the triplet colour charges and the "spinor" charge equal to twice the quark "spinor" charge, or the antitriplet colour charges and the anti "spinor" charge. They carry in addition the quantum

Table 9. Quantum numbers of the scalar gauge fields carrying the space index $t = (9, 10, \cdots, 14)$, appearing in Eq. (2), are presented. The space degrees of freedom contribute one of the triplets values to the colour charge of all these scalar fields. These scalars are with respect to the two $SU(2)$ charges, (τ^{13} and $\vec{\tau}^{2}$), and the two $\widetilde{SU}(2)$ charges, ($\vec{\tilde{\tau}}^{1}$ and $\vec{\tilde{\tau}}^{2}$), triplets (that is in the adjoint representations of the corresponding groups), and they all carry twice the "spinor" number (τ^4) of the quarks. The quantum numbers of the two vector gauge fields, the colour and the $U(1)_{II}$ ones, are added.

field	prop.	τ^4	τ^{13}	τ^{23}	(τ^{33}, τ^{38})	Y	Q	$\tilde{\tau}^4$	$\tilde{\tau}^{13}$	$\tilde{\tau}^{23}$	$\tilde{N}^3_L$	$\tilde{N}^3_R$
$A^{1\boxplus}_{9\,10}$ $(\oplus)$	scalar	$\oplus\,\frac{1}{3}$	$\boxplus\,1$	0	$(\oplus\,\frac{1}{2}, \oplus\,\frac{1}{2\sqrt{3}})$	$\oplus\,\frac{1}{3}$	$\oplus\,\frac{1}{3} + \boxplus\,1$	0	0	0	0	0
$A^{13}_{9\,10}$ $(\oplus)$	scalar	$\oplus\,\frac{1}{3}$	0	0	$(\oplus\,\frac{1}{2}, \oplus\,\frac{1}{2\sqrt{3}})$	$\oplus\,\frac{1}{3}$	$\oplus\,\frac{1}{3}$	0	0	0	0	0
$A^{1\boxplus}_{11\,12}$ $(\oplus)$	scalar	$\oplus\,\frac{1}{3}$	$\boxplus\,1$	0	$(\oplus\,\frac{1}{2}, \oplus\,\frac{1}{2\sqrt{3}})$	$\oplus\,\frac{1}{3}$	$\oplus\,\frac{1}{3} + \boxplus\,1$	0	0	0	0	0
$A^{13}_{11\,12}$ $(\oplus)$	scalar	$\oplus\,\frac{1}{3}$	0	0	$(\oplus\,\frac{1}{2}, \oplus\,\frac{1}{2\sqrt{3}})$	$\oplus\,\frac{1}{3}$	$\oplus\,\frac{1}{3}$	0	0	0	0	0
$A^{1\boxplus}_{13\,14}$ $(\oplus)$	scalar	$\oplus\,\frac{1}{3}$	$\boxplus\,1$	0	$(0, \oplus\,\frac{1}{\sqrt{3}})$	$\oplus\,\frac{1}{3}$	$\oplus\,\frac{1}{3} + \boxplus\,1$	0	0	0	0	0
$A^{13}_{13\,14}$ $(\oplus)$	scalar	$\oplus\,\frac{1}{3}$	0	0	$(0, \oplus\,\frac{1}{\sqrt{3}})$	$\oplus\,\frac{1}{3}$	$\oplus\,\frac{1}{3}$	0	0	0	0	0
$A^{2\boxplus}_{9\,10}$ $(\oplus)$	scalar	$\oplus\,\frac{1}{3}$	0	$\boxplus\,1$	$(\oplus\,\frac{1}{2}, \oplus\,\frac{1}{2\sqrt{3}})$	$\oplus\,\frac{1}{3} + \boxplus\,1$	$\oplus\,\frac{1}{3} + \boxplus\,1$	0	0	0	0	0
$A^{23}_{9\,10}$ $(\oplus)$ $\cdots$	scalar	$\oplus\,\frac{1}{3}$	0	0	$(\oplus\,\frac{1}{2}, \oplus\,\frac{1}{2\sqrt{3}})$	$\oplus\,\frac{1}{3}$	$\oplus\,\frac{1}{3}$	0	0	0	0	0
$\tilde{A}^{1\boxplus}_{910}$ $(\oplus)$	scalar	$\oplus\,\frac{1}{3}$	0	0	$(\oplus\,\frac{1}{2}, \oplus\,\frac{1}{2\sqrt{3}})$	$\oplus\,\frac{1}{3}$	$\oplus\,\frac{1}{3}$	0	$\boxplus\,1$	0	0	0
$\tilde{A}^{13}_{910}$ $(\oplus)$ $\cdots$	scalar	$\oplus\,\frac{1}{3}$	0	0	$(\oplus\,\frac{1}{2}, \oplus\,\frac{1}{2\sqrt{3}})$	$\oplus\,\frac{1}{3}$	$\oplus\,\frac{1}{3}$	0	0	0	0	0
$\tilde{A}^{2\boxplus}_{910}$ $(\oplus)$	scalar	$\oplus\,\frac{1}{3}$	0	0	$(\oplus\,\frac{1}{2}, \oplus\,\frac{1}{2\sqrt{3}})$	$\oplus\,\frac{1}{3}$	$\oplus\,\frac{1}{3}$	0	0	$\boxplus\,1$	0	0
$\tilde{A}^{23}_{910}$ $(\oplus)$ $\cdots$	scalar	$\oplus\,\frac{1}{3}$	0	0	$(\oplus\,\frac{1}{2}, \oplus\,\frac{1}{2\sqrt{3}})$	$\oplus\,\frac{1}{3}$	$\oplus\,\frac{1}{3}$	0	0	0	0	0
$\tilde{A}^{N_L\boxplus}_{910}$ $(\oplus)$	scalar	$\oplus\,\frac{1}{3}$	0	0	$(\oplus\,\frac{1}{2}, \oplus\,\frac{1}{2\sqrt{3}})$	$\oplus\,\frac{1}{3}$	$\oplus\,\frac{1}{3}$	0	0	0	$\boxplus\,1$	0
$\tilde{A}^{N_L 3}_{910}$ $(\oplus)$ $\cdots$	scalar	$\oplus\,\frac{1}{3}$	0	0	$(\oplus\,\frac{1}{2}, \oplus\,\frac{1}{2\sqrt{3}})$	$\oplus\,\frac{1}{3}$	$\oplus\,\frac{1}{3}$	0	0	0	0	0
$\tilde{A}^{N_R\boxplus}_{910}$ $(\oplus)$	scalar	$\oplus\,\frac{1}{3}$	0	0	$(\oplus\,\frac{1}{2}, \oplus\,\frac{1}{2\sqrt{3}})$	$\oplus\,\frac{1}{3}$	$\oplus\,\frac{1}{3}$	0	0	0	0	$\boxplus\,1$
$\tilde{A}^{N_R 3}_{910}$ $(\oplus)$ $\cdots$	scalar	$\oplus\,\frac{1}{3}$	0	0	$(\oplus\,\frac{1}{2}, \oplus\,\frac{1}{2\sqrt{3}})$	$\oplus\,\frac{1}{3}$	$\oplus\,\frac{1}{3}$	0	0	0	0	0
$A^{3i}_{9\,10}$ $(\oplus)$ $\cdots$	scalar	$\oplus\,\frac{1}{3}$	0	0	$(\boxplus\,1 + \oplus\,\frac{1}{2}, \oplus\,\frac{1}{2\sqrt{3}})$	$\oplus\,\frac{1}{3}$	$\oplus\,\frac{1}{3}$	0	0	0	0	0
$A^{4}_{9\,10}$ $(\oplus)$ $\cdots$	scalar	$\oplus\,\frac{1}{3}$	0	0	$(\oplus\,\frac{1}{2}, \oplus\,\frac{1}{2\sqrt{3}})$	$\oplus\,\frac{1}{3}$	$\oplus\,\frac{1}{3}$	0	0	0	0	0
$\vec{A}^{3}_{m}$	vector	0	0	0	octet	0	0	0	0	0	0	0
A^{4}_{m}	vector	0	0	0	0	0	0	0	0	0	0	0

numbers of the adjoint representations originating in S^{ab} or in $\tilde{S}^{ab}$ [17].

Let us choose the 57th line of Table 2, which represents in the spinor technique the left handed positron, $\bar{e}_L^{+}$, to see what do the scalar fields, appearing in Eq. (46) and in Table 9, do when applying on the left handed members of the Weyl representation presented on Table 2, containing quarks and leptons and antiquarks and antileptons [9, 10, 36].

[17] Although carrying the colour charge in one of the triplet or antitriplet states, these fields can not be interpreted as superpartners of the quarks since they do not have quantum numbers as required by, let say, the $N = 1$ supersymmetry. The hyper charges and the electromagnetic charges are namely not those required by the supersymmetric partners to the family members.

If we make, let say, the choice of the term $(\gamma^0 \stackrel{910}{(+)} \tau^{2\boxminus})\ A^{2\boxminus}_{\stackrel{9\,10}{(\oplus)}}$ (the scalar field $A^{2\boxminus}_{\stackrel{9\,10}{(\oplus)}}$ is presented in the 7th line in Table 9 and in the second line of Eq. (46)), the family quantum numbers will not be affected and they can be any. The state carries the "spinor" (lepton) number $\tau^4 = \frac{1}{2}$, the weak charge $\tau^{13} = 0$, the second $SU(2)_{II}$ charge $\tau^{23} = \frac{1}{2}$ and the colour charge $(\tau^{33}, \tau^{38}) = (0, 0)$. Correspondingly, its hyper charge $(Y(= \tau^4 + \tau^{23}))$ is 1 and the electromagnetic charge $Q(= Y + \tau^{13})$ is 1.

So, what does the term $\gamma^0 \stackrel{910}{(+)} \tau^{2\boxminus}\ A^{2\boxminus}_{\stackrel{9\,10}{(\oplus)}}$ make on this spinor $\bar{e}^{+}_{L}$? Making use of Eqs. (A.10, A.12, A.26) of Appendix A one easily finds that operator $\gamma^0 \stackrel{910}{(+)} \tau^{2-}$ transforms the left handed positron into $\stackrel{03}{(+i)}\stackrel{12}{(+)} \mid \stackrel{56}{[-]}\stackrel{78}{[-]} \mid\mid \stackrel{9\,10}{(+)}\ \ \stackrel{11\,12}{(-)}\ \ \stackrel{13\,14}{(-)}$, which is d^{c1}_{R}, presented on line 3 of Table 2. Namely, γ^0 transforms $\stackrel{03}{[-i]}$ into $\stackrel{03}{(+i)}$, $\stackrel{910}{(+)}$ transforms $\stackrel{9\,10}{[-]}$ into $\stackrel{9\,10}{(+)}$, while τ^{2-} $(= -\ \stackrel{56}{(-)}\ \stackrel{78}{(-)})$ transforms $\stackrel{56}{(+)}\ \stackrel{78}{(+)}$ into $\stackrel{56}{[-]}\ \stackrel{78}{[-]}$. The state d^{c1}_{R} carries the "spinor" (quark) number $\tau^4 = \frac{1}{6}$, the weak charge $\tau^{13} = 0$, the second $SU(2)_{II}$ charge $\tau^{23} = -\frac{1}{2}$ and the colour charge $(\tau^{33}, \tau^{38}) = (\frac{1}{2}, \frac{1}{2\sqrt{3}})$. Correspondingly its hyper charge is $(Y = \tau^4 + \tau^{23} =) -\frac{1}{3}$ and the electromagnetic charge $(Q = Y + \tau^{13} =) -\frac{1}{3}$. The scalar field $A^{2\boxminus}_{\stackrel{9\,10}{(\oplus)}}$ carries just the needed quantum numbers as we can see in the 7th line of Table 9.

If the antiquark $\bar{u}^{\bar{c}2}_{L}$, from the line 43 (it is not presented, but one can very easily construct it) in Table 2, with the "spinor" charge $\tau^4 = -\frac{1}{6}$, the weak charge $\tau^{13} = 0$, the second $SU(2)_{II}$ charge $\tau^{23} = -\frac{1}{2}$, the colour charge $(\tau^{33}, \tau^{38}) = (\frac{1}{2}, -\frac{1}{2\sqrt{3}})$, the hyper charge $Y(= \tau^4 + \tau^{23} =) -\frac{2}{3}$ and the electromagnetic charge $Q(= Y + \tau^{13} =) -\frac{2}{3}$ submits the $A^{2\boxminus}_{\stackrel{9\,10}{(\oplus)}}$ scalar field, it transforms into u^{c3}_{R} from the line 17 of Table 2, carrying the quantum numbers $\tau^4 = \frac{1}{6}$, $\tau^{13} = 0$, $\tau^{23} = \frac{1}{2}$, $(\tau^{33}, \tau^{38}) = (0, -\frac{1}{\sqrt{3}})$, $Y = \frac{2}{3}$ and $Q = \frac{2}{3}$. These two quarks, d^{c1}_{R} and u^{c3}_{R} can bind together with u^{c2}_{R} from the 9th line of the same table (at low enough energy, after the electroweak transition, and if they belong to a superposition with the left handed partners to the first family) into the colour chargeless baryon - a proton. This transition is presented in Figure 2.

The opposite transition at low energies would make the proton decay.

Similar transitions go also with other scalars from Eq. (46) and Table 9. The $\vec{\tilde{A}}^{1}_{\stackrel{t'\,t"}{(+)}}$, $\vec{\tilde{A}}^{2}_{\stackrel{t'\,t"}{(+)}}$, $\vec{\tilde{A}}^{N_L}_{\stackrel{t'\,t"}{(+)}}$ and $\vec{\tilde{A}}^{N_L}_{\stackrel{t'\,t"}{(+)}}$ fields cause transitions among the family members, changing a particular member into the antimember of another colour and of another family. The term $\gamma^0 \stackrel{910}{(+)}\ \tilde{N}^{-}_{R}\ A^{\tilde{N}_R -}_{\stackrel{9\,10}{(\oplus)}}$ transforms $\bar{e}^{+}_{R}$ into u^{c1}_{L}, changing the family quantum numbers.

The action from Eqs. (1, 2) manifests $\mathbb{C}_{\mathcal{N}} \cdot \mathcal{P}_{\mathcal{N}}$ invariance. All the vector and the spinor gauge fields are massless.

Since none of the scalar fields from Table 9 have been observed and also no vector gauge fields like $\vec{A}^{2}_{m}$, A^{4}_{m} and other scalar and vector fields, there must exist a mechanism, which makes the non observed scalar and vector gauge fields massive enough [18].

Scalar fields from Table 9 carry the colour and the electromagnetic charge. Therefore their nonzero vacuum expectation values would not be in agreement with the observed phenomena. One, however, notices that all the scalar gauge fields from Table 9 and several other scalar and

[18] I expect that the condensate (Table 1) appears on the scale of unification - $\geq\approx 10^{16}$ GeV. Interacting with these vector gauge fields the condensate make them massive.

IOP Conf. Series: Journal of Physics: Conf. Series **845** (2017) 012017 doi:10.1088/1742-6596/845/1/012017

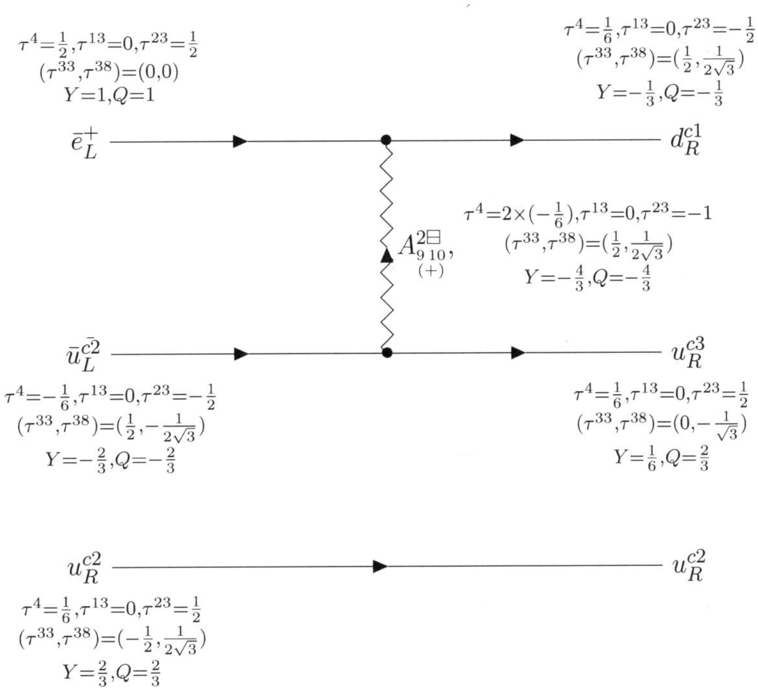

Figure 2. The birth of a "right handed proton" out of an positron $\bar{e}^+_L$, antiquark $\bar{u}^{\bar{c}2}_L$ and quark (spectator) u^{c2}_R. The family quantum number can be any.

vector gauge fields couple to the condensate with the nonzero quantum number τ^4 and τ^{23} and nonzero family quantum numbers.

It is not difficult to recognize that the desired condensate must have spin zero, $Y = \tau^4 + \tau^{23} = 0$, $Q = Y + \tau^{13} = 0$ and $\vec{\tau}^1 = 0$ in order that in the low energy limit the *spin-charge-family* theory would manifest effectively as the *standard model.*

I make a choice of the two right handed neutrinos of the VIIIth family coupled into a scalar, with $\tau^4 = -1$, $\tau^{23} = 1$, correspondingly $Y = 0$, $Q = 0$ and $\vec{\tau}^1 = 0$, and with family quantum numbers (Eqs. (A.24, A.23)) $\tilde{\tau}^4 = -1$, $\tilde{\tau}^{23} = 1$, $\tilde{N}^3_R = 1$, and correspondingly with $\tilde{Y} = \tilde{\tau}^4 + \tilde{\tau}^{23} = 0$, $\tilde{Q} = \tilde{Y} + \tilde{\tau}^{13} = 0$, and $\vec{\tilde{\tau}}^1 = 0$. The condensate carries the family quantum numbers of the upper four families, see Subsect. 3.4.

The condensate made out of spinors couples to spinors differently than to antispinors - "anticondensate" would namely carry $\tau^4 = 1$, and $\tau^{23} = -1$ - breaking correspondingly the $\mathbb{C}_{\mathcal{N}} \cdot \mathcal{P}_{\mathcal{N}}$ symmetry: The reactions creating particles from antiparticles are not any longer symmetric to those creating antiparticle from particles.

Such a condensate leaves the hyper field A^Y_m $(= \sin\vartheta_2\, A^{23}_m + \cos\vartheta_2\, A^4_m)$ (for the choice that $\sin\vartheta_2 = \cos\vartheta_2$ and $g^4 = g^2$, there is no justification for such a choice, $A^Y_m = \frac{1}{\sqrt{2}}\,(A^{23}_m + A^4_m)$ follows) massless, while it gives masses to $A^{2\pm}_m$ and $A^{Y'}_m$ $(= \frac{1}{\sqrt{2}}\,(A^4_m - A^{23}_m)$ for $\sin\vartheta_2 = \cos\vartheta_2)$ and it gives masses also to all the scalar gauge fields from Table 9, since they all couple to the condensate through τ^4.

The weak vector gauge fields, $\vec{A}^1_m$, the hyper charge vector gauge fields, A^Y_m, and the colour vector gauge fields, $\vec{A}^1_m$, remain massless.

The scalar fields with the scalar space index $s = (7, 8)$ (there are three singlets which couple to all eight families, two triplets which couple only to the upper four families and another two triplets which couple only to the lower four families) - carrying the weak and the hyper charges of the Higgs's scalar - wait for gaining nonzero vacuum expectation values to change their masses while causing the electroweak break.

The condensate does what is needed so that in the low energy regime the *spin-charge-family* manifests as an effective theory. This effective theory agrees with the *standard model* to such an extent that it is in agreement with the observed phenomena, explaining the *standard model* assumptions and predicting new fermion and boson fields.

It also may hopefully explain the observed matter-antimatter asymmetry if the conditions in the expanding universe would be appropriate, Ref. ([2], Sect. VI.). The work needed to check these conditions in the expanding universe within the *spin-charge-family* theory is very demanding. Although we do have some experience with following the history of the expanding universe [13], this study needs much more efforts, not only in calculations, but also in understanding the mechanism of the condensate appearance, relations among the velocity of the expansion, the temperature and the dimension of space-time in the period of the appearance of the condensate. This study has not yet been really started.

5. Conclusions

To better understand the history of the universe and also to make next step in understanding the dynamics of the elementary fermion fields and boson (vector and scalar) gauge fields it is needed to explain the assumptions of the *standard model*, as well as the phenomena like the existence of the dark matter, matter-antimatter asymmetry and dark energy.

We must understand the origin of: **A.** the family members quantum numbers, **B.** the family quantum numbers, **C.** the origin of vector gauge fields, **D.** the origin of the Higgs and Yukawa couplings.

One of the most urgent questions in the elementary particle physics is: *Where do the families originate?* Explaining the origin of families would answer the question about the number of families which are possibly observable at the low energy regime, about the origin of the scalar field(s) and the Yukawa couplings (the couplings of fermions to the scalar field(s)), about the differences in the fermions properties - the differences in the masses and mixing matrices among family members – quarks and leptons, as well as about the hierarchy in quark and lepton masses.

I demonstrated in this talk, that the *spin-charge-family* theory - starting with the simple action in $d = (13 + 1)$ for fermions and bosons - offers the explanation for all the assumptions of the *standard model*:

a. The theory explains all the properties of the family members - quarks and leptons, left and right handed, and their right and left handed antiquarks and antileptons [19], explaining why the left handed spinors carry the weak charge while the right handed do not (the right handed neutrino is the regular member of each family).

b. It explains the appearance and the properties of the families of family members.

c. It explains the existence of the gauge vector fields of the family members charges.

d. It explains the appearance and the properties of the scalar field (the Higgs) and the Yukawa couplings.

All the gauge fields, vector and scalar in $d = (3 + 1)$, origin in vielbeins and the two kinds of spin connection fields in $d = (13 + 1)$ - the gravity. The two spin connection fields are uniquely expressible with the vielbeins, if there are no spinors present [42].

[19] One Weyl representation of $SO(13 + 1)$ contains, if analyzed with respect to the *standard model* groups, all the members of one family, the coloured quarks and colourless leptons, and the anticoloured antiquarks and (anti)colourless antileptons, with the left handed spinors carrying the weak charge and the right handed ones weak chargeless, while the left handed antispinors are weak chargeless and the right handed ones carry the weak charge.

It also offers the explanation for the phenomena, which are not part of the *standard model*, like:
e. It explains the existence of the dark matter.
f. It explain the origin of the (ordinary) matter-antimatter asymmetry.

5.1. Predictions for the future experiments

The theory predicts:
g. There are twice two groups of four families of quarks and leptons at low energies.
g.i. The fourth family with masses above 1 TeV, weakly coupled to the observed three families, will be measured at the LHC.
g.ii. The quarks and leptons of the fifth family - that is of the stable one of the upper four families - form the dark matter. The family members, which form the chargeless clusters, manifest, due to their very heavy masses, a "new nuclear force".
h. The predicted scalar fields with the space index $(7, 8)$ manifest in $(d = (3+1))$ as the weak and hyper charges doublets (as required by the *standard model* Higgs) with respect to the space index. These scalars carry in addition:
h.i. Either they carry one of the three family members quantum numbers, (Q, Q', Y') - belonging correspondingly to one of three singlets.
h.ii. Or they carry family quantum numbers - belonging correspondingly to one of the twice two triplets.
h.iii. The three singlets and the two triplets determine mass matrices of the lower four families, contributing to masses of the heavy vector bosons.
h.iv. These scalars determine the observed Higgs and the Yukawa couplings.
i. The predicted scalar fields with the space index $(9, 10, .., 14)$ are triplets with respect to the space index. They cause the transitions from antileptons into quarks and antiquarks into quarks and back.
i.i. The condensate breaks the matter-antimatter symmetry, causing the asymmetry in the (ordinary) matter with respect to antimatter.
i.ii. These condensate is responsible also for the proton decay.
j. The condensate is a scalar of the two right handed neutrinos with the family quantum numbers of the upper four families.
k. The condensate gives masses to all the gauge fields with which it interacts.
k.i. It gives masses to all scalar fields and to vector fields, leaving massless only the colour, the weak, the hyper vector gauge fields and the gravity in $((3 + 1))$.
l. There is the $SU(2)$ (belonging together with the weak $SU(2)$ to $SO(4)$ gauge fields included in $SO(7, 1)$) vector gauge field, which gain masses of the order of the appearance of the condensate.
m. At the electroweak break the scalar fields with the space index $(7, 8)$ change their mutual interaction, and gaining nonzero vacuum expectation values, break the weak and the hyper charges and correspondingly the mass protection of fermions, making them massive.
n. The symmetry of mass matrices allow, in the case that the experimental data for the mixing submatrix 3×3 of the 4×4 mixing matrix would be accurate, to determine the mixing matrix and the masses of the fourth family quarks. The accuracy, with which the masses of the six lower families are measured so far, does not influence the results appreciably. Due to uncertainty of the experimental data for the 3×3 mixing submatrix we are only able to determine the 4×4 quark mixing matrix for a chosen masses of the the fourth family quarks. However, we also predict how will the 3×3 submatrix of the mixing matrix change with more accurate measurements.
n.i. The fourth family quarks mass matrices are for masses above 1 TeV closer and closer to the democratic matrices. The less the scalars with the family members quantum numbers contribute to masses of the fourth family quarks, the closer is m_{u_4} to m_{d_4}.

n.ii. The large contribution of the scalars with the family members quantum numbers (Q, Q', Y') to the masses of the lower four families manifests in the large differences of quarks masses of the lower four families.
n.iii. Although we have done calculations also for leptons, further analyses of their properties must wait for more accurate experimental data.
o. In the case that the u_4 and d_4 quarks have similar masses - determined mostly by the scalar fields carrying the family quantum numbers - they contribute mostly to the production of these scalars, while their contribution to the production of those scalars which carry the family members quantum numbers - to the Higgs in particular - is much weaker, which is in agreement with the experiment [20].
p. All the degrees of freedom discussed in this talk are already a part of the simple starting action Eq.(1).
p.i. The way of breaking symmetries (ordered by the conditions determining the history of our universe) is assumed so that it leads in $d = (3 + 1)$ to the observable symmetries, although we could in principle derive it from the starting action and boundary conditions.
p.ii. Also the effective interaction among scalar fields is assumed, although we could derive it in principle from the starting action and the boundary conditions.
r. The *spin-charge-family* theory easily explains what in the *standard model* seems like a miracle: no triangle anomalies.
s. A lot of efforts has been put in this theory to show that it could work as a next step below the *standard model* proving like:
s.i. There is possibility in the Kaluza-Klein-like theory that breaking symmetries can leave fermions massless.
s.ii. That vielbeins in the Kaluza-Klein theories and spin connections in the *spin-charge-family* theory represent the same vector gauge fields in $d = (3 + 1)$.

5.2. Open questions in the spin-charge-family theory

There are several open problems in the *spin-charge-family* theory:
t. Since this theory is, except that fermions carry two kinds of spins - one kind taking care of spin and charges, the second one taking care of families - a kind of the Kaluza-Klein theories, it shares at very high energy with these theories the quantization problem.
u. The dimension of space-time, $d = (13+1)$, is in the *spin-charge-family* theory chosen, since $SO(13, 1)$ contains all the members, assumed in the *standard model.*
It contains also the right handed neutrino (which carries the Y' quantum number).
u.i. It should be shown, however, how has nature "made the decision" in evolution to go through this dimension and what is indeed the dimension of space-time (infinite?).
v. There are many other open question, like:
v.i. What is the reason for the (so small) dark energy?
v.ii. At what energy the electroweak phase transition occurs? (Let me add that there is the answer within the *spin-charge-family* theory to this open problem.)
v.iii. Why do we have fermions and bosons? [43]

It is encouraging that the more work is done on this theory the more answers to the open questions is found.

[20] The coupling constants of the singlet scalar fields differ among themselves and also from the coupling constants of the two triplet scalar fields.

Acknowledgments
The author thanks the organizers of IARD - Martin Land and Matej Pavšič - to give her opportunity to present her work, which might have a real chance to show the next step beyond the standard models.

Appendix A. Short presentation of spinor technique [1, 7, 31, 32]

This appendix is a short review (taken from [4]) of the technique [7, 33, 31, 32], initiated and developed in Ref. [7], while proposing the *spin-charge-family* theory [5, 3, 6, 7, 8, 9, 10, 4, 11, 12, 13, 14, 15, 16, 17, 2, 1]. All the internal degrees of freedom of spinors, with family quantum numbers included, are describable in the space of d-anticommuting (Grassmann) coordinates [7], if the dimension of ordinary space is also d. There are two kinds of operators in the Grassmann space fulfilling the Clifford algebra and anticommuting with one another Eq.(A.1). The technique was further developed in the present shape together with H.B. Nielsen [33, 31, 32].

In this last stage we rewrite a spinor basis, written in Ref. [7] as products of polynomials of Grassmann coordinates of odd and even Grassmann character, chosen to be eigenstates of the Cartan subalgebra defined by the two kinds of the Clifford algebra objects, as products of nilpotents and projections, formed as odd and even objects of γ^a's, respectively, and chosen to be eigenstates of a Cartan subalgebra of the Lorentz groups defined by γ^a's and $\tilde{\gamma}^a$'s.

The technique can be used to construct a spinor basis for any dimension d and any signature in an easy and transparent way. Equipped with the graphic presentation of basic states, the technique offers an elegant way to see all the quantum numbers of states with respect to the two Lorentz groups, as well as transformation properties of the states under any Clifford algebra object.

Ref. [1], App. B, briefly represents the starting point [7] of this technique. There are two kinds of the Clifford algebra objects, γ^a's and $\tilde{\gamma}^a$'s.

These objects have properties,

$$\{\gamma^a, \gamma^b\}_+ = 2\eta^{ab}\,, \qquad \{\tilde{\gamma}^a, \tilde{\gamma}^b\}_+ = 2\eta^{ab}\,, \quad , \quad \{\gamma^a, \tilde{\gamma}^b\}_+ = 0\,. \tag{A.1}$$

If B is a Clifford algebra object, let say a polynomial of γ^a, $B = a_0 + a_a\,\gamma^a + a_{ab}\,\gamma^a\gamma^b + \cdots + a_{a_1 a_2 \ldots a_d}\,\gamma^{a_1}\gamma^{a_2}\ldots\gamma^{a_d}$, one finds

$$\begin{aligned}(\tilde{\gamma}^a B :&= i(-)^{n_B}\, B\gamma^a\,)\,|\psi_0>,\\ B &= a_0 + a_{a_0}\gamma^{a_0} + a_{a_1 a_2}\gamma^{a_1}\gamma^{a_2} + \cdots + a_{a_1\cdots a_d}\gamma^{a_1}\cdots\gamma^{a_d}\,,\end{aligned} \tag{A.2}$$

where $|\psi_0>$ is a vacuum state, defined in Eq. (A.16) and $(-)^{n_B}$ is equal to 1 for the term in the polynomial which has an even number of γ^b's, and to -1 for the term with an odd number of γ^b's, for any d, even or odd, and I is the unit element in the Clifford algebra.

It follows from Eq. (A.2) that the two kinds of the Clifford algebra objects are connected with the left and the right multiplication of any Clifford algebra objects B (Eq. (A.2)).

The Clifford algebra objects S^{ab} and $\tilde{S}^{ab}$ close the algebra of the Lorentz group

$$\begin{aligned}S^{ab} :&= (i/4)(\gamma^a\gamma^b - \gamma^b\gamma^a)\,,\\ \tilde{S}^{ab} :&= (i/4)(\tilde{\gamma}^a\tilde{\gamma}^b - \tilde{\gamma}^b\tilde{\gamma}^a)\,,\end{aligned} \tag{A.3}$$

$\{S^{ab}, \tilde{S}^{cd}\}_- = 0\,,\ \{S^{ab}, S^{cd}\}_- = i(\eta^{ad}S^{bc} + \eta^{bc}S^{ad} - \eta^{ac}S^{bd} - \eta^{bd}S^{ac})\,,\ \{\tilde{S}^{ab}, \tilde{S}^{cd}\}_- = i(\eta^{ad}\tilde{S}^{bc} + \eta^{bc}\tilde{S}^{ad} - \eta^{ac}\tilde{S}^{bd} - \eta^{bd}\tilde{S}^{ac})\,.$

We assume the "Hermiticity" property for γ^a's

$$\gamma^{a\dagger} = \eta^{aa}\gamma^a\,, \tag{A.4}$$

IOP Conf. Series: Journal of Physics: Conf. Series **845** (2017) 012017 doi:10.1088/1742-6596/845/1/012017

in order that γ^a are compatible with (A.1) and formally unitary, i.e. $\gamma^{a\,\dagger}\,\gamma^a = I$.

One finds from Eq. (A.4) that $(S^{ab})^\dagger = \eta^{aa}\eta^{bb}S^{ab}$.

Recognizing from Eq.(A.3) that the two Clifford algebra objects S^{ab}, S^{cd} with all indices different commute, and equivalently for $\tilde{S}^{ab}, \tilde{S}^{cd}$, we select the Cartan subalgebra of the algebra of the two groups, which form equivalent representations with respect to one another

$$\begin{aligned} S^{03}, S^{12}, S^{56}, \cdots, S^{d-1\,d}, \quad &\text{if} \quad d = 2n \geq 4, \\ S^{03}, S^{12}, \cdots, S^{d-2\,d-1}, \quad &\text{if} \quad d = (2n+1) > 4\,, \\ \tilde{S}^{03}, \tilde{S}^{12}, \tilde{S}^{56}, \cdots, \tilde{S}^{d-1\,d}, \quad &\text{if} \quad d = 2n \geq 4\,, \\ \tilde{S}^{03}, \tilde{S}^{12}, \cdots, \tilde{S}^{d-2\,d-1}, \quad &\text{if} \quad d = (2n+1) > 4\,. \end{aligned} \tag{A.5}$$

The choice for the Cartan subalgebra in $d < 4$ is straightforward. It is useful to define one of the Casimirs of the Lorentz group - the handedness Γ ($\{\Gamma, S^{ab}\}_- = 0$) in any d

$$\begin{aligned} \Gamma^{(d)} : &= (i)^{d/2} \prod_a (\sqrt{\eta^{aa}}\gamma^a), \quad \text{if} \quad d = 2n, \\ \Gamma^{(d)} : &= (i)^{(d-1)/2} \prod_a (\sqrt{\eta^{aa}}\gamma^a), \quad \text{if} \quad d = 2n+1\,. \end{aligned} \tag{A.6}$$

One proceeds equivalently for $\tilde{\Gamma}^{(d)}$, substituting γ^a's by $\tilde{\gamma}^a$'s. We understand the product of γ^a's in the ascending order with respect to the index a: $\gamma^0\gamma^1\cdots\gamma^d$. It follows from Eq.(A.4) for any choice of the signature η^{aa} that $\Gamma^\dagger = \Gamma$, $\Gamma^2 = I$. We also find that for d even the handedness anticommutes with the Clifford algebra objects γ^a ($\{\gamma^a, \Gamma\}_+ = 0$) , while for d odd it commutes with γ^a ($\{\gamma^a, \Gamma\}_- = 0$).

To make the technique simple we introduce the graphic presentation as follows

$$\stackrel{ab}{(k)}: = \frac{1}{2}(\gamma^a + \frac{\eta^{aa}}{ik}\gamma^b)\,, \qquad \stackrel{ab}{[k]}:= \frac{1}{2}(1 + \frac{i}{k}\gamma^a\gamma^b)\,, \tag{A.7}$$

where $k^2 = \eta^{aa}\eta^{bb}$. It follows then

$$\begin{aligned} \gamma^a &= \stackrel{ab}{(k)} + \stackrel{ab}{(-k)}\,, \qquad \gamma^b = ik\eta^{aa}\,(\stackrel{ab}{(k)} - \stackrel{ab}{(-k)})\,, \\ S^{ab} &= \frac{k}{2}(\stackrel{ab}{[k]} - \stackrel{ab}{[-k]})\,. \end{aligned} \tag{A.8}$$

One can easily check by taking into account the Clifford algebra relation (Eq. (A.1)) and the definition of S^{ab} and $\tilde{S}^{ab}$ (Eq. (A.3)) that the nilpotent $\stackrel{ab}{(k)}$ and the projector $\stackrel{ab}{[k]}$ are "eigenstates" of S^{ab} and $\tilde{S}^{ab}$

$$\begin{aligned} S^{ab}\,\stackrel{ab}{(k)} &= \frac{1}{2}k\,\stackrel{ab}{(k)}\,, \qquad S^{ab}\,\stackrel{ab}{[k]} = \frac{1}{2}k\,\stackrel{ab}{[k]}\,, \\ \tilde{S}^{ab}\,\stackrel{ab}{(k)} &= \frac{1}{2}k\,\stackrel{ab}{(k)}\,, \qquad \tilde{S}^{ab}\,\stackrel{ab}{[k]} = -\frac{1}{2}k\,\stackrel{ab}{[k]}\,, \end{aligned} \tag{A.9}$$

which means that we get the same objects back multiplied by the constant $\frac{1}{2}k$ in the case of S^{ab}, while $\tilde{S}^{ab}$ multiply $\stackrel{ab}{(k)}$ by k and $\stackrel{ab}{[k]}$ by $(-k)$ rather than (k). This also means that when $\stackrel{ab}{(k)}$ and $\stackrel{ab}{[k]}$ act from the left hand side on a vacuum state $|\psi_0\rangle$ the obtained states are the eigenvectors of S^{ab}. We further recognize that γ^a transform $\stackrel{ab}{(k)}$ into $\stackrel{ab}{[-k]}$, never to $\stackrel{ab}{[k]}$, while $\tilde{\gamma}^a$ transform $\stackrel{ab}{(k)}$ into $\stackrel{ab}{[k]}$, never to $\stackrel{ab}{[-k]}$

$$\gamma^a \stackrel{ab}{(k)} = \eta^{aa} \stackrel{ab}{[-k]},\; \gamma^b \stackrel{ab}{(k)} = -ik \stackrel{ab}{[-k]},\; \gamma^a \stackrel{ab}{[k]} = \stackrel{ab}{(-k)},\; \gamma^b \stackrel{ab}{[k]} = -ik\eta^{aa} \stackrel{ab}{(-k)},$$
$$\tilde{\gamma}^a \stackrel{ab}{(k)} = -i\eta^{aa} \stackrel{ab}{[k]},\; \tilde{\gamma}^b \stackrel{ab}{(k)} = -k \stackrel{ab}{[k]},\; \tilde{\gamma}^a \stackrel{ab}{[k]} = i \stackrel{ab}{(k)},\; \tilde{\gamma}^b \stackrel{ab}{[k]} = -k\eta^{aa} \stackrel{ab}{(k)}\,. \qquad \text{(A.10)}$$

From Eq.(A.10) it follows

$$\begin{aligned}
S^{ac} \stackrel{ab}{(k)}\stackrel{cd}{(k)} &= -\frac{i}{2}\eta^{aa}\eta^{cc} \stackrel{ab}{[-k]}\stackrel{cd}{[-k]}\,, & \tilde{S}^{ac} \stackrel{ab}{(k)}\stackrel{cd}{(k)} &= \frac{i}{2}\eta^{aa}\eta^{cc} \stackrel{ab}{[k]}\stackrel{cd}{[k]}\,,\\
S^{ac} \stackrel{ab}{[k]}\stackrel{cd}{[k]} &= \frac{i}{2} \stackrel{ab}{(-k)}\stackrel{cd}{(-k)}\,, & \tilde{S}^{ac} \stackrel{ab}{[k]}\stackrel{cd}{[k]} &= -\frac{i}{2} \stackrel{ab}{(k)}\stackrel{cd}{(k)}\,,\\
S^{ac} \stackrel{ab}{(k)}\stackrel{cd}{[k]} &= -\frac{i}{2}\eta^{aa} \stackrel{ab}{[-k]}\stackrel{cd}{(-k)}\,, & \tilde{S}^{ac} \stackrel{ab}{(k)}\stackrel{cd}{[k]} &= -\frac{i}{2}\eta^{aa} \stackrel{ab}{[k]}\stackrel{cd}{(k)}\,,\\
S^{ac} \stackrel{ab}{[k]}\stackrel{cd}{(k)} &= \frac{i}{2}\eta^{cc} \stackrel{ab}{(-k)}\stackrel{cd}{[-k]}\,, & \tilde{S}^{ac} \stackrel{ab}{[k]}\stackrel{cd}{(k)} &= \frac{i}{2}\eta^{cc} \stackrel{ab}{(k)}\stackrel{cd}{[k]}\,.
\end{aligned} \qquad \text{(A.11)}$$

From Eq. (A.11) we conclude that $\tilde{S}^{ab}$ generate the equivalent representations with respect to S^{ab} and opposite.

Let us deduce some useful relations

$$\begin{aligned}
\stackrel{ab}{(k)}\stackrel{ab}{(k)} &= 0\,, & \stackrel{ab}{(k)}\stackrel{ab}{(-k)} &= \eta^{aa} \stackrel{ab}{[k]}\,, & \stackrel{ab}{(-k)}\stackrel{ab}{(k)} &= \eta^{aa} \stackrel{ab}{[-k]}\,, & \stackrel{ab}{(-k)}\stackrel{ab}{(-k)} &= 0\,,\\
\stackrel{ab}{[k]}\stackrel{ab}{[k]} &= \stackrel{ab}{[k]}\,, & \stackrel{ab}{[k]}\stackrel{ab}{[-k]} &= 0\,, & \stackrel{ab}{[-k]}\stackrel{ab}{[k]} &= 0\,, & \stackrel{ab}{[-k]}\stackrel{ab}{[-k]} &= \stackrel{ab}{[-k]}\,,\\
\stackrel{ab}{(k)}\stackrel{ab}{[k]} &= 0\,, & \stackrel{ab}{[k]}\stackrel{ab}{(k)} &= \stackrel{ab}{(k)}\,, & \stackrel{ab}{(-k)}\stackrel{ab}{[k]} &= \stackrel{ab}{(-k)}\,, & \stackrel{ab}{(-k)}\stackrel{ab}{[-k]} &= 0\,,\\
\stackrel{ab}{(k)}\stackrel{ab}{[-k]} &= \stackrel{ab}{(k)}\,, & \stackrel{ab}{[k]}\stackrel{ab}{(-k)} &= 0, & \stackrel{ab}{[-k]}\stackrel{ab}{(k)} &= 0\,, & \stackrel{ab}{[-k]}\stackrel{ab}{(-k)} &= \stackrel{ab}{(-k)}\,.
\end{aligned}$$
$$\text{(A.12)}$$

We recognize in Eq. (A.12) the demonstration of the nilpotent and the projector character of the Clifford algebra objects $\stackrel{ab}{(k)}$ and $\stackrel{ab}{[k]}$, respectively. Defining

$$\stackrel{ab}{(\tilde{\pm} i)} = \frac{1}{2}\,(\tilde{\gamma}^a \mp \tilde{\gamma}^b)\,,\quad \stackrel{ab}{(\tilde{\pm} 1)} = \frac{1}{2}\,(\tilde{\gamma}^a \pm i\tilde{\gamma}^b)\,, \qquad \text{(A.13)}$$

one recognizes that

$$\stackrel{ab}{(\tilde{k})}\stackrel{ab}{(k)} = 0\,,\quad \stackrel{ab}{(\tilde{-k})}\stackrel{ab}{(k)} = -i\eta^{aa} \stackrel{ab}{[k]}\,,\quad \stackrel{ab}{(\tilde{k})}\stackrel{ab}{[k]} = i \stackrel{ab}{(k)}\,,\quad \stackrel{ab}{(\tilde{k})}\stackrel{ab}{[-k]} = 0\,. \qquad \text{(A.14)}$$

Recognizing that

$$\stackrel{ab}{(k)}^{\dagger} = \eta^{aa} \stackrel{ab}{(-k)}\,,\quad \stackrel{ab}{[k]}^{\dagger} = \stackrel{ab}{[k]}\,, \qquad \text{(A.15)}$$

we define a vacuum state $|\psi_0>$ so that one finds

$$< \stackrel{ab}{(k)}^{\dagger} \stackrel{ab}{(k)} >= 1\,,$$
$$< \stackrel{ab}{[k]}^{\dagger} \stackrel{ab}{[k]} >= 1\,. \qquad \text{(A.16)}$$

IOP Conf. Series: Journal of Physics: Conf. Series **845** (2017) 012017 doi:10.1088/1742-6596/845/1/012017

Taking into account the above equations it is easy to find a Weyl spinor irreducible representation for d-dimensional space, with d even or odd.

For d even we simply make a starting state as a product of $d/2$, let us say, only nilpotents $\stackrel{ab}{(k)}$, one for each S^{ab} of the Cartan subalgebra elements (Eq.(A.5)), applying it on an (unimportant) vacuum state. For d odd the basic states are products of $(d-1)/2$ nilpotents and a factor $(1 \pm \Gamma)$. Then the generators S^{ab}, which do not belong to the Cartan subalgebra, being applied on the starting state from the left, generate all the members of one Weyl spinor.

$$
\begin{array}{c}
\stackrel{0d}{(k_{0d})}\stackrel{12}{(k_{12})}\stackrel{35}{(k_{35})}\cdots\stackrel{d-1\,d-2}{(k_{d-1\,d-2})}\;|\psi_0>\\
\stackrel{0d}{[-k_{0d}]}\stackrel{12}{[-k_{12}]}\stackrel{35}{(k_{35})}\cdots\stackrel{d-1\,d-2}{(k_{d-1\,d-2})}\;|\psi_0>\\
\stackrel{0d}{[-k_{0d}]}\stackrel{12}{(k_{12})}\stackrel{35}{[-k_{35}]}\cdots\stackrel{d-1\,d-2}{(k_{d-1\,d-2})}\;|\psi_0>\\
\vdots\\
\stackrel{0d}{[-k_{0d}]}\stackrel{12}{(k_{12})}\stackrel{35}{(k_{35})}\cdots\stackrel{d-1\,d-2}{[-k_{d-1\,d-2}]}\;|\psi_0>\\
\stackrel{od}{(k_{0d})}\stackrel{12}{[-k_{12}]}\stackrel{35}{[-k_{35}]}\cdots\stackrel{d-1\,d-2}{(k_{d-1\,d-2})}\;|\psi_0>\\
\vdots
\end{array}
\qquad (A.17)
$$

All the states have the same handedness Γ, since $\{\Gamma, S^{ab}\}_- = 0$. States, belonging to one multiplet with respect to the group $SO(q, d-q)$, that is to one irreducible representation of spinors (one Weyl spinor), can have any phase. We made a choice of the simplest one, taking all phases equal to one.

The above graphic representation demonstrates that for d even all the states of one irreducible Weyl representation of a definite handedness follow from a starting state, which is, for example, a product of nilpotents $\stackrel{ab}{(k_{ab})}$, by transforming all possible pairs of $\stackrel{ab}{(k_{ab})}\stackrel{mn}{(k_{mn})}$ into $\stackrel{ab}{[-k_{ab}]}\stackrel{mn}{[-k_{mn}]}$. There are $S^{am}, S^{an}, S^{bm}, S^{bn}$, which do this. The procedure gives $2^{(d/2-1)}$ states. A Clifford algebra object γ^a being applied from the left hand side, transforms a Weyl spinor of one handedness into a Weyl spinor of the opposite handedness. Both Weyl spinors form a Dirac spinor.

We shall speak about left handedness when $\Gamma = -1$ and about right handedness when $\Gamma = 1$ for either d even or odd.

While S^{ab} which do not belong to the Cartan subalgebra (Eq. (A.5)) generate all the states of one representation, $\tilde{S}^{ab}$ which do not belong to the Cartan subalgebra (Eq. (A.5)) generate the states of $2^{d/2-1}$ equivalent representations.

Making a choice of the Cartan subalgebra set (Eq. (A.5)) of the algebra S^{ab} and $\tilde{S}^{ab}$ $S^{03}, S^{12}, S^{56}, S^{78}, S^{9\,10}, S^{11\,12}, S^{13\,14}$, $\tilde{S}^{03}, \tilde{S}^{12}, \tilde{S}^{56}, \tilde{S}^{78}, \tilde{S}^{9\,10}, \tilde{S}^{11\,12}, \tilde{S}^{13\,14}$, a left handed ($\Gamma^{(13,1)} = -1$) eigenstate of all the members of the Cartan subalgebra, representing a weak chargeless u_R-quark with spin up, hyper charge $(2/3)$ and colour $(1/2\,, 1/(2\sqrt{3}))$, for example, can be written as

$$
\begin{aligned}
&\stackrel{03}{(+i)}\stackrel{12}{(+)}\,|\,\stackrel{56}{(+)}\stackrel{78}{(+)}\,||\,\stackrel{9\,10}{(+)}\stackrel{11\,12}{(-)}\stackrel{13\,14}{(-)}\;|\psi_0\rangle = \\
&\frac{1}{2^7}(\gamma^0-\gamma^3)(\gamma^1+i\gamma^2)|(\gamma^5+i\gamma^6)(\gamma^7+i\gamma^8)|| \\
&(\gamma^9+i\gamma^{10})(\gamma^{11}-i\gamma^{12})(\gamma^{13}-i\gamma^{14})|\psi_0\rangle\,.
\end{aligned}
\qquad (A.18)
$$

This state is an eigenstate of all S^{ab} and $\tilde{S}^{ab}$ which are members of the Cartan subalgebra (Eq. (A.5)).

The operators $\tilde{S}^{ab}$, which do not belong to the Cartan subalgebra (Eq. (A.5)), generate families from the starting u_R quark, transforming the u_R quark from Eq. (A.18) to the u_R of another family, keeping all of the properties with respect to S^{ab} unchanged. In particular, $\tilde{S}^{01}$ applied on a right handed u_R-quark from Eq. (A.18) generates a state which is again a right handed u_R-quark, weak chargeless, with spin up, hyper charge (2/3) and the colour charge $(1/2\,,1/(2\sqrt{3}))$

$$\tilde{S}^{01} \stackrel{03}{(+i)}\stackrel{12}{(+)} \mid \stackrel{56}{(+)}\stackrel{78}{(+)} \mid\mid \stackrel{9\,10}{(+)}\stackrel{11\,12}{(-)}\stackrel{13\,14}{(-)} = -\frac{i}{2} \stackrel{03}{[+i]}\stackrel{12}{[+]} \mid \stackrel{56}{(+)}\stackrel{78}{(+)} \mid\mid \stackrel{9\,10}{(+)}\stackrel{11\,12}{(-)}\stackrel{13\,14}{(-)} \,. \tag{A.19}$$

Below some useful relations [9] are presented

$$\vec{N}_{\pm}(=\vec{N}_{(L,R)}): \;=\; \frac{1}{2}(S^{23} \pm iS^{01}, S^{31} \pm iS^{02}, S^{12} \pm iS^{03})\,, \tag{A.20}$$

$$\begin{aligned} \vec{\tau}^{1}: &= \frac{1}{2}(S^{58} - S^{67},\, S^{57} + S^{68},\, S^{56} - S^{78})\,, \\ \vec{\tau}^{2}: &= \frac{1}{2}(S^{58} + S^{67},\, S^{57} - S^{68},\, S^{56} + S^{78})\,, \end{aligned} \tag{A.21}$$

$$\begin{aligned} \vec{\tau}^{3} := &\; \frac{1}{2}\{S^{9\,12} - S^{10\,11}, S^{9\,11} + S^{10\,12}, S^{9\,10} - S^{11\,12}, \\ &\; S^{9\,14} - S^{10\,13}, S^{9\,13} + S^{10\,14}, S^{11\,14} - S^{12\,13}, \\ &\; S^{11\,13} + S^{12\,14}, \frac{1}{\sqrt{3}}(S^{9\,10} + S^{11\,12} - 2S^{13\,14})\}\,, \\ \tau^{4} := &\; -\frac{1}{3}(S^{9\,10} + S^{11\,12} + S^{13\,14})\,, \end{aligned} \tag{A.22}$$

$$\vec{\tilde{N}}_{L,R}: \;=\; \frac{1}{2}(\tilde{S}^{23} \pm i\tilde{S}^{01}, \tilde{S}^{31} \pm i\tilde{S}^{02}, \tilde{S}^{12} \pm i\tilde{S}^{03})\,, \tag{A.23}$$

$$\begin{aligned} \vec{\tilde{\tau}}^{1}: &= \frac{1}{2}(\tilde{S}^{58} - \tilde{S}^{67},\, \tilde{S}^{57} + \tilde{S}^{68},\, \tilde{S}^{56} - \tilde{S}^{78})\,, \\ \vec{\tilde{\tau}}^{2}: &= \frac{1}{2}(\tilde{S}^{58} + \tilde{S}^{67},\, \tilde{S}^{57} - \tilde{S}^{68},\, \tilde{S}^{56} + \tilde{S}^{78})\,, \end{aligned} \tag{A.24}$$

$$\tilde{\tau}^{4} := \;-\frac{1}{3}(\tilde{S}^{9\,10} + \tilde{S}^{11\,12} + \tilde{S}^{13\,14})\,. \tag{A.25}$$

$$\begin{aligned} N^{\pm}_{+} &= N^{1}_{+} \pm i\,N^{2}_{+} = -\stackrel{03}{(\mp i)}\stackrel{12}{(\pm)}\,, \quad N^{\pm}_{-} = N^{1}_{-} \pm i\,N^{2}_{-} = \stackrel{03}{(\pm i)}\stackrel{12}{(\pm)}\,, \\ \tilde{N}^{\pm}_{+} &= -\stackrel{03}{\widetilde{(\mp i)}}\stackrel{12}{\widetilde{(\pm)}}\,, \quad \tilde{N}^{\pm}_{-} = \stackrel{03}{\widetilde{(\pm i)}}\stackrel{12}{\widetilde{(\pm)}}\,, \\ \tau^{1\pm} &= (\mp)\stackrel{56}{(\pm)}\stackrel{78}{(\mp)}\,, \quad \tau^{2\mp} = (\mp)\stackrel{56}{(\mp)}\stackrel{78}{(\mp)}\,, \\ \tilde{\tau}^{1\pm} &= (\mp)\stackrel{56}{\widetilde{(\pm)}}\stackrel{78}{\widetilde{(\mp)}}\,, \quad \tilde{\tau}^{2\mp} = (\mp)\stackrel{56}{\widetilde{(\mp)}}\stackrel{78}{\widetilde{(\mp)}}\,. \end{aligned} \tag{A.26}$$

IOP Conf. Series: Journal of Physics: Conf. Series **845** (2017) 012017 doi:10.1088/1742-6596/845/1/012017

Appendix B. *Standard model* assumptions

More than 40 years ago the *standard model* offered an *elegant new step* in understanding the origin of fermions and bosons by postulating:

- The existence of *massless family members*: coloured quarks and colourless leptons, both left and right handed, the left handed members distinguishing from the right handed ones in the weak and hyper charges and correspondingly mass protected, Table B1.
- The existence of the *vector gauge fields* (massless before the electroweak break) to the observed charges of the family members, Table B2.
- The existence of a massive *scalar field* carrying the weak charge ($\pm\frac{1}{2}$) and the hyper charge ($\mp\frac{1}{2}$), which by its "nonzero vacuum expectation values" breaks the weak and the hyper charge and correspondingly breaks the mass protection of fermions and those vector bosons which interact with this vacuum, Table B3.
- The existence of the *Yukawa couplings* of fermions, which together with (the gluons and) the scalar take care of the properties of fermions after the electroweak break.

Table B1. Table represents the *standard model* assumptions for each of the three so far observed ($i = 1, 2, 3$) families of quarks and leptons, massless before the electroweak break. Each family contains the left handed weak charged quarks and right handed weak chargeless quarks, each quark belonging to the colour triplet $(1/2, 1/(2\sqrt{3}))$, $(-1/2, 1/(2\sqrt{3}))$, $(0, -1/(\sqrt{3}))$, and the left handed weak charged and right handed weak chargeless colourless leptons. τ^{13} defines the third component of the weak charge, Y is the hyper charge determining the electromagnetic charge $Q = Y + \tau^{13}$. The *standard model* assumes to each family member of each family the corresponding anti-fermions.

α name	hand-edness $-4iS^{03}S^{12}$	weak charge τ^{13}	hyper charge Y	colour charge	elm charge Q
u^i_L	-1	$\frac{1}{2}$	$\frac{1}{6}$	colour triplet	$\frac{2}{3}$
d^i_L	-1	$-\frac{1}{2}$	$\frac{1}{6}$	colour triplet	$-\frac{1}{3}$
ν^i_L	-1	$\frac{1}{2}$	$-\frac{1}{2}$	colourless	0
e^i_L	-1	$-\frac{1}{2}$	$-\frac{1}{2}$	colourless	-1
u^i_R	1	weakless	$\frac{2}{3}$	colour triplet	$\frac{2}{3}$
d^i_R	1	weakless	$-\frac{1}{3}$	colour triplet	$-\frac{1}{3}$
ν^i_R	1	weakless	0	colourless	0
e^i_R	1	weakless	-1	colourless	-1

The *standard model* assumptions have been confirmed without offering surprises. The last unobserved field, the Higgs, detected in June 2012, was confirmed in March 2013.

Table B2. Vector fields, the gauge fields of the hyper, weak and colour charges, all massless before the electroweak break. They all are vectors in the adjoint representations with respect to the weak, colour and hyper charges.

name	**hand-edness**	**weak charge**	**hyper charge**	**colour charge**	elm charge
hyper photon	0	0	0	colourless	0
weak bosons	0	**triplet**	0	colourless	**triplet**
gluons	0	0	0	**colour octet**	0

Table B3. Higgs is the scalar field with the weak charge and the hyper charge $\pm\frac{1}{2}$ and $\mp\frac{1}{2}$, respectively. The 0. Higgs_d and 0. Higgs_u are not assumed. In Table these two are added to manifest the fundamental representation of the charge groups. The two components, $<\text{Higgs}_u>$ and $<\text{Higgs}_d>$, "dress" in the *standard model* the right handed *u*-quarks and *d*-quarks, respectively, giving them the charges of the left partners.

name	**hand-edness**	**weak charge**	**hyper charge**	**colour charge**	**elm charge**
$0\cdot \text{Higgs}_u$	0	$\frac{1}{2}$	$\frac{1}{2}$	colourless	1
$<\text{Higgs}_d>$	0	$-\frac{1}{2}$	$\frac{1}{2}$	colourless	0
$<\text{Higgs}_u>$	0	$\frac{1}{2}$	$-\frac{1}{2}$	colourless	0
$0\cdot \text{Higgs}_d$	0	$-\frac{1}{2}$	$-\frac{1}{2}$	colourless	-1

References

[1] Mankoč Borštnik N S 2015 The explanation for the origin of the Higgs scalar and for the Yukawa couplings by the *spin-charge-family* theory *J. of Mod. Phys.* **6** 2244
[2] Mankoč Borštnik N S 2015 Can spin-charge-family theory explain baryon number non conservation? *Phys. Rev.* D **91** 6 (*Preprint* [arxiv:1409.7791, arXiv:1502.06786v1)
[3] Mankoč Borštnik N S 2012 *Do we have the explanation for the Higgs and Yukawa couplings of the* standard model (Bled workshops in physics vol 13 no 2) ed Mankoč Borštnik S N, Nielsen H B *et al* (Ljubljana: DMFA Založništvo) pp 56–71
[4] Mankoč Borštnik N S 2013 The spin-charge-family theory is explaining the origin of families, of the Higgs and the Yukawa Couplings *J. of Modern Phys.* **4** 823(*Preprint* arxiv:1312.1542)
[5] Mankoč Borštnik N S 2013 *Spin-charge-family theory is explaining appearance of families of quarks and leptons, of Higgs and Yukawa couplings* (Bled workshops in physics vol 14 no 2) ed Mankoč Borštnik S N, Nielsen H B *et al* (Ljubljana: DMFA Založništvo) pp 113–142
[6] Mankoč Borštnik N S 1992 *Phys. Lett.* B **292** 25–29
[7] Mankoč Borštnik N S 1993 *J. Math. Phys.* **34** 3731–3745
[8] Mankoč Borštnik 1994 *Int. J. Theor. Phys.* **40** 315–338
[9] Borštnik Bračič A and Mankoč Borštnik N S 2006 *Phys. Rev.* D **74** 073013
[10] Borštnik Bračič A and Mankoč Borštnik N S 2003 *The Approach Unifying Spins and Charges and Its Predictions* (Bled workshops in physics vol 4 no 2) ed Mankoč Borštnik S N, Nielsen H B *et al* (Ljubljana:

DMFA Založništvo) pp 31–57

[11] Mankoč Borštnik N S 1995 *Modern Phys. Lett.* A **10** 587–595

[12] Bregar G, Breskvar M, Lukman D and Mankoč Borštnik 2008 On the origin of families of quarks and leptons — predictions for four families *New J. Phys.* **10** 093002

[13] Bregar G and Mankoč Borštnik N S 2009 *Phys. Rev.* D **80**, 083534

[14] Bregar and Mankoč Borštnik N S 2013 *Can we predict the fourth family masses for quarks and leptons?* (Bled workshops in physics vol 14 no 2) ed Mankoč Borštnik S N, Nielsen H B *et al* (Ljubljana: DMFA Založništvo) pp 31–51

[15] Bregar G and Mankoč Borštnik N S 2014 *The new experimental data for the quarks mixing matrix are in better agreement with the* spin-charge-family theory predictions (Bled workshops in physics vol 15 no 2) ed Mankoč Borštnik S N, Nielsen H B *et al* (Ljubljana: DMFA Založništvo) pp 20–45

[16] Mankoč Borštnik N S 2012 *Do we have the explanation for the Higgs and Yukawa couplings of the* standard model (*Preprint* arxiv:1212.3184, arxiv:1011.5765)

[17] Mankoč Borštnik N S 2014 *The* spin-charge-family *theory explains why the scalar Higgs carries the weak charge* $\pm\frac{1}{2}$ *and the hyper charge* $\mp\frac{1}{2}$ (Bled workshops in physics vol 15 no 2) ed Mankoč Borštnik S N, Nielsen H B *et al* (Ljubljana: DMFA Založništvo) pp 163–182

[18] Lukman D and Mankoč Borštnik N S 2015 *Vector and Scalar Gauge Fields with Respect to* $d = (3+1)$ *in Kaluza-Klein Theories and in the* spin-charge-family theory (Bled workshops in physics vol 16 no 2) ed Mankoč Borštnik S N, Nielsen H B *et al* (Ljubljana: DMFA Založništvo) pp 158–164

[19] Ceccucci A (CERN), Ligeti Z (LBNL), Sakai Y (KEK), Particle Data Group, Aug. 29, 2014, [http://pdg.lbl.gov/2014/reviews/rpp2014-rev-ckm-matrix.pdf].

[20] Nakamura K *et al*, (Particle Data Group), J. Phys. G: **37** 075021 (2010); Z.Z. Xing, H. Zhang, S. Zhou, Phys. Rev. **D 77** (2008) 113016; Beringer et al, Phys. Rev. D 86 (2012) 010001, Particle Physics booklet, July 2012, PDG, APS physics.

[21] Pavšič M 2006 Spin Gauge Theory of Gravity in Clifford Space: A Realization of Kaluza-Klein Theory in 4-Dimensional Spacetime *Int.J.Mod.Phys.* A **21** 5905 (*Preprint* arXiv:gr-qc/0507053)

[22] L. Alvarez-Gaumé, "An Introduction to Anomalies", Erice School Math. Phys. 1985:0093.

[23] L. Alvarez-Gaumé, E. Witten, "Gravitational Anomalies", *Nucl. Phys.* **B 234** (1983)269-330.

[24] A. Bilal, "Lectures on anomalies", [arXiv:0802.0634].

[25] L. Alvarez-Gaumé, J.M. Gracia-Bondía, C.M. Martin, "Anomaly Cancellation and the Gauge Group of the Standard Model in NCG", [hep-th/9506115].

[26] Ali A and Neubert M 2016 private communication at the Singapore Conference on New Physics at the Large Hadron Collider, 29 February – 4 March 2016.

[27] Franceschini R *et al* 2015 (*Preprint* arXiv:1512.04933)

[28] CMS Collaboration 2012 CMS-PAS-EXO-12-045

[29] CMS Collaboration 2015 *Phys. Rev.* D **92** 032004

[30] Mankoč Borštnik N S 2016 *Do present experiments exclude the fourth family quarks?* to be sent to JHEP

[31] Mankoč Borštnik N S and Nielsen H B 2002 *J. of Math. Phys.* **43** 5782 (*Preprint* hep-th/0111257)

[32] Mankoč Borštnik N S and Nielsen H B 2003 *J. of Math. Phys.* **44** 4817(*Preprint* hep-th/0303224)

[33] Mankoč Borštnik N S and Nielsen H B 2000 *Phys. Rev.* D **62** 044010*Preprint* hep-th/9911032)

[34] Lukman D, Mankoč Borštnik N S and Nielsen H B 2011 An effective two dimensionality cases bring a new hope to the Kaluza-Klein-like theories *New J. Phys.* **13** 103027 (*Preprint* hep-th/1001.4679v5)

[35] Lukman D and Mankoč Borštnik N S 2012 *J. Phys. A: Math. Theor.* **45** 465401

[36] Mankoč Borštnik N S and Nielsen H B 2014 Discrete symmetries in the Kaluza-Klein theories *J. High Energy Phys.* JHEP04(2014)165 (*Preprint* arXiv:1212.2362) 28

[37] Troha T, Lukman D and Mankoč Borštnik N S 2014 Massless and massive representations in the *spinor technique Int. J. of Mod. Phys.* A **29** 1450124 (*Preprint* arXiv:1312.1541)

[38] Witten E 1981 Search for realistic Kaluza-Klein theory *Nucl. Phys.* B **186** 412; Fermion quantum numbers in Kaluza-Klein theories Princeton Technical Rep. PRINT -83-1056, October 1983.

[39] Blagojević M 2002 *Gravitation and gauge symmetries* (Bristol: IoP Publishing)

[40] Appelquist T, Chodos A and Freund P G O (1983) *An introduction to Kaluza-Klein theories* Ed Lee H C (Singapore: World Scientific)

[41] Mankoč Borštnik N S, Nielsen H B and Lukman D 2004 *An example of Kaluza-Klein-like theories leading after compactification to massless spinors coupled to a gauge field-derivations and proofs* (Bled workshops in physics vol 15 no 2) ed Mankoč Borštnik S N, Nielsen H B *et al* (Ljubljana: DMFA Založništvo) pp 64–84

[42] Mankoč Borštnik N S and Lukman D 2015 *Vector and scalar gauge fields with respect to* $d = (3+1)$ *in Kaluza-Klein theories and in the* spin-charge-family theory (Bled workshops in physics vol 16 no 2) ed Mankoč Borštnik S N, Nielsen H B *et al* (Ljubljana: DMFA Založništvo) pp 158–164 (*Preprint* arXiv:1604.00675)

[43] Mankoč Borštnik N S and Nielsen H B F 2015 *Fermionization in an arbitrary number of dimensions* (Bled workshops in physics vol 16 no 2) ed Mankoč Borštnik S N, Nielsen H B *et al* (Ljubljana: DMFA Založništvo) pp 111–128 (*Preprint* arXiv:1602.03175)

Branes and quantized fields

Matej Pavšič
Jožef Stefan Institute, Jamova 39, SI-1000 Ljubljana, Slovenia
E-mail: matej.pavsic@ijs.si

Abstract. It is shown that the Dirac-Nambu-Goto brane can be described as a point particle in an infinite dimensional space with a particular metric. This can be considered as a special case of a general theory in which branes are points in the brane space $\mathcal{M}$, whose metric is dynamical, just like in general relativity. Such a brane theory, amongst others, includes the flat brane space, whose metric is the infinite dimensional analog of the Minkowski space metric $\eta_{\mu\nu}$. A brane living in the latter space will be called "flat brane"; it is like a bunch of non interacting point particles. Quantization of the latter system leads to a system of non interacting quantum fields. Interactions can be included if we consider a non trivial metric in the space of fields. Then the effective classical brane is no longer a flat brane. For a particular choice of the metric in the field space we obtain the Dirac-Nambu-Goto brane. We also show how a Stueckelberg-like quantum field arises within the brane space formalism. With the Stueckelberg fields, we avoid certain well-known intricacies, especially those related to the position operator that is needed in our construction of effective classical branes from the systems of quantum fields.

1. Introduction

Relativistic membranes of arbitrary dimension (branes) [1]–[4], are very important objects in theoretical physics. An attractive possibility is the brane world scenario [5]–[26] in which our spacetime is a 4-dimensional surface embedded in a higher dimensions space. Quantization of gravity could then be achieved by quantizing the brane. Unfortunately, quantization of the Dirac-Nambu-Goto brane, satisfying the minimal surface action principle, is a tough problem that has not yet been solved in general. Although the quantization of the string, an extended object whose worldsheet has two dimensions, is rather well understood [27]–[29], this is not so in the case of branes with higher dimensional worldsheets (also called "worldvolumes").

We will show how to solve this problem by considering the brane as a point in an infinite-dimensional brane space $\mathcal{M}$ that in general can be curved. The idea is that the metric of $\mathcal{M}$ is dynamical, just like in general relativity [23, 30, 31]. In particular the metric of $\mathcal{M}$ can be such that it gives the *Dirac-Nambu-Goto brane*, which is just the usual "minimal surface" brane. For other choices of $\mathcal{M}$-space metric we have branes that differ from the Dirac-Nambu-Goto brane, i.e., they do not satisfy the minimal surface action principle, but some other action principle. In particular, the $\mathcal{M}$-space metric can be "flat", which means that at any point of $\mathcal{M}$ it can be cast into the diagonal form. Then we have a brane analogue of a point particle in flat spacetime. Such a brane, from now on called *flat brane*, sweeps a worldsheet that is a bunch of straight worldlines (Fig. 1).

A flat brane is thus just like a continuous system of point particles. Quantization of a flat brane then leads to a system of non interacting quantum fields, $\varphi^r(x)$. The index r distinguishes one field from another, and because in the classical theory we had a continuous set

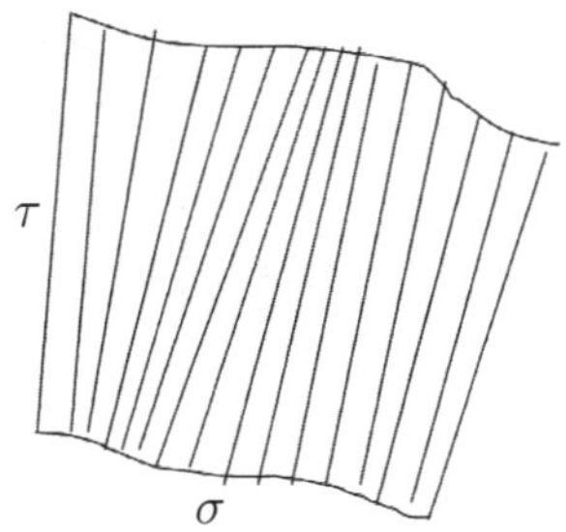

Figure 1. A schematic illustration of a "flat brane". For some exact plots see Fig. 2.

of point particles, r must be continuous. If we consider interactions among the quantum fields, $\lambda_{rs}\varphi^r(x)\varphi^s(x)$, then the effective classical theory gives a brane living in a curved brane space, which, in particular, can be the Dirac-Nambu-Goto brane [31].

Instead of one brane a system can consist of many branes. Within the framework of such an enlarged configuration space it is possible to formulate the Stueckelberg quantum field theory with an invariant evolution parameter. We show how the latter parameter is embedded in the system's configuration.

Each of those branes can be described with a finite number of degrees of freedom, namely, with the center of mass, and additional degrees of freedom that take into account finite extension of the brane. Such extra degrees of freedom can be the coordinates of the Clifford space [32]–[39] that includes scalars, oriented lengths, areas, volumes and 4- volumes (pseudoscalars). In describing a multi brane system we can choose one brane and sample it with the coordinates of Clifford space, while for the remaining branes we retain the description with embedding functions. Clifford space is a 16-dimensional ultrahyperbolic space with neutral signature (8,8). From the scalar and pseudoscalar coordinates that span a 2-dimensional subspace with signature $(+-)$, we can form, with a suitable superposition, the analog of the light cone coordinates. In ultrahyperbolic spaces the Cauchy problem in general cannot be well posed, unless we determine initial data on the "light cone". In such a way we obtain the generalized Stueckelberg description [40]–[51], [23] of particles and branes, both classical and quantum. The Stueckelberg theory is based on the introduction of an evolution parameter τ, which is invariant under Lorentz transformations. In the literature we can find various explanations about the physical origin of τ, but none is generally accepted. In the approach pursued in this and in a series of previous papers [23, 50], the evolution parameter τ is a superposition of the scalar and pseudoscalar coordinate of the Clifford space, and is thus embedded in the configuration of the chosen brane. The latter brane, which in fact need not be just a brane, but whatever extended object that can be sampled as a brane, thus serves as a clock with which we measure the motion of the remaining branes that form the considered system.

2. Brane as a point in the brane space $\mathcal{M}$

The Dirac-Nambu-Goto brane is described by the minimal surface action

$$I = \kappa \int \mathrm{d}^{p+1}\xi\,(-\gamma)^{1/2}, \tag{1}$$

where $\gamma \equiv \det \gamma_{ab}$, $\gamma_{ab} \equiv \partial_a X^\mu \partial_b X_\mu$ is the determinant of the worldsheet embedding functions $X^\mu(\xi^a)$, $\mu = 0, 1, 2, ..., D-1$, $a = 0, 1, 2, ..., p$.

An equivalent action is the Schild action [52]

$$I_{\text{Schild}} = \frac{\kappa}{2k} \int \mathrm{d}^{p+1}\xi\,(-\gamma), \tag{2}$$

IOP Conf. Series: Journal of Physics: Conf. Series **845** (2017) 012018 doi:10.1088/1742-6596/845/1/012018

from which as a consequence of the equations of motion we obtain

$$\partial_a(-\gamma) = 0. \tag{3}$$

The determinant of the induced metric γ is thus a constant whose choice determines a gauge. We will choose a gauge such that

$$-\gamma = k^2, \tag{4}$$

in which case the canonical momentum ${\pi_\mu}^c = \kappa\sqrt{-\gamma}\,\partial^c X_\mu$ derived from the action (1) coincides with the canonical momentum $\kappa(-\gamma)\partial^c X_\mu/k$ derived from the action (2). Additionally, we can choose a gauge so that the determinant factorizes according to

$$(-\gamma) = \dot{X}^2(-\bar{\gamma}), \tag{5}$$

where $\dot{X}^2 \equiv \dot{X}^\mu \dot{X}_\mu$, $\dot{X}^\mu \equiv \partial X^\mu/\partial\tau$, and $\bar{\gamma} \equiv \det \partial_{\bar{a}} X^\mu \partial_{\bar{b}} X_\mu$, $\bar{a}, \bar{b} = 1, 2, ..., p$, the worldsheet parameters being split as $\xi^a = (\tau, \xi^{\bar{a}})$. Then instead of (2) we have

$$I_{\text{Schild}} = \frac{\kappa}{2k}\int \mathrm{d}\tau\, \mathrm{d}^p\sigma\, \dot{X}^2(-\bar{\gamma}), \tag{6}$$

which can be written as

$$I_{\text{Schild}} = \frac{\kappa}{2k}\int \mathrm{d}\tau\, \mathrm{d}^p\sigma\, \mathrm{d}^p\sigma'\, (-\bar{\gamma})\, \rho_{\mu\nu}(\sigma,\sigma')\, \dot{X}^\mu(\tau,\sigma)\dot{X}^\nu(\tau,\sigma'), \tag{7}$$

which is the τ-integral of a quadratic form in an infinite dimensional space $\mathcal{M}$ with the metric

$$\rho_{\mu\nu}(\sigma,\sigma') = (-\bar{\gamma})\, \eta_{\mu\nu}\, \delta(\sigma - \sigma'). \tag{8}$$

At this point it is convenient to introduce a compact notation

$$\dot{X}^{\mu(\sigma)}(\tau) \equiv \dot{X}^\mu(\tau,\sigma)\,, \qquad \rho_{\mu(\sigma)\nu(\sigma)} \equiv \rho_{\mu\nu}(\sigma,\sigma'), \tag{9}$$

and write [31]

$$I_{\text{Schild}} = \frac{\kappa}{2k}\int \mathrm{d}\tau\, \rho_{\mu(\sigma)\nu(\sigma')}\dot{X}^{\mu(\sigma)}(\tau)\dot{X}^{\nu(\sigma')}(\tau). \tag{10}$$

Here we use the generalization of Einstein's summation convention, so that not only summation over the repeated indices μ, ν, but also the integration over the repeated continuous indices (σ), (σ') is assumed. Indices are lowered and raised, respectively, by $\rho_{\mu(\sigma)\nu(\sigma')}$ and its inverse $\rho^{\mu(\sigma)\nu(\sigma')}$. The infinite dimensional space $\mathcal{M}$ is called *brane space*, because its points $x^{\mu(\sigma)}$ represent kinematically possible branes [23, 30].

The quadratic form $\rho_{\mu(\sigma)\nu(\sigma')}\dot{X}^{\mu(\sigma)}(\tau)\dot{X}^{\nu(\sigma')}(\tau)$ is invariant under diffeomorphisms in the brane space $\mathcal{M}$. A curve in $\mathcal{M}$ is given by the parametric equation

$$x^{\mu(\sigma)} = X^{\mu(\sigma)}(\tau), \tag{11}$$

where $X^{\mu(\sigma)}(\tau)$ are τ-dependent functions. The velocity of a "point particle" in $\mathcal{M}$ is $\dot{X}^{\mu(\sigma)} \equiv \frac{\partial X^{\mu(\sigma)}}{\partial\tau}$.

The canonical momentum belonging to the action (10) is

$$p_{\mu(\sigma)} = \frac{\kappa}{k}\,\rho_{\mu(\sigma)\nu(\sigma')}\dot{X}^{\nu(\sigma')} = \frac{\kappa}{k}\dot{X}_{\mu(\sigma)} = \frac{\kappa}{k}\,(-\bar{\gamma})\dot{X}_\mu(\sigma). \tag{12}$$

IOP Conf. Series: Journal of Physics: Conf. Series **845** (2017) 012018 doi:10.1088/1742-6596/845/1/012018

Its contravariant components are

$$p^{\mu(\sigma)} = \rho^{\mu(\sigma)\nu(\sigma')} p_{\nu(\sigma')} = \frac{p^\mu(\sigma)}{(-\bar{\gamma})}, \tag{13}$$

where $p^\mu(\sigma) = \eta^{\mu\nu} p_\nu(\sigma)$.

The canonical momentum associated with the action (6) is

$$p_\mu(\sigma) = \frac{\kappa(-\bar{\gamma})\dot{X}_\mu}{k} = \frac{\kappa\sqrt{-\bar{\gamma}}\dot{X}_\mu}{\sqrt{\dot{X}^\mu \dot{X}_\mu}}, \tag{14}$$

where we have taken into account

$$k = \sqrt{\dot{X}^2}\sqrt{-\bar{\gamma}}, \tag{15}$$

which follows from (4) and (5). Using the latter equation (15), we verify, that both momenta, (12) and (14), are equal, as they should be. In our notation $p_{\mu(\sigma)} = p_\mu(\sigma)$, whereas $p^{\mu(\sigma)}$ is given by Eq. (13).

The momentum $p_\mu(\sigma)$ satisfies the following constraint:

$$p_\mu(\sigma)p^\mu(\sigma) = \eta^{\mu\nu} p_\mu(\sigma)p_\nu(\sigma) = \kappa^2(-\bar{\gamma}). \tag{16}$$

We can also form the quadratic form of the momenta in $\mathcal{M}$-space,

$$p_{\mu(\sigma)} p^{\mu(\sigma)} = \rho^{\mu(\sigma)\nu(\sigma')} p_{\mu(\sigma)} p_{\nu(\sigma')} = \tilde{\kappa}^2, \tag{17}$$

in which the integration over repeated indices (σ) and (σ') is assumed. Comparing (17) and (16), we obtain

$$\tilde{\kappa}^2 = \int \kappa^2 \, \mathrm{d}\sigma. \tag{18}$$

Let us introduce the quantity

$$\tilde{k}^2 = \rho_{\mu(\sigma)\nu(\sigma')} \dot{X}^{\mu(\sigma)} \dot{X}^{\nu(\sigma')} = \int \mathrm{d}\sigma (-\bar{\gamma}) \dot{X}^\mu \dot{X}^\nu = \int k^2 \mathrm{d}\sigma, \tag{19}$$

and take into account that Eqs. (18) and (19) imply $\tilde{\kappa}/\tilde{k} = \kappa/k$, i.e.,

$$\frac{\tilde{\kappa}}{\sqrt{\dot{X}^{\mu(\sigma)} \dot{X}_{\mu(\sigma)}}} = \frac{\kappa}{\sqrt{(-\bar{\gamma})\dot{X}^\mu \dot{X}_\mu}}. \tag{20}$$

Then we can write the Schild action (10) in terms of the quantities $\tilde{\kappa}$ and $\tilde{k}$:

$$I_{\text{Schild}} = \frac{\tilde{\kappa}}{2\tilde{k}} \int \mathrm{d}\tau \, \rho_{\mu(\sigma)\nu(\sigma')} \dot{X}^{\mu(\sigma)}(\tau) \dot{X}^{\nu(\sigma')}(\tau). \tag{21}$$

The latter action is just a gauge fixed action obtained from the action [23, 30, 31]

$$I = \tilde{\kappa} \int \mathrm{d}\tau \left(\rho_{\mu(\sigma)\nu(\sigma)} \dot{X}^{\mu(\sigma)}(\tau) \dot{X}^{\nu(\sigma')}(\tau) \right)^{1/2}, \tag{22}$$

which gives a minimal length worldline, i.e., a geodesic in $\mathcal{M}$-space. Indeed, the equations of motion derived from the action are [31]

$$\frac{\partial I}{\partial X^{\mu(\sigma)}} = \tilde{\kappa} \frac{\mathrm{d}}{\mathrm{d}\tau} \left(\frac{\dot{X}_{\mu(\sigma)}}{\sqrt{\dot{X}^2}} \right) - \frac{\tilde{\kappa}}{2} \partial_{\mu(\sigma)} \rho_{\alpha(\sigma')\beta(\sigma'')} \frac{\dot{X}^{\alpha(\sigma')} \dot{X}^{\beta(\sigma'')}}{\sqrt{\dot{X}^2}} = 0, \tag{23}$$

where

$$\dot{X}^2 \equiv \dot{X}^{\mu(\sigma)}\dot{X}_{\mu(\sigma)} = \rho_{\mu(\sigma)\nu(\sigma')}\dot{X}^{\mu(\sigma)}\dot{X}^{\nu(\sigma')}. \tag{24}$$

We use the following notation for functional derivatives:

$$\partial_{\mu(\sigma)} \equiv {}_{,\mu(\sigma)} \equiv \frac{\partial}{\partial x^{\mu(\sigma)}} \equiv \frac{\delta}{\delta x^{\mu}(\sigma)}. \tag{25}$$

Using $\dot{X}_{\mu(\sigma)} = \rho_{\mu(\sigma)\nu(\sigma)}\dot{X}^{\nu(\sigma)}$, and introducing the connection in $\mathcal{M}$,

$$\Gamma^{\mu(\sigma)}_{\alpha(\sigma')\beta(\sigma'')} = \frac{1}{2}\rho^{\mu(\sigma)\gamma(\sigma''')}(\rho_{\gamma(\sigma''')\alpha(\sigma'),\beta(\sigma'')} + \rho_{\gamma(\sigma''')\beta(\sigma''),\alpha(\sigma')} - \rho_{\alpha(\sigma')\beta(\sigma''),\gamma(\sigma''')}), \tag{26}$$

we can write Eq. (23) in the form

$$\frac{1}{\sqrt{\dot{X}^2}}\frac{\mathrm{d}}{\mathrm{d}\tau}\left(\frac{\dot{X}^{\mu(\sigma)}}{\sqrt{\dot{X}^2}}\right) + \frac{\Gamma^{\mu(\sigma)}_{\alpha(\sigma')\beta(\sigma'')}\dot{X}^{\alpha(\sigma')}\dot{X}^{\beta(\sigma'')}}{\dot{X}^2} = 0, \tag{27}$$

which is the equation of geodesic in $\mathcal{M}$. If we insert into the latter equation the metric (8), then we obtain the equation of motion for the Dirac-Nambu-Goto brane. Equivalently, if we insert the metric (8) into the action (22) we obtain

$$I = \tilde{\kappa}\int \mathrm{d}\tau\, \mathcal{L}[\dot{X}^{\mu}(\sigma), X^{\mu}(\sigma)], \tag{28}$$

where the Lagrangian

$$\mathcal{L}[\dot{X}^{\mu}(\sigma), X^{\mu}(\sigma)] = \left(\int \mathrm{d}^p\sigma\,(-\bar{\gamma})\dot{X}^2\right)^{1/2} \tag{29}$$

is a functional of infinite dimensional velocities and coordinates. From the Euler-Lagrange equations

$$\frac{\mathrm{d}}{\mathrm{d}\tau}\frac{\delta\mathcal{L}}{\delta\dot{X}^{\mu}(\sigma)} - \frac{\delta\mathcal{L}}{\delta X^{\mu}(\sigma)} = 0 \tag{30}$$

we obtain

$$\frac{\mathrm{d}}{\mathrm{d}\tau}\left(\frac{\tilde{\kappa}}{\sqrt{\dot{X}^2}}(-\bar{\gamma})\dot{X}_{\mu}\right) + \partial_{\bar{a}}\left(\frac{\tilde{\kappa}(-\bar{\gamma})\dot{X}^2\partial^{\bar{a}}X_{\mu}}{\sqrt{\dot{X}^2}}\right) = 0, \tag{31}$$

where $\dot{X}^2 \equiv \dot{X}^{\mu(\sigma)}\dot{X}_{\mu(\sigma)} = \int \mathrm{d}^p\sigma\,(-\bar{\gamma})\dot{X}^2$. Inserting Eq. (20) into the latter equation, we obtain

$$\frac{\mathrm{d}}{\mathrm{d}\tau}\left(\frac{\kappa\sqrt{-\bar{\gamma}}}{\sqrt{\dot{X}^2}}\dot{X}_{\mu}\right) + \partial_{\bar{a}}\left(\kappa\sqrt{-\bar{\gamma}}\sqrt{\dot{X}^2}\partial^{\bar{a}}X_{\mu}\right) = 0. \tag{32}$$

The same equation follows from the Dirac-Nambu-Goto action (1) in a gauge (5).

We have arrived at the minimal length action (22) by using a particular metric (8). However, once we have such an action, we can assume that the metric need not be of that particular form. We can generalize the validity of the action (22) and the corresponding geodesic equation to any metric. In fact, we can assume that the metric of $\mathcal{M}$ is dynamical, like in general relativity, and that to the action (22) we have to add a kinetic term for the metric $\rho_{\mu(\sigma)\nu(\sigma')}$. An approach along such lines was investigated in Ref. [23]. Within such a generalized theory the metric (8), leading to the usual Dirac-Nambu-Goto brane, is just one of many other possible metrics, including the metric that is the brane space analog of the flat spacetime metric $\eta_{\mu\nu}$.

3. Special case: flat brane space $\mathcal{M}$

The brane theory simplifies significantly if into the action (22) we plug the metric

$$\rho_{\mu(\sigma)\nu(\sigma')} = \eta_{\mu(\sigma)\nu(\sigma')} = \eta_{\mu\nu}\delta(\sigma - \sigma'). \tag{33}$$

Then we have [31]

$$I = \tilde{\kappa} \int \mathrm{d}\tau \left(\int \mathrm{d}^p\, \sigma\, \eta_{\mu\nu} \dot{X}^\mu(\tau, \sigma) \dot{X}^\nu(\tau, \sigma) \right)^{1/2}. \tag{34}$$

This is like an action for a point particle in a flat background space,

$$I = m \int \mathrm{d}\tau\, (\eta_{\mu\nu} \dot{X}^\mu \dot{X}^\nu)^{1/2}, \tag{35}$$

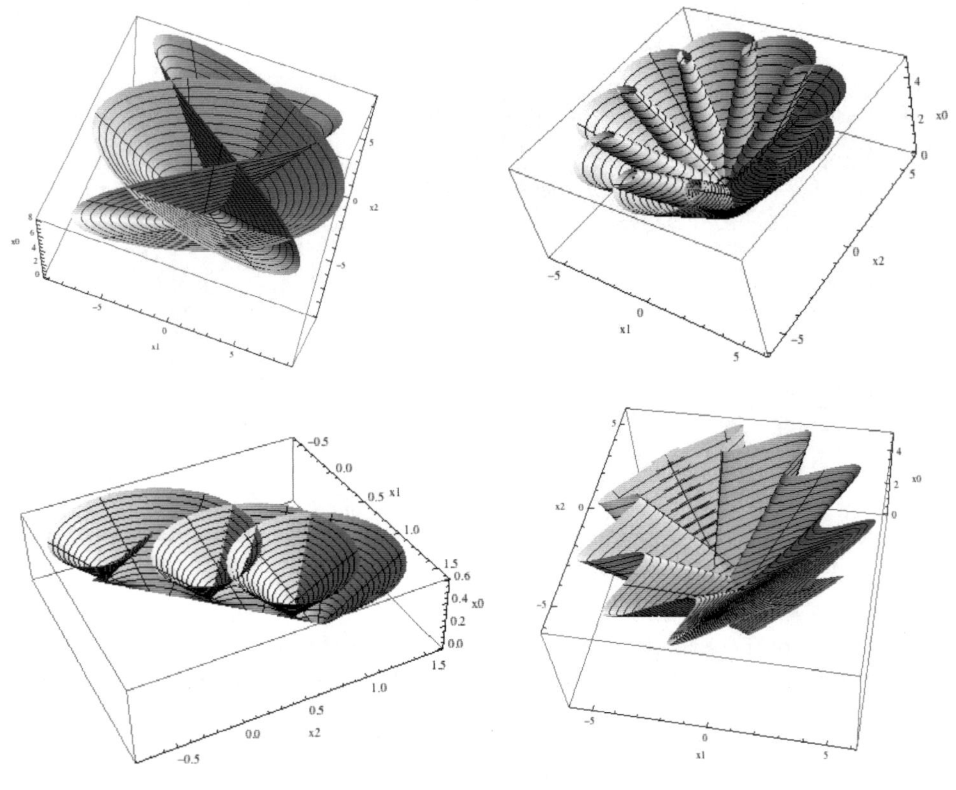

Figure 2. Examples of flat 1-branes for various choices of initial conditions.

but the background space is now infinite dimensional. Variation of (34) gives the following equations of motion

$$\frac{\mathrm{d}}{\mathrm{d}\tau} \left(\frac{\dot{X}^\mu(\tau, \sigma)}{\sqrt{\dot{X}^2}} \right) = 0, \tag{36}$$

where now we have $\dot{X}^2 \equiv \dot{X}^{\nu(\sigma)} \dot{X}_{\nu(\sigma)} = \int \mathrm{d}^p\sigma\, \dot{X}^\mu(\sigma) \dot{X}^\nu(\sigma) \eta_{\mu\nu}$. Choosing a gauge in which $\dot{X}^2 = 1$, we obtain the following simple equations of motion:

$$\ddot{X}^\mu(\tau, \sigma) = 0, \tag{37}$$

whose solution is

$$X^\mu(\tau,\sigma) = v^\mu(\sigma)\tau + X_0^\mu(\sigma). \tag{38}$$

This describes a bunch of straight worldlines that altogether form a special kind of brane's worldsheet, namely a worldsheet of a "flat brane". Equation (38) thus describes a continuum limit of a system of non-interacting point particles, tracing straight worldlines.

In Fig. 2 we give examples of flat 1-branes (i.e., strings) for various solutions of Eq. (38), i.e., for various choices of $v^\mu(\sigma)$. In Fig. 3 we illustrate how the situation looks in the case of a metric that differs from (33). For comparison, in Fig. 4 we show two examples of the usual Nambu-Goto strings.

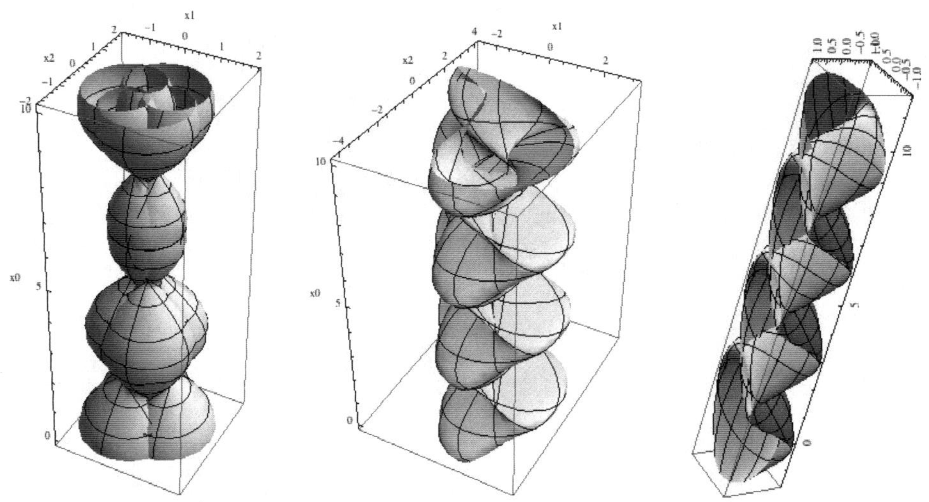

Figure 3. Examples of "curved" 1-branes for various choices of initial conditions. In all cases the brane space metric is $\rho_{\mu(\sigma)\nu(\sigma')} = \left(1 + \frac{X'^2}{\dot{X}^2}\right)\eta_{\mu\nu}\delta(\sigma - \sigma')$.

We see that flat branes can form involved self intersecting objects in spacetime. In the last example in Fig. 2 the worldsheet does not self intersect, which is a consequence of suitable boundary conditions.

Quantization of the system described by the action (34) can be performed in analogous way as the quantization of the point particle, described by (35).

In the case of the point particle (35), we have the constraint

$$p^\mu p_\mu - m^2 = 0\,, \qquad p_\mu = \frac{m\dot{X}_\mu}{(\dot{X}^2)^{1/2}}, \tag{39}$$

which upon quantization becomes the Klein-Gordon equation,

$$(\hat{p}^\mu \hat{p}_\mu - m^2)\phi(x^\mu) = 0\,, \qquad \hat{p}_\mu = -i\partial_\mu \equiv -i\frac{\partial}{\partial x^\mu}. \tag{40}$$

In the case of the brane (22) with the metric (33) we have the constraint

$$p^{\mu(\sigma)}p_{\mu(\sigma)} - \tilde{\kappa}^2 = 0\,, \qquad p_{\mu(\sigma)} = \frac{\tilde{\kappa}\dot{X}_{\mu(\sigma)}}{\sqrt{\dot{X}^2}}, \tag{41}$$

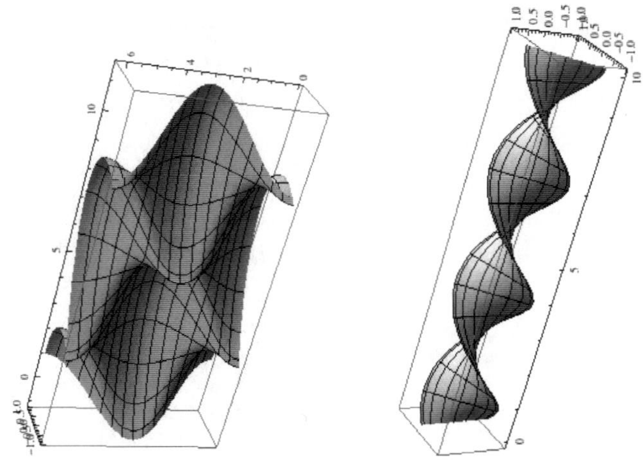

Figure 4. Examples of a special kind of curved 1-branes: the Nambu-Goto strings

which upon quantization becomes the generalized Klein-Gordon equation,

$$(\hat{p}^{\mu(\sigma)}\hat{p}_{\mu(\sigma)} - \tilde{\kappa}^2)\phi(x^{\nu(\sigma)}) = 0 \ , \tag{42}$$

$$\hat{p}_{\mu(\sigma)} \equiv -i\partial_{\mu(\sigma)} \equiv -i\frac{\partial}{\partial x^{\mu(\sigma)}} = -i\frac{\delta}{\delta x^\mu(\sigma)}. \tag{43}$$

Here

$$p^{\mu(\sigma)}p_{\mu(\sigma)} = \rho_{\mu(\sigma)\nu(\sigma')}p^{\mu(\sigma)}p^{\nu(\sigma')} = \int \mathrm{d}^p\sigma\, \eta_{\mu\nu}p^\mu(\sigma)p^\nu(\sigma), \tag{44}$$

and

$$\phi(x^{\nu(\sigma)}) \equiv \phi[x^\mu(\sigma)] \tag{45}$$

is a functional of the brane's embedding functions. The point particle equation (40) can be derived from the action

$$I[\phi(x^\mu)] = \frac{1}{2}\int \mathrm{d}^4x\, (\partial_\mu\phi\partial^\mu\phi - m^2\phi^2), \tag{46}$$

whereas the corresponding action for the brane equation (42) is

$$I[\phi(x^{\mu(\sigma)})] = \frac{1}{2}\int \mathcal{D}x^{\nu(\sigma)}(\partial_{\mu(\sigma)}\phi\, \partial^{\mu(\sigma)}\phi - \tilde{\kappa}^2\phi^2). \tag{47}$$

Explicitly, the equation of motion derived from the latter action is

$$\left(\partial_{\mu(\sigma)}\partial^{\mu(\sigma)} + \tilde{\kappa}^2\right)\phi = 0, \tag{48}$$

In ordinary notation this reads

$$\int \mathrm{d}^p\sigma \mathrm{d}^p\sigma'\, \eta^{\mu\nu}\delta(\sigma-\sigma')\left(\frac{\delta^2}{\delta x^\mu(\sigma)\delta x^\nu(\sigma')} + \tilde{\kappa}^2\right)\phi = 0. \tag{49}$$

As a classical flat brane is like a bunch of free point particles, so a ("first") quantized brane is like a "bunch", that is, a continuous set of "free", i.e., non interacting quantum fields. Therefore, we can write a solution of Eq. (48) as the product [31]

$$\phi(x^{\mu(\sigma)}) = \prod_{\sigma''}\varphi^{(\sigma'')}(x^{\mu(\sigma'')}), \tag{50}$$

where for every σ'' we have a field $\varphi^{(\sigma'')}$ which is a function of *four* spacetime coordinates $x^{\mu(\sigma'')}$ that bear a label σ''. This is just like a separation of variables that is commonly used in solutions of partial differential equations. We will now use Eq. (18) and introduce the mass

$$m = \kappa \Delta\sigma \tag{51}$$

within a region $\Delta\sigma \equiv \Delta^p\sigma \equiv \Delta\sigma^1\Delta\sigma^2...\Delta\sigma^p$. We will also use the following relation between the functional derivative and the partial derivative at a fixed point σ on the brane:

$$\partial_{\mu(\sigma)}\phi \equiv \frac{\delta}{\delta x^\mu(\sigma)}\phi = \lim_{\Delta\sigma\to 0}\frac{1}{\Delta\sigma}\frac{\partial_\mu\varphi^{(\sigma)}}{\partial x^{\mu(\sigma)}}\prod_{\sigma''\neq\sigma}\varphi^{(\sigma'')}(x^{\mu(\sigma'')}). \tag{52}$$

From (48),(50)–(52) we thus obtain

$$\left(\eta^{\mu\nu}\frac{\partial^2}{\partial x^{\mu(\sigma)}\partial x^{\nu(\sigma)}} + m^2\right)\varphi^{(\sigma)}(x^{\mu(\sigma)}) = 0. \tag{53}$$

Because σ is fixed, we can now rename the four spacetime coordinates $x^{\mu(\sigma)}$ into x^μ and write the latter equation simply as

$$\left(\eta^{\mu\nu}\frac{\partial^2}{\partial x^\mu\partial x^\nu} + m^2\right)\varphi^{(\sigma)}(x^\mu) = 0. \tag{54}$$

In our setup a segment of a classical flat brane around σ behaves as a free point particle, and after quantization it satisfies at each σ the Klein-Gordon equation. Because σ is any point on the brane, we have a continuous set of non interacting scalars fields $\varphi^{(\sigma)}$, every one of them satisfying the Klein-Gordon equation (54). In other words, we describe the flat brane by means of many particle non interacting field theory. Different segments of the brane behave as distinguishable particles, each being described by a different scalar field.

In the case of a discrete set of non interacting scalar fields $\varphi^r(x)$, the system is described by the action

$$I[\varphi^r(x)] = \frac{1}{2}\int \mathrm{d}^D x \sum_{r=1}^{N}(\partial_\mu\varphi^r\partial^\mu\varphi^r - m^2(\varphi^r)^2). \tag{55}$$

In the continuum limit, the discrete index r becomes the continuous index (σ) , and $\varphi^r(x)$ becomes $\varphi^{(\sigma)}(x) \equiv \varphi(\sigma, x)$, or shortly, $\varphi^{(\sigma)} \equiv \varphi(\sigma)$. The action is then

$$I[\varphi(\sigma)] = \frac{1}{2}\int \mathrm{d}^D x \int \mathrm{d}^p\sigma(\partial_\mu\varphi(\sigma)\partial^\mu\varphi(\sigma) - m^2\varphi^2(\sigma)). \tag{56}$$

A discrete system based on the action (55) can be straightforwardly second quantized, and so can be the continuous system (56). In the discrete case, the canonically conjugate variables, the fields $\varphi^r(t, \boldsymbol{x})$ and momenta $\Pi^r(t, \boldsymbol{x})$, become the operators satisfying the equal t commutation relations

$$\begin{aligned}&[\varphi^r(t,\boldsymbol{x}), \Pi^s(t,\boldsymbol{x}')] = i\delta^3(\boldsymbol{x}-\boldsymbol{x}')\delta^{rs},\\ &[\varphi^r(t,\boldsymbol{x}), \varphi^s(t,\boldsymbol{x}')] = 0, \qquad [\Pi^r(t,\boldsymbol{x}), \Pi^s(t,\boldsymbol{x}')] = 0.\end{aligned} \tag{57}$$

IOP Conf. Series: Journal of Physics: Conf. Series **845** (2017) 012018 doi:10.1088/1742-6596/845/1/012018

4. An interacting bunch of scalar fields

The action for our system of a continuous set of non interacting scalar fields (56) can be written in the form [31]

$$I[\varphi^{(\sigma)}] = \frac{1}{2}\int \mathrm{d}^D x \left(\partial_\mu \varphi^{(\sigma)} \partial^\mu \varphi^{(\sigma')} - m^2 \varphi^{\sigma)} \varphi^{(\sigma')}\right) s_{(\sigma)(\sigma')}, \tag{58}$$

where

$$s_{(\sigma)(\sigma')} = \delta_{(\sigma)(\sigma')} = \delta(\sigma - \sigma'). \tag{59}$$

The latter form of the action suggests its generalization to a continuous set of *interacting* fields. We see that $s_{(\sigma)(\sigma')}$ has the rôle of a metric in the space of the fields $\varphi^{(\sigma)}$. In principle it need not be the simple metric (59), but can be a generic metric. In such a way we introduce interactions among the fields, satisfying the action principle (58) in which now $s_{(\sigma)(\sigma')}$ is no longer the simple metric (59), but a more general metric.

The equations of motion derived from the action (58) are

$$\partial_\mu \partial^\mu \varphi_{(\sigma)} + m^2 \varphi_{(\sigma)} = 0, \tag{60}$$

where $\varphi_{(\sigma)} = s_{(\sigma)(\sigma')}\varphi^{(\sigma')}$. Assuming that $s_{(\sigma)(\sigma')}$ has the inverse $s^{(\sigma)(\sigma')}$, so that

$$s^{(\sigma)(\sigma'')} s_{(\sigma'')(\sigma')} = \delta^{(\sigma)}{}_{(\sigma')} \equiv \delta(\sigma - \sigma'), \tag{61}$$

then we also have $\varphi^{(\sigma)} = s^{(\sigma)(\sigma')}\varphi_{(\sigma')}$. Applying the latter relation on Eq. (60), we obtain

$$\partial_\mu \partial^\mu \varphi^{(\sigma)} + m^2 \varphi^{(\sigma)} = 0, \tag{62}$$

which is the equation of motion for $\varphi^{(\sigma)}$.

A peculiar property of the system so constructed is that even when the metric is non trivial so that there are interactions among the fields, a general solution of the equation of motion (60) has the familiar form

$$\varphi_{(\sigma)}(x) = \int \frac{\mathrm{d}^{\bar D}\boldsymbol{k}}{\sqrt{(2\pi)^{\bar D} 2\omega_{\boldsymbol{k}}}} \left(a_{(\sigma)}(\boldsymbol{k}) \mathrm{e}^{-ikx} + a^\dagger_{(\sigma)}(\boldsymbol{k}) \mathrm{e}^{ikx}\right), \tag{63}$$

where $\omega_{\boldsymbol{k}} = \sqrt{\boldsymbol{k}^2 + m^2}$, and $\bar D = D - 1$.. The quantities $\varphi_{(\sigma)}(x)$, $a_{(\sigma)}(\boldsymbol{k})$, and $a^\dagger_{(\sigma)}(\boldsymbol{k})$ can be raised by means of the inverse metric $s^{(\sigma)(\sigma')}$, so that we obtain

$$\varphi^{(\sigma)}(x) = \int \frac{\mathrm{d}^{\bar D}\boldsymbol{k}}{\sqrt{(2\pi)^{\bar D} 2\omega_{\boldsymbol{k}}}} \left(a^{(\sigma)}(\boldsymbol{k}) \mathrm{e}^{-ikx} + a^{(\sigma)\dagger}(\boldsymbol{k}) \mathrm{e}^{ikx}\right), \tag{64}$$

which is a solution of Eq. (62).

The canonically conjugated variables $\varphi^{(\sigma)}$ and $\Pi_{(\sigma)} = \partial\mathcal{L}/\partial\dot\varphi^{(\sigma)} = \dot\varphi_{(\sigma)}$ satisfy

$$[\varphi^{(\sigma)}(x), \Pi_{(\sigma')}(x')]|_{x^0 = x'^0} = \delta^{(\sigma)}{}_{(\sigma')} \delta^{\bar D}(\boldsymbol{x} - \boldsymbol{x}') \tag{65}$$

$$[\varphi^{(\sigma)}(x), \varphi^{(\sigma')}(x')]|_{x^0 = x'^0} = 0\ , \qquad [\Pi_{(\sigma)}(x), \Pi_{(\sigma')}(x')]|_{x^0 = x'^0} = 0. \tag{66}$$

From those quantities we construct the Hamiltonian as usual,

$$H = \int \mathrm{d}^{\bar D}\boldsymbol{x}\, (\Pi_{(\sigma)}\dot\varphi^{(\sigma)} - \mathcal{L}) = \frac{1}{2}\int \mathrm{d}^{\bar D}\boldsymbol{x}\, (\Pi_{\sigma)}\Pi^{(\sigma)} - \partial_i \varphi^{(\sigma)} \partial^i \varphi_{(\sigma)} + m^2 \varphi^{(\sigma)} \varphi_{(\sigma)}). \tag{67}$$

and rewrite it in terms the operators $a_{(\sigma)}(\boldsymbol{k})$, and $a^{\dagger}_{(\sigma)}(\boldsymbol{k})$. From Eqs. (63)–(66) we find that the latter operators must satisfy

$$[a^{(\sigma)}(\mathbf{p}), a^{\dagger}_{(\sigma')}(\mathbf{p}')] = \delta^{(\sigma)}{}_{(\sigma')}\delta^{\bar{D}}(\mathbf{p}-\mathbf{p}'). \tag{68}$$

$$[a^{(\sigma)}(\mathbf{p}), a_{(\sigma')}(\mathbf{p}')] = 0, \qquad [a^{(\sigma)\dagger}(\mathbf{p}), a^{\dagger}_{(\sigma')}(\mathbf{p}')] = 0. \tag{69}$$

The relation (68) can be written in the following equivalent forms:

$$[a_{(\sigma)}(\mathbf{p}), a^{\dagger}_{(\sigma')}(\mathbf{p}')] = s_{(\sigma)(\sigma')}\delta^{\bar{D}}(\mathbf{p}-\mathbf{p}'), \tag{70}$$

$$[a^{(\sigma)}(\mathbf{p}), a^{\dagger(\sigma')}(\mathbf{p}')] = s^{(\sigma)(\sigma')}\delta^{\bar{D}}(\mathbf{p}-\mathbf{p}'). \tag{71}$$

The Hamilton then becomes

$$\begin{aligned} H &= \frac{1}{2}\int \mathrm{d}^{\bar{D}}\boldsymbol{k}\,\omega_{\boldsymbol{k}}\left(a^{\dagger}_{(\sigma)}(\boldsymbol{k})a^{(\sigma)}(\boldsymbol{k}) + a^{(\sigma)}(\boldsymbol{k})a^{\dagger}_{(\sigma)}(\boldsymbol{k})\right) \\ &= \int \mathrm{d}^{\bar{D}}\boldsymbol{k}\,\omega_{\boldsymbol{k}}\,a^{\dagger}_{(\sigma)}(\boldsymbol{k})a^{(\sigma)}(\boldsymbol{k}) + H_{\text{z.p.}}, \end{aligned} \tag{72}$$

where $H_{\text{z.p.}}$ is the "zero point" Hamiltonian, and

$$a^{\dagger}_{(\sigma)}(\boldsymbol{k})a^{(\sigma)}(\boldsymbol{k}) = a^{(\sigma)\dagger}(\boldsymbol{k})a^{(\sigma')}(\boldsymbol{k})s_{(\sigma)(\sigma')} = a^{\dagger}_{(\sigma)}(\boldsymbol{k})a_{(\sigma')}(\boldsymbol{k})s^{(\sigma)(\sigma')}. \tag{73}$$

More generally, by using the standard field theoretic techniques that involve the Noether theorem, we obtain the stress-energy tensor

$$T^{\mu}{}_{\nu} = \frac{\partial\mathcal{L}}{\partial\partial_{\mu}\varphi(\sigma)}\partial_{\nu}\varphi^{(\sigma)} - \mathcal{L}\delta^{\mu}{}_{\nu}. \tag{74}$$

Integrating the latter tensor over a space like hypersurface, we obtain the D-momentum $P_{\nu} = \int \mathrm{d}\Sigma_{\mu}T^{\mu}{}_{\nu}$. In the reference frame in which the hypersurface has components $\mathrm{d}\Sigma_{\mu} = (\mathrm{d}\Sigma_0, 0, 0, ..., 0)$ with $\mathrm{d}\Sigma_0 = \mathrm{d}^{\bar{D}}\boldsymbol{x}$, the zero component of the D-momentum is the Hamiltonian (67), whilst the spatial components are $P_{\bar{\mu}} = \int \mathrm{d}^{\bar{D}}\boldsymbol{x}\,\dot{\varphi}_{(\sigma)}\partial_{\bar{\mu}}\varphi^{(\sigma)}$, where $\bar{\mu} = 1, 2, ..., \bar{D}$. After using the expansion (63),(64), we have

$$\begin{aligned} \hat{\mathbf{p}} &= \frac{1}{2}\int \mathrm{d}^{\bar{D}}\boldsymbol{k}\,\boldsymbol{k}\left(a^{\dagger}_{(\sigma)}(\boldsymbol{k})a^{(\sigma)}(\boldsymbol{k}) + a^{(\sigma)}(\boldsymbol{k})a^{\dagger}_{(\sigma)}(\boldsymbol{k})\right) \\ &= \int \mathrm{d}^{\bar{D}}\boldsymbol{k}\,\boldsymbol{k}\,a^{\dagger}_{(\sigma)}(\boldsymbol{k})a^{(\sigma)}(\boldsymbol{k}) + \hat{\mathbf{p}}_{\text{z.p.}}, \end{aligned} \tag{75}$$

where $\hat{\boldsymbol{p}}_{\text{z.p.}}$ is the "zero point" momentum.

By means of the operators $a_{(\sigma)}(\boldsymbol{k})$ and $a^{\dagger}_{(\sigma)}(\boldsymbol{k})$ we can construct the states of our system. Defining the vacuum state according to

$$a_{(\sigma)}(\boldsymbol{k})|0\rangle = 0, \tag{76}$$

the states with definite momenta are created by $a^{\dagger}_{(\sigma)}(\boldsymbol{k})$,

$$|\boldsymbol{k}_1\rangle = a^{\dagger}_{(\sigma)}(\boldsymbol{k}_1)|0\rangle\ , \quad |\boldsymbol{k}_1\boldsymbol{k}_2\rangle = a^{\dagger}_{(\sigma)}(\boldsymbol{k}_1)a^{\dagger}_{(\sigma)}(\boldsymbol{k}_2)|0\rangle\ , \ldots. \tag{77}$$

IOP Conf. Series: Journal of Physics: Conf. Series **845** (2017) 012018 doi:10.1088/1742-6596/845/1/012018

These are basis states, from which we can form various more general states. For instance, we can form single particle wave packet profile states at every σ, and sum (i.e., integrate) them over σ:

$$|\psi\rangle = \int \mathrm{d}\boldsymbol{p}\, g^{(\sigma)}(\boldsymbol{p}) a^\dagger_{(\sigma)}(\boldsymbol{p})|0\rangle, \tag{78}$$

where

$$g^{(\sigma)}(\boldsymbol{p}) a_{(\sigma)}(\boldsymbol{p}) = g^{(\sigma)}(\boldsymbol{p}) a^{\dagger(\sigma')}(\boldsymbol{p}) s_{(\sigma)(\sigma')}. \tag{79}$$

The action of an annihilation operator on such a state gives

$$a^{(\sigma')}(\boldsymbol{p}')|\psi\rangle = g^{(\sigma')}(\boldsymbol{p}')|0\rangle, \tag{80}$$

so that we have

$$\langle\psi|a^{\dagger(\sigma'')}(\boldsymbol{p}'') a^{(\sigma')}(\boldsymbol{p}')|\psi\rangle = g^{*(\sigma'')}(\boldsymbol{p}'') g^{(\sigma')}(\boldsymbol{p}')\langle 0|0\rangle. \tag{81}$$

We normalize the vacuum according to $\langle 0|0\rangle = 1$.

Let us now consider the state which is the product of "single particle" wave packet profiles [31]:

$$|\psi\rangle = \prod_\sigma \int \mathrm{d}^{\bar{D}}\boldsymbol{p}_{(\sigma)}\, g^{(\sigma)}(\boldsymbol{p}_{(\sigma)}) a^\dagger_{(\sigma)}(\boldsymbol{p}_{(\sigma)})|0\rangle \qquad \text{no integration over } (\sigma). \tag{82}$$

The action of an annihilation operator to the latter state gives

$$a^{(\sigma')}(\boldsymbol{p}'_{(\sigma')})|\psi\rangle = \int \mathrm{d}\boldsymbol{p}_{(\sigma)} \mathrm{d}\sigma' \delta(\sigma'-\sigma)\delta(\boldsymbol{p}'_{(\sigma')} - \boldsymbol{p}_{(\sigma)}) g^{(\sigma)}(\boldsymbol{p}_{(\sigma')})|\bar{\psi}\rangle = g^{(\sigma')}(\boldsymbol{p}'_{(\sigma')}|\bar{\psi}\rangle, \tag{83}$$

where $|\bar{\psi}\rangle$ is the product of all the single "particle" states, except the one picked up by $a^{(\sigma')}(\boldsymbol{p}'_{(\sigma')})$:

$$|\bar{\psi}\rangle = \left(\prod_{\sigma\neq\sigma'} \int \mathrm{d}\boldsymbol{p}_{(\sigma)} g^{(\sigma)}(\boldsymbol{p}_{(\sigma)}) a^\dagger_{(\sigma)}(\boldsymbol{p}_{(\sigma)})\right)|0\rangle. \tag{84}$$

We thus have

$$\langle\psi|a^{\dagger(\sigma'')}(\boldsymbol{p}''_{(\sigma'')}) a^{(\sigma')}(\boldsymbol{p}'_{(\sigma')})|\psi\rangle = g^{*(\sigma'')}(\boldsymbol{p}''_{(\sigma'')}) g^{(\sigma')}(\boldsymbol{p}'_{(\sigma')})\langle\bar{\psi}|\bar{\psi}\rangle, \tag{85}$$

where normalization can be such that $\langle\bar{\psi}|\bar{\psi}\rangle = 1$.

We are now going to calculate how the expectation value of the momentum operator changes with time $t \equiv x^0$. Using the Schrödinger equation we obtain [31]

$$\frac{\mathrm{d}}{\mathrm{d}t}\langle\psi|\hat{\boldsymbol{p}}|\psi\rangle = \left(\frac{\mathrm{d}}{\mathrm{d}t}\langle\psi|\right)\hat{\boldsymbol{p}}|\psi\rangle + \langle\psi|\hat{\boldsymbol{p}}\frac{\mathrm{d}}{\mathrm{d}t}|\psi\rangle = (-i)\langle\psi|\hat{\boldsymbol{p}}H - H^\dagger\hat{\boldsymbol{p}}|\psi\rangle. \tag{86}$$

In the above derivation we assumed that the Hamilton operator is not Hermitian. This is the case, if the mass $m = \kappa\Delta\sigma$ depends on position σ on the brane[1], so that also $\sqrt{m^2(\sigma) + \boldsymbol{k}^2} = \omega_{\boldsymbol{k}}(\sigma)$ is a function of σ. From the expression (72) for the Hamiltonian in which instead of a σ independent $\omega_{\boldsymbol{k}}$ stays $\omega_{\boldsymbol{k}}(\sigma)$, we then find $H^\dagger \neq H$.

If we insert into Eq. (86) either a state (72) or (82) we obtain [31]

$$\frac{\mathrm{d}}{\mathrm{d}t}\langle\psi|\hat{\boldsymbol{p}}|\psi\rangle = (-i)\int \mathrm{d}^{\bar{D}}\boldsymbol{p}\,\boldsymbol{p}\, g^*(\sigma,\boldsymbol{p}) g(\sigma',\boldsymbol{p}) s(\sigma,\sigma')(\omega_{\boldsymbol{p}}(\sigma) - \omega_{\boldsymbol{p}}(\sigma'))\mathrm{d}\sigma\mathrm{d}\sigma', \tag{87}$$

[1] In the discrete case this is equivalent to every particle (field) having a different mass m_r. In the continuous case this means that the brane's tension κ is σ dependent.

where we now write $g^{(\sigma)}(\boldsymbol{p}) \equiv g(\sigma, \boldsymbol{p})$, and $s_{(\sigma)(\sigma')} \equiv s(\sigma, \sigma')$ and explicitly denote the integration over σ and σ' In the case of a σ independent $\omega_{\boldsymbol{p}}$ the above expression vanishes, which means that the expectation values of the system's total momentum is conserved in time. This is indeed the case for an isolated system, whose tension κ, and thus $\omega_{\boldsymbol{k}}$ cannot change with σ. If the system is in interaction with another system, then in principle tension can depend on σ.

Let us now assume that there is the following local interaction between nearby brane segments [31]:

$$s_{(\sigma)(\sigma')} \equiv s(\sigma, \sigma') = (1 + \lambda_c \partial_{\bar{a}} \partial^{\bar{a}})\, \delta^p(\sigma - \sigma'). \tag{88}$$

Using the latter expression in Eq. (87), we obtain

$$\frac{\mathrm{d}}{\mathrm{d}t}\langle \boldsymbol{p} \rangle \equiv \frac{\mathrm{d}}{\mathrm{d}t}\langle \psi | \hat{\boldsymbol{p}} | \psi \rangle = (-i)\lambda_c \int \mathrm{d}^{\bar{D}}\boldsymbol{p}\, \mathrm{d}^p\sigma\, \boldsymbol{p}\, \omega_{\boldsymbol{p}}(\sigma)(g^* \partial_{\bar{a}}\partial^{\bar{a}} g - \partial_{\bar{a}}\partial^{\bar{a}} g^* g). \tag{89}$$

In the expression for the total brane's momentum,

$$\begin{aligned} \langle \psi | \hat{\boldsymbol{p}} | \psi \rangle &= \int \mathrm{d}\boldsymbol{p}\, \mathrm{d}\sigma\, \mathrm{d}\sigma'\, \boldsymbol{p}\, g^*(\sigma, \boldsymbol{p}) g(\sigma', \boldsymbol{p}) s(\sigma, \sigma') \\ &= \int \mathrm{d}\boldsymbol{p}\, \mathrm{d}\sigma\, \boldsymbol{p}(g^* g + g^* \partial_{\bar{a}}\partial^{\bar{a}} g) = \langle \hat{\boldsymbol{p}} \rangle = \int \mathrm{d}\sigma \langle \hat{\boldsymbol{p}} \rangle_\sigma, \end{aligned} \tag{90}$$

where there is the integrations over $\mathrm{d}\sigma \equiv \mathrm{d}^p\sigma$. If we omit this integration, then we have the expected momentum density of a brane's segment:

$$\langle \hat{\boldsymbol{p}} \rangle_\sigma = \int \mathrm{d}\boldsymbol{p}\, \boldsymbol{p}(g^* g + g^* \partial_{\bar{a}}\partial^{\bar{a}} g) = \langle \psi |_\sigma \hat{\boldsymbol{p}} | \psi \rangle_\sigma, \tag{91}$$

where

$$|\psi\rangle_\sigma = \int \mathrm{d}\boldsymbol{p}\, g(\sigma, \boldsymbol{p}) a^\dagger(\sigma, \boldsymbol{p}) |0\rangle, \tag{92}$$

is the state of the brane's element at $\sigma \equiv \sigma^{\bar{a}}$, i.e., the state (82) with the product over σ being omitted.

The time derivative of such an expected momentum density is obtained from Eq. (89), if we omit the integrations over $\mathrm{d}^p\sigma$:

$$\frac{\mathrm{d}}{\mathrm{d}t}\langle \hat{\boldsymbol{p}} \rangle_\sigma = (-i)\lambda_c \int \mathrm{d}^{\bar{D}}\boldsymbol{p}\, \boldsymbol{p}\, \omega_{\boldsymbol{p}}(g^* \partial_{\bar{a}}\partial^{\bar{a}} g - \partial_{\bar{a}}\partial^{\bar{a}} g^* g). \tag{93}$$

The latter expression can be different from zero even if $\omega_{\boldsymbol{p}}$ does not change with σ. In fact this is the continuity equation for the current density on the brane, isolated from its environment.

If, instead a wave packet profile in momentum space, we take a wave packet in coordinate space, the Fourier transformation being

$$g(\sigma, \boldsymbol{p}) = \frac{1}{(2\pi)^{\bar{D}/2}} \int \mathrm{e}^{-i\boldsymbol{p}\boldsymbol{x}} f(\sigma, \boldsymbol{x}) \mathrm{d}\boldsymbol{x}, \tag{94}$$

then Eq. (93) becomes

$$\begin{aligned} \frac{\mathrm{d}}{\mathrm{d}t}\langle \hat{\boldsymbol{p}} \rangle_\sigma = -\lambda_c \partial_{\bar{a}} \int \mathrm{d}^{\bar{D}}\boldsymbol{x} \Big[& f^*(\sigma, \boldsymbol{x}) \left(\nabla \sqrt{m^2 + (-i\nabla)^2} \right) \partial^{\bar{a}} f(\sigma, \boldsymbol{x}) \Big) \\ & - \left(\nabla \sqrt{m^2 + (-i\nabla)^2} \right) \partial^{\bar{a}} f^*(\sigma, \boldsymbol{x}) \Big) f(\sigma, \boldsymbol{x}) \Big]. \end{aligned} \tag{95}$$

IOP Conf. Series: Journal of Physics: Conf. Series **845** (2017) 012018 doi:10.1088/1742-6596/845/1/012018

Though we have not explicitly denoted so, the wave packet profiles $g(\sigma, \boldsymbol{p})$ and $f(\sigma, \boldsymbol{x})$ depend on time as well. Therefore a state such as (92) is time dependent and satisfies the time dependent Schrödinger equation with the Hamilton operator (72). The wave packet profile then satisfies [53]–[55],[56]

$$\sqrt{m^2 + (-i\nabla^2)}\, f = -i\frac{\partial}{\partial t} f. \tag{96}$$

Using the latter equation, we can express (95) in terms of the time derivative:

$$\frac{\mathrm{d}}{\mathrm{d}t}\langle \hat{\boldsymbol{p}}\rangle_\sigma = -\lambda_c \partial_{\bar{a}} \int \mathrm{d}^{\bar{D}}\boldsymbol{x} \left[f^*(\sigma,\boldsymbol{x}) \left(\nabla(-i)\frac{\partial}{\partial t}\partial^{\bar{a}} f(\sigma,\boldsymbol{x}) \right) - \left(\nabla(-i)\frac{\partial}{\partial t}\partial^{\bar{a}} f^*(\sigma,\boldsymbol{x}) \right) f(\sigma,\boldsymbol{x}) \right] \tag{97}$$

If we rewrite Eq. (97) in components,

$$\frac{\mathrm{d}}{\mathrm{d}t}\langle \hat{p}_{\bar{\mu}}\rangle_\sigma = -\lambda_c \partial_{\bar{a}} \int \mathrm{d}^{\bar{D}}\boldsymbol{x} \left[f^*(\sigma,\boldsymbol{x}) \left(-i\frac{\partial}{\partial t}\partial^{\bar{a}}\partial_{\bar{\mu}} f \right) - \left(-i\frac{\partial}{\partial t}\partial^{\bar{a}}\partial_{\bar{\mu}} f^* \right) f \right], \tag{98}$$

where $\nabla \equiv \partial_{\bar{\mu}}$, $\bar{\mu} = 1, 2, ...\bar{D}$, then we immediately recognize that the right-hand side of Eq. (98) is the divergence of the expectation value of the operator

$$\hat{\pi}^{\bar{a}}{}_{\bar{\mu}} = -i\lambda_c \frac{\partial}{\partial t} \overset{\leftrightarrow}{\partial}{}^{\bar{a}} \partial_{\bar{\mu}} \equiv -i\lambda_c \frac{\partial}{\partial t}\left(\overset{\leftarrow}{\partial}{}^{\bar{a}} - \overset{\rightarrow}{\partial}{}^{\bar{a}} \right)\partial_{\bar{\mu}}, \tag{99}$$

Close to the initial time $t = 0$ the solution of Eq. (96) for a minimal wave packet can be approximated with a Gaussian wave packet if its width is greater than the Compton wavelength:

$$f \approx A \mathrm{e}^{-\frac{(\boldsymbol{x} - \bar{\boldsymbol{X}}(\sigma))^2}{2\bar{\sigma}_0}} \mathrm{e}^{i\bar{\boldsymbol{p}}\boldsymbol{x}} \mathrm{e}^{i\bar{p}_0 t}, \tag{100}$$

where $\bar{\boldsymbol{X}}(\sigma)$, $\bar{\boldsymbol{p}}$ and $\bar{p}_0$ are the coordinates, momentum and energy of the wave packet center, respectively, whilst A is the normalization constant.

Inserting the wave packet (100) into (98), we obtain [31]

$$\frac{\langle \hat{p}_{\bar{\mu}}\rangle_\sigma}{\mathrm{d}t} = -\lambda_c \partial_{\bar{a}} \left(\frac{\bar{p}_0}{\Delta S} \frac{\partial^{\bar{a}} \bar{X}_{\bar{\mu}}}{\sigma_o} \right), \tag{101}$$

where $\Delta S = \int \mathrm{d}^p\sigma$. This is reminiscent of the brane equation of motion (32).

Let us now consider the following metric in the field space, covariant under reparamerizations of the brane parameters $\sigma^{\bar{a}}$:

$$s(\sigma, \sigma') = \sqrt{-\bar{\gamma}(\sigma)}\, \delta^p(\sigma - \sigma') + \lambda_c \partial_{\bar{a}} \left(\sqrt{-\bar{\gamma}(\sigma)} \gamma^{\bar{a}\bar{b}} \partial_{\bar{b}} \right) \delta^p(\sigma - \sigma') \tag{102}$$

where $\gamma \equiv \det \gamma_{\bar{a}\bar{b}}$. With such a metric, instead of (101) we obtain [31]

$$\frac{\langle \hat{p}_{\bar{\mu}}\rangle_\sigma}{\mathrm{d}t} = -\lambda_c \partial_{\bar{a}} \left(\frac{\bar{p}_0}{\Delta S} \frac{\sqrt{-\bar{\gamma}}\, \gamma^{\bar{a}\bar{b}} \partial_{\bar{b}} \bar{X}_{\bar{\mu}}}{\sigma_o} \right), \tag{103}$$

where now we have $\Delta S = \int \sqrt{-\bar{\gamma}(\sigma)} \mathrm{d}^p\sigma$. The latter equation is in fact the equation of motion (32) of a classical Dirac-Nambu-Goto brane if we make the following correspondence:

$$\langle \hat{p}_{\bar{\mu}}\rangle_\sigma \equiv \langle \hat{\boldsymbol{p}}\rangle_\sigma \longrightarrow p_{\bar{\mu}}(\sigma) = \frac{\kappa\sqrt{-\bar{\gamma}}\dot{X}_{\bar{\mu}}}{\sqrt{\dot{X}^2}} \tag{104}$$

$$\lambda_c \frac{\bar{p}_0}{\Delta S} \gamma^{\bar{a}\bar{b}} \partial^{\bar{b}} \bar{X}_{\bar{\mu}} \longrightarrow \kappa\sqrt{-\bar{\gamma}}\sqrt{\dot{X}^2}\partial^{\bar{a}} X_{\bar{\mu}} = p_0(\sigma)\dot{X}^2 \partial^{\bar{a}} X_{\bar{\mu}}, \tag{105}$$

and take $\dot{X}^2 = 1 - v^2 = 1$. The latter equality holds in a gauge in which $\tau = x^0$, if $v^2 = 0$. Recall that Eq. (103) has been calculated for the wave packet at $t \approx 0$, therefore $v^2 = 0$ is consistent with vanishing $\langle \hat{p}_{\bar{\mu}}\rangle \propto \dot{\bar{X}}_{\bar{\mu}}$, $\bar{\mu} = 1, 2, ..., \bar{D}$ at $t \approx 0$.

5. Generalization to arbitrary configurations

The exercises with the brane space were just a tip of an iceberg. Instead of one brane, a configuration can consist of many branes, or point particles, or both, as illustrated in Fig. 4. The action for such a system is a straightforward generalization of the brane action (22) to such an extended configurations space C:

$$I = \tilde{\kappa} \int \mathrm{d}\tau (\rho_{MN} \dot{X}^M \dot{X}^N)^{1/2} \tag{106}$$

Here we use the same compact indices M, N for coordinates in C in various cases:

$M = i\mu$	many point particles
$M = \mu(\sigma)$	a single brane
$M = k\mu(\sigma)$	many branes
$M = \mu_1\mu_2...\mu_r$	oriented r-volume associated with a brane,

where $\mu = 0, 1, 2, ..., \bar{D}$ denotes coordinates of D-dimensional spacetime, $i = 1, 2, 3, ...$ counts different particles, and $k = 1, 2, 3, ...$ different branes. The meaning of the last line will be explained shortly below.

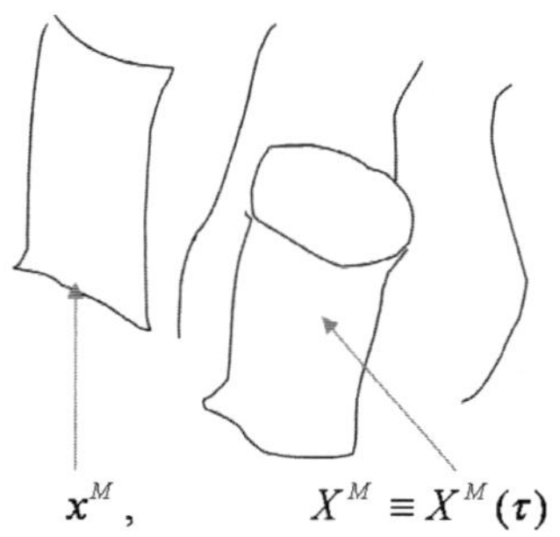

Figure 5. A configuration can consist of many branes, or point particles, or both.

We thus adopt a generic notation so that x^M and $X^M \equiv X^M(\tau)$ denotes, respectively, coordinates and τ-dependent functions in whatever configuration space, either a system of many particles, a single brane, or a system of many branes, or a Clifford space associated with a brane. Thus, depending on the considered physical system, $M = i\mu$, $M = \mu(\sigma)$, $M = k\mu(\sigma)$, or $M = \mu_1\mu_2...\mu_r$. Then Eq. (106) and derived equations are valid for all those cases of configuration spaces.

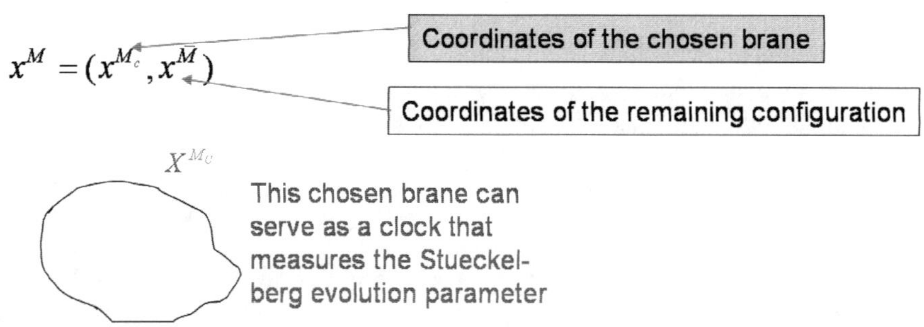

Figure 6. One of the branes within a configuration can be chosen to serve as a clock.

As a consequence of the invariance of the action (106) under reparametrizations of τ, the momenta $p^M = \frac{\tilde{\kappa}\dot{X}^M}{\sqrt{\dot{X}^N\dot{X}_N}}$ satisfy the constraint

$$p^M p_M - \tilde{\kappa}^2 = 0. \tag{107}$$

Let us consider a configuration which consists of many particles and/or branes. Let us choose one brane and denote its coordinates as X^{M_c} (Fig. 6).

A way to sample a brane is to describe it as a set of 16 oriented r-areas (or r-volumes) of all possible dimensionalities, $r = 0, 1, 2, ..., D$. We shall take $D = 4$. In Refs. [37] it has been shown how a brane, described by an infinite dimensional vector $x^{\mu(\sigma)}$, can be mapped into a vector of the space spanned by the basis elements of a Clifford algebra $Cl(1,3)$:

$$x^{\mu(\sigma)} \to x^{\mu_1\mu_2...\mu_r}\gamma_{\mu_a} \wedge \gamma_{\mu_2} \wedge ... \wedge \gamma_{\mu_r} \equiv x^{M_c}\gamma_{M_c} \tag{108}$$

To avoid multiple counting of the terms, it is convenient to order the indices according to $\mu_1 < \mu_2 < ... < \mu_r$, $r = 0, 1, 2, 3, 4$.

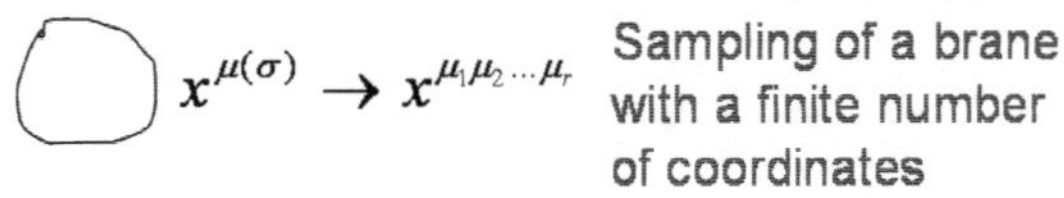

Figure 7. A brane can be sampled by coordinates of Clifford space.

If instead of one brane we consider two, three or more branes, such a system can also be described by 16 coordinates of the Clifford space [37] (see Fig. 8).

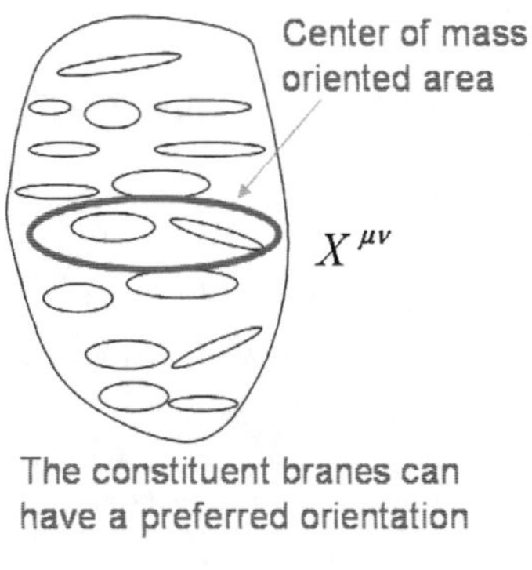

Figure 8. An effective ("center of mass") brane associated with a system of many branes.

Clifford algebra can be considered [32]–[39] as a tangent space to a manifold, called *Clifford space*, C. We will consider flat Clifford space, which is isomorphic to $Cl(1,3)$. Therefore, the points of C can be described by $x^{\mu_1\cdots\mu_r}$. In eq. (108) we have thus a mapping from the infinite dimensional brane space to the 16-dimensional Clifford space. A brane can be sampled by 16 coordinates $x^M \equiv x^{\mu_1\cdots\mu_r}$ of the Clifford space.

The metric of Clifford space is given by the scalar product of two basis elements:

$$\eta_{M_cN_c} = \gamma^{\ddagger}_{M_c} * \gamma_{N_c} = \langle\gamma^{\ddagger}_{M_c}\gamma_{N_c}\rangle_0, \tag{109}$$

where "‡" denotes reversion of the order of vectors in the product $\gamma_{M_c} = \gamma_{\mu_1}\gamma_{\mu_2}...\gamma_{\mu_r}$. The subscript "0" denotes the scalar part of the expression. Explicitly the metric (40) is [38]

$$\eta_{M_c N_c} = \text{diag}(1,1,1,1,1,1,1,1,-1,-1,-1,-1,-1,-1,-1,-1). \tag{110}$$

Clifford space is thus an ultrahyperbolic space.

The scalar product of $X^\ddagger = (x^{M_c}\gamma_{M_c})^\ddagger$ and $X = x^{M_c}\gamma_{M_c})$ gives

$$\begin{aligned} X^\ddagger * X &= \eta_{M_c N_c} x^{M_c} x^{N_c} \\ &= s^2 + \eta_{\mu\nu} x^\mu x^\nu + \tfrac{1}{4}(\eta_{\mu\beta}\eta_{\nu\alpha} - \eta_{\mu\alpha}\eta_{\nu\beta}) x^{\mu\alpha} x^{\nu\beta} + \eta_{\mu\nu}\tilde{x}^\mu \tilde{x}^\nu - \tilde{s}^2 \\ &= \eta_{\hat{\mu}\hat{\nu}} x^{\hat{\mu}} x^{\hat{\nu}} + s^2 - \tilde{s}^2, \end{aligned} \tag{111}$$

where s, $\tilde{s} = \frac{1}{4!}\epsilon_{\mu\nu\rho\sigma} x^{\mu\nu\rho\sigma}$ and $\tilde{x}^\mu = \frac{1}{3!}\epsilon^\mu{}_{\nu\rho\sigma} x^{\nu\rho\sigma}$ are the scalar, pseudoscalar and pseudovector coordinates, respectively. In the last expression we introduced $x^{\hat{\mu}} = (x^\mu, x^{\mu\nu}, \tilde{x}^\mu)$. We thus have $x^{M_c} = (s, \tilde{s}, x^{\hat{\mu}})$.

Upon ("first") quantization the constraint (107), associated with the action (106), becomes the Klein-Gordon equation in the configuration space:

$$\left(\rho^{MN} \frac{\partial^2}{\partial_M \partial_N} + \tilde{\kappa}^2\right)\phi. \tag{112}$$

The corresponding action for the scalar field $\phi(x^M)$ is

$$I = \int \mathcal{D}x^M \left(\rho^{MN} \frac{\partial\phi^*}{\partial x^M}\frac{\partial\phi}{\partial x^N} - \tilde{\kappa}^2 \phi^*\phi\right), \tag{113}$$

where $\mathcal{D}x^M \equiv \prod\limits_M \mathrm{d}x^M$ is a volume element in the configuration space. In the case of many branes, $\prod\limits_M \mathrm{d}x^M = \prod\limits_{k\mu(\sigma)} \mathrm{d}x^{k\mu(\sigma)}$, whereas in the case of many particles it is $\prod\limits_M \mathrm{d}x^M = \prod\limits_{i\mu} \mathrm{d}x^{i\mu}$.

5.1. Non interacting case

If the metric ρ^{MN} is a generalization of the Minkowski metric to the configuration space, then we have the Klein-Gordon equation in flat configuration space. We will now consider such a non interacting case.

By splitting the index M according to $M = (M_c, \bar{M})$, where M_c refers to one chosen brane, described in terms of the coordinates $x^{M_c} = (s, \tilde{s}, x^{\hat{\mu}})$ of the Clifford space, whereas $\bar{M}$ refers to the remaining particle and/or branes, and then renaming $\bar{M}$ back into M, the field action (113) becomes

$$I = \frac{1}{2}\int \mathcal{D}x^{M_c}\mathcal{D}x^M \left(\frac{\partial\phi^*}{\partial x^{M_c}}\frac{\partial\phi}{\partial x_{M_c}} + \frac{\partial\phi^*}{\partial x^M}\frac{\partial\phi}{\partial x_M} - \kappa^2\phi^*\phi\right), \tag{114}$$

Let us introduce the light-cone coordinates (where now the symbol τ has different meaning than in Sec. 2.)

$$\tau = \frac{1}{\sqrt{2}}(s + \tilde{s}) \ , \qquad \lambda = \frac{1}{\sqrt{2}}(s - \tilde{s}), \tag{115}$$

so that instead of the coordinates $x^{M_c} = (s, \tilde{s}, x^{\hat{\mu}})$, we have now the coordinates $x'^{M_c} = (\tau, \lambda, x^{\hat{\mu}})$. The field ϕ then depends on the light-cone coordinates τ, λ, the remaining 12 coordinates $x^{\hat{\mu}}$ of the Clifford space associated with the chosen brane, and on the coordinates x^M of the remaining objects (branes or particles) within the configurations.

Taking the ansatz

$$\phi(\tau,\lambda,x^{\hat{\mu}},x^M)=\mathrm{e}^{i\Lambda\lambda}\mathrm{e}^{ip_{\hat{\mu}}x^{\hat{\mu}}}\psi(\tau,x^M), \tag{116}$$

the action (114) becomes

$$I_0=\frac{1}{2}\int \mathrm{d}\tau\,\mathcal{D}x^M\left[i\lambda\left(\frac{\partial\psi^*}{\partial\tau}\psi-\psi^*\frac{\partial\psi}{\partial\tau}\right)+\partial_M\psi^*\partial^M\psi-(\kappa^2-p_{\hat{\mu}}p^{\hat{\mu}})\psi^*\psi\right]. \tag{117}$$

We have omitted the integration over λ and $x^{\hat{\mu}}$, because it gives a constant factor which can be absorbed into the redefinition of the action I_0. The equation of motion is the Stueckelberg equation in the configuration space:

$$i\frac{\partial\psi}{\partial\tau}=-\frac{1}{2\Lambda}(\partial_M\partial^M+M'^2)\psi, \tag{118}$$

where $M'^2=\kappa^2-p_{\hat{\mu}}p^{\hat{\mu}}$. The general solution is

$$\psi(\tau,x^M)=\int \mathcal{D}p^M\,c(p^M)\exp\left[ip_Mx^M-\frac{i}{2\Lambda}(p_Mp^M-M'^2)\tau\right], \tag{119}$$

in which there is no restriction on momenta p_M, therefore initial data at $\tau=0$ can be freely specified.

In particular it can be $c(p^M)=c(p^{1\mu(\sigma)})c(p^{2\mu(\sigma)})...c(p^{N\mu(\sigma)})$, for a multi brane configuration, or $c(p^M)=c(p^{1\mu})c(p^{2\mu})...c(p^{N\mu})$ for a multi particle configuration. Then the field $\phi(x^M)$ can be written as the product of sigle brane or single particle states. In the case of particles we have:

$$\psi(\tau,x^M)=\varphi(\tau,x^{1\mu})\varphi(\tau,x^{2\mu})...\varphi(\tau,x^{N\mu})\,\mathrm{e}^{\frac{i}{2\Lambda}M'^2\tau}, \tag{120}$$

where

$$\varphi(\tau,x^{1\mu})=\int \mathrm{d}^4p_1\,c(p^{1\mu})\exp\left[ip_{1\mu}x^{1\mu}-\frac{i}{2\Lambda}\,p_{1\mu}p^{1\mu}\tau\right], \tag{121}$$

and similarly for other particles labelled by $2,3,...,N$.

Writing now

$$\psi(\tau,x^M)=\varphi(\tau,x^\mu)\chi(x^{\bar{M}}), \tag{122}$$

where $\varphi(\tau,x^\mu)\equiv\varphi(\tau,x^{1\mu})$ is the field associated with a chosen particle (labelled by '1'), and

$$\chi(\tau,x^{\bar{M}})=\int \mathrm{d}\tau\,\mathrm{d}^{\mathcal{D}-4}\bar{p}\,c(\bar{p})\,\mathrm{e}^{ip_{\bar{M}}x^{\bar{M}}}\,\mathrm{e}^{-\frac{i}{2\Lambda}(p_{\bar{M}}p^{\bar{M}}-M'2)\tau} \tag{123}$$

is the field over the configuration of the remaining particles with coordinates $x^{\bar{M}}\equiv x^{\bar{i}\mu}$, $\bar{i}=2,3,...,N$, the action (117) becomes

$$I=Q\int \mathrm{d}\tau\,\mathrm{d}^4x\left[-i2\Lambda\varphi^*\frac{\partial\varphi}{\partial\tau}+\partial_\mu\varphi^*\partial^\mu\varphi-m^2\varphi^*\varphi\right], \tag{124}$$

where

$$m^2=\frac{(1-Q)}{Q}M'^2, \tag{125}$$

and

$$Q=\int \mathrm{d}^{\mathcal{D}-4}\bar{x}\,\chi^*\chi=\int \mathrm{d}^{D-4}\bar{p}\,c^*(\bar{p})c(\bar{p}). \tag{126}$$

We can normalize χ so that $Q = 1$. Then

$$I = \int \mathrm{d}^4 x \left[-i2\Lambda\varphi^* \frac{\partial\varphi}{\partial\tau} + \partial_\mu\varphi^*\partial^\mu\varphi \right], \tag{127}$$

which is the Stueckelberg action [40]–[51],[23] for a single particle field $\varphi(\tau, x^\mu)$. From (127) we obtain the Stueckelberg field equation

$$i\frac{\partial\varphi}{\partial\tau} = -\frac{1}{2\Lambda}\partial_\mu\partial^\mu\varphi. \tag{128}$$

The non interacting many particle Stueckelberg equation (118) thus contains the single particle Stueckelberg equation (128).

Upon quantization, $\varphi(\tau, x^\mu)$ becomes the operator that annihilates, and $\varphi^*(\tau, x^\mu)$ the operator $\varphi^\dagger(\tau, x^\mu)$ that creates a particle (more precisely, an 'instantonic' particle or and 'event') at x^μ. The evolution of the system is given in terms of the Stueckelberg evolution parameter τ, which in our setup is associated with the brane sampled by the coordinates x^{M_c} of the Clifford space. The latter brane[2] is a part of the overall considered configuration, and is given the role of a clock, which can be a "Stueckelberg clock". The Stueckelberg evolution parameter τ is thus embedded in the configuration.

In the Stueckelberg quantum field theory, the position operator is not considered as problematic[3]. It creates an *event* in spacetime.

5.2. Bunch of Stuckelberg fields interacting in a particular way

The procedure with branes and interacting quantized fields that we have performed in Sec. 4 can be done à la Stueckelberg as well. The Stueckelberg field action (127) or its more general form (124) refers to a single quantum field. Instead of one such a field we can have many fields, and even a continuous set of such fields, as in Sec. 4. But instead of the field action (58) we now have (upon quantization) the following action

$$I[\varphi^{(\xi)}] = \frac{1}{2}\int \mathrm{d}\tau \mathrm{d}^D x \left(-i2\Lambda\varphi^{\dagger(\xi)}\frac{\partial\varphi^{(\xi')}}{\partial\tau} + \partial_\mu\varphi^{\dagger(\xi)}\partial^\mu\varphi^{(\xi')} - m^2\varphi^{\dagger(\xi)}\varphi^{(\xi')} \right) s_{(\xi)(\xi')}, \tag{129}$$

where $\xi \equiv \xi^a$, $a = 1, 2, ..., d$, are d parameters. If the metric is $s_{(\xi)(\xi')} = \delta(\xi, \xi')$, then this is the action for a continuous set of non interacting Stueckelberg fields, otherwise it is an action for interacting Stueckelberg fields. The momentum, canonically conjugate to the field $\varphi^{(\xi)}(x)$ is $\Pi^{(\xi)} = -i\Lambda\varphi^{\dagger(\xi)}(x)$. We have the following commutation relations

$$[\varphi^{(\xi)}(\tau, x)\Pi_{(\xi')}(\tau, x')] = i\delta^{(\xi)}{}_{(\xi')}\delta^D(x - x'), \tag{130}$$

$$[\varphi^{(\xi)}(\tau, x), \varphi^{(\xi')}(\tau, x')] = 0\ , \qquad [\Pi^{(\xi)}(\tau, x), \Pi^{(\xi')}(\tau, x')] = 0. \tag{131}$$

The equation of motion derived from (129) is

$$i\frac{\partial\varphi_{(\xi')}}{\partial\tau} = -\frac{1}{2\Lambda}(\partial_\mu\partial^\mu + m^2)\varphi_{(\xi')}. \tag{132}$$

Its solution can be expanded according to

$$\varphi_{(\xi)}(\tau, x) = \frac{1}{\sqrt{(2\pi)^D\Lambda}}\int \mathrm{d}^D p\, a_{(\xi)}(p) \exp\left[ip_\mu x^\mu + \frac{i}{\Lambda}(p^\mu p_\mu - m^2)\tau \right], \tag{133}$$

[2] It need not be only one brane, there can be many branes, altogether sampled by x^{M_c} (see Ref. [37]).
[3] A careful analysis reveals that position operator is not problematic [31] even in the usual quantum field theory.

where the commutation relations (130) are satisfied provided that

$$[a^{(\xi)}(p), a^\dagger_{(\xi')}(p')] = \delta^{(\xi)}{}_{\xi')}\delta^D(p-p'), \tag{134}$$

while, as usually, the commutators of equal type operators, vanish.

An operator $a^\dagger_{(\xi)}(p)$ creates and $a^{(\xi)}(p)$ annihilates a (ξ)-type particle with momentum $p \equiv p^\mu$, $\mu = 0, 1, 2, ..., D$. Vacuum state is defined according to $a^{(\xi)}(p)|0\rangle = 0$. The Fourier transformed operators

$$a^\dagger_{(\xi)}(x) = \frac{1}{\sqrt{(2\pi)^D}} \int \mathrm{d}^D p\, a^\dagger_{(\xi)}(p) \mathrm{e}^{-ip_\mu x^\mu}, \tag{135}$$

$$a_{(\xi)}(x) = \frac{1}{\sqrt{(2\pi)^D}} \int \mathrm{d}^D p\, a_{(\xi)}(p) \mathrm{e}^{ip_\mu x^\mu}, \tag{136}$$

are creation and annihilation operators for a particle event at a spacetime point x^μ. Up to a factor $\sqrt{\Lambda}$ they coincide with the field operators $\varphi_{(\xi)}(\tau, x)$ and $\varphi^\dagger_{(\xi)}(\tau, x)$ at a fixed value of τ (say $\tau = 0$).

A many particle event state is obtained by successive action of creation operators on the vacuum. In the limit of infinitely many densely packed events such a configuration can be a brane (an extended event) in spacetime:

$$\prod_\xi a^\dagger_{(\xi)}(x_\xi)|0\rangle \equiv A^\dagger[X^\mu(\xi)]|0\rangle = |X^\mu(\xi)\rangle, \tag{137}$$

where $X^\mu(\xi)$ are a brane's embedding functions of d parameters $\xi \equiv \xi^a$, which now need not be all space like; one of them can be time like [23]. In such a case $X^\mu(\xi)$ describes a brane that extends into $d-1$ spacelike directions and into one time like direction of the embedding space. General states are superposition of the states (137) or their momentum space counterparts.

The Hamilton operator is

$$\begin{aligned} H &= \int \mathrm{d}^D x (\Pi_{(\xi)}\partial_\tau \varphi^{(\xi)} - \mathcal{L}) \\ &= \frac{1}{2\Lambda} \int \mathrm{d}^D x\, (\partial_\mu \varphi^{\dagger(\xi)} \partial^\mu \varphi^{(\xi')} - m^2 \varphi^{\dagger(\xi)} \partial^\mu \varphi^{(\xi')}) s_{(\xi)(\xi')} \\ &= \frac{1}{2\Lambda} \int \mathrm{d}^D p\, (p^2 - m^2)\, a^{\dagger(\xi)}(p) a^{(\xi')}(p) s_{(\xi)(\xi')}. \end{aligned} \tag{138}$$

Similarly, we obtain the momentum operator:

$$\hat{p}_\mu = \int \mathrm{d}^D p\, p_\mu\, a^{\dagger(\xi)}(p) a^{(\xi')}(p) s_{(\xi)(\xi')} \tag{139}$$

Let us now calculate how the expectation value of the momentum operator changes with the evolution parameter τ. The procedure is analogous to that in Sec. 4. Instead of the state (82) we now take

$$|\psi\rangle = \prod_\xi \int \mathrm{d}^D p_{(\xi)}\, g^{(\xi)}(p_{(\xi)}) a^\dagger_{(\xi)}(p_{(\xi)})|0\rangle \qquad \text{no integration over } (\xi). \tag{140}$$

Taking $m = m(\xi)$ and introducing

$$h(p, \xi) = \frac{\Lambda}{2}(p^2 - m^2), \tag{141}$$

we obtain

$$\frac{\mathrm{d}}{\mathrm{d}\tau}\langle\psi|\hat{p}_\mu|\psi\rangle = (-i)\int \mathrm{d}^D p\, p_\mu\, g^*(\xi,p)g(\xi',p)s(\xi,\xi')(h(p,\xi)-h(p,\xi'))\mathrm{d}\xi\mathrm{d}\xi', \tag{142}$$

where $g(\xi,p)\equiv g^{(\xi)}(p)$ and $s(\xi,\xi')\equiv s_{(\xi)(\xi')}$.

If we take the field space metric

$$s(\xi,\xi') = (1+\lambda_c\partial^a\partial_a)\,\delta^d(\xi-\xi'), \tag{143}$$

then

$$\frac{\mathrm{d}}{\mathrm{d}\tau}\langle p_\mu\rangle \equiv \frac{\mathrm{d}}{\mathrm{d}\tau}\langle\psi|\hat{p}_\mu|\psi\rangle = (-i)\lambda_c\int \mathrm{d}^D p\,\mathrm{d}^p\xi\, p_\mu\, h(p,\xi)(g^*\partial_a\partial^a g - \partial_a\partial^a g^* g). \tag{144}$$

This is the time derivative of the expectation value of the total momentum of the brane, and it vanishes if $h(p,\xi)$ does not change with ξ.

The expectation value of the total momentum is given by the integral over the momenta of the brane's segments:

$$\begin{aligned}\langle\psi|\hat{p}_\mu|\psi\rangle &= \int \mathrm{d}p\,\mathrm{d}\xi\,\mathrm{d}\sigma'\, p_\mu\, g^*(\xi,p)g(\xi',p)s(\xi,\xi')\\ &= \int \mathrm{d}p\,\mathrm{d}\xi\, p_\mu(g^*g + g^*\partial_a\partial^a g) = \langle\hat{p}_\mu\rangle = \int \mathrm{d}\xi\langle\hat{p}_\mu\rangle_\xi\end{aligned} \tag{145}$$

where

$$\langle\hat{p}_\mu\rangle_\xi = \int \mathrm{d}p\, p_\mu(g^*g + g^*\partial_a\partial^a g) = \langle\psi|_\xi\hat{p}_\mu|\psi\rangle_\xi \tag{146}$$

From Eq. (144) we then read the following expression for the τ derivative of the momentum of a brane segment:

$$\frac{\mathrm{d}}{\mathrm{d}\tau}\langle p_\mu\rangle_\xi = (-i)\lambda_c\int \mathrm{d}^D p\, p_\mu\, h(p,\xi)(g^*\partial_a\partial^a g - \partial_a\partial^a g^* g), \tag{147}$$

which in general is different from zero even if $h(p,\xi)$ does not change with ξ.

If in Eq. (147) we express the wave packet profile $g(\xi,p)$ in term of its position space counterpart $f(\xi,p)$,

$$g(\xi,p) = \frac{1}{(2\pi)^{D/2}}\int \mathrm{e}^{-ip_\mu x^\mu} f(\xi,x)\mathrm{d}x, \tag{148}$$

then we obtain

$$\frac{\mathrm{d}}{\mathrm{d}\tau}\langle\hat{p}_\mu\rangle_\xi = -\lambda_c\partial_a\int \mathrm{d}^D x\left[f^*(\xi,x)\left(\frac{1}{2\Lambda}(\partial_\mu\partial^\mu - m^2)\,\partial^a\partial_\mu f\right) - \left(\frac{1}{2\Lambda}(\partial_\mu\partial^\mu - m^2)\,\partial^a\partial_\mu f^*\right)f\right]. \tag{149}$$

Though not written explicitly, the wave packet profiles g and f depend on the evolution time τ. Using the Schrödinger equation with the Hamiltonina (138) for the state (140), we obtain the equation of motion for the wave packet profile f:

$$\frac{1}{2\Lambda}(\partial_\mu p^\mu - m^2)f = -i\frac{\partial}{\partial\tau}f. \tag{150}$$

Using the latter equation in Eq. (149) we obtain

$$\frac{\mathrm{d}}{\mathrm{d}\tau}\langle\hat{p}_\mu\rangle_\xi = -\lambda_c\partial_a\int \mathrm{d}^D x\left[f^*(\xi,x)\left(-i\frac{\partial}{\partial\tau}\,\partial^a\partial_\mu f\right) - \left(-i\frac{\partial}{\partial\tau}\,\partial^a\partial_\mu f^*\right)f\right], \tag{151}$$

IOP Conf. Series: Journal of Physics: Conf. Series **845** (2017) 012018 doi:10.1088/1742-6596/845/1/012018

For a Gaussian wave packet profile

$$f \approx A\mathrm{e}^{-\frac{(x-\bar{X}(\xi))^2}{2\tilde{\sigma}_0}} \mathrm{e}^{i\bar{p}_\mu x^\mu} \mathrm{e}^{i\bar{h},\tau}, \tag{152}$$

where

$$\bar{h} = \frac{1}{2\Lambda}(\bar{p}^2 - m^2), \tag{153}$$

equation (151) becomes

$$\frac{\langle \hat{p}_\mu \rangle_\xi}{\mathrm{d}\tau} = -\lambda_c \partial_a \left(\frac{\bar{h}}{\Delta S} \frac{\partial^a \bar{X}_\mu}{\tilde{\sigma}_0} \right). \tag{154}$$

The expressions with the metric (143) are not covariant with respect to arbitrary reparametrizations of ξ^a. If we take the metric

$$s(\xi,\xi') = \sqrt{-\gamma(\xi)}\delta^d(\xi-\xi') + \lambda_c \partial_a \left(\sqrt{-\gamma(\xi)}\gamma^{ab}\partial_b\right)\delta^d(\xi-\xi') \tag{155}$$

where $\gamma \equiv \det\gamma_{ab}$, then the expressions become covariant, and instead of (154) we obtain

$$\frac{\langle \hat{p}_\mu \rangle_\xi}{\mathrm{d}\tau} = -\lambda_c \partial_a \left(\frac{\bar{h}}{\Delta S} \frac{\sqrt{-\gamma(\xi)}\gamma^{ab}\partial_b \bar{X}_\mu}{\tilde{\sigma}_0} \right). \tag{156}$$

The latter equation tells how the expected momentum density $\langle \hat{p}_\mu \rangle_\xi$ changes with the evolution parameter τ, which in the Stueckelberg theory is the "true" time, whereas $x^0 \equiv t$ is just one of spacetime coordinates. In Appendix we show that Eq. (156) corresponds to the equation of motion of a classical Stueckelberg brane (see [23]), which is a generalization of the Stueckelberg point particle.

5.3. Self interacting Stueckelberg field in configuration space

In the absence of interactions, a field $\psi(\tau, x^M) \equiv \psi(\tau, x^{1\mu}, x^{2\mu}, ..., x^{N\mu})$ over a many particle configuration is the product (120) of the single particle fields. In the presence of interactions, in general this is no longer the case. An interacting field theory is described by the action (117) to which we add an interactive term I_{int}, so that the total action is

$$I = I_0 + I_{\text{int}}. \tag{157}$$

We will take $I_{\text{int}} = -\frac{G_0}{4!}(\psi^*\psi)^2$. Let us also assume that a particle, say, No. 1, can be singled out from the rest of the configuration according to (122). Inserting Eq. (122) into the action (157), we obtain the Stueckelberg action for the scalar field $\varphi(\tau, x^\mu)$ with the quartic self interaction:

$$I = \int \mathrm{d}\tau\, \mathrm{d}^4x \left[-i2\Lambda\varphi^* \frac{\partial\varphi}{\partial\tau} + \partial_\mu\varphi^*\partial^\mu\varphi + m_{\text{res}}\varphi^*\varphi - g_0(\varphi^*\varphi)^2 \right], \tag{158}$$

where $g_0 = G_0 \int \mathrm{d}^{D-4}\bar{x}(\chi^*\chi)^2$, and

$$m_{\text{res}}^2 = \int \mathrm{d}^{D-4}\bar{x} \left(\partial_{\bar{M}}\chi^*\partial^{\bar{M}}\chi - i2\Lambda\chi^*\frac{\partial\chi}{\partial\tau} \right), \tag{159}$$

is the residual mass that is determined by the presence of the field $\chi(\tau, x^{\bar{M}})$ due to all the other particles of the configuration. In general, m_{res}^2 is different from zero. In particular, in the absence of an interaction, χ is given by Eq. (123) and then $m_{\text{res}}^2 = 0$.

For an interacting field theory the factorization (120) of a field $\psi(\tau, x^M)$ is valid only if the particle No. 1 is not entangled with the other, mutually interacting, particles. If it is entangled, then (120) does not hold. We must then work with the field $\psi(\tau, x^M)$ without factoring out a single particle field.

We have thus arrived at the many particle analog of the brane theory, described by the classical action (22) or the first quantized action (47), in which now the metric $\rho_{\mu(\sigma)\nu(\sigma')}$ of the brane space is not flat. Then one cannot describe a brane as a bunch of point particles. Similarly, in general one cannot describe a many particle configuration as a bunch of point particles. Only if the metric is $g_{MN} \equiv g_{(i\mu)(j\nu)} = \delta_{ij} g_{\mu\nu}$ one has a bunch of point particles. In general, the metric need not be diagonal in the indices $(i\mu)$, $(j\nu)$. Then the particles are intertwined more than it is usually assumed. The physics, either classical or quantized, has to be done in a configuration space $\mathcal{C}$ of many particles/branes. The metric ρ_{MN} of $\mathcal{C}$ in general is curved. An interactive term such as $I_{\rm int} = -\frac{G_0}{4!}(\psi^*\psi)^2$ can be obtained from the dimensional reduction of the action of the form (117), along the lines similar to that of Ref. [57].

6. Conclusion

Within this approach configuration space $\mathcal{C}$ is primary even in classical physics, and the action principle must be formulated in $\mathcal{C}$, not in spacetime. In other words, physics, both classical and quantum, must be formulated in configuration space which can be a space of many point particles and/or branes. Space or spacetime is a subspace of a configuration space (Fig. 9). The concept of spacetime has to be revised by considering spacetime as a subspace of a configuration space, which ultimately is that of the whole universe.

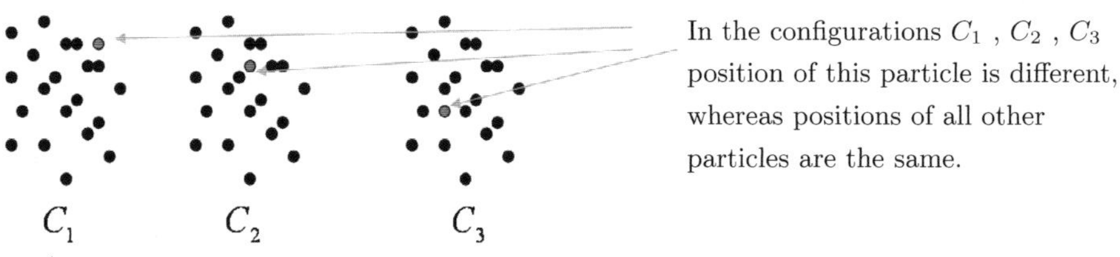

Figure 9. Space(time) as a subspace of configuration space.

We have arrived at such a conclusion by inspecting the action of a Dirac-Nambu-Goto brane. We have found that a brane can be considered as a point in an infinite dimensional brane space $\mathcal{M}$, moving along a geodesic in $\mathcal{M}$. The metric of $\mathcal{M}$ is not fixed, it is dynamical, like in general relativity. For a particular metric we obtain the usual Dirac-Nambu-Goto brane. More general metrics give us interesting fancy branes (Fig. 3) that might be useful in scenarios for quantum gravity in the presence of matter, where matter is given by the brane's self intersections [23, 24]. The simplest is the "flat" metric that gives us "flat branes" (Fig. 2). A flat brane can be straightforwardly quantized as a bunch of point particles. If we take suitable interactions between the quantum fields, we obtain as an "expectation value" the classical Dirac-Nambu-Goto brane.

We have thus found how to quantize branes: via flat brane space. Non flat branes are then objects of an effective classical theory that arises from the underlying QFT of many interacting fields.

The concept of configuration space is associated not only with branes, but with whatever physical systems, in classical and quantum theory. A configuration can be:

- a single brane, considered as a bunch of point particles,
- a discrete system of point particles,

- a mixed system of many branes and point particles,
- etc.

A closed brane or a system of closed branes (Fig. 8) can be approximately described by a finite number of degrees of freedom, which are coordinates of the 16-dimensional Clifford space. The latter space has signature (8,8), i.e., its points can be described by eight "time like" coordinates (associated with the plus sign of the metric) and eight "space like" coordinates (associated with the minus sign of the metric). By picking up one time like and one space like coordinate, and composing from them the analog of two light-cone coordinates, we have derived the Stueckelberg action for a scalar field. We have also shown how a continuous set of such locally interacting fields leads to the effective classical branes à la Stueckelberg. The latter objects satisfy the equations of motion that can be obtained by calculating the time derivative of the expectation value of the momentum operator with respect to certain "wave packet" like quantum states created by the Stueckelberg field operators.

Acknowledgments
This work has been supported by the Slovenian Research Agency.

Appendix: Stueckelberg point particle and its generalization to a brane

The phase space action for a point particle in a $(D+2)$-dimensional space with signature $(2, D)$ is

$$I = \int \mathrm{d}\tilde{\tau} \left[p_M \dot{x}^M - \frac{\alpha}{2}(p_M p^M - M^2) \right], \tag{160}$$

where $\tilde{\tau}$ is an arbitrary parameter[4] and α is a Lagrange multiplier whose variation gives the constrain $p_M p^M - M^2 = 0$. The signature of the extra two dimensions is $(+-)$, whilst the signature of the D-dimensional space is $(1, D-1)$. Now we take $D = 4$, so that we have an extra fifth and sixth dimension. If x^5 and x^6 are "light-cone" coordinates (also denoted as τ, λ), then the action reads

$$I = \int \mathrm{d}\tilde{\tau} \left[p_\mu \dot{x}^\mu + p_5 \dot{x}^5 + p_6 \dot{x}^6 - \frac{\alpha}{2}(p_\mu p^\mu - 2p_5 p_6 - M^2) \right]. \tag{161}$$

Taking a gauge in which $\tilde{\tau} = x^5 \equiv \tau$, we have

$$p_6 \equiv -\Lambda = \frac{\dot{x}_6}{\alpha}; \qquad x^5 \equiv \tau; \qquad \alpha = -\frac{\dot{x}_6}{\Lambda} = \frac{\dot{x}^5}{\Lambda} = \frac{1}{\Lambda}. \tag{162}$$

The action (161) can then be written as

$$I = \int \mathrm{d}\tau \left[p_\mu \dot{x}^\mu + p_5 - \frac{1}{2\Lambda}(p_\mu p^\mu - M^2) \right]. \tag{163}$$

Here Λ is not a Lagrange multiplier, but a fixed quantity, namely $\Lambda \equiv -p_6$.

But we can omit p_5 in the above action, because it does not contribute to the x^μ equations of motion. Then we have

$$I = \int \mathrm{d}\tau \left[p_\mu \dot{x}^\mu - \frac{1}{2\Lambda}(p_\mu p^\mu - M^2) \right]. \tag{164}$$

[4] Because, in accordance with the usual practice, we wish to use the symbol τ for the Stueckelberg evolution parameter, we now denote the arbitrary monotonically increasing parameter as $\tilde{\tau}$.

The corresponding Hamiltonian is

$$H = \frac{1}{2\Lambda}(p_\mu p^\mu - M^2) \tag{165}$$

The above action is the Stueckelberg action. It is derived from the higher dimensional action. The Stueckelberg evolution parameter occurring in (164) is identified with the extra coordinate x^5.

From the constraint $p_M p^M - M^2 = p_\mu p^\mu - 2p_5 p_6 - M^2$ we have

$$p_5 = \frac{1}{2p_6}(p_\mu p^\mu - M^2) = -H, \tag{166}$$

which means that the Hamiltonian is given by the fifth component of momentum, and is thus a generator of translations along $x^5 = \tau$, whilst the constant Λ is given by the sixth component of momentum.

Now let us do the same for a brane. Let $\xi \equiv \xi^a$, $a = 1, 2, ..., d$, be d parameters of a brane in $(D+2)$-dimensions. Now a brane need not be space like. It can extend either into space like or into time like directions, or both [23]. If $D = 4$ then the extra two dimensions are x^5 and x^6, but we may keep the same notation for the extra two dimensions even if $D > 4$. We then have $x^M = (x^\mu, x^5, x^6)$, $\mu = 0, 1, 2, 3, 7, 8, ..., D-3$. Again, x^5 and x^6 are assumed to be "light-cone" coordinates.

The phase space brane action is

$$\begin{aligned} I &= \int \mathrm{d}\tilde{\tau}\mathrm{d}^d\xi \left[p_M \dot{x}^M - \frac{\alpha}{2\kappa\sqrt{-\gamma}}(p_M p^M - \kappa^2(-\gamma) \right] \\ &= \int \mathrm{d}\tilde{\tau}\mathrm{d}^d\xi \left[p_\mu \dot{x}^\mu + p_5 \dot{x}^5 + p_6 \dot{x}^6 - \frac{\alpha}{2\kappa\sqrt{-\gamma}}(p_\mu p^\mu - 2p_5 p_6 - \kappa^2(-\gamma) \right]. \end{aligned} \tag{167}$$

Choosing a gauge $\tilde{\tau} = x^5 \equiv \tau$, and using

$$p_6 = -\sqrt{-\gamma}\tilde{\Lambda}; \quad \dot{x}_M = \frac{\alpha}{\kappa\sqrt{-\gamma}} p_M ; \quad \dot{x}_6 = \frac{\alpha}{\kappa\sqrt{-\gamma}}(-\sqrt{-\gamma}\tilde{\Lambda}) = -\dot{x}^5 = -1, \tag{168}$$

we obtain

$$I = \int \mathrm{d}\tau \mathrm{d}^d\xi \left[p_\mu \dot{x}^\mu - p_5 - \frac{1}{2\sqrt{-\gamma}\tilde{\Lambda}}(p_\mu p^\mu - \kappa^2(-\gamma)) \right]. \tag{169}$$

Let us omit p_5, because this term does no influence the x^μ equations of motion. Then we obtain the following unconstrained (Stueckelberg) action,

$$I = \int \mathrm{d}\tau \mathrm{d}^d\xi \left[p_\mu \dot{x}^\mu - \frac{1}{2\sqrt{-\gamma}\tilde{\Lambda}}(p_\mu p^\mu - \kappa^2(-\gamma)) \right] \tag{170}$$

which is a generalized of the Stueckelberg point particle action. The corresponding Hamiltonian is

$$H = \int \mathrm{d}^d\xi (p_\mu \dot{x}^\mu - L) = \int \mathrm{d}^d\xi \frac{1}{2\sqrt{-\gamma}\tilde{\Lambda}}(p_\mu p^\mu - \kappa^2(-\gamma)) = \int \mathrm{d}^d\xi \, \mathcal{H}. \tag{171}$$

From the constraint $p_\mu p^\mu - 2p_5 p_6 - \kappa^2(-\gamma) = 0$ we have

$$p_5 = \frac{1}{2p_6}(p_\mu p^\mu - \kappa^2(-\gamma)) = -\mathcal{H}. \tag{172}$$

The Hamiltonian of a brane segment is

$$h = \Delta\xi \frac{1}{2\sqrt{-\gamma}\tilde{\Lambda}}(p_\mu(\xi)p^\mu(\xi) - \kappa^2(-\gamma)). \tag{173}$$

Here $p_\mu(\xi)$ is the momentum density. We introduce the momentum and the mass of a brane segment

$$p_\mu = p_\mu(\xi)\Delta\xi \ , \qquad m = \kappa\sqrt{-\gamma}\Delta\xi \tag{174}$$

We also define

$$\frac{p_\mu(\xi)}{\tilde{\Lambda}} = \frac{p_\mu}{\Lambda} = \frac{p_\mu(\xi)\Delta\xi}{\Lambda} \tag{175}$$

from which it follows

$$\Lambda = \tilde{\Lambda}\Delta\xi \tag{176}$$

The Hamiltonian of a brane segment thus becomes

$$h = \frac{1}{2\Lambda}(p_\mu p^\mu - m^2) \tag{177}$$

Equation of motion derived from the Stueckelberg brane action (170) is

$$\frac{\mathrm{d}p_\mu(\xi)}{\mathrm{d}\tau} + \partial^a \left[\frac{1}{2\tilde{\Lambda}}\left(\frac{1}{\sqrt{-\gamma}}p^\mu p_\mu - \kappa^2\sqrt{-\gamma}\right)\partial^a x_\mu\right] = 0 \tag{178}$$

This corresponds to the quantum expectation value equation (156).

References

[1] Ne'eman Y and Elzenberg E 1995 *Membranes and Other Extendons ("p-Branes")* World Scientific Lecture Notes in Physics: Volume 39
[2] Papadopoulos G 1996 *Fortschr. Phys.* **44** 573
[3] West P 2012 *Introduction to Strings and Branes* (Cambridge: Univ. Press)
[4] Duff M J 2004 Benchmarks on the brane *Preprint* hep-th/0407175
[5] Rubakov V A and Shaposhnikov M E 1983 *Phys. Lett.* B **125** 136
[6] Akama K 1982 An Early Proposal of 'Brane World *Lect. Notes Phys.* **176** 267 (*Preprint* hep-th/0001113)
[7] Visser M 1985 *Phys. Lett.* **B159** 22
[8] Gibbons G W and Wiltshire D L 1987 *Nucl. Phys.* **B287** 717
[9] Pavšič M 1986 *Phys. Lett.* A **116** 1 (*Preprint* gr-qc/0101075)
[10] Pavšič M 1986 *Nuov. Cim.* A **95** 297
[11] Pavšič M 1985 *Class. Quant. Grav.* **2** 869
[12] Pavšič M 1985 *Phys. Lett.* A **107 66**
[13] Maia M D 1985 *Phys. Rev.* D **31** 262
[14] Maia M D 1989 *Class. Quant. Grav.* **6** 173
[15] Tapia V 1989 *Class. Quant. Grav.* **6** L49
[16] Hori T 1989 *Phys. Lett.* B **222** 188
[17] Davidson A and Karasik D 1998 *Mod. Phys. Lett.* A **13** 2187
[18] Davidson A 1999 *Class. Quant. Grav.* **16** 653
[19] Davidson A, Karasik D and Lederer Y 1999 *Class. Quant. Grav.* **16** 1349
[20] Pavšič M and Tapia V 2001 Resource letter on geometrical results for embeddings and branes *Preprint* gr-qc/0010045
[21] Gogberashvili M 2002 *Int. J. Mod. Phys.* D **11** 1639 (*Preprint* hep-ph/9908347)
[22] Randall L and Sundrum R 1999 *Phys. Rev. Lett.* **83** 4690 [arXiv:hep-th/9906064].
[23] Pavšič M 2001 *The Landscape of Theoretical Physics: A Global View; From Point Particles to the Brane World and Beyond, in Search of Unifying Principle* (Dordrecht: Kluwer Academic)
[24] Pavšič M 2001 *Phys. Lett.* A **283** 8 (*Preprint* hep-th/0006184)
[25] Pavšič M 1996 *Grav. Cosmol.* **2** 1 (*Preprint* gr-qc/9511020)
[26] Pavšič M 1994 *Found. Phys.* **24** 1495

[27] See, e.g., Green M B, Schwarz J H and Witten E 1987 *Superstring Theory* (Cambridge: Univ. Press).
[28] Kaku M 1988 *Introduction to Superstring* (New York: Springer).
[29] Danielson U 2001 *Rep. Progr. Phys.* **64**, 51
[30] Pavšič M 2003 General principles of brane kinematics and dynamics Bled Workshops Phys. **4** 150 (*Preprint* hep-th/0311060)
[31] Pavšič M 2016 *Int. J. Mod. Phys.* A **31** 1650115 (*Preprint* 1603.01405 [hep-th])
[32] Castro C 1999 *Chaos, Solitons and Fractals* **10** 295
[33] Castro C 2000 *Chaos, Solitons and Fractals* **11** 1663
[34] Castro C 2001 *Chaos, Solitons and Fractals* **10** 1585
[35] Castro C 2000 *Found. Phys.* **30** 1301
[36] Pavšič M 2001 *Found. Phys.* **31** 1185 (*Preprint* hep-th/0011216)
[37] Pavšič M 2003 Found. Phys. **33** 1277 (*Preprint* gr-qc/0211085)
[38] Pavšič M 2005 Found. Phys. **35** 1617 (*Preprint* hep-th/0501222)
[39] Castro C and Pavšič M 2006 Prog. Phys. **1** 31
[40] Fock V 1937 *Phys. Z. Sowj.* **12** 404
[41] Stueckelberg E C G 1941 *Helv. Phys. Acta*
[42] Stueckelberg E C G 1942 **15** 23
[43] Horwitz L P and C. Piron C 1973 *Helv. Phys. Acta* **46** 316
[44] Horwitz L P and Rohrlich F 1981 *Phys. Rev.* D **24** 1528
[45] Horwitz L P, Arshansky R I and Elitzur A C 1988 *Found. Phys* **18** 1159
[46] Feynman R P 1951 *Phys. Rev* **84** 108
[47] Fanchi J R 1993 *Found. Phys.* **23** 287, and many references therein
[48] Fanchi J R 1993 *Parametrized Relativistic Quantum Theory* (Dordrecht: Kluwer)
[49] Pavšič M 1991 *Found. Phys.* **21** 1005
[50] Pavšič M 2011 J. Phys. Conf. Ser. **330** 012011 (*Preprint* 1104.2462 [math-ph])
[51] Horwitz L P *Relativistic Quantum Mechanics* 2015 (Dordrecht: Springer)
[52] Schild A 1977 *Phys. Rev.* D **16** 1722
[53] Rosenstein B and Horwitz L P 1985 J. Phys. A: Math. Gen. **18** 2115
[54] Rosenstein B U and Usher M 1987 *Phys. Rev.* D **36** 2381
[55] Wagner E R, Shields B T, Ware M R, Su Q and Grobe R 2011 Phys. Rev. A **83** 062106
[56] Al-Hashimi M H, and Wiese U J 2009 *Annals of Physics* **324** 2599
[57] Pavšič M 2013 J. Phys. Conf. Ser. **437** 012006 (*Preprint* 1210.6820 [hep-th])

IOP Conf. Series: Journal of Physics: Conf. Series **845** (2017) 012019 doi:10.1088/1742-6596/845/1/012019

Electro-gravity via geometric chrononfield

Eytan H. Suchard

Even Yehuda, HaRotem 6, Israel

Email: eytansuchard@gmail.com, eytans@antbio.com

Abstract. In De Sitter / Anti De Sitter space-time and in other geometries, reference sub-manifolds from which proper time is measured along integral curves, are described as events. We introduce here a foliation with the help of a scalar field. The scalar field need not be unique but from the gradient of the scalar field, an intrinsic Reeb vector of the foliations perpendicular to the gradient vector is calculated. The Reeb vector describes the acceleration of a physical particle that moves along the integral curves that are formed by the gradient of the scalar field. The Reeb vector appears as a component of an anti-symmetric matrix which is a part of a rank-2, 2-Form. The 2-form is extended into a non-degenerate 4-form and into rank-4 matrix of a 2-form, which when multiplied by a velocity of a particle, becomes the acceleration of the particle. The matrix has one U(1) degree of freedom and an additional SU(2) degrees of freedom in two vectors that span the plane perpendicular to the gradient of the scalar field and to the Reeb vector. In total, there are U(1) x SU(2) degrees of freedom. SU(3) degrees of freedom arise from three dimensional foliations but require an additional symmetry to exist in order to have a valid covariant meaning.

Matter in the Einstein Grossmann equation is replaced by the action of the acceleration field, i.e. by a geometric action which is not anticipated by the metric alone. This idea leads to a new formalism that replaces the conventional stress-energy-momentum-tensor. The formalism will be mainly developed for classical physics but will also be discussed for quantized physics based on events instead of particles. The result is that a positive charge manifests small attracting gravity and a stronger but small repelling acceleration field that repels even uncharged particles that have a rest mass. Negative charge manifests a repelling anti-gravity but also a stronger acceleration field that attracts even uncharged particles that have rest mass.

Preliminary version: http://sciencedomain.org/abstract/9858

Keywords: *Time; general relativity; electro-gravity; Reeb class; Godbillon-Vey class, Symplectic Geometry.*

IOP Conf. Series: Journal of Physics: Conf. Series **845** (2017) 012019 doi:10.1088/1742-6596/845/1/012019

1. Introduction

The motivation of this theory is to show that matter can be put into correspondence with an acceleration field. There are two ways that measurement of proper time by a physical clock between events will be shortened: either by gravity in which the clock moves along geodesic curves but in curved space-time or by other interactions that prevent the clock from moving along a geodesic curve. These two approaches have to appear in the equation of gravity in order to describe Nature by a fully geometric model. The latter is not anticipated by the metric tensor alone and therefore it requires a new approach.

In this work, we will study the gradient of a real scalar field P which is $P_i \equiv \frac{dP}{dx^i}$. If a physical particle moves along the integral curves that are formed by P_i then its velocity is $\frac{V^i}{c} = \frac{P^i}{\sqrt{P^s P_s}} \Rightarrow V^i V_i = c^2$ where c is the speed of light. For convenience, throughout the paper we will use the notation $Z \equiv N^2 \equiv \|P_s\|^2 = P^s P_s$. The next step is to calculate the acceleration of such a particle, based on P_i. Taking the exterior derivative of $\frac{P_i}{\sqrt{Z}} dx^i$ as a 1-form, we will derive a 2-Form, while we continue with differential geometry conventions, comma as derivative and semi-colon as covariant derivative, $Z,_i \equiv Z_i \equiv \frac{dZ}{dx^i}$ and for a vector field V_k, $V_k,_i \equiv \frac{dV_k}{dx^i}$ where x^i are the contravariant coordinates. Also, the covariant derivatives are defined as in differential geometry, $V_k;_i \equiv V_k,_i - \Gamma^s{}_{ki} V_s$ and $V^k;_i \equiv V^k,_i + \Gamma^k{}_{si} V^s$ where $\Gamma^k{}_{si}$ are the affine connection, also known as second-type Christoffel symbols. We derive,

$$d\frac{P_i}{\sqrt{Z}}dx^i = (\frac{P_i}{\sqrt{Z}}),_j dx^i \wedge dx^j = ((\frac{P_i}{\sqrt{Z}}),_j - (\frac{P_j}{\sqrt{Z}},_i))dx^i \wedge dx^j \Rightarrow$$
$$(\frac{P_i}{\sqrt{Z}}),_j - (\frac{P_j}{\sqrt{Z}},_i) = (\frac{P_i,_j}{\sqrt{Z}} - \frac{P_i Z_j}{2Z^{3/2}}) - (\frac{P_j,_i}{\sqrt{Z}} - \frac{P_j Z_i}{2Z^{3/2}}) = \frac{P_j Z_i}{2Z^{3/2}} - \frac{P_i Z_j}{2Z^{3/2}} \tag{1}$$

From which

$$d(\frac{P_i}{\sqrt{Z}})dx^i = (\frac{P_j z_i}{2Z^{3/2}} - \frac{P_i z_j}{2Z^{3/2}})dx^i \wedge dx^j = A_{ij} dx^i \wedge dx^j = -A_{ji} dx^i \wedge dx^j$$

We now contract this anti-symmetric matrix with our original vector,

$$\frac{U_i}{2} \equiv A_{ij}\frac{P^j}{\sqrt{Z}} = \frac{P_j Z_i}{2Z^{3/2}}\frac{P^j}{\sqrt{Z}} - \frac{P_i Z_j}{2Z^{3/2}}\frac{P^j}{\sqrt{Z}} = \frac{Z_i}{2Z} - \frac{Z_j P^j}{2Z^2} P_i \tag{2}$$

It is immediately evident that the vector $\frac{U_i}{2}$ is perpendicular to $\frac{P_i}{\sqrt{Z}}$ both from

$$\frac{Z_i}{2Z}\frac{P^i}{\sqrt{Z}} - \frac{Z_j P^j}{2Z^2} P_i \frac{P^i}{\sqrt{Z}} = \frac{Z_i P^i}{2Z^{3/2}} - \frac{Z_j P^j}{2Z^{3/2}} = 0$$

and from a contraction of an anti-symmetric matrix A_{ij} by $\frac{P^i}{\sqrt{Z}}\frac{P^j}{\sqrt{Z}}=0$.

Physical meaning: A_{ij} transforms the vector $\frac{P_i}{\sqrt{Z}}$ to $\frac{U_i}{2}$ as a rotation and scaling transformation and is therefore, of rank 2. It can be extended to a non-degenerate matrix of rank 4, $\widetilde{A}_{ij}$ which defines a field of acceleration, i.e. $\widetilde{A}_{ij}\frac{V^j}{c}=\frac{a_i}{c^2}=g_{ij}\frac{1}{c^2}\frac{dV^j}{d\tau}$ where a_i is the covariant acceleration of the mass that interacts with the field, c is the speed of light, τ is proper time and g_{ij} is the metric tensor. The acceleration matrix $\widetilde{A}_{ij}$ will be discussed later and will appear as the sum of two matrices $A_{ij}+B_{ij}$. If the A_{ij} and B_{ij} are real then $Det(A_{ij}+B_{ij})=\pm Det(A_{ij}+\widetilde{B}_{ij})$ for other choices of $\widetilde{B}_{ij}\neq B_{ij}$.

$$A_{ij}=\frac{U_i}{2}\frac{P_j}{\sqrt{Z}}-\frac{U_j}{2}\frac{P_i}{\sqrt{Z}}=(\frac{Z_i}{2Z}\frac{P_j}{\sqrt{Z}}-\frac{Z_kP^kP_i}{2Z^2}\frac{P_j}{\sqrt{Z}})-(\frac{Z_j}{2Z}\frac{P_i}{\sqrt{Z}}-\frac{Z_kP^kP_j}{2Z^2}\frac{P_i}{\sqrt{Z}})= \frac{Z_i}{2Z}\frac{P_j}{\sqrt{Z}}-\frac{Z_j}{2Z}\frac{P_i}{\sqrt{Z}} \quad (3)$$

We identify this representation with foliation theory,(Reeb, 1948, 1952 [1] and Godbillon-Vey, 1971, [2],[3]). We can write $\omega\equiv\frac{P_i}{\sqrt{Z}}$ and $\eta\equiv\frac{U_i}{2}$ and we reach the Reeb representation $d\omega=\eta\wedge\omega$ where $\eta\equiv\frac{U_i}{2}$ is known as the Reeb vector [1] of the foliation that is perpendicular to the 1-Form ω.
The representation of the vector η leads to a far simpler term than the one represented by the Reinhart-Wood formula [2]. For this cohomology class, the following holds,

$$\omega\wedge d\omega=\frac{P_k}{\sqrt{Z}}(\frac{P_jz_i}{2Z^{3/2}}-\frac{P_iz_j}{2Z^{3/2}})dx^k\wedge dx^i\wedge dx^j=(\frac{P_jP_kz_i}{2Z^2}-\frac{P_iP_kz_j}{2Z^2})dx^k\wedge dx^i\wedge dx^j=0$$

The Godbillon-Vey 3-Form, of the foliation F that is perpendicular to the 1-Form ω is defined as $GV(F)=\eta\wedge d\eta$. Its De RhamCohomology class is $[\eta\wedge d\eta]\in H^3(M,R)$ where here R denotes the real numbers. This cohomology class is an invariant of the foliation F and is a closed 3-Form. An interesting property of the Reeb vector is that its restriction to the foliation F integrates to 0 on each closed curve on F.

A generalization of (2) to the complex numbers is easily defined

$$\frac{U_i}{2}\equiv A_{ij}\frac{P^{*j}}{\sqrt{Z}}=\frac{Z_i}{2Z}-\frac{Z_jP^{*j}}{2Z^2}P_i \quad (4)$$

And $Z=(P_iP^{*i}+P^*_iP^i)/2$ and because the metric tensor is symmetric and real, we can write $Z=P_iP^{*i}$. An interesting way to reach the Reeb vector $\frac{U_i}{2}$ is by the Lie derivative [4]

$$Lie\ (\frac{P^{*m}}{\sqrt{Z}}, \frac{P_i}{\sqrt{Z}}) = \frac{P^{*m}}{\sqrt{Z}}(\frac{P_i}{\sqrt{Z}}),_m + (\frac{P^{*m}}{\sqrt{Z}}),_i \frac{P_m}{\sqrt{Z}} \quad (5)$$

In which the second term is positive because the differentiated $\frac{P_i}{\sqrt{Z}}$ vector has a low index.

The first term becomes,

$$\frac{P^{*m}}{\sqrt{Z}}(\frac{P_i}{\sqrt{Z}}),_m = \frac{P^{*m}\ P_{i},_m}{Z} - \frac{P^{*m}}{\sqrt{Z}}\frac{P_i Z_m}{2Z^{3/2}} = \frac{P^{*m}\ P_{i},_m}{Z} - \frac{Z_m P^{*m}\ P_i}{2Z^2} \quad (6)$$

The second term is,

$$(\frac{P^{*m}}{\sqrt{Z}}),_i \frac{P_m}{\sqrt{Z}} = \frac{P^{*m},_i P_m}{Z} - \frac{P^{*m}\ P_m Z_i}{2Z^2} = \frac{P^{*m},_i P_m}{Z} - \frac{Z_i}{2Z} \quad (7)$$

We add (6) and (7) to get (5) and notice that $\frac{P^{*m}\ P_{i},_m}{Z} + \frac{P^{*m},_i P_m}{Z} = \frac{P^{*m}\ P_{m},_i}{Z} + \frac{P^{*m},_i P_m}{Z}$ from which (5) becomes

$$Lie\ (\frac{P^{*m}}{\sqrt{Z}}, \frac{P_i}{\sqrt{Z}}) = \frac{Z_i}{Z} - \frac{Z_i}{2Z} - \frac{Z_m P^{*m}\ P_i}{2Z^2} = \frac{Z_i}{2Z} - \frac{Z_m P^{*m}\ P_i}{2Z^2} = \frac{U_i}{2} \quad (8)$$

It is again the Reeb vector. It is important to say that the foliation F is covariant because its tangent vectors $T(F)$ consist of vectors which are perpendicular to $\omega \equiv \frac{P_i}{\sqrt{Z}}$ and orthogonality of vectors is invariant under local change of coordinates. The question that we should ask now, is how is the Reeb vector $\frac{U_i}{2}$ related to the curvature of the integral curves which are generated by ω? First of all, we have to notice that $Z = P_i P^{*i}$ is not constant and therefore $Z_k \equiv \frac{dZ}{dx^k} \neq 0$ unlike the case of a velocity of a particle $V_i V^i = c^2$. The squared curvature of the integral curves that are generated by P_i is expressible, according to differential geometry, by the measurement of how much the unit vector $\frac{P_i}{\sqrt{Z}}$ changes along an arc length parameterization t of the integral curves. Calculation of the second power of trajectory curvature of integral curves along a conserving field, can be left as an exercise to the reader but the author prefers to present its calculation. This calculation is valid for all integral curves that are generated by vector fields that are scalar gradients. In our case, the integral curves should not be geodesic if they pass through material fields.

Caution: The t parameterization may not be the time measured by any physical particle because the scalar field from which the vector field is derived may be the result of an intersection of multiple trajectories along which P is measured. However, a particle that follows the gradient curves will indeed measure t even if its trajectory is not geodesic. Let t be the arc length measured along the curves formed by the vector field P_μ. By differential geometry, we know that the second power of curvature along these curves is simply

$$Curv^2 = \frac{d}{dt}\frac{P_\lambda}{\sqrt{P_k P^k}}\frac{d}{dt}\frac{P_\mu}{\sqrt{P_k P^k}} g^{\lambda\mu} \tag{9}$$

such that $g^{\lambda\mu}$ is the metric tensor. (9) is an excellent candidate for an action operator. For convenience, we will write $Norm \equiv \sqrt{P^k P_k}$ and $\dot{P}_\lambda \equiv \frac{d}{dt}P_\lambda$. For the arc length parameter t. Here is the main trick, as was mentioned about $Z = Norm^2$, $Norm$ may not be constant because P_λ is not the 4-velocity of any particle, (to see an example of changing *Norm*, see "APPENDIX – The time field in the Schwarzschild solution"), An arc length parameterization along these curves is equivalent to proper time measured by a particle that moves along the curves, and in thereal numbers case, P can be indeed time. Unlike velocity's squared norms, Z is not constant.

Let W_λ denote: $W_\lambda = \frac{d}{dt}(\frac{P_\lambda}{\sqrt{P_k P_k}}) = \frac{\dot{P}_\lambda}{Norm} - \frac{P_\lambda}{Norm^3}P_k \dot{P}_v g^{kv}$

Obviously

$$W_\lambda P_k g^{\lambda k} = \frac{\dot{P}_\lambda P_k g^{\lambda k}}{Norm} - \frac{P_\lambda P_s g^{\lambda s}}{Norm^3}P_k \dot{P}_v g^{kv} = \frac{\dot{P}_\lambda P_k g^{\lambda k}}{Norm} - \frac{P_k \dot{P}_v g^{kv}}{Norm} = 0$$

Thus

$$Curv^2 = W_\lambda W^\lambda = \frac{\dot{P}_\lambda \dot{P}_v g^{\lambda v}}{Norm^2} - \frac{P_\lambda \dot{P}_s g^{\lambda s}}{Norm^4}P_k \dot{P}_v g^{kv} = \frac{\dot{P}_\lambda \dot{P}^\lambda}{Norm^2} - (\frac{P_\lambda \dot{P}^\lambda}{Norm^2})^2$$

Following the curves formed by $P_\lambda = P,_\lambda = \frac{dP}{dx^\lambda}$, the term $\frac{dx^r}{dt} = \frac{P_\lambda}{Norm}$ is the derivative of the normalized curve or normalized "velocity", using the upper Christoffel symbols, $P_\lambda;_r \equiv \frac{d}{dx^r}P_\lambda - P_s\Gamma^s_{\lambda r}$.

$\frac{d}{dt}P_\lambda = (\frac{d}{dx^r}P_\lambda - P_s\Gamma^s_{\lambda r})\frac{dx^r}{dt} = (P_\lambda;_r)\frac{P^r}{Norm}$ such that x^r denotes the local coordinates. If P_λ is a conserving field, then $P_\lambda;_r = P_r;_\lambda$ and thus $P_\lambda,_r P^r = \frac{1}{2}Norm^2,_\lambda$ and

$$Curv^2 = \frac{\dot{P}_\lambda \dot{P}^\lambda}{Norm^2} - (\frac{P_\lambda \dot{P}^\lambda}{Norm^2})^2 = \frac{1}{4}(\frac{Norm^2,_\lambda Norm^2,_k g^{\lambda k}}{Norm^4} - (\frac{Norm^2,_s P_r g^{sr}}{Norm^3})^2)$$

In the real case, we have achieved the Reeb vector,

$$U_m = \frac{(P^\lambda P_\lambda),_m}{P^i P_i} - \frac{(P^\lambda P_\lambda),_\mu P^\mu}{(P^i P_i)^2}P_m = \frac{Z_m}{Z} - \frac{Z_\mu P^\mu}{Z^2}P_m \tag{10}$$

And our candidate for a trajectory curvature action

$Action = \frac{1}{4} U_m U^m$ where in the complex case

$$Action = \frac{1}{8}(U^*_m U^m + U_m U^{*m}) \qquad (11)$$

Non-geodesic motion, as a result of interaction with a field, is not a geodesic motion in a gravitational field, i.e. it is not free fall. Moreover, material fields by this interpretation prohibit geodesic motion curves of particles moving at speeds less than the speed of light and by this, reduce the measurement of proper time. We return to the idea of acceleration by material fields.

We recall the work of TzviScarr and of Yaakov Friedman [5] which used an anti-symmetric matrix to map a 4-velocity vector V^μ to a 4-acceleration vector a_ν. Since (2),

$$\frac{U_\nu}{2} = \frac{a_\nu}{c^2} = A_{\mu\nu} \frac{V^\mu}{c} \qquad (12)$$

such that $c$is the speed of light, where V^μ is the 4-velocity of a material frame and $A_{\mu\nu}$ is the Scarr-Friedman matrix [5]. The known relation $a_\nu V^\nu = 0$ is obvious.
The real valued action above (11), will lead to a very different energy momentum tensor than that of a simple real valued scalar Klein Gordon energy momentum tensor, instead of $\frac{\hbar^2}{m}(2P_\mu P_\nu - P^\lambda P_\lambda g_{\mu\nu}) - mc^2 P^2 g_{\mu\nu}$ we will see $\frac{c^4}{8\pi K}\frac{1}{4}(U_\mu U_\nu - \frac{1}{2} g_{\mu\nu} U_\lambda U^\lambda - 2U^k{}_{;k} \frac{P_\mu P_\nu}{P_s P^s})$ where $U_\mu = \frac{Z_\mu}{Z} - \frac{Z_k P^k}{Z_2} P_\mu$ and where $Z_\mu = dZ / dx^\mu$ and $Z = P_s P^s$ and $P_\mu = dP / dx^\mu$. The term $2U^k{}_{;k} \frac{P_\mu P_\nu}{P_s P^s}$ was not expected by the author. In almost flat geometry, considering $|U_0| << Max\{|U_1|, |U_2|, |U_3|\}$ leads to the interpretation of electric charge up to multiplication by a constant. Another perhaps more illuminating way is to look at the action $\tilde{F}_{\mu\nu} = \frac{1}{4}(A_{\mu\nu} + B_{\mu\nu})$ where $B_{\mu\nu}$ is the anti-symmetric matrix that was mentioned just before (3) and $\frac{U_\mu U^\mu}{4}\sqrt{-g} = \frac{\tilde{F}_{\mu\nu}\tilde{F}^{\mu\nu}}{4}\sqrt{-g}$ such that $\sqrt{-g}$ is the volume coefficient term that appears in General Relativity. Unlike in the classical covariant electro-magnetic action, there is no mixed 4-current J^μ and vector potential A_μ component $A_\mu J^\mu \sqrt{-g}$ since such a term is redundant due to the unexpected $\frac{1}{4}(-2U^k{}_{;k} \frac{P_\mu P_\nu}{P_s P^s})$ term that will appear in the Euler Lagrange equations of the Lagrangian

$$\frac{U_\mu U^\mu}{4}\sqrt{-g} = \frac{\tilde{F}_{\mu\nu}\tilde{F}^{\mu\nu}}{4}\sqrt{-g}.$$

To summarize:

$$\frac{c^4}{8\pi K}\frac{\tilde{F}_{\mu\nu}\tilde{F}^{\mu\nu}}{4}\sqrt{-g} \neq -(\frac{F_{\mu\nu}F^{\mu\nu}}{4\mu_0} + A_\mu J^\mu)\sqrt{-g}$$

where μ_0 is the permeability constant of vacuum, $F_{\mu\nu}$ is the electro-magnetic tensor.

2. SU(2) X U(1) symmetries – partially symplectic space-time

There is, however, a problem with $A_{\mu\nu}$. There is a degree of freedom in the matrix $A_{\mu\nu}$ which is defined by two vectors, $\frac{P_i}{\sqrt{Z}}$ and by $\frac{U_i}{2}$. That means two additional vectors can be defined in order to express acceleration in the plane which is perpendicular to the local plane spanned by $\frac{P_i}{\sqrt{Z}}$ and $\frac{U_i}{2}$.
We continue with the TzviScarr and Yaakov Friedman acceleration representation matrix [5] and for simplicity, we restrict our discussion to the real case. $A_{\mu\nu}$ is singular and we can easily define a matrix that rotates vectors in a plane perpendicular to both U_μ and to P_ν in order to extend $A_{\mu\nu}$ to a regular matrix by adding to $A_{\mu\nu}$ a second singular matrix, denoted by $B_{\mu\nu}$. That is the matrix

$$B^{\alpha\beta} \equiv \frac{1}{\sqrt{2}}\varepsilon^{\mu\nu\alpha\beta}A_{\mu\nu} \tag{13}$$

where $\varepsilon^{\mu\nu\alpha\beta}$ is the Levi-Civita tensor (not symbol as the Levi-Civita symbol is a tensor density and not a tensor). It is easily verified that

$$(A^{\alpha\beta} + B^{\alpha\beta})(A_{\alpha\beta} + B_{\alpha\beta}) = A^{\alpha\beta}A_{\alpha\beta} + B^{\alpha\beta}B_{\alpha\beta}$$

and also

$$B^{\alpha\beta}B_{\alpha\beta} = \frac{1}{2}\varepsilon^{\mu\nu\alpha\beta}A_{\mu\nu}\varepsilon_{ij\alpha\beta}A^{ij} = \frac{1}{2}\varepsilon^{\mu\nu}{}_{ij}A_{\mu\nu}A^{ij} = \frac{1}{2}(\delta_i^\mu\delta_j^\nu - \delta_j^\mu\delta_i^\nu)A^{ij}A_{\mu\nu} =$$
$$\frac{1}{2}(A^{\mu\nu}A_{\mu\nu} - A^{\nu\mu}A_{\mu\nu}) = A^{\mu\nu}A_{\mu\nu}$$

Therefore $(A^{\alpha\beta} + B^{\alpha\beta})(A_{\alpha\beta} + B_{\alpha\beta}) = A^{\alpha\beta}A_{\alpha\beta} + B^{\alpha\beta}B_{\alpha\beta} = 2A^{\mu\nu}A_{\mu\nu}$ where δ_i^j is the Kroneckerdelta.In the real numbers case, there are two ways to extend $A_{\alpha\beta}$ to a regular matrix and to keep the norm of the acceleration vector after the extended matrix is multiplied by vectors perpendicular to both $\omega = \frac{P_\mu}{\sqrt{Z}}$ and to U_μ. These matrices are,

$$A_{\alpha\beta} + B_{\alpha\beta} \text{ and } A_{\alpha\beta} - B_{\alpha\beta} \tag{14}$$

and it is easy to see that $(A_{\alpha\beta} + B_{\alpha\beta})(A^{\alpha\beta} - B^{\alpha\beta}) = 0$ (14) is the matrix we have been looking for and it also results in an immediate degree of freedom in the representation of the acceleration matrix by two additional vectors to ω and U_μ but **not** in the matrix itself. (14) is quite similar to Dirac matrices but unlike them, it describes two acceleration planes and not a bi-spinor [6]. In particular, $\tilde{\gamma}_{\alpha}{}^{\mu}$ rotations in $SU(4)$, see [7], that do not affect $A_{\alpha\beta}$ may be applied to $B_{\alpha\beta}$. These rotations are in $SU(2)xU(1)$, $B_{\alpha\beta} = \pm\tilde{\gamma}_{\alpha}{}^{\mu} B_{\mu\nu} \tilde{\gamma}_{\beta}{}^{\nu}$ and $A_{\alpha\beta} = \tilde{\gamma}_{\alpha}{}^{\mu} A_{\mu\nu} \tilde{\gamma}_{\beta}{}^{\nu}$. There is no $SU(2)$ degree of freedom in $B_{\alpha\beta}$itself but only in its representation vectors, i.e., the normalized gradient of a scalar and its Reeb vector. As was suggested in (14), the singular $A_{\alpha\beta}$ acceleration matrix is replaced with $A_{\alpha\beta} \rightarrow A_{\alpha\beta} + \tilde{\gamma}_{\alpha}{}^{\mu} B_{\mu\nu} \tilde{\gamma}_{\beta}{}^{\nu}$

In the complex case, consider $e^{i\theta} B_{\alpha\beta}$ instead of $B_{\alpha\beta}$. $e^{i\theta}$ is a $U(1)$ member at angle θ.

2.1. Partially Symplectic space-time

For space-time to be symplectic, it is enough to show that for $A_{ij} = \frac{Z_i}{2Z}\frac{P_j}{\sqrt{Z}} - \frac{Z_j}{2Z}\frac{P_i}{\sqrt{Z}}$ there exists a $B_{ij} = \frac{W_i}{2W}\frac{Q_j}{\sqrt{W}} - \frac{W_j}{2W}\frac{Q_i}{\sqrt{W}}$ such that Q is a scalar function $W = Q_k Q^k$ such that $Q_k = dQ/dx^k$ is the Kth derivative of Q, and $W_j = dW/dx^j$ with local coordinates x^s and the following holds: $P_i Q^i = P_i W^i = Z_i Q^i = Z_i W^i = 0$ or in other words, the two forms A_{ij} and B_{ij} define transformations in perpendicular planes of the tangent space of space-time.

The result is that the following form is not zero $\frac{Z_i}{2Z}\frac{P_j}{\sqrt{Z}}\frac{W_s}{2W}\frac{Q_t}{\sqrt{W}} dx^i \wedge dx^j \wedge dx^s \wedge dx^t \neq 0$.

We define $\beta = (\frac{Z_i}{2Z}\frac{P_j}{\sqrt{Z}} + \frac{W_i}{2W}\frac{Q_j}{\sqrt{W}}) dx^i \wedge dx^j$. It is immediate that β is the exterior derivative of the 1-Form $\omega = (\frac{P_i}{\sqrt{Z}} + \frac{Q_i}{\sqrt{W}}) dx^i$ and that β is a closed form and obviously $\beta \wedge \beta = 2\frac{Z_i}{2Z}\frac{P_j}{\sqrt{Z}}\frac{W_s}{2W}\frac{Q_t}{\sqrt{W}} dx^i \wedge dx^j \wedge dx^s \wedge dx^t$ which is a 4th order non-degenerate form. The demand for a manifold to be symplectic, is that there will be a 1-Form ω such that $\beta = d\omega$ and such that $\beta^n = \beta \wedge \beta \wedge .. \beta$ will be of order $2n$ which is the dimension of the manifold and that β will be a closed form as in our case. By a theorem of Darboux [8], there exists a local basis (x^0, x^1, y^0, y^1) in which $\beta = x^0 \wedge y^0 + x^1 \wedge y^1$ or in a more illuminating way,

$$\begin{pmatrix} 0 & -1 & 0 & 0 \\ 1 & 0 & 0 & 0 \\ 0 & 0 & 0 & -1 \\ 0 & 0 & 1 & 0 \end{pmatrix}.$$

In a $2n$ dimensional Riemannian or Pseudo-Riemannian manifold, if there is a series of, $n$vectors, $\omega(1) = \frac{P(1)_\mu}{\sqrt{P(1)_\lambda P(1)^\lambda}} dx^\mu, \omega(2) = \frac{P(2)_\mu}{\sqrt{P(2)_\lambda P(2)^\lambda}} dx^\mu, ..., \omega(2) = \frac{P(n)_\mu}{\sqrt{P(n)_\lambda P(n)^\lambda}} dx^\mu$ which have non-degenerate exterior derivatives and therefore non-degenerate Reeb vectors $U(1)_\mu, ..., U(n)_\mu$ then

IOP Conf. Series: Journal of Physics: Conf. Series **845** (2017) 012019 doi:10.1088/1742-6596/845/1/012019

show that the nxn Gram determinant $Det(U(i)_\mu U(j)^\mu)$ is invariant for all such choices. In the real case, it can be shown that any other choice is reducible to $U(1)_\mu,...,U(n)_\mu$ by a composition of an $SO(n)$ transformation followed by a reflection which generalizes to $SU(n)$ and a reflection in the complex case.

3. SU(3) symmetry

We may want to express the acceleration matrix $A_{\alpha\beta}$ by three scalar fields that are defined in thefoliation F that is perpendicular to $\frac{P_i}{\sqrt{Z}}$. This is because P_i is a geometric object that defines foliations of space-time and can be conversely defined by the foliations. Another motivation is to show that $SU(3)$ that is seen in Quantum Chromo-Dynamics, may originate from geometry. By a theorem of Frobenius, necessary conditions for 3 vectors $h(j=1,2,3)$ to span the foliation F is that the vectors $h(s)$ areHolonomic if their Lie brackets depend on them $[h(i),h(k)]=\sum_{j=1}^{3} c_j h(j)$ for some coefficients c_j. The Lie brackets of each two vectors must depend on the vectors that span $T(F)$. We may write our 3 scalars a,b,c (here c is not the speed of light and not the previous coefficients but a scalar field) and their gradients that span the foliation's tangent space $T(F)$ as follows, $h_k(1)=\frac{d}{dx^k}a$, $h_k(2)=\frac{d}{dx^k}b$, $h_k(3)=\frac{d}{dx^k}c$.

We now express $A_{\alpha\beta}$ by $h_k(1)$, $h_k(2)$, $h_k(3)$ in a covariant formalism but we need some constraint on P_μ.

Condition: $P_\mu;^\lambda P*^\mu = P*_\mu;^\lambda P^\mu$

This condition is not trivial and in general, $P_\mu;^\lambda P*^\mu \neq P*_\mu;^\lambda P^\mu$.

Consider the following matrix:

$$D_{ij} = g_{ij} - a*_i a_j - b*_i b_j - c*_i c_j \tag{15}$$

and $a_i = \frac{\alpha_i}{\sqrt{(\alpha^\lambda \alpha*_\lambda + \alpha*^\lambda \alpha_\lambda)/2}}$ for some scalar function α whose gradient α_i is in the foliation perpendicular to P_μ etc. and in the same manner replace β_i by a normalized unit vector b_i and γ_i by a normalized $c_i = \frac{\gamma_i}{\sqrt{\gamma_\mu \gamma*^\mu + \gamma*_\mu \gamma^\mu}}$ vector. Also, we demand orthogonality, $a_i b*^i = a_i c*^i = b_i c*^i = 0$. Obviously $D_{ij} a*^j = D_{ij} b*^j = D_{ij} c*^j = 0$ and $D_{ij}\frac{P*^j}{\sqrt{Z}} = \frac{P*^i}{\sqrt{Z}}$ which then shows that $D_{ij} = \frac{P*_i P_j}{Z}$, $Z=(P_\lambda P*^\lambda + P*_\lambda P^\lambda)/2$. That is very interesting, because as was already said, we can represent $P*_i P_j$ by orthogonal fields that span the tangent space of the perpendicular foliation to P_j, namely T(F). Consider the following:

$$D_{ij;k} - D_{ik;j} = \left(\frac{P*_{i;k} P_j + P*_i P_{j;k}}{Z} - \frac{P*_i P_j Z_k}{Z^2}\right) - \left(\frac{P*_{i;j} P_k + P*_i P_{k;j}}{Z} - \frac{P*_i P_k Z_j}{Z^2}\right) = \left(\frac{P*_i P_k Z_j}{Z^2} - \frac{P*_i P_j Z_k}{Z^2}\right) + \left(\frac{P*_{i;k} P_j}{Z} - \frac{P*_{i;j} P_k}{Z}\right) \quad (16)$$

Now comes a little trick:

$$(D_{ij;k} - D_{ik;j})\frac{P^i P*^j}{Z} = \left(\frac{P*_i P_k Z_j}{Z^2} - \frac{P*_i P_j Z_k}{Z^2}\right)\frac{P^i P*^j}{Z} + \left(\frac{P*_{i;k} P_j}{Z} - \frac{P*_{i;j} P_k}{Z}\right)\frac{P^i P*^j}{Z} = \left(\frac{Z_j P*^j P_k}{Z^2} - \frac{Z_j}{Z}\right) + \left(\frac{P*_{i;k} P^i}{Z} - \frac{P^i P*^j P*_{i;j} P_k}{Z^2}\right)$$

By (4), it is obvious that the first two terms constitute minus twice the Reeb vector, $\frac{Z_j P*^j P_k}{Z^2} - \frac{Z_j}{Z} = -U_k = -2(\frac{U_k}{2})$. For the last two terms, we need a special condition $P_{\mu;}{}^{\lambda} P*^{\mu} = P*_{\mu;}{}^{\lambda} P^{\mu}$ although usually $P_{\mu;}{}^{\lambda} P*^{\mu} \neq P*_{\mu;}{}^{\lambda} P^{\mu}$. Then by this condition, $\frac{P*_{i;k} P^i}{Z} - \frac{P^i P*^j P*_{i;j} P_k}{Z^2} = \frac{U_k}{2}$ and therefore

$$(D_{ij;k} - D_{ik;j})D*^{ij} = (D_{ij;k} - D_{ik;j})\frac{P^i P*^j}{Z} = -\frac{U_k}{2} \quad (17)$$

Consider our assumption, $D_{ij} = g_{ij} - a*_i a_j - b*_i b_j - c*_i c_j$ and we have obtained an expression of the Reeb vector by the orthonormal vectors that represent the foliation. The additives of (15) are tensors. This leads us to an open question as follows: Is the condition $P_{\mu;}{}^{\lambda} P*^{\mu} = P*_{\mu;}{}^{\lambda} P^{\mu}$, the minimal condition which is needed for a representation of the Reeb vector by a_j, b_j, c_j as the sum of tensor terms ? In other words, is the condition $P_{\mu;}{}^{\lambda} P*^{\mu} = P*_{\mu;}{}^{\lambda} P^{\mu}$ a necessity for the tensor representation of the acceleration matrix by the foliation scalars, a, b, c ?

4. Invariance of the Reeb vector under different functions of P

Here we wish to explore another degree of freedom in the action operator of the "acceleration field" which results from the Reeb vector, as shown by a representative vector field $\frac{dP}{dx^i}$ which is tangent to a non-geodesic integral curve. We wish to show that P can be replaced with a smooth function $f(P)$ and that U_m is invariant under such a transformation.

We revisit our acceleration field and write $U_m = \frac{N^2{}_{,m}}{N^2} - \frac{N^2{}_{,\mu} P^{*\mu}}{N^4} P_m$ s.t. $Z = N^2 \equiv (P^{*i} P_i + P^{*i} P_i)/2$ (also found as Z in this paper) we can omit the comma for the sake of brevity the same way we write P_i instead of $P_{,i}$ for $\frac{dP}{dx^i}$ and write $U_m = \frac{N^2{}_m}{N^2} - \frac{N^2{}_\mu P^{*\mu}}{N^4} P_m$. We will prove the invariance of U_m where P is real, however, a similar proof is also valid where P is complex and where P is replaced with a smooth function of P.

Suppose that we replace P by $f(P)$ such that f is positive and increasing or decreasing, then $f(P)_i \equiv \frac{df(P)}{dx^i} = \frac{df(P)}{dP}\frac{dP}{dx^i} = f_p(P)P_i$. Let $N^2 \equiv \mathrm{P}^\lambda \mathrm{P}_\lambda$ then $\hat{N}^2 \equiv f(P)_\lambda f(P)^\lambda = N^2 f_p(P)^2$ and $\frac{\hat{N}^2{}_k}{\hat{N}^2} = \frac{N^2{}_k}{N^2} + \frac{2 f_{pp}(P)}{f_p(P)} P_k$ but also

$$\hat{U}_k = \frac{\hat{N}^2{}_k}{\hat{N}^2} - \frac{\hat{N}^2{}_s f_p(P) P^s f_p(P) P_k}{\hat{N}^2} =$$

$$\frac{N^2{}_k}{N^2} + \frac{2 f_{pp}(P)}{f_p(P)} P_k - \left(\frac{N^2{}_s}{N^2} + \frac{2 f_{pp}(P)}{f_p(P)} P_s\right) \frac{f_p(P) P^s f_p(P) P_k}{N^2 f_p(P)^2} = \tag{18}$$

$$\frac{N^2{}_k}{N^2} - \frac{N^2{}_\mu P^\mu}{N^4} P_k = U_k$$

Which proves the invariance of the Reeb vector $\frac{U_k}{2}$ under different parameterizations of the scalar field P.

5. Energy density by an acceleration field – Reeb vector at the classical non-covariant limit

We now show, how much energy density does this term $\frac{1}{4} U_k U^k$ represent.

For a clock that moves along the integral curves, formed by $\frac{P_\mu}{\sqrt{Z}}$, we have from (2) and (11)

$$\frac{a_\mu}{c^2} = \frac{U_\mu}{2} \tag{19}$$

IOP Conf. Series: Journal of Physics: Conf. Series **845** (2017) 012019 doi:10.1088/1742-6596/845/1/012019

In special relativity, the squared curvature of a trajectory of a particle is expressible by its 4-acceleration, divided by the squared speed of light, $\frac{dV_\mu / c}{cd\tau} = \frac{a_\mu}{c^2}$ where the proper time τ differentiates the velocity V_μ and where τ is an arc-length parameterization.

The classical limit of a gravitational field is not covariant and that even worse, the classical field is intrinsic to the body of mass M that generates gravity; however, it is valid tool for the assessment of a physical model. We consider small mass at rest in a Newtonian (obviously not covariant) gravitational field. By the principle of equivalence, this mass is accelerated, otherwise it would freely fall. So, if a force field can keep small mass from falling, the field's classical limit of energy, is the same as the energy of the classical non-covariant gravitational field. This results hints at the energy of an acceleration field that opposes weak gravity. Summation of the squared norm of non-covariant 3-acceleration, a^2 of clocks that are kept from falling in the weak gravity generated by the mass M is

$$\iiint_{V=Volume} a^2 dV = \int_{r0}^{\infty} \left(\frac{(KM)^2}{r^2} \right)^2 4\pi r^2 dr = K \frac{4\pi KM^2}{r_0} \tag{20}$$

Where K is Newton's gravity constant. Now we calculate the non-relativistic and non-covariant negative potential energy $-E_g$,

$$\int_0^M \left(\frac{Km}{r_0} \right) dm = \frac{KM^2}{2r_0} = -E_g \tag{21}$$

So from (20) and (21)

$$\frac{1}{8\pi K} \iiint_{V=Volume} a^2 dV = -E_g \tag{22}$$

(22) implies the following relation between energy and the non-gravitational acceleration field that prohibits geodesic motion, where ρc^2 is the energy density and ρ is the mass density.

$$\frac{a^2}{c^4} = \frac{8\pi K\rho}{c^2} \Rightarrow Energy - Density = \frac{1}{2}\frac{a^2}{4\pi K} = \frac{a^2}{8\pi K} \tag{23}$$

where c is the speed of light. (23) dictates in four dimensions,

$$\frac{U_\mu U^\mu}{4} = \frac{a_\mu a^\mu}{c^4} = -\frac{8\pi K}{c^4} * \text{Energy} - \text{Density} \tag{24}$$

Note that unlike (24), (23) is not a covariant expression. What does it mean in the non-covariant classical limit of the electro-static field E. Since an electric field is also a form of an energy density, $Energy - Density = \frac{\varepsilon_0}{2} E^2$ where ε_0 is the permittivity of vacuum and from (24) we can infer the

IOP Conf. Series: Journal of Physics: Conf. Series **845** (2017) 012019 doi:10.1088/1742-6596/845/1/012019

following non-covariant classical limit, $8\pi K \frac{\varepsilon_0}{2} E^2 = a^2$ where E^2 and a^2 are square norms of the 3-vectors $\vec{E}$ and $\vec{a}$. We can infer,

$$\sqrt{4\pi K \varepsilon_0}\, E = a \tag{25}$$

The acceleration in (25) is dauntingly small and very difficult to measure. It requires an immense field of 1 million volts over 1 millimetre to expose an acceleration of uncharged clocks, which is about 8.61cm/sec^2, less than 0.01 g, providing that there are no other fields that cancel out this acceleration. In fact, we will see below that charge also generates gravity and that for the choice $8\pi K$ in (23), the acceleration will be about 4.305 cm/sec^2. By the principle of parsimony, the fact that this acceleration field stores energy, i.e. $Energy\text{-}Density = \frac{a^2}{8\pi K}$ means that this acceleration is aligned with the electric charge, electro-static field curves because this can explain the electric charge attraction and repulsion by simply, increasing or decreasing the energy stored in such a weak acceleration field. As we shall see, if instead of $8\pi K$ we develop this theory such that $4\pi K$ divides the square norm of acceleration, no acceleration of neutral particles will be measured within a homogeneous electrostatic field. This is because, we will develop the Euler Lagrange equations of the Ricci scalar plus (11) and see that charge also generates gravity and not only inertial mass does.

We now use a covariant terminology of 4-acceleration a_μ and $a^2 \equiv a_\mu a^\mu$.

As a more general theory, we can write $Energy\_Density = -\frac{a^2}{\sigma K} = -\frac{a_\mu a^\mu}{\sigma K}$, such that $\sigma = 8\pi$.

Another important remark is that in the classical non-covariant limit, the divergence of the electric field can be written as,

$$\begin{aligned} &\frac{U_\mu;^\mu}{2} = Div\left(\frac{U_\mu}{2}\right) = Div\left(a_\mu / c^2\right) \approx \\ &\frac{\sqrt{4\pi K \varepsilon_0}}{c^2} Div(E) = \frac{\sqrt{4\pi K \varepsilon_0}}{c^2}\frac{\rho}{\varepsilon_0} = \sqrt{\frac{4\pi K}{\varepsilon_0}}\frac{\rho}{c^2} \end{aligned} \tag{26}$$

Such that ρ denotes charge density and not the previously defined energy density.

By experiments done by Hector Serrano, for NASA, the author believes that the acceleration field of even uncharged clocks in an electric field is towards the electron and out of the proton. A relation between charge and gravity can be developed, leading to unprecedented repercussions on the feasibility of Alcubierre Warp Drive, reference is given where it is discussed later.

There is a remark of Serrano [9], about a moving capacitor in vacuum, in a reply to Peter Liddicoat: *"Actually by the generally accepted definition of what constitutes high vacuum 10^-6 Torr is about in the middle. This pressure is about equal to low Earth orbit. More importantly at this pressure the 'Mean Free Path' of the molecules in the chamber is far too great to support Corona/Ion wind effects. We've tested from atmosphere to 10^-7 Torr with no change in performance either. However, I'm glad the results have you thinking. It looks simple, but trust me it's not"..*

Hector Serrano has mentioned in a patent [10], that a capacitor manifests weak thrust also in vacuum. Another indirect evidence is the Flyby Anomaly [11] which is possibly caused by ionosphere charge. For further evidence, see TimirDatta et. al. work as an elegant way to focus field lines by metal cone and plane and to observe an effect [12]. The author believes the acceleration of charge-less particles in an electric field is from positive to negative. In section 9 it is shown that there is an electro-gravitational effect opposite in direction to the acceleration of an uncharged particle in an electro-static

IOP Conf. Series: Journal of Physics: Conf. Series **845** (2017) 012019 doi:10.1088/1742-6596/845/1/012019

field. There is at least informal evidence that the elecro-gravitational effect shows thrust of the entire dipole towards the positive direction [9] and the author does not imply asymmetrical capacitors of 1 - 0.1 Pico-Farad with 45000 Volts. It is shown that such capacitors - according to the calculations in section 9, assuming a roughly approximated acceleration proportional to the gravitational field – will not manifest any measurable effect of at least 1 micro Newton thrust. Most likely is that any measurable thrust, using such small capacitors, will be solely based on ionic wind.

6. Experimental problems – electron mobility

The down side of the non-geodesic acceleration is that it is about 10 orders of magnitude smaller than the accepted and known electric field interaction. For example, negative charge suspended above the Earth will cause charge to move in the ground. This charge will have a much stronger effect than the interaction with the acceleration field as is, and will cause a shielding effect i.e. the fields will cancel out within the Earth. Even the almost ideal insulator, i.e. diamond crystals, have impurities such as Nitrogen Vacancies [13] that allow charge carriers to move in the lattice i.e. high electron mobility. In the purest diamonds, the NV impurities are about 10^{18} nodes per cm^{-3} comparing to $1.77\,x10^{23}$ carbon atoms per cm^{-3}. The donor electrons lie deep in the band gap of 5.47eV, at about 1.7 eV.

7. Vaknin's theory

We quote here one of the four models of Vaknin [14] as follows: This work contains a possible realization of space-time as an ideal geometric object that becomes physically accessible only where a wave function which is called "chronon" collapses. The physical model is therefore of events and not of particles. This paper offers the idea that matter occurs where the Reeb vector is not zero. Showing consistency of this model with Quantum Mechanics is a very difficult task although it is possible to show that the energy of an electric field is stored in an acceleration field by replacement of the electro-magnetic tensor with the anti-symmetric acceleration field.

Vaknin's description of the realization of event is as follows:*"Time as a wave function with observer-mediated collapse. Entanglement of allChronons at the exact "moment" of the Big Bang. A relativistic QFT with Chrononsas Field Quanta (excited states.) The integration is achieved via quantumsuperpositions".*

The main difference between Vaknin's approach and the author's approach is that Vaknin's approach is algebraic where the author's approach is geometric. Thus, the outcome is two different theories that discuss a similar idea. We now show the simplest implementation of Vaknin's model as a quantization idea of time by collapsible events, as an additional constraint to the action

$$\int_{\Omega^4} \frac{c^4}{\sigma K} Curv^2 \sqrt{-g}\, d\Omega^4 \text{ where } \sigma = 8\pi$$

Where $\sqrt{-g}$ is the root of the negative metric tensor determinant for the volume element, such that $P = \lim_{n \to \infty} \psi(1) + \psi(2) + \ldots + \psi(n)$ and such that: $\int_{\Omega^4} \psi(k)\psi^*(k)\sqrt{-g}\, d\Omega^4 = 1$ And

$$0 < j < k < \infty \Rightarrow \int_{\Omega^4} \psi(j)\psi^*(k)\sqrt{-g}\, d\Omega^4 = 0 .$$

The term c^4 is the speed of light to the power of 4 and K is Newton's gravity constant. $Curv^2$ is a generalization of the square norm of the Reeb vector to the complex numbers field,

$$Curv^2 = \frac{(U_j U^{*j} + U^*_j U^j)}{8} \text{ where } \frac{U_j}{2} = \frac{Z_j}{2Z} - \frac{Z_k P^{*k}}{2Z^2} P_j = \left(\frac{Z_j P_s}{2Z^{3/2}} - \frac{Z_s P_j}{2Z^{3/2}}\right) \frac{P^{s*}}{\sqrt{Z}}$$

and $A_{sj}=\frac{Z_jP_s}{2Z^{3/2}}-\frac{Z_sP_j}{2Z^{3/2}}$ can be extended to a regular matrix $A_{sj}+B_{sj}$ where $B^{\alpha\beta}\equiv r\frac{1}{\sqrt{2}}\varepsilon^{\mu\nu\alpha\beta}A_{\mu\nu}$ and $\varepsilon^{\mu\nu\alpha\beta}$ is the Levi Civita tensor (not symbol) and $rr^*=1$ due to a degree of freedom.

7.1. *Physical meaning:* The field $A_{js}+B_{js}$ will rotate and scale a scalar wave function φ of a particle, $\varphi_j=\frac{d\varphi}{dx^j}$ where $(A_{js}+B_{js})\varphi^{*s}=\frac{1}{c}\frac{d\varphi_j}{d\tau}$ and where τ measures proper time.

We generalize the acceleration field energy density from $-(\frac{a_ja^j}{8\pi K})=-\frac{U_jU^j}{4}\frac{c^4}{8\pi K}$ to $-(\frac{a_ja^{*j}+a^*_ja^j}{16\pi K})=-\frac{U_jU^{*j}+U^*_jU^j}{8}\frac{c^4}{8\pi K}$ and $\frac{a_j}{c^2}=\frac{U_j}{2}$.
As we saw in (18), P does not have to be the proper time measured along curves. Instead, it can be a function of such proper time. (18) motivates the decomposition of P into wave functions because P does not have to be a monotonically increasing function of the proper time measured along integral curves formed by P_k. The problem is that P is not any wave function of a particle. The simplest physical interpretation of the ψ wave function is that it describes events in space-time and not particles. Therefore, P becomes a sum of wave functions and $P=\lim_{n\to\infty}\psi(1)+\psi(2)+...+\psi(n)$ is a decomposition of the function P as a sum of wave functions.

As quantum states, these event wave functions $\psi(1)$, $\psi(2)$,... must be normalized to probability 1 on the space-time manifold and they should be independent of each other as was written in two integral constraints. The best motivation for the constraints $\int_{\Omega^4}\psi(k)\psi^*(k)\sqrt{-g}d\Omega^4=1$ and for $\int_{\Omega^4}\psi(k)\psi^*(j)\sqrt{-g}d\Omega^4=0$ s.t. $k\neq j$ is given by Vaknin [14] and Storkin [15] where they emphasize that physical events are discrete. Storkin considers decreasing probability functions of the number of events in a given -volume, e.g. Poisson distribution within 2+1 dimensions Minkowsky space-time. Causal sets are partially ordered graphs of events along paths in space-time. The approach of this paper is more robust than that of Storkin because causal sets are the result of the order of events along the integral curves that are naturally formed by P_k along with the mentioned constraints that induce a countable set of wave functions.

7.2. *Auto-rotation:* In this section, we will study the field of a particle, whose rest mass energy ispresumably stored in an acceleration field. The equation $(A_{js}+B_{js})\varphi^{*s}=\frac{1}{c}\frac{d\varphi_j}{d\tau}$ can be better understood, by recalling (8) and by replacing $\frac{P_k}{\sqrt{Z}}$ in (8) with φ_k, as in the following:

$A_{js}\varphi^{*s} = Lie(\frac{P^{*\lambda}}{\sqrt{Z}},\varphi_j) = \frac{P^{*\lambda}}{\sqrt{Z}}\varphi_{j},_{\lambda} + \left(\frac{P^{*s}}{\sqrt{Z}}\right),_j \varphi_s$ in which the right hand side describes the acceleration of φ_s along the vector $\frac{P^{*\lambda}}{\sqrt{Z}}$. On the other hand, the left hand side, $A_{js}\varphi^{*s}$ describes the acceleration of φ^{*s} along $\frac{P_s}{\sqrt{Z}}$ which does not include the portion of the acceleration $B_{js}\varphi^{*s}$, and which results in the equation above. By (8), If we replace φ_s with $\frac{P_s}{\sqrt{Z}}$, we get $\frac{U_j}{2}$ on both sides of the equation. What is the meaning of φ_k ? Dirac's equation consists of spinors [6], and of matrix blocks as basis elements of the Lie Algebra of a rotation group. In Dirac's equation, φ_k is not the gradient of a scalar function, the indices correspond to orthogonal unit vectors and $\varphi_k\varphi^{*k}$ or $(\varphi_k\varphi^{*k} + \varphi^{*}_k\varphi^k)/2$ can be reduced to a probability density. φ_k is not a vector in the usual sense of General Relativity because it transforms between different coordinate systems with the help of spin connections and not with the help of affine connections. The philosophy of the Dirac equation, stems from the motivation to represent spatial spin axes as three orthogonal quantum states and to predict their probabilities. Any measurement of a spin is either +spin or -spin, no matter from which angle the physical measurement is performed. This property of the spin is very different than that of the classical mechanics spin. The philosophy of this paper is very different than that of Dirac, because it is based on a fully geometric interpretation of matter as acceleration fields. We begin with a quest for φ_k that will be an ordinary vector, unlike in Dirac's equation. We know that if $\varphi_k\varphi^{*k}$ is a probability density, then by (24), if all the energy of the particle is in its acceleration field, then we must have $\frac{8\pi K}{c^4}\varphi_k\varphi^{*k}mc^2 = -\frac{U_kU^k}{4}$ which results in two equations, $A_{js}\varphi^{*s} = \frac{P^{*\lambda}}{\sqrt{Z}}\varphi_{j},_{\lambda} + \left(\frac{P^{*s}}{\sqrt{Z}}\right),_j \varphi_s$ and $\frac{8\pi K}{c^4}\varphi_k\varphi^{*k}mc^2 = -\frac{U_kU^k}{4}$ where mc^2 is the energy of the particle and $\varphi_k\varphi^{*k}mc^2$ is the energy density of the particle. There is a new open problem and a new ongoing research which is intended to answer whether φ_k, as presented here, offers a useful way to describe matter on the quantum level.

8. General Relativity for the deterministic limit

By General Relativity, we have to add the Hilbert-Einstein action [16][17][18] to the negative sign of the square curvature of the gradient of the scalar field in order to replace the energy-momentum tensor in the Einstein's field equations. Negative means that the curvature operator is mostly negative. As before, we assume $\sigma = 8\pi$ (from the previously discussed term, $-a_\mu a^\mu/8\pi K$ as an energy density).

$$Z = N^2 = P_\mu P^\mu \text{ and } U_\lambda = \frac{Z_\lambda}{Z} - \frac{Z_k P^k P_\lambda}{Z^2} \text{ and } L = \frac{1}{4} U^k U_k$$

$R = Ricci$ curvature.

$$Min\ Action = Min \int_\Omega \left(R - \frac{8\pi}{\sigma} L \right) \sqrt{-g}\, d\Omega = \tag{27}$$

$$Min \int_\Omega \left(R - \frac{1}{4} U^k U_k \right) . \sqrt{-g}\, d\Omega \text{ s.t. } \sigma = 8\pi$$

$\sqrt{-g}$ is a scalar density of the volume element, R is the Ricci curvature [16] and $\sqrt{-g}$ is the determinant of the metric tensor used for the 4-volume element as in tensor densities [17].

The variation of the Ricci scalar is well known. It uses the Platini identity and Stokes theorem to calculate the variation of the Ricci curvature and reaches the Einstein tensor [18], as follows, $\delta R = R_{\mu\nu} \delta g^{\mu\nu}$ and $\delta\sqrt{-g} = -\frac{1}{2} g_{\mu\nu} \delta g^{\mu\nu} \sqrt{-g}$ by which we infer $\delta(R\sqrt{-g}) = (R_{\mu\nu} - \frac{1}{2} R g_{\mu\nu}) \delta g^{\mu\nu}$ which will be later added to the variation of $\left(\frac{1}{2} R - \frac{8\pi}{\sigma} L \right) . \sqrt{-g}$ by $\delta g^{\mu\nu}$.

The following Euler Lagrange equations have to hold,

$$\left(\frac{\partial}{\partial g^{\mu\nu}} - \frac{d}{dx^m} \frac{\partial}{\partial (g^{\mu\nu}{}_{,m})} + \frac{d^2}{dx^m dx^s} \frac{\partial}{\partial (g^{\mu\nu}{}_{,m,s})} \right) \left(\left(\frac{1}{2} R - \frac{1}{4} U^k U_k \right) \sqrt{-g} \right) = 0$$

and

$$\left(\frac{\partial}{\partial p} - \frac{d}{dx^m} \frac{\partial}{\partial (P_m)} + \frac{d^2}{dx^m dx^s} \frac{\partial}{\partial (P_{,m,s})} \right) \left(\left(\frac{1}{2} R - \frac{1}{4} U^k U_k \right) \sqrt{-g} \right) = 0$$

$U^k U_k = \frac{Z_\mu Z^\mu}{Z^2} - \frac{(Z_s P^s)^2}{Z^3}$ which we obtain from the minimum Euler Lagrange equation because $U_\lambda P^\lambda = \frac{Z_\lambda P^\lambda}{Z} - \frac{Z_k P^k P_\lambda P^\lambda}{Z^2} = 0$. In order to calculate the minimum action Euler-Lagrange equations, we will separately treat the Lagrangians, $L = \frac{Z_\mu Z^\mu}{Z^2}$ and $L = \frac{(Z_s P^s)^2}{Z^3}$ to derive the Euler Lagrange equations of the Lagrangian $L = \frac{Z_\mu Z^\mu}{Z^2} - \frac{(Z_s P^s)^2}{Z^3} = U_\mu U^\mu$. The Euler Lagrange operator of the Ricci scalar $\left(\frac{\partial}{\partial g^{\mu\nu}} - \frac{d}{dx^m} \frac{\partial}{\partial (g^{\mu\nu}{}_{,m})} + \frac{d^2}{dx^m dx^s} \frac{\partial}{\partial (g^{\mu\nu}{}_{,m,s})} \right)$.

The reader may skip the following equations up to equation (33). Equations (33), (34) and (36) are however crucial.

IOP Conf. Series: Journal of Physics: Conf. Series **845** (2017) 012019 doi:10.1088/1742-6596/845/1/012019

$$L = \frac{(P^{\lambda} Z_{\lambda})^2}{Z^3} \quad s.t.\ Z = P_{\mu} P^{\mu}$$

$$\frac{\partial (L\sqrt{-g})}{\partial g^{\mu\nu}} - \frac{d}{dx^m}\frac{\partial (L\sqrt{-g})}{\partial (g^{\mu\nu}{}_{,m})} =$$

$$\left(\begin{array}{l} -2\left(\frac{(P^{\lambda}P_{\lambda})_{,s}P^s}{Z^3} P_{\mu}P_{\nu}P^m\right)_{;m} + \\ 2\frac{(P^{\lambda}P_{\lambda})_{,s}P^s}{Z^3}(\Gamma_{\mu}{}^{i}{}_{m}P_iP_{\nu}P^m + \Gamma_{\upsilon}{}^{i}{}_{m}P_{\mu}P_iP^m) + \\ +2\frac{(P^{\lambda}P_{\lambda})_{,s}P^s}{Z^3}(P_{\mu}P_{\nu})_{;m}P^m \\ -2\frac{(P^{\lambda}P_{\lambda})_{,s}P^s}{Z^3}(\Gamma_{\mu}{}^{i}{}_{m}P_iP_{\nu}P^m + \Gamma_{\upsilon}{}^{i}{}_{m}P_{\mu}P_iP^m) + \\ +2\left(\frac{(P^{\lambda}P_{\lambda})_{,s}P^s}{Z^3}\right)Z_{\mu}P_{\nu} - 3\left(\frac{((P^{\lambda}P_{\lambda})_{,s}P^s)^2}{Z^4}\right)P_{\mu}P_{\nu} \\ -\frac{1}{2}\frac{(P^{\lambda}Z_{\lambda})^2}{Z^3}g_{\mu\nu} \end{array} \right)\sqrt{-g} =$$

$$\left(\begin{array}{l} -2\left(\frac{(P^{\lambda}P_{\lambda})_{,m}P^m}{Z^3}P^k\right)_{;k}P_{\mu}P_{\nu} \\ -2\frac{(P^{\lambda}Z_{\lambda})^2}{Z^3}\frac{P_{\mu}P_{\nu}}{Z} + 2\left(\frac{(P^{\lambda}P_{\lambda})_{,s}P^s}{Z^3}\right)Z_{\mu}P_{\nu} + \\ -\frac{1}{2}\frac{(P^{\lambda}Z_{\lambda})^2}{Z^3}g_{\mu\nu} - \frac{(P^{\lambda}Z_{\lambda})^2}{Z^3}\frac{P_{\mu}P_{\nu}}{Z} \end{array} \right)\sqrt{-g} \qquad (28)$$

IOP Conf. Series: Journal of Physics: Conf. Series **845** (2017) 012019 doi:10.1088/1742-6596/845/1/012019

$$L = \frac{Z^{\lambda} Z_{\lambda}}{Z^{2}} \ s.t. Z = P_{\mu} P^{\mu}$$

$$\frac{\partial (L \sqrt{-g})}{\partial g^{\mu\upsilon}} - \frac{d}{dx^{m}} \frac{\partial (L \sqrt{-g})}{\partial g^{\mu\upsilon}{}_{,m}} =$$

$$\left(\begin{array}{l} -2 \left(\frac{Z^{m} P_{\mu} P_{\upsilon}}{Z^{2}} \right);_{m} \\ +2 \frac{(\Gamma_{\mu}{}^{i}{}_{m} P_{i} P_{v} Z^{m} + \Gamma_{\upsilon}{}^{i}{}_{m} P_{\mu} P_{i} Z^{m})}{Z^{2}})+ \\ +2 \frac{(P_{\mu} P_{\upsilon});_{m} Z^{m})}{Z^{2}} \\ -2 \frac{(\Gamma_{\mu}{}^{i}{}_{m} P_{i} P_{v} Z^{m} + \Gamma_{\upsilon}{}^{i}{}_{m} P_{\mu} P_{i} Z^{m})}{Z^{2}})+ \\ +\frac{Z_{\mu} Z_{v}}{Z^{2}} - 2 \frac{Z_{s} Z^{s}}{Z^{3}} P_{\mu} P_{v} - \frac{1}{2} \frac{Z_{m} Z^{m}}{(P^{i} P_{i})^{2}} g_{\mu v} \end{array} \right) \sqrt{-g} =$$

$$\left(-2 \left(\frac{Z^{m}}{Z^{2}} \right);_{m} P_{\mu} P_{v} - 2 \frac{Z^{\lambda} Z_{\lambda}}{Z^{2}} \frac{P_{\mu} P_{v}}{Z} - \frac{1}{2} \frac{Z_{k} Z^{k}}{Z^{2}} g_{\mu v} + \frac{Z_{\mu} Z_{v}}{Z^{2}} \right) \sqrt{-g}$$

$$L = \frac{Z^{\lambda} Z_{\lambda}}{Z^{2}} \ s.t. Z = P_{\mu} P^{\mu}$$

$$\frac{\partial (L \sqrt{-g})}{\partial g^{\mu\upsilon}} - \frac{d}{dx^{m}} \frac{\partial (L \sqrt{-g})}{\partial g^{\mu\upsilon}{}_{,m}} =$$

$$\left(\begin{array}{l} -2 \left(\frac{Z^{m} P_{\mu} P_{\upsilon}}{Z^{2}} \right);_{m} + 2 \frac{(\Gamma_{\mu}{}^{i}{}_{m} P_{i} P_{v} Z^{m} + \Gamma_{\upsilon}{}^{i}{}_{m} P_{\mu} P_{i} Z^{m})}{Z^{2}})+ \\ +2 \frac{(P_{\mu} P_{\upsilon});_{m} Z^{m})}{Z^{2}} - 2 \frac{(\Gamma_{\mu}{}^{i}{}_{m} P_{i} P_{v} Z^{m} + \Gamma_{\upsilon}{}^{i}{}_{m} P_{\mu} P_{i} Z^{m})}{Z^{2}})+ \\ +\frac{Z_{\mu} Z_{v}}{Z^{2}} - 2 \frac{Z_{s} Z^{s}}{Z^{3}} P_{\mu} P_{v} - \frac{1}{2} \frac{Z_{m} Z^{m}}{(P^{i} P_{i})^{2}} g_{\mu v} \end{array} \right) \sqrt{-g} = \tag{29}$$

$$\left(-2 \left(\frac{Z^{m}}{Z^{2}} \right);_{m} P_{\mu} P_{v} - 2 \frac{Z^{\lambda} Z_{\lambda}}{Z^{2}} \frac{P_{\mu} P_{v}}{Z} - \frac{1}{2} \frac{Z_{k} Z^{k}}{Z^{2}} g_{\mu v} + \frac{Z_{\mu} Z_{v}}{Z^{2}} \right) \sqrt{-g}$$

We subtract (28) from (29)

$$Z = P_\mu P^\mu \ and \ U_\lambda = \frac{Z_\lambda}{Z} - \frac{Z_k P^k P_\lambda}{Z^2} and \ L = U^\kappa U_k = \frac{Z_\lambda Z^\lambda}{Z^2} - \frac{(Z_k P^k)^2}{Z^3}$$

$$(\frac{\partial (L\sqrt{-g})}{\partial g^{\mu\upsilon}} - \frac{d}{dx^m}\frac{\partial (L\sqrt{-g})}{\partial g^{\mu\upsilon}{}_{,m}})U^\kappa U_k =$$

$$\left(\begin{array}{l} +2(\frac{(P^\lambda P_\lambda)_{,m} P^m}{Z^3} P^k)_{;k} P_\mu P_v + \\ 2\frac{(P^\lambda Z_\lambda)^2}{Z^3}\frac{P_\mu P_v}{Z} - 2(\frac{(P^\lambda P_\lambda)_{,s} P^s}{Z^3})Z_\mu P_v + \\ +\frac{1}{2}\frac{(P^\lambda Z_\lambda)^2}{Z^3} g_{\mu v} + \frac{(P^\lambda Z_\lambda)^2}{Z^3}\frac{P_\mu P_v}{Z} + \\ (-2(\frac{Z^m}{Z^2})_{;m} P_\mu P_v - 2\frac{Z^\lambda Z_\lambda}{Z^2}\frac{P_\mu P_v}{Z} - \frac{1}{2}\frac{Z_k Z^k}{Z^2} g_{\mu v} + \frac{Z_\mu Z_v}{Z^2}) \end{array} \right)\sqrt{-g} =$$

$$\left(\begin{array}{l} +2((\frac{(P^\lambda P_\lambda)_{,m} P^m}{Z^3} P^k)_{;k} - 2(\frac{Z^m}{Z^2})_{;m})P_\mu P_v + \\ +2\frac{(P^\lambda Z_\lambda)^2}{Z^3}\frac{P_\mu P_v}{Z} - 2\frac{Z^\lambda Z_\lambda}{Z^2}\frac{P_\mu P_v}{Z} + \\ +\frac{1}{2}\frac{(P^\lambda Z_\lambda)^2}{Z^3} g_{\mu v} - \frac{1}{2}\frac{Z_k Z^k}{Z^2} g_{\mu v} + \\ +\frac{Z_\mu Z_v}{Z^2} - 2(\frac{(P^\lambda P_\lambda)_{,s} P^s}{Z^3})Z_\mu P_v + \frac{(P^\lambda Z_\lambda)^2}{Z^3}\frac{P_\mu P_v}{Z} \end{array} \right)\sqrt{-g} =$$

$$\left(\begin{array}{l} +2((\frac{(P^\lambda P_\lambda)_{,m} P^m}{Z^3} P^k)_{;k} - 2(\frac{Z^m}{Z^2})_{;m})P_\mu P_v + \\ +2\frac{(P^\lambda Z_\lambda)^2}{Z^3}\frac{P_\mu P_v}{Z} - 2\frac{Z^\lambda Z_\lambda}{Z^2}\frac{P_\mu P_v}{Z} + \\ +U_\mu U_v - \frac{1}{2}U_k U^k g_{\mu v} \end{array} \right)\sqrt{-g} = \tag{30}$$

$$(U_\mu U_v - \frac{1}{2}U_k U^k g_{\mu v} - 2U^k{}_{;k}\frac{P_\mu P_v}{Z})\sqrt{-g}$$

IOP Conf. Series: Journal of Physics: Conf. Series **845** (2017) 012019 doi:10.1088/1742-6596/845/1/012019

$$L = \frac{(Z^s P_s)^2}{Z^3} \quad s.t.\ Z = P^\lambda P_\lambda \ and\ Z_m = (P^\lambda P_\lambda)_{,m}$$

$$\frac{\partial (L\sqrt{-g})}{\partial P_\mu} - \frac{d}{dx^v}\frac{\partial (L\sqrt{-g})}{\partial P_{\mu,v}} =$$

$$\begin{pmatrix} -4(\frac{(Z_s P^s)}{Z^3} P^\mu P^v);_v + 4\frac{(Z_s P^s)}{Z^3}\Gamma_i{}^\mu{}_v P^i P^v + \\ +4\frac{(Z_s P^s)}{Z^3} P^\mu;_v P^v - 4\frac{(Z_s P^s)}{Z^3}\Gamma_i{}^\mu{}_k P^i P^k + \\ +2\frac{Z_m P^m Z^\mu}{Z^3} - 6\frac{(Z_m P^m)^2}{Z^4} P^\mu \end{pmatrix}\sqrt{-g} = \tag{31}$$

$$(-4(\frac{(Z_s P^s)P^v}{Z^3});_v P^\mu + 2\frac{Z_m P^m Z^\mu}{Z^3} - 6\frac{(Z_m P^m)^2}{Z^4} P^\mu)\sqrt{-g}$$

$$L = \frac{Z^s Z_s}{Z^2} \quad s.t.\ Z = P^\lambda P_\lambda \ and\ Z_m = (P^\lambda P_\lambda)_{,m}$$

$$\frac{\partial (L\sqrt{-g})}{\partial P_\mu} - \frac{d}{dx^v}\frac{\partial (L\sqrt{-g})}{\partial P_{\mu,v}} =$$

$$\begin{pmatrix} -4(\frac{P^\mu Z^v}{Z^2});_v + \frac{4}{Z^2}\Gamma_i{}^\mu{}_k P^i Z^k + \\ +\frac{4}{Z^2} P^\mu;_v Z^v - \frac{4}{Z^2}\Gamma_i{}^\mu{}_k P^i Z^k + \\ -4\frac{Z_m Z^m}{Z^3} P^\mu \sqrt{-g} \end{pmatrix}\sqrt{-g} = \tag{32}$$

$$(-4(\frac{Z^v}{Z^2});_v - 4\frac{Z_m Z^m}{Z^3})P^\mu\sqrt{-g}$$

We subtracted the Euler Lagrange operators of $\frac{(Z^s P_s)^2}{Z^3}\sqrt{-g}$ in (28) from the Euler Lagrange operators of $\frac{Z^\lambda Z_\lambda}{Z^2}\sqrt{-g}$ in (29) and got (30) and we will subtract (31) from (32) to get two tensor equations of gravity, these will be (33), and (36).

Assuming $\sigma = 8\pi$, where the metric variation equations (27), (28), (29) and (30) yield

$$Z = N^2 = \mathrm{P}_\mu \mathrm{P}^\mu \, , \; U_\lambda = \frac{Z_\lambda}{Z} - \frac{Z_k P^k P_\lambda}{Z^2} \, , \; L = \frac{1}{4} U_i U^i \;\text{ and }\; Z = \mathrm{P}^k \mathrm{P}_k$$

$$\frac{8\pi}{\sigma}\frac{1}{4}\left(\begin{array}{l} + 2\left(\left(\frac{(\mathrm{P}^\lambda \mathrm{P}_\lambda),_{\mathrm{m}} P^m}{Z^3} P^k\right);_k - 2\left(\frac{Z^m}{Z^2}\right);_m\right) P_\mu P_\nu + \\ + 2\frac{(P^\lambda Z_\lambda)^2}{Z^3}\frac{P_\mu P_\nu}{Z} - 2\frac{Z^\lambda Z_\lambda}{Z^2}\frac{P_\mu P_\nu}{Z} + \\ + U_\mu U_\nu - \frac{1}{2} U_{\mathrm{k}} U^k g_{\mu\nu} \end{array} \right) =$$

$$\frac{8\pi}{\sigma}\frac{1}{4}\left(U_\mu U_\nu - \frac{1}{2} U_{\mathrm{k}} U^k g_{\mu\nu} - 2U^k;_k \frac{P_\mu P_\nu}{Z}\right) = R_{\mu\nu} - \frac{1}{2} R g_{\mu\nu} \tag{33}$$

$$\text{s.t.} \quad R = R_{\mu\upsilon} g^{\mu\upsilon}$$

$$\text{s.t.} \quad R_{kj} = (\Gamma_{jk}{}^P),_p - (\Gamma_{pk}{}^P),_j + \Gamma_{p\mu}{}^P \Gamma_{jk}{}^\mu - \Gamma_{pj}{}^\mu \Gamma_{k\mu}{}^p$$

$R_{\mu\nu}$ is the Ricci tensor and $R_{\mu\nu} - \frac{1}{2} R g_{\mu\nu}$ is the Einstein tensor [18].In general, by (27) and $\sigma = 8\pi$ (33) can be written as

$$\frac{1}{4}\left(U_\mu U_\nu - \frac{1}{2} U_{\mathrm{k}} U^k g_{\mu\nu} - 2U^k;_k \frac{P_\mu P_\nu}{Z}\right) = R_{\mu\nu} - \frac{1}{2} R g_{\mu\nu} \tag{34}$$

If we consider Vaknin's model [14] of realization of space-time by collapse events $\psi (i = 1,2,3,...., n \to \infty)$ then we have to add the lambda * constraint to the action operator and the resulting Euler Lagrange equations for a vanishing metric variation are:

$$\begin{array}{l} P = \lim\limits_{n\to\infty} \psi(1) + \psi(2) + ... + \psi(n) \\ \int\limits_{\Omega^4} \psi(k)\psi^*(k)\sqrt{-g}\, d\Omega^4 = 1 \\ 0 < j < k < \infty \Rightarrow \int\limits_{\Omega^4} \psi(j)\psi^*(k)\sqrt{-g}\, d\Omega^4 = 0 \\ \frac{1}{4}\left(U_\mu U_\nu - \frac{1}{2} U_{\mathrm{k}} U^k g_{\mu\nu} - 2U^k;_k \frac{P_\mu P_\nu}{Z}\right) - \frac{1}{2}\lambda P P^* g_{\mu\nu} = R_{\mu\nu} - \frac{1}{2} R g_{\mu\nu} \end{array} \tag{35}$$

for some cosmological constant λ . Also, note our choice $\sigma = 8\pi$.

We can also see that the ordinary local conservation laws are modified if $U^k;_k \neq 0$ unless the local average around charge $(-2\overline{U}^k;_k \frac{\overline{P}_\mu \overline{P}_\nu}{\overline{Z}});^\nu = 0$ which is expected due to symmetry around the charge.

IOP Conf. Series: Journal of Physics: Conf. Series **845** (2017) 012019 doi:10.1088/1742-6596/845/1/012019

Charge-less field: The term $-2U^k;_k \dfrac{P_\mu P_\nu}{Z}$ in (33) can be generalized to:

$-2((U^k;_k + U^{*k};_k)/2)\dfrac{(P_\mu P^*_\nu + P^*_\mu P_\nu)/2}{Z}$ and can be zero under the following condition:

$$4(A_{\mu\nu};^\mu \frac{P^{*\nu}}{\sqrt{Z}} + A^*_{\mu\nu};^\mu \frac{P^\nu}{\sqrt{Z}}) = U_\mu U^{*\mu} + U^*_\mu U^\mu \Rightarrow U^k;_k + U^{*k};_k = 0$$

The complimentary matrix $B^{\mu\nu} = \dfrac{1}{\sqrt{2}} A_{\alpha\beta}\varepsilon^{\alpha\beta\mu\nu}$ can be transformed to a real matrix due to the SU(2) x U(1) degrees of freedom and also be imaginary.From (31), (32) we have,

$$\frac{d}{dx^\mu}(\frac{\partial}{\partial P_\mu} - \frac{d}{dx^\nu}\frac{\partial}{\partial P_{\mu},_\nu})(U_k U^k \sqrt{-g}) = W^\mu;_\mu \sqrt{-g} = 0$$

We recall, $W^\mu = (\dfrac{\partial}{\partial P_\mu} - \dfrac{d}{dx^\nu}\dfrac{\partial}{\partial P_{\mu},_\nu})(U_k U^k \sqrt{-g})$

$$W^\mu =$$

$$(-4(\frac{Z^\nu}{Z^2});_\nu - 4\frac{Z_m Z^m}{Z^3})P^\mu + 4(\frac{(Z_s P^s)P^\nu}{Z^3});_\nu P^\mu - 2\frac{Z_m P^m Z^\mu}{Z^3} + 6\frac{(Z_m P^m)^2}{Z^4}P^\mu =$$

$$-4(\frac{Z^\nu}{Z^2});_\nu P^\mu - 4\frac{Z_m Z^m}{Z^3}P^\mu +$$

$$+4(\frac{(Z_s P^s)P^\nu}{Z^3});_\nu P^\mu + 4\frac{(Z_m P^m)^2}{Z^4}P^\mu$$

$$-2\frac{Z_m P^m}{Z^2}(\frac{Z^\mu}{Z} - \frac{Z_m P^m P^\mu}{Z^2}) =$$

$$-4((\frac{U^k}{Z});_k + \frac{U^k U_k}{Z})P^\mu - 2\frac{Z_m P^m}{Z^2}U^\mu = 0$$

$$W^\mu;_\mu = \left(-4U^\nu;_\nu \frac{P^\mu}{Z} - 2\frac{(Z_m P^m)}{Z^2}U^\mu\right);_\mu = 0 \qquad (36)$$

9. electro-gravity –unexpected gravity inducedby electric charge

We return to (35)

$$\frac{1}{4}(U_\mu U_\nu - \frac{1}{2}U_k U^k g_{\mu\nu} - 2U^k;_k \frac{P_\mu P_\nu}{Z}) - \frac{1}{2}\lambda PP^* g_{\mu\nu} = R_{\mu\nu} - \frac{1}{2}Rg_{\mu\nu}$$

and see a startling property of the term $\dfrac{1}{4}(-2U^k;_k \dfrac{P_\mu P_\nu}{Z})$. In comparison, $\dfrac{1}{4}(U_\mu U_\nu - \dfrac{1}{2}U_k U^k g_{\mu\nu})$ looks like an energy momentum tensor of a perfect fluid and in contrast, $\dfrac{P_\mu P_\nu}{Z}$ consists of the unit vector $\dfrac{P_\mu}{\sqrt{Z}} = \dfrac{P_\mu}{\sqrt{P_k P^k}}$ which points to a perpendicular direction to U_μ.

The "source" of the Reeb vector U_μ can be defined by a non-zero $U^k{}_{;k}$ and its velocity need not be parallel to $\frac{P_\mu}{\sqrt{Z}}$. This property means that this term does not behave like as expected from an ordinary Energy – Momentum tensor. From (26) and from Einstein – Grossman's equation in vacuo,

We have that $-\frac{U^k{}_{;k}}{2} = \frac{\sqrt{\frac{4\pi K}{\varepsilon_0}\frac{\rho(\text{Charge})}{c^2}}}{\frac{8\pi K}{c^4}} = \rho(\text{Charge-Gravitational-Mass})c^2$ from which we infer

$\frac{\rho(\text{Charge-Gravitational-Mass})}{\sqrt{16\pi K\varepsilon_0}} = \rho(\mathit{Gravitational-Mass})$ or in terms of charge Q and mass M instead of charge density and mass density,

$$M_{\text{Charge-Gravitational-Mass}} = \frac{\pm Q}{\sqrt{16\pi K\varepsilon_0}} \tag{37}$$

This means that charge can cause gravity or anti-gravity and its sign is opposite to the acceleration field around the charge. A more general form is

$M_{\text{Charge-Gravitational-Mass}} = \frac{\pm Q}{\sqrt{2\sigma K\varepsilon_0}}$ where $\sigma = 8\pi$, K is Newton's gravity constant and ε_0 is the permittivity of vacuum.

We will calculate $\frac{\pm Q}{\sqrt{16\pi K\varepsilon_0}}$ for ± 20Coulombs.

$$\frac{\pm 1\,Coulomb}{\sqrt{16\pi\varepsilon_0 K}} =\sim \pm 5.8023\ x10^{9}\ Kg\ .$$

Multiplied by 20 we have

$$\frac{\pm 20\,Coulomb}{\sqrt{16\pi\varepsilon_0 K}} =\sim \pm 1.1605\ x10^{11}\ Kg\ .$$

This renders Alcubierre Warp Drive [19] a feasible technology by charge separation. Capacitors of several pico-Farads will not yield any measurable thrust [20] because they do not separate enough charge. However, separation of virtual charge that appears during transition states of electrons as they interact with photons, and by a short lived vacuum charge, is not ruled out because they can explain the effect known as EMDrive [21] by Warp Drive [19] caused by (37).

10. Total acceleration around electric charge

In the classical non-relativistic limit, acceleration a around a charge Q at radius r will be the result of (37) and by the acceleration field that prohibits geodesic motion, see (25),

$$-\delta a \approx \frac{-KQ}{r^2\sqrt{2\sigma K\varepsilon_0}} + \sqrt{\frac{\sigma K\varepsilon_0}{2}}\frac{Q}{4\pi\varepsilon_0 r^2} = \sqrt{\frac{\sigma K}{2\varepsilon_0}}\frac{Q}{r^2}\left(\frac{1}{\sigma} - \frac{1}{4\pi}\right) = g_{Electro\_gravity} - a \tag{38}$$

If $\sigma = 8\pi$ then $-\delta a \approx \sqrt{\frac{\sigma K}{2\varepsilon_0}\frac{Q}{r^2}(\frac{1}{8\pi}-\frac{1}{4\pi})} = g_{Electro\_gravity} - a$

$$-\delta a \approx \sqrt{\frac{4\pi K}{\varepsilon_0}\frac{Q}{r^2}(\frac{1}{8\pi}-\frac{1}{4\pi})} = -\sqrt{\frac{K}{16\pi\varepsilon_0}\frac{Q}{r^2}}$$

which results in a sum of accelerations $\delta a \approx \sqrt{\frac{K}{16\pi\varepsilon_0}\frac{Q}{r^2}}$

11. Proof of conservation

Theorem: Conservation law of the real Reeb vector.

From the vanishing of the divergence of Einstein tensor and (33) in the paper, we have to prove the following:

$$\frac{1}{4}\left(U_\mu U_\nu - \frac{1}{2}U_k U^k g_{\mu\nu} - 2U^k{}_{;k}\frac{P_\mu P_\nu}{Z}\right)_{;}{}^{\mu} = G_{\mu\nu;}{}^{\mu} = (R_{\mu\nu} - \frac{1}{2}Rg_{\mu\nu})_{;}{}^{\mu} = 0 \quad (39)$$

Proof:

From the zero variation by the scalar time field (36)

$$W^\mu{}_{;\mu} = \left(-4U^\nu{}_{;\nu}\frac{P^\mu}{Z} - 2\frac{(Z_m P^m)}{Z^2}U^\mu\right)_{;\mu} = 0 \quad (40)$$

$$-\left(2U^\nu{}_{;\nu}\frac{P^\mu}{Z}\right)_{;\mu} = \left(\frac{(Z_m P^m)}{Z^2}U^\mu\right)_{;\mu} \quad (41)$$

$$\left(-2U^k{}_{;k}\frac{P^\mu P^\nu}{Z}\right)_{;\mu} = \left(\frac{(Z_m P^m)}{Z^2}U^\mu\right)_{;\mu}P^\nu - \left(2U^k{}_{;k}\frac{P^\mu}{Z}\right)P^\nu{}_{;\mu} = \left(\frac{(Z_m P^m)}{Z^2}U^\mu\right)_{;\mu}P^\nu - U^k{}_{;k}\frac{Z^\nu}{Z} \quad (42)$$

Now let $t \equiv Z_m P^m$

$$\left(\frac{t}{Z^2}U^\mu\right)_{;\mu}P^\nu - U^k{}_{;k}\frac{Z^\nu}{Z} = (\frac{t}{Z^2})_{;\mu}U^\mu P^\nu + \frac{t}{Z^2}U^\mu{}_{;\mu}P^\nu - U^k{}_{;k}\frac{Z^\nu}{Z} =$$

$$-U^\mu{}_{;\mu}U^\nu + (\frac{t}{Z^2})_{;\mu}U^\mu P^\nu$$

so,

$$(-2U^k{}_{;k}\frac{P^\mu P^\nu}{Z})_{;\mu} = -U^\mu{}_{;\mu}U^\nu + (\frac{t}{Z^2})_{;\mu}U^\mu P^\nu \quad (43)$$

$$\begin{aligned}
&\left(U^{\mu}U^{\nu}-\frac{1}{2}U_{k}U^{k}g^{\mu\nu}-2U^{k};_{k}\frac{P^{\mu}P^{\nu}}{Z}\right);^{\mu}= \\
&U^{\mu};_{\mu}U^{\nu}+U^{\mu}U^{\nu};_{\mu}-\frac{1}{2}(U_{k};_{\mu}U_{s}+U_{k}U_{s};_{\mu})g^{ks}g^{\mu\nu}- \\
&U^{\mu};_{\mu}U^{\nu}+(\frac{t}{Z^{2}});_{\mu}U^{\mu}P^{\nu}= \\
&U^{\mu}U^{\nu};_{\mu}-\frac{1}{2}(U^{s}U_{s});^{\nu}+(\frac{t}{Z^{2}});_{\mu}U^{\mu}P^{\nu}=0
\end{aligned} \tag{44}$$

Notice that

$$\begin{aligned}
&U^{\mu}U^{\nu};_{\mu}-\frac{1}{2}U^{s}U_{s};^{\nu}= \\
&U^{\mu}\left((\frac{Z_{k}}{Z});_{\mu}-(\frac{t}{Z^{2}});_{\mu}P_{k}-(\frac{t}{Z^{2}})P_{k};_{\mu}\right)g^{k\nu}- \\
&U^{s}\left((\frac{Z_{s}}{Z});_{k}-(\frac{t}{Z^{2}});_{k}P_{s}-(\frac{t}{Z^{2}})P_{s};_{k}\right)g^{k\nu}= \\
&-U^{\mu}(\frac{t}{Z^{2}});_{\mu}P^{\nu}
\end{aligned} \tag{45}$$

Since $-(\frac{t}{Z^{2}});_{k}P_{s}U^{s}=0$

$$U^{\mu}U^{\nu};_{\mu}-\frac{1}{2}(U^{s}U_{s});^{\nu}+(\frac{t}{Z^{2}});_{\mu}U^{\mu}P^{\nu}=-U^{\mu}(\frac{t}{Z^{2}});_{\mu}P^{\nu}+(\frac{t}{Z^{2}});_{\mu}U^{\mu}P^{\nu}=0 \tag{46}$$

and we are done.

Conclusion

An upper limit on measurable time from each event backwards to the "big bang" singularity as a limit or from a manifold of events as in de Sitter or anti - de Sitter, may exist only as a limit and is not a practical physical observable because it can only be theoretically measured. Since more than one curve on which such time can be virtually measured intersects the same event - as is the case in material fields which prohibit inertial motion, i.e. prohibit free fall - such a time can't be realized as a coordinate. Nevertheless, using such time as a scalar field, enables to describe matter as acceleration fields by using the gradient of the scalar field and it allows new physics to emerge by a replacement of the stress-energy-momentum tensor. One arrives at electro-gravity as a neat explanation of the Dark Matter effect and the advent of Sciama's Inertial Induction, which becomes realizable by separation of high electric charge. This paper totally rules out any measurable Biefeld Brown effect in vacuum on Pico-Farad or less, Ionocrafts due to insufficient amount of electric charge [20]. The electro-gravitational effect is due to field divergence and not directly due to intensity or gradient of the square norm. Inertial motion prohibition by material fields, e.g. intense electrostatic field, can be measured as a very small mass dependent force on neutral particles that have rest mass and thus can measure proper time. The non-gravitational acceleration should be from the positive to the negative charge. The electro-gravitational effect which is opposite in direction and half in intensity, requires large amounts of separated charge carriers and acts on the entire negative to positive dipole.

Acknowledgements

First, I should thank Dr. Sam Vaknin for his full support and for his ground breaking dissertation from 1982. Also thanks to Larry Horwitz from Tel Aviv university for his advice on an earlier work. His letter from Wed, Jul 23, 2008, encouraged me to continue my research which started in 2003 and led to this paper. Special thanks to professor David Lovelock for his online help in understanding the calculus of variations and in fixing an error several years ago. Thanks to electrical engineers, Elad Dayan, Ran Timar and Benny Versano for their help in understanding how profoundly difficult it is to separate charge as required by the idea of electro-gravity. Thanks to other involved engineers, AryeAldema, ErezMagen and to Dr.LiorHaviv from the Weizmann Institute of Science. Thanks to ZeevJabotinsky and to ErezOhana for his initiative of building a miniature warp drive engine with almost zero funds. Thanks to Martin Land for his assistance in the presentation of this paper because the author is not sufficiently acquainted with scientific writing of physics papers.

Thanks to Mr. AryehAldema for his recursive chromo-dynamics idea.

Thanks to Mr. Yossi Avni who approved my research in 2003 into minimum cost diffeomorphism as part of a pattern matching algorithm. This early research helped in acquiring special knowledge in geometry that was later used in the theory presented here.

Also, in the Book of Principals by the philosopher Rabbi Joseph Albo, essay 18 appears to be the first known historical account of what Measurable Time – In Hebrew "ZmanMeshoar" and Immeasurable Time "ZmanBiltiMeshoar" are. His Idea of the immeasurable time as a limit [22], is the very reason for 11 years of research and for this paper.

AppendixA– The time field in the Schwarzschild solution

Motivation: To make the reader familiar with the idea of maximal proper time from a sub-manifold and to calculate the background scalar time field of the Schwarzschild solution from that sub-manifold. We choose as a sub-manifold, a small 3 dimensional 3-spehere around the "Big Bang" singularity and therefore this example is limited to a "Big Bang" manifold. So, we want to connect each event in a Schwarzschild solution to a primordial sub-manifold a fraction of second after the presumed "Big Bang", with the longest possible curve under the assumption that no closed time-like curves occur.

In this limited case, the scalar field is uninteresting as it does not represent interactions with any charged particle or with other force fields and therefore, the Reeb vector is zero.

We would like to calculate $\frac{U_s U^s}{4} = \frac{1}{4}\left(\frac{(P_\lambda P^\lambda),_m (P_\lambda P^\lambda),_k g^{mk}}{(P_i P^i)^2} - \frac{((P_\lambda P^\lambda),_m P^m)^2}{(P_i P^i)^3}\right)$ in Schwarzschild coordinates for a freely falling particle. This theory predicts that where there is no matter, the result must be zero. The speed U of a falling particle from very far away, as measured by an observer in the gravitational field is

$$V^2 = \frac{U^2}{c^2} = \frac{R}{r} = \frac{2GM}{rc^2} \tag{A.1}$$

Where R is the Schwarzschild radius. If speed V is normalized in relation to the speed of light then $V = \frac{U}{c}$. For a far observer, the deltas are denoted by dt', dr' and,

$$\dot{r}^2 = \left(\frac{dr}{dt}\right)^2 = V^2\left(1 - \frac{R}{r}\right) \tag{A.2}$$

because $dr = dr' / \sqrt{1 - R/r}$ and $dt = dt'\sqrt{1 - R/r}$.

IOP Conf. Series: Journal of Physics: Conf. Series **845** (2017) 012019 doi:10.1088/1742-6596/845/1/012019

$$P=\int_0^t\left((1-\frac{R}{r})-\frac{\dot{r}^2}{(1-\frac{R}{r})}\right)^{\frac{1}{2}}dt=\int_0^t\left((1-\frac{R}{r})-\frac{\frac{R}{r}(1-\frac{R}{r})^2}{(1-\frac{R}{r})}\right)^{\frac{1}{2}}dt=$$

$$\int_0^t\left((1-\frac{R}{r})^2\right)^{\frac{1}{2}}dt=\int_0^t(1-\frac{R}{r})dt$$

which results in,

$$P_t=\frac{dP}{dt}=(1-\frac{R}{r}) \tag{A.3}$$

Here t is not a tensor index and it denotes derivative by t!
On the other hand

$$P=\int_0^r\left((1-\frac{R}{r})\frac{1}{\dot{r}^2}-\frac{1}{(1-\frac{R}{r})}\right)^{\frac{1}{2}}dr=\int_0^r\left(\frac{(1-\frac{R}{r})\frac{r}{R}}{(1-\frac{R}{r})^2}-\frac{1}{(1-\frac{R}{r})}\right)^{\frac{1}{2}}dr=\int_0^r\left(\frac{\frac{r-R}{R}}{\frac{r-R}{r}}\right)^{\frac{1}{2}}dr=$$

$$\int_0^r\sqrt{\frac{r}{R}}dr$$

Which results in

$$P_r=\frac{dP}{dr}=\sqrt{\frac{r}{R}} \tag{A.4}$$

Here, r is not a tensor index and it denotes derivative by r !
For the square norms of gradients, we use the inverse of the metric tensor,

So, we have $(1-\frac{R}{r})\rightarrow(1-\frac{R}{r})^{-1}$ and $(1-\frac{R}{r})^{-1}\rightarrow(1-\frac{R}{r})$

So, we can write

$$N^2=P_\lambda P^\lambda=(1-\frac{R}{r})P_r^{\ 2}-(1-\frac{R}{r})^{-1}P_t^{\ 2}=(1-\frac{R}{r})(\frac{r}{R}-1)=\frac{r}{R}+\frac{R}{r}-2$$

$$N^2=\frac{r}{R}+\frac{R}{r}-2 \tag{A.5}$$

$N^2{}_\lambda=\frac{dN^2}{dx^\lambda}$ And we can calculate

$$\frac{N^2{}_\lambda N^{2\lambda}}{(N^2)^2}=\frac{(1-\frac{R}{r})^2(\frac{1}{R}-\frac{R}{r^2})^2}{(\frac{r}{R}+\frac{R}{r}-2)^2} \tag{A.6}$$

We continue to calculate

$$N^2{}_t P_t=(1-\frac{R}{r})^2(\frac{1}{R}-\frac{R}{r^2})\sqrt{\frac{R}{r}} \text{ and } \frac{N^2{}_t P_t}{(1-\frac{R}{r})}=(1-\frac{R}{r})(\frac{1}{R}-\frac{R}{r^2})\sqrt{\frac{R}{r}} \tag{A.7}$$

Note that here t is not a tensor index and it denotes derivative by t !

$$(1-\frac{R}{r})N^2{}_r P_r=(1-\frac{R}{r})(\frac{1}{R}-\frac{R}{r^2})\sqrt{\frac{r}{R}} \tag{A.8}$$

Please note, here r is not a tensor index and it denotes derivative by r !

$$N^2{}_\lambda P^\lambda=(1-\frac{R}{r})(\frac{1}{R}-\frac{R}{r^2})(\sqrt{\frac{r}{R}}-\sqrt{\frac{R}{r}}) \text{ and }$$

$$(N^2{}_\lambda P^\lambda)^2=(1-\frac{R}{r})^2(\frac{1}{R}-\frac{R}{r^2})^2(\frac{r}{R}+\frac{R}{r}-2) \tag{A.9}$$

So

$$\frac{(N^2{}_\lambda P^\lambda)^2}{(N^2)^3}=\frac{(1-\frac{R}{r})^2(\frac{1}{R}-\frac{R}{r^2})^2}{(\frac{r}{R}+\frac{R}{r}-2)^2} \tag{A.10}$$

And finally, from (A.6) and (A.10) we have,

$$\frac{(P^\lambda P_\lambda)_{,m}(P^s P_s)_{,k}\, g^{mk}}{(P_i P^i)^2}-\frac{(P^\lambda P_\lambda)_{,m}P^m}{(P_i P^i)^3}=$$

$$\frac{N^2{}_\lambda N^{2\lambda}}{(N^2)^2}-\frac{N^2{}_\lambda P^\lambda}{(N^2)^3}=\frac{(1-\frac{R}{r})^2(\frac{1}{R}-\frac{R}{r^2})^2}{(\frac{r}{R}+\frac{R}{r}-2)^2}-\frac{(1-\frac{R}{r})^2(\frac{1}{R}-\frac{R}{r^2})^2}{(\frac{r}{R}+\frac{R}{r}-2)^2}=0 \tag{A.11}$$

which shows that indeed the gradient of time measured, by a falling particle until it hits an event in the gravitational field, has zero curvature as expected.

Appendix B – Planck Area Gravity – Based on a lecture by professor Seth Lloyd of the M.I.T combined with the Geometric Chronon model and its correlation with sub-atomic particles

Suppose we have an atomic length L , The speed of light is c so the maximal acceleration will be $\frac{c}{\frac{L}{c}}=\frac{c^2}{L}$. By the real case (33), $\frac{1}{4}(U_\mu U_\nu-\frac{1}{2}U_k U^k g_{\mu\nu}-2U^k{}_{;k}\frac{P_\mu P_\nu}{Z})=R_{\mu\nu}-\frac{1}{2}Rg_{\mu\nu}$

Then

$$\frac{1}{4}(U_\mu U_\nu - \frac{1}{2}U_k U^k g_{\mu\nu} - 2U^k{}_{;k}\frac{P_\mu P_\nu}{Z})\frac{P^\mu P^\nu}{Z} = (R_{\mu\nu} - \frac{1}{2}Rg_{\mu\nu})\frac{P^\mu P^\nu}{Z}$$

which becomes

$$-\frac{1}{8}U_k U^k g_{\mu\nu} - \frac{1}{2}U^k{}_{;k} = (R_{\mu\nu} - \frac{1}{2}Rg_{\mu\nu})\frac{P^\mu P^\nu}{Z} \quad \text{(B.1)}$$

If the right-hand side if multiplied by $\frac{1}{2}$ and then by $\frac{\pi}{12}L^4$ then it yields the missing or added area to the sphere perpendicular to the unit vector $\frac{P^\mu}{\sqrt{Z}}$ [23], [24]. The term $\frac{1}{2}$ is required because $\frac{1}{2}(R_{\mu\nu} - \frac{1}{2}Rg_{\mu\nu})\frac{P^\mu P^\nu}{Z}$ is the sum of sectional curvatures of the infinitesimal 3-volume ball which is perpendicular to the vector P^μ. When this sum of sectional curvatures is multiplied by $\frac{\pi}{12}L^4$, it yields the area that is subtracted or added due to gravity or anti-gravity.

$$\frac{\pi}{24}(R_{\mu\nu} - \frac{1}{2}Rg_{\mu\nu})\frac{P^\mu P^\nu}{Z}L^4 = -Area \quad \text{(B.2)}$$

Then by (B.1)

$$\frac{\pi}{24}(-\frac{1}{8}U_k U^k g_{\mu\nu} - \frac{1}{2}U^k{}_{;k})L^4 = -Area \quad \text{(B.3)}$$

We replace $-\frac{1}{2}U_k$ by 4-acceleration divided by c^2 and we have $-\frac{1}{2}U_k = \frac{c^2}{Lc^2} = \frac{a_k}{c^2} = \frac{1}{L}$ so (B.3) becomes

$$\frac{\pi}{24}(\frac{1}{2}\frac{1}{L^2} \pm \frac{1}{c^2}\frac{c^2}{L}\frac{1}{L})L^4 = \frac{\pi}{24}L^2(\frac{1}{2} \pm 1) = -Area \quad \text{(B.4)}$$

Because $-\frac{1}{8}U_k U^k g_{\mu\nu} = \frac{1}{2}\frac{1}{L}\frac{1}{L} = \frac{1}{2}\frac{1}{L^2}$ and also $-\frac{1}{2}U^k{}_{;k} = \pm\frac{1}{c^2}\frac{c^2}{L^2}$

Which is either an addition to the area or subtraction from the area due to the divergence term $\frac{1}{4}(-2U^k{}_{;k}) = -\frac{1}{2}U^k{}_{;k}$ so, $-\frac{\pi}{24}L^2\frac{3}{2} = Area$ or $+\frac{\pi}{24}L^2\frac{1}{2} = Area$

Divide these areas by the area of the two-dimensional sphere $4\pi L^2$ and we have ratios,

$$\frac{-\frac{\pi}{24}L^2\frac{3}{2}}{4\pi L^2} = \frac{-\frac{1}{24}\frac{3}{2}}{4} = -\frac{1}{64} \quad \text{or} \quad \frac{+\frac{1}{24}\frac{1}{2}}{4} = +\frac{1}{192} \quad \text{(B.5)}$$

But we didn't take into account that the geodesic motion prohibition field i.e. acceleration field changes its density on the sphere in accordance with increased or decreased area ratio α.
(B.4) can be rewritten as a more enlightening term,

$$1 + \frac{\frac{\pi}{24}(-\frac{1}{2}\frac{1}{\alpha^2 L^2} \pm \frac{1}{c^2}\frac{c^2}{L\alpha}\frac{1}{L})L^4}{4\pi L^2} = 1 + \frac{1}{96}(-\frac{1}{2}\alpha^{-2} \pm \alpha^{-1}) = \frac{4\pi L^2 + Area}{4\pi L^2} = \alpha \quad \text{(B.6)}$$

Such that α is either bigger than 1 or smaller than 1 and denotes the increase or decrease in area. Note that the term α measures how much the square acceleration field changes as the area grows or dwindles.

The resulting equation is a cubic equation: $1+\frac{1}{96}(-\frac{1}{2}\alpha^{-2}\pm\alpha^{-1})=\alpha$ that can be easily solved numerically.

$$1+\frac{1}{96}(-\frac{1}{2}\alpha^{-2}\pm\alpha^{-1})=\alpha\Rightarrow$$
$$192\alpha^3=192\alpha^2\pm2\alpha^{-1}-1\Rightarrow\alpha^3=\frac{192\alpha^2\pm2\alpha^{-1}-1}{192}\Rightarrow\alpha=(\frac{192\alpha^2\pm2\alpha^{-1}-1}{192})^{\frac{1}{3}} \quad (B.7)$$

The area is increased or decreased by α and the portion of the area that changes is

$$\alpha \sim= 1.00520819\ \ 3610747100 \ \Rightarrow (\frac{1}{1-\alpha})^{-1} \sim= +(192.005150\ \ 87160028..\ \ .)^{-1}$$

around a negative charge or

$$\alpha \sim= 0.99207267\ \ 636432284 \ \Rightarrow (\frac{1}{\alpha-1})^{-1} \sim= -(62.6395393\ \ 39674555..\ \ .)^{-1}$$

around a positive charge. The problem is that there is no stable charged particle without spin and therefore our discussion could mean a temporary decomposition of electrically neutral Bosons into two energy states, one temporarily behaving like a negative charge and one like a positive one. The reasoning behind such a claim is that if matter is expressible by a weak acceleration field and the weak acceleration field energy is the energy of an electric field, then elementary neutral particles, even with zero magnetic momentum and with zero electric dipole, should have an internal electric field.
The question is how to infer such a structure. The idea is that area changes are relative to energy ratios even if they are changes due to charge electro-gravity and not due to inertial mass. It is a manifestation of a holographic principle [23], [24]. Our modest test will be to divide the Higgs energy by 2 and then either by 192.005150..,or by 62.6395393.... That is by Beta = **384.010301743200560** or by Alpha =**125.279078679349110**.

For example: 125 GeV / 125.279078679349110 ~= **0.9977 GeV** which should be a Baryonic energy state. Another energy is 125 GeV / 384.01030174320056 ~= **325.5 MeV**
This energy is the model dependent vacuum constituent Quark energy according to Zhao Zhanget. al. [25].

According to this paper, no neutral particle can avoid having an internal structure, otherwise, the particle would not be able to manifest an acceleration field as energy. This leads to the possible model of BS Meson, Z Boson and Higgs Boson as either oscillating + and – charge such that both the magnetic and electric dipoles are zero, or as spinning + and - charge such that both magnetic and electric dipoles vanish. The problem is the Z and the Higgs bosons which are considered elementary particles. The Z boson mass is $91.1876 \pm 0.0021\ GeV/C^2$. If we split this mass into two charges, then 1/ 192.00515087160028 of area around the negative charge will be added, which is considered as proportional to mass [23]. But that portion is of half of the mass that splits to two charges, so we seek1/384.00258393161619 of the mass of the Z boson as having a physical meaning.

$$91.1876\ GeV\ /(384.0103) = 237.4613\ \text{MeV} \quad (B.8)$$

which is the energy difference between the Phi (previously Eta) and Omega Mesons !

By the Checkered Board Model and EMS [26] the mass of the up Quark deviates from the Standard Model's ~ 2.3 MeV/C ^{2} and is 237.31 MeV/C ^{2} according to that very same model, the down Quark is 42.39Mev/C ^{2} unlike the S.M. ~ 4.8 MeV/C ^{2}. The Z boson can contribute to mass fluctuations through half of its mass by area fluctuations around a positive charge too but that yields 727.87572MeV and there is no known 727.876 MeV resonance in the particles world.

Here is a summary of the electro-gravity energy in the Planck scale around a positive and a negative charge that split an elementary boson and by this, these energies are beyond the Standard Model.

Table 1. Presumed Beta and Alpha energy ratios due to a splitting of an Elementary charge-less boson into positive and negative charge – a supposed process which is beyond the Standard Model. A second assumption is that the magnetic moments and the electric dipoles of such bosons are zero, therefore split charge should be fluctuating and so is the area around positive and negative charge.

	Energy delta (GeV)	**Energy sum (GeV)**	Area expansion ratio around e- r2=1.005208194 Beta=2/(r2-1) **1/384.0103087 of the energy**	Area contraction ration around e+ r1=0.984035643 Alpha=2/(1-r1) **1/125.27908 of the energy**
Higgs Boson energy portion	125 GeV *(1/Alpha-1/Beta)	125 GeV *(1/Alpha+1/Beta)	125 GeV/Beta	125 GeV/Alpha
Higgs Boson electro-gravitationa l energy	0.67226024 Delta of the last two cells in this row	1.323284428 About **1.323GeV** Sum of the next two cells	Gravitational energy delta around e- 0.325512095 GeV About **325.5 MeV**	Gravitational energy delta around e+ 0.997772333 GeV About **1.0 GeV**
Z Boson Energy portion	91.1876GeV *(1/Alpha-1/Beta)	91.1876GeV *(1/Alpha+1/Beta)	91.1876 GeV/Beta	91.1876 GeV/Alpha
Z Boson electro-gravitationa-l energies	0.49041438 Delta of the last two cells in this row	0.965337049 About **0.965GeV** Sum of the next two cells	Gravitational energy delta around e- 0.237461334 GeV About **237.46 MeV**	Gravitational energy delta around e+ 0.727875715 GeV About **728 MeV**

References

[1] G. Reeb. Sur certaines propri´t´s topologiques des vari´et´es feuillet´ees. Actualit´e Sci. Indust. 1183, Hermann, Paris (1952).

[2] B. Reinhart and J. Wood. "A Metric Formula for the Godbillon–Vey Invariant for Foliations". Proceedings of the American Mathematical Society 38: 427-430, 1973.

[3] S. Hurder and R. Langevin. Dynamics and the Godbillon-Vey Class of C^1 Foliations. preprint, September 2000.
[4] D. Lovelock and H. Rund, *Tensors, Differential Forms and Variational Principles*, Dover Publications Inc. Mineola, N.Y., 1989, ISBN 0-486-65840-6, p. 121-126, 4.4 The Lie Derivative.
[5] Y. Friedman and T. Scarr, "Making the relativistic dynamics equation covariant: explicit solutions for motion under a constant force", Volume 86 Number 6, 2012 PhysicaScripta 065008 doi:10.1088/0031-8949/86/06/065008.
[6] Elie Cartan, The Theory of Spinors, Dover Publications Inc. Mineola, N.Y., ISBN 0-486-64070-1, p. 145,149 (14), 1981.
[7] Walter Pfeifer, "The Lie Algebra su(N), an Introduction" , Chapter 5, The Lie Algebra SU(4), pp 87-106, 2003 (revised 2008), ISBN 3-7643-2418-X.
[8] Ana Cannas da Silva , "Lectures on Symplectic Geometry", Page 7, Darboux Theorem, Revision 2006, Original ISBN 978-3-540-45330-7, Springer-Verlag.
[9] Hector Luis Serrano, (President of Gravitec Inc.) capacitor at 1.72 x 10^-6 Torr, Experiment by NASA and Gravitec, Video from July 3, 2003.
[10] Hector Luis Serrano, Patent WO 00/58623, Page 12, Line 28, "such a vehicle can operate in any dielectric environment such as air or vaccum of space", granted on November 2001.
[11] L Acebo. The Flyby anomaly: A multivariate analysis approach, 20/Jan/2017, arXiv: 1701.05735
[12] T. Datta, M. Yin, A. Dimofte, M. C. Bleiweissand Z. Cai, "Experimental Indications of Electro-Gravity", 8 Sep 2005 arXiv:physics/0509068[physics.gen-ph]
[13] Y. Doi, T. Makino, H. Kato et al. D. Takeuchi, "Deterministic Electrical Charge-State Initialization of Single Nitrogen-Vacancy Center in Diamond", Published 31 March 2014, Physical Review X 4, 011057, DOI: 10.1103/PhysRevX.4.011057, Publisher: Aerican Physical Society, March/2014.
[14] S. Vaknin, "Time Asymmetry Re-visited", LC Classification: Microfilm 85/871 (Q). [microform], Library of Congress.LC Control Number: 85133690, Thesis (Ph. D.)--Pacific Western University, 1982, c1984, Ann Arbor, MI : University Microfilms International, 1984.
[15] Rafael D. Storkin, "Indications of causal set cosmology", 10/Mar/2000, arXiv: gr-qc/0003043.
[16] D. Lovelock and H. Rund, *Tensors, Differential Forms and Variational Principles*, Dover Publications Inc. Mineola, N.Y., ISBN 0-486-65840-6, p. 261, 3.19 and 3.26, for Ricci tensor and Ricci scalar.
[17] D. Lovelock and H. Rund, "The Numerical Relative Tensors", *Tensors, Differential Forms and Variational Principles*, 4.2, Dover Publications Inc. Mineola, N.Y. , ISBN 0-486-65840-6, p. 113, 2.18, p. 114, 2.30.
[18] D. Lovelock and H. Rund, *Tensors, Differential Forms and Variational Principles*, Dover Publications Inc. Mineola, N.Y., 1989 ,ISBN 0-486-65840-6, p. 262, 3.27 is the Einstein tensor.
[19] M. Alcubierre, "The warp drive: hyper-fast travel within general relativity", Class.Quant.Grav.11:L73-L77, 1994, DOI: 10.1088/0264-9381/11/5/001.
[20] Thomas B. Bahder and Chris Fazi, "Force on an Asymmetric Capacitor", Army Research Laboratory ARL, March 2003, site: http://arxiv.org/ftp/physics/papers/0211/0211001.pdf
[21] Harold White et. al. "Measurement of impulsive thrust from a closed radio frequency cavity in vacuum", 17/Nov/2016, Aerospace Research Central, http://dx.doi.org/10.2514/1.B36120.
[22] J. Albo, Book of Principles (Sefer Ha-ikarim), "Immeasurable time - Maamar*18",* "measurable time by movement". (Circa 1380-1444, unknown), The Jewish publication Society of America (1946), ASIN: B001EBBSIC, Chapter 2, Chapter 13.
[23] Seth Lloyd - "Deriving general relativity from quantum measurement", Institute For Quantum Computing - IQC, lecture loaded to YouTube on 16/August/2013,

http://www.youtube.com/watch?v=t9zcBKoFrME

[24] Ted Jacobson, Thermodynamics of Spacetime: The Einstein Equation of State Physical Review Letters. 75, 1260 - Published 14 August 1995, DOI:https://doi.org/10.1103/PhysRevLett.75.1260

[25] Zhao Zhang et. Al., "Number of the QCD critical points with neutral color superconductivity" 2008,arXiv:0808.3371v3 (Remark: See also Phys.Rev.D79:014004,2009, 10.1103/PhysRevD.79.014004).

[26] Theodore M. Lach, Masses of Sub-Nuclear Particles arXiv:nucl-th/0008026v1, August 2000.

Quantum dynamics of bound states under spacetime fluctuations

Teodora Oniga and Charles H-T Wang

Department of Physics, University of Aberdeen, King's College, Aberdeen AB24 3UE, United Kingdom

E-mail: t.oniga@abdn.ac.uk, c.wang@abdn.ac.uk

Abstract. With recent developments in high-precision quantum measurements, the question of whether observations of decoherence from spacetime fluctuations are accessible experimentally arises. Here we investigate the dynamics of bound states interacting with an environment of gravitons under the Markov approximation. The corresponding Lindblad master equation is presented that enables gravitational decoherence and dissipation due to zero-point spacetime fluctuations to be analyzed. Specifically, we consider a one-dimensional cavity of massless scalar particles that models a light beam with negligible spin polarizations being reflected between two free masses. Numerical simulations have been performed to illustrate the wave-modal dependent decoherence and dissipation of such a configuration. We further demonstrate the existence of nontrivial collective effects akin to superradiance, providing amplifications of gravitational decoherence for a large number of identical bosonic particles.

1. Introduction

Over the past few decades, quantum decoherence has become an ever more important field of physics. Given the universal nature of the matter-gravity interaction, spacetime fluctuations constitute an inescapable environment which may cause decoherence to occur in physical systems. Such effects may have an impact of high-precision measurements [1, 2, 3, 4, 5] and astronomical observations [6] and, in addition, may also play a role in the classical appearance of macroscopic objects [7, 8]. Thus it is important to understand the implications of environmentally induced gravitational decoherence. In the existing literature, this has been addressed mostly through phenomenological models, with different causes of decoherence, such as semiclassical metric fluctuations [9, 10, 11], quantized Newtonian gravity [12], a thermal spacetime foam [13], weak gravity [14, 15] etc. These approaches have typically been limited by their specific requirements, including the Markov approximation, the use of single non-relativistic particles and assuming classical stochastic spacetime fluctuations.

Recently, a new theory of decoherence due to the entanglement of bosonic matter with spacetime fluctuations was developed from first principles [16, 17]. This approach is based on Dirac's constraint quantization of linearized gravity coupled with matter that allows to derive a gauge invariant influence functional technique used to obtain a general gravitational master equation for the reduced density matrix of the matter system by averaging over the environmental degrees of freedom. It was found that when this master equation is applied to a broad class of free bosons including scalar particles, for long times and with the Markov approximation, the nonunitary part of the master equation vanishes, suppressing completely decoherence. However, while the decoherence terms in the master equation disappear for free fields, for bound fields this may not be the case. Furthermore, it has been suggested that when a large number of identical particles is considered, nontrivial collective decoherence phenomena occur. Motivated by the above developments, in this paper we will apply the formalism in [16, 17] to scalar particles confined in a finite box, with discrete modes and zero boundary conditions, where we are indeed able to observe decoherence. The scalar particles provide a useful initial description of the behaviours of other bosonic particles in a similar configuration, which is physically analogous to photons inside a reflective cavity when spin polarizations are negligible.

In the following we will use the relativistic units with the speed of light $c = 1$ and rescale the gravitational constant $G \to Gc^4$ and the reduced Planck constant $\hbar \to \hbar c$ as well as other physical quantities accordingly. Time derivatives will be denoted by an overdot $\dot{}$ and Hermitian conjugates by an superscript dagger †. Spatial indices running from 1 to 3 are denoted by Latin letters and spatial and temporal indices running from 0 to 3 are denoted by Greek alphabets. Coordinates are given by $x^\mu = (x^0, x^1, x^2, x^3) = (x^0, \boldsymbol{x}) = (x)$ where the zeroth coordinate is time. When indices are repeated, a summation is implied. The background metric is given by the Minkowski metric $\eta_{\mu\nu} = \mathrm{diag}(-1, 1, 1, 1)$.

2. The gravitational master equation

Let us consider an ensemble of environmental gravitons in a Gaussian state weakly interacting with a system of bosonic particles. We will further assume that the gravitational field follows a Planck distribution $N(\omega) = 1/(e^{\hbar\omega/k_B T} - 1)$ at temperature T, with graviton frequency ω and the Boltzmann constant k_B, although in principle it is not required that the gravitational field state be thermal and as such, the same formalism can be used to investigate decoherence for a Gaussian state environment with a general distribution function $N(\omega)$. The interaction picture evolution of the reduced density matrix of the bosonic system is then given by the exact master equation [16]:

$$\begin{aligned} \frac{d}{dt}\rho(t) &= -\frac{i}{\hbar}[H_{\mathrm{self}}(x), \rho(t)] - \frac{8\pi G}{\hbar}\int \frac{\mathrm{d}^3k}{2(2\pi)^3 k} \\ &\quad \Big\{\big[\tau^\dagger_{ij}(\boldsymbol{k}, t),\, \tilde{\tau}_{ij}(\boldsymbol{k}, t)\rho(t)\big] + N(\omega_k)\big[\tau^\dagger_{ij}(\boldsymbol{k}, t),\, \big[\tilde{\tau}_{ij}(\boldsymbol{k}, t),\, \rho(t)\big]\big] + \mathrm{h.c.}\Big\}. \end{aligned} \tag{1}$$

The gravitational self-interaction of the matter field is described by the Hamiltonian

$$H_{\mathrm{self}} = -2G \int \mathrm{d}^3x\, \mathrm{d}^3x' \frac{T^{\mu\nu}(\vec{x}, x^0)\bar{T}_{\mu\nu}(\vec{x}', x'^0)}{|\vec{x} - \vec{x}'|} \tag{2}$$

where $T^{\mu\nu}$ is the stress energy tensor of the system of interest and $\bar{T}_{\mu\nu}$ is the trace-reversed stress energy tensor. Here we have denoted by $\tau_{ij}(\boldsymbol{k}, t)$ the normal ordered and particle number preserving stress energy tensor of the system of interest, in momentum space and in the transverse-traceless (TT) gauge. The TT projection operator

$$P_{ijkl} = \frac{1}{2}P_{ik}P_{jl} + \frac{1}{2}P_{il}P_{jk} - \frac{1}{2}P_{ij}P_{kl} \tag{3}$$

where we have the transverse projection operator $P_{ij} = \delta_{ij} - \frac{\partial_i \partial_j}{\partial_k \partial_k}$ gives the TT stress tensor $\tau_{ij}(\boldsymbol{k}, t) = P_{ijkl}(\boldsymbol{k})T^{kl}(\boldsymbol{k}, t)$. The time integration from the initial time taken to be 0 to a final time t is denoted by:

$$\tilde{\tau}_{ij}(\boldsymbol{k}, t) = \int_0^t dt' e^{-i\omega_k(t-t')}\tau_{ij}(\boldsymbol{k}, t'). \tag{4}$$

As the gravitational self-interaction Hamiltonian part of the master equation corresponds to a unitary time evolution, it does not cause decoherence. Therefore, from now on we will only consider the dissipative part of the master equation.

3. Discrete mode scalar field master equation

A massive scalar field with dispersion relation $\omega_k^2 = k^2 + \mu^2$ and nonnegative ω_k has the following expression for its field operator in position space:

$$\phi(\boldsymbol{x},t) = \int \mathrm{d}^3k \sqrt{\frac{\hbar}{2(2\pi)^3\omega_k}} \left[a_{\boldsymbol{k}} e^{ikx} + a_{\boldsymbol{k}}^\dagger e^{-ikx}\right]. \tag{5}$$

The spatial components of the stress energy tensor

$$T_{ij} = \phi_{,i}\phi_{,j} + \frac{1}{2}\delta_{ij}(\phi_{,0}\phi_{,0} - \phi_{,k}\phi_{,k} - \mu^2\phi^2) \tag{6}$$

lead to the transverse-traceless stress energy tensor and its Hermitian conjugate

$$\begin{aligned}\tau_{ij}(\boldsymbol{k},t) &= P_{ijkl}(\boldsymbol{k}) \int \mathrm{d}^3k' \frac{\hbar}{2\sqrt{\omega_{k'}}} k'_k k'_l \\ &\left[\frac{1}{\sqrt{\omega_{\boldsymbol{k}'-\boldsymbol{k}}}} a^\dagger_{\boldsymbol{k}'-\boldsymbol{k}} a_{\boldsymbol{k}'} e^{-i(\omega_{\boldsymbol{k}'}-\omega_{\boldsymbol{k}'-\boldsymbol{k}})t} + \frac{1}{\sqrt{\omega_{\boldsymbol{k}'+\boldsymbol{k}}}} a^\dagger_{\boldsymbol{k}'} a_{\boldsymbol{k}'+\boldsymbol{k}} e^{-i(\omega_{\boldsymbol{k}'+\boldsymbol{k}}-\omega_{\boldsymbol{k}'})t}\right] \end{aligned} \tag{7}$$

$$\begin{aligned}\tau^\dagger_{ij}(\boldsymbol{k},t) &= P_{ijkl}(\boldsymbol{k}) \int \mathrm{d}^3k' \frac{\hbar}{2\sqrt{\omega_{k'}}} k'_k k'_l \\ &\left[\frac{1}{\sqrt{\omega_{\boldsymbol{k}'-\boldsymbol{k}}}} a^\dagger_{\boldsymbol{k}'} a_{\boldsymbol{k}'-\boldsymbol{k}} e^{i(\omega_{\boldsymbol{k}'}-\omega_{\boldsymbol{k}'-\boldsymbol{k}})t} + \frac{1}{\sqrt{\omega_{\boldsymbol{k}'+\boldsymbol{k}}}} a^\dagger_{\boldsymbol{k}'+\boldsymbol{k}} a_{\boldsymbol{k}'} e^{i(\omega_{\boldsymbol{k}'+\boldsymbol{k}}-\omega_{\boldsymbol{k}'})t}\right]. \end{aligned} \tag{8}$$

The operators $a_{\boldsymbol{k}}$ and $a^\dagger_{\boldsymbol{k}}$ are time-independent and satisfy the following commutation relations:

$$\left[a_{\boldsymbol{k}}, a^\dagger_{\boldsymbol{k}'}\right] = \delta(\boldsymbol{k}-\boldsymbol{k}') \tag{9}$$

$$\left[a_{\boldsymbol{k}}, a_{\boldsymbol{k}'}\right] = \left[a^\dagger_{\boldsymbol{k}}, a^\dagger_{\boldsymbol{k}'}\right] = 0. \tag{10}$$

In a box of volume V with zero boundary conditions, such that particles are reflected at the boundary, with $\tau_{ij}(\boldsymbol{x},t) = 0$ if $\boldsymbol{x}$ is outside the box, in terms of discrete momenta $\mathbf{k}$:

$$\tau_{ij}(\mathbf{k},t) = \int_V \mathrm{d}^3x\, \tau_{ij}(\boldsymbol{x},t) e^{-i\mathbf{k}\cdot\boldsymbol{x}}. \tag{11}$$

We can express the stress energy tensor in terms of continuous momenta $\boldsymbol{k}$ as

$$\begin{aligned}\tau_{ij}(\boldsymbol{k},t) &= \int_V \mathrm{d}^3x\, \tau_{ij}(\boldsymbol{x},t) e^{-i\boldsymbol{k}\cdot\boldsymbol{x}} \\ &= \sum_{\mathbf{k}} \Delta(\boldsymbol{k}-\mathbf{k})\tau_{ij}(\mathbf{k},t) \end{aligned} \tag{12}$$

where

$$\Delta(\boldsymbol{k}) = \frac{1}{V}\int_V \mathrm{d}^3x\, e^{-i\boldsymbol{k}\boldsymbol{x}} \tag{13}$$

with the limit $\lim_{V\to\infty} V\Delta(\boldsymbol{k}) = (2\pi)^3\delta(\boldsymbol{k})$.

IOP Conf. Series: Journal of Physics: Conf. Series **845** (2017) 012020 doi:10.1088/1742-6596/845/1/012020

4. Decoherence of an idealized light beam cavity

As a one-dimensional reduction of the above discussion, it is interesting to consider next in some detail a situation analogous to a light beam bouncing between two end mirrors. Physically, these mirrors may be considered to be attached to classical free masses that average out metric fluctuations and so do not move in the TT coordinates currently adopted. The spin polarizations of the light are not taken into account by modelling the light as a massless scalar field. We then have the following field operator inside the light beam with $0 < x < L$, $0 < y < \Delta y$ and $0 < z < \Delta z$:

$$\phi(x,t) = \sum_n \sqrt{\frac{\hbar}{L\omega_n}} \left(a_n e^{-i\omega_n t} + a_n^\dagger e^{i\omega_n t}\right) \sin\frac{\pi n x}{L} \tag{14}$$

where $V = L^3$ and $n_1, n_2, n_3 = 1, 2, \cdots$ such that

$$\mathbf{k} = \left(\frac{\pi n_1}{L}, \frac{\pi n_2}{L}, \frac{\pi n_3}{L}\right). \tag{15}$$

This leads to the following creation and annihilation operators a_n and $a_{n'}^\dagger$:

$$[a_n, a_{n'}^\dagger] = \delta_{n,n'}, \quad [a_n, a_{n'}] = 0 = [a_n^\dagger, a_{n'}^\dagger]. \tag{16}$$

Inside the light beam, the stress energy tensor takes the following effective 3-dimensional form:

$$T_{ij}^{\mathrm{TT}}(\boldsymbol{x},t) = \frac{1}{\Delta y \Delta z} P_{ij11} \phi_{,x}(\boldsymbol{x},t)\phi_{,x}(\boldsymbol{x},t).$$

For a small beam width limit with $\Delta y, \Delta z \ll L \sim 1/k$ and by normal ordering and neglecting the terms of $T_{ij}^{\mathrm{TT}}(\boldsymbol{k},t)$ not conserving particle numbers, we obtain $\tau_{ij}(\boldsymbol{k},t)$:

$$\tau_{ij}(\boldsymbol{k},t) = P_{ij11}(\boldsymbol{k}) \sum_{n,n'} F(n,n',\boldsymbol{k})\, a_{n'}^\dagger a_n e^{-i(\omega_n - \omega_{n'})t} \tag{17}$$

$$\tau_{ij}^\dagger(\boldsymbol{k},t) = P_{ij11}(\boldsymbol{k}) \sum_{n,n'} F^*(n,n',\boldsymbol{k})\, a_n^\dagger a_{n'} e^{i(\omega_n - \omega_{n'})t} \tag{18}$$

with

$$F(n,n',\boldsymbol{k}) := \frac{2\pi^2 n\, n' \hbar}{L^3\sqrt{\omega_n \omega_{n'}}} \int_0^L \mathrm{d}x \cos\frac{\pi n x}{L} \cos\frac{\pi n' x}{L}\, e^{-ikx\cos\theta} \tag{19}$$

where the continuous wave vector $\boldsymbol{k}$ of modulus k forms a small angle θ with the x-axis. For the time evolution of this system neglecting memory effects, we can apply the Markov approximation. The time variable in the integration is changed from t' to s using $t' = t - s$, taking the limit $\int_0^t \mathrm{d}s \to \int_0^\infty \mathrm{d}s$. After applying the Sokhotski-Plemelj theorem $\int_0^\infty \mathrm{d}s\, e^{-i\epsilon s} = \pi\delta(\epsilon) - i\,\mathrm{P}\frac{1}{\epsilon}$ where P denotes the Cauchy principal value, we can neglect the terms containing P which do not give rise to decoherence to obtain:

$$\begin{aligned}\tilde{\tau}_{ij}(\boldsymbol{k},t) &= \int_0^\infty \mathrm{d}s\, \tau_{ij}(\boldsymbol{k},t')e^{-iks} \\ &= \pi P_{ij11}(\boldsymbol{k}) \sum_{n \geq n'} F(n,n',\boldsymbol{k})\, a_{n'}^\dagger a_n e^{-i(\omega_n - \omega_{n'})t}\, \delta(k - \omega_n + \omega_{n'}).\end{aligned} \tag{20}$$

IOP Conf. Series: Journal of Physics: Conf. Series **845** (2017) 012020 doi:10.1088/1742-6596/845/1/012020

After applying the customary rotating wave approximation and adopting the following notation:

$$\begin{aligned} \omega(n,\Delta n) &= \omega_{n+\Delta n} - \omega_n \\ \underline{\sum_{n,m,\Delta n,\Delta m}} &= \sum_{\substack{n,m,\Delta n,\Delta m \\ \omega(n,\Delta n)=\omega(m,\Delta m)}} \end{aligned}$$

where $\boldsymbol{k}(n,\Delta n)$ denotes $\boldsymbol{k}$ with $k = \omega(n,\Delta n)$ and $\Delta n, \Delta m \geq 0$, we arrive at the master equation in Lindblad form [18]:

$$\begin{aligned} \frac{\mathrm{d}}{\mathrm{d}t}\rho(t) &= \underline{\sum_{n,n',\Delta n,\Delta n'}} \Gamma(n,n',\Delta n,\Delta n') \\ &\Big[(1+N(\omega(n,\Delta n)))\big(A(n',\Delta n')\rho A^\dagger(n,\Delta n) - \frac{1}{2}\{A^\dagger(n,\Delta n)A(n',\Delta n'),\rho\}\big) \\ &+N(\omega(n,\Delta n)))\big(A^\dagger(n,\Delta n)\rho A(n',\Delta n') - \frac{1}{2}\{A(n',\Delta n')A^\dagger(n,\Delta n)),\rho\}\big)\Big]. \end{aligned} \tag{21}$$

in terms of the operators $A(n,\Delta n) = a_n^\dagger a_{n+\Delta n}$ and with the transition rate coefficient

$$\begin{aligned} \Gamma(n,n',\Delta n,\Delta n') &= \frac{G}{2\pi\hbar}\omega(n,\Delta n) \\ &\int \mathrm{d}\Omega(\boldsymbol{k}(n,\Delta n))P_{11}^2(\boldsymbol{k}(n,\Delta n))F^*(n+\Delta n,n,\boldsymbol{k}(n,\Delta n))F(n'+\Delta n',n',\boldsymbol{k}(n,\Delta n)) \end{aligned} \tag{22}$$

related directly to the rate of decoherence and dissipation.

To investigate the influence of zero-point spacetime fluctuations on a photon reflected by the boundaries of this light beam, let us consider a massless scalar master equation at zero gravitational temperature with $N(\omega) = 0$. For massless particle $\Delta n = \Delta n'$ and the resulting transition rate coefficient $\Gamma(n,n',\Delta n)$ takes the form

$$\Gamma(n,n',\Delta n) = \frac{4\pi^3\hbar G}{L^3}\hat{\Gamma}(n,n',\Delta n) \tag{23}$$

in terms of the dimensionless coefficients

$$\begin{aligned} \hat{\Gamma}(n,n',\Delta n) &= \Delta n\sqrt{n(n+\Delta n)}\,\sqrt{n'(n'+\Delta n)}\int_{-1}^{1}\mathrm{d}\sigma\,(1-\sigma^2)^2 \\ &\int_0^1 \mathrm{d}s\,\cos(\pi n s)\,\cos(\pi(n+\Delta n)s)\,e^{i\pi\sigma\Delta n s}\int_0^1 \mathrm{d}s'\,\cos(\pi n' s')\,\cos(\pi(n'+\Delta n)s')\,e^{-i\pi\sigma\Delta n s'}. \end{aligned} \tag{24}$$

As an illustrative numerical example in units where $c = 1$ and $4\pi^3\hbar G/L^3 = 1$ displayed in the Fig. 1, we take a one-photon system with density matrix $\rho_{n,n'}(t) = \langle n|\rho|n'\rangle(t)$ truncated for modes with $1 \leq (n,n') \leq 20$ and hence $1 \leq \Delta n \leq 19$. As shown, the initial density matrix at $t = 0$ has a chosen 2-dimensional Gaussian profile centred at $(n,n') = (10,10)$. Under the nonunitary evolution, its centre moves towards the ground state with $(n,n') = (1,1)$ due to energy dissipation through spontaneous emissions of gravitons, with a gradually reduced off-diagonal width as a result of decoherence.

Fig. 2 shows a similar evolution where the photon density matrix is initially diagonal, having no superposition between different n-modes, with constant diagonal elements up to the truncation mode $n = 20$. The diagonality is dynamically preserved. It can clearly be seen that the higher decay rates for larger n-modes cause the formation of a peak off the smaller n-modes while this peak settles towards the ground state at a decelerated pace.

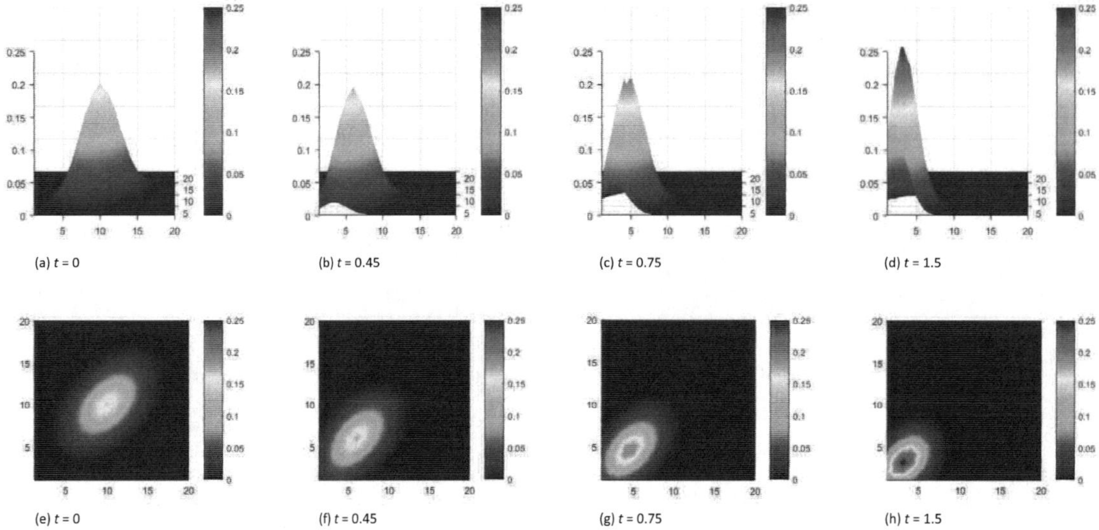

Figure 1. As explained in the main text, under the nonunitary evolution, the centre of the density matrix profile with an initial Gaussian shape moves towards the ground state due to energy dissipation, with a gradually reduced off-diagonal width as a result of decoherence. The evolution of the cavity photon beam density matrix $\rho_{n,n'}(t)$ is represented in side views (a)–(d) and in top views (e)–(h) where the matrix value is the height and matrix indices (n, n') are along the horizontal axes.

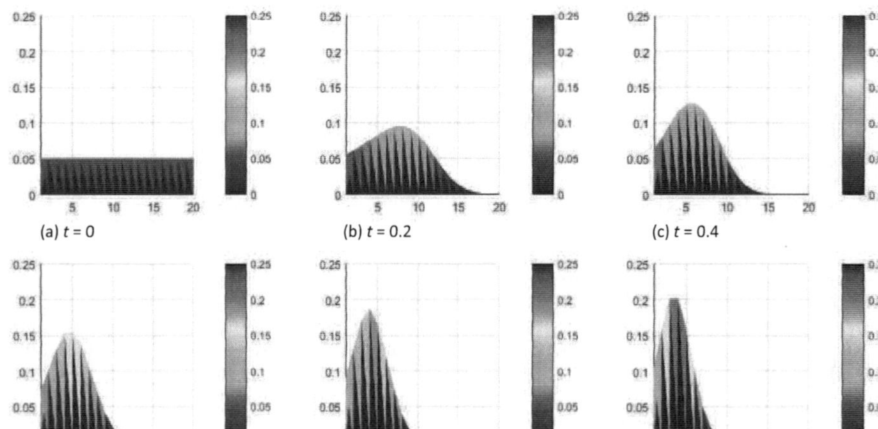

Figure 2. Here plots (a)–(f) demonstrate the evolution of the photon density matrix, which is initially diagonal, with constant diagonal elements $= 0.05$ for $1 \leq n \leq 20$. It can clearly be seen that the higher decay rates for larger n-modes cause the formation of a peak (rather like traffic congestion) off the smaller n-modes while this peak settles towards the ground state in a slowing fashion.

5. Collective decoherence and radiance processes

We would now like to consider the collective quantum dynamics of massless scalar particles at zero gravitational temperature using the master equation (21) with $N(\omega) = 0$ and transition rate coefficient $\Gamma(n, n', \Delta n)$ given by (23). To start with, for one particle states given by $|m\rangle = a_m^\dagger|0\rangle$, the density matrix will decay into lower energy states as $\Gamma(n, n, \Delta n) > 0$. Let us look at a diagonal density matrix that at some point in time takes the form $\rho = \sum_m \rho(m)|m\rangle\langle m|$ normalized so that $\sum_m \rho(m) = 1$. Then we have

$$\dot{\rho} = \sum_m \sum_{n<m} \rho(m)\Gamma(n, n, m-n)\big[|n\rangle\langle n| - |m\rangle\langle m|\big]$$

with the matrix elements given by

$$\begin{aligned}
\langle m|\dot{\rho}|m\rangle &= \sum_{m'} \sum_{n<m'} \rho(m')\Gamma(n, n, m'-n)\langle m|\big[|n\rangle\langle n| - |m'\rangle\langle m'|\big]|m\rangle \\
&= -\sum_{m'} \sum_{n<m'} \rho(m')\Gamma(n, n, m'-n)\delta_{m,m'} + \sum_{m'} \sum_{n<m'} \rho(m')\Gamma(n, n, m'-n)\delta_{m,n} \\
&= -\sum_{n<m} \rho(m)\Gamma(n, n, m-n) + \sum_{n>m} \rho(n)\Gamma(m, m, n-m).
\end{aligned}$$

Note that the density matrix stays diagonal throughout the time evolution. For comparison, let us also consider a many particle state, given by N particles found in the $|m\rangle$, normalized such that $|N_m\rangle = \frac{1}{\sqrt{N_m!}}(a_m^\dagger)^N|0\rangle$, with a multiple particle diagonal density matrix $\rho = |N_m\rangle\langle N_m|$ at a given time. Then, at that time, the master equation becomes:

$$\begin{aligned}
\dot{\rho} &= \frac{N_m^2}{N_m!}\sum_{n<m} \Gamma(n, n, m-n)a_n^\dagger(a_m^\dagger)^{N_m-1}|0\rangle\langle 0|(a_m)^{N_m-1}a_n \\
&\quad - \frac{N_m}{N_m!}\sum_{n<m} \Gamma(n, n, m-n)(a_m^\dagger)^{N_m}|0\rangle\langle 0|(a_m)^{N_m} \\
&\quad - \frac{N_m(N_m-1)}{2N_m!}\sum_{n<m} \Gamma(m, n, m-n)a_{2m-n}^\dagger a_n^\dagger(a_m^\dagger)^{N_m-2}|0\rangle\langle 0|(a_m)^{N_m} \\
&\quad - \frac{N_m(N_m-1)}{2N_m!}\sum_{n<m} \Gamma(n, m, m-n)(a_m^\dagger)^{N_m}|0\rangle\langle 0|(a_m)^{N_m-2}a_n a_{2m-n}.
\end{aligned}$$

This reduces to the above one particle case when $\rho = |m\rangle\langle m|$. However, we can see that for many particle states we have addition factors equal to the number of particles in the state, so that we have a collective, i.e. superradiant-like, decay of elements of the density matrix of the form:

$$\langle N_m|\dot{\rho}|N_m\rangle = -N_m \sum_{n<m} \Gamma(n, n, m-n). \tag{25}$$

More specifically, as with the description of electromagnetic superradiance [19, 20], the above relation shows the rate of spontaneous emission of gravitons by N identical particles in the same state $|m\rangle$ increases by a factor of the particle number N while the radiation intensity is increased by a further factor of N resulting in an N^2 intensity amplification.

To investigate collective decoherence, we now consider a superposition state

$$|\psi\rangle = \frac{1}{\sqrt{2}}(|N_m\rangle + |N_{m'}\rangle)$$

with $m \neq m'$ and

$$\begin{aligned} \rho &= |\psi\rangle\langle\psi| \\ &= \frac{1}{2}\big(|N_m\rangle\langle N_m| + |N_m\rangle\langle N_{m'}| + |N_{m'}\rangle\langle N_m| + |N_{m'}\rangle\langle N_{m'}|\big). \end{aligned} \tag{26}$$

Therefore, we have arrived at a collective effect on the decoherence of the superposition of states $|N_m\rangle$ and $|N_m\rangle$ described by

$$\langle N_{m'}|\dot{\rho}|N_m\rangle = -\frac{1}{4}\big[N_{m'}\sum_{n<m'}\Gamma(n,n,m'-n) + N_m\sum_{n<m}\Gamma(n,n,m-n)\big] \tag{27}$$

where the decay rate of the off-diagonal element $\langle N_{m'}|\rho|N_m\rangle$, increases with the particle numbers of the superposition states.

6. Conclusion

Motivated by recent theoretical and experimental developments in gravitational decoherence and quantum measurements, we have carried out an investigation of the open quantum dynamics of certain bound states subject to a fluctuating gravitational environment. Their non-free nature allows such systems to be formulated in terms of relatively well-understood Lindblad master equations, through which the gravitational decoherence and dissipation due to zero-point spacetime fluctuations have been analyzed. In a simplified resemblance to a (gravitationally noisy) laser interferometer, we then consider a one-dimensional cavity of massless scalar particles that models a light beam with negligible spin polarizations being reflected between two free masses. Numerical simulations have been performed to illustrate the wave-modal dependent decoherence and dissipation of such a configuration where a strong wave-modal dependence has been identified. Finally, we demonstrate the existence of novel collective quantum effects akin to superradiance originally formulated with electromagnetic interactions. This new collective process may eventually provide important amplification of normally weak gravitational decoherence for a large number of identical bosonic particles.

Acknowledgments

The authors are grateful for hospitality to Martin Land and other Organizers of the IARD 2016 Conference, where a related lecture was delivered by TO. This work was supported by the Carnegie Trust for the Universities of Scotland (TO) and by the EPSRC GG-Top Project and the Cruickshank Trust (CW).

References

[1] Amelino-Camelia G 2000 *Phys. Rev.* D **62** 024015
[2] Schiller S, Lammerzahl C, Muller H, Braxmaier C, Herrmann S and Peters A 2004 *Phys. Rev.* D **69** 027504
[3] Pikovski I, Vanner M R, Aspelmeyer M, Kim M S and Brukner C 2012 *Nature Phys.* **8** 393
[4] Lamine B, Herve R, Lambrecht A and Reynaud S 2006 *Phys. Rev. Lett.* **96** 050405
[5] Pfister C *et al.* 2016 *Nat. Commun.* **7** 13022
[6] Amelino-Camelia G 2005 Introduction to Quantum-Gravity Phenomenology *Planck Scale Effects in Astrophysics and Cosmology*, ed Amelino-Camelia G and Kowalski-Glikman J (Berlin: Springer)
[7] Schlosshauer M 2008 *Decoherence and the Quantum-to-Classical Transition* (Berlin: Springer)
[8] Joos E, Zeh H D, Kiefer C, Giulini D, Kupsch J and Stamatescu I-O 2003 *Decoherence and the Appearance of a Classical World in Quantum Theory* (Berlin: Springer)
[9] Wang C H-T, Bingham R and Mendonca J T 2006 *Class. Quantum Grav.* **23** L59
[10] Breuer H-P, Goklu E and Lammerzahl C 2009 *Class. Quantum Grav.* **26** 105012
[11] Power W L and Percival I C 1999 *Proc. R. Soc. Lond.* A **456** 955
[12] Kay B S 1998 *Class. Quantum Grav.* **15** L59

IOP Conf. Series: Journal of Physics: Conf. Series **845** (2017) 012020 doi:10.1088/1742-6596/845/1/012020

[13] Garay L 1998 *Phys. Rev.* D **58** 124015
[14] Anastopoulos C and Hu B L 2013 *Class. Quantum Grav.* **30** 165007
[15] Blencowe M P 2013 *Phys. Rev. Lett.* **111** 021302
[16] Oniga T and Wang C H-T 2016 *Phys. Rev.* D **93** 044027
[17] Oniga T and Wang C H-T 2016 *Phys. Rev.* D **94** 061501 (R)
[18] Breuer H-P and Petruccione F 2002 *The Theory of Open Quantum Systems* (New York: Oxford University Press)
[19] Dicke R H 1954 *Phys. Rev.* **93** 99
[20] Gross M and Haroche S 1982 *Phys. Rep.* **93** 301

IOP Conf. Series: Journal of Physics: Conf. Series **845** (2017) 012021 doi:10.1088/1742-6596/845/1/012021

Complete integrability of geodesics in Sasaki-Einstein space $Y^{p,q}$ via action-angle variables

Mihai Visinescu

Department of Theoretical Physics, National Institute for Physics and Nuclear Engineering, Magurele, P.O.Box M.G.-6, Romania

E-mail: mvisin@theory.nipne.ro

Abstract. We describe the construction of the integrals of geodesic motions in the five-dimensional Sasaki-Einstein space $Y^{p,q}$. The complete integrability of geodesics makes possible the construction of the action-angle variables. We find that the Hamiltonian involves only four action variables which have the corresponding frequencies different from zero.

1. Introduction

Sasaki geometry, as an odd-dimensional counterpart of the Kähler geometry, has become of renewed interest in some modern developments in mathematics and physics [1, 2]. There has been particular interest in Sasaki-Einstein (SE) manifolds in connection with the AdS/CFT correspondence. The AdS/CFT correspondence conjectures that for a five-dimensional SE manifold Y_5, type IIB string theory on $AdS_5 \times Y_5$ is dual to a four-dimensional $\mathcal{N} = 1$ superconformal field theory. In dimension five, the SE $Y^{p,q}$ metrics have played a central role as they provide an infinite class of dualities.

The purpose of this paper is to describe the construction of the action-angle variables for the geodesics in SE space $Y^{p,q}$. The description of the integrability of geodesics in $Y^{p,q}$ space in terms of action-angle variables gives us a comprehensive geometric description of the dynamics. We find that one of the fundamental frequencies is zero giving way to chaotic behavior when the system is perturbed.

The organization of the paper is as follows. In the next Section we give a brief presentation of well-known results concerning Killing tensors and SE geometry. In Section 3 we describe the complete integrability of geodesic motions in $Y^{p,q}$ space. In Section 4 we construct the action-angle variables and the frequencies of the motions. The paper ends with conclusions in Section 5.

2. Preliminaries regarding Killing tensors and Sasaki-Einstein geometry

We consider a Riemannian manifold M with the metric g_M and let ∇ be its Levi-Civita connection.

A vector field field K_μ, associated with an isometry, satisfies the Killing equation

$$\nabla_{(\mu}K_{\nu)} = 0\,, \tag{1}$$

where a round bracket denotes a symmetrization over the indices within.

IOP Conf. Series: Journal of Physics: Conf. Series **845** (2017) 012021 doi:10.1088/1742-6596/845/1/012021

A Stäckel-Killing (SK) tensor is a rank r symmetric tensor defined on M such that

$$\nabla_{(\mu} K_{\nu_1 \cdots \nu_r)} = 0\,. \tag{2}$$

In the presence of a SK tensor the system of a free particle with the Hamiltonian

$$H = \frac{1}{2} g^{\mu\nu} P_\mu P_\nu\,, \tag{3}$$

admits the conserved quantity

$$K = K^{\mu_1 \cdots \mu_r} P_{\mu_1} \cdots P_{\mu_r}\,, \tag{4}$$

commuting with Hamiltonian (3) in the sense of Poisson brackets. Here P_μ are canonical momenta conjugate to the coordinates x^μ, $P_\mu = g_{\mu\nu}\dot{x}^\nu$ with overdot denoting proper time derivative.

Another important generalization of Killing vector fields is given by Killing-Yano (KY) tensors which are antisymmetric tensors and satisfy the equation

$$\nabla_{(\mu} \Psi_{\nu_1)\nu_2 \dots \nu_r} = 0\,. \tag{5}$$

There is a remarkable connection between these two generalizations of the Killing vectors. Namely, the partially contracted product of two KY tensors $\Psi^{i_1,\dots,i_r}$ and $\Sigma^{i_1,\dots,i_r}$ generates a SK tensor of rank 2:

$$K_{ij}^{(\Psi,\Sigma)} = \Psi_{i i_2 \dots i_r} \Sigma_j^{\;\; i_2 \dots i_r} + \Sigma_{i i_2 \dots i_r} \Psi_j^{\;\; i_2 \dots i_r}\,. \tag{6}$$

In order to find the conserved quantities it is necessary to know the Killing vectors and SK tensors. However the generalized Killing equations (2) and (5) are difficult to be solved. Sometimes it is possible to find the KY tensors using the geometrical properties of the manifold. Using (6) we are able to generate SK tensors and implicitly the conserved quantities (4). That is the case of the SE manifolds for which it is possible to construct their complete set of KY tensors.

A $(2n-1)$-dimensional manifold M is a *contact manifold* if there exists a 1-form η (called a contact 1-form) on M such that

$$\eta \wedge (d\eta)^{n-1} \neq 0\,. \tag{7}$$

The *Reeb vector field* ξ dual to η satisfies:

$$\eta(\xi) = 1 \quad \text{and} \quad \xi \lrcorner\, d\eta = 0\,, \tag{8}$$

where $\lrcorner$ is the operator dual to the wedge product.

A contact Riemannian manifold (M, g_M) is *Sasakian* if its metric cone is Kähler [1]

$$C(M) \cong \mathbb{R}_+ \times M \quad , \quad g_{C(M)} = dr^2 + r^2\, g_M\,. \tag{9}$$

Here $r \in (0, \infty)$ may be regarded as a coordinate on the positive real line $\mathbb{R}_+$.

If the Sasaki space is Einstein

$$\mathrm{Ric}\, g_M = 2(n-1) g_M\,, \tag{10}$$

then the metric cone is Ricci-flat ($\mathrm{Ric}\, g = 0$), i.e. Calabi-Yau manifold.

On a $(2n-1)$-dimensional SE manifold with the contact 1-form η there are the following KY tensors:

$$\Psi_k = \eta \wedge (d\eta)^k \quad , \quad k = 0, 1, \cdots, n-1\,. \tag{11}$$

Besides these KY tensors, there are two additional KY tensors connected with the real and imaginary parts of the complex volume form of the Calabi-Yau metric cone $C(M)$ [3].

IOP Conf. Series: Journal of Physics: Conf. Series **845** (2017) 012021 doi:10.1088/1742-6596/845/1/012021

3. Complete integrability of geodesics in Sasaki-Einstein space $Y^{p,q}$

The metric of the Sasaki-Einstein space $Y^{p,q}$ is given by the line element [4]

$$\begin{aligned} ds^2 = & \frac{1-y}{6}(d\theta^2 + \sin^2\theta\, d\phi^2) + \frac{1}{w(y)q(y)}dy^2 + \frac{q(y)}{9}(d\psi - \cos\theta\, d\phi)^2 \\ & + w(y)\left[d\alpha + \frac{a - 2y + y^2}{6(a-y^2)}[d\psi - \cos\theta\, d\phi]\right]^2 , \end{aligned} \tag{12}$$

where

$$\begin{aligned} w(y) &= \frac{2(a-y^2)}{1-y} , \\ q(y) &= \frac{a - 3y^2 + 2y^3}{a - y^2} . \end{aligned} \tag{13}$$

A detailed analysis of the SE metric $Y^{p,q}$ [5] showed that for

$$0 < a < 1 , \tag{14}$$

one can take the range of the angular coordinates (θ, ϕ, ψ) to be $[0, 2\pi]$, while y lies between two zeros of $q(y)$, i.e. $y_1 \leq y \leq y_2$ with $q(y_i) = 0$. To be more specific, the roots y_i of the cubic equation

$$a - 3y^2 + 2y^3 = 0 , \tag{15}$$

are real, one negative (y_1) and two positive, the smallest being y_2.

Finally, the period of α is chosen so as to describe a principal S^1 bundle over $B_4 = S^2 \times S^2$. For any p and q coprime, the space $Y^{p,q}$ is topologically $S^2 \times S^3$ and one may take [4, 5]

$$0 \leq \alpha \leq 2\pi\ell , \tag{16}$$

where

$$\ell = \frac{q}{3q^2 - 2p^2 + p(4p^2 - 3q^2)^{1/2}} . \tag{17}$$

The contact 1-form η is

$$\eta = -2y d\alpha + \frac{1-y}{3}(d\psi - \cos\theta d\phi) , \tag{18}$$

and the Reeb vector field is [4]

$$K_\eta = 3\frac{\partial}{\partial\psi} - \frac{1}{2}\frac{\partial}{\partial\alpha} . \tag{19}$$

In order to describe the conserved quantities for the geodesic motions we list the conjugate momenta to the coordinates $(\theta, \phi, y, \alpha, \psi)$:

$$\begin{aligned}
P_\theta &= \frac{1-y}{6}\dot{\theta}\,, \\
P_y &= \frac{1}{6p(y)}\dot{y}\,, \\
P_\alpha &= w(y)\left(\dot{\alpha} + f(y)\left(\dot{\psi} - \cos\theta\dot{\phi}\right)\right)\,, \\
P_\psi &= w(y)f(y)\dot{\alpha} + \left[\frac{q(y)}{9} + w(y)f^2(y)\right]\left(\dot{\psi} - \cos\theta\dot{\phi}\right)\,, \\
P_\phi &= \frac{1-y}{6}\sin^2\theta\dot{\phi} - \cos\theta P_\psi \\
&= \frac{1-y}{6}\sin^2\theta\dot{\phi} - \cos\theta w(y)f(y)\dot{\alpha} - \cos\theta\left[\frac{q(y)}{9} + w(y)f^2(y)\right]\dot{\psi} \\
&+ \cos^2\theta\left[\frac{q(y)}{9} + w(y)f^2(y)\right]\dot{\phi}\,,
\end{aligned} \tag{20}$$

where we introduced the functions

$$f(y) = \frac{a - 2y + y^2}{6(a - y^2)}\,, \tag{21}$$

$$p(y) = \frac{w(y)q(y)}{6} = \frac{a - 3y^2 + 2y^3}{3(1-y)}\,, \tag{22}$$

which simplify writing formulas.

Using the conjugate momenta (20), the Hamiltonian (3) governing the geodesic motions in $Y^{p,q}$ space is:

$$\begin{aligned}
H =& \frac{1}{2}\left\{6p(y)P_y^2 + \frac{6}{1-y}\left(P_\theta^2 + \frac{1}{\sin^2\theta}(P_\phi + \cos\theta P_\psi)^2\right) + \frac{1-y}{2(a-y^2)}P_\alpha^2\right. \\
&\left.+ \frac{9(a-y^2)}{a - 3y^2 + 2y^3}\left(P_\psi - \frac{a - 2y + y^2}{6(a-y^2)}P_\alpha\right)^2\right\}.
\end{aligned} \tag{23}$$

The integrals of motions associated with the isometry $SU(2) \times U(1) \times U(1)$ of the metric (23) are easily to find. First of all we observe that the angles (ϕ, ψ, α) are cyclic coordinates and consequently the corresponding momenta are conserved:

$$\begin{aligned}
P_\phi &= c_\phi\,, \\
P_\psi &= c_\psi\,, \\
P_\alpha &= c_\alpha\,,
\end{aligned} \tag{24}$$

where $(c_\phi\,, c_\psi\,, c_\alpha)$ are some constants. P_ϕ corresponds to the third component of the $SU(2)$ angular momentum and P_ψ, P_α are associated with the $U(1)$ factors. Moreover the total $SU(2)$ angular momentum

$$\vec{J}^{\,2} = P_\theta^2 + \frac{1}{\sin^2\theta}\left(P_\phi + \cos\theta P_\psi\right)^2 + P_\psi^2\,, \tag{25}$$

is also conserved.

Other conserved quantities can be constructed using the KY forms of the SE space $Y^{p,q}$ [6, 7, 8]. However the maximum number of conserved quantities which are functionally independent and in involution (their Poisson brackets are equal to zero) is *five* [6]. In what follows we choose these functionally independent integrals to be the the set $\mathbf{F} = (H, P_\phi, P_\psi, P_\alpha, \vec{J}^{\,2})$. The system is complete integrable and we can proceed to construct the action-angle variables.

IOP Conf. Series: Journal of Physics: Conf. Series **845** (2017) 012021 doi:10.1088/1742-6596/845/1/012021

4. Action-angle variables

The construction of the action-angle variables starts with fixing a level surface $\mathbf{F} = \mathbf{c}$ [9, 10]. We search a generating function S for the canonical transformation $(\mathbf{p}, \mathbf{q}) \to (\mathbf{J}, \mathbf{w})$ where $\mathbf{p}$ is represented by the momenta (20) and $\mathbf{q}$ is the set of coordinates $(\theta, \phi, y, \alpha, \psi)$.

The generating function S is

$$S(\mathbf{q}, \mathbf{c}) = \int_{\mathbf{F}=\mathbf{c}} \mathbf{p}\, d\mathbf{q}\,, \tag{26}$$

where $\mathbf{p} = \mathbf{p}(\mathbf{q}, \mathbf{c})$ by use of the equations of motion.

Since the Hamiltonian (23) has no explicit time dependence, we can write

$$S(\mathbf{q}, \mathbf{c}) = W(\mathbf{q}, \mathbf{c}) - Et\,, \tag{27}$$

where $W(\mathbf{q}, \mathbf{c})$ is the Hamilton characteristic function

$$W = \sum_i \int p_i dq_i\,, \tag{28}$$

and E is the energy $H = E$ corresponding to the chosen level surface $\mathbf{F} = \mathbf{c}$.

In the case of the geodesics in SE space $Y^{p,q}$ the variables in Hamilton-Jacobi equation are separable and consequently we seek a solution of the form

$$W(y, \theta, \phi, \psi, \alpha) = W_y(y) + W_\theta(\theta) + W_\phi(\phi) + W_\psi(\psi) + W_\alpha(\alpha)\,. \tag{29}$$

The action variables $\mathbf{J}$ are given by

$$J_i = \oint p_i dq_i = \oint \frac{\partial W_i(q_i; \mathbf{c})}{\partial q_i} dq_i \qquad \text{(no summation)}\,, \tag{30}$$

where $\oint$ means an integration over one cycle. The J_i's form n independent functions of the constants $\mathbf{c}$ and can be taken as a set of new constant momenta.

Finally, the angle variables can be found from the equation

$$w_i = \frac{\partial W}{\partial J_i} = \sum_{j=1}^{n} \frac{\partial W_j(q_j; J_1, \cdots, J_n)}{\partial J_i}\,, \tag{31}$$

having a linear evolution in time

$$w_i = \omega_i t + \beta_i\,, \tag{32}$$

with β_i other constants of integration and ω_i are frequencies associated with the periodic motion of q_i.

Applying the general prescriptions, we can manage to evaluate the action-angle variables for geodesics in $Y^{p,q}$. For the cyclic variables the task is easy as the Hamilton characteristic functions (28) related to cyclic variables are

$$\begin{aligned} W_\phi &= P_\phi \phi = c_\phi \phi\,, \\ W_\psi &= P_\psi \psi = c_\psi \psi\,, \\ W_\alpha &= P_\alpha \alpha = c_\alpha \alpha\,, \end{aligned} \tag{33}$$

with the constants c_ϕ, c_ψ, c_α introduced in (24). The corresponding action variables are

$$\begin{aligned} J_\phi &= 2\pi c_\phi\,, \\ J_\psi &= 2\pi c_\psi\,, \\ J_\alpha &= 2\pi \ell c_\alpha\,. \end{aligned} \tag{34}$$

IOP Conf. Series: Journal of Physics: Conf. Series **845** (2017) 012021 doi:10.1088/1742-6596/845/1/012021

For a fixed energy E and using (29), the Hamilton (23) can be written in the form:

$$\begin{aligned}&\left(\frac{\partial W_\theta}{\partial\theta}\right)^2+\frac{1}{\sin^2\theta}(c_\phi+\cos\theta_\psi)^2\\&=\frac{1-y}{3}E-p(y)(1-y)\left(\frac{\partial W_y}{\partial y}\right)^2-\frac{(1-y)^2}{12(a-y^2)}c_\alpha^2\\&-\frac{3(a-y^2)(1-y)}{2(a-3y^2+2y^3)}\left[c_\psi-\frac{a-2y+y^2}{6(a-y^2)}c_\alpha\right]^2.\end{aligned}\tag{35}$$

We separated in the LHS the terms which depend only on θ, while the RHS depends only on y. Therefore we may set

$$\left(\frac{\partial W_\theta}{\partial\theta}\right)^2+\frac{1}{\sin^2\theta}(c_\phi+\cos\theta c_\psi)^2=c_\theta^2\,,\tag{36}$$

with c_θ another constant.

The action variable J_θ is given by the integral

$$J_\theta=\oint d\theta\sqrt{c_\theta^2-\frac{(c_\phi+c_\psi\cos\theta)^2}{\sin^2\theta}}\,.\tag{37}$$

Denoting by $\theta_\pm$ the roots of the function appearing in the square root, the complete cycle of θ involves going from θ_- to θ_+ and back to θ_-. The most efficient way to evaluate the integral (37) is to put $\cos\theta=t$, extend t to a complex variable z and interpret the integral as a closed contour integral in the complex z-plane [11]. The turning points of the t-motions are

$$t_\pm=\frac{-c_\phi c_\psi\pm c_\theta\sqrt{c_\theta^2+c_\psi^2-c_\phi^2}}{c_\theta^2+c_\psi^2}\,.\tag{38}$$

They are real for

$$c_\theta^2+c_\psi^2-c_\phi^2\geq 0\,,\tag{39}$$

and situated in the interval $(-1,+1)$.

The contour of integration in the complex z-plane can be deformed to a large circular contour plus two contour integrals about the poles at $z=\pm1$. The standard evaluation of the residues and taking into account the contribution of the large contour integral finally gives

$$J_\theta=2\pi\left[\sqrt{c_\theta^2+c_\psi^2}-c_\phi\right]\,.\tag{40}$$

Concerning the angular variable w_ϕ we have

$$w_\phi=\frac{1}{2\pi}J_\phi+\frac{\partial W_\theta}{\partial J_\phi}\,.\tag{41}$$

Putting $\cos\theta=t$, and using (34), (36), (40), the second term is

$$\begin{aligned}\frac{\partial W_\theta}{\partial J_\phi}&=-\frac{1}{2\pi}\int dt\frac{(J_\phi+J_\theta)t^2+J_\psi t}{(1-t^2)\sqrt{-(J_\phi+J_\theta)^2t^2-2J_\phi J_\psi t+(J_\theta^2+2J_\theta J_\phi-J_\psi^2)}}\\&=\frac{1}{2\pi}\int\frac{dt}{1-t^2}\frac{dt^2+et}{\sqrt{a+bt+ct^2}}\,,\end{aligned}\tag{42}$$

IOP Conf. Series: Journal of Physics: Conf. Series **845** (2017) 012021 doi:10.1088/1742-6596/845/1/012021

where

$$\begin{aligned} a =& J_\theta^2 + 2J_\theta J_\phi - J_\psi^2\,, \\ b =& -2J_\theta J_\psi\,, \\ c =& -(J_\theta + J_\phi)^2\,, \\ d =& J_\theta + J_\phi\,, \\ e =& J_\psi\,. \end{aligned} \tag{43}$$

We need the following integrals [12]

$$\begin{aligned} I_1(a,b,c;t) &= \int \frac{dt}{\sqrt{a+bt+ct^2}} = \frac{-1}{\sqrt{-c}} \arcsin\left(\frac{2ct+b}{\sqrt{-\Delta}}\right), \\ I_2(a,b,c;t) &= \int \frac{dt}{t\sqrt{a+bt+ct^2}} = \frac{1}{\sqrt{-a}} \arctan\left(\frac{2a+bt}{2\sqrt{-a}\sqrt{a+bt+ct^2}}\right), \end{aligned} \tag{44}$$

evaluated for $c < 0$, $\Delta = 4ac - b^2 < 0$ and for $a < 0$ respectively. That is the case of the parameters a, b, c taking into account the constraint (39).

Using these integrals we finally get for the angular variable w_ϕ

$$\begin{aligned} w_\phi =& \frac{1}{2\pi} J_\phi - \frac{d}{2\pi} I_1(a,b,c;\cos\theta) \\ & - \frac{d+e}{4\pi} I_2(a+b+c, b+2c, c; \cos\theta - 1) \\ & - \frac{e-d}{4\pi} I_2(a-b+c, b-2c, c; \cos\theta + 1)\,. \end{aligned} \tag{45}$$

For the action variable corresponding to y coordinate we have from (35)

$$\begin{aligned} \frac{\partial W_y}{\partial y} =& \left\{ \frac{1-y}{a-3y^2+2y^3} E - \frac{3}{a-3y^2+2y^3} c_\theta^2 \right. \\ & - \frac{9(a-y^2)(1-y)}{2(a-3y^2+2y^3)^2} c_\psi^2 + \frac{3(a-2y+y^2)(1-y)}{2(a-3y^2+2y^3)^2} c_\psi c_\alpha \\ & \left. - \frac{(1-y)(2a+a^2-6ay-2y^2+2ay^2+6y^3-3y^4)}{8(a-3y^2+2y^3)^2(a-y^2)} c_\alpha^2 \right\}^{\frac{1}{2}}. \end{aligned} \tag{46}$$

In contrast with the action variable J_θ (37), the evaluation of the action variable J_y involves an intricate integral as can be seen from (46). After all, the closed-form of J_y is not at all illuminating. For our purposes, it is enough to observe that J_y depends only on four constants of motion: $E, J_\theta, J_\alpha, J_\psi$. In consequence, the energy depends only on four action variables $J_y, J_\theta, J_\alpha, J_\psi$ representing a reduction of the number of action variables entering the expression of the energy of the system.

The explicit evaluation of the angular variables $w_\theta, w_\psi, w_\alpha, w_y$ is again intricate due to the absence of a simple closed-form for the action variable J_y. However, it is remarkable the fact that one of the fundamental frequencies (32)

$$\omega_i = \frac{\partial H}{\partial J_i}\,, \tag{47}$$

is zero, namely

$$\omega_\phi = \frac{\partial H}{\partial J_\phi} = 0\,, \tag{48}$$

since the action J_ϕ does not enter the expression of the energy.

Therefore the Hamiltonian (23) depends on that action variables for which the corresponding frequencies are different from zero. The presence of a vanishing frequency refers to the degeneracy of the system giving way to chaotic behavior when the system is perturbed [9] This fact was observed in several recent studies of non-integrability and chaotic behavior of some classical configurations of strings in the context of AdS/CFT correspondence (see e. g. [13, 14, 15]).

5. Conclusions

The existence of the conserved quantities are very important not only for the particle motions, but also leading to the separation of Hamilton-Jacobi and quantum Klein-Gordon, Dirac equations. The action-angle approach to the integrable geodesics represents a useful tool in the study of near-integrable systems (Kolmogorov-Arnold-Moser (KAM) theory) and for the quantization of integrable systems (Bohr-Sommerfeld rule).

It would be interesting to extend the action angle formulation to higher dimensional SE metrics relevant for the predictions of the AdS/CFT correspondence.

Acknowledgments

This work has been partly supported by the joint Romanian-LIT, JINR, Dubna Research Project, theme no. 05-6-1119-2014/2016 and partly by the program NUCLEU 16 42 01 01/2016, Romania.

References

[1] Boyer C and Galicki K 2008 *Sasakian geometry* (Oxford Mathematical Monographs, Oxford University Press, Oxford)
[2] Sparks J 2011 Sasaki-Einstein manifolds *Surv. Diff. Geom.* **16** 265 (*Preprint* 1004.2461)
[3] Semmelmann U 2003 Conformal Killing forms on Riemannian manifolds *Math. Z.* **245** 503 (*Preprint* math/0206117)
[4] Martelli D and Sparks J 2006 Toric geometry, Sasaki-Einstein manifolds and a new infinite class of AdS/CFT duals *Commun. Math. Phys.* **262** 51 (*Preprint* hep-th/0411238)
[5] Gauntlett J P, Martelli D, Sparks J and Waldram D 2004 Sasaki-Einstein metrics on $S^2 \times S^3$ *Adv. Theor. Math. Phys.* **8** 711 (*Preprint* hep-th/0403002)
[6] Babalic E M and Visinescu M 2015 Complete integrability of geodesic motion in Sasaki-Einstein toric $Y^{p,q}$ spaces *Mod. Phys. Lett. A* **30** 1550180 (*Preprint* 1505.03976)
[7] Slesar V, Visinescu M and Vîlcu G E 2014 Special Killing forms on toric Sasaki-Einstein manifolds *Phys. Scr.* **89** 125205 (*Preprint* 1403.1015)
[8] Slesar V, Visinescu M and Vîlcu G E 2015 Toric data, Killing and complete integrability of geodesics in Sasaki-Einstein spaces $Y^{p,q}$ *Annals of Physics* **361** 548
[9] Arnold V I 1989 *Mathematical Methods of Classical Mechanics* (Springer-Verlag, New York, 2nd edition)
[10] Goldstein H, Poole C and Safko J 2002 *Classical Mechanics* (Addison-Wesley, San Francisco, 3rd edition)
[11] Visinescu M 2016 Integrability of geodesics and action-angle variables in Sasaki-Einstein space $T^{1,1}$ *Eur. Phys. J. C* **76** 498 (*Preprint* 1604.03705)
[12] Gradshteyn I S and Ryzhik I M 2007 *Table of Integrals, Series, and Products* (Edited by A. Jeffrey and D. Zwillinger; Academic Press, New York)
[13] Pando Zayas L A and Terrero-Escalante C A 2010 Chaos in the gauge/gravity correspondence *JHEP* **1009**, 094 (*Preprint* 1007.0277)
[14] Basu P and Pando Zayas L A 2011 Chaos rules out integrability of strings on $AdS_5 \times T^{1,1}$ *Phys. Lett. B* **700**, 243 (*Preprint* 1103.4107)
[15] Basu P and Pando Zayas L A 2011 Analytic non-integrability in string theory *Phys. Rev. D* **84**, 046006 (*Preprint* 1105.2540)

IOP Conf. Series: Journal of Physics: Conf. Series **845** (2017) 012022 doi:10.1088/1742-6596/845/1/012022

Quantum models as classical cellular automata

Hans-Thomas Elze

Dipartimento di Fisica "Enrico Fermi", Università di Pisa, Largo Pontecorvo 3, I-56127 Pisa, Italia

E-mail: elze@df.unipi.it

Abstract. A synopsis is offered of the properties of discrete and integer-valued, hence "natural", cellular automata (CA). A particular class comprises the "Hamiltonian CA" with discrete updating rules that resemble Hamilton's equations. The resulting dynamics is *linear* like the unitary evolution described by the Schrödinger equation. Employing Shannon's Sampling Theorem, we construct an invertible *map* between such CA and continuous quantum mechanical models which incorporate a fundamental *discreteness* scale l. Consequently, there is a one-to-one correspondence of quantum mechanical and CA conservation laws. We discuss the important issue of linearity, recalling that nonlinearities imply nonlocal effects in the continuous quantum mechanical description of intrinsically local discrete CA – requiring locality entails linearity. The admissible CA observables and the existence of solutions of the l-dependent dispersion relation for stationary states are mentioned, besides the construction of multipartite CA obeying the Superposition Principle. We point out problems when trying to match the deterministic CA here to those envisioned in 't Hooft's *CA Interpretation of Quantum Mechanics*.

1. Introduction

A novel interpretation of quantum mechanics (QM) – which amounts to redesigning the foundations of quantum theory in accordance with "classical" concepts, foremost with determinism – has recently been laid out by G. 't Hooft [1]. The hope for a comprehensive theory expressed there is founded on the observation that quantum mechanical features arise in a large variety of deterministic "mechanical" models. While practically all of these models have been singular cases, i.e., which cannot easily be generalized to cover a realistic range of phenomena incorporating interactions, cellular automata (CA) promise to provide the necessary versatility, as we shall discuss [2, 3]. For an incomplete list of various earlier attempts in this field, see, for example, Refs. [4, 5, 6, 7, 8, 9, 10, 11, 12] and further references therein.

The linearity of quantum mechanics (QM) is a fundamental feature most notably embodied in the Schrödinger equation. This linearity does not depend on the particular object under study, provided it is sufficiently isolated from anything else. This is reflected in the Superposition Principle and entails the "quantum essentials" of interference and entanglement.

Nevertheless, the linearity of QM has been questioned repeatedly and nonlinear modifications have been proposed, *e.g.* in Ref. [13, 14] – not only as suitable approximations for complicated many-body dynamics, but especially in order to test experimentally the robustness of QM against such *nonlinear deformations*, as well as in attempts to address the measurement problem in dynamical schemes (none of which has been generally accepted). This has been thoroughly discussed by T.F. Jordan who presented a proof 'from within' quantum theory that the theory has to be linear, given the essential *separability* assumption "... that the system we are considering can be described as part of a larger system without interaction with the rest of the larger system." [15].

Recently, we have considered a seemingly unrelated *discrete* dynamical theory which deviates drastically from quantum theory, at first sight. However, we have shown with the help of sampling theory that the deterministic mechanics of the class of Hamiltonian CA can be related to QM in the presence of a fundamental time scale. This relation demonstrates that consistency of the action principle of the underlying discrete dynamics implies, in particular, the linearity of both theories.

Our CA approach may offer additional insight into interference and entanglement, in the limit when the discreteness scale can be considered as small.

In Section 2., we summarize results obtained so far. While in Section 3., we present some considerations about the exact solutions of the CA equations of motion. In particular, we will comment about the relation to ontological CA and their states, the existence of which was introduced as the working hypothesis in Ref. [1].

2. Natural Hamiltonian CA – from the action to multipartite automata

Reasoning about the linearity of QM has led us to the consideration of CA models, which are based on three essential ingredients [2, 3]: **i)** Deterministic discrete mechanics as proposed by T.D. Lee in his program aimed to arrive at a manageable discretization for quantum gravity [16, 17] (and references therein); it presumes the existence of a *minimal time* scale l and *discrete updating rules* for dynamics. **ii)** Sampling Theory [18, 19] for discrete structures on or of spacetime [20]; we employ this to construct a *map* between CA and QM models in the continuum, which obtain finite-l corrections. **iii)** The "oscillator representation" of QM based on decomposing wave functions into real and imaginary parts, $\psi \equiv x + ip$ [21]; this is suggestive of a relation between apparently classical CA and quantum mechanical models.

To summarize, the following results hold for a particular class of CA (to be specified):

- Evolution of these CA can be described by a *continuous time* Schrödinger equation modified by l-dependent *higher-order derivatives* with respect to time.
- There is a l-dependent *dispersion relation* for stationary states.
- There are l-dependent *conservation laws* in one-to-one correspondence with those of the corresponding QM models in the continuum.
- There are *multipartite CA* obeying the Superposition Principle, *i.e.* which show the tensorial structures of QM.
- If space is discrete as well (assuming the same scale l for simplicity), a generalized Uncertainty Principle can be derived from Robertson's inequality, $\Delta A \Delta B \geq |\langle [A, B] \rangle|/2$, with $\Delta A := (\langle A^2 \rangle - \langle A \rangle^2)^{1/2}$, *etc.*; let $X_{rs} := lr\delta_{r,s}$ (r, s numbering spatial sites) and $P_{rs} := -i(\delta_{r,s-1} - \delta_{r,s+1})/2l$, then: $\Delta X \Delta P \geq |1 + l^2 \langle P^2 \rangle /2|$, which yields a *minimum uncertainty* $\Delta X_{min} = l/\sqrt{2}$ [22].

Last not least, we find:

- The known QM results are obtained correctly in the *continuum limit*, $l \rightarrow 0$.

2.1. The CA Action Principle

We describe natural Hamiltonian CA with countably many degrees of freedom presently in terms of *complex integer-valued* state variables ψ_n^α, *i.e.* by *Gaussian integers* ($\{z|z \equiv m + in,\ m, n \in \mathbf{Z}\}$), where $\alpha \in \mathbf{N}_0$ denote different degrees of freedom and $n \in \mathbf{Z}$ different states labelled by this discrete *clock variable*. Various equivalent forms of the action for such CA exist, as indicated earlier [2]. – In particular, making use of the mentioned decomposition into real and imaginary parts, $\psi \equiv x + ip$, our theory describes a perfectly classical looking discrete mechanical system, cf. Section 2.2. However, we will employ here the more compact description in terms of complex variables, which is useful for the construction of composite CA in analogy with multipartite QM systems.

Let $\hat{H} := \{H^{\alpha\beta}\}$ denote a self-adjoint matrix of Gaussian integers that will play the role of Hamilton operator. Furthermore, let us introduce $\dot{O}_n := O_{n+1} - O_{n-1}$, for any quantity O_n depending on the clock variable n. With a summation convention for Greek indices, $r^\alpha s^\alpha \equiv \sum_\alpha r^\alpha s^\alpha$, we often simplify notation further by writing $\psi_n^{*\alpha} H^{\alpha\beta} \psi_n^\beta \equiv \psi_n^* \hat{H} \psi_n$. – Then, with ψ_n^α and $\psi_n^{*\alpha}$ as independent variables, the CA action $\mathcal{S}$ is defined by:

$$\mathcal{S}[\psi, \psi^*] := \sum_n [\frac{1}{2i}(\psi_n^* \dot{\psi}_n - \dot{\psi}_n^* \psi_n) + \psi_n^* \hat{H} \psi_n] \equiv \psi^* \hat{\mathcal{S}} \psi \ , \tag{1}$$

with $\hat{\mathcal{S}}$ an useful abbreviation later on. In order to set up the variational principle, we introduce *integer-valued* variations δf applied to a polynomial g:

$$\delta_f g(f) := [g(f + \delta f) - g(f - \delta f)]/2\delta f \ , \tag{2}$$

and $\delta_f g \equiv 0$, if $\delta f = 0$. – Remarkably, variations of terms that are *constant, linear, or quadratic* in integer-valued variables yield analogous results as standard infinitesimal variations of corresponding expressions in the continuum.

With the help of these ingredients, the variational principle is postulated:

[*CA Action Principle*] The discrete evolution of a CA is determined by stationarity of the action under arbitrary integer-valued variations of all dynamical variables, $\delta\mathcal{S} = 0$. •

We emphasize the following characteristics of this CA Action Principle:

- While infinitesimal variations do not conform with integer valuedness, there is *a priori* no restriction of integer variations. Hence *arbitrary* integer-valued variations are admitted.
- One could imagine higher order terms in the action (1), *i.e.* of higher than second order in ψ_n or ψ_n^*. However, given arbitrary variations $\delta\psi_n^\alpha$ and $\delta\psi_n^{*\alpha}$, such additional contributions cannot be admitted. Otherwise the number of equations of motion generated by variation of the action, according to Eq. (2), would exceed the number of variables. (A suitable small number of such terms, which are nonzero only for some fixed values of n, could encode *initial conditions* for the evolution.)

These features of the CA Action Principle are essential in constructing a map between Hamiltonian CA and equivalent quantum mechanical continuum models [2]. – For curiosity, generalizations of the variations defined in Eq. (2) have been considered, in order to allow higher than second order polynomial terms in the action. While leading to consistent discrete equations of motion, however, these nonlinear equations generally are beset with undesirable nonlocal features in the corresponding continuum description [23].

2.2. The equations of motion

Applying the CA Action Principle to the action $\mathcal{S}$ of Eq. (1) with variations $\delta\psi_n^*$ and $\delta\psi_n$, *cf.* Eq. (2), gives discrete analogues of the Schrödinger equation and its adjoint, respectively:

$$\dot{\psi}_n = \frac{1}{i}\hat{H}\psi_n \ , \tag{3}$$

$$\dot{\psi}_n^* = -\frac{1}{i}(\hat{H}\psi_n)^* \ , \tag{4}$$

recalling that $\hat{H} = \hat{H}^\dagger$ and $\dot{\psi}_n = \psi_{n+1} - \psi_{n-1}$, *etc.* – Note that the action $\mathcal{S}$ vanishes when evaluated for solutions of these finite difference equations.

Setting $\psi_n^\alpha =: x_n^\alpha + ip_n^\alpha$, with real integer-valued variables x_n^α and p_n^α, and suitably separating real and imaginary parts of Eqs. (3)–(4), leads to discrete equations that superficially resemble Hamilton's equations for a network of coupled classical oscillators:

$$\dot{x}_n^\alpha = h_S^{\alpha\beta} p_n^\beta + h_A^{\alpha\beta} x_n^\beta \ , \quad \dot{p}_n^\alpha = -h_S^{\alpha\beta} x_n^\beta + h_A^{\alpha\beta} p_n^\beta \ , \tag{5}$$

where we also split the self-adjoint matrix $\hat{H}$ into real integer-valued symmetric and antisymmetric parts, respectively, $H^{\alpha\beta} =: h_S^{\alpha\beta} + ih_A^{\alpha\beta}$. – Notwithstanding their appearance, all finite difference equations here are of second order, necessitating *two initial values*, unlike in the continuum [21, 24]. We will address this point in Section 3. In any case, the form of equations (5) suggested the name *Hamiltonian CA*, which is also justified by the fact that analogues of Poisson brackets and classical like observables can be introduced here [25].

2.3. Conservation laws

The time-reversal invariant equations of motion of Section 2.2. give rise to conservation laws which are in *one-to-one correspondence* with those of the Schrödinger equation in the continuum. In fact, the following theorem can be easily verified:

[*Theorem A*] For any matrix $\hat{G}$ that commutes with $\hat{H}$, $[\hat{G}, \hat{H}] = 0$, there is a *discrete conservation law*:

$$\psi_n^{*\alpha} G^{\alpha\beta} \dot{\psi}_n^\beta + \dot{\psi}_n^{*\alpha} G^{\alpha\beta} \psi_n^\beta = 0 \ . \tag{6}$$

For self-adjoint $\hat{G}$, defined by Gaussian integers, this relation is about real integer quantities. •

A rearrangement of Eq. (6) yields the corresponding conserved quantity $q_{\hat{G}}$:

$$q_{\hat{G}} := \psi_n^* \hat{G} \psi_{n-1} + \psi_{n-1}^* \hat{G} \psi_n = \psi_{n+1}^* \hat{G} \psi_n + \psi_n^* \hat{G} \psi_{n+1} \ , \tag{7}$$

i.e. a complex (real for $\hat{G} = \hat{G}^\dagger$) integer-valued two-'time' *correlation function* which is invariant under a shift $n \to n+m$, $m \in \mathbf{Z}$. – In particular, for $\hat{G} := \hat{1}$, the conservation law amounts to a constraint on the state variables:

$$q_{\hat{1}} = 2\mathrm{Re}\ \psi_n^* \psi_{n-1} = 2\mathrm{Re}\ \psi_{n+1}^* \psi_n = \mathrm{const} \ . \tag{8}$$

This plays a similar role for discrete CA as the familiar *normalization* of state vectors in continuum QM; *cf.* Eqs. (13)–(14) below. – We also define the symmetrized conserved quantity:

$$\psi_n^* \hat{\mathcal{Q}} \psi_n := \frac{1}{2}\mathrm{Re}\ \psi_n^*(\psi_{n+1} + \psi_{n-1}) \equiv \frac{1}{2}\mathrm{Re}\ \psi_n^{*\alpha}(\psi_{n+1}^\alpha + \psi_{n-1}^\alpha) \ , \tag{9}$$

for later use.

2.4. Continuum representation

There exists an one-to-one invertible *map* between the dynamics of discrete Hamiltonian CA and continuum QM in presence of a fundamental time scale l [2, 3, 23]. Due to the finite discreteness scale, the continuous time wave functions are *bandlimited*, *i.e.*, their Fourier transforms have only finite support in frequency space, $\omega \in [-\pi/l, \pi/l]$. Hence, Sampling Theory allows one to reconstruct continuous time signals, wave functions $\psi^\alpha(t)$, from their discrete samples, the CA state variables ψ_n^α, and *vice versa* [18, 19, 20]. – Further aspects relating discrete and continuous dynamics, in particular concerning models that do not belong to the class of Hamiltonian CA, have also been discussed in Ref. [1].

The resulting mapping rules obtained through the reconstruction formula of Shannon's Sampling Theorem [18, 19] can be summarized as follows:

$$\psi_n^\alpha \longmapsto \psi^\alpha(t) \;, \tag{10}$$

$$\psi_{n\pm 1}^\alpha \longmapsto \exp\,[\,\pm\,l\frac{\mathrm{d}}{\mathrm{d}t}]\psi^\alpha(t) = \psi^\alpha(t\pm l) \;, \tag{11}$$

$$\psi^\alpha(nl) \longmapsto \psi_n^\alpha \;, \tag{12}$$

keeping in mind that the continuum wave function is bandlimited. – We remark that kernels that differ from the original *sinc* function can be used to modify the reconstruction formula, turning the sharp bandlimit into other types of cut-off. However, as long as the reconstruction formular remains a linear map, the form of the equations of motion and of the conservation laws does not change.

With the help of these results, all CA equations are mapped to their continuum versions. For example, corresponding to Eqs. (6)–(9), there exist analogous conservation laws and conserved quantities, which are obtained by applying the mapping rules separately to all wave function factors that appear. Thus, for example, the Eq. (9) yields the conserved quantity:

$$\mathrm{const} = \psi_n^*\hat{\mathcal{Q}}\psi_n \longmapsto \psi^*(t)\hat{\mathcal{Q}}\psi(t) = \mathrm{Re}\;\psi^*(t)\cosh\,[l\frac{\mathrm{d}}{\mathrm{d}t}]\psi(t) \tag{13}$$

$$= \psi^{*\alpha}(t)\psi^\alpha(t) + \frac{l^2}{2}\mathrm{Re}\;\psi^{*\alpha}(t)\frac{\mathrm{d}^2}{\mathrm{d}t^2}\psi^\alpha(t) + \mathrm{O}(l^4) \;, \tag{14}$$

with l-dependent corrections to the continuum limit ($l \to 0$), namely the usually conserved normalization $\psi^{*\alpha}\psi^\alpha = \mathrm{const}$.

Similarly, the Schrödinger equation and with its finite-l correction terms are obtained from Eq. (3) [2]:

$$2\sinh(l\partial_t)\psi(t) = -i\hat{H}\psi(t) \;. \tag{15}$$

This leads, in particular, to the l-dependent dispersion relation for stationary states:

$$lE_\alpha = \arcsin(\tfrac{l\epsilon_\alpha}{2}) = \tfrac{l\epsilon_\alpha}{2}[1 + (\tfrac{l\epsilon_\alpha}{2})^2/6 + \mathrm{O}((l\epsilon_\alpha)^4)] \;, \tag{16}$$

giving their energies E_α in terms of the eigenvalues of the diagonalized *dimensionless* Hamiltonian, $\hat{H} \to \{l\epsilon_\alpha\}$. We observe an effect of the band limit, $|E_\alpha| \leq \pi/2l \equiv \omega_{max}/2$, as expected. Interestingly, by inverting Eq. (16), we find that the eigenvalues $l\epsilon_\alpha$ must be constrained by $-2 \leq l\epsilon_\alpha \leq 2$, for all α. A complete classification of such Hamiltonians represented by *symmetric matrices* has been provided recently [26] – indeed there are classes of them with infinite numbers of members besides irregular ones . For self-adjoint matrices instead, analogous results are only partially known.

2.5. From single to multipartite Hamiltonian CA [27]

So far, we have been concerned with isolated single Hamiltonian CA. Of course, we may ask: Can discrete CA form composite multipartite systems? – Only if the answer is 'Yes', the quantum features of CA can possibly cover the Superposition Principle to its fullest extent.

Therefore, we wonder whether not only the *linearity* of the evolution law but also the *tensor product structure* of composite wave functions finds its analogue here. These are the fundamental ingredients of the usual continuum theory reflected in interference and entanglement of states or observables. Which should be recovered in the continuum limit ($l \to 0$) of the CA picture, at least. Furthermore, when the discreteness scale l is finite, the dynamics of composites of

IOP Conf. Series: Journal of Physics: Conf. Series **845** (2017) 012022 doi:10.1088/1742-6596/845/1/012022

CA which do not interact with each other should lead to *no spurious correlations* among them – the principle of "no correlations without interactions", which holds in all of known physics. Naturally, this principle does not rule out the possibility of entangled states, presenting *the* prototype of quantum correlations, which may have been formed during interactions among parts of a multipartite system or have been present already from initial conditions for its evolution.

Let us first discuss in more detail the obstructions encountered when trying to formulate composites of Hamiltonian CA along the lines familiar from QM. While the way to overcome them is described in the next Subsection 2.5.1.

The want-to-be discrete time derivative, $\dot{O}_n := O_{n+1} - O_{n-1}$, for any quantity O_n depending on the clock variable n, which appears all-over in CA equations of motion and conservation laws, does not obey the *Leibniz rule*:

$$[A_n\dot{B}_n] = \dot{A}_n\tfrac{B_{n+1}+B_{n-1}}{2} + \tfrac{A_{n+1}+A_{n-1}}{2}\dot{B}_n \neq \dot{A}_nB_n + A_n\dot{B}_n \ . \quad (17)$$

Similar observations can be expected for other definitions, in particular first-order ones, one might come up with.

The failure of the Leibniz rule for the above 'derivative' implies that a multi-CA equation of motion analogous to the single-CA Eq. (3):

$$\dot{\Psi}_n = \frac{1}{i}\hat{H}_0\Psi_n \ , \quad (18)$$

where $\hat{H}_0$ describes a block-diagonal Hamiltonian in the absence of interactions among CA, cannot be decomposed in the usual way. Because of Eq. (17), *factorization* of Eq. (18) is hindered on the left-hand side, since unphysical correlations will be produced among the components of a factorized wave function, such as

$$\Psi_n^{\alpha\beta\gamma\cdots} = \psi_n^\alpha\phi_n^\beta\kappa_n^\gamma\cdots \ , \quad (19)$$

or for superpositions of such factorized terms.

Furthermore, applying the mapping rules of Section 2.4., we find that bilinear terms, $\dot{\overline{\psi_n\phi_n}}$, for example, do not converge to the correct QM expression, when taking the limit $l \to 0$. Which should be $\partial_t(\psi\phi) = (\partial_t\psi)\phi + \psi\partial_t\phi$, in order to allow the decoupling of two subsystems that do not interact. Even on the right-hand side of Eq. (18) we encounter this kind of obstruction.

The latter is a generic problem of nonlinear terms in the equations of motion of discrete CA: *The linear map provided by Shannon's Theorem does not commute with the multiplication implied by the nonlinearities.* This follows from the explicit reconstruction formula (or any variant thereof that is linear) [2, 18, 19, 23].

2.5.1. The many-time formulation The difficulties just pointed out can be traced to the implicit assumption that components of a multipartite CA are *synchronized* to the extent that they share a common clock variable n. A a radical way out of the impasse encountered is to resort to a *many-time* formalism [27]. This means giving up synchronization among parts of the composite CA by introducing a set of clock variables, $\{n(1), \ \dots, \ n(m)\}$, one for each one out of m components.

It may be surprising to find this in the present nonrelativistic context, since the many-time formalism has been introduced by Dirac, Tomonaga, and Schwinger in their respective formulations of relativistically covariant many-particle QM or quantum field theory, where a global synchronization cannot be maintained [28, 29, 30].

We replace here the single-CA action of Eq. (1) by an integer-valued multipartite-CA action:

$$\mathcal{S}[\Psi,\Psi^*] := \Psi^*(\sum_{k=1}^{m}\hat{\mathcal{S}}_{(k)} + \hat{\mathcal{I}})\Psi \ , \quad (20)$$

with $\Psi := \Psi^{\alpha_1 \ldots \alpha_m}_{n_1 \ldots n_m}$ and, correspondingly, Ψ^* as independent *Gaussian integer* variables; the self-adjoint operator $\hat{\mathcal{I}}$ incorporates interactions between different CA; whereas $\hat{\mathcal{S}}_{(k)}$ is as introduced in Eq. (1), with subscript $_{(k)}$ indicating that it concerns *exclusively* the pair of indices pertaining to the k-th single-CA subsystem:

$$\Psi^* \hat{\mathcal{S}}_{(k)} \Psi := \sum_{\{n_k\}} [(\mathrm{Im}\ \Psi^{*\ldots\alpha_k\ldots}_{\ldots n_k \ldots}\ \dot{\Psi}^{\ldots\alpha_k\ldots}_{\ldots n_k\ldots} + \Psi^{*\ldots\alpha_k,\ldots}_{\ldots n_k\ldots}\ H^{\alpha_k\beta_k}_{(k)} \Psi^{\ldots\beta_k\ldots}_{\ldots n_k\ldots}] \ , \tag{21}$$

with summation over *all* clock variables and over Greek indices appearing twice ; the $\dot{}$-operation, however, acts only with respect to the explicitly indicated n_k, $\dot{f}(n_k) := f(n_k+1) - f(n_k-1)$, while the single-CA Hamiltonian, $\hat{H}_{(k)}$, requires a matrix multiplication, as before.

Obviously, we can apply the *CA Action Principle* also to the present situation with the generalized action of Eq. (20). This results in the following discrete equations of motion:

$$\sum_{k=1}^{m} \dot{\Psi}^{\ldots\alpha_k\ldots}_{\ldots n_k\ldots} = \frac{1}{i}(\sum_{k=1}^{m} H^{\alpha_k\beta_k}_{(k)} \Psi^{\ldots\beta_k\ldots}_{\ldots n_k\ldots} + \mathcal{I}^{\ldots\alpha_k\ldots\ \beta_1\ldots\beta_m} \Psi^{\beta_1\ldots\beta_m}_{\ldots n_k\ldots}) \ , \tag{22}$$

together with the adjoint equations; here the interaction $\hat{\mathcal{I}}$, like $\hat{H}_{(k)}$, is assumed to be independent of the clock variables and the $\dot{}$-operation acts only with respect to n_k in the k-th term on the left-hand side.

We have verified that this many-time formulation avoids the problems of a single-time multi-CA equation, *cf.* Eq. (18) [27]. – In particular, in the absence of interactions with each other, between CA subsystems, $\hat{\mathcal{I}} \equiv 0$, *no unphysical correlations* are introduced among independent CA subsystems.

Furthermore, *continuous multi-time equations* corresponding to Eqs. (22) are obtained by applying the mapping rules given in Section 2.3. to the discrete equations. We find no problem of incompatibility between multiplication according to nonlinear terms *vs.* linear mapping according to *Shannon's Theorem*, since a separate mapping is applied for each one of the clock variables. This effectively replaces $n_k \to t_k$, $k = 1, \ldots, m$, where t_k is a continuous real time variable. In this way, a *modified multi-time Schrödinger equation* is obtained:

$$\sum_{k=1}^{m} \sinh [l \frac{\mathrm{d}}{\mathrm{d}t_k}] \Psi^{\ldots\alpha_k\ldots}_{\ldots t_k\ldots} = \frac{1}{i}(\sum_{k=1}^{m} H^{\alpha_k\beta_k}_{(k)} \Psi^{\ldots\beta_k\ldots}_{\ldots t_k\ldots} + \mathcal{I}^{\ldots\alpha_k\ldots\ \beta_1\ldots\beta_m} \Psi^{\beta_1\ldots\beta_m}_{\ldots t_k\ldots}) \ , \tag{23}$$

where an overall factor of two from the left-hand side has been absorbed into the matrices on the right. By construction, here Ψ is bandlimited with respect to each variable t_k.

Performing the continuum limit, $l \to 0$, we arrive at the multi-time Schrödinger equation (one power of l^{-1} providing the physical dimension of $\hat{H}_{(k)}$ and $\hat{\mathcal{I}}$) considered by Dirac and Tomonaga [28, 29]. However, when l is fixed and finite, modifications in the form of powers of $l\mathrm{d}/\mathrm{d}t_k$ arise on its left-hand side, similarly as in single CA case before.

In the present nonrelativistic context, it may be appropriate to identify $t_k \equiv t$, $k = 1, ..., m$, in which case the operator on the left-hand side of Eq. (23), for $l \to 0$, can be simply replaced by $\mathrm{d}/\mathrm{d}t$. This results in the usual (single-time) *many-body Schrödinger equation.*

Finally, the study of the *conservation laws* of the multipartite CA equations of motion can be performed along the lines of Section 2.3. and analogous results have been obtained [27].

2.5.2. On the Superposition Principle in composite CA The equivalent discrete or continuous many-time equations (22) and (23) are both linear in the CA wave function Ψ. Therefore, superpositions of solutions of these equations also present solutions and the *Superposition Principle* does indeed hold for multipartite Hamiltonian CA.

IOP Conf. Series: Journal of Physics: Conf. Series **845** (2017) 012022 doi:10.1088/1742-6596/845/1/012022

As in the case of single CA, this entails the fact that these discrete systems – with all variables, parameters, *etc.* presented by Gaussian integers – can produce *interference* effects as in quantum mechanics. Even more interesting, their composites can also show *entanglement*, which is deemed an essential feature of QM. This follows from the form of the equations of motion, which allow for superpositions of factorized states.

A warning is in order, concerning expressions borrowed from QM which we used freely here, such as "wave functions" and "states". They usually invoke the notion of vectors in a *Hilbert space*, which turns into a complex projective space upon normalization of the vectors. However, as has become obvious in Section 2.3., see Eqs. (8)–(9), and which can be seen similarly in the multipartite case, as long as the CA are truly discrete ($l \neq 0$), the normalization (squared) of vectors is not among the conserved quantities, hence not applicable, but is replaced by a conserved two-time correlation function.

Furthermore, the space of states presently is *not* a Hilbert space, since it fails in two respects: the vector-space and completeness properties are absent. Instead, the space of states in the presented CA theory can be classified as a *pre-Hilbert module over the commutative ring of Gaussian integers* [27].

We conclude here that superpositions of states, interference effects, and entanglement, as in QM, all can be found already on the "primitive" level of the presently considered natural Hamiltonian CA, discrete single or multipartite systems which are characterized by (complex) integer-valued variables and couplings.

3. The formal solution of discrete CA equations of motion

We recall that the Schrödinger equation, $\partial_t\psi(t) = -i\hat{H}\psi(t)$, can be formally solved by exponentiating, $\psi(t) = \exp(-i\hat{H}t)\psi(0)$. Here we present the analogous formal solution of the CA equation of motion (3), *i.e.*, of the second-order finite difference equation $\dot{\psi}_n = \psi_{n+1} - \psi_{n-1} = -i\hat{H}\psi_n$, which can be solved by elementary means. We obtain:

$$\psi_n = (2\cos\hat{\phi})^{-1}\left(\mathrm{e}^{-in\hat{\phi}}[\mathrm{e}^{i\hat{\phi}}\psi_0 + \psi_1] + (-1)^n\mathrm{e}^{in\hat{\phi}}[\mathrm{e}^{-i\hat{\phi}}\psi_0 - \psi_1]\right) \; , \tag{24}$$

with $2\sin\hat{\phi} := \hat{H}$. From this implicit definition of the operator $\hat{\phi}$, we read off once again[1] that the eigenvalues of an admissible dimensionless Hamiltonian $\hat{H}$ must be constrained by $-2 \leq l\epsilon_\alpha \leq 2$, *i.e.*, in order to have solutions that neither grow nor decay exponentially in n.

We observe that the general solution of our second-order equation is determined by *two* initial values, ψ_0 and ψ_1, unlike the case of the first-order Schrödinger equation. In order to better understand this, we consider the continuum limit, letting $l \approx 0$, but keeping $t := nl$ fixed and the *dimensionful* eigenvalues ϵ_α as well. Thus, we have $2\sin\phi_\alpha = \epsilon_\alpha l \approx 0$, *i.e.* $\phi_\alpha \approx 0$. In this limit, the solution (24) correctly exponentiates, if and only if $\psi_1 \equiv \psi_0$.

Therefore, we have to restrict the set of solutions by allowing only the subset of initial values which satisfy the condition $\psi_1 \equiv \psi_0$, in order to assure quantum mechanical behaviour in the continuum limit. We speculate about the role of solutions with $\psi_1 \neq \psi_0$ shortly.

Rewriting Eq. (24) symbolically as $\psi_n = \hat{T}(n+1)\psi_1 + \hat{T}(n)\psi_0$, the evolution fulfills the composition law $\psi_n = \hat{T}(n-m+1)\psi_{m+1} + \hat{T}(n-m)\psi_m$, even for general initial conditions. This can be demonstrated by induction and corresponds to the semigroup property of unitary evolution in quantum mechanics.

3.1. Are Hamiltonian CA of the ontological kind?

In order to address this question, we have to define what is meant by 'ontological'. Here, we refer to the fundamental hypothesis of the *CA interpretation of quantum mechanics* proposed

[1] *Cf.* the remarks following Eq. (16).

and elaborated by G. 't Hooft [1]. Namely, there are so-called *ontological states which evolve deterministically.* They constitute the physical reality of nature that we wish to explore. [2]

We presently consider a CA characterized by a finite number of degrees of freedom. Its corresponding discrete states may serve to define the basis of a Hilbert space. However, reflecting their ontological character, formal superpositions among these states – which we do not hesitate to work with in QM – are not admitted. Superpositions of ontological states are *not* ontological, they are not "out there" in the Universe. As a consequence, a finite number of ontological states can only evolve by *permutations* among themselves! [3]

The question in the title of this section can now be asked more precisely: Is the Hamiltonian CA dynamics, as described, compatible with the permutation dynamics of ontological states? We will find some partial answers by resorting to very simple indicative examples.

Let ψ_n^α, $\alpha = 1, 2$ denote the state variables and $\hat{H}$ a 2x2 self-adjoint Hamiltonian matrix, all (complex) integer-valued, as before in Section 2. It is convenient to collect the state variables into a two-dimensional vector and write the evolution equation (3) as:

$$\psi_n = \psi_{n-2} - i\hat{H}\psi_{n-1} \;\; , \;\;\; \psi_n \equiv \begin{pmatrix} \psi_n^1 \\ \psi_n^2 \end{pmatrix} \; . \tag{25}$$

We have argued that the formal solution of such an equation, given in Eq. (24), approaches an exponential as in QM for the continuum limit ($l \to 0$), if and only if the necessary two inital values coincide. Assuming this, *e.g.* $\psi_1 = \psi_0 := (1, 0)^t$, we find that the evolution generally takes place in the two-dimensional space given; however, it cannot be avoided that in the sequence $\{\psi_n\}$ appear regularly states which are superpositions of any two linearly independent states, say ψ_m and $\psi_{m'}$, which we may select as ontological basis states, no matter how we choose $\hat{H}$. In this case, evolution does *not* consist of permutations only.

In other words, in this two-dimensional example, we learn that the discrete evolution which conforms with QM in the continuum limit passes unavoidably through superposition states as intermediates. Which are not ontological. We leave it as a *conjecture* that this happens independently of the dimensionality of the state space. Which seems to disqualify such a Hamiltonian CA as an example of a strictly ontological model.

It is tempting speculate now that an ontological model must deviate in one way or another from the Hamiltonian CA under consideration and from a QM model in its continuum limit. – One way could be to give up self-adjointness of the Hamiltonian, which we do not follow here. – Another is to abandon the condition of coinciding initial values, $\psi_1 \equiv \psi_0$, thereby renouncing the requirement that such a model have an obvious QM continuum limit, *cf.* Section 2.4.

For example, we may choose one of the Pauli matrices as Hamiltonian:

$$\hat{H} := \begin{pmatrix} 0 & 1 \\ 1 & 0 \end{pmatrix} \; , \tag{26}$$

with $\hat{H}^2 = \mathbf{1}$. Then, considering the initial values $\psi_0 = (1, 0)^t$ and $\psi_1 = (0, 1)^t$, we simply evaluate the first few steps of the evolution according to Eq. (25) to obtain:

$$\psi_2 = (1-i)\psi_0 \; , \; \psi_3 = -i\psi_1 \; , \; \psi_4 = -i\psi_0 \; , \; \psi_5 = -(1+i)\psi_1 \; , \; \psi_6 = -\psi_0 \; , \; \psi_7 = -\psi_1 \; , \; \dots \; . \tag{27}$$

It is obvious how the sequence continues and arrives at the initial values within the next 6 steps, thus becoming *periodic*. We could have used the general solution (24) instead, with $\hat{\phi} = \frac{\pi}{6}\hat{H}$,

[2] The essence of this hypothesis is summarized in Figure 6 of the book [1] and detailed in the ensuing discussion there, pointing towards astonishing implications.

[3] At present, we are not concerned with the important relations and distinctions between ontological and either quantum mechanical or classical states, which are specified in Ref. [1].

in the present example. – The rescaling in the intermediate states ψ_2 and ψ_5 would violate the conserved normalization of a state vector in QM. However, it is perfectly in accordance with the corresponding discrete conservation laws discussed in Section 2.3. and applicable here. – In this way, we have found an example of an elementary permutation dynamics realized by a Hamiltonian CA.

It will be interesting to see whether or how the findings here generalize to more complex higher dimensional ontological models.

4. Conclusion

This presents a brief review of earlier work which has demonstrated surprising quantum features arising in integer-valued, hence "natural", *Hamiltonian cellular automata* [2, 3, 23, 25, 27].

The study of this particular class of CA is motivated by 't Hooft's *CA interpretation of quantum mechanics* [1] and various recent attempts to construct models which may eventually lead to demonstrating that the essential features of QM can all be understood to emerge from pre-quantum deterministic dynamics and that its puzzles, such as the measurement problem, can be satisfactorily resolved after all.

The single CA we have considered allow practically for the first time to reconstruct quantum mechanical models with nontrivial Hamiltonians in terms of such systems with a *finite discreteness scale*. – Furthermore, we have extended this study by describing *multipartite systems*, analogous to many-body QM. Not only is this useful for the construction of more complex models *per se* (especially with a richer structure of energy spectra), but it is also necessary, in order to extend the Superposition Principle of QM to a description at the CA level. We find that it can be introduced already there to the fullest extent, compatible with a tensor product structure of multipartite states, which entails the possibilities of their *interference* and *entanglement*.

Surprisingly, we have been forced – in our approach employing Sampling Theory to construct the map between CA and an equivalent continuum picture – to introduce a many-time formulation, which only appeared in relativistic quantum mechanics before, as introduced by Dirac, Tomonaga, and Schwinger [28, 29, 30]. This points towards a crucial further step in these developments, which is still missing, namely a relativistic CA model of *interacting quantum fields*. Without the possibility of interacting multipartite CA with quantumlike features, as described here, it is hard to envisage a CA picture of dynamical fields spread out in spacetime.

Last not least, we have presented a primitive attempt to see the relation, if any, between the Hamiltonian CA here and ontological models as advocated in Ref. [1]. Which resulted in the suggestion that it might depend on initial conditions, whether such a CA behaves either as a QM model (with corrections due to discreteness) or is of the ontological kind.

Acknowledgments

It is a pleasure to thank Jack Ng for a discussion, the organizers of the 10th Biennial Conference on Classical and Quantum Relativistic Dynamics of Particles and Fields IARD 2016 (Ljubljana, June 2016) for the invitation to this very nice meeting, and Martin Land and Matej Pavsic for their kind hospitality.

References

[1] 't Hooft G 2014 The Cellular Automaton Interpretation of Quantum Mechanics. A View on the Quantum Nature of our Universe, Compulsory or Impossible? *Preprint* arXiv:1405.1548

[2] Elze H-T 2014 Action principle for cellular automata and the linearity of quantum mechanics *Phys. Rev.* A **89** 012111

[3] Elze H-T 2016 Quantum features of natural cellular automata *J. Phys.: Conf. Ser.* **701** 01201; *do.* 2014 The linearity of quantum mechanics from the perspective of Hamiltonian cellular automata *J. Phys.: Conf. Ser.* **504** 012004

IOP Conf. Series: Journal of Physics: Conf. Series **845** (2017) 012022 doi:10.1088/1742-6596/845/1/012022

[4] 't Hooft G 1990 Quantization of discrete deterministic theories by Hilbert space extension *Nucl. Phys.* B **342** 471
[5] 't Hooft G, Isler K and Kalitzin S 1992 Quantum field theoretic behavior of a deterministic cellular automaton *Nucl. Phys.* B **386** 495
[6] 't Hooft G 1997 Quantummechanical behaviour in a deterministic model *Found. Phys. Lett.* **10** 105
[7] Haba Z and Kleinert H 2002 Towards a simulation of quantum computers by classical systems *Phys. Lett.* A **294** 139
[8] Elze H-T and Schipper O 2002 Time without time: a stochastic clock model *Phys. Rev.* D **66** 044020
[9] Grössing G 2004 From classical Hamiltonian flow to quantum theory: derivation of the Schrödinger equation *Found. Phys. Lett.* **17** 343
[10] Blasone M, Jizba P, Scardigli F and Vitiello G 2009 Dissipation and quantization for composite systems *Phys. Lett.* A **373** 4106
[11] Sakellariadou M, Stabile A and Vitiello G 2011 Noncommutative spectral geometry, algebra doubling and the seeds of quantization *Phys. Rev.* D **84** 045026
[12] Acosta D, Fernandez de Cordoba P, Isidro J M and Santander J L G 2012 An entropic picture of emergent quantum mechanics *Int. J. Geom. Meth. Mod. Phys.* **9** 1250048
[13] Weinberg S 1989 Precision tests of quantum mechanics *Phys. Rev. Lett.* **62** 485; *do.* 1989 Testing quantum mechanics *Annals of Physics* **194** 336
[14] Elze H-T 2008 A relativistic gauge theory of nonlinear quantum mechanics and Newtonian gravity *Int. J. Theor. Phys.* **47** 455
[15] Jordan T F 2006 Assumptions that imply quantum dynamics is linear *Phys. Rev.* A **73** 022101; *do.* 2009 Why quantum dynamics is linear *J. Phys.: Conf. Ser.* **196** 012010
[16] Lee T D 1983 Can time be a discrete dynamical variable? *Phys. Lett.* **122**B 217
[17] Elze H-T 2013 Discrete mechanics, time machines and hybrid systems *EPJ Web of Conferences* **58** 01013
[18] Shannon C E 1949 Communications in the presence of noise *Proc. IRE* **37** 10
[19] Jerri A J 1977 The Shannon Sampling Theorem — Its Various Extensions and Applications: A Tutorial Review *Proc. IEEE* **65** 1565
[20] Kempf A 2010 Spacetime could be simultaneously continuous and discrete in the same way that information can *New J. Phys.* **12** 115001
[21] Heslot A 1985 Quantum mechanics as a classical theory *Phys. Rev.* D **31** 1341
[22] Gigli D 2014 Application of Shannon's Sampling Theorem in Quantum Mechanics *Master Thesis* (University of Pisa, December 2014)
[23] Elze H-T 2015 Are nonlinear discrete cellular automata compatible with quantum mechanics? *J. Phys.: Conf. Ser.* **631** 012069
[24] Skinner T E 2013 An exact mapping between the states of arbitrary N-level quantum systems and the positions of classical coupled oscillators *Phys. Rev.* A **88** 012110
[25] Elze H-T 2014 Quantumness of discrete Hamiltonian cellular automata *EPJ Web of Conferences* **78** 02005
[26] McKee J F and Smyth C J 2007 Integer symmetric matrices having all their eigenvalues in the interval [-2,2] *J. Algebra* **317** (1) 260
[27] Elze H-T 2016 Multipartite cellular automata and the superposition principle *Int. J. Qu. Info. (IJQI)* **14** (4) 1640001
[28] Dirac P A M 1932 Relativistic Quantum Mechanics *Proc. Roy. Soc. London* A **136** 453
[29] Tomonaga S 1946 On a Relativistically Invariant Formulation of the Quantum Theory of Wave Fields *Prog. Theor. Phys.* **1** (2) 27
[30] Schwinger J 1948 Quantum Electrodynamics. I. A covariant Formulation *Phys. Rev.* **74** 1439

IOP Conf. Series: Journal of Physics: Conf. Series **845** (2017) 012023 doi:10.1088/1742-6596/845/1/012023

The Feynman-Dyson view

Tepper L. Gill [1]

[1] Department of Mathematics, Physics and E&CE, Howard University Washington DC 20059 USA

E-mail: tgill@howard.edu

Abstract. This paper is a survey of our work on the mathematical foundations for the Feynman-Dyson program in quantum electrodynamics (QED). After a brief discussion of the history, we provide a representation theory for the Feynman operator calculus. This allows us to solve the general initial-value problem and construct the Dyson series. We show that the series is asymptotic, thus proving Dyson's second conjecture for quantum electrodynamics. In addition, we show that the expansion may be considered exact to any finite order by producing the remainder term. This implies that every nonperturbative solution has a perturbative expansion. Using a physical analysis of information from experiment versus that implied by our models, we reformulate our theory as a sum over paths. This allows us to relate our theory to Feynman's path integral, and to prove Dyson's first conjecture that the divergences are in part due to a violation of Heisenberg's uncertainly relations. As a by-product, we also prove Feynman's conjecture about the relationship between the operator calculus and has path integral. Thus, providing the first rigorous justification for the Feynman formulation of quantum mechanics.

Introduction

Following Dirac's quantization of the electromagnetic field in 1927, and his relativistic electron theory in 1928, the equations for quantum electrodynamics (QED) were developed by Heisenberg and Pauli in the years 1929-30 (see Miller [1] and Schweber [2]). From the beginning, when researchers attempted to use the straightforward and physically intuitive time-dependent perturbation expansion to compute physical observables, a number of divergent expressions appeared. Although it was known that the same problems also existed in classical electrodynamics, it was noted by Oppenheimer that there was a fundamental difference in the quantum problem when compared to the classical one. Dirac [3] had shown that, in the classical case, one could account for the problem of radiation reaction without directly dealing with the self-energy divergence by using both advanced and retarded fields and a particular limiting procedure.

Early attempts to develop subtraction procedures for the divergent expressions were very discouraging because they depended on both the gauge and the Lorentz frame, making them appear ambiguous. Although the equations of QED were both Lorentz and gauge covariant, it was generally believed that, in a strict sense, they had no solutions expandable in powers of the charge. The thinking of the times was clearly expressed by Oppenheimer in his 1948 report to the Solvay Conference, "If one wishes to explore these solutions, bearing in mind that certain infinite terms will, in a later theory, no longer be infinite, one needs a covariant way of identifying these terms; and for that, not merely the field equations themselves, but the whole method of approximation and solution must at all stages preserve covariance."

The solution to the problem posed by Oppenheimer was made independently by Tomonaga [4], Schwinger [5] and Feynman [6],[7], [8]. These papers may be found in Schwinger [5]. Tomonaga introduced what is now known as the interaction representation and showed how the approximation process could be carried out in a covariant manner. Schwinger developed the general theory and applied it to many of the important problems. Feynman took a holistic view of physical reality in his development. He suggested that we view a physical event as occurring on a film which exposes more and more of the outcome as the film unfolds. His idea was to deal directly with the solutions to the equations describing the physical system, rather than the equations themselves. In addition to solving the problem posed by Oppenheimer, Feynman's approach led to a new perturbation series, which provided an easy, intuitive, and computationally simple method to study interacting particles while giving physical meaning to each term in his expansion.

Since Feynman's method and approach was so different, it was not clear how it related to that of Schwinger and Tomonaga. Dyson [9], [10] made a major contribution. Dyson realized that Feynman and Schwinger were both dealing with different versions of Heisenberg's S-matrix. He then formally introduced time-ordering and provided a unified approach by demonstrating the equivalence of the Feynman and Schwinger-Tomonaga theories. This approach also allowed him to show how the Schwinger theory could be greatly simplified and extended to all orders of the perturbation expansion. Dyson's time-ordering idea was actually obtained from discussions with Feynman, who later explored and fully developed it into his time-ordered operator calculus.

After the problem proposed by Oppenheimer was resolved, attitudes toward the renormalization program and quantum field theory could be classified into three basic groups. The first group consisted of those who were totally dissatisfied with the renormalization program. The second group considered the renormalization program an interim step and believed that the divergences were an indication of additional physics, which could not be reached by present formulations. The first two groups will not be extensively discussed here. However, we can associate the names of Dirac and Landau with the first group, and Sakata and Schwinger with the second. (See Dirac,[11] Sakata,[12] Schwinger, [13] and also Schweber.[14].

The third group was more positive, and directed its attention toward investigating the mathematical foundations of quantum field theory with the hope of providing a more orderly approach to the renormalization program (assuming that the theory proved consistent). This direction was clearly justified since part of the problem had been consistently blamed on a mathematical issue, the perturbation expansion. Indeed, the whole renormalization program critically depended on the expansion of the S-matrix in powers of the coupling constant. This concern was further supported since attempts to use the expansion when the coupling constant was large led to meaningless results. Additional unease could be attributed to the fact that, at that time, not much was actually known about the physically important cases where one was dealing with unbounded operator-valued distributions.

Researchers working on the mathematical foundations of quantum electrodynamics and quantum field theory adopted the name axiomatic field theory starting in the 1950s. These researchers focused on trying to find out what could be learned about the existence of local relativistic quantum field theories based on certain natural assumptions which included the postulates of quantum mechanics, locality, Poincare invariance, and a reasonable spectrum. This approach was initiated by the work of Wightman [15], and Lehmann, Symanzik, and Zimmermann [16]. In this approach, the quantized field is interpreted mathematically as an operator-valued Schwartz distribution. Explicit use of the theory of distributions was a major step, which helped to partially make the theory mathematically sound by smoothing out the fields locally.

The axiomatic approach proved very fruitful, providing the first rigorous proofs of a number of important general results, and attracted many able researchers. The favored name 1960-1970 era was algebraic quantum field theory. The books by Jost [17], Streater and Wightman [18], and Bogolubov and Shirkov [19] are the classics, while more recent work can be found in Haag [20]. (See also the book by Bogolubov, Logunov, and Todorov,[21].)

Although a great deal of work was done in constructive field theory between 1950 and 1990, many difficult problems still remain. For example, the appearance of difficulties with the constructive approach to polynomial types of field theories is discussed in the paper by Sokal [22]. He identified some mathematical difficulties and conjectured that the preliminary successes in two and three space-time dimensions might not work as expected in four space-time. His conjecture was independently verified by Aizenman and Graham [23] and Fröhlich [24]. Since then, research on this and related problems has almost vanished.

Purpose

During his Gibbs Lecture [25] Dyson, suggested that providing the mathematical foundations for Feynman's sum over histories, or path integral approach would also provide the foundations for quantum field theory within the framework of the C^*-algebra approach due to Haag and Kastler [26].

At the end of his book on path integrals with Hibbs [27], Feynman states: "Nevertheless, many of the results and formulations of path integrals can be reexpressed by another mathematical system, a kind of ordered operator calculus. In this form many of the results of the preceding chapters find an analogous but more general representation ... involving noncommuting variables." Thus, Feynman conjectured that mathematical meaning for the time-ordered operator calculus would automatically solve the path integral problem and lead to wider applications.

A number of mathematicians have formally investigated the time-ordering problem: Miranker and Weiss [28], Nelson [29], Araki [30] and Maslov [31]. The recent comprehensive work by Johnson and Lapidus contains references to almost all contributions to the field (see [32]). However, none were able to provide any insight into the physical problems associated with quantum electrodynamics and quantum field theory in general, nor were they able to connect time-ordering to Feynman's path integral.

In this paper, we discuss a rigorous implementation of Feynman's program (see also [33]). First, von Neumann's infinite tensor product Hilbert space [34] is used to construct the mathematical version of Feynman's film. Then, the basic methods of analysis are extended to this setting, making it possible to prove Feynman's conjecture and the remaining two conjectures of Dyson. (It is of historical note that many of the tools required for this work were developed by von Neumann for other purposes in the 1930's [34].)

This paper is part of a new investigation into the physical and mathematical foundations of relativistic quantum theory. Our overall goal is to construct a self-consistent relativistic quantum theory of particles and fields. For this paper, we have two specific objectives. Our first and major objective is to construct a physically simple and computationally useful representation theory for the Feynman time-ordered operator calculus. A correct formulation and representation theory for the Feynman time-ordered operator calculus should at least have the following desirable features:

(i) It should provide a transparent generalization of current analytic methods without sacrificing the physically intuitive and computationally useful ideas of Feynman.

(ii) It should provide a clear approach to some of the mathematical problems of relativistic quantum theory.

(iii) It should explain the connection with path integrals.

In the course of his analysis, unification, and simplification of the Feynman, Schwinger Tomonaga theory, Dyson made two important suggestions (conjectures). The first conjecture concerned the divergences in QED, while the second was concerned with the convergence of the renormalized perturbation series. In addressing the problem of divergences, Dyson conjectured that they may be due to an idealized conception of measurability resulting from the infinitely precise knowledge of the space-time positions of particles implied by our Hamiltonian formulation which leads to a violation of the Heisenberg uncertainty principle. This point of view can be traced directly to the Bohr-Rosenfeld theory of measurability for field operators. In addressing the renormalized S-matrix, Dyson suggested that it might be more reasonable to expect the expansion to be asymptotic rather than convergent and gave physical arguments to support his claim. The lack of a clear mathematical framework made it impossible to formulate and investigate his suggestions. Schweber [2] notes that Dyson made two other well-known conjectures. The overlapping divergences conjecture was proved by Salam. Dyson's conjecture that a certain Feynman integral converges, necessary for showing that the ultraviolet divergences cancel to all orders, was proved by Weinberg. Our second objective is to provide proofs of the above two conjectures under general conditions that should apply to any formulation of quantum field theory which does not abandon Hamiltonian generators for unitary solution operators. The proof of the first conjecture is, to some extent expected, and is a partial vindication of our belief in the consistency of quantum electrodynamics in the sense that the ultraviolet problem is caused by an effect that is basically simple. Such a result is partly anticipated since the effect can be made to disappear via appropriate cutoffs. We also identify special conditions under which the renormalized perturbation series may actually converge.

1. Time-Ordering

In the Feynman view, particles live and die on a photograph of their motion, in which time acts as a director. If the time interval (for convenience) is $I = [-T, T]$, physical analysis shows that the actual model can be represented as a curve on $\mathbf{X} = \prod_{-T}^{T} \mathbb{R}_p^3(t)$, where $\mathbb{R}_p^3(t)$ is a physical coordinate system, envisioned as $\mathbb{R}^3$ attached to an observer, including measuring devices and any background effects (either local or distant), which affect the observer's ability to obtain precise information about the physical evolution at time t.

Thus, true physical events occur on $\mathbf{X}$, where actual experimental information is modified by fluctuations in $\mathbb{R}_p^3(t)$ and by the interaction of the micro-system with the measuring equipment. However, in terms of theoretical representations, mathematical models of the evolution are represented by wave functions, amplitudes, and/or operator-valued distributions, etc, on a Hilbert space. It follows that there are three spaces, the physical space of evolution for the micro-system, the observer's space of obtainable information concerning this evolution and the mathematical modeling space.

1.1. The Film

The modeling space of the obtainable information or film, is a Hilbert bundle over the time interval I. Fix a separable space $\mathcal{H}$ and let $\hat{\otimes}_{t\in I}\mathcal{H}(t) =: \mathcal{H}_\otimes^2$ be the infinite tensor product Hilbert space of von Neumann [34]. In order to gain a feel for this space, we need to defined uncountable products.

In order to avoid trivialities, we always assume that, in any product, all terms are nonzero.

Definition 1.1. *If $\{z_t\}$ is a sequence of complex numbers indexed by $t \in I$,*

(i) We say that the product $\prod_{t\in I} z_t$ is convergent with limit z if, for every $\varepsilon > 0$, there is a finite set $J(\varepsilon)$ such that, for all finite sets $J \subset I$, with $J(\varepsilon) \subset J$, we have $\left|\prod_{t\in J} z_t - z\right| < \varepsilon$.

(ii) We say that the product $\prod_{t\in I} z_t$ is quasi-convergent if $\prod_{t\in I} |z_t|$ is convergent. (If the product is quasi-convergent, but not convergent, we assign it the value zero.)

IOP Conf. Series: Journal of Physics: Conf. Series **845** (2017) 012023 doi:10.1088/1742-6596/845/1/012023

To see why the second definition is natural, let $z_t = e^{i\theta(t)}$, where $\theta(t)$ is an arbitrary real-valued function. Since I is not countable, we note that

$$0 < \left|\prod\nolimits_{t\in I} z_t\right| < \infty \text{ if and only if } \sum\nolimits_{t\in I} |1 - z_t| < \infty. \tag{1.1}$$

Thus, it follows that convergence implies that at most a countable number of the $z_t \neq 1$.

Definition 1.2. *Let $\phi = \underset{t\in I}{\otimes} \phi_t$ and $\psi = \underset{t\in I}{\otimes} \psi_t$ be in $\mathcal{H}^2_{\otimes}$.*

(i) We say that ϕ is strongly equivalent to ψ ($\phi \equiv^s \psi$), if and only if $\sum\limits_{t\in I} |1 - \langle\phi_t, \psi_t\rangle_t| < \infty$.

(ii) We say that ϕ is weakly equivalent to ψ ($\phi \equiv^w \psi$), if and only if $\sum\limits_{t\in I} |1 - |\langle\phi_t, \psi_t\rangle_t|\,| < \infty$.

We note the slight difference between strong and weak, $\langle\phi_t, \psi_t\rangle_t$ verses $|\langle\phi_t, \psi_t\rangle_t|$.

The proofs of results in this section are due to von Neumann [34].

Theorem 1.3. *The relations defined above are equivalence relations on $\mathcal{H}^2_{\otimes}$, which decomposes $\mathcal{H}^2_{\otimes}$ into disjoint equivalence classes.*

Definition 1.4. *For $\varphi = \underset{t\in I}{\otimes} \varphi_t \in \mathcal{H}^2_{\otimes}$, we define $\mathcal{H}^2_{\otimes}(\varphi)$ to be the closed subspace generated by the span of all $\psi \equiv^s \varphi$ and we call it the strong partial tensor product space generated by the vector φ. (von Neumann called them incomplete tensor product spaces.)*

Theorem 1.5. *For the partial tensor product spaces, we have the following:*

(i) If $\psi_t \neq \varphi_t$ occurs for at most a finite number of t, then $\psi = \underset{t\in I}{\otimes} \psi_t \equiv^s \varphi = \underset{t\in I}{\otimes} \varphi_t$.

(ii) The space $\mathcal{H}^2_{\otimes}(\varphi)$ is the closure of the linear span of $\psi = \underset{t\in I}{\otimes} \psi_t$ such that $\psi_t \neq \varphi_t$ occurs for at most a finite number of t.

(iii) If $\Phi = \otimes_{t\in I}\varphi_t$ and $\Psi = \otimes_{t\in I}\psi_t$ are in different equivalence classes of $\mathcal{H}^2_{\otimes}$, then $(\Phi, \Psi)_{\otimes} = \prod_{t\in I} \langle\varphi_t, \psi_t\rangle_t = 0$.

(iv) $\mathcal{H}^2_{\otimes}(\varphi)^w = \underset{\psi\equiv^w\phi}{\oplus} \left[\mathcal{H}^2_{\otimes}(\psi)^s\right]$.

If $L[\mathcal{H}^2_{\otimes}]$ is the set of bounded linear operators on $\mathcal{H}^2_{\otimes}$, for each $t \in I$, define $L[\mathcal{H}(t)] \subset L[\mathcal{H}^2_{\otimes}]$ by:

$$L[\mathcal{H}(t)] = \left\{\mathbb{H}(t) = (\underset{T\geqslant s>t}{\hat{\otimes}} \mathrm{I}_s) \otimes H(t) \otimes (\underset{t>s\geqslant -T}{\otimes} \mathrm{I}_s), \forall H(t) \in L[\mathcal{H}]\right\}, \tag{1.2}$$

where I_s is the identity operator. Let $L^{\#}[\mathcal{H}^2_{\otimes}]$ be the uniform closure of the algebra generated by $\{L[\mathcal{H}(t)],\ t \in I\}$. If the family $\{H(t), t \in I\}$ is in $L[\mathcal{H}]$, then the operators $\{\mathbb{H}(t), t \in I\} \in L^{\#}[\mathcal{H}^2_{\otimes}]$ commute when acting at different times:

$$\mathbb{H}(t)\mathbb{H}(\tau) = \mathbb{H}(\tau)\mathbb{H}(t) \text{ for } t \neq \tau.$$

Let $E = \underset{t\in I}{\otimes} e_t$ be a unit vector, let $\mathcal{H}^2_{\otimes}(E)$ be the partial tensor product space generated by E and let $\mathbf{P}_E$ denote the projection from $\mathcal{H}^2_{\otimes}$ onto $\mathcal{H}^2_{\otimes}(E)$. If $\mathbf{T}^t_\theta$ is used to represent the mapping from $L(\mathcal{H}) \to L(\mathcal{H}(t))$ in (2), von Neumann [34] proved that:

Theorem 1.6. *The set $L(\mathcal{H}(t))$ is a uniformly closed algebra and $\mathbf{T}^t_\theta : L(\mathcal{H}) \to L(\mathcal{H}(t))$ is an isometric isomorphism of algebras. (We call $\mathbf{T}^t_\theta$ the time-ordering morphism.)*

Theorem 1.7. *If $\mathbf{T} \in L^{\#}[\mathcal{H}^2_{\otimes}]$, then $\mathbf{P}_E\mathbf{T} = \mathbf{T}\mathbf{P}_E$.*

Thus, every operator in $L^{\#}[\mathcal{H}^2_{\otimes}]$ is the uniform limit of operators from the algebra generated by the family $\{L[\mathcal{H}(t)],\ t \in I\}$ and is invariant on $\mathcal{H}^2_{\otimes}(E)$ (for any E).

Definition 1.8. *An exchange operator* $\mathbf{E}[t,t']$ *is a linear map defined for pairs* t, t' *such that:*

(i) $\mathbf{E}[t,t'] : L[\mathcal{H}(t)] \to L[\mathcal{H}(t')]$, *(isometric isomorphism),*
(ii) $\mathbf{E}[s,t']\mathbf{E}[t,s] = \mathbf{E}[t,t']$,
(iii) $\mathbf{E}[t,t']\mathbf{E}[t',t] = I$,
(iv) for $s \neq t,\ t'$, $\mathbf{E}[t,t']\mathcal{H}(s) = \mathcal{H}(s)$, *for all* $\mathcal{H}(s) \in L[\mathcal{H}(s)]$.

The exchange operator acts to interchange the time positions of a pair of operators in an expression. We can now construct our film. Let $\{e^i\}$ be an orthonormal basis for $\mathcal{H}$ and set $E^i = \otimes e^i_t$, where $e^i_t = e^i$ for each i. Let $\mathcal{FD}^i_2 = \mathcal{H}^2_{\otimes}(E^i)$ be the partial tensor product space generated by the vector E^{i} and set $\mathcal{FD}^2_{\otimes} = \overset{\infty}{\underset{i=1}{\oplus}} \mathcal{FD}^i_2$. It is not hard to see that the fiber at each time-slice is isomorphic to $\mathcal{H}$.

Definition 1.9. *We call* $\mathcal{FD}^2_{\otimes}$ *the Feynman-Dyson space (FD-space) on* $\mathcal{H}$ *over* I.

In order to understand why $\mathcal{FD}^2_{\otimes}$ is the correct structure for our film, we need an explicit basis for each $\mathcal{FD}^i_2$. Let $\mathbf{F}$ be the set of all functions $f(\,\cdot\,) : I \to \mathbb{N} \cup \{0\}$ such that $f(t)$ is zero for all but a finite number of t, and let $F(f)$ denote the image of the function $f(\cdot)$. Set $E^i_{F(f)} = \otimes_{t \in I} e^i_{t,f(t)}$ with $e^i_{t,0} = e^i$, and $f(t) = k$ implies $e^i_{t,k} = e^k$.

It is shown by von Neumann that the set of vectors $\{E^i_{F(f)} \,|F(f) \in \mathbf{F}\}$ is an orthonormal basis for $\mathcal{FD}^i_2$ (so that $\left\langle E^i_{F(f)}, E^i_{F(g)} \right\rangle = \prod_{t \in I} \left\langle e^{\mathrm{i}}_{t,f(t)}, e^{\mathrm{i}}_{t,g(t))} \right\rangle = 0$ unless $f(t) = g(t)$ for all $t \in I$.)

1.2. *Integrals and Evolutions*

Recall that a unitary group $U(t)$ is said to be a C_0-unitary group if it is strongly continuous, $U(0) = I$, and $\lim\limits_{|t| \to 0} U(t)\phi = \phi$ for all $\phi \in \mathcal{H}$.

We now assume that $H(t)$ is strongly continuous in t and is the generator of a C_0-unitary group on $\mathcal{H}$, with domain $D(H(t))$, independent of $t \in I$. It follows that $Q[T,-T] = \int_{-T}^{T} H(t)dt \in C(\mathcal{H})$ (the closed densely defined linear operators on $\mathcal{H}$) as a strong Riemann integral.

If $\Phi \in \mathcal{FD}^2_{\otimes}$, with $\Phi = \sum_{\mathrm{i}} \sum_{F(f)} a^{\mathrm{i}}_{F(f)} E^{\mathrm{i}}_{F(\mathrm{f})}$ and if $\sum\limits_{F(f)}^{K} \left| a^{\mathrm{i}}_{F(f)} \right|^2 = \left| b^i \right|^2$, we have: (using the basis)

Theorem 1.10. *(Fundamental Theorem for Time-Ordered Integrals)*
Let $\mathbb{H}(t)$ *be the corresponding time-ordered version of the family* $\{H(t),\ t \in I\}$. *If* $\Phi \in D(H(t))$, $-T < t \leq T$, *then*

(i) The operator $\mathbf{Q}[t,-T] = \int_{-T}^{t} \mathbb{H}(s)ds \in C(\mathcal{FD}^2_{\otimes})$ *is a strong Riemann integral, which generates the* C_0*-unitary group* $exp\{i\mathbf{Q}[t,-T]\}$, *on* $\mathcal{FD}^2_{\otimes}$ *and, for each* $t \leq T$.
(ii) $\mathbf{Q}[t,s]\Phi + \mathbf{Q}[s,-T]\Phi = \mathbf{Q}[t,-T]\Phi$ *(a.s.).*
(iii) $\|\mathbf{Q}[t,-T]\Phi\|^2_{\otimes} = \sum\limits_{\mathrm{i}}^{J} \left| b^i \right|^2 \left| \langle Q[t,-T]e^i, e^i \rangle \right|^2$ *(a.s).*
(iv) If $\Phi(t) = exp\{i\mathbf{Q}[t,-T]\}\Phi$, *then*

$$i\frac{\partial \Phi(t)}{\partial t} = \mathbb{H}(t)\Phi(t), \quad \Phi(-T) = \Phi$$

IOP Conf. Series: Journal of Physics: Conf. Series **845** (2017) 012023 doi:10.1088/1742-6596/845/1/012023

The proof can be found in [33]. In (2) and (3) of the above theorem, the exceptional set is at most countable. This theorem shows that using time to order operators offers much more than just an index. It is shown in [33] that we need only assume weak continuity of $H(t)$, along with a growth condition, in order to obtain the same result. This theorem implies that all of analysis can be lifted to this setting under less conditions with no losses.

In physics, the interesting models represent interacting particles with $\mathbb{H}(t) = \mathbb{H}_0(t) + w\mathbb{H}_1(t)$, where $w\mathbb{H}_1(t)$ is small in some sense. (For example, $\|\mathbb{H}_1\Phi\|/\|\mathbb{H}_0\Phi\| \ll w$ with a few mathematical conditions will suffice.) In the most useful cases, everything is calculated using perturbation theory. The relevant computational algorithm starts with:

$$\mathbf{U}_0[t,-T] = exp\left\{-\frac{i}{\hbar}\mathbf{Q}_0[t,-T]\right\},\ \mathbf{Q}_0[t,-T] = \int_{-T}^{t} \mathbb{H}_0(\tau)d\tau.$$

The interaction representation is then defined by $\mathbb{H}_\mathbf{I}(t) = \mathbf{U}_0[-T,t]\mathbb{H}_1(t)\mathbf{U}_0[t,-T]$ and ($v = -\frac{iw}{\hbar}$)

$$\mathbf{U}_\mathbf{I}^w[t,-T] = exp\left\{v\mathbf{Q}_\mathbf{I}[t,-T]\right\},\ \mathbf{Q}_\mathbf{I}[t,-T] = \int_{-T}^{t} \mathbb{H}_\mathbf{I}(\tau)d\tau.$$

One then writes

$$\Phi(t) = \Phi + \sum_{k=1}^{\infty} v^k \int_{-T}^{t} ds_1 \int_{-T}^{s_1} ds_2 \cdots \int_{-T}^{s_{k-1}} ds_k \mathbb{H}_\mathbf{I}(s_1)\mathbb{H}_\mathbf{I}(s_2)\cdots\mathbb{H}_\mathbf{I}(s_k)\Phi,$$

where $\Phi(t) = \mathbf{U}_\mathbf{I}^w[t,-T]\Phi$ and w is a small constant (charge).

An important theorem of Haag [35] shows that the interaction representation in sharp time (as above) does not exist in a rigorous sense (see Streater and Wightman [36], pg. 161). Haag shows that the equal time commutation relations for the canonical variables of an interacting field are equivalent to those of a free field. In trying to explain this unfortunate result, Streater and Wightman point out that (see [36], p. 168) "... What is even more likely in physically interesting quantum field theories is that equal-time commutation relations will make no sense at all; the field might not be an operator unless smeared in time as well as space." It has now been experimentally confirmed that there is quantum interference in time (see [37]). Thus, Hagg's assumption of sharp time is not physically valid.

The mathematical problem is to provide some justification for the amazingly accurate results obtained by Feynman and others. To address this problem, we begin with the notion of an asymptotic expansion. We follow Hille and Phillips [38]. Let $\mathbf{U}_\mathbf{I}^w[t,-T] = \exp\left\{v\mathbf{Q}_\mathbf{I}[t,a]\right\}$.

Definition 1.11. *We say that* $\mathbf{U}_\mathbf{I}^w[t,-T]$ *is asymptotic in the sense of Poincaré if, for each* n *and each* $\Phi \in D\left[(\mathbf{Q}_\mathbf{I}[t,-T])^{n+1}\right]$, *we have*

$$\lim_{w\to 0} v^{-(n+1)}\left\{\mathbf{U}_\mathbf{I}^w[t,-T] - \sum_{k=1}^{n} \frac{(v\mathbf{Q}_\mathbf{I}[t,-T])^k}{k!}\right\}\Phi = \frac{\mathbf{Q}_\mathbf{I}[t,-T]^{n+1}}{(n+1)!}\Phi.$$

This is the operator version of an asymptotic expansion in the classical sense due to Poincaré (see Coddington and Levinson [39]). In order to get time smearing, we use the exchange operator $\mathbf{E}[t,s]$ of Definition 3, and let $\bar{\mathbf{U}}_0[t,-T]$ be defined by:

$$\bar{\mathbf{U}}_0[t,-T] = \exp\{(-i/\hbar)\int_{-T}^{t} \mathbf{E}[t,s]\mathbb{H}_0(s)ds\}. \tag{1.3}$$

In this case the interaction representation for $\mathbb{H}_{\mathbf{I}}(t)$ is given by:

$$\mathbb{H}_{\mathbf{I}}(t) = \bar{\mathbf{U}}_0[-T, t]\mathbb{H}_1(t)\bar{\mathbf{U}}_0[t, -T].$$

Theorem 1.12. *The time-ordered integral $\mathbf{Q_I}[t, -T]$ generates the C_0 unitary group $\mathbf{U}_{\mathbf{I}}^{w}[t, -T]$ on $\mathcal{FD}_{\otimes}^{2}$ and, for $\Phi(t) = \mathbf{U}_{\mathbf{I}}^{w}[t, -T]\Phi$, we have:*

(i)

$$i\hbar\frac{\partial}{\partial t}\Phi(t) = w\mathbb{H}_{\mathbf{I}}(t)\Phi(t), \ \Phi(-T) = \Phi. \tag{1.4}$$

(ii) The operator $\mathbf{U}_{\mathbf{I}}^{w}[t, a] = exp\{v\mathbf{Q_I}[t, a]\}$ is asymptotic in the sense of Poincaré.

(iii) For each n and each $\Phi \in D\left[(\mathbf{Q_I}[t, -T])^{n+1}\right]$, we have

$$\begin{aligned}\Phi(t) = \Phi + \sum_{k=1}^{n}\nu^k \int_{-T}^{t} ds_1 \int_{-T}^{s_1} ds_2 \cdots \int_{-T}^{s_{k-1}} ds_k \mathbb{H}_{\mathbf{I}}(s_1)\mathbb{H}_{\mathbf{I}}(s_2)\cdots\mathbb{H}_{\mathbf{I}}(s_k)\Phi \\ + \int_{0}^{w} (\nu-\xi)^n d\xi \int_{-T}^{t} ds_1 \int_{-T}^{s_1} ds_2 \cdots \int_{-T}^{s_n} ds_{n+1}\mathbb{H}_{\mathbf{I}}(s_1)\mathbb{H}_{\mathbf{I}}(s_2)\cdots\mathbb{H}_{\mathbf{I}}(s_{n+1})\mathbf{U}_{\mathbf{I}}^{\xi}\left[s_{n+1}, -T\right]\Phi.\end{aligned} \tag{1.5}$$

where $\Phi(t) = \mathbf{U}_{\mathbf{I}}^{w}[t, -T]\Phi$.

Remark 1.13. *The above theorem includes all generators of C_0-unitary groups and provides a precise formulation and proof of Dyson's second conjecture for quantum electrodynamics that, in general, we can only expect the expansion to be asymptotic. Actually, more is proved, in that the remainder term above makes the perturbation expansion exact for all finite n. The theory is more general and applies to all of applied analysis.*

The existence of a remainder should not lead one to conclude that it is useful for all problems. In the case of bound states, all important information resides in the remainder term, independent of n. It was this problem that led to work on nonperturbative methods. An early approach was developed by Salpeter and Bethe [40].

2. Sum Over Paths

As noted earlier, Feynman stated in his book with Hibbs [27] that the operator calculus is more general than the path integral, and includes it. We now show that his expectation was indeed warranted. First we construct what we call the experimental evolution operator. This allows us to rewrite our theory as a sum over paths. We use a general argument so that the ideas apply to all cases.

Assume that the family $\{\tau_1, \tau_2, \cdots, \tau_n\}$ represents the time positions of n possible measurements of a micro system trajectory, as it appears on a film of its history. We assume that information is available beginning at time $-T$ and ends at time t. Define $\mathbf{Q}_{ex}[\tau_1, \tau_2, \cdots, \tau_n]$ by

$$\mathbf{Q}_{ex}[\tau_1, \tau_2, \cdots, \tau_n] = \sum_{j=1}^{n}\int_{t_{j-1}}^{t_j} E[\tau_j, s]\mathbb{H}(s)ds. \tag{2.1}$$

Here, $t_0 = \tau_0 = -T$, $t_j = (1/2)[\tau_j + \tau_{j+1}]$ (for $1 \leqslant j \leqslant n$), and $E[\tau_j, s]$ is the exchange operator. The effect of $E[\tau_j, s]$ is to concentrate all information contained in $[t_{j-1}, t_j]$ at τ_j, the midpoint of the time interval around τ_j relative to τ_{j-1} and τ_{j+1}. We can rewrite $\mathbf{Q}_{ex}[\tau_1, \tau_2, \cdots, \tau_n]$ as

$$\mathbf{Q}_{ex}[\tau_1, \tau_2, \cdots, \tau_n] = \sum_{j=1}^{n}\Delta t_j\left[\frac{1}{\Delta t_j}\int_{t_{j-1}}^{t_j} E[\tau_j, s]\mathbb{H}(s)ds\right]. \tag{2.2}$$

Thus, we have an average over each adjacent interval, with information concentrated at the midpoint. The evolution operator is given by

$$U[\tau_1, \tau_2, \cdots, \tau_n] = \exp\left\{-\frac{i}{\hbar}\sum_{j=1}^{n} \Delta t_j \left[\frac{1}{\Delta t_j}\int_{t_{j-1}}^{t_j} E[\tau_j, s]\mathbb{H}(s)ds\right]\right\}.$$

For $\Phi \in \mathcal{FD}^2_\otimes$, we define the function $\mathbf{U}[N(t), -T]\Phi$ by:

$$\mathbf{U}[N(t), -T]\Phi = U[\tau_1, \tau_2, \cdots, \tau_n]\Phi. \tag{2.3}$$

$\mathbf{U}[N(t), -T]\Phi$ is a $\mathcal{FD}^2_\otimes$-valued random variable, which represents the distribution of the number of measurements, $N(t)$, that are possible up to time t. In order to relate $\mathbf{U}[N(t), -T]\Phi$ to actual experimental results, we must compute its expected value. Let λ^{-1} denote the smallest time interval in which a measurement can be made, and define $\bar{\mathbf{U}}_\lambda[t, -T]\Phi$ by:

$$\bar{\mathbf{U}}_\lambda[t, -T]\Phi = \mathcal{E}\left[\mathbf{U}[N(t), -T]\Phi\right] = \sum_{n=0}^{\infty} \mathcal{E}\left\{\mathbf{U}[N(t), -T]\Phi \,|N(t) = n\right\} \Pr\left[N(t) = n\right],$$

where

$$\begin{aligned} &\mathcal{E}\{\mathbf{U}[N(t), -T]\Phi\,|N(t) = n\} \\ &= \int_0^t \frac{d\tau_1}{t}\int_0^t \frac{d\tau_2}{t-\tau_1}\cdots\int_0^t \frac{d\tau_n}{t-\tau_{n-1}}\mathbf{U}[\tau_n, \cdots, \tau_1]\Phi = \mathbf{U}_n[t, 0]\Phi. \end{aligned}$$

We make the natural assumption that: (see Gill and Zachary [33])

$$\Pr\left[N(t) = n\right] = (n!)^{-1}\left(\lambda t\right)^n \exp\{-\lambda t\}.$$

Since we are only interested in what happens when $\lambda \to \infty$, and, as the mean number of possible measurements up to time t is λt, we can take $\tau_j = (jt/n), 1 \leqslant j \leqslant n$, ($\Delta t_j = t/n$ for each n). We now replace $\bar{\mathbf{U}}_n[t, -T]\Phi$ by $\mathbf{U}_n[t, -T]\Phi$ and, with this understanding, we continue to use τ_j, so that

$$\mathbf{U}_n[t, -T]\Phi = \exp\left\{-\frac{i}{\hbar}\sum_{j=1}^{n}\int_{t_{j-1}}^{t_j} E[\tau_j, s]\mathbb{H}(s)ds\right\}\Phi. \tag{2.4}$$

We define the experimental evolution operator $\mathbf{U}_\lambda[t, -T]\Phi$ by

$$\mathbf{U}_\lambda[t, 0]\Phi = \sum_{n=0}^{[\![\lambda t]\!]} \frac{(\lambda t)^n}{n!}\exp\{-\lambda t\}\mathbf{U}_n[t, -T]\Phi. \tag{2.5}$$

The following result is a consequence of the fact that Borel summability is regular.

Theorem 2.1. *Assume that $\mathbb{H}(t)$ is strongly continuous in t and the generator of a C_0-unitary group. Then*

$$\lim_{\lambda\to\infty} \bar{\mathbf{U}}_\lambda[t, -T]\Phi = \lim_{\lambda\to\infty} \mathbf{U}_\lambda[t, -T]\Phi = \mathbf{U}[t, -T]\Phi. \tag{2.6}$$

Since $\lambda \to \infty$ implies $\lambda^{-1} \to 0$, this means that the average time between measurements is zero (in the limit) so that we obtain a continuous path. It should be observed that this continuous path arises from averaging the sum over an infinite number of (discrete) paths. The first term in (3.5) corresponds to the path of a system that created no information (i.e., the film is blank). This event has probability proportional to $\exp\{-\lambda t\}$ (which approaches zero as $\lambda \to \infty$). The n-th term corresponds to the path that creates n possible measurements, (with probability $[(\lambda t)^n/n!]\exp\{-\lambda t\}$) etc.

IOP Conf. Series: Journal of Physics: Conf. Series **845** (2017) 012023 doi:10.1088/1742-6596/845/1/012023

3. The S-matrix

We are now in a position to investigate the sense in which we can believe Dyson's first conjecture. At the end of his second paper on the relationship between the Feynman and Schwinger-Tomonaga theories, he explored the difference between the divergent Hamiltonian formalism that one must begin with and the finite S-matrix that results from renormalization. He took the view that it is a contrast between a real observer and a fictitious (ideal) observer. The real observer can only determine particle positions with limited accuracy and always gets finite results from his measurements. Dyson then suggests that "... The ideal observer, however, using non-atomic apparatus whose location in space and time is known with infinite precision, is imagined to be able to disentangle a single field from its interactions with others, and to measure the interaction. In conformity with the Heisenberg uncertainty principle, it can perhaps be considered a physical consequence of the infinitely precise knowledge of particle location allowed to the ideal observer, that the value obtained when he measures the interaction is infinite." He goes on to remark that, if his analysis is correct, the problem of divergences is attributable to an idealized concept of measurability.

In order to explore this idea, we work in the interaction representation with an obvious change in notation, replace $\mathcal{H}(t)$ by $-\frac{i}{\hbar}\mathcal{H}(t)$ and $\mathbf{U}_\lambda[t,-T]$ by

$$\mathbf{S}_\lambda[T,-T]\Phi = \sum_{n=0}^{\infty} \frac{(2\lambda T)}{n!} \exp[-2\lambda T]\mathbf{S}_n[T,-T]\Phi,$$

$$\mathbf{S}_n[T,-T]\Phi = \exp\left\{-\frac{iw}{\hbar}\sum_{j=1}^{n}\int_{t_{j-1}}^{t_j} E[\tau_j,s]\mathbb{H}_\mathbf{I}(s)ds\right\}\Phi,$$

and $\mathbb{H}_\mathbf{I}(t) = \int_{\mathbf{R}^3} \mathbb{H}_\mathbf{I}(\mathbf{x}(t),t)d\mathbf{x}(t)$ is the interaction energy. We now give a physical interpretation of our formalism. If we now note that the term $exp\{-2\lambda T\}$ may be expressed as

$$\exp\{-2\lambda T\} = \exp\left\{-\frac{i}{\hbar}\left[\int_{-T}^{T}(-i\lambda\hbar)ds\right]\right\},$$

we can rewrite $\mathbf{S}_\lambda[T,-T]\Phi$ as:

$$\mathbf{S}_\lambda[T,-T]\Phi = \sum\nolimits_{n=0}^{\infty} \frac{(2\lambda T)}{n!} \exp\left\{-\frac{i}{\hbar}\sum_{j=1}^{n}\int_{t_{j-1}}^{t_j} \left[wE[\tau_j,s]\mathbb{H}_\mathbf{I}(s) - i\lambda\hbar I_\otimes\right]ds\right\}\Phi.$$

In this form, the term $-i\lambda\hbar I_\otimes$ has the physical interpretation as the absorption of photon energy of amount $\lambda\hbar$ in each subinterval $[t_j, t_{j-1}]$ (see Mott and Massey [41]). When we compute the $lim_{\lambda\to\infty}$, we get the standard $\mathbf{S}$-matrix on $[-T,T]$. It follows that we must add an infinite amount of photon energy to the mathematical description of the experimental picture at each point in time. This is the ultraviolet divergence and shows explicitly that the transition from the experimental to the ideal scattering operator requires that we illuminate the particle throughout its entire path. Thus, it appears that we have violated the uncertainty relation. This is further supported if we look at the form of the standard S-matrix:

$$\mathbf{S}[T,-T]\Phi = \exp\left\{-\frac{iw}{\hbar}\int_{-T}^{T}\mathbb{H}_\mathbf{I}(s)ds\right\}\Phi$$

and note that the differential ds in the exponent implies perfect infinitesimal time knowledge at each point, so that the energy should be totally undetermined. If violation of the Heisenberg

IOP Conf. Series: Journal of Physics: Conf. Series **845** (2017) 012023 doi:10.1088/1742-6596/845/1/012023

uncertainty relation is the cause for the ultraviolet divergence then, since it is a variance relation, it will not appear in any first order perturbation calculation. However, it should show up in all higher-order terms (as it does). On the other hand, if the divergent terms are eliminated in second order, we would expect the method to prevent them from appearing in any higher order term of the expansion (as it does). The fact that this is precisely the case in quantum electrodynamics is a clear verification of Dyson's conjecture. (If we allow T to become infinite, we once again introduce an infinite amount of energy into the mathematical description of the experimental picture, as this is also equivalent to precise time knowledge at infinity. Of course, this is the well-known infrared divergence and can be eliminated by keeping T large but finite.)

3.1. Path Integrals

We now investigate the relationship of the operator calculus and the path integral. Suppose that the unitary group $U[t,-T]$ for the generator $H(t)$ (defined on $\mathcal{H}$) has an associated kernel $\mathbf{K}[\mathbf{x}(t),t\,;\,\mathbf{x}(s),s]$ such that:

$$\mathbf{K}\left[\mathbf{x}(t),t;\mathbf{x}(s),s\right]=\int_{\mathbf{R}^3}\mathbf{K}\left[\mathbf{x}(t),t;d\mathbf{x}(\tau),\tau\right]\mathbf{K}\left[\mathbf{x}(\tau),\tau;\mathbf{x}(s),s\right],$$

$$U[t,s]\varphi(s)=\int_{\mathbf{R}^3}\mathbf{K}\left[\mathbf{x}(t),t;d\mathbf{x}(s),s\right]\varphi(s).$$

Let $\mathbf{U}[t,s]$ be the corresponding time-ordered version, with kernel $\mathbb{K}_{\mathbf{f}}\left[\mathbf{x}(t),t;\mathbf{x}(s),s\right]$, on $\mathcal{FD}^2_{\otimes}$. Then, since $\mathbf{U}[t,\tau]\mathbf{U}[\tau,s]=\mathbf{U}[t,s]$, we have:

$$\mathbb{K}_{\mathbf{f}}\left[\mathbf{x}(t),t;\mathbf{x}(s),s\right]=\int_{\mathbf{R}^3}\mathbb{K}_{\mathbf{f}}\left[\mathbf{x}(t),t;d\mathbf{x}(\tau),\tau\right]\mathbb{K}_{\mathbf{f}}\left[\mathbf{x}(\tau),\tau;\mathbf{x}(s),s\right].$$

From our sum over paths representation for $\mathbf{U}[t,s]$, we see that:

$$\mathbf{U}_k[t,s]\Phi(s)=\exp\left\{(-i/\hbar)\sum_{j=1}^{k}\int_{t_{j-1}}^{t_j}\mathbf{E}[(j/\lambda),\tau]\mathbb{H}(\tau)d\tau\right\}\Phi(s)$$

corresponds to

$$\prod_{j=1}^{k}\int_{\mathbf{R}^3}\mathbb{K}_{\mathbf{f}}[t_j,\mathbf{x}(t_j);d\mathbf{x}(t_{j-1}),t_{j-1}]\Big|^{(j/\lambda)}\Phi(s),$$

where $|^{(j/\lambda)}$ denotes the fact that the integration is performed in time slot (j/λ). We define $\mathbb{K}_{\mathbf{f}}[\mathcal{D}_\lambda\mathbf{x}(\tau);\mathbf{x}(\mathrm{s})]$ by:

$$\int_{\mathbf{R}^{3[t,s]}}\mathbb{K}_{\mathbf{f}}[\mathcal{D}_\lambda\mathbf{x}(\tau);\mathbf{x}(\mathrm{s})]$$
$$=e^{-\lambda(t-s)}\sum_{k=0}^{[\![\lambda(t-s)]\!]}\frac{[\lambda(t-s)]^k}{k!}\left\{\prod_{j=1}^{k}\int_{\mathbf{R}^3}\mathbb{K}_{\mathbf{f}}[t_j,\mathbf{x}(t_j)\,;\,d\mathbf{x}(t_{j-1}),t_{j-1}]|^{(j/\lambda)}\right\},$$

where $[\![\lambda(t-s)]\!]$ is the greatest integer in $\lambda(t-s)$.

Definition 3.1. *The Feynman path integral associated with* $\mathbf{U}[t,s]$ *is defined by:*

$$\mathbf{U}[t,s]=\int_{\mathbf{R}^{3[t,s]}}\mathbb{K}_{\mathbf{f}}[\mathcal{D}\mathbf{x}(\tau);\mathbf{x}(\mathrm{s})]=\lim_{\lambda\to\infty}\int_{\mathbf{R}^{3[t,s]}}\mathbb{K}_{\mathbf{f}}[\mathcal{D}_\lambda\mathbf{x}(\tau);\mathbf{x}(\mathrm{s})].$$

Theorem 3.2. *For the time-ordered theory, whenever a kernel exists, we have that:*

$$\lim_{\lambda\to\infty} \mathbf{U}_\lambda[t,s]\Phi(s) = \mathbf{U}[t,s]\Phi(s) = \int_{\mathbf{R}^{3[t,s]}} \mathbb{K}_{\mathbf{f}}[\mathcal{D}\mathbf{x}(\tau);\mathbf{x}(s)]\Phi[\mathbf{x}(s)].$$

Our approach is independent of the space of continuous functions and makes it clear that the need for a measure is more of a natural expectation based on past experience than a death blow to the foundations for the Feynman path integral.

For an intrinsic approach to the path integral in the manner intended by Feynman, see [42].

3.2. Feynman-Kac

We now assume that $H = H_0(t) + H_1(t)$, where $H_0(t)$ and $H_1(t)$ are strongly continuous in t and generators of C_0-unitary groups on I, and let $\mathbb{H}_{1,\rho}(t) = \rho\mathbb{H}_1(t)\mathbf{R}(\rho,\mathbb{H}_1(t))$ be the Yosida approximator for the time-ordered version of $H_1(t)$, where $\mathbf{R}(\rho,\mathbb{H}_1(t))$ is the corresponding resolvent operator. (It is a standard result that $\mathbb{H}_{1,\rho}(t)$ is bounded and $\lim_{\rho\nearrow\infty}\mathbb{H}_{1,\rho}(t)\Phi = \mathbb{H}_1(t)\Phi$, for $\Phi\in D(\mathbb{H}_1(t))$.) Define $\mathbf{U}_\rho[t,-T]$ and $\mathbf{U}_0[t,-T]$ by:

$$\mathbf{U}_\rho[t,-T] = \exp\{(-i/\hbar)\int_{-T}^{t}[\mathbb{H}_0(s)+\mathbb{H}_{1,\rho}(s)]ds\},$$

$$\mathbf{U}_0[t,-T] = \exp\{(-i/\hbar)\int_{-T}^{t}\mathbb{H}_0(s)ds\}.$$

If $\mathbf{U}_0[t,-T]$ has an associated kernel, so that $\mathbf{U}_0[t,s] = \int_{\mathbf{R}^{3[t,s]}}\mathbb{K}_{\mathbf{f}}[\mathcal{D}\mathbf{x}(\tau);\mathbf{x}(-T)]$, we have the following generalization of the famous Feynman-Kac Theorem, which is independent of the space of continuous functions or the concept of measure.

Theorem 3.3. *(Feynman-Kac)* If $\mathbb{H}(s) = \mathbb{H}_0(s)+\mathbb{H}_1(s)$ is the generator of a C_0-unitary group, then*

$$\lim_{\rho\to\infty}\mathbf{U}^\rho[t,-T]\Phi = \mathbf{U}[t,-T]\Phi$$

$$= \int_{\mathbf{R}^{3[t,-T]}}\mathbb{K}_{\mathbf{f}}[\mathcal{D}\mathbf{x}(\tau);\mathbf{x}(-\mathrm{T})]\exp\{(-i/\hbar)\int_{-T}^{\tau}\mathbb{H}_1(s)ds]\}\Phi[\mathbf{x}(-T)].$$

This result can be extended to all C_0-semigroup evolution operators that have a kernel. Thus, this result is a very strong generalization and shows that Feynman's conjecture was indeed correct.

4. Conclusion

In this paper we have provided a mathematically rigorous development of Feynman's operator calculus, in which the time index determines when operators operate (rather than their position on paper). This allows operators acting at different times to commute. This approach also allows time to assume it's natural role in the representation of physical reality. We have constructed the mathematical film that allows us to view physical reality as a three-dimensional motion picture, on which one becomes more and more aware of the future as more and more of the film unfolds. The actual physical path traveled between any two space-time points is seen as an average over all possible paths with the same beginning and end point. This leads to the first mathematically justified theory for Feynman's path integral formulation of quantum mechanics. As an application, we proved the last two remaining conjectures of Dyson's concerning the foundations of quantum electrodynamics (QED). First, we showed that the perturbation expansion is asymptotic in a well-defined sense. Secondly, we proved explicitly that the ultraviolet divergence is caused by a violation of the uncertainty relation as conjectured by Dyson.

Acknowledgments
The work reported here represents part of a program begun by the author and W. W. Zachary at Howard University over thirty five years ago.

This paper is an extended version of a talk given at the The 10th Biennial Conference on Classical and Quantum Relativistic Dynamics of Particles and Fields held in Ljubljana, Slovenia, June 2016.

References

[1] A. I. Miller, *Early Quantum Electrodynamics*, Cambridge University Press, Cambridge, (1994).
[2] S. S. Schweber, *QED and the Men Who Made It: Dyson, Feynman, Schwinger, and Tomonaga*, Princeton University Press, Princeton, NJ, 1994.
[3] P. A. M. Dirac, in *The Birth of Particle Physics*, edited by L. M. Brown and L. Hodderson Cambridge University Press, Cambridge, 1983, p. 39.
[4] S. Tomonaga, *On a relativistically invariant formulation of the quantum theory of wave fields I,* Prog. Theor. Phys. **1** (1946), 1-13.
[5] J. Schwinger, *Selected Papers on Quantum Electrodynamics*, Dover, New York, 1958.
[6] R. P. Feynman, *The theory of positrons*, Phys. Rev. **76** (1949), 749-759.
[7] R. P. Feynman, *Space-time approach to quantum electrodynamics*, Phys. Rev. **76** (1949), 769-789.
[8] R. P. Feynman, *Mathematical formulation of the quantum theory of electromagnetic interaction*, Phys. Rev. **80** (1950), 440-457.
[9] F. Dyson, *The radiation theories of Tomonaga, Schwinger, and Feynman*, Phys. Rev. **75** (1949), 486-502.
[10] F. Dyson, *The S-matrix in quantum electrodynamics*, Phys. Rev. **75** (1949), 1736-1755.
[11] P. A. M. Dirac, Proc. R. Soc. London, Ser. A **167**, (1938) 148.
[12] S. Sakata, Prog. Theor. Phys. **16**, (1956), 686.
[13] J. Schwinger, *On quantum electrodynamics and the magnetic moment the electron,* Phys. Rev. **73** (1948), 416-417.
[14] S. S. Schweber, *An Introduction to Relativistic Quantum Field Theory,* Harper & Row, New York, 1961.
[15] A. S. Wightman, Phys. Rev. **101** (1956), 860.
[16] H. Lehmann, K. Symanzik, and W. Zimmermann, Nuovo Cimento 6, 319 1957 .
[17] R. Jost, *The General Theory of Quantized Fields*, American Mathematical Society, Providence, RI, (1965).
[18] R. F. Streater and A. S. Wightman, *PCT, Spin and Statistics and All That* Benjamin, New York, 1964.
[19] N. N. Bogolubov and D. V. Shirkov, Introduction to the Theory of Quantized Fields Interscience, London, 1958.
[20] R. Haag, *Local Quantum Physics*, 2nd ed. Springer, New York, (1996).
[21] N. N. Bogolubov, A. A. Logunov, and I. T. Todorov, Introduction to Axiomatic Quantum Field Theory Benjamin, New York, 1975 .
[22] A.D. Sokal, Ann. Inst. Henri Poincare Sect. A 37, 13 (1982).
[23] M. Aizenman, R. Graham, Nucl. Phys. B 225, 261 (1983).
[24] J. Fröhlich, Nucl. Phys. B 200, 281 (1982).
[25] F. Dyson, *Missed opportunities*, Bull. Amer. Math. Soc. **78** (1972), 635-652.
[26] R. Haag and D. Kastler, *An algebraic approach to quantum field theory*, J. Math. Phys. **5** (1964), 848-861.
[27] R. P. Feynman and A. R. Hibbs, *Quantum Mechanics and Path Integrals*, McGraw-Hill, New York, (1965).
[28] W. L. Miranker and B. Weiss, *The Feynman operator calculus*, SIAM Rev. **8**, (1966) 224-232.
[29] E. Nelson, Operants: a functional calculus for non-commuting operators, in Functional Analysis and Related Fields, ed. by F. Browder (Springer, Berlin, 1970)
[30] H. Araki, *Expansional in Banach algebras*, Ann. scient. Éc. Norm. Sup. Math. **4** (1973), 67-84.

[31] V. P. Maslov, *Operational Methods*, Mir, Moscow, (1973). English translation 1976 (revised from the Russian edition).

[32] G. W. Johnson and M. L. Lapidus, *The Feynman Integral and Feynman's Operational Calculus*, Oxford U. Press, New York, (2000).

[33] T. L. Gill and W. W. Zachary, *Foundations for relativistic quantum theory I: Feynman's operator calculus and the Dyson conjectures*, J. M. Phys. **43** (2002), 69-93.

[34] J. von Neumann, *On infinite direct products,* Compositio Math. **6** (1938), 1-77.

[35] R. Haag, *On quantum field theories.* Dan Mat. Fys. Medd. **29**, 1937 (1955).

[36] R. F. Streater and A. S. Wightman, *PCT, Spin and Statistics and All That*, Benjamin, New York, (1964).

[37] L.P. Horwitz, *On the significance of a recent experiment demonstrating quantum interference in time.* Phys. Lett. A. **355**, 1?6 (2006).

[38] E. Hille and R. S. Phillips, *Functional Analysis and Semigroups*, Amer. Math. Soc. Colloq. Pub. 31, Amer. Math. Soc. Providence, RI, (1957).

[39] C. A. Coddington and N. Levinson, *Theory of Ordinary Differential Equations*, McGraw-Hill, New York N. Y., (1955).

[40] E. Salpeter and H. A. Bethe, *A relativistic equation for bound-state problems,* Phys. Rev. **84** (1951), 1232-1242.

[41] N. F. Mott, H. S. W. Massey, *The Theory of Atomic Collisions*, (The International Series of Monographs on Physics.) Oxford, Clarendon Press, and New York, Oxford University Press, (1933).

[42] T. L. Gill and W. W. Zachary, *Banach Spaces for the Feynman integral*, Real Analysis Exchange **34**(2) (2008)/(2009), 267-310.

IOP Conf. Series: Journal of Physics: Conf. Series **845** (2017) 012024 doi:10.1088/1742-6596/845/1/012024

Speeds of light in Stueckelberg-Horwitz-Piron electrodynamics

Martin Land

Department of Computer Science, Hadassah College, 37 HaNeviim Street, Jerusalem

E-mail: martin@hadassah.ac.il

Abstract. Stueckelberg-Horwitz-Piron (SHP) electrodynamics formalizes the distinction between coordinate time (measured by laboratory clocks) and chronology (temporal ordering) by defining 4D spacetime events x^μ as functions of an external evolution parameter τ. As τ grows monotonically, the spacetime evolution of classical events $x^\mu(\tau)$ trace out particle worldlines dynamically and induce the *five* U(1) gauge potentials through which events interact.

In analogy with the constant c that associates a unit of length x^0 with intervals of time t in standard relativity, we introduce a constant c_5 associated with the external time τ. Whereas the nonrelativistic limit of special relativity can be found by taking $c \to \infty$, we show that 5D SHP goes over to an equilibrium state of Maxwell theory in the limit $c_5 \to 0$. Thus, the dimensionless ratio c_5/c parameterizes the deviation of SHP from standard electrodynamics, in particular the coupling of events. Put another way, Maxwell theory can be understood as currents and fields relaxing to an equilibrium independent of chronological time as $c_5\tau$ slows to zero. We find that taking $0 < c_5/c < 1$ enables the resolution of several longstanding difficulties in SHP theory.

1. Introduction

The description of an antiparticle as a particle propagating backward in time was given by Stueckelberg [1] in the context of a relativistic Hamiltonian classical mechanics. In his model, a pair process is represented by a single worldline, generated dynamically by the interaction of a spacetime event $x^\mu(\tau)$ with an electromagnetic field as the external parameter τ grows monotonically from $\tau = -\infty$ to $\tau = \infty$. By explicitly distinguishing [2] the coordinate time x^0 (measured by laboratory clocks) from chronological time τ (temporal ordering), the parameter τ is given a role formally similar to the Galilean time in Newtonian theory, serving Stueckelberg's broader goal of generalizing the techniques of nonrelativistic classical and quantum mechanics to covariant form. This formalism has been extended by Piron, Horwitz, *et al.* [1] - [18], so that in its present form Stueckelberg-Horwitz-Piron (SHP) electrodynamics describes a U(1) gauge theory in which the τ-dependence of gauge transformations requires *five* compensation potentials. The resulting theory can be understood as a microscopic event dynamics in which the chronological increment $\tau \to \tau' = \tau + d\tau$ generates, by way of a covariant Hamiltonian, evolution of a 4D block universe defined at τ to an infinitesimally close 4D block universe defined at τ'. Standard Maxwell theory emerges as an equilibrium limit in which the system becomes τ-independent, and the 4D block universe remains static.

In a formal approach to special relativity that takes Minkowski geometry as its starting point, the constant c is introduced as a means of measuring time in units of spatial distance, and the notion of a speed of light emerges from the role played by c in wave equations for U(1) gauge fields. Defining the flat metric in Minkowski space as

$$c^2 ds^2 = -g_{\mu\nu} dx^\mu dx^\nu = c^2 dt^2 - \mathbf{dx}^2 \qquad g_{\mu\nu} = diag(-1,1,1,1) \tag{1}$$

and parameterizing by proper time s imposes a constraint on four-velocity

$$-c^2 = \frac{dx^\mu}{ds}\frac{dx_\mu}{ds} = -\left(\frac{dx^0}{ds}\right)^2 + \left(\frac{d\mathbf{x}}{ds}\right)^2 = -\left(\frac{dx^0}{ds}\right)^2 \left[1 - \left(\frac{d\mathbf{x}}{dx^0}\right)^2\right] \tag{2}$$

so that

$$\frac{dt}{ds} = \gamma = \frac{1}{\sqrt{1 - \dfrac{\mathbf{v}^2}{c^2}}} \ , \tag{3}$$

equivalent to the mass-shell constraint

$$c^2 m^2 = -\left(m\frac{dx^\mu}{ds}\right)\left(m\frac{dx_\mu}{ds}\right) = -p^2 = \frac{E^2}{c^2} - \mathbf{p}^2 \ . \tag{4}$$

Given this constraint, the nonrelativistic limit can be recovered by taking $c \to \infty$, expressing instantaneous action at a distance.

Using natural units ($\hbar = c = 1$) in the development of SHP electrodynamics, no explicit constant was assigned to the external time τ, implicitly associating c with both times. In this paper we associate a new constant c_5 with the invariant time τ, identify the expressions in which it must appear and study its role in the classical electromagnetic theory. We will show that 5D SHP goes over to Maxwell theory as an equilibrium state in the limit $c_5 \to 0$, expressing the suppression of τ-dependent microscopic events as $c_5 \Delta\tau \to 0$ and restoring the static 4D block universe. Thus, the dimensionless ratio c_5/c parameterizes the deviation of SHP from standard electrodynamics, in particular the coupling of events. We find that taking $0 \le c_5/c < 1$ enables the resolution of longstanding difficulties in SHP theory.

2. Stueckelberg-Horwitz-Piron (SHP) electrodynamics as gauge theory

2.1. *Quantum foundation*

The generalized Stueckelberg-Schrodinger equation

$$\left(i\hbar\partial_\tau + \frac{e}{c}\phi\right)\psi(x,\tau) = \frac{1}{2M}\left(p^\mu - \frac{e}{c}a^\mu\right)\left(p_\mu - \frac{e}{c}a_\mu\right)\psi(x,\tau) \tag{5}$$

describes the interaction of an event characterized by the wavefunction $\psi(x,\tau)$ with five gauge fields $a_\mu(x,\tau)$ and $\phi(x,\tau)$. Equation (5) is invariant under the local gauge transformations

$$\begin{array}{llcl} \text{Wavefunction} & \psi(x,\tau) & \to & \exp\left[\dfrac{ie}{\hbar c}\Lambda(x,\tau)\right]\psi(x,\tau) \\ \text{Vector potential} & a_\mu(x,\tau) & \to & a_\mu(x,\tau) + \partial_\mu\Lambda(x,\tau) \\ \text{Scalar potential} & \phi(x,\tau) & \to & \phi(x,\tau) + \partial_\tau\Lambda(x,\tau) \end{array} \tag{6}$$

whose τ-dependence is the essential departure from Stueckelberg's work, and determines the structure of the resulting theory [6]. The corresponding global gauge invariance leads to the conserved Noether current

$$\partial_\mu j^\mu + \partial_\tau \rho = 0 \tag{7}$$

where

$$j^\mu = -\frac{i\hbar}{2M}\left\{\psi^*\left(\partial^\mu - \frac{ie}{c}a^\mu\right)\psi - \psi\left(\partial^\mu + \frac{ie}{c}a^\mu\right)\psi^*\right\} \qquad \rho = |\psi(x,\tau)|^2 \ . \tag{8}$$

The scalar event density $\rho(x,\tau)$ describes the probability of finding the event at the spacetime point x^μ at the chronological time τ. Although this four-vector current is not divergenceless, the five-current conservation can be integrated

$$\partial_\mu \int \frac{d\tau}{\lambda}\, j^\mu(x,\tau) + \int \frac{d\tau}{\lambda}\, \partial_\tau \rho(x,\tau) = \partial_\mu J^\mu(x) = 0 \tag{9}$$

under the boundary condition

$$\rho(x,\tau) \xrightarrow[\tau\to\pm\infty]{} 0 \tag{10}$$

where λ is a new parameter with units of time. Thus, $J^\mu(x)$ can be identified with the Maxwell four-current. In earlier work, the parameter λ appeared as an explicit coefficient multiplying the potentials — here it is absorbed into the definitions of the potentials. We will see that λ plays the role of a correlation length, characterizing the range of the electromagnetic interaction.

In analogy with the notation $x^0 = ct$ we adopt the formal designations

$$x^5 = c_5\tau \qquad \partial_5 = \frac{1}{c_5}\partial_\tau \qquad j^5 = c_5\rho \qquad a_5 = \frac{1}{c_5}\phi \tag{11}$$

and the index convention

$$\mu,\nu = 0,1,2,3 \qquad \alpha,\beta,\gamma = 0,1,2,3,5 \tag{12}$$

so that the gauge and current conditions (6) and (7) can be written

$$a_\alpha \to a_\alpha + \partial_\alpha \Lambda \qquad \partial_\alpha j^\alpha = 0 \ . \tag{13}$$

We will see that the values of λ and c_5 are not entirely independent. It is convenient to choose the factor g_{55} and $g^{55} = 1/g_{55}$ to apply when raising and lowering the 5-index, so that

$$\partial_\alpha j^\alpha = g^{\mu\nu}\partial_\mu j_\mu + g^{55}\partial_5 j_5 \ . \tag{14}$$

These expressions suggest a higher dimensional symmetry that breaks to O(3,1) in the presence of interactions — it may be O(4,1) for $g_{55} = 1$ or O(3,2) for $g_{55} = -1$.

2.2. *Classical event dynamics*

The classical mechanics of a relativistic event is found by rewriting the Stueckelberg-Schrodinger equation in the form

$$i\hbar\partial_\tau\psi(x,\tau) = \left[\frac{1}{2M}\left(p^\mu - \frac{e}{c}a^\mu\right)\left(p_\mu - \frac{e}{c}a_\mu\right) - \frac{e}{c}\phi\right]\psi(x,\tau) = K\psi(x,\tau) \tag{15}$$

and transforming the classical Hamiltonian to Lagrangian form as

$$L = \dot{x}^\mu p_\mu - K = \frac{1}{2} M \dot{x}^\mu \dot{x}_\mu + \frac{e}{c} \dot{x}^\alpha a_\alpha \tag{16}$$

where

$$\dot{x}^\mu = \frac{dx^\mu}{d\tau} \qquad \dot{x}^5 = \frac{dx^5}{d\tau} \equiv c_5 \ . \tag{17}$$

The Euler-Lagrange equations

$$\frac{d}{d\tau} \frac{\partial L}{\partial \dot{x}_\mu} - \frac{\partial L}{\partial x_\mu} = 0 \tag{18}$$

are

$$\frac{d}{d\tau} \left(M \, \dot{x}^\mu + \frac{e}{c} a^\mu \right) - \partial^\mu \left(\frac{e}{c} \dot{x}^a a_a \right) = 0 \tag{19}$$

leading to the Lorentz force

$$\begin{aligned} M\ddot{x}^\mu &= \frac{e}{c} \left[\dot{x}^\alpha \partial^\mu a_\alpha - \dot{x}^\alpha \partial_\alpha a^\mu \right] = \frac{e}{c} f^\mu{}_\alpha(x,\tau) \dot{x}^\alpha \\ &= \frac{e}{c} f^\mu{}_\nu(x,\tau) \dot{x}^\nu + \frac{e}{c} f^\mu{}_5(x,\tau) \dot{x}^5 \\ &= \frac{e}{c} f^\mu{}_\nu(x,\tau) \dot{x}^\nu - g_{55} \frac{ec_5}{c} f^{5\mu}(x,\tau) \end{aligned} \tag{20}$$

where

$$f^\mu{}_\alpha = \partial^\mu a_\alpha - \partial_\alpha a^\mu \ . \tag{21}$$

Because the four components of $\dot{x}^\mu$ are independent, the event evolution may be off-shell. In this context, on-shell evolution obeys the mass-shell constraint $\dot{x}^2 = -c^2$ of standard relativity. In SHP electrodynamics

$$\dot{x}^2 = \left(c\frac{dt}{d\tau}, \frac{d\mathbf{x}}{d\tau} \right)^2 = \left(c\frac{dt}{d\tau} \right)^2 \left(1, \frac{1}{c} \left(\frac{d\mathbf{x}}{dt} \right) \right)^2 = -c^2 \dot{t}^2 \left(1 - \frac{\mathbf{v}^2}{c^2} \right) \tag{22}$$

so that an event evolves on-shell when

$$\left| \frac{dt}{d\tau} \right| = |\dot{t}| = \frac{1}{\sqrt{1 - \frac{\mathbf{v}^2}{c^2}}} \tag{23}$$

and is said to be off-shell when $|\dot{t}|$ takes any other value. In the SHP formalism, particles may exchange mass with fields through

$$\frac{d}{d\tau}(-\tfrac{1}{2} M\dot{x}^2) = -M\dot{x}^\mu \ddot{x}_\mu = -\frac{e}{c} \, \dot{x}^\mu (c_5 f_{\mu 5} + f_{\mu\nu} \dot{x}^\nu) = \frac{ec_5}{c} \, \dot{x}^\mu f_{5\mu} = g_{55} \frac{ec_5}{c} \, f^{5\mu} \dot{x}_\mu \tag{24}$$

and the mass shell relation is demoted from the status of constraint to that of conservation law for interactions in which $\dot{x}^\mu f_{5\mu} = 0$ (which usually entails $f_{5\mu} = 0$). However, if the scale of the fields $f_{5\mu}$ is small compared to the Maxwell fields $f_{\mu\nu}$ then the exchange of mass will be correspondingly small. It would be convenient to find that

$$c_5 \to 0 \quad \Rightarrow \quad f_{5\mu} \to 0 \tag{25}$$

so that c_5/c can be understood as the scale of dynamic evolution in the microscopic system, approaching an equilibrium equivalent to standard Maxwell theory as $c_5 \to 0$. With this expectation in mind we examine the fields produced by the motions of charged events.

IOP Conf. Series: Journal of Physics: Conf. Series **845** (2017) 012024 doi:10.1088/1742-6596/845/1/012024

2.3. Electromagnetic interaction

To specify an electromagnetic action, we expand the interaction term as

$$\dot{X}^\alpha a_\alpha \to \int d^4x \; \dot{X}^\alpha(\tau)\delta^4\left(x - X(\tau)\right) a_\alpha(x,\tau) = \frac{1}{c}\int d^4x \; j^\alpha(x,\tau)a_\alpha(x,\tau) \tag{26}$$

and choose a kinetic term for the gauge field; this term must be both gauge and O(3,1) invariant. We write the action as

$$S_{\text{em}} = \int d^4x d\tau \left\{ \frac{e}{c^2} \, j^\alpha(x,\tau)a_\alpha(x,\tau) - \int \frac{ds}{\lambda}\frac{1}{4c}\left[f^{\alpha\beta}(x,\tau)\Phi(\tau - s)f_{\alpha\beta}(x,s)\right]\right\} \tag{27}$$

where λ is the parameter introduced in (9). The five components of the local event current

$$j^\alpha(x,\tau) = c\dot{X}^\alpha(\tau)\delta^4\left(x - X(\tau)\right) \tag{28}$$

have support at the spacetime location $X^\mu(\tau)$ of the corresponding event, where again $\dot{X}^5 = c_5$. As in (9) the τ-integral of (28) along the worldline concatenates the instantaneous event current into the Maxwell particle current

$$J^\mu(x) = c\int d\tau \dot{X}^\mu(\tau)\delta^4\left(x - X(\tau)\right) \tag{29}$$

in the standard form. The field interaction kernel is taken to be

$$\Phi(\tau) = \delta\left(\tau\right) - (\xi\lambda)^2\delta''\left(\tau\right) = \int \frac{d\kappa}{2\pi}\left[1 + (\xi\lambda\kappa)^2\right] e^{-i\kappa\tau} \tag{30}$$

where

$$\xi = \frac{1}{2}\left[1 + \left(\frac{c_5}{c}\right)^2\right] \tag{31}$$

is chosen so that the low energy Lorentz force agrees with Coulomb's law. The second term in $\Phi(\tau)$ is nonlocal in τ and breaks any higher dimensional symmetry associated with the fields to O(3,1). The inverse function of the interaction kernel

$$\Phi^{-1}(\tau) = \int \frac{d\kappa}{2\pi}\frac{e^{-i\kappa\tau}}{1 + (\xi\lambda\kappa)^2} = \frac{1}{2\xi\lambda}e^{-|\tau|/\xi\lambda} \tag{32}$$

satisfies

$$\int ds \; \Phi^{-1}\left(\tau - s\right)\Phi\left(s\right) = \delta(\tau) \qquad\qquad \int d\tau \; \Phi^{-1}\left(\tau\right) = 1 \tag{33}$$

and appears in the field equations as a smoothing of the particle current over a neighborhood of width λ (in time) around the exact τ-synchronization of each event. For convenience we will write

$$\varphi(\tau) = \lambda\Phi^{-1} = \frac{1}{2\xi}e^{-|\tau|/\xi\lambda} \tag{34}$$

so that

$$\int \frac{d\tau}{\lambda}\;\varphi\left(\tau\right) = 1 \; . \tag{35}$$

This definition of φ differs by a factor of λ from earlier work.

Varying the action (27) with respect to the fields, and using (34) to remove the kernel Φ, leads to the field equations

$$\partial_\beta f^{\alpha\beta}(x,\tau) = \frac{e}{c}\int ds\ \varphi(\tau - s)\, j^\alpha(x,s) = \frac{e}{c}\, j^\alpha_\varphi(x,\tau) \tag{36}$$

$$\partial_\alpha f_{\beta\gamma} + \partial_\gamma f_{\alpha\beta} + \partial_\beta f_{\gamma\alpha} = 0 \tag{37}$$

which are formally similar to Maxwell's equations in 5D, and are called pre-Maxwell equations. Rewriting the field equations in four-vector and scalar components, they take the form

$$\begin{array}{ll} \partial_\nu f^{\mu\nu} - \dfrac{1}{c_5}\partial_\tau f^{5\mu} = \dfrac{e}{c}\, j^\mu_\varphi & \partial_\mu f^{5\mu} = \dfrac{e}{c}\, j^5_\varphi = \dfrac{c_5}{c}\, e\rho_\varphi \\ \partial_\mu f_{\nu\rho} + \partial_\nu f_{\rho\mu} + \partial_\rho f_{\mu\nu} = 0 & \partial_\nu f_{5\mu} - \partial_\mu f_{5\nu} + \dfrac{1}{c_5}\partial_\tau f_{\mu\nu} = 0 \end{array} \tag{38}$$

which may be compared with the 3-vector form of Maxwell equations

$$\begin{array}{ll} \nabla \times \mathbf{B} - \dfrac{1}{c}\partial_t \mathbf{E} = \dfrac{e}{c}\,\mathbf{J} & \nabla \cdot \mathbf{E} = \dfrac{e}{c}\, J^0 \\ \nabla \cdot \mathbf{B} = 0 & \nabla \times \mathbf{E} + \dfrac{1}{c}\partial_t \mathbf{B} = 0 \end{array} \tag{39}$$

with $f_{5\mu}$ playing the role of the vector electric field and $f^{\mu\nu}$ playing the role of the magnetic field.

2.4. *Wave equations and induced fields*

The pre-Maxwell equations in Lorenz gauge lead to the wave equation

$$\partial_\beta \partial^\beta a^\alpha = (\partial_\mu \partial^\mu + \partial_\tau \partial^\tau) a^\alpha = (\partial_\mu \partial^\mu + \frac{g_{55}}{c_5^2}\,\partial_\tau^2) a^\alpha = -\frac{e}{c}\, j^\alpha_\varphi(x,\tau) \tag{40}$$

whose solutions may respect 5D symmetries broken by the O(3,1) symmetry of the event dynamics. A Green's function solution to

$$(\partial_\mu \partial^\mu + \frac{g_{55}}{c_5^2}\,\partial_\tau^2) G(x,\tau) = -\delta^4(x)\,\delta(\tau) \tag{41}$$

can be used to obtain potentials in the form

$$\begin{aligned} a^\alpha(x,\tau) &= -\frac{e}{c}\int d^4x' d\tau'\ G(x-x',\tau-\tau')\, j^\alpha_\varphi(x',\tau') \\ &= -e\int d^4x' d\tau' ds\ G(x-x',\tau-\tau')\,\varphi(\tau'-s)\dot{X}^\alpha(s)\delta^4(x'-X(s)) \\ &= -e\int ds\left[\int d\tau'\ G(x-X(s),\tau-\tau')\,\varphi(\tau'-s)\right]\dot{X}^\alpha(s)\,. \end{aligned}$$

Comparing $\dot{X}^5(s) = c_5$ with the 4-vector $\dot{X}(s) = \dot{X}^0(s)(c,\mathbf{v})$, where $|\mathbf{v}| < c$, we see that the fifth potential $a^5(x,\tau)$ is in general scaled by a factor of c_5/c with respect to $a^\mu(x,\tau)$.

The principal part Green's function was found (in natural units, writing $c = c_5 = 1$) in [7] by adapting Schwinger's derivation of the Klein-Gordon propagator. Working through the derivation and replacing τ with $x^5 = c_5\tau$ the result takes the form

$$G_P(x,\tau) = -\frac{1}{2\pi}\delta(x^2)\delta(\tau) - \frac{c_5}{2\pi^2}\frac{\partial}{\partial x^2}\theta(-g_{55}g_{\alpha\beta}x^\alpha x^\beta)\frac{1}{\sqrt{-g_{55}g_{\alpha\beta}x^\alpha x^\beta}} \tag{42}$$

$$= G_{Maxwell} + G_{Correlation} \tag{43}$$

so that both terms have units of distance^{-2} $\times$ time^{-1}. The support of $G_{Correlation}$ is

$$-g_{55}g_{\alpha\beta}x^\alpha x^\beta = \begin{cases} -\left(x^2 + c_5^2\tau^2\right) = c^2t^2 - \mathbf{x}^2 - c_5^2\tau^2 > 0 \quad , \quad g_{55} = 1 \\ \left(x^2 - c_5^2\tau^2\right) = \mathbf{x}^2 - c^2t^2 - c_5^2\tau^2 > 0 \quad , \quad g_{55} = -1 \end{cases} \tag{44}$$

with causality properties discussed in [7]. In particular, we see that for $g_{55} = 1$, $G_{Correlation}$ has timelike support with respect to the event trajectory, opening the possibility of a self-interaction of a type not present in standard Maxwell theory. Since Schwinger's method involves singling out a direction for τ in the 5D space (x,τ), we see that the first term $G_{Maxwell}$ breaks O(4,1) or O(3,2) symmetry to O(3,1).

To avoid singularities, we must take particular care in handling the distribution functions and the order of integration. Evaluating the derivative in (42) we find for $g_{55} = 1$,

$$G_{correlation}(x,\tau) = -\frac{c_5}{2\pi^2}\frac{\partial}{\partial x^2}\theta(-g_{55}g_{\alpha\beta}x^\alpha x^\beta)\frac{c_5}{\sqrt{-g_{55}g_{\alpha\beta}x^\alpha x^\beta}}$$

$$= -\frac{c_5}{2\pi^2}\frac{\partial}{\partial x^2}\frac{\theta(-x^2 - c_5^2\tau^2)}{\left(-x^2 - c_5^2\tau^2\right)^{1/2}}$$

$$G_{correlation}(x,\tau) = -\frac{c_5}{2\pi^2}\left(\frac{1}{2}\frac{\theta(-x^2 - c_5^2\tau^2)}{\left(-x^2 - c_5^2\tau^2\right)^{3/2}} - \frac{\delta\left(-x^2 - c_5^2\tau^2\right)}{\left(-x^2 - c_5^2\tau^2\right)^{1/2}}\right) . \tag{45}$$

Although the second term appears highly singular, we will see that when calculating potentials, singularities in the terms of $G_{correlation}$ cancel when the subtraction is performed before applying the limits of integration, as required by Schwinger's prescription.

The 'static' Coulomb potential in this framework is induced by an isolated event moving uniformly along the t axis. Writing the event as

$$X(\tau) = (c\tau, 0, 0, 0) \tag{46}$$

produces the currents

$$j^0(x,\tau) = c^2\delta(t-\tau)\,\delta^3(\mathbf{x}) \qquad \mathbf{j}(x,\tau) = 0 \qquad j^5(x,\tau) = cc_5\delta(t-\tau)\,\delta^3(\mathbf{x}) \tag{47}$$

$$j^0_\varphi(x,\tau) = c^2\varphi(t-\tau)\,\delta^3(\mathbf{x}) \qquad \mathbf{j}_\varphi(x,\tau) = 0 \qquad j^5_\varphi(x,\tau) = cc_5\varphi(t-\tau)\,\delta^3(\mathbf{x}) \tag{48}$$

and the Maxwell part of the Green's function induces

$$a^0(x,\tau) = \frac{e}{4\pi|\mathbf{x}|}\varphi\left(\tau - \left(t - \frac{|\mathbf{x}|}{c}\right)\right) \qquad \mathbf{a} = 0 \qquad a^5(x,\tau) = \frac{c_5}{c}a^0(x,\tau) . \tag{49}$$

In Appendix A we show that the contribution from $G_{Correlation}$ is smaller than the $G_{Maxwell}$ contribution by c_5/c and drops off as $1/\left|\mathbf{x}\right|^2$, so that it may be neglected at at low energy. Using (34) for $\varphi(\tau)$, a test event evolving in parallel at $x(\tau) = (c\tau, \mathbf{x})$ will see the Yukawa-type potential

$$a^0(x,\tau) = \frac{e}{4\pi|\mathbf{x}|}\frac{1}{2\xi}e^{-|\mathbf{x}|/\xi\lambda c} \tag{50}$$

with photon mass spectrum $m_\gamma \sim \hbar/\xi\lambda c^2$. If λ is small (so that $\frac{1}{\lambda}\varphi$ approaches a delta function and the current narrows to a small neighborhood around the event) the mass spectrum becomes wide. If λ is large, the support of the current spreads along the worldline, the potential becomes Coulomb-like and the photon mass is small.

The field strength components found from (50) are

$$f^{k0}(x,\tau) = \partial^k \frac{e}{4\pi|\mathbf{x}|}\frac{1}{2\xi}e^{-|\mathbf{x}|/\xi\lambda c} \qquad f^{k5}(x,\tau) = \frac{c_5}{c}f^{k0}(x,\tau)$$
$$f^{ij}(x,\tau) = f^{50}(x,\tau) = 0 \tag{51}$$

and the test event will experience the Coulomb force through the Lorentz force (20) as

$$M\ddot{x}^k = \frac{e}{c}f^k{}_\nu \dot{x}^\nu - g_{55}\frac{ec_5}{c}f^{5k} = -\frac{e}{c}f^{k0}\left(\dot{x}^0 - g_{55}c_5\right) \ . \tag{52}$$

Since the test event velocity is $\dot{x}(\tau) = (c, \mathbf{0})$ this becomes

$$M\ddot{\mathbf{x}} = -\frac{e^2}{2\xi}\left(1 - g_{55}\frac{c_5}{c}\right)\nabla\left(\frac{e^{-|\mathbf{x}|/\xi\lambda c}}{4\pi\left|\mathbf{x}\right|}\right) = -e^2\frac{1 - g_{55}\frac{c_5}{c}}{1 + \left(\frac{c_5}{c}\right)^2}\nabla\left(\frac{e^{-|\mathbf{x}|/\xi\lambda c}}{4\pi\left|\mathbf{x}\right|}\right) \tag{53}$$

where we used (31) for ξ. For an antiparticle event, the source evolves backward in time with $\dot{x}^0 = -c$ so that the Coulomb force induced by a particle event (upper sign) or antiparticle event (lower sign) can be written

$$\mathbf{F} = \mp e^2\frac{1 \mp g_{55}\frac{c_5}{c}}{1 + \left(\frac{c_5}{c}\right)^2}\nabla\left(\frac{e^{-|\mathbf{x}|/\xi\lambda c}}{4\pi\left|\mathbf{x}\right|}\right) \ . \tag{54}$$

This expression permits the resolution a longstanding difficulty in SHP theory. Under the assumption that $c_5/c = 1$ there is no choice for $g_{55} = \pm 1$ that can describe classical elastic scattering for both e^-/e^- and e^+/e^-. However, with $c_5 < c$ either signature for g_{55} is permitted, and (54) provides an experimental bound on c_5/c associated with the discrepancy between particle/particle and particle/antiparticle scattering at very low energy.

The form of $G_{Maxwell}$ allows us apply standard techniques associated with the Liénard-Wiechert potential. Writing

$$\begin{aligned} a^\alpha(x,\tau) &= -\frac{e}{c}\int d^4x' d\tau' \; G_{Maxwell}\left(x - x', \tau - \tau'\right) j^\alpha_\varphi\left(x', \tau'\right) \\ &= \frac{e}{2\pi}\int ds \; \varphi(\tau - s)\,\dot{X}^\alpha(s)\,\delta\left(\left(x - X(s)\right)^2\right)\theta^{ret} \end{aligned} \tag{55}$$

and using the identity

$$\int d\tau f(\tau)\,\delta\left[g(\tau)\right] = \frac{f(\tau_R)}{\left|g'(\tau_R)\right|} \ , \tag{56}$$

IOP Conf. Series: Journal of Physics: Conf. Series **845** (2017) 012024 doi:10.1088/1742-6596/845/1/012024

where τ_R is the retarded time found from

$$g(\tau) = (x - X(\tau_R))^2 = 0 \qquad \theta^{ret} = \theta\left(x^0 - X^0(\tau_R)\right) , \tag{57}$$

provides

$$a^\alpha(x,\tau) = \frac{e}{4\pi}\varphi(\tau - \tau_R)\frac{\dot{X}^\alpha(\tau_R)}{\left|\left(x^\mu - X^\mu(\tau_R)\right)\dot{X}_\mu(\tau_R)\right|} . \tag{58}$$

Using this potential, where we write the event velocity and line of observation as

$$u^\mu = \dot{X}^\mu(\tau) \qquad z^\mu = x^\mu - X^\mu(\tau) \tag{59}$$

the potential takes the form

$$a^\mu(x,\tau) = \frac{e}{4\pi}\varphi(\tau - \tau_R)\frac{u^\mu}{|u\cdot z|} \qquad a^5(x,\tau) = \frac{e}{4\pi}\varphi(\tau - \tau_R)\frac{c_5}{|u\cdot z|} . \tag{60}$$

By a similar procedure we find the field strengths as

$$f^{\mu\nu}(x,\tau) = -\frac{e}{4\pi}\varphi(\tau - \tau_R)\left\{\frac{(z^\mu u^\nu - z^\nu u^\mu)\,u^2}{(u\cdot z)^3} + \frac{\epsilon(\tau-\tau_R)}{\lambda}\frac{z^\mu u^\nu - z^\nu u^\mu}{(u\cdot z)^2}\right.$$
$$\left.\frac{(z^\mu \dot{u}^\nu - z^\nu \dot{u}^\mu)(u\cdot z) - (z^\mu u^\nu - z^\nu u^\mu)(\dot{u}\cdot z)}{(u\cdot z)^3}\right\} \tag{61}$$

$$f^{5\mu}(x,\tau) = c_5\frac{e}{4\pi}\varphi(\tau - \tau_R)\left\{\frac{z^\mu u^2 - u^\mu(u\cdot z)}{(u\cdot z)^3} + \frac{\epsilon(\tau-\tau_R)}{\lambda}\frac{z^\mu - u^\mu(u\cdot z)}{(u\cdot z)^2}\right.$$
$$\left. - \frac{(\dot{u}\cdot z)\,z^\mu}{(u\cdot z)^3}\right\} \tag{62}$$

where we used $|u\cdot z| = -(u\cdot z)$ (easily seen in a co-moving frame) and

$$\frac{d}{d\tau_R}\varphi(\tau - \tau_R) = -\frac{1}{2\xi}\frac{d}{d\tau}e^{-|\tau-\tau_R|/\xi\lambda} = -\frac{\epsilon(\tau-\tau_R)}{\xi\lambda}\varphi(\tau - \tau_R) \tag{63}$$

where $\epsilon(\tau) = \text{signum}(\tau)$. Notice that the explicit τ-dependence in these expressions is limited to the smoothing function $\varphi(\tau - \tau_R)$ and its derivative.

3. Recovering Maxwell theory from SHP

The classical SHP framework, described by the pre-Maxwell equations (36) and (37) and the Lorentz force (20), is a theory of interacting events. Each event $x^\mu(\tau)$ induces an instantaneous five-component current density $j^\alpha_\varphi(x,\tau)$ with support in a neighborhood $x^\mu(\tau \pm \lambda)$ along its dynamically evolving worldline. The five gauge potentials $a^\alpha(x,\tau)$ locally induced by these currents moderate interactions between individual events. The complete worldlines characterized by the Maxwell currents $J^\mu(x)$ can only be known *a posteriori* by concatenation of these instantaneous event currents. We now compare two approaches for associating SHP with the standard Maxwell theory of interacting particle worldlines, as an equilibrium state of this microscopic event dynamics.

3.1. *Concatenation by τ-integration*

In (29) it was seen that the divergenceless 4D Maxwell current is obtained by integrating the instantaneous event current over τ, concatenating the events into a particle worldline. The source of the pre-Maxwell fields is

$$j_{\varphi}^{\alpha}(x,\tau) = \int ds\ \varphi(\tau - s) j^{\alpha}(x,s) \tag{64}$$

formed by smoothing the delta function support of the instantaneous current defined in (28) by convolution with $\varphi(\tau) = \lambda\Phi^{-1}(\tau)$. It follows from (35) that

$$\int \frac{d\tau}{\lambda}\ j_{\varphi}^{\mu}(x,\tau) = \int ds\ j^{\mu}(x,s) = J^{\mu}(x) \tag{65}$$

so that smoothing the local event current leaves the concatenation unchanged. Following the Stueckelberg boundary condition (10), one may impose the additional requirement

$$f^{5\mu}(x,\tau) \xrightarrow[\tau\to\pm\infty]{} 0 \tag{66}$$

so that integration of the field equations provide

$$\left.\begin{array}{c} \partial_{\beta} f^{\alpha\beta}(x,\tau) = \frac{e}{c} j_{\varphi}^{\alpha}(x,\tau) \\ \\ \partial_{[\alpha} f_{\beta\gamma]} = 0 \\ \\ \partial_{\alpha} j^{\alpha} = 0 \end{array}\right\} \xrightarrow[\int \frac{d\tau}{\lambda}]{} \left\{\begin{array}{c} \partial_{\nu} F^{\mu\nu}(x) = \frac{e}{c} J^{\mu}(x) \\ \\ \partial_{\nu} F^{5\nu}(x) = \frac{e}{c} J^{5}(x) \\ \\ \partial_{[\mu} F_{\nu\rho]} = 0 \\ \\ \partial_{\mu} J^{\mu}(x) = 0 \end{array}\right. \tag{67}$$

where we use (65) and

$$A^{\alpha}(x) = \int \frac{d\tau}{\lambda}\ a^{\alpha}(x,\tau) \qquad\qquad F^{\alpha\nu}(x) = \int \frac{d\tau}{\lambda}\ f^{\alpha\nu}(x,\tau)\ . \tag{68}$$

Integration of (42)

$$\int d\tau\ G_{Maxwell} = D(x) = -\frac{1}{2\pi}\delta(x^2) \qquad\qquad \int d\tau\ G_{Correlation} = 0 \tag{69}$$

similarly recovers the 4D Maxwell Green's function. As seen in equations (60) — (62), the τ-dependence of the potentials and fields resides in multiplicative factors of the type $\varphi(\tau - \tau_R)$, so that integration along the worldline recovers the standard Maxwell fields in their explicit form. In particular, integration of (50) yields the standard Coulomb potential.

In the resulting picture, the microscopic dynamics approach an equilibrium state because the boundary conditions (10) and (66) hold pointwise in x as $\tau \to \infty$, asymptotically eliminating interactions that cannot be described in Maxwell theory. The Maxwell-type description recovered by concatenating the microscopic dynamics may thus be understood as a self-consistent summary constructed *a posteriori* from the complete worldlines.

3.2. The limit $c_5 \to 0$

We have assumed that $0 \leq c_5 < c$ and we must check that SHP theory remains finite for $c_5 \to 0$.

First we notice that c_5 appears explicitly three times in the pre-Maxwell equations (38)

$$\partial_\nu f^{\mu\nu} - \frac{1}{c_5}\partial_\tau f^{5\mu} = \frac{e}{c} j^\mu_\varphi \qquad \partial_\mu f^{5\mu} = \frac{e}{c} j^5_\varphi = \frac{c_5}{c} e\rho_\varphi$$

$$\partial_\mu f_{\nu\rho} + \partial_\nu f_{\rho\mu} + \partial_\rho f_{\mu\nu} = 0 \qquad \partial_\nu f_{5\mu} - \partial_\mu f_{5\nu} + \frac{1}{c_5}\partial_\tau f_{\mu\nu} = 0$$

twice in the form $\frac{1}{c_5}\partial_\tau$ and once multiplying the event density ρ_φ. The derivative term poses no problem in the homogeneous pre-Maxwell equation, which is satisfied identically for fields derived from potentials. Specifically, the fields $f_{5\mu}$ contain terms of the type $\partial_5 a_\mu = \frac{1}{c_5}\partial_\tau a_\mu$ that cancel the explicit τ-derivative of $f_{\mu\nu}$, evaluated before passing to the limit $c_5 \to 0$. However, the homogeneous equation does impose a new condition through

$$c_5 \left(\partial_\nu f_{5\mu} - \partial_\mu f_{5\nu}\right) + \partial_\tau f_{\mu\nu} = 0 \quad \xrightarrow[c_5 \to 0]{} \quad \partial_\tau f_{\mu\nu} = 0 \tag{70}$$

requiring that the field strength $f^{\mu\nu}$ become τ-independent in this limit. From equation (61) we see that this condition is violated by the multiplicative factor $\varphi(\tau - \tau_R)$ unless we simultaneously require $c_5 \to 0 \Rightarrow \lambda \to \infty$, in which case $\varphi(x,\tau) \to 1/2\xi = 1$, where we use (31) for ξ. This requirement effectively spreads the event current j^α_φ uniformly along the particle worldline, recovering the τ-independent particle current

$$\begin{aligned} j^\mu_\varphi(x,\tau) &= \int ds\, \varphi(\tau - s)\, j^\mu(x,s) \quad \longrightarrow \quad \int ds\, 1 \cdot j^\mu(x,s) = J^\mu(x) \\ j^5_\varphi(x,\tau) &= \int ds\, \varphi(\tau - s)\, j^5(x,s) \quad \longrightarrow \quad \int ds\, j^5(x,s) \\ \partial_\mu j^\mu_\varphi(x,\tau) &+ \frac{1}{c_5}\partial_\tau j^5_\varphi(x,\tau) \quad \longrightarrow \quad \partial_\mu J^\mu(x) = 0 \end{aligned} \tag{71}$$

associated with Maxwell theory. Generally, because the τ-dependence of the potentials and fields is contained in φ, the condition $\lambda \to \infty$ eliminates all the terms in the pre-Maxwell equations containing ∂_τ. Similarly, the photon mass $m_\gamma \sim \hbar/\xi\lambda c^2$ must vanish.

Equation (62) shows that $f^{5\mu}$ is generally proportional to c_5 for fields of the Liénard-Wiechert type. Therefore we can write the inhomogeneous pre-Maxwell equations in the finite form

$$\partial_\nu f^{\mu\nu} = \frac{e}{c} j^\mu_\varphi \qquad \partial_\mu \left(\frac{1}{c_5} f^{5\mu}\right) = \frac{e}{c}\rho_\varphi \tag{72}$$

where we see that $f^{5\mu}$ decouples from the field $f^{\mu\nu}$ that now satisfies Maxwell's equations.

To find the limiting form of the electromagnetic interactions, we consider an arbitrary event $X^\mu(\tau)$, which induces the current

$$j^\alpha_\varphi(x,\tau) = c\int ds\, \varphi(\tau - s)\, \dot{X}^\alpha(s)\, \delta^4\left[x - X(s)\right] \ . \tag{73}$$

IOP Conf. Series: Journal of Physics: Conf. Series **845** (2017) 012024 doi:10.1088/1742-6596/845/1/012024

Using (20), (31), (32), (61) and (62) the Lorentz force on a test event moving in the field induced by this current can be written

$$\begin{aligned}
M\ddot{x}^{\mu} &= \frac{e}{c}\left[f^{\mu}{}_{\nu}(x,\tau)\dot{x}^{\nu} + f^{5\mu}(x,\tau)\dot{x}^{5}\right] \\
&= \frac{e}{c}\frac{e}{4\pi}\varphi\left(\tau-\tau_R\right)\left[\mathcal{F}^{\mu}{}_{\nu}(x,\tau)\dot{x}^{\nu} + c_5^2\,\mathcal{F}^{5\mu}(x,\tau)\right] \\
&= \frac{e}{c}\frac{e}{4\pi}\frac{1}{2\xi}e^{-|\tau-\tau_R|/\xi\lambda}\left[\mathcal{F}^{\mu}{}_{\nu}(x,\tau)\dot{x}^{\nu} + c_5^2\,\mathcal{F}^{5\mu}(x,\tau)\right] \\
&= \frac{e^2}{4\pi c}\,e^{-|\tau-\tau_R|/\xi\lambda}\,\frac{1}{2\xi}\left[\mathcal{F}^{\mu}{}_{\nu}(x,\tau)\dot{x}^{\nu} + c_5^2\,\mathcal{F}^{5\mu}(x,\tau)\right] \\
&= \frac{e^2}{4\pi c}\,e^{-|\tau-\tau_R|/\xi\lambda}\,\frac{\mathcal{F}^{\mu}{}_{\nu}(x,\tau)\dot{x}^{\nu} + c_5^2\,\mathcal{F}^{5\mu}(x,\tau)}{1+\left(c_5/c\right)^2}
\end{aligned} \tag{74}$$

where

$$\begin{aligned}
\mathcal{F}^{\mu\nu}(x,\tau) = &-\frac{\left(z^{\mu}u^{\nu}-z^{\nu}u^{\mu}\right)u^2}{(u\cdot z)^3} - \frac{\left(z^{\mu}\dot{u}^{\nu}-z^{\nu}\dot{u}^{\mu}\right)(u\cdot z) - \left(z^{\mu}u^{\nu}-z^{\nu}u^{\mu}\right)(\dot{u}\cdot z)}{(u\cdot z)^3} \\
&+\frac{\epsilon\left(\tau-\tau_R\right)}{\lambda}\frac{z^{\mu}u^{\nu}-z^{\nu}u^{\mu}}{(u\cdot z)^2}
\end{aligned} \tag{75}$$

$$\mathcal{F}^{5\mu}(x,\tau) = \frac{z^{\mu}u^2 - u^{\mu}\left(u\cdot z\right)}{(u\cdot z)^3} - \frac{\left(\dot{u}\cdot z\right)z^{\mu}}{(u\cdot z)^3} + \frac{\epsilon\left(\tau-\tau_R\right)}{\lambda}\frac{z^{\mu}-u^{\mu}\left(u\cdot z\right)}{(u\cdot z)^2} \tag{76}$$

are independent of c_5. In the limit $c_5 \to 0$ and $\lambda \to \infty$, all $\mathcal{F}^{\alpha\mu}(x,\tau)$ become τ-independent and the Lorentz force interaction reduces to

$$M\ddot{x}^{\mu} = \frac{e^2}{4\pi c}\,\mathcal{F}^{\mu}{}_{\nu}(x)\dot{x}^{\nu} \tag{77}$$

decoupling from the field $\mathcal{F}^{5\mu}$. We thus see that in addition to recovering Maxwell's equations, the parameter c_5/c provides a continuous scaling of the Lorentz force to the standard form in Maxwell theory. That is, the combined limit $\lambda \to \infty$ and $c_5 \to 0$ restricts the possible dynamics in SHP to those of Maxwell theory.

The contribution associated with $G_{Correlation}$ is less straightforward, but in Appendix B we gain some insight by studying the δ-function term and assuming that the structure of the θ-function term must be sufficiently similar to permit cancellation of singularities.

4. Conclusions

We have reviewed classical Stueckelberg-Horwitz-Piron electrodynamics, inserting where appropriate a new constant c_5 associated with the chronological time τ in much the way that c is associated with the coordinate time t. Although it has been previously assumed (implicitly) that $c_5 = c$, we now consider the possibility that $0 \leq c_5/c < 1$ and find that this resolves longstanding difficulties in consistently formulating elastic particle/particle scattering and particle/antiparticle scattering in classical electrodynamics.

We showed that SHP remains finite for all values of $0 \leq c_5/c < 1$, where consistency of the pre-Maxwell equations requires that $\lambda \to \infty$ when $c_5 \to 0$. This limit reproduces standard Maxwell dynamics, as the pre-Maxwell equations and SHP Lorentz equation go over to Maxwell's equations and their associated Lorentz equation. We note that while a Maxwell-like description can be obtained from SHP as an *a posteriori* equilibrium limit by integration over the worldlines, the parameter c_5/c provides a smooth transition, so that its value can be determined from experimental deviations of SHP from standard Maxwell theory. In the resulting picture, the Maxwell approximation to pre-Maxwell theory can be understood as an event dynamics evolving so slowly in τ that the system is essentially in equilibrium, with vanishing event density and no mass exchange.

As noted in the introduction, off-shell event evolution involves mass exchange between particles and fields. Another longstanding issue in SHP is finding an adequate explanation for the fixed masses of the observed particles. In a subsequent paper [18] we discuss a mechanism by which the field produced through $G_{Correlation}$ for $g_{55} = 1$ leads to a self-interaction with timelike support whose effect is to restore on-shell evolution and stabilize the particle mass.

Acknowledgments

The author gratefully acknowledges the inspiration and guidance of Professor Lawrence P. Horwitz, whose work has opened many fields of inquiry and whose direction has set many students on a productive paths in physics.

Appendix A. Coulomb potential from $\mathbf{G_{Correlation}}$

We are interested in an event moving as

$$X = (c\tau, \mathbf{0}) \qquad u^2 = -c^2 \tag{A.1}$$

where we approximate

$$\varphi(\tau' - s) = \lambda\delta(\tau' - s) \tag{A.2}$$

so that

$$a^\alpha(x,\tau) = -\lambda e \int ds\, G(x - X(s), \tau - s)\ \dot{X}^\alpha(s) = \frac{\lambda e}{2\pi^2}\dot{X}^\alpha(s) \int ds\, G(x - X(s), \tau - s)\ . \tag{A.3}$$

We introduce the function $g(s)$ to express terms of the type

$$-\left((x - X(s))^2 + c_5^2(\tau - s)^2\right) = -\left(((ct, \mathbf{x}) - (cs, \mathbf{0}))^2 + c_5^2(\tau - s)^2\right) = c^2 g(s) \tag{A.4}$$

where

$$g(s) = (t - s)^2 - \frac{R^2}{c^2} - \frac{c_5^2}{c^2}(\tau - s)^2 = Cs^2 + Bs + A \tag{A.5}$$

and

$$\zeta^2 = \frac{c_5^2}{c^2} \qquad C = \left(1 - \zeta^2\right) \qquad B = -2\left(t - \zeta^2\tau\right) \qquad A = t^2 - \frac{R^2}{c^2} - \zeta^2\tau^2 \tag{A.6}$$

so that the potential can be written as

$$a(x,\tau) = \frac{\lambda e c_5}{2\pi^2 c^3}(c, \mathbf{0}, c_5) \int ds \left[\frac{1}{2}\frac{\theta(g(s))}{g^{3/2}(s)} - \frac{\delta(g(s))}{g^{1/2}(s)}\right]\theta(t - s)\ . \tag{A.7}$$

IOP Conf. Series: Journal of Physics: Conf. Series **845** (2017) 012024 doi:10.1088/1742-6596/845/1/012024

The zeros of $g(s)$ are found to be

$$s_{\pm} = \frac{-B \pm \sqrt{B^2 - 4AC}}{2C} = \frac{(t - \zeta^2\tau) \pm \sqrt{\frac{R^2}{c^2}(1-\zeta^2) + \zeta^2 (t-\tau)^2}}{(1-\zeta^2)} \tag{A.8}$$

and since we assume $\zeta^2 < 1$ there will be roots for any values of t and R. In addition, the condition $\theta^{ret} = \theta(t-s)$ requires $t > s$.

If $t < s_-$ then

$$t < \frac{(t - \zeta^2\tau) - \sqrt{\frac{R^2}{c^2}(1-\zeta^2) + \zeta^2 (t-\tau)^2}}{(1-\zeta^2)} \qquad \Rightarrow \qquad -\zeta^2 (t-\tau)^2 > \frac{R^2}{c^2} \tag{A.9}$$

and so $t \geq s_-$ becomes a condition of integration for the θ term. Similarly, if $t > s_+$ then

$$t > \frac{(t - \zeta^2\tau) + \sqrt{\frac{R^2}{c^2}(1-\zeta^2) + \zeta^2 (t-\tau)^2}}{(1-\zeta^2)} \qquad \Rightarrow \qquad -\zeta^2 (t-\tau) > \frac{R^2}{c^2} \tag{A.10}$$

leading to the condition

$$\theta(g(s))\,\theta(t-s) \neq 0 \qquad \Rightarrow \qquad s_- \leq s \leq t \leq s_+ \tag{A.11}$$

from which

$$a(x,\tau) = \frac{\lambda e c_5}{2\pi^2 c^3}\left(1, \mathbf{0}, \frac{c_5}{c}\right)\left(\frac{1}{2}\int_{-\infty}^{s_-} ds \frac{1}{g^{3/2}(s)} - \int_{-\infty}^{\infty} ds \frac{\delta(g(s))}{g^{1/2}(s)} \theta(t-s)\right) . \tag{A.12}$$

Using the well-known form [19]

$$\int \frac{dx}{(Cx^2 + Bx + A)^{3/2}} = \frac{2(2Cs + B)}{q(Cx^2 + Bx + A)^{1/2}} \tag{A.13}$$

where

$$q = 4AC - B^2 \tag{A.14}$$

we notice from (A.8) that

$$s_- = \frac{-B - \sqrt{B^2 - 4AC}}{2C} = \frac{-B - \sqrt{-q}}{2C} \qquad \Rightarrow \qquad -\sqrt{-q} = 2Cs_- + B \tag{A.15}$$

and so

$$\begin{aligned} \frac{1}{2}\int_{-\infty}^{s_-} ds \frac{1}{g^{3/2}(s)} &= \frac{2Cs_- + B}{qg^{1/2}(s_-)} - \left.\frac{2Cs + B}{qg^{1/2}(s)}\right|_{-\infty} \\ &= \frac{-\sqrt{-q}}{qg^{1/2}(s_-)} + \frac{2\sqrt{C}}{(2Cs_- + B)^2} \\ &= \frac{1}{\sqrt{-q}g^{1/2}(s_-)} + \frac{1}{2}\frac{\sqrt{1-\zeta^2}}{\frac{R^2}{c^2}(1-\zeta^2) + \zeta^2 (t-\tau)^2} \end{aligned} \tag{A.16}$$

IOP Conf. Series: Journal of Physics: Conf. Series **845** (2017) 012024 doi:10.1088/1742-6596/845/1/012024

The second term is

$$\int_{-\infty}^{\infty} ds \frac{\delta\left(g\left(s\right)\right)}{g^{1/2}\left(s\right)} \theta\left(t-s\right) \tag{A.17}$$

and using the identity

$$\int ds\; f\left(s\right)\; \delta\left(g\left(s\right)\right) = \left.\frac{f\left(s^{-}\right)}{\left|g'\left(s^{-}\right)\right|}\right|_{s^{-}=g^{-1}(0)} \tag{A.18}$$

we can evaluate

$$\int_{-\infty}^{\infty} ds \frac{\delta\left(g\left(s\right)\right)}{g^{1/2}\left(s\right)} \theta\left(t-s\right) = \frac{\theta\left(t-s_{-}\right)}{\left|g'\left(s_{-}\right)\right| g^{1/2}\left(s_{-}\right)} = \frac{1}{\left|g'\left(s_{-}\right)\right| g^{1/2}\left(s_{-}\right)} . \tag{A.19}$$

Since

$$g'\left(s_{-}\right) = \left(Cs_{-}^{2} + Bs_{-} + A\right)' = 2Cs_{-} + B = -\sqrt{-q} \tag{A.20}$$

we see that this term cancels the singularity in the first term, leaving

$$\frac{1}{2}\int_{-\infty}^{s_{-}} ds \frac{1}{g^{3/2}\left(s\right)} - \int_{-\infty}^{\infty} ds \frac{\delta\left(g\left(s\right)\right)}{g^{1/2}\left(s\right)} \theta\left(t-s\right) = \frac{1}{2} \frac{\sqrt{1-\zeta^{2}}}{\frac{R^{2}}{c^{2}}\left(1-\zeta^{2}\right) + \zeta^{2}\left(t-\tau\right)^{2}} \tag{A.21}$$

and

$$a\left(x,\tau\right) = \frac{\lambda e}{4\pi^{2}}\left(c,\mathbf{0},c_{5}\right)\frac{c_{5}}{c}\frac{\sqrt{1-\frac{c_{5}}{c}}}{R^{2}\left(1-\frac{c_{5}}{c}\right)+\frac{c_{5}}{c}c^{2}\left(t-\tau\right)^{2}} . \tag{A.22}$$

We notice that the potential has units of $\lambda c/\text{distance}^2 = 1/\text{distance}$, as does the potential associated with $G_{Maxwell}$. This contribution to the potential is smaller by a factor of c_5/c than the Yukawa potential found in (50), and drops off faster with distance.

Appendix B. c_5-dependence of general potential from $G_{\text{Correlation}}$

We are interested in an arbitrary event moving as

$$X\left(\tau\right) = \left(ct\left(\tau\right), \mathbf{x}\left(\tau\right)\right) \qquad\qquad X^{5} = c_{5}\tau \tag{B.1}$$

and the induced field

$$a^{\alpha}\left(x,\tau\right) = -e\int ds\; G_{\varphi}\left(x - X(s), \tau - s\right)\; \dot{X}^{\alpha}(s) \; . \tag{B.2}$$

Making the approximation

$$\varphi(\tau' - s) = \lambda\delta(\tau' - s) \tag{B.3}$$

leads to

$$a^{\alpha}\left(x,\tau\right) = \frac{\lambda e c_{5}}{2\pi^{2}}\int ds\; \dot{X}^{\alpha}\left(s\right)\left(\frac{1}{2}\frac{\theta\left(-\left(x-X\left(s\right)\right)^{2} - c_{5}^{2}(\tau-s)^{2}\right)}{\left[-\left(x-X\left(s\right)\right)^{2} - c_{5}^{2}(\tau-s)^{2}\right]^{3/2}}\right.$$
$$\left. - \frac{\delta\left(-\left(x-X\left(s\right)\right)^{2} - c_{5}^{2}(\tau-s)^{2}\right)}{\left(-\left(x-X\left(s\right)\right)^{2} - c_{5}^{2}(\tau-s)^{2}\right)^{1/2}}\right)\theta^{ret} \tag{B.4}$$

IOP Conf. Series: Journal of Physics: Conf. Series **845** (2017) 012024 doi:10.1088/1742-6596/845/1/012024

We designate

$$g\left(s\right)=-\left(x-X\left(s\right)\right)^{2}-c_{5}^{2}(\tau-s)^{2}\qquad\qquad s^{\pm}=g^{-1}\left(0\right)\tag{B.5}$$

and assume that the observation point x is in the timelike future of $X\left(-\infty\right)$ so that

$$\begin{aligned}a^{\alpha}\left(x,\tau\right)=&\ \frac{\lambda e}{2\pi^{2}}\int_{-\infty}^{s^{\pm}}ds\,\frac{1}{2}\dot{X}^{\alpha}\left(s\right)\frac{\theta\left(-\left(x-X\left(s\right)\right)^{2}-c_{5}^{2}(\tau-s)^{2}\right)}{\left[-\left(x-X\left(s\right)\right)^{2}-c_{5}^{2}(\tau-s)^{2}\right]^{3/2}}\theta\left(ct-X^{0}\left(s\right)\right)\\&-\frac{\lambda e}{2\pi^{2}}\int_{-\infty}^{\infty}ds\,\dot{X}^{\alpha}\left(s\right)\frac{\delta\left(-\left(x-X\left(s\right)\right)^{2}-c_{5}^{2}(\tau-s)^{2}\right)}{\left(-\left(x-X\left(s\right)\right)^{2}-c_{5}^{2}(\tau-s)^{2}\right)^{1/2}}\theta\left(ct-X^{0}\left(s\right)\right)\ .\end{aligned}\tag{B.6}$$

Using the identity

$$\int ds\,f\left(s\right)\delta\left[g\left(s\right)\right]=\sum_{s^{\pm}=g^{-1}(0)}\frac{f\left(s\right)}{\left|g'\left(s\right)\right|}\tag{B.7}$$

the second term in the integral becomes

$$\int_{-\infty}^{\infty}ds\,\dot{X}\left(s\right)\frac{\delta\left(g\left(s\right)\right)}{\left(g\left(s\right)\right)^{1/2}}\theta\left(ct-X^{0}\left(s\right)\right)=\frac{\dot{X}(s^{\pm})\theta\left(ct-X^{0}\left(s^{\pm}\right)\right)}{\left(g\left(s^{\pm}\right)\right)^{1/2}\left|g'\left(s^{\pm}\right)\right|}\ .\tag{B.8}$$

At the observation point $(x,x^{5})=(x,c_{5}\tau)$ we define a 5D line of observation as

$$Z=\left(z,z^{5}\right)=\left(x,x^{5}\right)-\left(X\left(s\right),X^{5}\right)=\left(x-X(s),c_{5}\tau-c_{5}s\right)\tag{B.9}$$

and a 5-velocity

$$U=\left(u,u^{5}\right)\qquad\qquad u^{\mu}=\dot{X}^{\mu}\left(s\right)\qquad\qquad u^{5}=\dot{X}^{5}\tag{B.10}$$

leading to a generalization of the denominator of (60) in the form

$$g'\left(s\right)=2U\cdot Z\tag{B.11}$$

so that (B.8) becomes

$$\left.\frac{\dot{X}(s)\theta\left(ct-X^{0}\left(s\right)\right)}{\left(g\left(s\right)\right)^{1/2}\left|2U\cdot Z\right|}\right|_{s\longrightarrow s^{\pm}}\tag{B.12}$$

which is singular as $s\rightarrow s^{\pm}$. As seen in Appendix A, we expect that this singularity is canceled by a corresponding singularity in the θ-term of (B.6). Nevertheless, for $s\neq s^{\pm}$ this expression remains finite if we take $c_{5}\rightarrow 0$. Since we expect the δ-term to have a similar structure to the θ-term, it seems that the contribution of $G_{correlation}$ to the field induced by a general event will split as

$$\frac{\mathcal{F}^{\mu}{}_{\nu}(x,\tau)\dot{x}^{\nu}+c_{5}^{2}\,\mathcal{F}^{5\mu}(x,\tau)}{1+\left(c_{5}/c\right)^{2}}\tag{B.13}$$

where $\mathcal{F}^{\mu}{}_{\nu}(x,\tau)$ and $\mathcal{F}^{5\mu}(x,\tau)$ remain finite as $c_{5}\rightarrow 0$.

IOP Conf. Series: Journal of Physics: Conf. Series **845** (2017) 012024 doi:10.1088/1742-6596/845/1/012024

References

[1] Stueckelberg E C G 1941 *Helv. Phys. Acta* **14** 322
Stueckelberg E C G 1941 *Helv. Phys. Acta* **14** 588
[2] Horwitz L P, Arshansky R I and Elitzur A C 1988 *Found. of Phys.* **18** 1159.
[3] Horwitz L P and Piron C 1973 *Helv. Phys. Acta* **48** 316
[4] Arshansky R L, Horwitz P and Lavie Y 1983 *Found. of Phys.* **13** 1167
[5] Horwitz L P, Arshansky R I and Elitzur A C 1988 *Found. of Phys.* **18** 1159.
[6] Saad D, Horwitz L P and Arshansky R I 1989 *Found. of Phys.* **19** 1126
[7] Land M C and Horwitz L P 1991 *Found. of Phys.* **21** 299
[8] Land M C and Horwitz L P 1991 *Found. of Phys. Lett.* **4** 61
[9] Land M C, Shnerb N and Horwitz L P 1995 *J. Math. Phys.* **36** 3263
[10] Land M C 1996 *Found. of Phys.* **27** 19
[11] Land M C and Horwitz L P 1998 Land M C **A239** 135
[12] Land M C 2001 *Found. of Phys.* **31** 967
[13] Land M C 2003 *Found. of Phys.* **33** 1157
[14] Land M C 2011 *J. Phys.: Conf. Ser.* **330** 012015
[15] Land M 2013 *J. Phys.: Conf. Ser.* **437** 012012
[16] Aharonovich I and Horwitz L P 2012 *J. Math. Phys.* **53** 032902
[17] Horwitz L P 2015 *Relativistic Quantum Mechanics* (Dordrecht: Springer Science+Business Media)
[18] Land M, to appear 2017 *J. Phys.: Conf. Ser.*
[19] Pierce P O 1899 *A short table of integrals* (New York: Ginn and Company) 24.

Mass stability in classical Stueckelberg-Horwitz-Piron electrodynamics

Martin Land

Department of Computer Science, Hadassah College, 37 HaNevi'im Street, Jerusalem

E-mail: martin@hadassah.ac.il

Abstract. It is well-known that the 5D gauge structure of Stueckelberg-Horwitz-Piron (SHP) electrodynamics permits the exchange of mass between particles and the electromagnetic fields induced by their motion, even at the classical level. This phenomenon presents two closely related problems: (1) Under what circumstances can real particles evolve sufficiently off-shell to account for mass changing phenomena such as flavor-changing neutrino interactions and low energy nuclear reactions? (2) What accounts for the stability of the measured masses of the known particles? To approach these questions, we first propose a toy model in which a particle evolving through a complex charged environment can acquire a significant mass shift for a short time. We then consider a classical self-interaction that tends to restore on-shell propagation.

1. Introduction

Stueckelberg-Horwitz-Piron (SHP) electrodynamics (see [1] to [20] and references therein) is a theory of interacting events that has been shown [8] to be the most general dynamical system consistent with the quantum commutation relations

$$[x^\mu, x^\nu] = 0 \qquad\qquad m[x^\mu, \dot{x}^\nu] = -i\hbar g^{\mu\nu}(x) \tag{1}$$

defined on an unconstrained 8D phase space. The velocity $\dot{x}^\mu(\tau)$ is the derivative of the event $x^\mu(\tau)$ with respect to the evolution parameter τ. The absence of constraints among the four velocities has three immediate and closely related consequences. First, τ cannot be identified with the proper time of the motion, but is proportional to it through

$$ds^2 = -g_{\mu\nu}dx^\mu dx^\nu = -\dot{x}^2(\tau)\, d\tau^2 \tag{2}$$

where $\dot{x}^2(\tau)$ is a dynamical quantity. The parameter τ must be understood as a monotonically increasing chronological time, independent of the spacetime coordinates and thus similar to the external time in nonrelativistic Newtonian mechanics. Second, the system is invariant under $\tau \to \tau + a$ for constant a but is not generally reparametrization invariant. Therefore, the dynamical particle evolution is not restricted to a mass shell. In particular, events and the gauge fields they induce may exchange mass, although the total mass of the particles and fields is conserved. Third, an event may reverse direction in coordinate time $x^0 = ct$, implementing the Stueckelberg model in which a pair process is represented by a single worldline, generated dynamically by the interaction of a spacetime event $x^\mu(\tau)$ with an electromagnetic field, as τ grows monotonically from $\tau = -\infty$ to $\tau = \infty$.

Pair creation/annihilation at the classical level is an example of a phenomenon that cannot occur without mass exchange. In the continuous evolution of an event $x^\mu(\tau)$ from a future trajectory $\dot{x}^0 \geq c$ to the past $\dot{x}^0 \leq -c$ the event must pass through the spacelike region around $\dot{x}^0 = 0$, where $\dot{x}^2$ changes sign. It was shown in [17] that under certain conditions SHP electrodynamics permits these classical pair processes, with no *a priori* guarantee that the outgoing particles will have identical mass. Small changes in particle mass are similarly relevant to neutrino oscillations [16] and low energy nuclear reactions [21]. As a practical matter such shifts in the asymptotic mass will be small [19], but even small mass shifts might be observable in a way that poses empirical difficulties for the theory.

Horwitz has recently shown [20] that by modeling a particle as an ensemble of events, the total particle mass is determined by a chemical potential. Following collisions governed by a general class of interactions that includes pair processes, particles return to their equilibrium mass values. This result may be related to another mass-related feature of SHP electrodynamics. The invariance of the action under $\tau \to \tau + a$ leads to singularities in classical currents that have been treated by regarding the particle as an ensemble of events $x^\mu(\tau + \Delta\tau)$ along the worldline, where $\Delta\tau$ is distributed as $\exp(-|\Delta\tau|/\lambda)$ and λ is parameter with dimensions of time. The low-energy interaction in the theory then takes the form of a Yukawa potential where the photon possesses a mass spectrum of width $m_\gamma \sim \hbar/\lambda c^2$.

In this paper we propose a specific model that suggests a deterministic relaxation of off-shell to on-shell propagation, along with restoration of the fixed mass. In section 2 we present a quick overview of SHP interactions, by way of field equations and the Lorentz force. In section 3 we motivate the discussion of mass shift by showing a toy model in which a particle evolving through a complex charged environment can acquire a significant mass shift for a short time. In section 4 we consider a classical self-interaction in SHP that tends to restore on-shell propagation.

2. Interactions in SHP

In its present form, SHP electrodynamics is a U(1) gauge theory in which the τ-dependence of gauge transformations requires *five* compensation potentials. In analogy with the notation $x^0 = ct$ we associate a constant c_5 with the invariant time τ, and adopt the formal designations

$$x^5 = c_5\tau \qquad \partial_5 = \frac{1}{c_5}\partial_\tau \qquad g_{55} = g^{55} = \pm 1 \tag{3}$$

and the index convention

$$\mu, \nu = 0,1,2,3 \qquad \alpha, \beta, \gamma = 0,1,2,3,5\ . \tag{4}$$

Because the low energy Coulomb particle-particle interaction differs [19] from the particle-antiparticle interaction by a factor $1 \mp g_{55}(c_5/c)$ we take c_5/c to be small.

A classical event $X^\mu(\tau)$ induces an instantaneous five-component current density

$$j^\alpha_\varphi(x,\tau) = c\int ds\, \varphi(\tau - s)\, \dot{X}^\alpha(s)\, \delta^4\left[x - X(s)\right] \tag{5}$$

IOP Conf. Series: Journal of Physics: Conf. Series **845** (2017) 012025 doi:10.1088/1742-6596/845/1/012025

where

$$\varphi(\tau) = \frac{1}{2\xi} e^{-|\tau|/\xi\lambda} \qquad \xi = \frac{1}{2}\left[1 + \left(\frac{c_5}{c}\right)^2\right] \tag{6}$$

and λ is a parameter with units of time. These currents induce five gauge potentials $a^\alpha(x,\tau)$ and field strengths

$$f^{\alpha\beta} = \partial^\alpha a^\beta - \partial^\beta a^\alpha \tag{7}$$

leading to τ-dependent field equations

$$\partial_\beta f^{\alpha\beta}(x,\tau) = \frac{e}{c} j^\alpha_\varphi(x,\tau) \qquad \partial_\alpha f_{\beta\gamma} + \partial_\gamma f_{\alpha\beta} + \partial_\beta f_{\gamma\alpha} = 0 \tag{8}$$

which are formally similar to Maxwell's equations in 5D, and are called pre-Maxwell equations. The fields act on the events through the Lorentz force

$$M\ddot{x}^\mu = \frac{e}{c} f^\mu{}_\alpha(x,\tau)\dot{x}^\alpha \tag{9}$$

which entails mass exchange through

$$\frac{d}{d\tau}\left(-\tfrac{1}{2}M\dot{x}^2\right) = -M\dot{x}^\mu\ddot{x}_\mu = -\frac{e}{c}\,\dot{x}^\mu(c_5 f_{\mu 5} + f_{\mu\nu}\dot{x}^\nu) = \frac{ec_5}{c}\,\dot{x}^\mu f_{5\mu} = g_{55}\frac{ec_5}{c}\,f^{5\mu}\dot{x}_\mu\ . \tag{10}$$

The pre-Maxwell equations in Lorenz gauge lead to the wave equation

$$\partial_\beta\partial^\beta a^\alpha = (\partial_\mu\partial^\mu + \partial_\tau\partial^\tau)a^\alpha = (\partial_\mu\partial^\mu + \frac{g_{55}}{c_5^2}\,\partial_\tau^2)a^\alpha = -\frac{e}{c}\,j^\alpha_\varphi(x,\tau) \tag{11}$$

for which the principal part Green's function was found in [6] to be

$$G_P(x,\tau) = -\frac{1}{2\pi}\delta(x^2)\delta(\tau) - \frac{c_5}{2\pi^2}\frac{\partial}{\partial x^2}\theta(-g_{55}g_{\alpha\beta}x^\alpha x^\beta)\frac{1}{\sqrt{-g_{55}g_{\alpha\beta}x^\alpha x^\beta}} \tag{12}$$

$$= G_{Maxwell} + G_{Correlation}\ . \tag{13}$$

The term $G_{Maxwell}$ imposes τ-instantaneous lightlike causality in spacetime, while the support of $G_{Correlation}$

$$-g_{55}g_{\alpha\beta}x^\alpha x^\beta = \begin{cases} -\left(x^2 + c_5^2\tau^2\right) = c^2t^2 - \mathbf{x}^2 - c_5^2\tau^2 > 0 \ , & g_{55} = 1 \\ \left(x^2 - c_5^2\tau^2\right) = \mathbf{x}^2 - c^2t^2 - c_5^2\tau^2 > 0 \ , & g_{55} = -1 \end{cases} \tag{14}$$

depends on the sign of g_{55}. If $g_{55} = 1$, $G_{Correlation}$ has timelike support with respect to the event trajectory, opening the possibility of a self-interaction of a type not present in standard Maxwell theory. We take $g_{55} = 1$ in the remaining sections of this paper. Evaluating the derivative in (12) for this case,

$$G_{correlation}(x,\tau) = -\frac{c_5}{2\pi^2}\left(\frac{1}{2}\frac{\theta(-x^2 - c_5^2\tau^2)}{\left(-x^2 - c_5^2\tau^2\right)^{3/2}} - \frac{\delta\left(-x^2 - c_5^2\tau^2\right)}{\left(-x^2 - c_5^2\tau^2\right)^{1/2}}\right)\ . \tag{15}$$

Although this expression appears highly singular, the two terms cancel when calculating potentials, as long as the subtraction is performed before applying the limits of integration.

IOP Conf. Series: Journal of Physics: Conf. Series **845** (2017) 012025 doi:10.1088/1742-6596/845/1/012025

The form of $G_{Maxwell}$ allows us apply standard techniques associated with the Liénard-Wiechert potential. Writing

$$\begin{aligned} a^{\alpha}(x,\tau) &= -\frac{e}{c}\int d^4x'd\tau'\, G_{Maxwell}(x-x',\tau-\tau')\, j^{\alpha}_{\varphi}(x',\tau') \\ &= \frac{e}{2\pi}\int ds\, \varphi(\tau-s)\, \dot{X}^{\alpha}(s)\, \delta\left((x-X(s))^2\right)\theta^{ret} \end{aligned} \tag{16}$$

and using the identity

$$\int d\tau f(\tau)\,\delta[g(\tau)] = \frac{f(\tau_R)}{|g'(\tau_R)|}\ , \tag{17}$$

where τ_R is the retarded time found from

$$g(\tau) = (x-X(\tau_R))^2 = 0 \qquad\qquad \theta^{ret} = \theta\left(x^0 - X^0(\tau_R)\right)\ , \tag{18}$$

provides

$$a^{\mu}(x,\tau) = \frac{e}{4\pi}\varphi(\tau-\tau_R)\frac{u^{\mu}}{|u\cdot z|} \qquad\qquad a^5(x,\tau) = \frac{e}{4\pi}\varphi(\tau-\tau_R)\frac{c_5}{|u\cdot z|} \tag{19}$$

where we write the event velocity and line of observation as

$$u^{\mu} = \dot{X}^{\mu}(\tau) \qquad\qquad z^{\mu} = x^{\mu} - X^{\mu}(\tau)\ . \tag{20}$$

By a similar procedure we find the field strengths as

$$\begin{aligned} f^{\mu\nu}(x,\tau) = -\frac{e}{4\pi}\varphi(\tau-\tau_R)\Bigg\{ &\frac{(z^{\mu}u^{\nu}-z^{\nu}u^{\mu})\,u^2}{(u\cdot z)^3} + \frac{\epsilon(\tau-\tau_R)}{\lambda}\frac{z^{\mu}u^{\nu}-z^{\nu}u^{\mu}}{(u\cdot z)^2} \\ &\frac{(z^{\mu}\dot{u}^{\nu}-z^{\nu}\dot{u}^{\mu})(u\cdot z) - (z^{\mu}u^{\nu}-z^{\nu}u^{\mu})(\dot{u}\cdot z)}{(u\cdot z)^3}\Bigg\} \end{aligned} \tag{21}$$

$$\begin{aligned} f^{5\mu}(x,\tau) = c_5\frac{e}{4\pi}\varphi(\tau-\tau_R)\Bigg\{ &\frac{z^{\mu}u^2 - u^{\mu}(u\cdot z)}{(u\cdot z)^3} + \frac{\epsilon(\tau-\tau_R)}{\lambda}\frac{z^{\mu}-u^{\mu}(u\cdot z)}{(u\cdot z)^2} \\ &- \frac{(\dot{u}\cdot z)\,z^{\mu}}{(u\cdot z)^3}\Bigg\} \end{aligned} \tag{22}$$

where we used

$$\frac{d}{d\tau_R}\varphi(\tau-\tau_R) = -\frac{1}{2\xi}\frac{d}{d\tau}e^{-|\tau-\tau_R|/\xi\lambda} = -\frac{\epsilon(\tau-\tau_R)}{\xi\lambda}\varphi(\tau-\tau_R) \tag{23}$$

and $\epsilon(\tau) = \text{signum}(\tau)$. The contribution from $G_{Correlation}$ is smaller than the $G_{Maxwell}$ contribution by c_5/c and drops off as $1/|\mathbf{x}|^2$. It may thus be neglected at low energy [19] compared to the fields (19), (21) and (22).

IOP Conf. Series: Journal of Physics: Conf. Series **845** (2017) 012025 doi:10.1088/1742-6596/845/1/012025

3. A toy model for mass shift

We consider an event propagating uniformly on-shell as

$$x(\tau) = u\tau = \left(u^0, \mathbf{u}\right) \qquad\qquad u^2 = -c^2 \tag{24}$$

until it passes through a dense region of charged particles that induce a small stochastic perturbation $X(\tau)$ such that

$$x(\tau) = u\tau + X(\tau) \quad . \tag{25}$$

If the typical distance scale between force centers is d then the perturbation will be roughly periodic with characteristic period

$$\frac{d}{|\mathbf{u}|} = \frac{\text{very short distance}}{\text{moderate velocity}} = \text{very short time}, \tag{26}$$

fundamental frequency

$$\omega_0 = 2\pi\frac{|\mathbf{u}|}{d} = \text{very high frequency}, \tag{27}$$

and amplitude on the order of

$$|X^\mu(\tau)| \sim \alpha d \tag{28}$$

for some macroscopic factor $\alpha < 1$. We may expand the perturbation in a Fourier series

$$X(\tau) = \text{Re}\sum_n a_n\, e^{in\omega_0\tau} \tag{29}$$

and write the four-vector coefficients as

$$a_n = \alpha d s_n = \alpha d\left(s_n^0, \mathbf{s}_n\right) = \alpha d\left(cs_n^t, \mathbf{s}_n\right) \tag{30}$$

where the s_n represent a normalized Fourier series ($s_0^\mu \sim 1$). The perturbed motion

$$X(\tau) = \alpha d\ \text{Re}\sum_n s_n^\mu\, e^{in\omega_0\tau} \tag{31}$$

is seen to be of scale d, but the perturbed velocity

$$\begin{aligned}
\dot{x}^\mu(\tau) &= u^\mu + \dot{X}^\mu(\tau) \\
&= u^\mu + \alpha d\ \text{Re}\sum_n n\omega_0\, s_n^\mu\, ie^{in\omega_0\tau} \\
&= u^\mu + \alpha d\ \text{Re}\sum_n n\left(2\pi\frac{|\mathbf{u}|}{d}\right) s_n^\mu\, ie^{in\omega_0\tau} \\
&= u^\mu + \alpha\,|\mathbf{u}|\ \text{Re}\sum_n 2\pi n\, s_n^\mu\, ie^{in\omega_0\tau}
\end{aligned} \tag{32}$$

is of macroscopic scale $\alpha\,|\mathbf{u}|$. The unperturbed on-shell mass is

$$m = -\frac{M\dot{x}^2(\tau)}{c^2} = M \tag{33}$$

IOP Conf. Series: Journal of Physics: Conf. Series **845** (2017) 012025 doi:10.1088/1742-6596/845/1/012025

and the perturbed mass is

$$
\begin{aligned}
m &= -\frac{M\dot{x}^2(\tau)}{c^2} = -\frac{M}{c^2}\left(u + \alpha\,|\mathbf{u}|\ \mathrm{Re}\sum_n 2\pi n\, s_n\, ie^{in\omega_0\tau}\right)^2 \\
&= -\frac{M}{c^2}\left(u^2 + \left(\alpha\,|\mathbf{u}|\ \mathrm{Re}\sum_n 2\pi n\, s_n\, ie^{in\omega_0\tau}\right)^2 + 2\alpha\,|\mathbf{u}|\ \mathrm{Re}\sum_n 2\pi n\ (u\cdot s_n)\ ie^{in\omega_0\tau}\right) \\
&= -\frac{M}{c^2}\left(-c^2 + 2\alpha\,|\mathbf{u}|\ \mathrm{Re}\sum_n 2\pi n\ (u\cdot s_n)\ ie^{in\omega_0\tau}\right. \\
&\qquad\qquad \left. - (\alpha\,|\mathbf{u}|)^2\,\mathrm{Re}\sum_{n,m}(2\pi)^2\, nm\, s_n\cdot s_m\, e^{i(n+m)\omega_0\tau}\right) \\
m &= M\left(1 - \frac{2\alpha\,|\mathbf{u}|}{c^2}\mathrm{Re}\sum_n 2\pi n\ (u\cdot s_n)\ ie^{in\omega_0\tau}\right. \\
&\qquad\qquad \left. + \frac{\alpha^2\mathbf{u}^2}{c^2}\mathrm{Re}\sum_{n,m}(2\pi)^2\, nm\, s_n\cdot s_m\, e^{i(n+m)\omega_0\tau}\right) \quad . \qquad (34)
\end{aligned}
$$

Evaluating the typical coefficients in the rest frame of the unperturbed motion

$$\frac{2\alpha\,|\mathbf{u}|}{c^2}2\pi n\ (u\cdot s_n) = \frac{4\pi\alpha\,|\mathbf{u}|\,n}{c^2}\ (c,\mathbf{0})\cdot\left(cs_n^t,\mathbf{s}_n\right) = -4\pi\alpha\,|\mathbf{u}|\ ns_n^t \qquad (35)$$

$$\frac{\alpha^2\mathbf{u}^2}{c^2}(2\pi)^2\, nm\, s_n\cdot s_m = (2\pi)^2\,\alpha^2\mathbf{u}^2\, nm\left(s_n^t s_m^t - \frac{\mathbf{s}_n\cdot\mathbf{s}_m}{c^2}\right) \qquad (36)$$

and neglecting the α^2 term, we find

$$m \simeq M\left(1 + 4\pi\alpha\,|\mathbf{u}|\,\mathrm{Re}\sum_n n\, s_n^t\, ie^{in\omega_0\tau}\right) \qquad (37)$$

which expresses a macroscopic mass shift

$$m \longrightarrow m\left(1 + \frac{\Delta m}{m}\right) \qquad \frac{\Delta m}{m} = 4\pi\alpha\,|\mathbf{u}|\,\mathrm{Re}\sum_n n\, s_n^t\, ie^{in\omega_0\tau} \ . \qquad (38)$$

Larger mass shifts can be observed if $\alpha > 1$ and the second order term in α^2 becomes significant.

4. A self-interaction

It was seen in (10) that particles may exchange mass with the fifth electromagnetic field through

$$\frac{d}{d\tau}(-\tfrac{1}{2}M\dot{x}^2) = \frac{e}{c}\, f^{5\mu}\dot{x}_\mu$$

and despite the scaling of $f^{5\mu}$ by c_5/c this effect cannot be assumed to be insignificant. In order to account for the observed stability of particle masses, we must find some mechanism that tends to enforce on-shell evolution, perhaps by damping off-shell behavior in the manner

IOP Conf. Series: Journal of Physics: Conf. Series **845** (2017) 012025 doi:10.1088/1742-6596/845/1/012025

of air friction producing a terminal velocity. If, for example, some circumstance were to provide a field of the form $f^{5\mu} = \sigma \dot{x}^\mu$ then

$$\frac{d}{d\tau}(-\tfrac{1}{2}M\dot{x}^2) = \frac{e}{c}\,\sigma\dot{x}^\mu\dot{x}_\mu = -\frac{2e\sigma}{Mc}\,(-\tfrac{1}{2}M\dot{x}^2) \tag{39}$$

producing mass decay.

In this section, we propose a model for a self-interaction between a moving event and its electromagnetic field that produces a mass decay but vanishes for on-shell propagation. Unlike the self-interaction between a particle and its radiation field, associated with the Abraham-Lorentz-Dirac equation, this model involves the influence of the field induced through $G_{Correlation}$ with retarded timelike support. The event experiences a force along its worldline produced by its earlier motion along that worldline.

4.1. Framework

We study the motion of an arbitrarily moving event $X^\mu(\tau)$ in a co-moving frame so that

$$X(\tau) = (ct(\tau), \mathbf{0}) \qquad \dot{X}(\tau) = (c\dot{t}(\tau), \mathbf{0}) \ . \tag{40}$$

In this frame

$$\dot{X}^2 = -c^2\dot{t}^2 \tag{41}$$

and so off-shell propagation is characterized by $\dot{t} \neq 1$ in the rest frame. We are interested in the self-force on the event at time τ^* and write the observation point as a later event along the worldline

$$X(\tau^*) = (ct(\tau^*), \mathbf{x}(\tau^*)) \tag{42}$$

so that

$$X(\tau^*) - X(s) = (ct(\tau^*), \mathbf{x}(\tau^*)) - (ct(s), \mathbf{x}(s)) = c(t(\tau^*) - t(s), \mathbf{0}) \ . \tag{43}$$

Because $G_{Maxwell} = 0$ on this timelike separation, the sole contribution comes from the previously neglected term $G_{Correlation}$. We approximate $\varphi(\tau' - s) = \lambda\delta(\tau' - s)$, which becomes exact in the limit $\lambda \to 0$, introduce the function $g(s)$ to express terms of the type

$$c^2 g(s) = -\left((X(\tau^*) - X(s))^2 + c_5^2(\tau^* - s)^2\right) = c^2\left((t(\tau^*) - t(s))^2 - \frac{c_5^2}{c^2}(\tau^* - s)^2\right) \ , \tag{44}$$

and write

$$a^\alpha(X(\tau^*), \tau^*) = \frac{\lambda e c_5}{2\pi^2 c^3}\int ds\, \dot{X}^\alpha(s)\left(\frac{1}{2}\frac{\theta(g(s))}{(g(s))^{3/2}} - \frac{\delta(g(s))}{(g(s))^{1/2}}\right)\theta^{ret} \tag{45}$$

for the self-field experienced by the event. We designate the two terms as

$$a^\alpha(X(\tau^*), \tau^*) = a^\alpha_\theta + a^\alpha_\delta \tag{46}$$

IOP Conf. Series: Journal of Physics: Conf. Series **845** (2017) 012025 doi:10.1088/1742-6596/845/1/012025

4.2. *Uniform on-shell motion*

For an event evolving uniformly on-shell we have

$$t(\tau^*) = \tau^* \qquad g(s) = \left(1 - \frac{c_5^2}{c^2}\right)(\tau^* - s)^2 \tag{47}$$

and using identity (17) are led to

$$a(X(\tau^*), \tau^*) = \frac{\lambda e c_5}{2\pi^2 c^3}(c, \mathbf{0}, c_5) \int ds\, \theta(\tau^* - s) \left(\frac{1}{2} \frac{\theta\left(\left(1 - \frac{c_5^2}{c^2}\right)(\tau^* - s)^2\right)}{\left(\left(1 - \frac{c_5^2}{c^2}\right)(\tau^* - s)^2\right)^{3/2}} - \frac{\delta\left(\left(1 - \frac{c_5^2}{c^2}\right)(\tau^* - s)^2\right)}{\left(\left(1 - \frac{c_5^2}{c^2}\right)(\tau^* - s)^2\right)^{1/2}} \right)$$

$$= \frac{\lambda e c_5 (c, \mathbf{0}, c_5)}{2\pi^2 c^3 \left(1 - \frac{c_5^2}{c^2}\right)^{3/2}} \int_{-\infty}^{\tau^*} ds \left(\frac{1}{2} \frac{1}{2(\tau^* - s)^3} - \frac{\delta(\tau^* - s)\,\theta(\tau^* - s)}{\left|(\tau^* - s)^2\right|} \right). \tag{48}$$

Since

$$\int_{-\infty}^{\tau^*} ds\, \frac{1}{(\tau^* - s)^3} = \left. \frac{1}{2(\tau^* - s)^2} \right|_{-\infty}^{\tau^*} = \lim_{s \to \tau^*} \frac{1}{2(\tau^* - s)^2} \tag{49}$$

and

$$\int_{-\infty}^{\tau^*} ds\, \frac{\delta(\tau^* - s)\,\theta(\tau^* - s)}{(\tau^* - s)^2} = \lim_{s \to \tau^*} \frac{\theta(\tau^* - s)}{(\tau^* - s)^2} = \lim_{s \to \tau^*} \frac{\frac{1}{2}}{(\tau^* - s)^2} \tag{50}$$

we find that for uniform on-shell motion

$$a(X(\tau^*), \tau^*) = \frac{\lambda e c_5}{2\pi^2 c^3}(c, \mathbf{0}, c_5) \lim_{s \to \tau^*} \left(\frac{1}{2(\tau^* - s)^2} - \frac{\frac{1}{2}}{(\tau^* - s)^2} \right) = 0 \tag{51}$$

the self-force vanishes.

4.3. *Field strengths*

From $\dot{X}^i = 0$ and the form of (45)

$$a^i = 0 \qquad \partial_i a^0 = \partial_i a^5 = 0 \qquad \Rightarrow \qquad f^{\mu\nu} = f^{5i} = 0 \tag{52}$$

and so the field reduces to

$$f^{50} = \partial^5 a^0 - \partial^0 a^5 = g^{55} \frac{1}{c_5} \partial_{\tau^*} a^0 - g^{00} \frac{1}{c} \partial_t a^5 = \frac{1}{c_5} \partial_{\tau^*} a^0 + \frac{1}{c} \partial_t a^5 \tag{53}$$

where the partial derivative ∂_{τ^*} only acts on the explicit variable (not on $t(\tau^*)$ or θ^{ret}). Similarly, the velocity appears as $\dot{X}^\alpha(s)$ and is constant with respect to ∂_{τ^*}.

Working piece-by-piece

$$\partial^5 a_\theta^0 = \frac{\lambda e c_5}{4\pi^2 c^3} \frac{1}{c_5} \partial_{\tau^*} \int ds\, c\dot{t}(s) \frac{\theta\left((t(\tau^*) - t(s))^2 - \frac{c_5^2}{c^2}(\tau^* - s)^2\right)}{\left[(t(\tau^*) - t(s))^2 - \frac{c_5^2}{c^2}(\tau^* - s)^2\right]^{3/2}} \theta^{ret} \tag{54}$$

IOP Conf. Series: Journal of Physics: Conf. Series **845** (2017) 012025 doi:10.1088/1742-6596/845/1/012025

contains

$$\partial_{\tau^*}\theta\left(\left(t\left(\tau^*\right)-t\left(s\right)\right)^2-\frac{c_5^2}{c^2}(\tau^*-s)^2\right)=-2\frac{c_5^2}{c^2}\delta\left(\left(t\left(\tau^*\right)-t\left(s\right)\right)^2-\frac{c_5^2}{c^2}(\tau^*-s)^2\right)(\tau^*-s) \quad (55)$$

and

$$\partial_{\tau^*}\frac{1}{\left[\left(t\left(\tau^*\right)-t\left(s\right)\right)^2-\frac{c_5^2}{c^2}(\tau^*-s)^2\right]^{3/2}}=3\frac{c_5^2}{c^2}\frac{\tau^*-s}{\left[\left(t\left(\tau^*\right)-t\left(s\right)\right)^2-\frac{c_5^2}{c^2}(\tau^*-s)^2\right]^{5/2}}\ . \quad (56)$$

Similarly,

$$\frac{1}{c}\partial_t a_\theta^5=\frac{\lambda e c_5}{4\pi^2c^3}\frac{1}{c}\partial_{t(\tau^*)}\int ds\ c_5\frac{\theta\left(\left(t\left(\tau^*\right)-t\left(s\right)\right)^2-\frac{c_5^2}{c^2}(\tau^*-s)^2\right)}{\left(\left(t\left(\tau^*\right)-t\left(s\right)\right)^2-\frac{c_5^2}{c^2}(\tau^*-s)^2\right)^{3/2}}\theta^{ret} \quad (57)$$

contains

$$\partial_{t(\tau^*)}\theta\left(\left(t\left(\tau^*\right)-t\left(s\right)\right)^2-\frac{c_5^2}{c^2}(\tau^*-s)^2\right)=2\left(t\left(\tau^*\right)-t\left(s\right)\right)\times$$
$$\delta\left(\left(t\left(\tau^*\right)-t\left(s\right)\right)^2-\frac{c_5^2}{c^2}(\tau^*-s)^2\right) \quad (58)$$

$$\partial_{t(\tau^*)}\frac{1}{\left[\left(t\left(\tau^*\right)-t\left(s\right)\right)^2-\frac{c_5^2}{c^2}(\tau^*-\tau')^2\right]^{3/2}}=-3\frac{t\left(\tau^*\right)-t\left(s\right)}{\left[\left(t\left(\tau^*\right)-t\left(s\right)\right)^2-\frac{c_5^2}{c^2}(\tau^*-\tau')^2\right]^{5/2}} \quad (59)$$

and

$$\partial_{t(\tau^*)}\theta^{ret}=\partial_{t(\tau^*)}\theta\left(t\left(\tau^*\right)-t\left(s\right)\right)=\delta\left(t\left(\tau^*\right)-t\left(s\right)\right)=0 \quad (60)$$

where the last expression vanishes because $t\left(\tau^*\right) = t\left(s\right)$ makes the argument of $\theta(g(s))$ negative. Putting the pieces together we find

$$\partial^5 a_\theta^0-\partial^0 a_\theta^5=\frac{3\lambda e c_5}{4\pi^2c^3}\frac{c_5}{c}\int ds\ \frac{\theta\left(\left(t\left(\tau^*\right)-t\left(s\right)\right)^2-\frac{c_5^2}{c^2}(\tau^*-s)^2\right)}{\left[\left(t\left(\tau^*\right)-t\left(s\right)\right)^2-\frac{c_5^2}{c^2}(\tau^*-s)^2\right]^{5/2}}\theta^{ret}\,\Delta\left(\tau^*,s\right)$$
$$-\frac{\lambda e c_5}{2\pi^2c^3}\frac{c_5}{c}\int ds\ \frac{\delta\left(\left(t\left(\tau^*\right)-t\left(s\right)\right)^2-\frac{c_5^2}{c^2}(\tau^*-s)^2\right)}{\left[\left(t\left(\tau^*\right)-t\left(s\right)\right)^2-\frac{c_5^2}{c^2}(\tau^*-s)^2\right]^{3/2}}\theta^{ret}\,\Delta\left(\tau^*,s\right) \quad (61)$$

where

$$\Delta\left(\tau^*,s\right)=\dot{t}(s)(\tau^*-s)-\left(t\left(\tau^*\right)-t\left(s\right)\right) \quad (62)$$

characterizes the energy acceleration in the rest frame, which will be associated with mass shift. Similarly, the derivatives of a_δ produce

$$\partial^5 a_\delta^0-\partial^0 a_\delta^5=-\frac{\lambda e c_5}{2\pi^2c^3}\frac{c_5}{c}\int ds\ \frac{\delta\left(\left(t\left(\tau^*\right)-t\left(s\right)\right)^2-\frac{c_5^2}{c^2}(\tau^*-s)^2\right)}{\left(\left(t\left(\tau^*\right)-t\left(s\right)\right)^2-\frac{c_5^2}{c^2}(\tau^*-s)^2\right)^{3/2}}\theta^{ret}\,\Delta\left(\tau^*,s\right)$$
$$-\frac{\lambda e c_5}{2\pi^2c^3}\frac{c_5}{c}\int ds\ \frac{2\delta'\left(\left(t\left(\tau^*\right)-t\left(s\right)\right)^2-\frac{c_5^2}{c^2}(\tau^*-s)^2\right)}{\left(\left(t\left(\tau^*\right)-t\left(s\right)\right)^2-\frac{c_5^2}{c^2}(\tau^*-s)^2\right)^{1/2}}\theta^{ret}\,\Delta\left(\tau^*,s\right) \quad (63)$$

IOP Conf. Series: Journal of Physics: Conf. Series **845** (2017) 012025 doi:10.1088/1742-6596/845/1/012025

and combining terms we find

$$f^{50} = f^{50}_{\theta} + f^{50}_{\delta} + f^{50}_{\delta'} \tag{64}$$

where

$$f^{50}_{\theta} = \frac{3\lambda e}{4\pi^2} \frac{c_5^2}{c^4} \int ds \, \frac{\theta\left(\left(t\left(\tau^*\right) - t\left(s\right)\right)^2 - \frac{c_5^2}{c^2}\left(\tau^* - s\right)^2\right)}{\left[\left(t\left(\tau^*\right) - t\left(s\right)\right)^2 - \frac{c_5^2}{c^2}\left(\tau^* - s\right)^2\right]^{5/2}} \, \theta^{ret} \, \Delta\left(\tau^*, s\right) \tag{65}$$

$$f^{50}_{\delta} = -\frac{\lambda e}{\pi^2} \frac{c_5^2}{c^4} \int ds \, \frac{\delta\left(\left(t\left(\tau^*\right) - t\left(s\right)\right)^2 - \frac{c_5^2}{c^2}\left(\tau^* - s\right)^2\right)}{\left[\left(t\left(\tau^*\right) - t\left(s\right)\right)^2 - \frac{c_5^2}{c^2}\left(\tau^* - s\right)^2\right]^{3/2}} \, \theta^{ret} \, \Delta\left(\tau^*, s\right) \tag{66}$$

$$f^{50}_{\delta'} = -\frac{\lambda e}{\pi^2} \frac{c_5^2}{c^4} \int ds \, \frac{\delta'\left(\left(t\left(\tau^*\right) - t\left(s\right)\right)^2 - \frac{c_5^2}{c^2}\left(\tau^* - s\right)^2\right)}{\left(\left(t\left(\tau^*\right) - t\left(s\right)\right)^2 - \frac{c_5^2}{c^2}\left(\tau^* - s\right)^2\right)^{1/2}} \, \theta^{ret} \, \Delta\left(\tau^*, s\right) \tag{67}$$

Notice that if the particle remains at constant velocity (in any uniform frame), then

$$x^0\left(\tau\right) = u^0 \tau \Rightarrow \Delta\left(\tau^*, s\right) = \frac{u^0}{c}\left(\tau^* - s\right) - \left(\frac{u^0}{c}\tau^* - \frac{u^0}{c}s\right) = 0 \tag{68}$$

and so the self-force f^{50} vanishes. For any smooth $t\left(\tau\right)$, we may approximate

$$\begin{aligned} t\left(\tau^*\right) - t\left(s\right) &= t\left(s\right) + \dot{t}(s)(\tau^* - s) + \frac{1}{2}\ddot{t}(s)(\tau^* - s)^2 + o\left((\tau^* - s)^3\right) - t\left(s\right) \\ &= \dot{t}(s)(\tau^* - s) + \frac{1}{2}\ddot{t}(s)(\tau^* - s)^2 + o\left((\tau^* - s)^3\right) \end{aligned} \tag{69}$$

so the function

$$\Delta\left(\tau^*, s\right) = \dot{t}(s)(\tau^* - s) - \left(t\left(\tau^*\right) - t\left(s\right)\right) = -\frac{1}{2}\ddot{t}(s)(\tau^* - s)^2 + o\left((\tau^* - s)^3\right) \tag{70}$$

is nonzero only when the time coordinate accelerates in the rest frame, equivalent to a shift in the particle mass.

4.4. *Mass jump*

As a first order example, we consider a small, sudden jump in mass at $\tau = 0$ characterized by

$$t\left(\tau\right) = \begin{cases} \tau & , \quad \tau < 0 \\ \left(1 + \beta\right)\tau & , \quad \tau > 0 \end{cases} \qquad \Rightarrow \qquad \dot{t}\left(\tau\right) = \begin{cases} 1 & , \quad \tau < 0 \\ 1 + \beta & , \quad \tau > 0 \end{cases} \tag{71}$$

and calculate the self-interaction. Since θ^{ret} enforces $t(\tau^*) > t(s)$, it follows that

$$\tau^* < 0 \quad \Rightarrow \quad s < 0 \quad \Rightarrow \quad \dot{t}(\tau^*) = \dot{t}(s) = 1 \quad \Rightarrow \quad \Delta(\tau^*, s) = 0 \ . \tag{72}$$

Similarly,

$$\tau^* > 0 \text{ and } s > 0 \quad \Rightarrow \quad \dot{t}(\tau^*) = \dot{t}(s) = 1 + \beta \quad \Rightarrow \quad \Delta(\tau^*, s) = 0 \ . \tag{73}$$

IOP Conf. Series: Journal of Physics: Conf. Series **845** (2017) 012025 doi:10.1088/1742-6596/845/1/012025

But when $\tau^* > 0$ and $s < 0$,

$$\Delta(\tau^*,s) = \dot{t}(s)(\tau^* - s) - (t(\tau^*) - t(s)) = (\tau^* - s) - [(1+\beta)(\tau^*) - s] = -\beta\tau^* \tag{74}$$

and f^{50} can be found from the contributions (65) – (67). Writing

$$g(s) = (t(\tau^*) - t(s))^2 - \frac{c_5^2}{c^2}(\tau^* - s)^2 = ((1+\beta)\tau^* - s)^2 - \frac{c_5^2}{c^2}(\tau^* - s)^2 \tag{75}$$

and solving for $g(s^*) = 0$, we find

$$s^* = \left(1 + \frac{\beta}{1 - \frac{c_5}{c}}\right)\tau^* > \tau^* \tag{76}$$

so that $g(s) > 0$ in the region of interest $s < 0 < \tau^*$ and there will be no contribution from the terms (66) or (67). Thus,

$$\begin{aligned} f^{50} = f^{50}_\theta &= (-\beta\tau^*)\frac{3\lambda e}{4\pi^2}\frac{c_5^2}{c^4}\int_{-\infty}^0 ds \frac{1}{\left[(t(\tau^*) - t(s))^2 - \frac{c_5^2}{c^2}(\tau^* - s)^2\right]^{5/2}} \\ &= (-\beta\tau^*)\frac{3\lambda e}{4\pi^2}\frac{c_5^2}{c^4}\int_{-\infty}^0 ds \frac{1}{\left[((1+\beta)\tau^* - s)^2 - \frac{c_5^2}{c^2}(\tau^* - s)^2\right]^{5/2}} . \end{aligned} \tag{77}$$

Shifting the integration variable as $x = \tau^* - s$ the integral becomes

$$\int_{-\infty}^0 ds \frac{1}{\left[((1+\beta)\tau^* - s)^2 - \frac{c_5^2}{c^2}(\tau^* - s)^2\right]^{5/2}} = -\int_\infty^{\tau^*} \frac{dx}{(Cx^2 + Bx + A)^{5/2}} \tag{78}$$

where

$$C = 1 - \frac{c_5^2}{c^2} \qquad B = 2\beta\tau^* \qquad A = (\beta\tau^*)^2 \ . \tag{79}$$

This integral can be evaluated using the well-known form [22]

$$\int \frac{dx}{(Cx^2 + Bx + A)^{5/2}} = \frac{2(2Cx + B)}{3q\sqrt{Cx^2 + Bx + A}}\left(\frac{1}{Cx^2 + Bx + A} + \frac{8C}{q}\right) \tag{80}$$

where

$$q = 4AC - B^2 \tag{81}$$

IOP Conf. Series: Journal of Physics: Conf. Series **845** (2017) 012025 doi:10.1088/1742-6596/845/1/012025

leading to

$$-\int_{\infty}^{\tau^*}\frac{dx}{\left(Cx^2+Bx+A\right)^{5/2}}=-\frac{1}{3\left(\beta\tau^*\right)^4}\times$$

$$\left[2\frac{c^4}{c_5^4}\left(1-\frac{c_5^2}{c^2}\right)^{3/2}\left(1-\frac{\left(1-\frac{c_5^2}{c^2}\right)^{1/2}\left(1+\frac{\beta}{\left(1-\frac{c_5^2}{c^2}\right)}\right)}{\left[1+\frac{2\beta}{1-\frac{c_5^2}{c^2}}+\frac{\beta^2}{1-\frac{c_5^2}{c^2}}\right]^{1/2}}\right)\right.$$

$$\left.+\frac{c^2}{c_5^2}\frac{\beta^2\left(1+\frac{c_5^2}{c^2}\frac{\beta}{1-\frac{c_5^2}{c^2}}\right)}{\left(1-\frac{c_5^2}{c^2}\right)^{1/2}\left[1+\frac{2\beta}{1-\frac{c_5^2}{c^2}}+\frac{\beta^2}{1-\frac{c_5^2}{c^2}}\right]^{3/2}}\right] \tag{82}$$

and the field strength in the form

$$f^{50}=\frac{\lambda e}{4\pi^2}\frac{1}{c_5^2\left(\beta\tau^*\right)^3}\,Q\left(\beta,\frac{c_5^2}{c^2}\right) \tag{83}$$

where $Q\left(\beta,\frac{c_5^2}{c^2}\right)$ is the positive, dimensionless factor

IOP Conf. Series: Journal of Physics: Conf. Series **845** (2017) 012025 doi:10.1088/1742-6596/845/1/012025

$$Q\left(\beta,\frac{c_5^2}{c^2}\right)=\left[2\left(1-\frac{c_5^2}{c^2}\right)^{3/2}\left(1-\frac{\left(1-\frac{c_5^2}{c^2}\right)^{1/2}\left(1+\frac{\beta}{\left(1-\frac{c_5^2}{c^2}\right)}\right)}{\left[1+\frac{2\beta}{1-\frac{c_5^2}{c^2}}+\frac{\beta^2}{1-\frac{c_5^2}{c^2}}\right]^{1/2}}\right)\right.$$
$$\left.+\frac{\beta^2\frac{c_5^2}{c^2}\left(1+\frac{c_5^2}{c^2}\frac{\beta}{1-\frac{c_5^2}{c^2}}\right)}{\left(1-\frac{c_5^2}{c^2}\right)^{1/2}\left[1+\frac{2\beta}{1-\frac{c_5^2}{c^2}}+\frac{\beta^2}{1-\frac{c_5^2}{c^2}}\right]^{3/2}}\right] \tag{84}$$

which is seen to be finite for $c_5 < c$, with

$$Q\left(\beta,\frac{c_5^2}{c^2}\right)\xrightarrow[c_5\to 0]{}2\left(1-\frac{1+\beta}{[1+2\beta+\beta^2]^{1/2}}\right)=0\ . \tag{85}$$

Since $f^{\mu\nu}=0$, the Lorentz force induced by this field strength is then

$$M\ddot{x}^\mu = ef^{\mu\alpha}\dot{x}_\alpha = ef^{\mu 5}\dot{x}_5 = -ef^{5\mu}\dot{x}_5 = -g_{55}ef^{5\mu}\dot{x}^5 = -ef^{5\mu}c_5 \tag{86}$$

and since $f^{5i}=0$

$$M\ddot{x}^i = 0 \tag{87}$$

$$M\ddot{x}^0 = -c_5 e f^{50} = \begin{cases} 0 & , \quad \tau^* < 0 \\ -\dfrac{\lambda e^2}{4\pi^2}\dfrac{1}{c_5\,(\beta\tau^*)^3}\,Q\left(\beta,\dfrac{c_5^2}{c^2}\right) & , \quad \tau^* > 0 \end{cases} \tag{88}$$

which causes the 0-coordinate to decelerate. When the event returns to on-shell propagation the function $\Delta(\tau^*,s)$ and field strength f^{50} again vanish. The mass decay can also be seen in the Lorentz force for the mass

$$\frac{d}{d\tau}\left(-\frac{1}{2}M\dot{x}^2\right)=ef^{5\mu}\dot{x}_\mu = ef^{50}\dot{x} = -ecf^{50}\dot{t} = -\frac{\lambda e^2}{4\pi^2}\frac{c}{c_5^2\,(\beta\tau^*)^3}\,Q\left(\beta,\frac{c_5^2}{c^2}\right)\dot{t}\ . \tag{89}$$

We notice that if $\beta < 0$ then f^{50} changes sign so that the self-interaction results in damping or anti-damping to push the trajectory toward on-shell behavior. Although this model is approximate, it seems to indicate that the self-interaction of the event with the field generated by its mass shift will restore the event to on-shell propagation.

5. Summary

In SHP theory, the mass shell relation $\dot{x}^2 = 1$ is demoted from the status of *a priori* constraint on the phase space to that of first integral, identifying by way of Noether's theorem a constant of the motion for interactions mediated by τ-independent potentials. Exchange of particle mass with the electromagnetic field can occur in various elastic scattering scenarios, including pair processes and the stochastic interaction with a region of densely packed charges described in section 3. Although the mass exchange may be small and total mass of particles and fields is conserved, the theory must account for the empirical observation of fixed asymptotic masses.

In this paper we have considered a classical self-interaction that acts on an off-shell event trajectory in a way that tends to restore on-shell propagation. In this model, an event that undergoes a mass increase, equivalent to acceleration of the time coordinate in the particle rest frame, experiences a force as it passes through the electromagnetic field induced by its earlier motion along its worldline. Such an interaction is prohibited by the lightlike support of $G_{Maxwell}$, the Maxwell part of the SHP Green's function, and by the spacelike support of $G_{Correlation}$ for $g_{55} = -1$. However, for $g_{55} = +1$ the Green's function has timelike support and includes the particle's own future worldline. We found that for uniform motion this self-interaction vanishes, so that on-shell propagation is unaffected. But the Lorentz force acting on a particle that undergoes a discrete jump in $\dot{x}^0$ in its rest frame was found to oppose the time acceleration and restore the particle to on-shell propagation (for which the interaction again vanishes). This self-interaction would appear to provide an underlying mechanism for the asymptotic on-shell behavior found by Aharonovich and Horwitz in numerical solutions [14], and perhaps an explanation of the observed mass stability of the known particles.

To complete the investigation begun in this paper involves solving the Lorentz force equations under realistic conditions, for example a uniformly evolving event approaching a charge sheet, which was shown in [15] to induce a small mass shift. Because the interaction itself is non-linear and the self-interaction term must be evaluated iteratively at each point along the worldline by integrating with the Green's function up to that point on the worldline, these configurations require careful numerical solution. Such solutions will be presented in a subsequent paper.

Acknowledgments

The author gratefully acknowledges the inspiration and guidance of Professor Lawrence P. Horwitz, whose work has opened many fields of inquiry and whose direction has set many students on a productive paths in physics. The author is also grateful to Mark Davidson for useful conversations on the nature mass variation in SHP.

References

[1] Stueckelberg E C G 1941 *Helv. Phys. Acta* **14** 322
Stueckelberg E C G 1941 *Helv. Phys. Acta* **14** 588
[2] Horwitz L P and Piron C 1973 *Helv. Phys. Acta* **48** 316
[3] Arshansky R L, Horwitz P and Lavie Y 1983 *Found. of Phys.* **13** 1167
[4] Horwitz L P, Arshansky R I and Elitzur A C 1988 *Found. of Phys.* **18** 1159.
[5] Saad D, Horwitz L P and Arshansky R I 1989 *Found. of Phys.* **19** 1126
[6] Land M C and Horwitz L P 1991 *Found. of Phys.* **21** 299
[7] Land M C and Horwitz L P 1991 *Found. of Phys. Lett.* **4** 61
[8] Land M C, Shnerb N and Horwitz L P 1995 *J. Math. Phys.* **36** 3263
[9] Land M C 1996 *Found. of Phys.* **27** 19

[10] Land M C and Horwitz L P 1998 Land M C **A239** 135
[11] Land M C 2001 *Found. of Phys.* **31** 967
[12] Land M C 2003 *Found. of Phys.* **33** 1157
[13] Land M C 2011 *J. Phys.: Conf. Ser.* **330** 012015
[14] Aharonovich I and Horwitz L P 2012 *J. Math. Phys.* **53** 032902
[15] Land M 2013 *J. Phys.: Conf. Ser.* **437** 012012
[16] Horwitz L P and Aharonovich I 2013 *J. Phys.: Conf. Ser.* **437** 012021
[17] Land M 2015 *J. Phys.: Conf. Ser.* **615** 012007
[18] Horwitz L P 2015 *Relativistic Quantum Mechanics* (Dordrecht: Springer Science+Business Media)
[19] Land M, to appear 2017 *J. Phys.: Conf. Ser.*
[20] Horwitz LP http://arxiv.org/abs/1607.03742v2
[21] Davidson M 2015 J. Phys.: Conf. Ser. 615 012016
[22] Pierce P O 1899 *A short table of integrals* (New York: Ginn and Company) 24.

IOP Conf. Series: Journal of Physics: Conf. Series **845** (2017) 012026 doi:10.1088/1742-6596/845/1/012026

A statistical mechanical model for mass stability in the SHP theory

Lawrence P. Horwitz
School of Physics, Tel Aviv University Ramat Aviv, Israel
Department of Physics, Bar Ilan University, Ramat Gan, Israel
Department of Physics, Ariel University, Ariel, Israel

Abstract. We construct a model for a particle in the framework of the theory of Stueckelberg, Horwitz and Piron (SHP) as an ensemble of events subject to the laws of covariant classical equilibrium statistical mechanics. The canonical and grand canonical ensembles are constructed without an *a priori* constraint on the total mass of the system. We show that the total mass of the system, corresponding the mass of this particle is determined by a chemical potential. This model has the property that under perturbation, such as collisions in the SHP theory for which the final asymptotic mass of an elementary event is not constrained by the basic theory, the particle returns to its equilibrium mass value. A mechanism similar to the Maxwell construction for more than one equilibrium mass state may result in several possible masses in the final state.

1. Introduction

The relativistic classical and quantum mechanics based on the theory of Stueckelberg, Horwitz and Piron [SHP] [1],[2],[3] admits a wide range of values after collision for the quantity $p_\mu p^\mu = \mathbf{p}^2 - E^2 = -m^2$ for each of the particles, bounded only by kinematics and the conserved value of, for example, the Hamiltonian of a system of two particles in a potential type interaction,

$$K = \frac{p_{1\mu} p_1{}^\mu}{2M_1} + \frac{p_{2\mu} p_2{}^\mu}{2M_2} + V(x_1 - x_2), \tag{1.1}$$

where M_1 and M_2 are intrinsic parameters with dimension mass for each of the particles (these quantities may be thought of as the Galilean limiting masses, as discussed in ref. [4]), and $x \equiv \{x^\mu\}$, the spacetime coordinates of the *events* that are considered the basic dynamical objects of the theory (we shall call these *particles*, although the identification of these objects with particles in the usual sense, as discussed in [5] and in Jackson's book [6] involves the construction of conserved currents). The equations of motion, according to the Hamilton equations derived from this Hamiltonian, describe evolution in an invariant parameter τ [1][2][3]. Solving the equations of motion for an asymptotically free ($V \cong 0$) incoming state with well-defined incoming mass values, in the outgoing asymptotic state the particles may have different mass values. In order to maintain the identity of particles in such asymptotic states, it has been assumed in many applications, that there is a mechanism, such as relaxation to a minimum free energy, which restores the particle masses (*e.g.*, at the ionization point for the bound state problem [7]). For example, an electron suffering many collisions, even though

IOP Conf. Series: Journal of Physics: Conf. Series **845** (2017) 012026 doi:10.1088/1742-6596/845/1/012026

the mass changes may be small in each collision, could drift measurably, in contradiction with the observed stability of the electron mass under a wide range of conditions. Although the assumption of some relaxation process has been adequate to obtain quantitative results in agreement with experiment, the availability of an effective physical model could have important predictive consequences [8].

In this paper, I formulate a model for an "elementary" Stueckelberg particle by constructing a statistical ensemble of the type studied by Horwitz, Schieve and Piron [4], where the relativistic Gibbs ensembles for a system of events in thermal equilibrium were worked out. In order to achieve a structure for which the nonrelativistic limit provided the correct *a priori* nonrelativistic limit for the total mass of the system, they assumed that the value of the many-body Hamiltonian was restricted, in the microcanonical ensemble, to this value of total nonrelativistic mass. In the model I construct here, the total mass value of the ensemble is not restricted; it may vary according to the dynamical Hamiltonian laws of motion as described above, but in the asymptotically free state, the mass value is determined, as for the average energy, by a chemical potential analogous to a temperature (one could think of the associated relaxation as accompanied by emission or absorption of radiation). The statistical ensemble therefore has the property that its mean spacetime value behaves under interaction as the usual Stueckelberg particle, but contains a mechanism for stability of its mass after interaction, with the possibility of transitions to other equilibrium states according to the Maxwell construction, as a phase transition induced by the interactions. The fundamental origin of such an ensemble may lie in the vacuum fluctuations of a quantum field theory, but we leave this for future study.

2. SHP Dynamics and Statistical Mechanics: The Model

For a single particle, Stueckelberg [1] considered the possibility that the world line of a particle might turn from advancing in t to a state in which the motion in t is reversed, and recognized that this situation describes pair annihilation in classical mechanics. Since this motion is not single valued in t, he defined a parameter τ along the trajectory. In terms of this parameter, he was able to define a canonical Hamiltonian formalism in spacetime. Horwitz and Piron [2] then generalized the structure of the theory, postulating that τ is universal (as for t in Newtonian-Galilean mechanics), in order to deal with the many-body problem (see ref. 3 for a discussion of applications and consequences of this idea). The notion of τ evolution of a many body system permits us to think of equilibrium ensembles for a many body system [4]. In the following we formulate the statistical mechanics of a system of many degrees of freedom that we imagine to be associated with the structure of a *single Stueckelberg particle* to find equilibrium conditions for the stability of its mass.

The relativistically invariant Hamiltonian for an N-body system of particles interacting through a Poincaré invariant potential is defined [2] by

$$K = \Sigma_{i=1}^{N} \frac{p_{i\mu} p_i{}^{\mu}}{2M_i} + V(x_1.x_2, \ldots, x_N) \tag{2.1}$$

We imagine that this set of particles forms an ensemble which plays the dynamical role of a *single* Stueckelberg particle with dynamical coordinates P^μ, X^μ in phase space, corresponding to the center-of-mass properties of the ensemble. We then define the microcanonical ensemble [4] as

$$\Gamma(\kappa, E) = \int d\Omega \delta(K-\kappa)\delta(\Sigma E_i - E), \tag{2.2}$$

where $d\Omega$ is the infinitesimal volume element in the phase space of the many-body system. We do not choose *a priori*, however, the parameter κ to correspond to the total Galilean mass of the system, but leave its eventual value to equilibrium conditions imposed, as for E, by a chemical potential (a "mass temperature").

The canonical ensemble is then constructed by extracting a small subensemble for which the energy and mass parameter are assumed to be additive as in the usual statistical mechanics. This procedure distinguishes what we shall think of as the "particle" (the subensemble) from its "environment" (the "bath"). We write the microcanonical ensemble in terms of a sum over all possible partitions of energy and mass parameter between the system and the bath (one usually thinks of an exchange of energy between the subsystems and the bath; in this case, we include also an exchange of mass, reflecting not the flow of heat only, but also of off-shell mass as an intrinsic degree of freedom):

$$\begin{aligned}
\Gamma(\kappa, E) &= \int d\Omega_b d\Omega_s d\kappa_b d\kappa_s \\
&\quad \delta(K_b - \kappa_b)\delta(K_s - \kappa_s)\delta(E_s + E_b - E)\delta(\kappa_s + \kappa_b - \kappa) \\
&= \int d\kappa_s d\Omega_s \Gamma_b(\kappa - \kappa_s, E - E_s)\delta(K_s - \kappa_s) \\
&= \int d\kappa' dE' d\kappa_s d\Omega_s \Gamma_b(\kappa - \kappa', E - E')\delta(\kappa_s - \kappa')\delta(E_s - E') \\
&= \int d\kappa' dE' \Gamma_b(\kappa - \kappa', E - E')\Gamma_s(\kappa', E'),
\end{aligned} \tag{2.3}$$

where the subscript b corresponds to the bath, and s to the subsystem.[1]

We then suppose that the integrand has a maximum over both variables κ', E'. In case there is more than one local maximum in either variable, the usual Maxwell construction will yield, in addition to several equilibrium energy states, possibly several equilibrium mass states (as, for example, in the muon-electron system). The conditions for a maximum in two variables contain the necessary condition that there is a vanishing derivative in each of the variables. This condition is sufficient to construct the associated thermodynamics. The remaining requirement for a maximum on the two variables is consistently automatically satisfied in the neighborhood of this maximum by the additivity property of the entropy, as we demonstrate in the following.

In addition to the necessary condition of the vanishing of the derivative of the function in each of the two variables (in our case, E' and κ'), the eigenvalues of the matrix $f_{ij} = \frac{\partial^2 f}{\partial x_i \partial x_j}$, where

$$f(x_1, x_2) = g(x_1, x_2)h(x_1, x_2), \tag{2.4}$$

and

$$\begin{aligned}
g(E', \kappa') &= \Gamma_s(E', \kappa') \\
h(E', \kappa') &= \Gamma_b(E - E', \kappa - \kappa'),
\end{aligned} \tag{2.5}$$

must be non-positive. The quadratic equation for these eigenvalues yields non-positive roots only if for the first term of the solution of the quadratic,

$$f_{11} + f_{22} \leq 0, \tag{2.6}$$

[1] We remark that the integration $\int d\kappa \Gamma(\kappa, E)$ is convergent and maximizing the remaining integrand over E' would provide just a single temperature T.

IOP Conf. Series: Journal of Physics: Conf. Series **845** (2017) 012026 doi:10.1088/1742-6596/845/1/012026

from which one obtains

$$(\ln g)_{11} + (\ln h)_{11} + (\ln g)_{22} + (\ln h)_{22} \leq 0. \tag{2.7}$$

For the second, square root term, with a little manipulation, one finds for the condition that it cannot, with the positive sign, outweigh the first term in magnitude,

$$\begin{aligned}\left((\ln g)_{12} + (\ln h)_{12}\right)^2 \leq \big((\ln g)_{11} &+ (\ln h)_{11}\big) \\ \times \big((\ln g)_{22} &+ (\ln h)_{22}\big)\end{aligned} \tag{2.8}$$

Now, as we shall discuss in Section 3 (see (3.3)), the (constant in the neighborhood of the maximum) total entropy is, by additivity,

$$S \cong S_b + S_s = \ln h + \ln g, \tag{2.9}$$

so that

$$(\ln h)_{ij} = -(\ln g)_{ij}, \tag{2.10}$$

and therefore (2.7) and (2.8) are identically satisfied. Therefore, in the neighborhood of the assumed maximum, the condition for the absence of a saddle point configuration is consistently satisfied.

We emphasize again that our conditions are local, and that there may be more than one such maximum in the microcanonical distribution, as in the formulation of the usual statistical mechanics, with more than one mass state comprising different phases. The conditions for equilibrium can therefore be written

$$\frac{1}{\Gamma_b(\kappa-\kappa', E-E')}\frac{\partial\Gamma_b}{\partial E}(\kappa-\kappa', E-E')|_{max} = \frac{1}{\Gamma_s(\kappa', E')}\frac{\partial\Gamma_s}{\partial E}(\kappa', E')|_{max} \equiv \frac{1}{T} \tag{2.11}$$

and

$$\frac{1}{\Gamma_b(\kappa-\kappa', E-E')}\frac{\partial\Gamma_b}{\partial \kappa}(\kappa-\kappa', E-E')|_{max} = \frac{1}{\Gamma_s(\kappa', E')}\frac{\partial\Gamma_s}{\partial \kappa}(\kappa', E)|_{max} \equiv \frac{1}{T_\kappa}, \tag{2.12}$$

defining a new effective "mass temperature".

3. Canonical Thermodynamics

Let us now define the entropies of the bath and the subsystem,

$$\begin{aligned} S_b(\kappa, E) &= \ln\Gamma_b(\kappa, E) \\ S_s(\kappa, E) &= \ln\Gamma_s(\kappa, E), \end{aligned} \tag{3.1}$$

so that (we take the Boltzmann constant k_B, which we set equal to unity, the same for both definitions to assure additivity) at maximum, according to (2.4) and (2.5),[2]

$$\begin{aligned} \frac{\partial S_b}{\partial E} &= \frac{\partial S_s}{\partial E} = \frac{1}{T} \\ \frac{\partial S_b}{\partial \kappa} &= \frac{\partial S_s}{\partial \kappa} = \frac{1}{T_\kappa}. \end{aligned} \tag{3.2}$$

[2] Note that K is actually proportional to a negative mass in our metric, so that $-T_\kappa$ is a positive number, to be identified with a "mass temperature".

IOP Conf. Series: Journal of Physics: Conf. Series **845** (2017) 012026 doi:10.1088/1742-6596/845/1/012026

The total entropy of the system is then, independent of κ', E' in the neighborhood of the maximum,

$$S(\kappa, E) \cong \ln\Gamma_b(\kappa-\kappa', E-E') + \ln\Gamma_s(\kappa', E') \tag{3.3}$$

In this neighborhood we may define

$$\Gamma_b(\kappa-\kappa', E-E') = e^{S_b(\kappa-\kappa', E-E')}. \tag{3.4}$$

For κ' and E' small compared to κ and E,

$$\begin{aligned}\Gamma_b(\kappa-\kappa', E-E') &= e^{S_b(\kappa-\kappa', E-E')}\\ &\cong e^{S_b(\kappa,E)-\kappa'\frac{\partial S_b}{\partial\kappa}-E'\frac{\partial S_b}{\partial E}}\\ &= e^{S_b(\kappa,E)}e^{-\frac{\kappa'}{T_\kappa}}e^{-\frac{E'}{T}};\end{aligned} \tag{3.5}$$

We then have, integrating over κ' and E' (essentially restricted to the neighborhood of the maximum)

$$\Gamma(\kappa, E) = \int d\kappa' dE' e^{S_b(\kappa,E)}e^{-\kappa'\frac{\partial S_b}{\partial\kappa}}e^{-E'\frac{\partial S_b}{\partial E}}\Gamma_s(\kappa', E'), \tag{3.6}$$

Recall that

$$\Gamma_s(\kappa', E') = \int d\Omega_s\delta(K_s-\kappa')\delta(E_s-E') \tag{3.7}$$

and therefore we may write

$$\begin{aligned}\Gamma(\kappa, E) &= \int d\Omega_s d\kappa' dE'\delta(K_s-\kappa')\delta(E_s-E')\\ &\quad e^{S_b(\kappa,E)}e^{-\frac{\kappa'}{T_\kappa}}e^{-\frac{E'}{T}}\\ &= e^{S_b(\kappa,E)}\int d\Omega_s e^{-\frac{K_s}{T_\kappa}}e^{-\frac{E_s}{T}}.\end{aligned} \tag{3.8}$$

Since $S_b(\kappa, E)$ is an overall factor, it cancels out in any computation of average values. The distribution function for the N-body canonical ensemble is then (we drop the subscript s)

$$d_N(q,p) = e^{-\frac{K}{T_\kappa}}e^{-\frac{E}{T}} \tag{3.9}$$

and define the partition function as

$$Q_N(T_\kappa, T) = \int d\Omega e^{-\frac{K}{T_\kappa}}e^{-\frac{E}{T}}, \tag{3.10}$$

a very different result from [4], since K is considered here (as for E) as a dynamical variable.

Let us now define the Helmholtz free energy A by

$$Q_N(T_\kappa, T) = e^{-A(T_\kappa,T)\beta}, \tag{3.11}$$

so that (following Huang [9]), we write

$$\int d\Omega e^{-\beta_\kappa K}e^{(A-E)\beta} = 1, \tag{3.12}$$

IOP Conf. Series: Journal of Physics: Conf. Series **845** (2017) 012026 doi:10.1088/1742-6596/845/1/012026

where $\beta = 1/T$ and $\beta_\kappa = 1/T_\kappa$. Differentiating with respect to β one obtains

$$0 = \int d\Omega e^{-\beta_\kappa K}(A - E + \beta\frac{\partial A}{\partial \beta})e^{(A-E)\beta}. \tag{3.13}$$

Using (3.10) for the definition of Q_N, we see that

$$\begin{aligned} A &=< E > -\beta\frac{\partial A}{\partial \beta} \\ &=< E > +T\frac{\partial A}{\partial T}. \end{aligned} \tag{3.14}$$

Comparing this result with what we know of thermodynamics, if we define

$$S = -\frac{\partial A}{\partial T}, \tag{3.15}$$

we have

$$A =< E > -TS. \tag{3.16}$$

The formula (3.15) can be derived, as we shall see, from the grand canonical ensemble.

There is another relation, however, that we can obtain by taking the derivative of (3.12) with respect to β_κ; one obtains in this way

$$\begin{aligned} 0 &= \int d\Omega\{-Ke^{-\beta_\kappa K} + e^{-\beta_\kappa K}\beta\frac{\partial A}{\beta_\kappa}\}e^{(A-E)\beta} \\ &= - < K > +\beta\frac{\partial A}{\partial \beta_\kappa}, \end{aligned} \tag{3.17}$$

so that

$$< K >= \frac{1}{T}\frac{\partial A}{\partial \beta_\kappa} = -\frac{T_\kappa^2}{T}\frac{\partial A}{\partial T_\kappa}, \tag{3.18}$$

which approaches a limiting value of $-\frac{1}{2}Mc^2$ in the nonrelativistic limit, as required in [4].

We therefore obtain a mean value for $< K >$, the effective center-of-mass mass of the subensemble, which is determined by T_κ and T, under the canonical distribution, corresponding to an equilibrium of both heat and mass, without exchange of particles with the bath.[3]

We further remark that from (3.14),

$$\frac{\partial A}{\partial \beta_\kappa} = \frac{\partial < E >}{\partial \beta_\kappa} - \beta\frac{\partial^2 A}{\partial \beta \partial \eta_\kappa}; \tag{3.19}$$

so that, by (3.18),

$$< K >= \beta\frac{\partial < E >}{\partial \beta_\kappa} - \beta^2\frac{\partial^2 A}{\partial \beta \partial \eta_\kappa} \tag{3.20}$$

This result emphasizes the relation between the dynamical variables E and K.

[3] Note that the Particle Physics Booklet [10], in the discussion of big bang cosmology, page 236, summarizes deconfinement temperatures for several quarks and particles.

IOP Conf. Series: Journal of Physics: Conf. Series **845** (2017) 012026 doi:10.1088/1742-6596/845/1/012026

4. Fluctuations in Mass and Energy

To compute the fluctuations in energy, we again follow the efficient method used in Huang [9], and take the derivative with respect to β of (from (3.12))

$$0 = \int d\Omega e^{-\beta_\kappa K}(E- < E >)e^{(A-E)\beta}. \tag{4.1}$$

Using, from (3.14), the result that $A + \frac{\partial A}{\partial \beta}\beta =< E >$, one obtains

$$< (E- < E >)^2 >= -\frac{\partial < E >}{\partial \beta} = T^2 \frac{\partial < E >}{\partial T} \tag{4.2}$$

As for the usual statistical mechanics, we can define at constant space volume $V^{(3)}$, implicitly assumed here,

$$C_V = (\frac{\partial < E >}{\partial T})_V \tag{4.3}$$

In the same way, the mass fluctuations can be computed by considering

$$0 = \int d\Omega e^{-\beta_\kappa K}(K- < K >)e^{(A-E)\beta}. \tag{4.4}$$

Differentiating with respect to β_κ, and using (3.18), one obtains a result analogous to (4.2),

$$< (K- < K >)^2 >= -\frac{\partial < K >}{\partial \beta_\kappa}, \tag{4.5}$$

implying that $\frac{\partial <K>}{\partial \beta_\kappa} < 0$, or, since the right side is $+{T_\kappa}^2 \frac{\partial <K>}{\partial T_\kappa}$, that

$$\frac{\partial < K >}{\partial T_\kappa} > 0, \tag{4.6}$$

*i.e.*the mean mass (negative K)rises with the "mass temperature"(negative T_κ). We note that, furthermore, by (3.18),

$$\frac{\partial < K >}{\partial \beta_\kappa} = \beta \frac{\partial^2 A}{\partial {\beta_\kappa}^2}. \tag{4.7}$$

These relations provide an interesting starting point for the development of the thermodynamics of the mass of our model particle, which will be further developed elsewhere.

In addition to the mean square mass and energy fluctuations, it is interesting to compute the cross correlation. This can be achieved by differentiating (4.4) with respect to β (or (4.1) with respect to β_κ), and yields in the same way the relation

$$\begin{aligned} < (E- < E >)(K- < K >) > &=< EK > - < E >< K > \\ &= -\frac{\partial < K >}{\partial \beta} = -\frac{\partial < E >}{\partial \beta_\kappa}, \end{aligned} \tag{4.8}$$

so that we have the symmetrical result

$$\frac{\partial < K >}{\partial \beta} = \frac{\partial < E >}{\partial \beta_\kappa}. \tag{4.9}$$

This last result also follows from the continuity of the free energy in β and β_K by differentiating (3.14) with respect to β_K and (3.18) with respect to β.

We remark that the bracket type formula

$$\begin{aligned}\{<E>,<K>\} &= \frac{\partial<E>}{\partial\beta}\frac{\partial<K>}{\partial\beta_K} - \frac{\partial<K>}{\partial\beta}\frac{\partial<E>}{\partial\beta_K} \\ &=<(E-<E>)^2><(K-<K>)^2> \\ &-<(E-<E>)(K-<K>)>^2\end{aligned} \tag{4.10}$$

vanishes in the absence of correlation[4] and thus carries a measure of the correlation of fluctuations in E and K in the relativistic canonical ensemble.

5. Grand Canonical Ensemble

We recognize that the results above were obtained for a given number N of events in the ensemble, a picture that is not very realistic in view of our interpretation that the source of the ensemble might lie in the vacuum fluctuations of a quantum field theory. In this section, we allow the number of events in the ensemble to vary (exchange with the bath) as well as its volume, and construct the grand canonical ensemble. We write the decomposition of the full microcanonical in terms of its canonical subsets

$$\begin{aligned}Q_N(V,T,T_K) &= \Sigma_{N_s=0}^{N}\int d\Omega_s e^{-\beta_K K_s}e^{-\beta E_s} \\ &\times Q_{N-N_s}(V-V_s,T,T_K),\end{aligned} \tag{5.1}$$

where s refers to the N_s body subsystem (T and T_K are equilibrium parameters for the whole system), and, for the bath at each N_s,

$$Q_{N-N_s}(V-V_s,T,T_K) = \int d\Omega_b e^{-\beta_K K_b}e^{-\beta E_b}, \tag{5.2}$$

where $K_b = K - K_s$ and $E_b = E - E_s$. The normalized phase space density is then

$$\begin{aligned}\rho(\Omega_s,N_s) &= \frac{1}{Q_N(V,T,T_K)}e^{-\beta_K K_s}e^{-\beta E_s} \\ &\times Q_{N-N_s}(V-V_s,T,T_K),\end{aligned} \tag{5.3}$$

We can now write (for N_s small compared to N, which can be arbitrarily large),

$$\begin{aligned}Q_{N-N_s}(V-V_s,T,T_K) &= e^{-\beta A(V-V_s,T,T_K,K-K_s,N-N_s)} \\ \cong Q_N(V,T,T_K)e^{\beta V_s\frac{\partial A}{\partial V}+\frac{\partial A}{\partial K}K_s+\frac{\partial A}{\partial N}N_s}\end{aligned} \tag{5.4}$$

We then use the usual identifications with the pressure and chemical potential for the number

$$\begin{aligned}\frac{\partial A}{\partial V} &= -P \\ \frac{\partial A}{\partial N} &= \mu\end{aligned} \tag{5.5}$$

[4] It clearly vanishes in the nonrelativistic limit, but also in the relativistic on-shell theory, *i.e.*, for which $E=\sqrt{\mathbf{p}^2+M^2}$.

and define the new mass chemical potential

$$\frac{\partial A}{\partial K} = -\mu_K \tag{5.6}$$

The microcanonical partition function can now be written as (with N arbitrarily large)

$$Q_N(V,T,T_K) = \Sigma_{N_s=0}^{N} \int \Omega_s e^{-\beta_K K_s} e^{-\beta E_s} \times e^{-\beta V_s E - \mu_K \beta K_s + \beta \mu N_s}. \tag{5.7}$$

We note that K_s appears multiplied by a linear combination of $\mu_K\beta$ and β_K, arising from an interplay between the equilibrium requirements of the canonical ensemble (allowing variable mass) and the structure of the grand canonical ensemble (allowing varying particle number). Let us define

$$\hat{\mu}_K = \mu_K + \frac{\beta_K}{\beta} = \mu_K + \frac{T}{T_K}. \tag{5.8}$$

We shall, furthermore, assume that V_s has the common value V for every subensemble. The grand partition function can then be defined as

$$\mathcal{Q}(V,T,T_K) = e^{\beta VP} = \Sigma_{N_s 0}^{N} z^{N_s} Q_{N_s}(T,K_s,E_s), \tag{5.9}$$

where

$$Q_{N_s}(T,K_s,E_s) = \int d\Omega_s \zeta^{K_s} e^{-\beta E_s} \tag{5.10}$$

and

$$z = e^{\mu\beta}, \quad \zeta = e^{-\hat{\mu}_K \beta}. \tag{5.11}$$

It then follows that

$$< N >= z\frac{\partial}{\partial z} \ln \mathcal{Q} \tag{5.12}$$

and, for the mean grand canonical mass,

$$< K >= \zeta\frac{\partial}{\partial \zeta} \ln \mathcal{Q}. \tag{5.13}$$

6. Grand Canonical Thermodynamics

In this section, we make a connection with entropy and thermodynamics, and show consistency with our earlier definition of entropy in the canonical ensemble,

We define the Helmholtz free energy for the grand canonical ensemble, modified from the usual definition by taking into account the additional mass degree of freedom,

$$A = \frac{< N >}{\beta} \ln z + \frac{< K >}{\beta} \ln \zeta - \frac{1}{\beta} \ln \mathcal{Q} \tag{6.1}$$

so that

$$\mathcal{Q} = e^{-\beta A} z^{<N>} \zeta^{<K>} \tag{6.2}$$

IOP Conf. Series: Journal of Physics: Conf. Series **845** (2017) 012026 doi:10.1088/1742-6596/845/1/012026

¿From (5.9), (5.10), we then see, replacing N_s by N and taking the sum to infinity, that

$$1 = \Sigma_{N=0}^{\infty} z^{N-<N>} \int d\Omega \zeta^{K-<K>} e^{-\beta(E-A)}. \tag{6.3}$$

Differentiating with respect to β (holding z, ζ, V fixed), one obtains

$$\frac{\partial}{\partial\beta}(\beta A) =< E > + \frac{\partial < N >}{\partial\beta} \ln z + \frac{\partial < K >}{\partial\beta} \ln \zeta. \tag{6.4}$$

Now, by (6.1),

$$\beta A =< N > \ln z + < K > \ln \zeta - \ln Q, \tag{6.5}$$

so that

$$\frac{\partial}{\partial\beta}(\beta A) = \frac{\partial < N >}{\partial\beta} \ln z + \frac{\partial < K >}{\partial\beta} \ln \zeta - \frac{\partial}{\partial\beta} \ln Q. \tag{6.6}$$

With (6.4), we find the result for the internal energy

$$U \equiv< E >= -\frac{\partial}{\partial\beta} \ln Q. \tag{6.7}$$

Now, add and subtract (6.5) (divided by β) to make the dependence on A explicit, we have, with the definitions of z and ζ in terms of the chemical potentials,

$$U = A- < N > \mu + \hat{\mu}_\kappa < K > + \frac{1}{\beta} \ln Q - \frac{\partial}{\beta} \ln Q. \tag{6.8}$$

Using

$$-\frac{\partial}{\beta} \ln Q = +T^2 \frac{\partial}{\partial T} \ln Q \tag{6.9}$$

and using the thermodynamic relation

$$U = A + TS, \tag{6.10}$$

one finds

$$S = \frac{\partial}{\partial T}(T \ln Q) + \frac{\hat{\mu}_\kappa}{T} < K > - \frac{\mu}{T} < N > . \tag{6.11}$$

With the definition (5.8), we can write this as

$$S = \frac{\partial}{\partial T}(T \ln Q + (\frac{\mu_\kappa}{T} + \frac{1}{T_\kappa}) < K > - \frac{\mu}{T} < N > . \tag{6.12}$$

We can now obtain the Maxwell relations by taking the differential of (6.1) and using

$$\frac{\partial}{\partial V} \ln Q = \frac{P}{T} \tag{6.13}$$

to obtain

$$\begin{aligned} dA = T \ln z d < N > + T \ln \zeta d < E > + \{\frac{< N >}{T}\mu \\ - \frac{< K >}{T}\hat{\mu}_\kappa - \frac{\partial}{\partial T}(T \ln Q)\} dT - P dV. \end{aligned} \tag{6.14}$$

The terms in brackets are the negative of our definition of entropy (6.11), so that we obtain the Maxwell relations

$$\begin{aligned} S &= -(\frac{\partial A}{\partial T})|_{V,<N>,<K>} \\ P &= -(\frac{\partial A}{\partial V})|_{<N>,<K>,T} \end{aligned} \tag{6.15}$$

as well as

$$\begin{aligned} \frac{\partial A}{\partial <N>} &= \mu \\ \frac{\partial A}{\partial <K>} &= -\hat{\mu}_\kappa = -(\mu_\kappa + \frac{T}{T_\kappa}) \end{aligned} \tag{6.16}$$

When the free energy reaches a critical point in $< K >$, *i.e.*,

$$\frac{\partial A}{\partial <K>} = 0, \tag{6.17}$$

then (since T_κ is negative, this corresponds to a positive μ_κ).

$$\frac{T}{T_\kappa} = -\mu_\kappa. \tag{6.18}$$

7. Conclusions

In the classical SHP theory, the fundamental dynamical object is an *event*, moving in spacetime according to laws of motion generated, in the case of our study here, by a Hamiltonian according to Hamilton's equations. The usual notion of a particle is associated with the world line of this event [3][5][6], but for simplicity we have called this fundamental object a "particle" here. The mass (squared)of the particle, the negative of $p_\mu p^\mu$ (in our metric), after a collision, may asymptotically take on a wide range of values, limited by kinematics and the conserved quantity K, the (invariant) Hamiltonian. One sees that in experiment, the mass of a particle, such as the electron, is generally to be found very close to a definite value even after many collisions have taken place, and the question naturally arises of how this stability can be understood in the framework of the theory.

We have constructed, in this paper, a model for which the mass of the particle is controlled by a chemical potential, so that asymptotic variations in the mass can be restored to a given value by relaxation to satisfy the equilibrium conditions. In this model, we have taken the "particle" to be a statistical ensemble which has both an equilibrium energy and an equilibrium mass, controlled by the temperature and chemical potentials, thus assuring asymptotic states with the correct mass. The thermodynamic properties of such a system, involving the maximization of the integrand in the microcanonical ensemble, was worked out for such a system, where both the energy and the mass are considered to be parameters of the distribution. As we have seen in Section 6, a critical point in the free energy is made available by the interplay of the equilibrium requirements of the canonical ensemble, where the total mass of the system is considered variable (off-shell), as for the energy, and the equilibrium requirements of the grand canonical ensemble, where a chemical potential arises for the particle number.

In case there is more than one such maximum under the integrand of the microcanonical distribution, there may be more than one equilibrium mass state, such as in the electron muon system, with transitions between these states corresponding to phase transitions.

The study of how such a system behaves under the dynamical perturbations of collisions may be studied, in this model, through the (covariant) Boltzmann equation [11]. Such a study is left for future research.

Acknowledgments

I would like to thank Mark Davidson and M. Wagman for their critical reading of the manuscript, and Y. Strauss, J. Levitan, A. Yahalom, G. Elgressy, Y. Bachar and A. Kremer for discussions.

References

[1] E.C.G. Stueckeleberg, Helv. Phys. Acta **14**, 588 (1961); **15** , 23 (1942).
[2] L.P. Horwitz and C. Piron, Helv. Phys. Acta **46**, 316 (1973).
[3] Lawrence P. Horwitz,*Relativistic Quantum Mechanics*, Fundamental Theories in Physics **180**, (Springer, Dordrecht, 2015).
[4] L.P. Horwitz, W.C. Schieve and C.Piron, Ann. of Phys. **137**, 306 (1981).
[5] L.P. Horwitz and Y. Lavie, Phys. Rev. **D 26**819 (1982).
[6] J.D. Jackson, *Classical Electrodynamics*, 2nd edn,(Wiley, New York 1974).
[7] R.I. Arshansky and L.P. Horwitz, Jour. Math. Phys. **30**, 66, 213,380 (1989).
[8] I thank Don Reed and Mark Davidson for emphasizing this question, and to C. Piron, for first qualitatively referring to such a mechanism.
[9] K. Huang, *Statistical Mechanics*, Wiley, New York (1967).
[10] Particle Physics Booklet, July (2014), extracted from *Review of Particle Physics*, K.A. Olive *et al*, Chin. Physics C **38**, 090001 (2014).
[11] . L.P. Horwitz, S. Shashoua and W.C. Schieve, Physica **A 161**, 300 (1989).

Parametrized relativistic dynamical framework for neutrino oscillations

John R. Fanchi
Department of Engineering and TCU Energy Institute
Texas Christian University
Fort Worth, Texas, 76129 USA

E-mail:`j.r.fanchi@tcu.edu`

Abstract. Mass state transitions are a key feature of parametrized relativistic dynamics (PRD). PRD is a manifestly covariant quantum theory with invariant evolution parameter. The theory has been applied to neutrino flavor oscillations between two mass states. It is generalized to transitions between three mass states and applied to the survival of electron neutrinos. The analysis shows that significant differences exist between theoretical results of the conventional model and the PRD model.

1. Introduction

Experiments with solar neutrinos, atmospheric neutrinos, reactor neutrinos, and accelerator neutrinos have demonstrated that flavor mixing can occur between two or three neutrino flavors composed of up to three neutrino mass states [1,2]. Mass state transitions are a key feature of parametrized relativistic dynamics (PRD). Introductions to PRD are presented by Fanchi [3, 4], Pavsic [5], and Horwitz [6]. A review of the highlights of the history of neutrinos is presented to establish a framework for analysing neutrino oscillations in the context of PRD. We then extend the PRD formalism for two-state flavor mixing [7] to three-state flavor mixing. The formalism is illustrated by applying it to the neutrino transition $\nu_e \rightarrow \nu_\mu$ from an electron neutrino ν_e to a muon neutrino ν_μ in vacuum. In this application we assume the transition $\nu_e \rightarrow \nu_\tau$ from an electron neutrino ν_e to a tau neutrino ν_τ is negligible and that only two flavor states are involved. Calculation of electron neutrino survival probabilities using both the conventional analysis and PRD shows that there are significant differences between the two theories.

1.1 The Neutrino Hypothesis

We begin our review of some highlights of the history of neutrinos with the observation that particles can be emitted by radioactive elements. Wilhelm Roentgen [8] observed a new kind of radiation called x-rays. Roentgen's x-rays could pass through a human body and create a shadowy image of bone. Henri Becquerel learned about x-rays and began to look for a connection between x-rays and the phosphorescence of a uranium salt. Becquerel [9] published a series of articles about his discovery ofa

IOP Conf. Series: Journal of Physics: Conf. Series **845** (2017) 012027 doi:10.1088/1742-6596/845/1/012027

radiation that appeared without the stimulation of sunlight and could penetrate matter. The radiation discovered by Becquerel was not the same as Roentgen's x-rays and was later named radioactivity by Pierre and Marie Curie.

J.J. Thomson [10] recognized that "cathode rays" are negatively charged particles. Thomson used a magnet to alter the trajectory of the cathode rays and measure the mass-to-charge ratio. Cathode rays are now known as electrons. Robert Millikan [11, 12] used the oil drop experiment to measure the charge of the cathode ray. The mass of the electron was determined by combining Millikan's charge measurement with Thomson's mass-to-charge ratio.

Ernest Rutherford [13] was the first to observe that uranium emitted at least two distinct types of rays that he called alpha radiation and beta radiation. Shortly thereafter Paul Villard [14] studied the decay of radium and discovered a third type of electrically neutral radiation that Rutherford [15] later named gamma radiation. Today we know that alpha radiation is the helium nucleus, beta radiation is an electron, and gamma radiation is an energetic photon.

Rutherford [16] bombarded thin gold foils with alpha particles. The scattering of alpha particles by atoms in the foil showed that most of the atomic mass must occupy a small space inside the atom. This mass was called the nucleus. The constituents of the nucleus were later identified as the proton and a new, electrically neutral particle called the neutron. At this point in time, the atom was thought to contain a massive, positively charged nucleus surrounded by a distribution of negatively charged particles.

Experimental studies of radioactive elements that emitted alpha particles or gamma particles showed that the decays satisfied conservation of energy. By contrast, studies of radioactive elements that emitted beta particles did not appear to conserve energy. During beta decay, the positive charge of the nucleus increases by one positive charge in association with the emission of a negatively charged beta particle. James Chadwick [17] measured kinetic energy of emitted beta particles by passing the negatively charged beta particles through a magnetic field and calculating the kinetic energy from the deflection of the beta particle trajectory in the magnetic field. Chadwick showed that the kinetic energy of beta particles from beta decay experiments displayed a broad, continuous distribution of kinetic energy. The beta particles did not have enough kinetic energy to account for all of the energy needed to conserve energy in the beta decay process. Was energy conservation being violated?

Neils Bohr was a proponent of the idea that energy conservation was being violated [18, page 63 ff).Wolfgang Pauli was not prepared to abandon energy conservation. In a 1930 letter to a conference organizer, Pauli suggested "the possibility that inside the nuclei there are particles electrically neutral, that I will call neutrons, which have spin ½ and follow the exclusion principle" [19, pages 16-17]. Pauli's neutron was later named neutrino (Italian for "little neutral one") by Enrico Fermi [20] to distinguish it from the more massive neutron discovered by Chadwick [21].

The concept of a neutron was introduced by Rutherford in 1920. He suggested that "the nuclei of atoms contain electrons as well as positively charged bodies, and that the positive charge of the nucleus represents the excess positive charge." [22, page 377]. The basic building blocks were the positively charged nucleus of the hydrogen atom (the proton) and the negatively charged electron. In the case of a single electron and a single hydrogen nucleus, "it may be possible for an electron to combine much more closely with the H nucleus to form a kind of neutral doublet" [22, page 396]. The neutron was discovered in 1932 [21]. Was the neutron the bound state of an electron and a proton, or was it a new particle?

IOP Conf. Series: Journal of Physics: Conf. Series **845** (2017) 012027 doi:10.1088/1742-6596/845/1/012027

The beta decay process was evidence that the neutron was not a stable particle. Fermi [20] adopted Pauli's "neutron" concept, which he renamed "neutrino," and hypothesized a 4-point beta decay in which the neutron decayed into a proton, an electron, and a neutrino. The study of cosmic rays led Sakata and Inoue [23] to suggest that there may be two neutrinos: the electron neutrino $\nu_e$associated with electrons, and the muon neutrino $\nu_\mu$associated with muons discovered in cosmic rays.

1.2 Detecting the Neutrino

It was reasonable to infer from Fermi's 4-point theory that the neutrino could interact with other particles in such a way that it could be detectable. One interaction is beta decay in which the neutron decays into a proton, an electron, and an electron antineutrino: $n^0 \to p^+ + e^- + \overline{\nu_e}$. Another possible interaction is inverse beta decay in which an electron antineutrino interacts with a proton to form a neutrino and a positron: $\overline{\nu_e} + p^+ \to n^0 + e^+$.The electron capture process occurs when an electron in a lower energy level of an atom is captured by a proton in the nucleus to produce a neutron and a neutrino: $e^- + p^+ \to n^0 + \nu_e$.

It was not clear at the time if the neutrino and the antineutrino behaved differently when interacting with matter. Two possible beta emission processes were positive beta emission $p^+ \to n^0 + e^+ + \nu_e$ in which a proton transformed into a neutron, a positron and a neutrino, and negative beta emission $n^0 \to p^+ + e^- + \overline{\nu_e}$ in which a neutron transformed into a proton, an electron and an antineutrino. More generally, positive beta emission for a nucleus *X* with mass number *A* and atomic number *Z* can be written as ${}^A_ZX \to {}_{Z-1}^{\ \ A}Y + e^+ + \nu_e$ where the symbol *Y* represents the resulting nucleus. Positive beta emission may be viewed as the transformation of a proton inside the nucleus to a neutron: $p^+ \to n^0 + e^+ + \nu_e$. Similarly, negative beta emission can be written as ${}^A_ZX \to {}_{Z+1}^{\ \ A}Y + e^- + \overline{\nu_e}$ which may be viewed as the transformation of a neutron inside the nucleus to a proton: $n^0 \to p^+ + e^- + \overline{\nu_e}$.

Bruno Pontecorvo [24] proposed detecting neutrinos by allowing the neutrino to interact with a proton or neutron of a nucleus heavier than hydrogen so that Z > 1. The interaction would change a proton into a neutron and yield an isotope of a new element. Neutrino detection would be achieved by starting with a large amount of a pure isotope and look for the formation of a neutrino-generated element as a function of time. One reaction proposed by Pontecorvo was the reaction in which a neutrino interacts with a proton in a chlorine nucleus to produce an electron and an argon nucleus: $\nu_e + {}^{37}_{17}Cl \to e^- + {}^{37}_{18}Ar$. Ray Davis performed the experiment using radiation emitted by different reactors and detectors containing carbon tetrachloride.

Nuclear reactors emit antineutrinos from negative beta decays of fission products. Pontecorvo's reaction $\nu_e + {}^{37}_{17}Cl \to e^- + {}^{37}_{18}Ar$ assumed a neutrino would interact with a proton. Davis was aware that his experiment was testing the relative performance of neutrinos and antineutrinos: "If neutrinos and antineutrinos areidentical in their interactions with nucleons one shouldbe able to observe the process upon carrying the experimentto the required sensitivity. However, if neutrinosand antineutrinos differ in their interactions withnucleons one would not expect to induce the reaction${}^{37}_{17}Cl(\overline{\nu_e}, e^-){}^{37}_{18}Ar$." [25, pg. 766] Assuming the experiment had adequate sensitivity, Davis argued that a null result using chlorine in the detector implied that the interaction between neutrinos and nucleons is not the same as the interaction between antineutrinos and nucleons.His early results were not conclusive [25,26].

IOP Conf. Series: Journal of Physics: Conf. Series **845** (2017) 012027 doi:10.1088/1742-6596/845/1/012027

The first tentative observation of neutrinos was made by Reines and Cowan in 1953 at the Hanford nuclear reactor [27] on the Columbia River in the state of Washington. Electron neutrinos from nuclear fission reactors have since been identified as electron antineutrinos. Reines and Cowan observed the inverse beta decay process in which an electron antineutrino interacted with a proton to from a neutrino and a positron $\overline{\nu_e} + p^+ \rightarrow n^0 + e^+$.The positron produced by the reaction interacted with an electron in a pair-annihilation process to yield two gamma rays in the reaction $e^+ + e^- \rightarrow 2\gamma$.

The Hanford work identified gamma rays from pair annihilation, but this was not considered sufficiently conclusive because the gamma ray signal could not be distinguished from cosmic ray signals. The experiment was moved to the Savannah River Nuclear Reactor facility near Augusta, Georgia where it could be shielded from cosmic rays. The experiment was also modified to include a neutron absorber (cadmium chloride) in the scintillation tank. The neutron produced in the inverse beta decay process was absorbed by a cadmium nucleus in the neutron capture process $n^0 + {}^{108}_{48}Cd \rightarrow {}^{109}_{48}Cd + \gamma$. Photomultiplier tubes in the scintillation tank should be able to detect gamma rays from the pair-annihilation process approximately 5 microseconds before gamma ray signals from the neutron capture process.The neutrino signature was the combination of the detection of gamma rays from pair-annihilation followed by the detection of gamma rays from the neutron capture process. Cowan, Reines and collaborators[28] reported the direct detection of the electron neutrino in 1956.

1.3 Neutrino Oscillations

Ray Davis continued to improve his experimental techniques. Together with his colleagues, Davis increased the volume of ${}^{37}_{17}Cl$ in his detector and moved the detector underground to the Homestake gold mine in South Dakota. His goal was to minimize extraneous signals from cosmic rays as he sought to measure neutrino production by fusion reactions in the interior of the sun. Davis [29] found that solar neutrino flux was much lower than predicted by John Bahcall [30]. Bahcall and colleagues [31, 32] refined their theoretical predictions, but additional measurements by Davis and colleagues [32] showed that observed solar neutrino flux wasstill lower than predicted. Where were the missing solar neutrinos?

Work with kaons had shown that kaons could oscillate between neutral kaon $K^0$and neutral anti-kaon $\overline{K^0}$flavor states [33]. The observable kaons were the long-lived kaon state (symmetric superposition $K_L^0 = \frac{1}{\sqrt{2}}(K^0 + \overline{K^0})$) and short-lived kaon state (anti-symmetric superposition $K_S^0 = \frac{1}{\sqrt{2}}(K^0 - \overline{K^0})$). By analogy, Pontecorvo suggested that the electron neutrino ν_e, and the muon neutrino ν_μ could oscillate between states [34,35]. Danby, et al. [36] showedthat there is experimental evidence that there are at least two different types of neutrinos.Maki, Nakagawa and Sakata [37] presented a theory of 2-neutrino mixing between the muon neutrino and electron neutrino.

A third type of neutrino was proposed by Perl, et al. [38]. They were the first to report evidence of the tau lepton and suggested that a third neutrino, the tau neutrino, may be associated with the tau lepton. Thetau neutrino was observed in 2000 by the DONUT(Direct Observation of Nu Tau) collaboration at Fermilab in the USA and Super-Kamiokande in Japan [39]. Super-Kamiokandebegan operation in 1996. The collaboration detected oscillations in atmospheric neutrinos by 1998 [40], which implies that neutrinos have mass.

Non-technical reviews of the history of neutrinos and additional references are provided by Solomey [19], Franklin [18], Close [41], and Jayawardhana [42].

2. Mass Basis and Flavor Basis

We are interested in developing a formalism within the context of PRD that can describe transitions between three neutrino flavor states $\{|\nu_\alpha\rangle; \alpha = e, \mu, \tau\}$ given the assumption that neutrinos are composed of up to three mass states $\{|\nu_j\rangle; j = 1, 2, 3\}$. The mass and flavor states can be written as 3-component column vectors:

$$|\nu_j\rangle = \begin{bmatrix} |\nu_1\rangle \\ |\nu_2\rangle \\ |\nu_3\rangle \end{bmatrix} \tag{2.1}$$

and

$$|\nu_\alpha\rangle = \begin{bmatrix} |\nu_e\rangle \\ |\nu_\mu\rangle \\ |\nu_\tau\rangle \end{bmatrix} \tag{2.2}$$

The mass basis $\{|\nu_j\rangle; j = 1, 2, 3\}$ is related to the flavor basis $\{|\nu_\alpha\rangle; \alpha = e, \mu, \tau\}$ by a unitary transformation:

$$\begin{bmatrix} |\nu_e\rangle \\ |\nu_\mu\rangle \\ |\nu_\tau\rangle \end{bmatrix} = U \begin{bmatrix} |\nu_1\rangle \\ |\nu_2\rangle \\ |\nu_3\rangle \end{bmatrix} \tag{2.3}$$

where U is the unitary matrix

$$U = \begin{bmatrix} u_{e1} & u_{e2} & u_{e3} \\ u_{\mu 1} & u_{\mu 2} & u_{\mu 3} \\ u_{\tau 1} & u_{\tau 2} & u_{\tau 3} \end{bmatrix} \tag{2.4}$$

satisfying

$$U^{-1} = \left(U^*\right)^T \tag{2.5}$$

The elements of the unitary matrix are

$$u_{\alpha j}^{-1} = u_{j\alpha}^{*} \; ; \; j = 1,2,3 \; and \; \alpha = e, \mu, \tau \tag{2.6}$$

The expanded form of the unitary transformation is

$$\begin{aligned} |\nu_e\rangle &= u_{e1}|\nu_1\rangle + u_{e2}|\nu_2\rangle + u_{e3}|\nu_3\rangle \\ |\nu_\mu\rangle &= u_{\mu 1}|\nu_1\rangle + u_{\mu 2}|\nu_2\rangle + u_{\mu 3}|\nu_3\rangle \\ |\nu_\tau\rangle &= u_{\tau 1}|\nu_1\rangle + u_{\tau 2}|\nu_2\rangle + u_{\tau 3}|\nu_3\rangle \end{aligned} \tag{2.7}$$

A mass basis state satisfies the temporal evolution equation

$$\begin{aligned} T\left|\nu_j\right\rangle &= i\hbar\frac{\partial}{\partial s}\left|\nu_j\right\rangle \\ &= T_j\left|\nu_j\right\rangle \\ &= \hbar^2\frac{k_j^\mu k_{j\mu}}{2m_j}\left|\nu_j\right\rangle \end{aligned} \quad (2.8)$$

where T_j is the eigenvalue of the temporal operator T, s is the scalar evolution parameter, k_j^μ is the energy-momentum of state j and m_j is the mass of state j. Equation (2.8) has the formal solution

$$\begin{bmatrix} \left|\nu_1\right\rangle \\ \left|\nu_2\right\rangle \\ \left|\nu_3\right\rangle \end{bmatrix} = \begin{bmatrix} e^{-i\frac{T_1 s}{\hbar}} & 0 & 0 \\ 0 & e^{-i\frac{T_2 s}{\hbar}} & 0 \\ 0 & 0 & e^{-i\frac{T_3 s}{\hbar}} \end{bmatrix} \begin{bmatrix} \left|\nu_1(0)\right\rangle \\ \left|\nu_2(0)\right\rangle \\ \left|\nu_3(0)\right\rangle \end{bmatrix} \quad (2.9)$$

where $\nu_j(0)$ is mass state j at $s = 0$.

3. Transitions between Flavor States: Electron Neutrino Disappearance

The formalism for three mass states presented in the previous section is illustrated by applying the formalism to the disappearance of electron neutrinos. We begin with a pure beam of electron neutrinos in flavor state $\left|\nu_e\right\rangle$. The probability of forming ν_μ and ν_τ is

$$P(\nu_e \to \nu_\mu) = \left|\left\langle \nu_\mu \middle| \nu_e(s)\right\rangle\right|^2 \quad (3.1)$$

and

$$P(\nu_e \to \nu_\tau) = \left|\left\langle \nu_\tau \middle| \nu_e(s)\right\rangle\right|^2 \quad (3.2)$$

respectively. The matrix element for $\beta = \mu \ or \ \tau$ is

$$\left\langle \nu_\beta \middle| \nu_e(s)\right\rangle = \left[u_{\beta1}^*\nu_1^* + u_{\beta2}^*\nu_2^* + u_{\beta3}^*\nu_3^*\right] \times \left[u_{e1}\nu_1(s) + u_{e2}\nu_2(s) + u_{e3}\nu_3(s)\right] \quad (3.3)$$

or, in expanded form,

$$\begin{aligned} \left\langle \nu_\beta \middle| \nu_e(s)\right\rangle &= u_{\beta1}^*\left[u_{e1}\nu_1^*\nu_1(s) + u_{e2}\nu_1^*\nu_2(s) + u_{e3}\nu_1^*\nu_3(s)\right] \\ &+ u_{\beta2}^*\left[u_{e1}\nu_2^*\nu_1(s) + u_{e2}\nu_2^*\nu_2(s) + u_{e3}\nu_2^*\nu_3(s)\right] \\ &+ u_{\beta3}^*\left[u_{e1}\nu_3^*\nu_1(s) + u_{e2}\nu_3^*\nu_2(s) + u_{e3}\nu_3^*\nu_3(s)\right] \end{aligned} \quad (3.4)$$

Equation (3.4) is simplified by applying the orthonormality condition $\left\langle \nu_i \middle| \nu_j\right\rangle = \delta_{ij}$ to obtain

$$\begin{aligned} \left\langle \nu_\beta \middle| \nu_e(s)\right\rangle &= u_{\beta1}^* u_{e1}\nu_1^*\nu_1(s) \\ &+ u_{\beta2}^* u_{e2}\nu_2^*\nu_2(s) \\ &+ u_{\beta3}^* u_{e3}\nu_3^*\nu_3(s) \end{aligned} \quad (3.5)$$

The temporal dependence is obtained by expressing Eq. (3.5) in terms of the mass state at $s = 0$:

$$\begin{aligned}\langle \nu_\beta | \nu_e(s) \rangle &= u^*_{\beta 1} u_{e1} \exp\left(-i\frac{T_1 s}{\hbar}\right) \\ &+ u^*_{\beta 2} u_{e2} \exp\left(-i\frac{T_2 s}{\hbar}\right) \\ &+ u^*_{\beta 3} u_{e3} \exp\left(-i\frac{T_3 s}{\hbar}\right)\end{aligned} \tag{3.6}$$

4. Application to the $\nu_e \to \nu_\mu$ Transition

We apply the 3 flavor state-formalism to the transition $\nu_e \to \nu_\mu$ in vacuum between two flavor states. In this application we assume the transition $\nu_e \to \nu_\tau$ is negligible and that only 2 flavor states are involved. A 3-state unitary matrix that effectively simplifies the problem so that we only need to consider flavor states 1 and 2 is

$$\begin{aligned}U(3) &= \begin{bmatrix} U_{12}(2) & 0 \\ 0 & 1 \end{bmatrix} \\ &= \begin{bmatrix} \cos\theta_{12} & \sin\theta_{12} & 0 \\ -\sin\theta_{12} & \cos\theta_{12} & 0 \\ 0 & 0 & 1 \end{bmatrix}\end{aligned} \tag{4.1}$$

where θ_{12} refers to the mixing angle between mass states 1 and 2 in vacuum.

The transition probability amplitude is

$$\begin{aligned}\langle \nu_\mu | \nu_e(s) \rangle &= u^*_{\mu 1} u_{e1} \exp\left(-i\frac{T_1 s}{\hbar}\right) \\ &+ u^*_{\mu 2} u_{e2} \exp\left(-i\frac{T_2 s}{\hbar}\right)\end{aligned} \tag{4.2}$$

where $\{T_j, j = 1,2\}$ are the eigenvalues of the temporal evolution operator, and

$$\begin{aligned}u^*_{\mu 1} &= -\sin\theta_{12} \\ u_{e1} &= \cos\theta_{12} \\ u^*_{\mu 2} &= \cos\theta_{12} \\ u_{e2} &= \sin\theta_{12}\end{aligned} \tag{4.3}$$

In the conventional theory, T_j is the energy E_j of state j, while in PRD it is the eigenvalue K_j of the mass operator for mass state j. Substituting Eq. (4.3) into Eq. (4.2) gives

IOP Conf. Series: Journal of Physics: Conf. Series **845** (2017) 012027 doi:10.1088/1742-6596/845/1/012027

$$\begin{aligned}\langle\nu_\mu|\nu_e(s)\rangle &= -\sin\theta_{12}\cos\theta_{12}\exp\left(-i\frac{T_1 s}{\hbar}\right)\\ &+\cos\theta_{12}\sin\theta_{12}\exp\left(-i\frac{T_2 s}{\hbar}\right)\\ &=\sin\theta_{12}\cos\theta_{12}\left[\exp\left(-i\frac{T_2 s}{\hbar}\right)-\exp\left(-i\frac{T_1 s}{\hbar}\right)\right]\end{aligned} \quad (4.4)$$

and

$$\langle\nu_\tau|\nu_e(s)\rangle = 0 \quad (4.5)$$

The transition probability is

$$P(\nu_e\to\nu_\mu) = |\langle\nu_\mu|\nu_e(s)\rangle|^2 \quad (4.6)$$

5. Application to Neutrino Oscillation Experiments

The evolution equation in PRD for a state may be written in terms of the evolution parameter s as

$$i\hbar\frac{\partial}{\partial s}|\nu_j\rangle = K_j|\nu_j\rangle \quad (5.1)$$

where K_j is the eigenvalue of the mass operator for mass state j. The evolution parameter dependent solution of Eq. (5.1) in the mass basis for two mass states is

$$\begin{bmatrix}|\nu_1(s)\rangle\\ |\nu_2(s)\rangle\end{bmatrix} = \begin{bmatrix}e^{-iK_1 s/\hbar} & 0\\ 0 & e^{-iK_2 s/\hbar}\end{bmatrix}\begin{bmatrix}|\nu_1(0)\rangle\\ |\nu_2(0)\rangle\end{bmatrix} \quad (5.2)$$

where

$$K_j = \hbar^2 k_j^\mu k_{j\mu}/2m_j = \hbar^2\left[(\omega_j/c)^2 - k_j\cdot k_j\right]/2m_j \quad (5.3)$$

In PRD, the components of the energy-momentum four-vector k_j^μ are observables and the mass m_j is a function of statistical values of k_j^μ.

In the flavor oscillation process $\nu_e\to\nu_\mu$, we begin with a pure beam of electron neutrino ν_e particles and calculate the probability for formation of muon neutrino ν_μ particles. The PRD result for the probability of forming the final state ν_μ from initial state ν_e is

$$\begin{aligned}P_{PRD}(\nu_e\to\nu_\mu) &= sin^2 2\theta\, sin^2\left\{\frac{(m_2-m_1)c^2}{4\hbar}s\right\}\\ &\equiv sin^2 2\theta\, sin^2\alpha_{PRD}\end{aligned} \quad (5.4)$$

where s is temporal duration measured by an evolution parameter clock [4].Dynamical factors are collected in the term α_{PRD}.

Flavor oscillations may be described by quantifying the behavior of two particles. One particle propagates without interaction or oscillation from the source to the detector and serves as a "clock" for the scalar evolution parameter s. The other particle is the oscillating particle. In this application, the source and detector are separated by a distance L.

IOP Conf. Series: Journal of Physics: Conf. Series **845** (2017) 012027 doi:10.1088/1742-6596/845/1/012027

The most probable trajectory of the non-interacting s-clock particle is

$$s^2 = (\delta t)^2 - \frac{(\delta x)^2}{c^2} = (\delta t)^2\left[1-\beta^2\right], \beta = \frac{v}{c}, v = \frac{\delta x}{\delta t} \tag{5.5}$$

The distance δx traveled by the s-clock particle in the interval δt is L, so we obtain

$$s = \frac{L}{c}\frac{\left[1-\beta^2\right]^{1/2}}{\beta}, \delta x = L \tag{5.6}$$

Substituting Eq. (5.6) into Eq. (5.4) gives

$$P_{PRD}(v_e \to v_\mu) = \sin^2 2\theta \sin^2 \alpha_{PRD},$$
$$\alpha_{PRD} = \frac{(m_2 - m_1)c^2}{4\hbar}\frac{L}{c}\frac{\left[1-\beta^2\right]^{1/2}}{\beta} \tag{5.7}$$

The result for the conventional theory denoted by subscript Std is

$$P_{Std}(v_e \to v_\mu) = \sin^2 2\theta \sin^2 \alpha_{Std}$$
$$\alpha_{Std} = \frac{(m_2^2 - m_1^2)c^4}{4\hbar E_v}\frac{L}{c} \tag{5.8}$$

where E_v is the energy of the ultrarelativistic incident neutrino

$$E_v = \frac{m_v c^2}{\left[1-\beta^2\right]^{1/2}} \tag{5.9}$$

We combine Eqs. (5.7) and (5.9) and rearrange to simplify comparison with Eq. (5.8):

$$P_{PRD}(v_e \to v_\mu) = \sin^2 2\theta \sin^2 \alpha_{PRD},$$
$$\alpha_{PRD} = \frac{(m_2 - m_1)c^2}{4\hbar}\frac{L}{c}\frac{m_v c^2}{E_v \beta} = \frac{m_v (m_2 - m_1)c^4}{4\hbar E_v}\frac{L}{c}\frac{1}{\beta} \tag{5.10}$$

The ratio of the dynamical factors $\alpha_{PRD}, \alpha_{Std}$ is

$$\frac{\alpha_{Std}}{\alpha_{PRD}} = \frac{m_2^2 - m_1^2}{m_v (m_2 - m_1)}\beta = \frac{m_1 + m_2}{m_v}\beta \tag{5.11}$$

and the ratio of probabilities in Eqs. (5.8) and (5.10) is

$$\frac{P_{Std}}{P_{PRD}} = \frac{\sin^2 \alpha_{Std}}{\sin^2 \alpha_{PRD}} \tag{5.12}$$

Comparing P_{PRD}, P_{Std} and the dynamical factors $\alpha_{PRD}, \alpha_{Std}$ shows that the PRD model and the conventional theory have the same dependence on the flavor mixing angle θ, but their dependence on dynamical factors differs significantly.If the mass difference between neutrino mass and flavor states is very small and the neutrinos are ultrarelativistic, then $(m_1 + m_2)/m_v \approx 2$ and $\beta \approx 1$.The ratio of dynamical factors $\alpha_{Std}/\alpha_{PRD} \approx 2$ in this case.

The survival probability of the electron neutrino is

$$P_{PRD}(v_e \to v_e) = 1 - P_{PRD}(v_e \to v_\mu) = 1 - \sin^2 2\theta \sin^2 \alpha_{PRD} \tag{5.13}$$

in the PRD model, and

$$P_{Std}(v_e \to v_e) = 1 - P_{Std}(v_e \to v_\mu) = 1 - \sin^2 2\theta \sin^2 \alpha_{Std} \qquad (5.14)$$

in the conventional Std model. These probabilities are compared in Figure 1. The angle α_{Std} is calculated using $L = 180\,km$, $\Delta m^2 = 7.0\times10^{-5} eV^2$ and $\sin^2 2\theta = 0.84$ as a function of neutrino energy that varies from 0.1 MeV to 15 MeV [1, Fig. 14.1, pg. 63]. The angle α_{PRD} is calculated using $\alpha_{Std} / \alpha_{PRD} \approx 2$ which is based on the assumption that the neutrinos are ultrarelativistic in vacuum and $(m_1 + m_2)/ m_v \approx 2$.

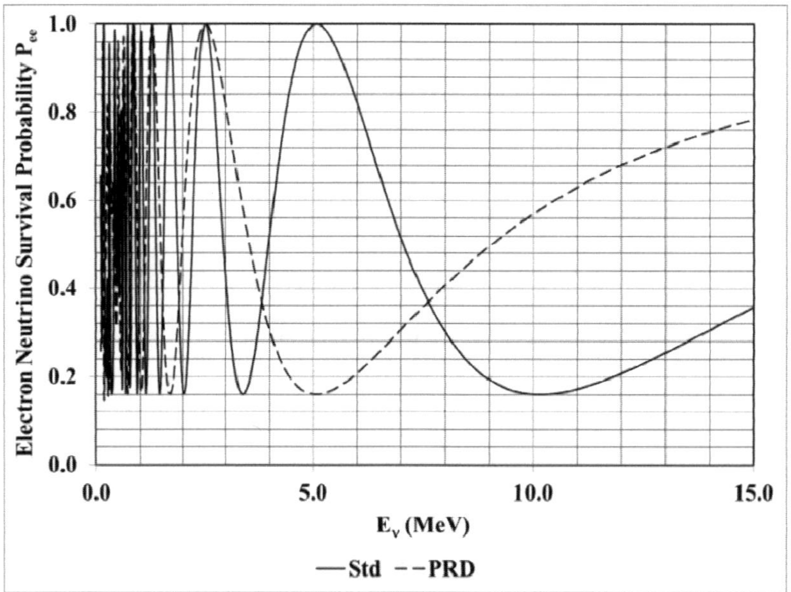

Figure 1. Comparison of Conventional (Std) and PRD Survival Probabilities as a Function of Neutrino Energy

Figure 2 shows survival probability of the electron neutrino as a function of L at a neutrino energy of 10 MeV. The angle α_{Std} is calculated using $\Delta m^2 = 7.0\times10^{-5} eV^2$ and $\sin^2 2\theta = 0.84$. It is clear from the figures that there are significant differences between conventional (Std) and PRD theoretical results. Figures 1 and 2 are especially useful displays of experimental resultsbecause they provide theory-independent information that can be used to determine the probability of disappearance $P(v_e \to v_\mu)$ or survival $P(v_e \to v_e)$ of electron neutrinos.

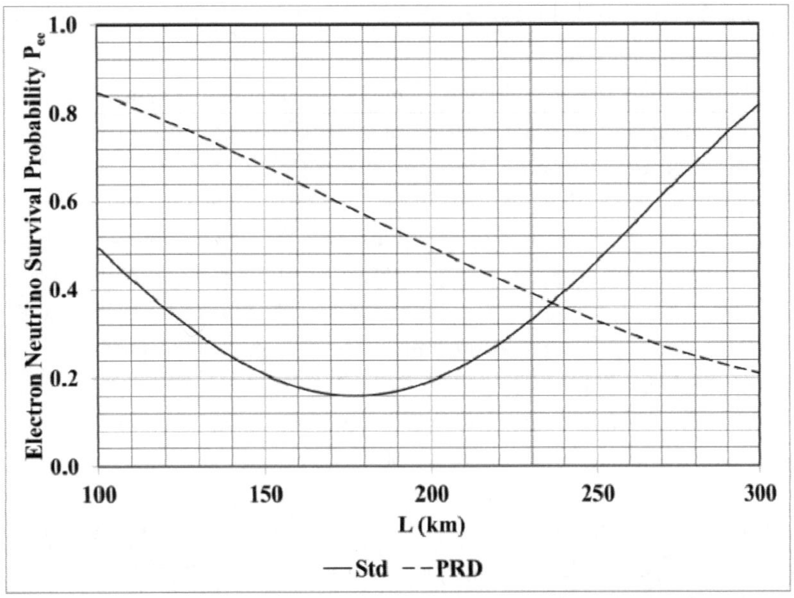

Figure 2. Comparison of Conventional (Std) and PRD Survival Probabilities as a Function of the Distance L between Source and Detector

6. Conclusions

A framework for studying mass state transitions between neutrino flavor states with up to three flavors was presented in the context of parametrized relativistic dynamics (PRD) and applied to the survival of electron neutrinos. The analysis shows that significant differences exist between theoretical results of the conventional model and the PRD model. Current estimates of neutrino masses depend on the mass dependence $\Delta m^2 \equiv m_2^2 - m_1^2$ of the conventional theory. Further work will be needed to examine experimental results within the context of PRD. The result may be a set of neutrino masses that is consistent with the experimental results but differs from the conventional analysis.

References

[1] Olive, K.A., et al. (2014): "14. Neutrino Mass, Mixing, and Oscillations," Particle Data Group, *Chin. Phys.***C38**, 090001 (http://pdg.lbl.gov) ; as of August 29, 2014 ; updated May 2014 by K. Nakamura (Kavli IPMU (WPI), U. Tokyo, KEK), and S.T. Petcov (SISSA/INFN Trieste, Kavli IPMU (WPI), U. Tokyo, Bulgarian Academy of Sciences); http://www-pdg.lbl.gov/2015/reviews/rpp2015-rev-neutrino-mixing.pdf accessed March 25, 2016.

[2] Gonzalez-Garcia, M.C., M. Maltoni, and T. Schwetz (2014) : "Updated fit to three neutrino mixing: status of leptonic CP violation,"*J. High Energy Physics***11**, 52; arXiv:1409.5439 [hep-ph].

[3] Fanchi, J.R. (1993): *Parametrized Relativistic Quantum Theory*, Kluwer: Dordrecht.

[4] Fanchi, J.R. (2011): "Manifestly Covariant Quantum Theory with Invariant Evolution Parameter in Relativistic Dynamics," *Found. Phys.***41**, 4-32.

[5] Pavsic, M. (2001): *The Landscape of Theoretical Physics: A Global View*, Kluwer: Dordrecht.

[6] Horwitz, L.P. [2015]: *Relativistic Quantum Mechanics*, Springer: Dordrecht.

[7] Fanchi, J.R. (1998): "The Mass Operator and Neutrino Oscillations," *Found. Phys.***28**, 1521-1528.

[8] Roentgen, W.K. (1895): "On a New Kind of Rays," A. Stanton translation in *Nature* (Volume 53, pages 274-276, 1896).

[9] Becquerel, H. (1896): "Sur les Radiations Emisés par Phosphorescence," *Comptes Rendus des Séances de L'Académie des Sciences*, Volume 122, pages 420-421; "Sur les Radiations Emisés Invisibles par les corps Phosphorescents," *Comptes Rendus des Séances de L'Académie des Sciences*, Volume 122, pages 501-503; "Sur Quelques Propriétés Nouvelles des Radiations Invisibles Emisés par Divers Corps Phosphorescents," *Comptes Rendus des Séances de L'Académie des Sciences*, Volume 122, pages 559-564; "Sur les Radiations Invisibles Emisés par les Sels d'Uranium," *Comptes Rendus des Séances de L'Académie des Sciences*, Volume 122, pages 689-694; and "Sur les Propriétés Differentes des Radiations Invisibles Emisés par les Sels d'Uranium, et du Rayonnement de la Paroi Anticathodique d'un Tube de Crookes," *Comptes Rendus des Séances de L'Académie des Sciences*, Volume 122, pages 762-767; references from Franklin (2001).

[10] Thomson, J.J. (1897): "Cathode Rays," *Philosophical Magazine*, Volume 44, pages 293-316.

[11] Millikan, R.A. (1911) : "The Isolation of an Ion, a Precision Measurement of Its Charge, and the Correction of Stoke's Law," *Physical Review*, Volume 32, pages 349-397.

[12] Millikan, R.A. (1913) : "On the Elementary Electrical Charge and the Avogadro Constant," *Physical Review*, Volume II, Number 2, pages 109-143.

[13] Rutherford, E. (1899): "Uranium Radiation and the Electrical Conduction Produced by It," *Philosophical Magazine*, Volume 24, pages 453-462.

[14] Villard, P. (1900): "Sur la réflexion et la réfraction des rayons cathodiques et des rayons déviables du radium," *Comptes Rendus des Séances de L'Académie des Sciences*, Volume 130, pages 1010-1012; see also Villard, P. (1900): "Sur le rayonnement du radium," *Comptes Rendus des Séances de L'Académie des Sciences*, Volume 130, pages 1178-1179.

[15] Rutherford, E. (1903): ""The magnetic and electric deviation of the easily absorbed rays from radium," *Philosophical Magazine*, Series 6, Volume 5, Number 26, pages 177-187.

[16] Rutherford, E. (1911): "The Scattering of α and β Particles by Matter and the Structure of Matter," *Philosophical Magazine*, Volume 21, pages 669–688.

[17] Chadwick, J. (1914): "Intensitätsverteilung im magnetischen Spektrum der β-Strahlen von Radium B+C," *Verhandlungen der Deutschen Physikalischen Gesellschaft*, Volume 16, pages 383–391.

[18] Franklin, A. (2001) : *Are There Really Neutrinos ? An Evidential History*, Perseus Books : Cambridge, Massachusetts.

[19] Solomey, N. (1997): *The Elusive Neutrino: A Subatomic Detective Story*, Scientific American Library: New York

[20] Fermi, E. (1934): "Attempt at a Theory of β-rays," *Il Nuovo Cimento*, Volume 11, pages 1-21; and "Versuch einer Theorie der β-Strahlen," *Zeitschrift fur Physik*, Volume 88, pages 161-177.

[21] Chadwick, J. (1932): "Possible Existence of a Neutron," *Nature*, Volume 129, page 312; and "The Existence of a Neutron," *Proceedings of the Royal Society* (London), Volume A136, pages 692-708.

[22] Rutherford, E. (1920): "Bakerian Lecture: Nuclear Constitution of Atoms," downloaded from http://rspa.royalsocietypublishing.org/ on September 23, 2016.

[23] Sakata, S. and T. Inoue (1946): "On the correlations between mesons and Yukawa particles,"

IOP Conf. Series: Journal of Physics: Conf. Series **845** (2017) 012027 doi:10.1088/1742-6596/845/1/012027

Progress of Theoretical Physics, Volume 1, Number 4, Nov-Dec, Pages 143-150.

[24] Pontecorvo, B. (1946): "Inverse beta process," Chalk River Report, Pd-205, http://pontecorvo.jinr.ru/work/ibp.html accessed September 25, 2016.

[25] Davis, R. (1955): "Attempt to Detect the Antineutrinos from a Nuclear Reactor by the $Cl^{37}(\nu, e^-)A^{37}$ Reaction," *Physical Review*, Volume 97, pages 766-769.

[26] Davis, R. (1956): "Attempt to Detect the Antineutrinos from a Nuclear Reactor by the $Cl^{37}(\nu, e^-)A^{37}$ Reaction," *Bulletin of the American Physical Society*, Volume 1, page 219.

[27] Reines, F. and C. L. Cowan, Jr. (1953): "A Proposed Experiment to Detect the Free Neutrino," *Physical Review*, Volume 90, pages 492-493 (1953); and "Detection of the Free Neutrino," *Physical Review*, Volume 92, pages 830-831 (1953).

[28] Cowan, C.L. Jr., F. Reines, F. B. Harrison, H. W. Kruse and A. D. McGuire (1956): "Detection of the Free Neutrino: A Confirmation", *Science*, Volume 124, pages 103-104.

[29] Davis, R. (1964): "Solar Neutrinos II. Experimental," *Physical Review Letters*, Volume 12, pages 303-305.

[30] Bahcall, J.N. (1964): "Solar Neutrinos I. Theoretical," *Physical Review Letters*, Volume 12, pages 300-302.

[31] Bahcall, J.N. and G. Shaviv (1968): "Solar Models and Neutrino Fluxes," *Astrophysical Journal*, Volume 153, pages 113-125.

[32] Bahcall, J.N., N.A. Bahcall, W.A. Fowler, and G. Shaviv (1968): "Solar Neutrinos and Low-Energy Nuclear Cross sections," *Physics Letter*, Volume 26B, pages 359-361.

[32] Davis, R., D.S. Harmer and K.C. Hoffman (1968): "Search for Neutrinos from the Sun," *Physical Review Letters*, Volume 20, pages 1205-1209.

[33] Gell-Mann, M. and A. Pais (1955) : "Behavior of Neutral Particles under Charge Conjugation," *Physical Review*, Volume 97, pages 1387-1389.

[34] Pontecorvo, B. (1957-1958): "Mesonium and antimesonium," *Zhurnal Eksperimental'noi i Teoreticheskoi Fiziki*, Volume 33, pages 549 ff, 1957; *Soviet Physics—JETP*, Volume 6, pages 429-431, 1958.

[35] Pontecorvo, B. (1959-1960): "Electron and Muon Neutrinos," *Zhurnal Eksperimental'noi i Teoreticheskoi Fiziki*, Volume 37, pages 1751 ff, 1959; *Soviet Physics—JETP*, Volume 10, pages 1236-1240, 1960.

[36] Danby, G., J.-M. Gaillard, K. Goulianos, L.M. Lederman, N. Mistry, M. Schwartz and J. Steinberger (1962): "Observation of High-Energy Neutrino Interactions and the Existence of Two Kinds of Neutrinos", *Physical Review Letters*, Volume 9, pages 36-44.

[37] Maki, Z., M. Nakagawa and S. Sakata (1962): "Remarks on the Unified Model of Elementary Particles," *Progress of Theoretical Physics*, Volume 28, pages 870-880.

[38] Perl, M.L., et al. (1975): "Evidence for Anomalous Lepton Production in e^+–e^- Annihilation," *Physical Review Letters*, Volume 35, pages 1489-1492.

[39] DONUT (2001): "Observation of tau neutrino interactions," *Physics Letters B*, Volume 504, 12 April 2001, pages 218–224.

[40] Fukuda Y., et al. (Super-Kamiokande collaboration) (1998) : "Evidence for oscillation of atmospheric neutrinos,"*Physical Review Letters*, Volume 81, pages 1562–1567

[41] Close, F. (2010): *Neutrino*, Oxford University Press: Oxford, U.K.

[42] Jayawardhana, R. (2013): *Neutrino Hunters*, Scientific American/Farrar, Strauss and Giroux: New York.

IOP Conf. Series: Journal of Physics: Conf. Series **845** (2017) 012028 doi:10.1088/1742-6596/845/1/012028

Relativistic Newtonian dynamics

Yaakov Friedman and Joseph Mendel Steiner

Jerusalem College of Technology
Departments of Mathematics and Physics
P.O.B. 16031 Jerusalem 9116001, Israel
email: friedman@jct.ac.il

E-mail: `friedman@jct.ac.il`

Abstract. A new Relativistic Newtonian Dynamics (RND) for motion under a conservative force capable to describe non-classical behavior in astronomy is proposed. The rotor experiments using Mössbauer spectroscopy with synchrotron radiation, described in the paper, indicate the influence of non-gravitational acceleration or potential energy on time. Similarly, the observed precession of Mercury and the periastron advance of binaries can be explained by the influence of gravitational potential energy on spacetime. The proposed RND incorporates the influence of potential energy on spacetime in Newton's dynamics. The effect of this influence on time intervals, space increments and velocities is described explicitly by the use of the concept of escape trajectory. For an attracting conservative static potential we derived the RND energy conservation and the dynamics equation for motion of objects with non-zero mass and for massless particles. These equations are subsequently simplified for motion under a central force. Without the need to curve spacetime, this model predicts accurately the four non-classical observations in astronomy used to test the General Relativity.

1. Introduction

Newton's laws of motion are still relevant today and continue to be used as an excellent approximation in everyday experience. At the end of the 19th century, most problems in physics could be addressed with classical Newton theory except:

- non-covariance of Maxwell equations under Galilean transformations
- Michelson-Morley experiment
- gravitational redshift
- the anomalous precession of Mercury's orbit

These indicate a logical incompleteness of Newtons laws of motion.

The first of these problems was resolved by Einstein with his theory of Special Relativity (SR), in which the observed rate of a moving clock differs from the rate of a stationary clock and depends on the clock's velocity. This implies that the Lorentz transformations are the correct spacetime transformations between inertial frames. Einstein's new velocity addition formula of SR also explained the Michelson-Morley experiment.

The gravitational redshift and the anomalous precession of Mercury's orbit, however, cannot be explained within the framework of SR. These problems were solved using Einstein's Theory of Gravitation commonly known as General Relativity (GR), in which he basically accounted for gravity as an attribute of curved spacetime. As Princeton physicist Wheeler later described it: Spacetime tells matter how to move, matter tells spacetime how to curve.

Although Einstein's theories of relativity originate in the logical incompleteness of Newton's laws of motion, that incompleteness itself has not been understood completely, as yet, and therefore there are still global and chronic problems which still need further clarification. We propose that this incompleteness arises from the fact that *Newton's theory does not consider the influence of energy on spacetime.*

SR considers the influence of kinetic energy on spacetime. The curving of spacetime in GR can be regarded as an expression of the the influence of the gravitational potential on spacetime and the weak equivalence principle [29]. It is known [25] that the gravitational redshift (time dilation due to gravitational potential), can be derived solely from energy conservation and Planck's equation. But also for non-gravitational fields energy conservation and Planck's equation predict a time dilation depending on the position in space, However, none of the relativity theories considers the influence of *non-gravitational potential energy* on spacetime.

In our search for such an influence, we looked for the effect of (non-gravitational) acceleration on time. Our recent rotor experiments using Mössbauer spectroscopy and synchrotron radiation at PETRAIII, Hamburg and at ESRF, Grenoble indicate the existence of non-gravitational time dilation due to the acceleration.

All the above suggested to us to derive a new *Relativistic Newtonian Dynamics* (RND) defined as:

Definition 1. *RND is a dynamics incorporating the effects of any energy (gravitational or non-gravitational) on spacetime into classical Newtonian dynamics.*

We conjecture:

Conjecture 1. *RND provides the necessary relativistic corrections to Newtonian dynamics with the capability to explain all the above mentioned problems without the need for radical changes.*

Our derivation incorporates well established principles of relativity into Newtonian dynamics without any a priori assumptions.

Currently, RND incorporates the influence of potential energy on spacetime by invoking an extensions of Einstein's Equivalence Principle and the Clock Hypothesis (CH) to describe this influence. For an attracting conservative static potential with zero potential at infinity we derived the RND energy conservation and the dynamics equation for the motion of both objects with non-zero mass and for massless particles. The relativistic corrections of RND dynamics equation depend on the reduced potential $\hat{U}$- the potential per unit mass, and on $\nabla\hat{U}$ - the classical acceleration generated by this potential.

Specialising further these equations to a central inverse square force, predict accurately both the anomalous precession of Mercury [14] as well as the periastron advance of any binary [17], without the need of curving spacetime. RND predicts [16], [7] a chaotic behavior in the microscopic region similar to the probabilistic description in quantum mechanics.

In IARD2012 we presented [8] an Extended Relativistic Dynamics (ERD), based on the assumption that the CH is not valid (acceleration influences time in addition to the influence of the velocity of the comoving frame) which predicts [4] the existence of a maximal acceleration, explored in [5]. The relativistic correction of the dynamic equation in this model depends only on $\nabla\hat{U}$ the classical acceleration generated by this potential and on the value of the maximal acceleration. ERD predicts [10], [8] the discreteness of the radiation spectrum (compatible with Planck's Hypothesis for appearance of electromagnetic energy in quanta) for the harmonic oscillator in the microscopic region. However, ERD does not predicts correctly the astronomical observations mentioned above.

This suggests:

Conjecture 2. *The final RND model should incorporate the relativistic corrections from both RND and ERD to describe non-classical behavior in both astronomical and microscopic regions.*

2. Testing Einstein's Clock Hypothesis at PETRAIII

A cornerstone of Relativity Theory is the CH, stating that the rate of an accelerated clock is equal to that of a co-moving unaccelerated clock [3]. As it was shown [4] and [8], if the CH is false, (meaning that the rates of two clocks moving at different accelerations but with instantaneously equal velocities are different) then there exists a maximal acceleration a_m and an additional Doppler-type shift $\Delta f/f = a/a_m$ is predicted for an accelerated source of radiation [5] and [8].

The aim of the experiment performed in 2013 under the leadership of Ralf Rölsberger at the Dynamics Beamline P01 of the synchrotron PETRAIII at DESY, Hamburg, Germany was to test the CH by looking for this Doppler type shift between two sources with different acceleration. Time-domain interferometry based on nuclear resonant scattering was used to measure this shift.

Nuclear resonant absorption experiments with samples subject to high accelerations have been performed in the past to test special relativity via measurement of the transverse Doppler effect. The most prominent one is the experiment by Kündig, [22], who employed a resonant absorber mounted on a fast rotating disk. A reanalysis [20] of this experiment claimed a 20% deviation between the observed shift and the one predicted by SR. This deviation can be explained [4] by the failure of the CH and the existence of a maximal acceleration of $a_m = 10^{19} m/s^2$. However, since the current observation [15] revealed that non-random vibrations (not measured in the above experiment) also contribute to the shift, this estimate of a_m is questionable.

Since the expected shift is very small one must use tools very sensitive to such shift. Such an effect could be detected via the ^{57}Fe Mössbauer nuclear resonance that exhibits a relative linewidth of $\Gamma_0/E_0 = 3*10^{-13}$. High radial accelerations **a** were obtained with the rotors used in the magic-angle spinning technique of nuclear magnetic resonance (NMR). The rotor diameter was $d = 3mm$ with rotational frequencies of up to $15kHz$. The rotor is a hollow cylinder in which the Mössbauer foil was placed to cover its full inner circumference.

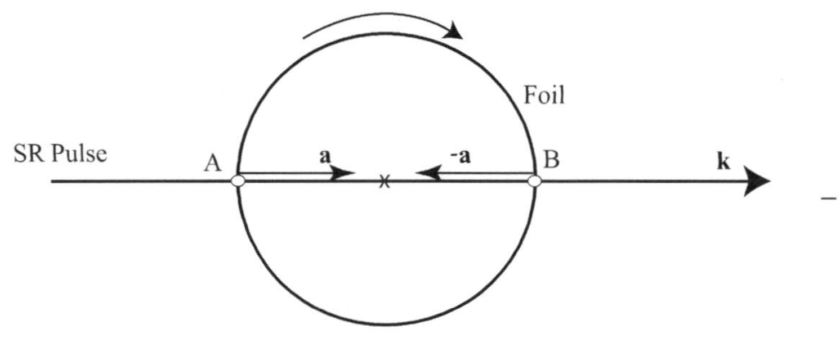

Figure 1. PETRAIII experimental geometry

The experimental geometry is shown in Figure 1. A focused beam of x-ray pulses (cross section $5 \times 5\mu m$ from the KB optics [19] at beamline P01, $1meV$ bandwidth) traversing the cylinder along its diameter perpendicular to its rotation axis meets the rotating ^{57}Fe foil twice: first, at the point of incidence A and then at the point of incidence B. The radial acceleration **a** is directed parallel at A and $-$**a** antiparallel at B to the photon wavevector **k**. Assuming that acceleration causes a frequency shift, we expect a spectral shift between the resonant lines of the foil at the two points of incidence of the rotating cylinder. This spectral shift cannot be measured directly at this facility.

What could be measured at this facility is the so called *time spectrum*. A coherent pulse with frequency close to the resonant frequency of the sample, propagating through a sample, excites the nuclei in the sample thereby generating a delayed secondary radiation. The time spectrum of the sample is the intensity of this delayed radiation exiting the sample as a function of the delay time. In our case, the sample was the ^{57}Fe foil at the two points of incidence A and B. The formula for the time spectrum of a single resonant sample from a coherent synchrotron radiation pulse propagating through the sample obtained in [13] based on [1] is

$$I(\tau) \propto e^{-\tau} T^2 \left(\frac{J_1^2(\sqrt{T\tau})}{T\tau} + \alpha^2 J_0^2(\sqrt{T\tau}) \right), \tag{1}$$

where τ is the dimensionless time delay (the time delay divided by the lifetime of the excited level of the nuclei), T is the *effective thickness* of the sample, J_0, J_1 are the Bessel function of first kind of order zero and one, respectively and the constant $\alpha \ll 1$. In runs using an effectively thin foil, the only statistically significant parameter of the time spectrum is the effective thickness T. The measured effective thickness of the time spectrum of this rotating sample at different rotational frequencies is given in Figure 2.

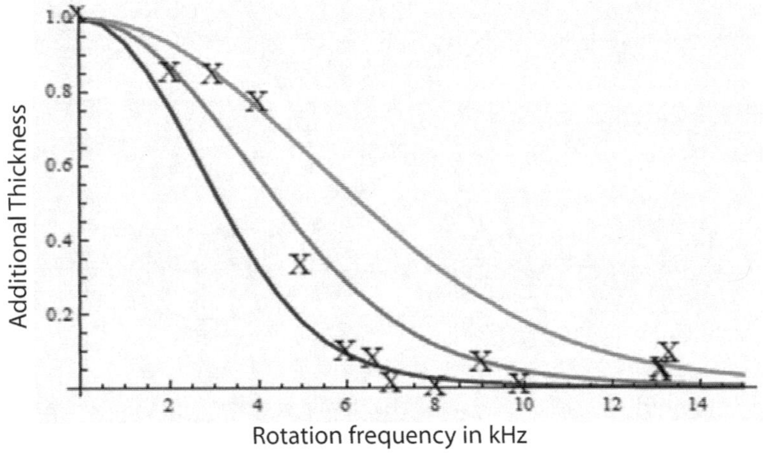

Figure 2. The measured $\times$ excess in the effective thickness of the time spectrum of the thin rotating sample at different rotational frequencies. Theoretical predictions for different values of a_m, Blue:$5 \cdot 10^{18}$, Red: 10^{19}, Green:$2 \cdot 10^{19} m/s^2$

The unit of thickness was chosen to be one for our foil and we reported the excess measured effective thickness relative to one foil. At rest, the measured effective thickness of the delayed radiation was twice the effective thickness of the foil (excess thickness one) because the radiation came from both points of incidence. The rotation of the sample will cause only a transverse Doppler shift of the sample spectrum, if the CH is valid, and this shift will be the same for both points of incidents. Such a shift does not affect the time spectrum, and one would not expect to get any effect on the time spectrum caused by rotation.

In the experiment we observed a gradual change from the spectrum at rest to the spectrum of one foil with the increase of the rotation frequency. Assuming that acceleration influences time, this will cause a gradually increasing shift (similar to the usual Doppler shift) between resonant lines of the ^{57}Fe foil at the two points of incidence A and B. We observed that with rotations larger than 6kHz the effective thickness is the same as for one foil.

If there exists a strong shift between the absorption spectra at the two points of incidence at these rotational frequencies, as we expected, we may assume that their resonant frequencies are disjoint. Thus, the foil at B is transparent to the delayed radiation exiting A because of the difference in their resonant frequencies. Similarly, the radiation delayed at B was not delayed at A. Thus, the combined delayed radiation exiting the cylinder is a superposition of two rays, each of which was delayed only once. If the coherence between these two rays is lost, then the combined time spectrum will be that of a one foil spectrum. Thus, the results of the experiment indicate that the acceleration does effect time, contradicting the CH.

For an enriched sample the picture gets more complicated, indicating that probably there are additional unknown factors during the rotation affecting the time spectrum.

3. Testing acceleration time dilation at ESRF

Additional two experiments were performed at the European Synchrotron Radiation Facility (ESRF), Grenoble, France in 2014 and 2015. These experiments follow the outline presented in [8], see Figure 3. In [6] we predicted that the absorption line of a rotating Mössbauer absorber

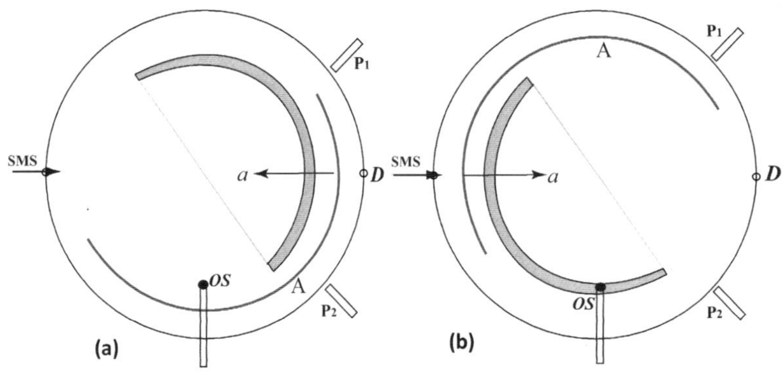

Figure 3. The setup and two states (a) and (b), SMS source, semicircular absorber A, optical sensor OS, proximity sensors P_1, P_2 and detector D

gets broader during the rotation. This is due to the fact that the beam must have a finite width. Therefore, the velocity of the absorber is not always perpendicular to all individual rays. Hence, these rays undergo a longitudinal Doppler shift in addition to the expected transversal one. This shift is very significant even for very narrow beams.

In 2014 experiment we confirmed experimentally [11] that, for a source outside the rotating disk, the absorption line of a rotating Mössbauer absorber is broadened during the rotation and that this broadening is linearly proportional to the rotation frequency. This necessitates the use of a strong Mössbauer source with the capability to focus the beam to the center of the disk. The Synchrotron Mössbauer Source (SMS) [26] at the Nuclear Resonance Beamline ID18 [27] of ESRF together with the KB-optics [19] to focus this beam, was the ideal choice for our experiments. The beam emitted by the SMS is almost fully resonant and fully polarized, has high brilliance and can be focused by use of KB-optics to a $10\mu m \times 5\mu m$ spot size at the center of the rotating disk. The beam from the SMS passes the KB-optics, hits the fast rotating semicircular absorber of radius $R = 0.05m$ and is finally detected by the APD detector D

diametrically opposed to the SMS (Figure 3). In this experiment we obtained for the first time the *entire resonant line* of a rotating absorber.

Based on the experiences of this experiment, a follow-up improved experiment was performed in 2015 at the same facility with the same rotor system. The main objective of this experiment was to explore the effect of acceleration on time dilation. The time dilation is measured by the shift in the Mössbauer resonant line of an accelerated absorber. The acceleration of the absorber was generated by rotating it using a rotor system. Using the setup diagrammatically presented in Figure 3 we separated the absorption of a rotating absorber into two states (a) and (b) when the acceleration is anti-parallel and parallel to the SMS radiation, respectively, and obtained the absorption lines for each state. Since the absorption line for each of the states are generated by the same source and absorber, and are equally affected by the velocity of the absorber with respect to the source, these two spectra should be identical without the influence of acceleration on time.

The experiment revealed [18] that within the given setup and rotation frequencies of 200Hz and higher, the resonant shifts of the two states are not the same. We observed a stable statistically significant non-zero relative shift between the spectral lines in the two states. In particular, for two long runs at 200 Hz in which the effect of vibrations was negligible, the almost constant relative shift of about $0.41(14)mm/s$, indicates the effect of acceleration on time dilation (see Figure 4 and Table 1). Existence of the statistically significant, almost

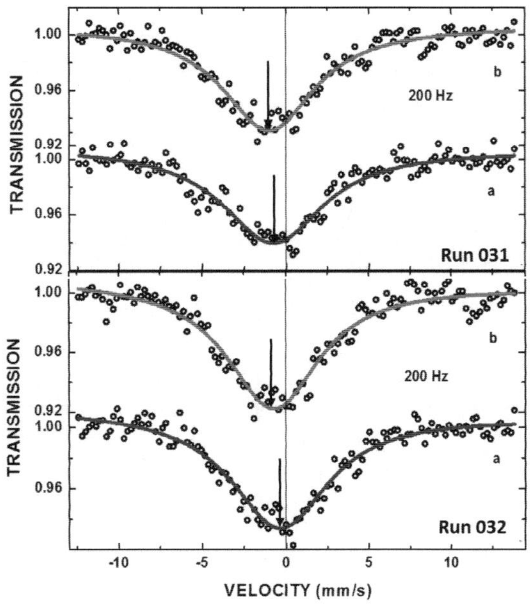

Figure 4. The absorption spectra, converted to average linear velocity form, of run 31 (top) and 32 (bottom) for states (a) and (b) at 200Hz rotation

constant, non-zero actual relative shift for two runs at 200 Hz indicates the effect of acceleration on the absorption resonant line and hence on time dilation.

Since non-gravitational accelerations were found to have an effect on time dilation, it would be interesting to compare this effect with that due to the redshift. To do this, one can imagine a massive object of mass M positioned at the center of the disk producing

Run	Effect %	γ [mm/s]	Shift x_0 [mm/s]	Vibr.sh.[mm/s]
31a	6.68(18)	3.29(24)	-0.58(11)	-0.007
31b	8.01(27)	3.14(22)	-1.01 (09)	0.007
32a	7.29(20)	3.25(20)	-0.22(12)	-0.01
32b	8.31(29)	3.13(21)	-0.61(09)	0.01

Table 1. Parameters of resonant spectra and the vibration shifts for states (a) and (b) at 200Hz rotation

a gravitational acceleration on the rim equal to centripetal acceleration in the experiment $a = R\omega^2 = 0.05(2\pi \cdot 200)^2 = 8 \cdot 10^4 m/s^2$. To achieve this acceleration on the rim $GM/R^2 = a$ requires that $GM = 197 m^3/s^2$. The redshift in units of velocity due to such potential is $\frac{2GM}{Rc} = 0.013 mm/s$. This value is more then one order of magnitude less then the observed relative shift. This is not surprising, since the rotation generate a non-conservative field, while gravitation is conservative. Moreover, since in both states the absorber is at the same gravitational potential level, one should not expect a relative redshift.

Currently, the only relativistic model that predicts a non-zero relative shift in such experiment is the ERD model [4],[5] which incorporates the influences of kinetic energy and acceleration assuming that Einstein's Clock Hypothesis is not valid and predicts a maximal acceleration. If we use the maximal acceleration $a_m = 10^{19} m/s^2$ based on Kündig's experiment [22] an an estimate, the observed relative shift is larger then expected. We plan to repeat the experiment with an improved system and study the dependence of the shift on frequency and direction of rotation.

A new RND theoretical model capable to explain the observed influence of the acceleration on time, as well as other phenomena in astronomy and microscopic region, is developed.

4. Relativistic Newtonian Dynamics

In this section we will derive the Relativistic Newtonian Dynamics (RND) for an attracting force with negative potential and apply it to astronomical predictions. This dynamics incorporates the influence of *any potential* energy vanishing at infinity into Newtonian dynamics. We begin by recalling some facts from Newtonian dynamics for a conservative force.

4.1. Newtonian dynamics under conservative force

We recall the description of the motion of a mass particle under a conservative force in Newtonian dynamics. Pick an inertial lab frame K with coordinates $\mathbf{x} = (x^1, x^2, x^3)$ and time t. We denote the mass of the particle by m, which will be assumed to be constant (rest mass), and denote the force by $\mathbf{F}(\mathbf{x})$ and its potential by $U(\mathbf{x})$. The trajectory $\mathbf{x}(t)$ of the motion of the particle is described by Newton's second law

$$m\frac{d^2\mathbf{x}(t)}{dt^2} = \mathbf{F}(\mathbf{x}(t)) = -\nabla U(\mathbf{x}(t)), \tag{2}$$

where ∇U denotes the gradient of U. Introduce a vector field $\mathbf{n}(\mathbf{x})$ which at any space point $\mathbf{x}$ is the normalized vector in the direction of the gradient $\nabla U(\mathbf{x})$, i.e.

$$\mathbf{n}(\mathbf{x}) = \nabla U(\mathbf{x})/|\nabla U(\mathbf{x})|. \tag{3}$$

Taking the dot product of this equation with $\mathbf{v} = d\mathbf{x}/dt$ we obtain

$$m\frac{d^2\mathbf{x}(t)}{dt^2} \cdot \frac{d\mathbf{x}}{dt} + \nabla U(\mathbf{x}) \cdot \frac{d\mathbf{x}}{dt} = \frac{d}{dt}\left(\frac{m}{2}\mathbf{v}^2 + U(\mathbf{x})\right) = 0$$

implying that the total energy E, which is a function of the state, defined by

$$E(\mathbf{x},\mathbf{v}) = \frac{m}{2}\mathbf{v}^2 + U(\mathbf{x}) \tag{4}$$

is conserved during the motion. The first term on the right-hand side is the kinetic energy and the second one is the potential energy.

From (2), the potential U is defined up to a constant. We will need a uniquely defined U satisfying the condition $U(\infty) = 0$. For an attractive force such U must be negative. Requiring the potential to vanish at infinity excludes polynomial potentials (e.g., oscillator) and other unbounded potentials and in its current status limits the applicability of the theory.

The *escape velocity* $v_e(\mathbf{x})$ is defined as the minimum speed in direction of $\mathbf{n}$ needed for a mass particle at $\mathbf{x}$ to "break free" from the attraction defined by the potential. More particularly, it is the velocity (speed traveled away from the starting point) at which the sum of a particle's kinetic and potential energies is equal to zero. Thus,

$$\beta_e^2 := \frac{v_e^2(\mathbf{x})}{c^2} = -\frac{2U(\mathbf{x})}{mc^2}. \tag{5}$$

We further define the *Schwarzschild region*, the region S defined by

$$S = \{\mathbf{x}:\ \beta_e^2(\mathbf{x}) \le 1\}. \tag{6}$$

Any mass particle positioned in S cannot escape the attraction of the potential and will be absorbed. We will consider only motion outside S.

4.2. Influence of the potential energy on time intervals, space increments and velocity

Let $U(\mathbf{x})$ be the gravitational potential of an attracting static force field at any space point $\mathbf{x}$ in an inertial lab frame K. We assume that $U(\mathbf{x})$ is zero at infinity, where the attraction tends to zero. We will describe the *influence of the potential energy* $U(\mathbf{x})$ on spacetime in the neighbourhood of any space point $\mathbf{x}_0$.

Define the *escape trajectory* $\mathbf{x}(\tau)$ for some parameter τ as the (de)accelerated trajectory of the test object with velocity $\mathbf{v}_e(\mathbf{x}_0)$ defined by (5) at $\mathbf{x}_0$ and progressing freely to the ultimate point O with zero potential. Denote by $\tilde{K}(\mathbf{x}(\tau))$ the Frenet frames [12] defined by this escape trajectory, by K' the comoving frame to the accelerated frame $\tilde{K}(\mathbf{x}_0)$, and by $\tilde{K}(O)$ the final frame on this trajectory. Note that since $\tilde{K}(O)$ is an inertial frame in which the potential is time independent, it can be viewed as a translation of the lab frame K. Since the time intervals and space increments are not affected by a translation, the increments in $\tilde{K}(O)$ are the same as in the lab frame K.

On the escape trajectory the total energy E is zero, implying the equality of the kinetic and potential energies on this trajectory. Moreover, since there is a continual exchange between these two energies, it is natural to postulate a conjecture similar to the Equivalence Principle:

Conjecture 3. *The influence of the potential energy on spacetime in the neighbourhood of* $\mathbf{x}_0$ *is the same as the influence of the acceleration and velocity on rest points in the accelerated frame* $\tilde{K}(\mathbf{x}(\tau))$ *defined by the escape trajectory at* $\mathbf{x}_0$.

This conjecture allows one to express the effect of the potential energy on the time intervals and space increments, by their corresponding changes from $\tilde{K}(\mathbf{x}_0)$ to $\tilde{K}(O)$. Then, using an extension of Einstein's CH, introduced in [9, 12] and [23, 24], one can quantify these changes by the use of the Lorentz spacetime transformation between the frames K'-the comoving frame to $\tilde{K}(\mathbf{x}_0)$ and $\tilde{K}(O)$. Finally, since since increments in $\tilde{K}(O)$ are the same as in K, one can express the effect of the potential energy on the time and space increments by a map $\Phi : \tilde{K}(\mathbf{x}_0) \to K$ based on the Lorentz transformation from K' to K. Using (5) this map is

$$\Phi(dt) = \tilde{\gamma} dt,\ \Phi(d\mathbf{x}_n) = \tilde{\gamma} d\mathbf{x}_n,\ \Phi(d\mathbf{x}_{tr)} = d\mathbf{x}_{tr}\,, \tag{7}$$

where

$$\tilde{\gamma} = \frac{1}{\sqrt{1 - v_e^2(\mathbf{x}_0)/c^2}} = \frac{1}{\sqrt{1 + 2U(\mathbf{x}_0)/mc^2}} \tag{8}$$

and the components of the space increments in the direction $\mathbf{n}$ and transverse to $\mathbf{n}$ are

$$d\mathbf{x}_n = (d\mathbf{x} \circ \mathbf{n})\mathbf{n},\ \ d\mathbf{x}_{tr} = d\mathbf{x} - (d\mathbf{x} \circ \mathbf{n})\mathbf{n} \tag{9}$$

with $\circ$ denoting the Euclidean dot product and $\mathbf{n}$ defined by (3) is the direction of the escape velocity.

Note that $\tilde{\gamma}$ defined by (8) is the known gravitational *time dilation factor*, if $U(\mathbf{x})$ represents a gravitational potential. From (7) it follow that the influence of potential energy at $\mathbf{x}_0$ on any 3D velocity $\mathbf{v} = \frac{d\mathbf{x}}{dt}$, expressed by the map Φ, is

$$\Phi(\mathbf{v}_n) = \mathbf{v}_n,\ \Phi(\mathbf{v}_{tr}) = \tilde{\gamma}^{-1}\mathbf{v}_{tr}. \tag{10}$$

This means, that the velocity of an object at $\mathbf{x}_0$ is spacetime influenced by the potential energy is transformed to the velocity in the lab frame K by multiplying the component of the velocity transverse to $\mathbf{n}$ by $\tilde{\gamma}^{-1}$, where the time dilation factor $\tilde{\gamma}$ is defined by (8).

4.3. RND energy conservation

As shown above, the potential energy influences time intervals and space increments in the direction of $\mathbf{n}(\mathbf{x})$, while the space increments transverse to $\mathbf{n}(\mathbf{x})$ are not influenced by this potential energy. We obtain the RND energy conservation formula, by analogy to the principle of least action, a variational principle that defines the path of motion as the path with the least value of some action. In an analogous way the motion in such an influenced spacetime can be viewed as the motion along a geodesic (path with least distance) with respect to some metric. By the above observation this metric is of the form

$$ds^2 = f(\mathbf{x})(cdt)^2 - g(\mathbf{x})d\mathbf{x}_n^2 - d\mathbf{x}_{tr}^2, \tag{11}$$

where the two vector projections of the space increments $d\mathbf{x}_n$ and $d\mathbf{x}_{tr}$ parallel and transverse to $\mathbf{n}(\mathbf{x})$ are defined by (9) and the positive valued functions $f(\mathbf{x}), g(\mathbf{x})$ describe the influence of the potential energy on the time intervals and space increments in the $\mathbf{n}(\mathbf{x})$ direction.

Consider now the time dilation due to the influence of potential energy of a clock resting at some point $\mathbf{x}$ in K. Since for such a clock $d\mathbf{x} = 0$, from (11) $ds = \sqrt{f(\mathbf{x})}cdt$ implying that the time dilation of such a clock is $c\frac{dt}{ds} = \frac{1}{\sqrt{f(\mathbf{x})}}$. From (7), this time dilation is defined by the gamma-factor $\tilde{\gamma}$ given by (8), implying that $\sqrt{f(\mathbf{x})} = \sqrt{1 + 2U(\mathbf{x})/mc^2}$. Introducing the *dimensionless potential energy*

$$u(\mathbf{x}) = \frac{-2U(\mathbf{x})}{mc^2}, \tag{12}$$

one obtains immediately

$$f(\mathbf{x}) = 1 - u(\mathbf{x}). \tag{13}$$

This formula reduces to the gravitational time dilation in the vicinity of a non-rotating massive spherically symmetric object, as derived from the Schwarzschild metric.

IOP Conf. Series: Journal of Physics: Conf. Series **845** (2017) 012028 doi:10.1088/1742-6596/845/1/012028

The motion of an object of mass $m > 0$ along a time-like geodesic $\mathbf{X}(\lambda) = (ct(\lambda), \mathbf{x}(\lambda))$ in spacetime is parameterized by an *affine parameter* λ on the trajectory, chosen to be the arc length s defined by (11). For massless objects (photons) moving along light-like geodesics $ds = 0$, this affine parameter is redefined as in [25] p.575 with the normalization condition

$$\lim_{x^2 \to \infty} c\frac{dt}{d\lambda} = 1. \tag{14}$$

The four-velocity on the trajectory is $\dot{\mathbf{X}} = \frac{d\mathbf{X}}{d\lambda} = (c\dot{t}, \dot{\mathbf{x}})$, where *dot* denotes differentiation with respect to λ.

Since the metric (11) is independent of t, the vector $\mathbf{K}_0 = (1, 0, 0, 0)$ is a Killing vector ([25], p.651) and its scalar product $\dot{\mathbf{X}} \cdot \mathbf{K}_0$ with respect to this metric with the four-velocity is conserved on the trajectory. Thus, on any trajectory there exists a constant k related to the total energy on it such that

$$cf(\mathbf{x})\dot{t} = k, \quad c\dot{t} = \frac{k}{1 - u(\mathbf{x})}. \tag{15}$$

Using (11),(13) and (15), the four-velocity norm is

$$\dot{\mathbf{X}}^2 = \frac{k^2}{1 - u(\mathbf{x})} - g(\mathbf{x})\dot{\mathbf{x}}_n^2 - \dot{\mathbf{x}}_{tr}^2 = \epsilon, \tag{16}$$

where $\epsilon = 1$ for objects with non-zero mass and $\epsilon = 0$ for massless particles. For objects with non-zero mass multiply this equation by $(1 - u(\mathbf{x}))$ to obtain

$$(1 - u(\mathbf{x}))g(\mathbf{x})\dot{\mathbf{x}}_n^2 + (1 - u(\mathbf{x}))\dot{\mathbf{x}}_{tr}^2 - u(\mathbf{x}) = k^2 - 1\,. \tag{17}$$

This formula can be considered as the RND dimensionless energy conservation equation on the trajectory. The first two terms of the left side are the relativistically corrected dimensionless kinetic energy (the square of the influenced dimensionless velocity $\Phi(\dot{\mathbf{x}})^2$) and the third term is the dimensionless potential energy. The right side is the dimensionless total energy $k^2 - 1$.

To calculate the influenced dimensionless kinetic energy, note that $\dot{\mathbf{x}} = \dot{t}\mathbf{v}$, where $\mathbf{v} = \frac{d\mathbf{x}}{dt}$ and using (10) and that $\tilde{\gamma}^{-2} = 1 - u$ one obtains

$$\Phi(\dot{\mathbf{x}})^2 = \dot{t}^2\Phi(\mathbf{v})^2 = \dot{t}^2(\mathbf{v}_n^2 + (1 - u(\mathbf{x}))\mathbf{v}_{tr}^2) = \dot{\mathbf{x}}_n^2 + (1 - u(\mathbf{x}))\dot{\mathbf{x}}_{tr}^2.$$

Comparing this to the first two terms in (17) one obtains

$$g(\mathbf{x}) = \frac{1}{1 - u(\mathbf{x})}, \tag{18}$$

hence, the RND dimensionless energy conservation equation for an object with non-zero mass becomes

$$\dot{\mathbf{x}}_n^2 + (1 - u(\mathbf{x}))\dot{\mathbf{x}}_{tr}^2 - u(\mathbf{x}) = \mathcal{E}, \tag{19}$$

where the dimensionless total energy on the trajectory is $\mathcal{E} = 2E/mc^2 = k^2 - 1$.

For a massless particle, the analysis is only applicable when restricted to a gravitational potential for which the dimensionless potential energy $u(\mathbf{x})$ is independent of the mass m. For such a potential one can define a *reduced potential* or potential per unit mass as

$$\hat{U} = \frac{U}{m}. \tag{20}$$

For such potential, the dimensionless potential defined by (12) is $u(\mathbf{x}) = -2\hat{U}/c^2$. For such particle ($\epsilon = 0$), multiplying (16) by $(1 - u(\mathbf{x}))$ and using (18) one obtains

$$\dot{\mathbf{x}}_n^2 + (1 - u(\mathbf{x}))\dot{\mathbf{x}}_{tr}^2 = k^2. \tag{21}$$

Even though gravitation does not act directly on the photon as a force since its mass is zero, which is expressed in missing term $-u(\mathbf{x})$ in (19), its momentum and angular momentum are not zero and its motion is affected by the influence of the gravitational potential on spacetime expressed by the relativistic term in this equation. Moreover, from (14) and (15) one obtains $k = 1$.

Thus, the RND dimensionless energy conservation equation for both objects with non-zero mass and massless particles is

$$\dot{\mathbf{x}}_n^2 + (1 - u(\mathbf{x}))\dot{\mathbf{x}}_{tr}^2 - \epsilon u(\mathbf{x}) = 1 + \epsilon(\mathcal{E} - 1). \tag{22}$$

Unlike in GR, in RND the reduced energy (obtained by multiplying the dimensionless energy by $c^2/2$)

$$H(\mathbf{x}, \dot{\mathbf{x}}) = \frac{c^2\dot{\mathbf{x}}^2}{2} + \hat{U}(\mathbf{x})(\dot{\mathbf{x}}^2 - (\dot{\mathbf{x}} \circ \mathbf{n})^2) + \epsilon\hat{U}(\mathbf{x}) \tag{23}$$

is conserved on the trajectory. Note that the relativistic addition to a classical reduced total energy, expressed in the middle term of right-hand side, depends both on the potential energy at that point and the velocity of the moving object.

4.4. RND Equation of Motion under a Conservative Force

By use of (9) equation (22) can be rewritten as

$$\dot{\mathbf{x}}^2 - \epsilon u(\mathbf{x}) - u(\mathbf{x})(\dot{\mathbf{x}}^2 - (\dot{\mathbf{x}} \circ \mathbf{n})^2) = 1 + \epsilon(\mathcal{E} - 1). \tag{24}$$

The equation of motion is obtained by differentiating (24) by λ which leads to

$$2\ddot{\mathbf{x}} \circ \dot{\mathbf{x}} - \epsilon\dot{u} - \dot{u}\dot{\mathbf{x}}_{tr}^2 - 2u(\ddot{\mathbf{x}} \circ \dot{\mathbf{x}} - (\dot{\mathbf{x}} \circ \mathbf{n})(\ddot{\mathbf{x}} \circ \mathbf{n} + \dot{\mathbf{x}} \circ \dot{\mathbf{n}})) = 0,$$

Substituting $u(\mathbf{x}) = -\hat{2}U/c^2$ and its derivative $\dot{u} = -\frac{2\nabla\hat{U}\circ\dot{\mathbf{x}}}{c^2}$, one obtains

$$\ddot{\mathbf{x}} + \frac{2\hat{U}(\mathbf{x})}{c^2}\ddot{\mathbf{x}}_{tr} = -\frac{\epsilon\nabla\hat{U}}{c^2} - \frac{\nabla\hat{U}\dot{\mathbf{x}}_{tr}^2}{c^2} + \frac{2\hat{U}}{c^2}(\dot{\mathbf{x}} \circ \dot{\mathbf{n}})\mathbf{n}. \tag{25}$$

Decomposing this vector equation into the parallel and perpendicular components to $\mathbf{n}$, the perpendicular component is

$$\left(1 + \frac{2U(\mathbf{x})}{mc^2}\right)\ddot{\mathbf{x}}_{tr} = 0. \tag{26}$$

implying that $\ddot{\mathbf{x}}_{tr} = 0$ for $\mathbf{x}$ outside the Schwarzschild region defined by(6).

Thus, for motion outside the Schwarzschild region the *RND equation of motion* for any object/particle is

$$c^2\ddot{\mathbf{x}} = -\epsilon\nabla\hat{U} - \dot{\mathbf{x}}_{tr}^2\nabla\hat{U} + 2\hat{U}(\mathbf{x})(\dot{\mathbf{x}} \circ \dot{\mathbf{n}})\mathbf{n}. \tag{27}$$

Noting that $\frac{d}{d\lambda} = \frac{k}{c(1-u)}\frac{d}{dt}$ from (15) and remembering that $\mathbf{v} = \frac{d\mathbf{x}}{dt}$ one obtains the RND equation of motion in the space and time of the lab frame K as

$$\frac{k^2}{1-u}\frac{d}{dt}\left(\frac{\mathbf{v}}{1-u}\right) = -\epsilon\nabla\hat{U} - \frac{k^2\mathbf{v}_{tr}^2\nabla\hat{U}}{c^2(1-u)^2} + \frac{2k^2\hat{U}}{c^2(1-u)^2}(\mathbf{v} \circ \frac{d\mathbf{n}}{dt})\mathbf{n}. \tag{28}$$

Note that the acceleration $\frac{d^2\mathbf{x}}{dt^2}$ is not in the direction of $\mathbf{n}$, but $\ddot{\mathbf{x}}$ is.

The first relativistic correction depends on $-\nabla\hat{U}$- the classical acceleration caused by the force, and the second one depends on the reduced potential $\hat{U}$. Unlike in the classical Newtonian equation of motion involving only the force $\mathbf{F}$, the RND equation involves both the potential U (as in Schrodinger quantum dynamics equation) as well as ∇U.

4.5. Central force motion in RND

Let $U(r)$ be a potential of a central force which is attractive and static in an inertial lab frame K with origin at the center of the force. Assume that $U(r)$ is negative and vanishes at infinity. In this case $\mathbf{n}$ defined by (3) is the radial direction and from (7) it follows that the time intervals and the radial spacial displacements are influenced by the potential energy, while the spacial displacements transverse to the radial direction are not affected.

Introduce spherical coordinates r, φ, θ in K. Using the results of Section 4.3, the influence of the potential energy on spacetime in the neighbourhood of any point is described by the metric

$$ds^2 = (1 - u(r))(cdt)^2 - \frac{1}{1-u(r)}(dr)^2 - r^2((d\theta)^2 + \sin^2\theta(d\varphi)^2). \tag{29}$$

As above, assume that the motion of an object/particle is along the geodesic in spacetime $\mathbf{X}(\lambda) = (t(\lambda), r(\lambda), \varphi(\lambda), \theta(\lambda))$ parameterised by the affine parameter λ. Then the four-velocity is $\dot{\mathbf{X}} = \frac{d\mathbf{X}}{d\lambda} = (c\dot{t}, \dot{r}, \dot{\varphi}, \dot{\theta})$. From symmetry considerations one may assume that if initially the position and velocity of the object/particle are in the plane $\theta = \pi/2$, they will remain in this plane during the motion. Thus, one may assume that $\theta = \pi/2$ and $\dot{\theta} = 0$.

Using that the metric is independent of φ, the vector $\mathbf{K}_3 = (0,0,1,0)$ is a Killing vector implying that

$$J = r^2\dot{\varphi} \tag{30}$$

remains constant on the trajectory, where cJ has the meaning of angular momentum per unit mass for objects with non-zero mass and J has units of length.

Using that $\dot{\mathbf{x}}_n = \dot{r}$ and that $\dot{\mathbf{x}}_{tr}^2 = \frac{J^2}{r^2}$ one can rewrite the RND dimensionless energy conservation equation (22) as

$$\dot{r}^2 + (1-u(r))\frac{J^2}{r^2} - \epsilon u(r) = 1 + \epsilon(\mathcal{E} - 1). \tag{31}$$

Using (30) and (31) one obtains the parameter-free equation for the trajectory $r(\varphi)$

$$\left(\frac{J}{r^2}\frac{dr}{d\varphi}\right)^2 + (1-u(r))\frac{J^2}{r^2} - \epsilon u(r) = 1 + \epsilon(\mathcal{E} - 1). \tag{32}$$

This equation with $\epsilon = 1$ was used in the previous papers [14], [16] and [17] for objects with non-zero mass.

Introducing polar coordinates and complex numbers in the plane $\theta = \pi/2$, $\mathbf{r} = re^{i\varphi}$. Using (30),

$$\dot{\mathbf{r}} = \dot{r}e^{i\varphi} + ri\dot{\varphi}e^{i\varphi} = \dot{r}e^{i\varphi} + i\frac{J}{r}e^{i\varphi}. \tag{33}$$

and

$$\ddot{\mathbf{r}} = \ddot{r}e^{i\varphi} + \dot{r}\frac{J}{r^2}ie^{i\varphi} - \dot{r}\frac{J}{r^2}ie^{i\varphi} - \frac{J^2}{r^3}e^{i\varphi} = \left(\ddot{r} - \frac{J^2}{r^3}\right)e^{i\varphi}.$$

For a central force $\mathbf{n} = e^{i\varphi}$ and using (33) and (30), $\dot{\mathbf{r}} \cdot \dot{\mathbf{n}} = \frac{J^2}{r^3}$. Substituting this into (27), the *RND equation of motion for a central force* becomes

$$c^2\left(\ddot{r} - \frac{J^2}{r^3}\right) = -\epsilon\nabla\hat{U} - \frac{J^2}{r^2}\nabla\hat{U} + \frac{2\hat{U}(\mathbf{r})J^2}{r^3}. \tag{34}$$

This equation defines $r(\lambda)$. Then, equation (30) defines $\varphi(\lambda)$ and hence the trajectory of any object/particle.

4.6. Inverse square law in RND

Consider now a inverse square law force $\mathbf{F}(\mathbf{r})/m = -\frac{w}{r^3}\mathbf{r}$ with potential $\hat{U}(\mathbf{r}) = -\frac{w}{r}$. Substituting this into (34) one obtains

$$\left(\ddot{r} - \frac{J^2}{r^3}\right) = -\frac{w}{c^2r^2} - \frac{wJ^2}{c^2r^4} - \frac{2wJ^2}{c^2r^4},$$

or

$$\ddot{r} = \frac{J^2}{r^3} - \frac{w}{c^2r^2} - \frac{3J^2w}{c^2r^4}. \tag{35}$$

Note that for this case the two relativistic correction reinforce each other.

This equation is further simplified if one expresses it in terms of $u(\varphi)$ defined by (8) as $u(r) = \frac{2w}{c^2r} = \frac{r_s}{r}$ with $r_s = \frac{2w}{c^2}$. Denoting by $u' = \frac{du}{d\varphi}$ it can be shown by use of (30) that $\dot{r} = -\frac{J}{r_s}u'$ and $\ddot{r} = -\frac{J^2}{r_sr^2}u''$. With these substitutions (35) becomes

$$-\frac{J^2}{r_s^3}u^2u'' = \frac{J^2}{r_s^3}u^3 - \frac{w}{r_s^2}u^2 - \frac{3J^2w}{c^2r_s^4}u^4, \tag{36}$$

which simplifies to

$$u'' + u = \frac{r_sw}{J^2} + \frac{3w}{c^2r_s}u^2.$$

Using the definition of r_s one obtains

$$u'' + u = \frac{2w^2}{c^2J^2} + \frac{3}{2}u^2. \tag{37}$$

The relativistic correction turns the linear classical Newtonian equation into a non-linear one.

Defining the orbit parameter $\mu = \frac{2w^2}{c^2J^2}$, equation (37) reproduces equation (25) in [14] for planetary motion under RND which predicts the known anomalous precession $3\pi\mu$ of Mercury [2], [21]. Note that the first relativistic correction term of (34) depending on the force contributes $\pi\mu$ to the precession, while the second correction term depending on the potential contributes $2\pi\mu$ to the precession.

Furthermore, the two body problem has also a solution in RND. Using a similar analysis as above, we have shown in [17] that RND is able to handle effectively the periastron advance of any binary. For a hydrogen-like atom RND predicts [16] a chaotic behavior compatible to the probabilistic model of quantum mechanics.

The RND model also handles motion with unbounded trajectories, predicts accurately gravitational lensing and the Shapiro time delay for a light ray passing in vicinity of a spherically symmetric massive object. These will appear in future publications.

5. Discussion and conclusions

The observed anomalous precession of Mercury and periastron advance of the Hulse-Taylor pulsar and other binaries were predicted by the authors [14], [17] by considering the influence of potential energy on spacetime. Our recent experiments [18] indicate that non-gravitational acceleration also influences time. Suchard [28] also predicts that the potential of an electromagnetic field influences spacetime. These suggest that any potential energy, gravitational or not, influences spacetime.

In order to understand how the laws of dynamics should be modified by this influence, one must express explicitly the relativistic corrections due to it. A new RND model is proposed to describe non-classical behavior. So far we have shown that this model passes successfully all the non-classical tests in astronomy used for the justification of GR. But this model is not limited to gravitation and could be applied to other conservative fields, like the electric field. So, it has a potential to describe the non-classical behavior of the microscopic world.

In Sections 2 and 3, we presented our recent experiments using Mössbauer spectroscopy of a rotating absorber and synchrotron radiation which indicate the influence of non-gravitational acceleration on time.

The RND was then derived in Section 4 for a conservative static attracting force with zero potential at infinity. The influence of the potential energy on spacetime in the neighbourhood of a point $\mathbf{x}_0$ is described by the use of the escape trajectory and conjecturing that this influence is the same as the influence of velocity and acceleration on the escape trajectory initiating at this point. Then, this influence can be described by the transformation from the the initial accelerated frame at $\mathbf{x}_0$ to the final inertial (lab) frame on the escape trajectory. The transformation in the model is bases on the validity of the CH for an accelerated system in flat spacetime.

The RND reduced total energy (23) for objects of non-zero mass and massless particles introduce an additional term depending on both the kinetic and potential energies into the classical energy equation (4). The RND equation of motion (27) introduce two relativistic correction terms of order 2 in v/c into. The first relativistic correction depends on $-\nabla\hat{U}$- the classical acceleration caused by the force, and the second one depends on the reduced potential $\hat{U}$- potential per unit mass.

For a central force, using angular momentum conservation (30), the RND equation of motion is simplified (34) and even more for an inverse square law (35) leading to a single additional relativistic term to the classical equation. This equation predicts accurately the anomalous precession of Mercury and periastron advance of any binary. These deviations from Newtonian behavior in these astronomical observations were predicted by GR and experimentally confirm its validity for weak gravitational fields only [29]. Recently we have shown that RND predicts accurately gravitational lensing and Shapiro time delay of light as it moves over a finite distance through a change in the gravitational potential of a massive spherically symmetric object M. Thus, RND passes successfully all four tests of GR.

As mentioned above, one of the limitations of the current RND model is the requirement of zero potential at infinity - zero point of the force. Even though the force generated by the harmonic oscillator is conservative and its potential could be defined to be zero at the point where the force is zero, this zero point is not at infinity and an escape trajectory cannot be defined properly in this model. So, it is not surprising that when the RND dynamics equation of motion is applied to the oscillator, the relativistic correction terms cancel out. So RND in its current form is not able to describe the change of behavior of the oscillator in the microscopic region involving strong fields. The needed modifications for strong fields can be obtained by use of Extended Relativistic Dynamics (ERD) [5]. In ERD the speed of any moving object is limited by c and the magnitude of its acceleration is limited by a_m. The relativistic correction in ERD depends only on the classical acceleration caused by the force $-\nabla\hat{U}$. It was shown [10] that the vibrations of the ERD harmonic oscillator poses an identical spectrum to that of the quantum oscillator in the microscopic region, which implies Planck's assumption.

We plan to extend RND also for non-conservative forces by use of a four-potential instead of the scalar potential. We plan to study the implications of this model to the electromagnetic field and experimentally test these implications.

In order to predict accurately both astronomical observations and the dynamics in the microscopic region, both RND and ERD corrections together with an extension of the model to non-conservative forces is probably needed.

References

[1] Burnham D C and Chiao R Y *Phys. Rev.* **188** (1969) 667-675

[2] Grøn Ø, Hervik S. 2007 *Einstein's General Theory of Relativity: With Modern Applications in Cosmology* (Springer)
[3] Einstein A 1907 *Jahrb. Radioakt. Elektron.* **4** 411
[4] Friedman Y, Gofman Yu 2010 *Phys. Scr.* **82** 015004
[5] Friedman Y 2011 *Ann. Phys. (Berlin)*, **523 523** 408-416.
[6] Friedman Y, Nowik I 2012 *Phys. Scr.* **85** 065702
[7] Friedman Y, Resin E 2012 *Phys. Scr.* **86** 015002
[8] Friedman Y 2013 *Proc. of IARD 2012, J. of Phys.: Conf. Ser.* **437** 012017
[9] Friedman Y, Scarr T 2013 *Phys. Scr.* **87** 055004
[10] Friedman Y 2013 *Phys. Scr.* **87** 065702
[11] Friedman Y *et al.* 2015 *J. Synch. Rad.* **22** 723
[12] Friedman Y, Scarr T 2015 *Gen. Relat. and Grav.* **47** 121
[13] Friedman Y, Steiner J M, Yudkin E 2015 *Eur. Phys. Lett.* **111** 34003
[14] Friedman Y, Steiner J M 2016 *Eur. Phys. Lett.***113** 39001
[15] Friedman Y *et al.* 2016 *Eur. Phys. Lett.* **114** 50010
[16] Friedman Y 2016 *Eur. Phys. Lett.* **116** 19001
[17] Friedman Y, Livshitz S, Steiner J M 2016 *European Physics Letters* **116** 59001
[18] Friedman Y *et al.* 2017 *Advances in testing the effect of acceleration on time dilation using a Synchrotron Mössbauer Source*, *J. Synch. Rad.*, to appear
[19] Kirkpatrick P, Baez A V 1948 *J. Opt. Soc. Am.* **38** 766
[20] Kholmetskii A L, Yarman T and Missevitch O V, 2008 *Phys. Scr.* **77** 035302
[21] Kopeikin S, Efroimsky M, Kaplan G 2011 *Relativistic Celestial Mechanics of the Solar System* (Wiley-VCH, Berlin)
[22] Kündig W 1963 *Phys. Rev.* **126** 2371
[23] Mashhoon B 1990 *Phys. Lett. A* **143** 176-182
[24] Mashhoon B 1990 *Phys. Lett. A* **145** 147-153
[25] Misner C W, Thorne K S and Wheeler J A *Gravitation* Freeman and co.(1973)
[26] Potapkin V, Chumakov A I, Smirnov G V, Celse J-P, Rüffer R, McCammon C and Dubrovinsky L 2012 *J. Synch. Rad.* **19** 559
[27] Rüffer R,Chumakov A I 1996 *Hyp. Int.*, **97/98** 589-604
[28] Suchard E H 2015 *Physical Science International Journal* **7** 152
[29] Will C M 2014 *Liv. Rev. Relat.* **17** 4

IOP Conf. Series: Journal of Physics: Conf. Series **845** (2017) 012029 doi:10.1088/1742-6596/845/1/012029

Free massless particle Dirac like wave equations of any spin in curved space-time

Amnon Moalem and **Alexander Gersten**
Ben Gurion University of the Negev, Beer Sheva, Israel
E-mail: Amnon.moalem@gmail.com

Abstract. We derived Dirac like equations for massless particles of any spin in a global curved space-time. In an axially symmetric space-time the wave functions factorize in two parts. A spin-connection independent function which resembles the solution in a flat space-time, and a second spin connection dependent function which is calculated analytically for static Friedman-Robertson-Walker and Schwarzschild metrics.

1. Introduction

In previous contributions [1] it was demonstrated that in a Minkowski flat space-time, wave equations for massless particles of any spin can be written as a matrix equation,

$$\left(\left(\frac{\hat{E}}{c}\right)\gamma^0 - \gamma_1\hat{p}_1 - \gamma_2\hat{p}_2 - \gamma_3\hat{p}_3\right)\Phi^{(4s)} = 0, \quad (1)$$

where s is the spin, $\hat{E} = i\hbar\partial_0$, $\hat{p}_i = -i\hbar\partial_i$; $(i = 1,2,3)$ are energy and momentum operators, γ_i are (4s)× $(4s)$ hermitian matrices and γ_0 is the $(4s) \times (4s)$ unit matrix. The wave functions $\Phi^{(4s)}$ form bases for $D^{(s-1/2,1/2)}$ representations of the Lorentz group, which we may write as a column matrix with (2s-1) components of spin (s-1) and (2s+1) components of spin s, altogether $(4s)$ components. Explicitly,

$$\Phi^{(4s)} = \begin{pmatrix} \psi^{s-1}_{s-1} \\ \cdot \\ \cdot \\ \psi^{s-1}_{-s+1} \\ \psi^{s}_{s} \\ \cdot \\ \cdot \\ \psi^{s}_{-s} \end{pmatrix}. \quad (2)$$

Here we note that the Hermitian conjugate of $\Phi^{(4s)}$ is defined as $\left[\Phi^{(4s)}\right]^H = \left[\Phi^{*(4s)}\right]^T$ and that $\left[\Phi^{(4s)}\right]^H \gamma^0 \left[\Phi^{(4s)}\right]$ is a scalar. Likewise, $\left[\Phi^{(4s)}\right]^H \gamma^a \left[\Phi^{(4s)}\right]$ form a vector. We shall use these observations latter on to calculate the spin connection for the wave function $\Phi^{(4s)}$. For free particles the spin (s-1) components of the wave function are set to be zero. By doing so, the subsidiary conditions which restrict helicity values to forward helicity and backward helicity only are satisfied.

IOP Conf. Series: Journal of Physics: Conf. Series **845** (2017) 012029 doi:10.1088/1742-6596/845/1/012029

For spin 1 particles (a photon) Eq. (1) is equivalent to the Maxwell equations. For spin 2 Eq. (1) describes a free graviton with no sources. Our main objective in the present note is to reformulate these equations in a curved space-time. To this aim we make use of the Principle of Equivalence. Namely, Curved space-time is locally a Minkowski flat space-time. Accordingly, Eq. (1) is valid in within sufficiently small region around any point in a curved space-time. With this in mind, we calculate in Section 2 the appropriate spin connection for the function $\Phi^{(4s)}$and rewrite Eq. (1) for a general global space-time. In section 3 we consider the equations derived in the vicinity of spherically symmetric massive body such as a neutron star or a black hole. Following Refs. [2,3] we factorize the wave function into a spin connection dependent term which is rather easy to calculate in axis-symmetric space-time and a spin connection independent term which can be obtained from solving an equation rather similar to Eq. (1). We conclude in section 4.

2. Covariant formulation

In what follows we reformulate Eq. (1) in a general global curved space-time. We first rewrite Eq. (1) using natural units with $\hbar = c = 1$ and a metric signature $(+,-,-,-)$. Greek indices refer to the general world index, whilst Latin indices refer to the flat Minkowski tangent space. We denote the metric of the curved space by $g_{\mu\nu} = e^a_\mu e^b_\nu \eta_{ab}$, where $\eta_{ab} = diag(+1,-1,-1,-1)$ stands for the Minkowski metric. With these notations we write Eq. (1) as,

$$\gamma^b \eta_{ab} \partial_a \Phi = 0 \ . \tag{3}$$

The Gamma matrices in Eqs. (1,3) form presentations of bi-quaternions, they factorize the d'Alembertian operator and have eigenvalues $+1$, and -1only. This is a reflection of the fact that the subsidiary conditions are encoded in the gamma matrices [1]. The gamma matrices satisfy the following relations,

$$(\gamma^a)^\dagger = (\gamma^a) \ ; \ (\gamma^a)^2 = \gamma^0, \tag{4.1}$$
$$\{\gamma^a, \gamma^b\} = 2I^{(0)}\delta^{ab} \ ; \ a, b = 0,1,2,3, \tag{4.2}$$
$$[\gamma^a, \gamma^b] = i\varepsilon^{abc}\gamma_c \ ; \ a, b, c = 0,1,2,3, \tag{4.3}$$
$$trace(\gamma^a)=0 \ ; \ a=1,2,3. \tag{4.4}$$

The gamma matrices above carry tangent space indices so that they maintain a flat space-time form. As shown below, the spinor representations of the Lorentz generators are commutators of the gamma matrices, an observation that simplifies the calculation of the affine connection for the wave function Φ.

In order to rewrite Eq. (3) in a curved space-time one must replace the Minkowski metric η_{ab} by a curved space-time metric tensor $g_{\mu\nu}$and replace the partial local derivatives by covariant derivatives. In addition, one must transform the local gamma matrices by global gamma matrices by means of vierbein fields [4]. One obtains,

$$E^\mu_a \gamma^a g_{\mu\nu} [\partial_\nu + \Omega_\nu(x)] \Phi(x) = 0. \tag{5}$$

In the above equation E^μ_a are inverse veirbein fields, and Ω_ν is the connection coefficient for the wave function. A parallel transport of the wave function should then satisfy the relation,

$$\Phi(x + dx) = \Phi(x) - \Omega_\mu(x)\Phi(x)dx^\mu. \tag{6}$$

To evaluate the connection $\Omega_\mu(x)$ we recall that $S(x) = \Phi^H(x)\gamma^0\Phi(x)$is a scalar and that $V^a(x) = \Phi^H(x)\gamma^a\Phi(x)$ is a vector and therefore must obey the transformation rules for scalars and vectors,

IOP Conf. Series: Journal of Physics: Conf. Series **845** (2017) 012029 doi:10.1088/1742-6596/845/1/012029

respectively. Consider first $S(x)$ which must remain unchanged under parallel transport. Using Eq. (6) one obtains,

$$S(x+dx) = S(x) - \Phi^H(x)[\gamma^0 \Omega^H_\mu(x) + \Omega_\mu(x)\gamma^0]\Phi(x)dx^\mu, \tag{7}$$

where terms of order $dx^\mu dx^\nu$have been neglected. For Eq. (8) to satisfy scalar parallel transport rule the quantity in the square bracket must vanish, i.e.,

$$\Omega^H_\mu(x) = -\Omega_\mu(x) \tag{8}$$

Next we consider the vector $V(x)$ which transports as a local vector. Again using Eq. (7) one obtains,

$$V^a(x+dx) = V^a(x) - \Phi^H(x)[\gamma^a, \Omega_\mu(x)]\Phi(x)dx^\mu. \tag{9}$$

Then the second term must satisfy,

$$\Phi^H(x)[\gamma^a, \Omega_\mu(x)]\Phi(x) = \omega^a_{\mu b}V^b, \tag{10}$$

where $\omega^a_{\mu b}$ is the spin connection [4]. This last expression indicates that the commutator in the expression above must be proportional to γ^b and that Ω_μ is related to multiplications of gamma matrices. Let us assume that,

$$\Omega_\mu = C\omega_{\mu bc}\gamma^b\gamma^c, \tag{11}$$

where$C$ is a constant. It is straightforward to show using properties of the gamma matrices Eqs. (4) above that,

$$[\gamma^a, \Omega_\mu(x)] = 4C\omega^a_{\mu b}\gamma^b. \tag{12}$$

Comparing this last expression with Eq. (11) we choose $C = 1/4$ so that $\Omega_\mu = (1/4)\omega_{\mu bc}\gamma^b\gamma^c$. Then the spin connection term in Eq. (5) can be written explicitly as,

$$E^\mu_a\gamma^a\Omega_\mu = \frac{1}{2}\gamma^b[E^\sigma_b\partial_\sigma lne + \partial_\sigma E^\sigma_b], \tag{13}$$

where we have substitute the well-known expression [4] for the spin connection, *i.e.*,

$$\omega^a_{\mu b} = e^a_\mu E^\sigma_a\Gamma^\nu_{\sigma\mu} + e^a_\nu\partial_\mu E^\nu_b .$$

In the expression above$g = det\,(g_{\mu\nu})$ stands for the determinant of the metric tensor and $e = \sqrt{-g}$. Substituting Eq. (14) in Eq. (5) gives the covariant form of Eq. (1), *i.e.*,

$$E^\mu_a\gamma^a\frac{1}{\sqrt{e}}\partial_\mu(\sqrt{e}\Phi) = -\frac{1}{2}\gamma^b\partial_\sigma E^\sigma_b\Phi. \tag{14}$$

For a Minkowski space-time the terms on the right vanish so that in the limit of the correspondence principle, Eq. (1) that was derived in a flat space-time is recovered.

IOP Conf. Series: Journal of Physics: Conf. Series **845** (2017) 012029 doi:10.1088/1742-6596/845/1/012029

3. Covariant Massless Particle Equation In Axis-symmetric Space-time

It may not be easy to solve Eq. (15) as it is for general global metric tensor. In what follows we attempt to define a scheme based on variable separation similar to that of Refs. [2,3] that may ease the solution in axially symmetric space-time. Black holes are stationary axis-symmetric solutions of the Einstein equation [5] and such a scheme can be useful to study the behaviour of free massless particles such as the photons and gravitons in the environment of massive black holes. Axial symmetry restricts the form of the metric tensor and consequently the corresponding vierbein and inverse vierbein fields. Following Ref. [2] we assume that Φ factorize as,

$$\Phi(x^0,x^1,x^2,x^3) = \psi(x^0,x^1,x^2,x^3)\, f(x^1,x^2), \tag{15}$$

and require that the function ψsatisfies an equation independent of the spin connection, *i.e.*,

$$E_a^\mu \gamma^a \partial_\mu \psi = 0. \tag{16}$$

This expression does not depend on the spin connection. This is absorbed into the function f which as we shall see below can be calculated for axially symmetric space-time analytically. Define $E_0^0\partial_\mu = \partial_t$, $E_a^k\partial_k = -\partial_a$ the equation for ψ becomes $\gamma^0\partial_0 - \gamma^a\partial_a\psi = 0$, rather similar to Eq. (3) and can be solved as shown in Ref. [1]. The function $f(x^1,x^2)$ satisfies the equation,

$$E_a^\mu \gamma^a \partial_\mu f = -\frac{1}{2}\left(\gamma^b E_b^\sigma \partial_\sigma lne + \gamma^b \partial_\sigma e_b^\sigma \right) f. \tag{17}$$

Multiplying the expression by γ^aon the left and take the matrix trace on both sides of the above expression the equation for f resumes a compact form,

$$\partial_\mu \ln(f\sqrt{e}) = -\frac{1}{2}\left(e_\mu^b\, \partial_\sigma E_b^\sigma \right). \tag{18}$$

Eq. (18) depends on the metric tensor. For the Minkowski, FRW and Schwarzschild metrics Eq. (18) is integrable. A list of the function f for these is given in Table 1.

Table 1. The function $f(x^1,x^2)$ for axially symmetric space-times

Metric	$e_{\mu\nu}$	$f(x^1,x^2)$
Minkowski	$diag\,(1,1,r,r\,sin\theta)$	$1/rsin^{1/2}\,\theta$
FWR	$diag\,(1,a/F,ar,arsin\theta)$; $F=\sqrt{1-kr^2}$ *)	$1/a^{3/2}rsin^{1/2}\,\theta$
Schwarzschild	$diag\,(S,1/S,r,rsin\theta)$; $S=\sqrt{1-2M/r}$ **)	$1/\sqrt{S}\,r\,sin^{1/2}\,\theta$

*) a is the size of the universe and has the dimension of length; r is dimensionless parameter; and k= $\{-1, 0, 1\}$.

**) For $2M/r \ll 1$, the calculated f reduces, as expected, to that of the Minkowski space-time.

IOP Conf. Series: Journal of Physics: Conf. Series **845** (2017) 012029 doi:10.1088/1742-6596/845/1/012029

4. Conclusions

Dirac like equations for free massless particles of any spin were derived in a curved space- time. These can serve to study the behaviour of free photons and gravitons in near compact massive objects such as neutron stars or black holes. All black holes are stationary axis-symmetric solutions of the Einstein's equation [5] and our results focus on this class of space-time. The free particle wave function was written in terms of a scaler function $f(x^1, x^2)$ and a spinor $\psi(x^0, x^1, x^2, x^3)$ which satisfies an equation similar to that derived originally in Minkowski flat space. Axial symmetry imposes restrictions which are encoded in the vierbein fields and spin connection and renders the evaluation of $f(x^1, x^2)$ to be rather simple. We believe that the equations presented above can serve as a platform to study eigenvalues and degeneracies of photons and gravitons on the 3-sphere.

References

[1] A. Gersten & A. Moalem, Jour. Phys., Conference Series; **330**(2011), 012010, **437**(2013)012019 , **615**(2015)012011; see also A. Gersten and A. moalem (2016) in the present volume.

[2] Dag Morten Sjostrom, "Bosons and Fermions in Curved Space-time", Thesis, Norwegian University of Science and Technology, 2013.

[3] Seyd Alwi B. Ahmad , Proceedings of the Conference in Honor of 90th Birthday of Freeman Dyson, Edited by Phua K K et al. Published by World Scientific Publishing Co. Pte. Ltd., 2014. pp. 419-424.

[4] See for example Jeffery Yepez, arXiv:1106.3037, 10 (2011).

[5] See for example, Robert M. Wald, General Relativity, University of Chicago Press (1984) and F. De Felice and C.I. S. Clarke, Relativity on Curved Manifolds, Cambridge University Press (1990).

A generalized spin statistics theorem

Paul O'Hara

Istituto Universitario Sophia,Via San Vito, 28 - Loppiano, 50064 Figline e Incisa Valdarno (FI), Italy

E-mail: paul.ohara@iu-sophia.org

Abstract. In this article we generalize the spin statistics theorem and show that a state obeys Fermi-Dirac statistics if and only if the state is invariant under the action of $SL(n, \mathcal{C})$. We also briefly discuss the experimental evidence and how the theorem relates to spin entanglement.

Key Words: spin statistics theorem, $SL(n, \mathcal{C})$ invariance, entanglement.

1. Introduction

The origin of quantum statistics seems to have begun in 1920 when S.K. Bose sent a paper to Einstein seeking his help in getting it published. Einstein recommended it to Zeitschrift but later also published his own version in which the notion of indistinguishable photon states were introduced [12]. This was the beginning of what is now referred to as Bose-Einstein statistics. Another development took place in 1925 with the formulation of the Pauli exclusion principle which asserts that no two electrons in an atom could be in the same quantum state. In the 1930's this was subsequently generalized by Fermi and Dirac into what is now referred to as Fermi-Dirac statistics [1]. At about the same time Jordan and Wigner second quantized the Schrodinger equation and showed that Bose-Einstein statistics and Fermi-Dirac statistics respectively obeyed a set of commutator relations and anti-commutator relations applied to creation and annihilation operators [12]. This was the precursor of a connection between spin and statistics, first formulated by Markus Fiertz in 1939 [5], and then further developed by Pauli a year later [10]. In his paper Pauli claims *the necessity of Fermi-Dirac statistics for particles with arbitrary half-integral spin*, and of *the necessity of Einstein-Bose statistics for particles with arbitrary integral spin.* Also by invoking relativistic invariance he shows that bosons cannot be quantized as fermions and vice-versa.

Most subsequent work on spin-statistics takes for granted Pauli's conclusions but also struggles to understand the physical (as opposed to the mathematical) principles involved. For example, Feynman in his Lecture in Physics series states: *...An explanation has been worked out by Pauli from complicated arguments of QFT and relativity...but we haven't found a way of reproducing his arguments on an elementary level...this probably means that we do not have a complete understanding of the fundamental principle involved...*[4] Indeed within the context of Feynman's obsrvation, Duck and Sudarshan give a comprehensive analysis of the many different approaches to spin-statistics including work by De Wet, Wightman, Schwinger, Feynman, Hall, Luder and Zumino and conclude that the various proofs, including their own, *[were] not completely free from the complications of relativistic quantum field theory* [3]. Also

IOP Conf. Series: Journal of Physics: Conf. Series **845** (2017) 012030 doi:10.1088/1742-6596/845/1/012030

Berry and Robbins article on the subject published in 1997 cannot be considered *elementary* in Feynman's sense [11], [2]. With this in mind, in Theorem 1 we prove another version of the spin-statistics theorem which is free of quantum field theory complications. We show that Fermi-Dirac statistics is directly related to $SL(n, \mathcal{C})$ invariance.

However, before doing so, we note that this result is itself a generalization of Theorem 2 in [7] where it was previously shown that the rotational invariance associated with the existence of pairwise entangled states was sufficient for the Pauli exclusion principle. It is also suggested both in [7] and [8] that pairwise entanglement can be used to explain the stability of spin-$\frac{1}{2}$ baryons. In other words, " spin-$\frac{3}{2}$ baryons may be viewed as excited states of spin-$\frac{1}{2}$ baryons" [7] which will decay into a stable spin-$\frac{1}{2}$ proton.

2. A spin statistics theorem

The importance of this paper is not the discussion about entanglement per se but rather the proof of Theorem 1 which states that a necessary and sufficient condition to have Fermi-Dirac statistics is invariance under the action of $SL(n, \mathcal{C})$. The rotational invariance is embedded in the observation that $SU(2, \mathcal{C}) \subset SL(n, \mathcal{C})$. The theorem is very general. It applies to any tensor product vector space of the form $V = V_1 \otimes \cdots \otimes V_n$. In particular if we choose a vector of the form

$$v \equiv v_1 \wedge v_2 \wedge \ldots \wedge v_n, \quad v_i = (v_{ij}), 1 \leq j \leq n$$

where the wedge product indicates an anti-symmetric vector then

$$\begin{aligned} v &= \begin{pmatrix} v_{11} \\ \vdots \\ v_{n1} \end{pmatrix} \wedge \begin{pmatrix} v_{12} \\ \vdots \\ v_{n2} \end{pmatrix} \wedge \ldots \wedge \begin{pmatrix} v_{n1} \\ \vdots \\ v_{nn} \end{pmatrix} \\ &= \begin{vmatrix} v_{11} & v_{12} & \cdots & v_{1n} \\ v_{21} & v_{22} & \cdots & v_{2n} \\ \vdots & \vdots & \vdots & \vdots \\ v_{n1} & v_{n2} & \cdots & v_{nn} \end{vmatrix} \mathbf{e}_1 \wedge \mathbf{e}_2 \wedge \ldots \wedge \mathbf{e}_n \\ &= |v| \mathbf{e}_1 \wedge \mathbf{e}_2 \wedge \ldots \wedge \mathbf{e}_n, \text{ where } |v| = \begin{vmatrix} v_{11} & v_{12} & \cdots & v_{1n} \\ v_{21} & v_{22} & \cdots & v_{2n} \\ \vdots & \vdots & \vdots & \vdots \\ v_{n1} & v_{n2} & \cdots & v_{nn} \end{vmatrix} \end{aligned}$$

$|v|$ is usually called the Slater determinant and it remains invariant for any choice of orthonormal basis $\{\mathbf{e}_1, \mathbf{e}_2 \ldots \mathbf{e}_n\}$ of V_i. Historically, in the physics literature, the Slater determinant has always been associated with Fermi-Dirac statistics, and used to characterize the anti-symmetric nature of the wavefunction.

As a consequence of the invariance of the Slater determinant defined with respect to an orthonormal basis, it follows that if we take any matrix element, T of the group $SL(n, \mathcal{C})$ (which by definition is the group of all elements with determinant 1), and apply this operator to each component of the antisymmetric vector v then

$$Tv = |Tv| \mathbf{e}_1 \wedge \mathbf{e}_2 \wedge \ldots \wedge \mathbf{e}_n = |T||v| |\mathbf{e}_1 \wedge \mathbf{e}_2 \wedge \ldots \wedge \mathbf{e}_n = v.$$

Moreover, as the second theorem notes, the antisymmetric tensor v is the only vector with this property. Consequently, the two theorems taken together suggest that Fermi-Dirac statistics for n indistingushable particles be formally defined as any statistic that is invariant under the action of $SL(n, \mathcal{C})$. The formal proofs are presented below.

IOP Conf. Series: Journal of Physics: Conf. Series **845** (2017) 012030 doi:10.1088/1742-6596/845/1/012030

Theorem 1 *Let $V = V_1 \otimes \cdots \otimes V_n$, where for all i, j, each $V_i \cong V_j$ and V_i is an n-dimensional vector space. Let $T = T_1 \otimes \cdots \otimes T_n$ where for each i, j, $T_i = T_j$ and T_i is a linear operator on V_i. Let*

$$\begin{aligned} v &\equiv v_1 \wedge v_2 \wedge \ldots \wedge v_n \\ &= \begin{pmatrix} v_{11} \\ \vdots \\ v_{n1} \end{pmatrix} \wedge \begin{pmatrix} v_{12} \\ \vdots \\ v_{n2} \end{pmatrix} \wedge \ldots \wedge \begin{pmatrix} v_{1n} \\ \vdots \\ v_{nn} \end{pmatrix} \end{aligned}$$

then for $v \neq 0$

$$Tv = v \iff T \in \bigotimes_1^n SL(n, \mathcal{C}).$$

In other words, Fermi-Dirac statistics is invariant under the action of $SL(n, \mathcal{C})$. Note by definition $v_1 \wedge v_2 \wedge \ldots \wedge v_n = \frac{1}{n!}\delta^{i_1 \ldots i_n}_{1 \ldots n} v_{i_1} \otimes \ldots \otimes v_{i_n}$

Proof: Let $\{\mathbf{e}_1, \mathbf{e}_2 \ldots \mathbf{e}_n\}$ be an orthonormal basis of V_i, then

$$\begin{aligned} v &= \begin{pmatrix} v_{11} \\ \vdots \\ v_{n1} \end{pmatrix} \wedge \begin{pmatrix} v_{12} \\ \vdots \\ v_{n2} \end{pmatrix} \wedge \ldots \wedge \begin{pmatrix} v_{n1} \\ \vdots \\ v_{nn} \end{pmatrix} \\ &= \begin{vmatrix} v_{11} & v_{12} & \cdots & v_{1n} \\ v_{21} & v_{22} & \cdots & v_{2n} \\ \vdots & \vdots & \vdots & \vdots \\ v_{n1} & v_{n2} & \cdots & v_{nn} \end{vmatrix} \mathbf{e}_1 \wedge \mathbf{e}_2 \wedge \ldots \wedge \mathbf{e}_n \\ &= |v| \mathbf{e}_1 \wedge \mathbf{e}_2 \wedge \ldots \wedge \mathbf{e}_n, \text{ where } |v| = \begin{vmatrix} v_{11} & v_{12} & \cdots & v_{1n} \\ v_{21} & v_{22} & \cdots & v_{2n} \\ \vdots & \vdots & \vdots & \vdots \\ v_{n1} & v_{n2} & \cdots & v_{nn} \end{vmatrix} \end{aligned}$$

The linearity of T gives

$$\begin{aligned} Tv &= |v| T_1\mathbf{e}_1 \wedge T_2\mathbf{e}_2 \wedge \ldots \wedge T_n\mathbf{e}_n \\ &= |v| \begin{pmatrix} t_{11} \\ \vdots \\ t_{n1} \end{pmatrix} \wedge \begin{pmatrix} t_{12} \\ \vdots \\ t_{n2} \end{pmatrix} \wedge \ldots \wedge \begin{pmatrix} t_{n1} \\ \vdots \\ t_{nn} \end{pmatrix} \\ &= |v||T_1| \mathbf{e}_1 \wedge \mathbf{e}_2 \wedge \ldots \wedge \mathbf{e}_n, \qquad T_1 = T_2 = \ldots = T_n \end{aligned}$$

Therefore, since $v \neq 0$ implies $|v| \neq 0$ then

$$Tv = v \Rightarrow |T_1| = 1 \text{ and } T_1 \in SL(n, \mathcal{C})$$

Conversely

$$T_1 \in SL(n, \mathcal{C}) \Rightarrow Tv = v$$

This proves the theorem.

As mentioned in the introduction, this result can be seen as a generalization of a theorem where Fermi-Dirac statistics can be derived using rotational invariance [7]. The first thing to note is that $SU(n,\mathcal{C}) \subset SL(n,\mathcal{C})$ and therefore the Fermi-Dirac statistic is automatically rotationally invariant. In itself this already gives us a deeper insight into Fermi-Dirac statistics. The fact is $SU(2,\mathcal{C})$ and $SO(2,\mathcal{C})$ groups are subgroups of $SL(n,\mathcal{C})$ and consequently particles which are invariant under the action of these groups are pairwise entangled. This means that singlet states become the building blocks of Fermi-Dirac statistics. For example, in two dimensions if we let

$$R(\theta) = \begin{pmatrix} \cos c\theta & \sin c\theta \\ -\sin c\theta & \cos c\theta \end{pmatrix} \in SL(2,\mathcal{C})$$

then direct calculation shows that $R(\theta)\mathbf{e}_1 \wedge R(\theta)\mathbf{e}_2 = \mathbf{e}_1 \wedge \mathbf{e}_2$.

It is also important to note that $2\mathbf{e}_1 \wedge \mathbf{e}_2 = \mathbf{e}_1 \otimes \mathbf{e}_2 - \mathbf{e}_2 \otimes \mathbf{e}_1$ represents a singlet state and is therefore entangled by definition. In the case of n dimensions if

$$R_{ij}(\theta) = \begin{array}{c} \\ \\ \\ i \\ \\ j \\ \\ n \end{array} \begin{pmatrix} & & i & & j & & n \\ 1 & \cdots & 0 & \cdots & 0 & \cdots & 0 \\ \vdots & \ddots & \vdots & \ddots & \vdots & \ddots & \vdots \\ 0 & \cdots & \cos(c\theta) & \cdots & \sin(c\theta) & \cdots & 0 \\ \vdots & \ddots & \vdots & \ddots & \vdots & \ddots & \vdots \\ 0 & \cdots & -\sin(c\theta) & \cdots & \cos(c\theta) & \cdots & 0 \\ \vdots & \ddots & \vdots & \ddots & \vdots & \ddots & \vdots \\ 0 & \cdots & 0 & \cdots & 0 & \cdots & 1 \end{pmatrix} \in SL(n,\mathcal{C}) \tag{1}$$

then

$$\mathbf{e}_1 \wedge \ldots R_{ij}(\theta)\mathbf{e}_i \wedge \ldots \wedge R_{ij}(\theta)\mathbf{e}_j \wedge \mathbf{e}_n = \mathbf{e}_1 \wedge \ldots \mathbf{e}_i \wedge \ldots \wedge \mathbf{e}_j \wedge \mathbf{e}_n \ .$$

This captures the pairwise rotational invariance associated with the ij singlet state represented by the wedge product. Moreover, since $\mathbf{e}_1 \wedge \mathbf{e}_2 \ldots \wedge \mathbf{e}_n$ is associative this n-fold state can be interpreted as being built from pairwise entangled states.

There is a second theorem closely related to the first. It is a uniqueness theorem affirming that only Fermi-Dirac states are invariant under the action of $SL(n,\mathcal{C})$. Its proof requires the following lemma:

Lemma 1 *Let $V = V_1 \otimes \ldots \otimes V_n$ as in Theorem 1. Let $\{e_1, \ldots, e_n\}$ be an orthonormal basis of V_i and R_{ij} be as in equation (1). If $R_{ij}u = u$ where*

$$u = \sum_{\sigma(1)\cdots\sigma(n)\in S_n} c_{\sigma(1)\cdots\sigma(n)} e_{\sigma(1)} \otimes \cdots \otimes e_{\sigma(n)} \ ,$$

S_n is the permutation group and $c_{\sigma(1)\cdots\sigma(n)}$ are constants then

$$c_{\sigma(1)\cdots i\cdots j\cdots\sigma(n)} = -c_{\sigma(1)\cdots j\cdots i\cdots\sigma(n)}$$

Proof: If $u = 0$ then the lemma follows trivially, Assume $u \neq 0$. Note that in the case $n = 2$

$$\begin{aligned} & R_{12}(c_{12}\mathbf{e_1} \otimes \mathbf{e_2} + c_{21}\mathbf{e_2} \otimes \mathbf{e_1}) \\ = \; & (c_{12} + c_{21}) \cos\theta \sin\theta \mathbf{e_1} \otimes \mathbf{e_1} + (c_{12}\cos^2\theta - c_{21}\sin^2\theta)\mathbf{e_1} \otimes \mathbf{e_2} \\ & -(c_{12}\sin^2\theta - c_{21}\cos^2\theta)\mathbf{e_2} \otimes \mathbf{e_1} - (c_{12} + c_{21})\cos\theta\sin\theta\mathbf{e_2} \otimes \mathbf{e_2} \\ = \; & c_{12}\mathbf{e_1} \otimes \mathbf{e_2} + c_{21}\mathbf{e_2} \otimes \mathbf{e_1} \qquad \text{by assumption.} \end{aligned}$$

It follows by the linear independence of $\mathbf{e_1}$ and $\mathbf{e_2}$ that $c_{12} = -c_{21}$.

To extend this to the n dimensional case, note that $R_{ij}\mathbf{e_k} = \mathbf{e_k}$ in the case of $k \neq i$ and $k \neq j$. Direct calculation gives

$$\begin{aligned} & (R(ij) \otimes \cdots \otimes R(ij))u \\ = & \sum_{\sigma(1)\cdots i \cdots j \cdots \sigma(n) \in S_n} c_{\sigma(1)\ldots\sigma(n)} Re_{\sigma(1)} \otimes \cdots Re_i \cdots \otimes Re_j \cdots \otimes Re_{\sigma(n)} \\ = & \sum_{\sigma(1)\cdots i \cdots j \cdots \sigma(n) \in S_n} c_{\sigma(1)\ldots\sigma(n)} e_{\sigma(1)} \otimes \cdots Re_i \cdots \otimes Re_j \cdots \otimes e_{\sigma(n)} \\ = & \; u \qquad \text{since } u \text{ is an eigenvector with eigenvalue 1.} \end{aligned}$$

Mathematically, this is equivalent to the $n = 2$ case already worked out above. It follows from linear independence that

$$c_{\sigma(1)\cdots i \cdots j \cdots \sigma(n)} = -c_{\sigma(1)\cdots j \cdots i \cdots \sigma(n)}$$

The result has been proven.

We now state and prove the theorem:

Theorem 2 *Let $V = V_1 \otimes \cdots \otimes V_n$ as in Theorem 1, and $T = T_1 \otimes \cdots \otimes T_n$ where for each i, j, $T_i = T_j$ and $T_i(\theta) \in SL(n, \mathcal{C})$. If for all $T \in \bigotimes^n SL(n, \mathcal{C})$, $Tv = v$, $v \neq 0$ then*

$$v = \kappa(v_1 \wedge v_2 \wedge \ldots \wedge v_n), \;\; \kappa \text{ an arbitrary constant.} \tag{2}$$

This means that if $v \neq 0$ is invariant under the action of $SL(n, \mathcal{C})$ and $\kappa = 1$ then it must be a Fermi-Dirac statistic.

Proof: We need to show that if v is invariant under the action of any operator $T \in \bigotimes^n SL(n, \mathcal{C})$ then v is given as in (2). Indeed, from Theorem 1, we know that the Fermi-Dirac state is invariant under the action of $SL(n, \mathcal{C})$. It remains to show that it is unique upto a multiplicative constant.

In general if $u \in V \otimes \cdots \otimes V$ then

$$u = \sum_{i_1 \ldots i_n \in \aleph_n} c_{i_1 \ldots i_n} e_{i_1} \otimes \cdots \otimes e_{i_n} \tag{3}$$

where the $\aleph_n = \{1, \cdots, n\}$ and $\{e_{i_1} \otimes \cdots \otimes e_{i_n} | i_j \in \aleph_n\}$ forms a basis for the space. Note that there are n^n summed terms in equation (3). It remains to show that if $Tu = u$ for an arbitrary T then $u = v$. This is achieved by showing that the action of suitably chosen elements of $SL(n, \mathcal{C})$ on u impose restrictions on equation (3) until only v remains.

In particular, for the Lie group $\{\exp(\theta J) | tr(J) = 0\} \in SL(n, \mathcal{C})$, if we let $L(\theta) = (L_1(\theta) \otimes \ldots \otimes L_n(\theta))$, where each $L_i(\theta) = \exp(\theta J)$ there exists a complete set of eigenvectors $\{\mathbf{e}_1, \mathbf{e}_2 \ldots \mathbf{e}_n\}$ of L_i, forming a basis for V_i, with eigenvalues $e^{\lambda_1 \theta}, e^{\lambda_2 \theta}, \ldots e^{\lambda_n \theta}$ such that [6]

$$\lambda_1 + \lambda_2 + \ldots + \lambda_n = 0 \tag{4}$$

and $L_i = diag\{\mathbf{e}_1, \mathbf{e}_2 \ldots \mathbf{e}_n\}$. It is clear that for every permutation $\sigma \in S_n$, where S_n is the permutation group, the set of tensor products

$$\{\mathbf{e}_{\sigma(1)} \otimes \mathbf{e}_{\sigma(2)} \otimes \ldots \otimes \mathbf{e}_{\sigma(n)}\}$$

characterize a basis for all independent eigenvectors of $L(\theta) = L_1 \otimes \ldots \otimes L_n$ with eigenvalue 1. Indeed, all other linearly independent eigenvectors of $L(\theta)$ can be expressed in the form

$$\mathbf{e}_{\sigma(1)} \cdots \otimes \mathbf{e}_i \otimes \ldots \mathbf{e}_i \ldots \otimes \mathbf{e}_{\sigma(n)}, \qquad \mathbf{e}_i \neq \mathbf{e}_j$$

IOP Conf. Series: Journal of Physics: Conf. Series **845** (2017) 012030 doi:10.1088/1742-6596/845/1/012030

with

$$\lambda_1 + \ldots \lambda_i \ldots \lambda_i \ldots + \lambda_n \neq \lambda_1 + \ldots \lambda_i \ldots \lambda_j \ldots + \lambda_n = 0$$

unless $\lambda_i = \lambda_j$. However we have chosen L such that each λ_i is a distinct n-th root of unity and therefore $\lambda_i \neq \lambda_j$. It follows that equation (3) under the action of $L(\theta)$ reduces to

$$u = \sum_{\sigma(1)\ldots\sigma(n)\in S_n} c_{\sigma(1)\ldots\sigma(n)} e_{\sigma(1)} \otimes \cdots \otimes e_{\sigma(n)} \tag{5}$$

To conclude the proof, we turn to the lemma. Let $T = R_{ij}$ be as above. Note $R(ij) \in SL(n, \mathcal{C})$. Invoking the lemma now requires that

$$c_{\sigma(1)\ldots\sigma(i)\ldots\sigma(j)\ldots\sigma(n)} = -c_{\sigma(1)\ldots\sigma(j)\ldots\sigma(i)\ldots\sigma(n)}$$

for every $i \neq j$. This gives $u = v$. The theorem has been proven.

The above theorem applies to any n-dimensional vector space with an n-fold tensor product defined on it. We now extend this to include an n-fold vector space with an m-fold tensor product $(n \geq m)$

Corollary 1 *Let $V = V_1 \otimes \cdots \otimes V_m$, where each V_i is an n-dimensional vector space $(m \leq n)$, and $W_i \subset V_i$ an m-dimensional subspace. Let $T = T_1 \otimes \cdots \otimes T_m$ where for each i, j, $T_i = T_j$ and T_i is a linear operator on V_i leaving W_i invariant, i.e. $T_i = T_{iW} \oplus T_{i(V-W)}$, with the understanding that T_{iW} is the operator T restricted to the subspace W. If*

$$v \equiv v_1 \wedge v_2 \wedge \ldots \wedge v_m \in W \text{ and } v \neq 0$$

then

$$T_W v = v \iff T_W \in \bigotimes_1^m SL(n, \mathcal{C}).$$

In other words, Fermi-Dirac statistics restricted to a subspace is invariant under the action of $SL(n, \mathcal{C})$ restricted to the same subspace.

Proof: $T = T_W \oplus T_{V-W}$ The proof then follows by applying Theorem 1 to T_W and noting that T_W is restricted to W.

3. Bose-Einstein statistics

Based on the above, an alternative definition of a Fermi-Dirac statistics can be given:

Definition 1 *In a tensor product space of the form $V_1 \otimes \ldots \otimes V_n$, where each $V_i \cong V_j$, a Fermi-Dirac statistic is a state which is invariant under the action of the group $SL(n, \mathcal{C})$.*

Theorem 2 affirms that once a normalization is chosen such a state is unique. Moreover, in order to generate non Fermi-Dirac statistics, it is sufficient to relax the conditions specified by the definition. Specifically, in keeping with the usual definition we can define Bose-Einstein statistics as follows:

Definition 2 *In a tensor product space of the form $V_1 \otimes \ldots \otimes V_n$, where each $V_i \cong V_j$, a Bose-Einstein statistic is a state which is invariant under the action of the permutation group S_n.*

IOP Conf. Series: Journal of Physics: Conf. Series **845** (2017) 012030 doi:10.1088/1742-6596/845/1/012030

It is important to note that both the Fermi-Dirac and Bose-Einstein states are invariant under the action of the set of even permutations $A_n \subset S_n$. However, in the case of the Fermi-Dirac statistic the invariance under the action of A_n is not per se sufficient to have such states. We also require the invariance under the action of $SL(n, \mathcal{C})$, which as we have already previously noted is connected with the presence of spin singlet states. This means that from the perspective of physics, Fermi-Dirac statististics can be understood as the statistics of n-indistinguisable particles forming spin singlet states, while Bose-Einstein statistics can be understood as the statistics of n-indistinguisable particles where the spin singlet state dependency has been broken. In the case of Bose-Einstein statistics the spin states of indistinguisable particles are independent of each other. We now express this observation in the following lemma and corollary. It is also worth pointing out that in the case of a statistic which is neither Fermi-Dirac or Bose-Einstein, the invariance under A_n is violated. An example of this is also given below.

Lemma 2 *Let $\sigma \in S_n$ be a permutaion of (1,...,n), with the identity permutation denoted by id. If*

$$\mathbf{v} = \sum_{\sigma} c_\sigma (v_{\sigma(1)} \ldots \otimes v_{\sigma(n)}) \qquad \textit{such that } c_{id} = \frac{1}{n!} \text{ and } c_\sigma = \pm\frac{1}{n!} \textit{ otherwise}$$

is defined on the space $V_1 \otimes \ldots \otimes V_n$, and is invariant under the action of A_n then $\mathbf{v}$ obeys either the Fermi-Dirac or the Bose-Einstein statistic.

Proof: Let

$$\begin{aligned} \mathbf{v}_0 &= \sum_{\sigma \in A_n} \sigma(c_{id}(v_1 \otimes v_2 \ldots \otimes v_n)) \\ \mathbf{v}_1 &= \sum_{\sigma \in A_n} \sigma(c_{id}(v_2 \otimes v_1 \ldots \otimes v_n)) \end{aligned}$$

where $\sigma(c_{id}(v_1 \otimes \ldots \otimes v_n) \equiv c_{id}(v_{\sigma_i(1)} \otimes \ldots \otimes v_{\sigma_i(n)})$. This means that $\mathbf{v}_0$ and $\mathbf{v}_1$ are invariant by construction under the action of A_n, since they are respectively the sum of all even and odd permutations of $v_1 \otimes \ldots \otimes v_n$. Therefore, the invariance of $\mathbf{v}$ with respect to A_n requires that

$$\mathbf{v} - (\mathbf{v}_0 \pm \mathbf{v}_1)$$

is also invariant. By using linear independence we find that this can only occur if

$$\mathbf{v} = (\mathbf{v}_0 + \mathbf{v}_1) \qquad \text{or} \qquad \mathbf{v} = (\mathbf{v}_0 - \mathbf{v}_1)$$

which define the Bose-Einstein and Fermi-Dirac statistics respectively. The result follows.

Corollary 2 *Let $\mathbf{v}$ be as above such that no two particles are in a singlet state then $\mathbf{v} = (\mathbf{v}_0 + \mathbf{v}_1)$, which means this system of particles obeys the Bose-Einstein statistics.*

Proof: Since $\mathbf{v}$ is invaraint under the action of A_n then it must be either a Fermi-Dirac or Bose-Einstein statistic. However, there are no singlets, and so it cannot be invariant under $SL(2, C) \subset SL(n, C)$. Therefore, it cannot be a Fermi-Dirac statistic by definition. Therefore, it obeys Bose-Einstein statistics.

Inherent in this lemma and its corollary is the fact that Fermi-Dirac statistics requires not only indistinguishability but also that the particles form singlet states. In other words, Fermi-Dirac statistics presupposes that particles are entangled and consequently dependent on each other while Bose-Einstein statistics is a consequence of breaking the entanglement. Within the context of atoms or molecules this entanglement can be associated with the electron orbitals.

It might be instructive to apply the above theorem to a three particle wave function that is *not* of the above type. Consider:

$$\mathbf{v} = v_1 \otimes (v_2 \otimes v_3 + v_3 \otimes v_2) + v_2 \otimes (v_3 \otimes v_1 + v_1 \otimes v_3) + v_3 \otimes (v_1 \otimes v_2 - v_2 \otimes v_1)$$

On putting $v_1 = v_2$,

$$\mathbf{v} = v_1 \otimes (v_2 \otimes v_3 + v_3 \otimes v_2) + v_2 \otimes (v_3 \otimes v_1 + v_1 \otimes v_3)$$

which is not invariant under A_3 and a fortiori $SL(3, \mathcal{C})$. It is also not invariant under S_3.

The above theorems and lemma also implicitely explain how to construct various types of parastatistics. For example the five electrons in the boron atom obey the Fermi-Dirac statistics associated with $SL(5, \mathcal{C})$ invariance. On the other hand, if we consider the two electrons of the helium atom together with the three electrons of the lithium atom then theses electrons obey $SL(2, \mathcal{C}) \otimes SL(3, \mathcal{C})$ statistics. The process of assigning electrons to different atoms partially distinguishes them.

4. Relationship to Special Relativity and QFT

The above theorems and corollary suggest a general criteria for classifying Fermi-Dirac statistics and Bose-Einstein statistics. In order to complete the transition, we first need to establish some algebraic connections between the tensor formalism and matrix representations. Indeed, if we impose some further structure on the tensor products, we can relate the above theorem to relativity and quantum field theory.

4.1. A Clifford Algebra Approach

Given the relationship between the Pauli spin matrices and a Clifford Algebra, we begin with a 2-component spinor of the group $SL(2, \mathcal{C})$ such that $\phi' = S(l)\phi$, where $S(l) = e^{\omega_{ab}\sigma_{ab}} \in SL(2, \mathcal{C})$, $\sigma_a = (1, \vec{\sigma})$ forms a basis for the Clifford algebra and the vector $x_a = \phi^\dagger \sigma_a \phi$ is a Lorentz 4-vector of a massless particle [9]. Moreover, any vector in this space can be expressed as $X = x^a \sigma_a$. It should be noted that the restriction $a, b \in \{1, 2, 3\}$ means $\{\sigma_a, \sigma_b\} = 0$ and $[\sigma_a, \sigma_b] = 2i\sigma_c$. Oftentimes, physicists prefer to work with the matrices $\mathbf{S}_i = (\hbar/2)\sigma_i$, which are called the Pauli spin matrices. However, for this paper it is more convenient to work with σ_i, and we will call these the Pauli matrices.

In general if $S(l) \in SL(2, \mathcal{C})$ with adjoint $S^\dagger$ and X is a hermitian (non-singular) 2×2 matrix then

$$X' = SXS^\dagger \tag{6}$$

is a transformation mapping the vector X into the vector X' [6] and $det(X') = det(X)$. We now prove the following lemma.

Lemma 3 *Let X and X^* represent a (hermitian) 4 vector and its conjugate defined respectively by*

$$X = \begin{pmatrix} x_0 - x_1 & x_2 + ix_3 \\ x_2 - ix_3 & x_0 + x_1 \end{pmatrix}, \; X^* = \begin{pmatrix} x_0 + x_1 & -x_2 - ix_3 \\ -x_2 + ix_3 & x_0 - x_1 \end{pmatrix}$$

If $S(l) \in SL(2, \mathcal{C})$ and $T = (S^\dagger)^{-1}$ then S and T preserve conjugacy. In other words,

$$X' = SXS^\dagger = \begin{pmatrix} x'_0 - x'_1 & x'_2 + ix'_3 \\ x'_2 - ix'_3 & x'_0 + x'_1 \end{pmatrix}$$

IOP Conf. Series: Journal of Physics: Conf. Series **845** (2017) 012030 doi:10.1088/1742-6596/845/1/012030

implies

$$X'^* = \begin{pmatrix} x'_0 + x'_1 & -x'_2 - ix'_3 \\ -x'_2 + ix'_3 & x'_0 - x'_1 \end{pmatrix} = TX^*T^\dagger$$

Proof: A simple calculation shows $X^* = X^{-1}det(X)$. Therefore

$$X'^* = det(X')(X')^{-1} = det(X')(SXS^\dagger)^{-1} = det(X)(S^\dagger)^{-1}X^{-1}S^{-1} = TX^*T^\dagger$$

The lemma is proven.
Remark:

- It immediately follows from the lemma that

$$X'X'^* = XX^* = x_0^2 - x_1^2 - x_2^2 - x_3^2$$

  is Lorentz invariant.
- In general $X'^* \neq X^{*\prime}$. As a counter example consider

$$X^{*\prime} = SX^*S^\dagger \text{ where } S = \begin{pmatrix} e^\omega & 0 \\ 0 & e^{-\omega} \end{pmatrix}$$

This means that while $X^{*\prime}$ is a vector it is **not** necessarily the conjugate vector of X'. Indeed, in order for $X'^* = X^{*\prime}$, S needs to be an element of $SU(2)$ which is a subgroup of $SL(2, \mathcal{C})$. We express this as a corollary.

Corollary 3 *If $S \in SU(n)$ then $T = S$*

Proof: By definition of $SU(n)$, $S^\dagger = S^{-1}$ and therefore $T = (S^\dagger)^{-1} = S$. The result follows.

4.2. Conjugate Solutions and Majorana Fields
There is a remarkable connection between the conjugate states X and X^* and the solutions to the Majorana equations. Recall that for a free particle the Majorana equations are given by

$$\mathcal{D}\eta = \sigma_\mu \partial^\mu \eta = -m\chi \qquad \text{and} \qquad \mathcal{D}^*\chi = \sigma_\mu^* \partial^\mu \chi = -m\eta \tag{7}$$

where

$$\mathcal{D} = \begin{pmatrix} \partial_0 - \partial_1 & \partial_2 + i\partial_3 \\ \partial_2 - i\partial_3 & \partial_0 + \partial_1 \end{pmatrix}, \ \mathcal{D}^* = \begin{pmatrix} \partial_0 + \partial_1 & -\partial_2 - i\partial_3 \\ -\partial_2 + i\partial_3 & \partial_0 - \partial_1 \end{pmatrix}$$

Clearly $\frac{1}{2}\{\mathcal{D}, \mathcal{D}^*\}_M = \mathcal{D}\mathcal{D}^* = \partial_0^2 - \nabla^2$. Moreover, if we define $\chi' = S(\Lambda)\chi$ and $\eta' = T(\Lambda)\eta = (S^\dagger)^{-1}(\Lambda)\eta$ then equations (7) are covariant under $SL(2, \mathcal{C})$ which means

$$\sigma^\mu \partial'_\mu \eta' = -m\chi' \qquad \text{and} \qquad \sigma_\mu^* \partial'^\mu \chi' = -m\eta' \tag{8}$$

This is equivalent to the Lorentz invariance of the Klein-Gordan equation

$$(\partial_0^2 - \nabla^2)\psi = m^2\psi$$

which can be factored into (7).

Note also that $[\mathcal{D}, \mathcal{D}^*] = 0$. We refer to the pair (X, X^*) as Majorana fields.

IOP Conf. Series: Journal of Physics: Conf. Series **845** (2017) 012030 doi:10.1088/1742-6596/845/1/012030

4.3. Pauli Products

Noting that

$$X'X'^* = XX^* = x_0^2 - x_1^2 - x_2^2 - x_3^2$$

and that for any (pseudo) inner product $\langle X+Y, X+Y\rangle = (X+Y)(X^*+Y^*)$ is a scalar, it follows from the bi-linearity and symmetry that

$$\langle X+Y, X+Y\rangle = \langle X,X\rangle + 2\langle X,Y\rangle + \langle Y,Y\rangle$$

and therefore

$$2\langle X,Y\rangle = XY^* + YX^*$$

Equivalently, we can define a Pauli inner product by[1]

$$\begin{aligned}
\langle X,Y\rangle_P &\equiv \frac{1}{4}(\{X,Y^*\}+\{Y,X^*\}) && (9)\\
&= \frac{1}{4}(XY^*+Y^*X+YX^*+X^*Y) && (10)\\
&= x_0y_0 - x_1y_1 - x_2y_2 - x_3y_3 = \langle x,y\rangle && (11)
\end{aligned}$$

which is clearly Lorentz invariant. Indeed, it is invariant under $SL(2,\mathcal{C})$ (and not just $SU(2,\mathcal{C})$), provided we agree that conjugacy is preserved according to the rule established in the previous lemma. In other words if $T = (S^\dagger)^{-1}$,

$$\begin{aligned}
\{X',Y'\}_P &\equiv \frac{1}{4}(SXS^\dagger TY^*T^\dagger + TY^*T^\dagger SXS^\dagger)\\
&\quad + \frac{1}{4}(SYS^\dagger TX^*T^\dagger + TX^*T^\dagger SYS^\dagger)\\
&= \frac{1}{4}(SXY^*T^\dagger + TY^*XS^\dagger) + \frac{1}{4}(TX^*YS^\dagger + SYX^*T^\dagger)\\
&= \frac{1}{4}(S(XY^*+YX^*)T^\dagger + T(X^*Y+Y^*X)S^\dagger)\\
&= \frac{1}{2}S(x_0y_0 - x_1y_1 - x_2y_2 - x_3y_3)T^\dagger\\
&\quad + \frac{1}{2}T(x_0y_0 - x_1y_1 - x_2y_2 - x_3y_3)S^\dagger\\
&= \{X,Y\}_P
\end{aligned}$$

Therefore the Pauli inner product preserves conjugacy.

By making a slight modification[2] of equation (20) of article ([10]), we can also define a Pauli outer product by

$$[X,Y]_P \equiv \frac{1}{2}([X,Y^*]+[X^*,Y]) = \frac{1}{2}(XY^* - Y^*X + X^*Y - YX^*) \qquad (12)$$

This reduces with a little algebra to

$$[X,Y]_P = -[\tilde{X},\tilde{Y}] \qquad (13)$$

[1] Apart from a notation change, this expression is identical with equation (20) of Pauli's original paper [10], applied to what Pauli refers to as the Jordan and Wigner bracket.

[2] Pauli's equation for commutators is symmetric and contains the expression $[X,Y^*]+[Y,X^*]$. He failed to note that it is always 0 for all X and Y regardless of the statistics. In contrast, we write $[X,Y^*]+[X^*,Y]$ which is anti-symmetric.

where $\tilde{X} = x_1\sigma_1 + x_2\sigma_2 + x_3\sigma_3$ is a vector in three dimensional space. Note that this also means $[X, X^*] = [X, X] = 0$.

In particular, in the case of a singlet state, $\tilde{X}$ and $\tilde{Y}$, $[\tilde{X}, \tilde{Y}] \neq 0$. It also follows from equations (9) - (11) that in the case of singlet states

$$\{\tilde{X}, \tilde{Y}\}_P = 0 \ ,$$

while in the case of non-singlet states

$$\{\tilde{X}, \tilde{Y}\}_P < 0 \ .$$

In other words, two particles cannot simultaneously obey Fermi-Dirac and Bose-Einstein statistics.

4.4. Restricted tensor products

In Theorem 1, we let $V = V_1 \otimes \ldots \otimes V_n$, where each V_i is an n-dimensional vector space, which means we chose the dimension of each vector to be the same value as the number of tensor products of the space itself. Specifically in the case of a two dimensional Euclidean space, the wedge product of two vectors $\mathbf{x} = (x_1, x_2)$ and $\mathbf{y} = (y_1, y_2)$ generates a singlet state:

$$2\mathbf{x} \wedge \mathbf{y} = \mathbf{x} \otimes \mathbf{y} - \mathbf{y} \otimes \mathbf{x} = (x_1 y_2 - y_1 x_2)\mathbf{e}_1 \wedge \mathbf{e}_2$$

where $\mathbf{e}_1$ and $\mathbf{e}_2$ represent respectively the unit vectors $(1, 0)$ and $(0, 1)$. On the other hand, if we switch to a spinor formulation and use the Pauli spin matrices as a basis then we can identify

$$\mathbf{x} \leftrightarrow X = \begin{pmatrix} x_1 & x_2 \\ x_2 & -x_1 \end{pmatrix} \qquad \text{and} \qquad \mathbf{y} \leftrightarrow Y = \begin{pmatrix} y_1 & y_2 \\ y_2 & -y_1 \end{pmatrix}$$

A direct calculation gives $[X, Y] = 2(x_1 y_2 - y_1 x_2)\sigma_1\sigma_2$. It is worth noting that in the case of two dimensional Majorana fields $[X, X^*] = 0$. In other words, Majorana fields can never form a singlet state and consequently never obey (by Theorem 1) Femi-Dirac statistics.

To complete the theory, we need to extend our results to Minkowski space. With this in mind, let $x,\ y \in \mathcal{R}_3^1$ and consider the tensor product $x \otimes y = x^i y^j e_i \otimes e_j$ and a linear map $\phi(x \otimes y) = x^i y^j \sigma_i \sigma_j$. This is equaivalent to identifying for $i \neq j$

$$e_0 \otimes e_j = e_j \otimes e_0 \ and \ e_i \otimes e_j = -e_j \otimes e_i$$

In keeping with the Pauli outer product defined above, we define the conjugate wedge product by

$$\mathbf{x} \wedge_c \mathbf{y} \equiv \frac{1}{2}(\mathbf{x} \wedge \mathbf{y}^* + \mathbf{x}^* \wedge \mathbf{y}) \ .$$

Now let $\mathbf{x} = x^j e_k \equiv x_0 + \tilde{x}$, where $\tilde{x} = x_1 e_1 + x_2 e_2 + x_3 e_3$ then this reduces to the bi-vector

$$\mathbf{x} \wedge_C \mathbf{y} = -\tilde{x} \wedge \tilde{y} \qquad (14)$$

Clearly we can identify $4(\mathbf{x} \wedge_C \mathbf{y}) \leftrightarrow [X, Y]_P$

Moreover, from basic geometry, we can see that any linear transformation with an eigenvector $\tilde{z}$ that is orthogonal to the plane spanned by $\tilde{x}$ and $\tilde{y}$ will be such that $\tilde{x} \wedge \tilde{y}$ remains invariant in accordance with Corollary 1. On the other hand those rotations that shift the eigenvector will not remain invariant. However, the norm of the bivector will remain invariant under $SL(2, \mathcal{C})$.

IOP Conf. Series: Journal of Physics: Conf. Series **845** (2017) 012030 doi:10.1088/1742-6596/845/1/012030

5. Conclusion

Based on the above, it should be clear that a necessary and sufficient condition for the Pauli exclusion principle to be valid is the requirement that the quantum state of a system of n particles be invariant under the action of the $SL(n, \mathcal{C})$ group. As we have already noted this requires the existence of spin singlet states, which means that spin entanglement is a necessary requirement to exhibit Fermi-Dirac statistics. Indeed, Fermi-Dirac statistics can be defined as the statistics of n indistinguishable singlet states [7].

References

[1] *Encyclopedia Brittanica: The Pauli Exclusion Principle*

[2] Berry, M.V and J.M.Robbins, *Indistinguishability for quantum particles: spin, statistics and the geometric phase*, Proc. R. Soc,. London Ser. A 453, 1771-1790(1997).

[3] Duck,I and E.C.G. Sudarshan, *Pauli and the Spin-Statistics Theorem*, World Scientific, Singapore (1997), p.485.

[4] Feynman, R., *Lectures of Physics*, Addison-Wesley, Massachusetts (1963), 3(Chap. 4).

[5] Fierz,M. Helv. Phys. Acta 12, 3(1939).

[6] Fässler, A. and E. Stiefler, *Group Theoretical Methods and Their Applications*,Birkhauser, Chaps. 4.2, 7.1(1992)

[7] O'Hara,P., *Rotational Invariance and the Spin-Statistics Theorem*, Foun. Phys., 33(9), 1360-1362(2003).

[8] O'Hara, P.,Bell's Inequality, the Pauli Exclusion Principle and Baryonic Structure, *Spin 96 Proceedings*, World Scientific (1997). arXiv.hep-th/9701089.

[9] O'Raifeartaigh, L., *The Dawning of Gauge Theory*, Princeton Press, Princeton(1997),p 109.

[10] Pauli,W., *The Connection Between Spin and Statistics* Phy. Rev., 716-722(1940).

[11] Romer,R., *The Spin Sytatistics Theorem*, Am.J. Phys 70(8), 791(2002).

[12] Streater,R.F., *Spin and Statistics*, scholarpedia.org/article/Spin_and_Statistics .

The transformations of spinors[1]

Marco Budinich
University of Trieste and INFN, Trieste, Italy
E-mail: `mbh@ts.infn.it`

Abstract. We begin showing that for even dimensional vector spaces V all automorphisms of their Clifford algebras are inner. So all orthogonal transformations of V are restrictions to V of inner automorphisms of the algebra. Thus under orthogonal transformations P and T — space and time reversal — all algebra elements, including vectors v and spinors φ, transform as $v \to xvx^{-1}$ and $\varphi \to x\varphi x^{-1}$ for some algebra element x. We show that while under combined PT spinor $\varphi \to x\varphi x^{-1}$ remain in its spinor space, under P or T separately φ goes to a *different* spinor space and may have opposite chirality. We conclude with a preliminary characterization of inner automorphisms with respect to their property to change, or not, spinor spaces.

1. Introduction

Élie Cartan introduced spinors in 1913 [7, 8] and, after more than a century, this field still yields good harvests. Spinors were later thoroughly investigated by Claude Chevalley [10] in the mathematical frame of Clifford algebras where they were identified as elements of Minimal Left Ideals (MLI) of the algebra. Many years later Benn and Tucker [1] and Porteous [14] wrote books with many of these results easier to be assimilated by physicists.

In this paper we address the transformation properties of spinors under certain inner automorphisms of Clifford algebra exploiting the Extended Fock Basis (EFB) of Clifford algebra [2, 3, 4]. With this formalism one can write vectors as linear superpositions of simple spinors, thus supporting the well-known Penrose twistor program [13] that spinor structure is the underlying — more fundamental — structure of Minkowski spacetime.

In section 2 we review some quite general properties of a simple Clifford algebra and in particular the fact that it contains many different MLI, namely many different spinor spaces, that are completely equivalent in the sense that each of them can be the carrier of a representation; moreover the algebra, as a vector space, is the direct sum of these spinor spaces. These properties are known and recently Pavsic suggested that multiple spinor spaces play a role in physics [12].

One of the most important properties of Clifford algebra is that it establishes a deep connection between the orthogonal transformations of vector space V with scalar product g (more precisely: its image in the algebra) and the automorphisms of Clifford algebra $\mathcal{C}\ell(g)$. In section 3 we show first that if the vector space is even dimensional then all $\mathcal{C}\ell(g)$ automorphisms are *inner* automorphisms and thus that all orthogonal transformations on V lift to inner automorphisms of $\mathcal{C}\ell(g)$. We then examine in detail the so called discrete orthogonal transformations of V, namely $\mathbb{1}_V, P, T$ and PT (V identity, space and time reversal

[1] an older version of this paper [4] had been submitted before IARD 2016 and was accepted after the conference. This version is newer with some parts removed and with some new results.

IOP Conf. Series: Journal of Physics: Conf. Series **845** (2017) 012031 doi:10.1088/1742-6596/845/1/012031

and their composition) and we focus on the inner algebra automorphism they induce. This study takes advantage of the properties of the EFB that allow to remain within the algebra without using representations. At the same time we exhibit the elements of the algebra that generate these inner automorphisms. It follows that we can look at $\mathbb{1}_V, P, T$ and PT as at restrictions of automorphisms of the entire algebra to V, thus unifying the treatment of the discrete transformations of V with those of the continuous ones of the $\mathrm{Pin}\,(g)$ group.

The same approach was taken also by Varlamov [16, 17] to study the hierarchies of $\mathrm{Pin}\,(g)$ and $\mathrm{O}\,(g)$ groups and he succesfully classified the automorphisms of $\mathcal{C}\ell(g)$ showing that the eight double coverings of $\mathrm{O}\,(g)$, the Dabrowski groups [11], correspond to the eight types of real Clifford algebras: the so called "spinorial clock" [5].

Here we exploit the same unification to investigate a different subject: given an inner automorphism

$$\alpha : \mathcal{C}\ell(g) \to \mathcal{C}\ell(g); \;\; \alpha(\mu) = x\mu x^{-1} \qquad x \in \mathcal{C}\ell(g)$$

it is manifest that all algebra elements must transform accordingly and in particular that the typical physics equations $v\varphi = 0$, where $v \in V$ and φ is a spinor, must go to $\alpha(v\varphi) = 0$. We remark that φ is *both* a carrier of the regular representation *and* an element of $\mathcal{C}\ell(g)$ so the equation $\alpha(v\varphi) = 0$ is justified. Since the automorphism is inner it follows that both v and spinor φ must transform as $\alpha(\varphi) = x\varphi x^{-1}$ thus adding an "extra" x^{-1} to the "traditional rule" stating that vectors transform as $v \to xvx^{-1}$ while spinors as $\varphi \to x\varphi$. This consequence is unavoidable if we accept that spinors are part of the Clifford algebra and not elements of some "external" linear space, a point of view that, even if historical, is rarely taken nowadays.

We examine in detail the spinors transformations $\alpha(\varphi) = x\varphi x^{-1}$ proving that if on one side they can not alter in any way the solutions of $v\varphi = 0$, on the other hand, in some cases, they "move" φ to a different spinor space, one of the many equivalent ones in $\mathcal{C}\ell(g)$. In particular we show that while the automorphisms corresponding to $\mathbb{1}_V$ and PT do not move spinors, those corresponding to P and T move them, thus populating other spinor spaces.

In section 4 we begin the characterization of these automorphisms: those that keep the spinor space constant, like $\mathbb{1}_V$ and PT and those that do not, like P and T, and we show that the latter transformations can also invert spinor chiralities. This is the first part of the study of these transformations that will be completed examining also continuous transformations.

2. Multiple spinor spaces

We consider Clifford algebras $\mathcal{C}\ell(g)$ [10, 14, 5] over the fields $\mathbb{F} = \mathbb{R}$ or $\mathbb{C}$ and an *even* dimensional vector space V equipped with a non degenerate scalar product g; any base $e_1, e_2, \ldots, e_n$ with $n = 2m$ generates the algebra that results: simple, central and of dimension 2^n.

EFB formalism is fully developed for *neutral* spaces: $V = \mathbb{C}^{2m}$ or $\mathbb{R}^{m,m}$, spaces for which Witt decomposition is the direct sum of two totally null (isotropic) subspaces of dimension m; when we refer to this case we indicate the corresponding Clifford algebra $\mathcal{C}\ell(m,m)$. This choice allows to treat a simpler case, avoiding the many intricacies brought in by other signatures (the extension of the formalism to other cases is under development). At the same time this restriction is much milder than it may seem since the following results apply also to the complexification of the Clifford algebras of even dimensional real spaces of *any* signature.

In neutral spaces the e_i's form an orthonormal basis of V with e.g.

$$2g(e_i, e_j) = e_i e_j + e_j e_i := \{e_i, e_j\} := g_{ij} = 2\delta_{ij}(-1)^{i+1}$$

while $\{e^i, e_j\} = 2\delta^i_j$ and

$$\begin{cases} e_{2i-1}^2 & = & 1 \\ e_{2i}^2 & = & -1 \end{cases} \qquad i = 1, \ldots, m \; . \tag{1}$$

The Witt, or null, basis of the vector space V is defined, for both fields:

$$\begin{cases} p_i &= \frac{1}{2}(e_{2i-1}+e_{2i}) \\ q_i &= \frac{1}{2}(e_{2i-1}-e_{2i}) \end{cases} \Rightarrow \begin{cases} e_{2i-1} &= p_i+q_i \\ e_{2i} &= p_i-q_i \end{cases} \qquad i=1,2,\ldots,m \ . \tag{2}$$

The EFB of $\mathcal{C}\ell(m,m)$ is given by the 2^{2m} different sequences

$$\psi_1\psi_2\cdots\psi_m := \Psi \qquad \psi_i \in \{q_ip_i, p_iq_i, p_i, q_i\} \qquad i=1,\ldots,m \tag{3}$$

in which each ψ_i can take four different values and we reserve Ψ for EFB elements and ψ_i for its components. The main characteristics of EFB is that all its 2^{2m} elements Ψ are simple spinors [6]. Each element of this basis of the algebra is completely determined by two binary signatures, h and $h\circ g$ signatures, a and b, [4], that represent also row and column indexes of the matrix algebra; we will thus indicate one element of this basis with Ψ_{ab}.

$\mathcal{C}\ell(m,m)$, as a vector space, is the direct sum of subspaces of different $h\circ g$ signatures [3]. Given the Clifford product properties in EFB formalism [4] these subspaces are also MLI of $\mathcal{C}\ell(m,m)$ and thus coincide with 2^m different spinor spaces S_{hg} that in turn correspond to different columns of the isomorphic matrix algebra $\mathbb{F}(2^m)$. To establish a further link between EFB and the familiar definition of MLI in $\mathcal{C}\ell(m,m)$ [1] we remark that in EFB the 2^m elements with identical h and $h\circ g$ signatures [4], namely Ψ_{aa}, are primitive idempotents and the MLI S_a can thus be written as $S_a = \mathcal{C}\ell(m,m)\Psi_{aa}$.

Each of the 2^m spinor spaces supports a regular, faithful and irreducible representation of $\mathcal{C}\ell(m,m)$ and since the algebra is simple there exist isomorphisms intertwining the representations. This has been known for a long time but recently mirror particles [12] have been proposed as a possible realization of multiple spinor spaces. Here we show how, under certain transformations, e.g. P and T, a spinor moves to a different spinor space.

We choose a particular spinor space, e.g. $h\circ g = f$ so that when speaking of a generic spinor we will refer to spinor space S_f (used to build the Fock basis in [6]). Its generic element $\varphi \in S_f$ can thus be expanded in the Fock basis

$$\varphi = \sum_a \xi_{af}\Psi_{af} \tag{4}$$

and, since the second index f is constant, in principle it could be omitted.

Let this $\varphi \in S_f$ be a solution of the Weyl equation $v\varphi = 0$ where $v \in V$; we remark that the equation is solved also by all $\varphi' = \sum_a \xi_a \Psi_{af'} \in S_{f'}$ for any f'; we will return to this point in paragraph 4.

2.1. Representations of Clifford algebra $\mathcal{C}\ell(g)$

We resume some quite general properties we need in the sequel: let $\gamma : \mathcal{C}\ell(g) \to \text{End}S$ be a faithful irreducible representation of $\mathcal{C}\ell(g)$ and let β be the so called main antiautomorphism

$$\begin{cases} \beta(\mu\nu) &= \nu\mu & \forall\, \mu,\nu \in \mathcal{C}\ell(g) \\ \beta(v) &= v & \forall\, v \in V \\ \beta(\mathbb{1}) &= \mathbb{1} \end{cases} \tag{5}$$

that reverses the order of multiplication and that is involutive. With β it is possible to define the contragredient representation in S^*, the dual of S, $\check{\gamma} : \mathcal{C}\ell(g) \to \text{End}S^*$ given by $\check{\gamma}(\mu) = \gamma(\beta(\mu))^*$ and, since in our case V is even dimensional, $\mathcal{C}\ell(g)$ is simple and central and thus there exists an isomorphism $B : S \to S^*$ intertwining the two representations: $\check{\gamma}B = B\gamma$ which is either

IOP Conf. Series: Journal of Physics: Conf. Series **845** (2017) 012031 doi:10.1088/1742-6596/845/1/012031

symmetric or antisymmetric $B = \pm B^*$ [5, 15, 9] and that also defines on S the structure of an inner product space ($\langle\cdot,\cdot\rangle$ represents the bilinear product or contraction)

$$S \times S \to \mathbb{F} \qquad B(\varphi, \phi) := \langle B\varphi, \phi\rangle \in \mathbb{F} \ .$$

This structure extends to End$S$: there is a symmetric isomorphism $B \otimes B^{-1}$: End$S \to$ (End$S)^*$ = EndS^* given, for every $\gamma \in$ End$S$, by $(B \otimes B^{-1})(\gamma) = B\gamma B^{-1}$. These results are fully general and hold thus also when γ is the regular representation $\gamma(\mu) = \mu$, End$S = \mathcal{C}\ell(g)$ and S is one of its MLI.

3. Automorphisms of Clifford algebra $\mathcal{C}\ell(g)$

We begin with a general proposition and thus in this section there are no restrictions on the dimensions of the vector space V.

Proposition 1. *For a Clifford algebra over fields $\mathbb{R}$ and $\mathbb{C}$ all its automorphisms are inner if and only if the dimension of the vector space is even.*

Proof. For any non degenerate, even dimensional, vector space V its Clifford algebra is central and simple [5] and, by Skolem – Noether theorem, all its automorphisms are inner. To prove the converse we take an odd dimensional vector space and we examine the so called main automorphism of its Clifford algebra, the automorphism that reverses all vectors (7). In this case the volume element $\omega = e_1 e_2 \cdots e_n$ is formed by an odd number of vectors and thus the main automorphism sends $\omega \to -\omega$. But in this case ω belongs also to the center of the algebra and thus for any inner automorphism $x\omega x^{-1} = \omega$, thus the main automorphism is not inner. □

A simple example is $\mathcal{C}\ell_{\mathbb{R}}(0,1) \cong \mathbb{C}$ where the main automorphism coincides with complex conjugation and is not inner. In general the Clifford algebra of an odd dimensional vector space has the form $\mathcal{C}\ell(g) \oplus \mathcal{C}\ell(g)$ and have no inner elements giving the main automorphism, nevertheless the main automorphism is obtained by the "swap" automorphism $\begin{pmatrix} 0 & \mathbb{1} \\ \mathbb{1} & 0 \end{pmatrix}$ that *is not* an element of the algebra but *is* an element of the matrix representations of the algebra.

A corollary that follows from the universality of Clifford algebra is that all orthogonal transformations on an even dimensional V lift to inner automorphisms of $\mathcal{C}\ell(g)$. This corollary gives a simpler proof, but only for even dimensional spaces, of the result quoted in [16] that all "fundamental automorphisms", even discrete ones like P and T, are inner automorphisms.

So in even dimensional spaces

$$\text{Aut}\left(\mathcal{C}\ell(g)\right) = \{x \in \mathcal{C}\ell(g) : \exists x^{-1}\} := C_g^\star \tag{6}$$

and the Clifford Lipschitz group is its subgroup that stabilizes vectors that in turn, when restricted on vector space, is the orthogonal group $\mathrm{O}\,(g)$.

3.1. Fundamental automorphisms of $\mathcal{C}\ell(g)$

In general in $\mathcal{C}\ell(g)$ there are four automorphisms corresponding to the two involutions and to the two antinvolutions induced by the orthogonal transformations $\mathbb{1}_V$ and $-\mathbb{1}_V$ of vector space V [14, Theorem 15.32]. They are called fundamental or discrete automorphisms and form a finite group, isomorphic to the Klein four group $\mathbb{Z}_2 \otimes \mathbb{Z}_2$ [16]. We review them briefly to show how they appear in EFB formalism and to exhibit the elements of $C_g^\star$ realizing the inner automorphisms in even dimensional vector spaces.

From now on we restrict again to even dimensional, neutral, spaces to fully exploit EFB properties.

3.2. Identity automorphism of $\mathcal{C}\ell(g)$

The just quoted theorem proves also that $\mathbb{1}_V$ induces the algebra identity automorphism, its internal element in $\mathcal{C}\ell(m,m)$ is $\mathbb{1}$.

3.3. Main automorphism of $\mathcal{C}\ell(g)$

The main automorphism α of $\mathcal{C}\ell(g)$ (main involution in [14]) is induced by V orthogonal transformation $-\mathbb{1}_V$, namely

$$\alpha(v) = -v \quad \forall\, v \in V \tag{7}$$

it is involutive and defines the basic $\mathbb{Z}_2$ grading of $\mathcal{C}\ell(g)$.

It's easy to see that given the volume element $\omega = e_1 \cdots e_n$ we obtain $\omega e_i = (-1)^{(n-1)} e_i \omega$ and so, for even dimensional spaces, we have $\omega v \omega^{-1} = -v$ for any $v \in V$, where $\omega^{-1} = \omega^3 = \pm\omega$ and thus in this case the main automorphism on the entire algebra may be written as:

$$\alpha(\mu) = \omega \mu \omega^{-1} \ .$$

For the EFB expansion $\mu = \sum_{ab} \xi_{ab} \Psi_{ab}$ we find

$$\alpha(\Psi_{ab}) = \omega \Psi_{ab} \omega^{-1} = \theta_{ab} \Psi_{ab} = \epsilon(a)\epsilon(b)\Psi_{ab} \tag{8}$$

where $\theta_{ab} = \pm 1$ is the global parity of the EFB element Ψ_{ab} that can be expressed, defining the function $\epsilon : \{\pm 1\}^m \to \{\pm 1\}, \epsilon(h) = \prod_{i=1}^m h_i$, from the h and g signatures of EFB, namely a and b [4].

3.4. Reversion automorphism of $\mathcal{C}\ell(g)$

The main antiautomorphism β (5) is the antiautomorphism induced by the orthogonal transformation $\mathbb{1}_V$ of V (reversion antiautomorphism in [14]) and gives an automorphism when "transposed to the dual space". If S_f is a MLI of $\mathcal{C}\ell(g)$, the space of spinors, and γ the regular representation $\gamma(\mu) = \mu$, then $\check{\gamma}(\mu) = (\beta(\mu))^*$ is the contragredient representation that defines also the reversion automorphism; with (5) we get its main property:

$$\check{\gamma}(e_{i_1} \cdots e_{i_k}) = e^*_{i_1} \cdots e^*_{i_k} \ .$$

Since it is an automorphism it must be inner, thus there exists $\tau \in \mathcal{C}\ell(g)$ such that $\check{\gamma}(\mu) = \tau\mu\tau^{-1}$ and τ is fully defined by its action on the generators $\check{\gamma}(e_i) = e^*_i = \tau e_i \tau^{-1}$ and since $\{e^*_i, e_j\} = g^i_j = 2\delta^i_j$ it follows that

$$\tau e_i \tau^{-1} = e^*_i = e_i^{-1} = e_i^3 \ .$$

With this result and remembering (1) and (2) it is simple to get the explicit form of τ that depends on the parity of m

$$\begin{aligned} \tau &= \begin{cases} e_2 e_4 \cdots e_{2m} & \text{for } m \text{ even} \\ e_1 e_3 \cdots e_{2m-1} & \text{for } m \text{ odd} \end{cases} \\ &= (p_1 + s\, q_1)(p_2 + s\, q_2) \cdots (p_m + s\, q_m) \qquad s = (-1)^{m+1} \ . \end{aligned}$$

To evaluate the reversion automorphism on EFB elements we easily get

$$\check{\gamma}(\Psi_{ab}) = \beta(\Psi_{ab})^* = \beta(\psi_1\psi_2 \cdots \psi_m)^* = \beta(\psi_1)^* \beta(\psi_2)^* \cdots \beta(\psi_m)^* \ . \tag{9}$$

By (5) $\beta(p_i) = p_i$, $\beta(q_i) = q_i$ so that $\beta(p_i q_i) = q_i p_i$ and $\beta(q_i p_i) = p_i q_i$. Since $e^*_i = e_i^{-1}$ by (2) we obtain that $(p_i)^* = q_i$, $(p_i q_i)^* = p_i q_i$ and $(q_i p_i)^* = q_i p_i$. We can resume saying that the g_i and

h_i signatures of $\beta(\psi_i)^*$ are respectively equal to g_i and $-h_i$ of that of ψ_i so that the effect of reversion automorphism is to change sign to both h and $h \circ g$ signatures. We can thus conclude that for the reversion automorphism we have

$$\beta(\Psi_{ab})^* = \tau \Psi_{ab} \tau^{-1} = \Psi_{-a-b} \tag{10}$$

and we remark that while Ψ_{ab} belongs to spinor space S_b, $\beta(\Psi_{ab})^*$ belongs to S_{-b}, always a *different* spinor space, the main result of this paper.

For completeness we report the results of similar exercises:

$$\beta(\Psi_{ab}) = s'(a,b)\Psi_{-b-a} \qquad s'(a,b) = \pm 1 \tag{11}$$

where the sign $s'(a,b)$, straightforward, if slightly tedious to calculate, depends on the indices; it is easy to double check that it satisfies the properties of the main antiautomorphism (5). We also obtain

$$\Psi^*_{ab} = s'(a,b)\Psi_{ba} \qquad s'(a,b) = \pm 1 \tag{12}$$

that could also be deduced directly from the natural matrix structure of the EFB since a and b are just the matrix indexes written in binary form; combining both these formulas we reobtain (10). Since both (11) and (12) are involutive we have

$$s'(a,b) = s'(b,a) = s'(-b,-a) = s'(-a,-b) \ .$$

3.5. Conjugation automorphism of $\mathcal{C}\ell(g)$
The composition of the main (8) and reversion automorphisms (10) is called conjugation and results in

$$\alpha\left(\beta(\Psi_{ab})^*\right) = \omega\tau\Psi_{ab}(\omega\tau)^{-1} = \theta_{ab}\Psi_{-a-b} \tag{13}$$

since $\theta_{-a-b} = \theta_{ab}$ given that also this automorphism is involutive; clearly $\alpha\left(\beta(e_i)^*\right) = -e_i^{-1}$.

3.6. A simple example in $\mathcal{C}\ell(1,1)$
We conclude with a simple example in $\mathcal{C}\ell(1,1) \cong \mathbb{R}(2)$ where the EFB is formed by 4 elements: $\{qp_{++}, pq_{--}, p_{-+}, q_{+-}\}$ with the subscripts indicating respectively h and $h \circ g$ signatures; its EFB matrix is

$$\begin{array}{c c} & \begin{array}{cc} + & - \end{array} \\ \begin{array}{c} + \\ - \end{array} & \begin{pmatrix} qp & q \\ p & pq \end{pmatrix} \end{array}$$

and the generic element $\mu \in \mathcal{C}\ell(1,1)$ can be written in the EFB as

$$\mu = \xi_{++}qp_{++} + \xi_{--}pq_{--} + \xi_{-+}p_{-+} + \xi_{+-}q_{+-} \qquad \xi \in \mathbb{F}$$

and the application of the three inner automorphisms gives

$$\begin{array}{rcl} \omega\mu\omega^{-1} & = & \xi_{++}qp_{++} + \xi_{--}pq_{--} - \xi_{-+}p_{-+} - \xi_{+-}q_{+-} \\ \tau\mu\tau^{-1} & = & \xi_{--}qp_{++} + \xi_{++}pq_{--} + \xi_{+-}p_{-+} + \xi_{-+}q_{+-} \\ \omega\tau\mu(\omega\tau)^{-1} & = & \xi_{--}qp_{++} + \xi_{++}pq_{--} - \xi_{+-}p_{-+} - \xi_{-+}q_{+-} \end{array}$$

and $\omega = e_1e_2 = [q,p]$, $\tau = e_1 = p + q$, $\omega\tau = -e_2 = q - p$ and $\mathbb{1} = e_1^2 = \{q,p\}$. For comparison, the same automorphisms applied to the standard e_i formulation gives the ordinary results

$$\begin{array}{rcl} \mu & = & \xi_0\mathbb{1} + \xi_1e_1 + \xi_2e_2 + \xi_{12}e_1e_2 \qquad \xi \in \mathbb{F} \\ \omega\mu\omega^{-1} & = & \xi_0\mathbb{1} - \xi_1e_1 - \xi_2e_2 + \xi_{12}e_1e_2 \\ \tau\mu\tau^{-1} & = & \xi_0\mathbb{1} + \xi_1e_1 - \xi_2e_2 - \xi_{12}e_1e_2 \\ \omega\tau\mu(\omega\tau)^{-1} & = & \xi_0\mathbb{1} - \xi_1e_1 + \xi_2e_2 - \xi_{12}e_1e_2 \ . \end{array}$$

IOP Conf. Series: Journal of Physics: Conf. Series **845** (2017) 012031 doi:10.1088/1742-6596/845/1/012031

4. Spinors transformations

We begin by observing that in the complex case, and also in complex representations of the real case, also complex conjugation is responsible for an automorphism and things get more complicated: the finite group of fundamental automorphism doubles its size and has been examined in detail in [17]. We leave aside for the moment complex conjugation and examine what's going on in our simpler case since it is enough for our purpose of introducing general properties of spinor transformations.

The inner automorphisms of section 3 are fully general and their restrictions to V correspond to the V transformations: $\mathbb{1}_V$, P, T and PT. We already saw that the restriction of the algebra identity to V is $\mathbb{1}_V$ while we identified the main automorphism (7), (8) with PT. If we accept that T changes sign to timelike e_{2i} than reversion (10) and conjugation (13) restricted to V correspond respectively to T and P but other identifications are possible. A word of caution on this point: when we deal with complex representation of real algebras, how it is customary to do for Dirac spinors, a Wick rotation can easily swap timelike and spacelike vectors, e.g. going from $\mathbb{R}^{3,1}$ to $\mathbb{R}^{1,3}$. Things would be different for real Clifford algebras but in this case our formalism take us to consider neutral spaces, $\mathbb{R}^{m,m}$ and again the identification of timelike and spacelike coordinates is ambiguous. On top of that there is the fact that the automorphism group of $\{\mathbb{1}_V, P, T, PT\} \cong \mathbb{Z}_2 \otimes \mathbb{Z}_2$ is the group of permutations of $\{P, T, PT\}$ that thus can be freely permuted, and so there are no indications neither from this side. Anyhow whatever the precise identification of P and T their corresponding automorphisms both move also the spinorial space supporting the regular representation of $\mathcal{C\ell}(g)$ both sending Ψ_{ab} to Ψ_{-a-b} and in any case $b \neq -b$.

It is evocative to write the general form of these elements in EFB

$$\begin{aligned}
\mathbb{1} &= \{q_1, p_1\}\{q_2, p_2\}\cdots\{q_m, p_m\} \\
\omega &= [q_1, p_1][q_2, p_2]\cdots[q_m, p_m] \\
\tau &= (p_1 + s\,q_1)(p_2 + s\,q_2)\cdots(p_m + s\,q_m) \qquad s = (-1)^{m+1} \\
\omega\tau &= (-1)^m (p_1 - s\,q_1)(p_2 - s\,q_2)\cdots(p_m - s\,q_m)
\end{aligned}$$

and the first two result: even under the main automorphism, they do not move spinor spaces and form a group isomorphic to $\mathbb{Z}_2$ that provides $\mathcal{C\ell}(g)$ grading. The last two, have parity $(-1)^m$ under the main automorphism, move spinor spaces and form, together with the first two, the group of discrete automorphisms isomorphic to $\mathbb{Z}_2 \otimes \mathbb{Z}_2$ group. Moreover since $\eta(\Psi_{ab}) = \epsilon(a)$ and $\theta(\Psi_{ab}) = \epsilon(a)\epsilon(b)$ [4] and observing that $\epsilon(-x) = (-1)^m \epsilon(x)$ we find

$$\begin{aligned}
\eta(\Psi_{-a-b}) &= (-1)^m \eta(\Psi_{ab}) \\
\theta(\Psi_{-a-b}) &= \theta(\Psi_{ab})
\end{aligned}$$

showing that, when m is odd, the chirality is reversed by automorphisms that move spinor spaces; a subtler study is needed to generalize this result from neutral spaces to real spaces of different signature.

To investigate how these inner automorphisms behave on generic spinors (4) and not only on EFB elements we give a simple result:

Proposition 2. *for any inner automorphism* $\alpha \in C_g^\star$ *the image of a MLI is a MLI, moreover* $x S_{hg} x^{-1} = S_{hg}$ *if and only if* $S_{hg} x^{-1} = S_{hg}$.

Proof. The first part follows immediately from the fact that a MLI must have rank 1; for the second part let $x S_{hg} x^{-1} = S_{hg}$, then since S_{hg} is a MLI we have also $S_{hg} x^{-1} = S_{hg}$; vice versa let $S_{hg} x^{-1} = S_{hg}$, since S_{hg} is a MLI $x S_{hg} = S_{hg}$ and thus $x S_{hg} x^{-1} = S_{hg}$. □

Thus since for any spinor $\varphi = \varphi \mathbb{1}$ the identity do not change spinor space. Going to PT since $\Psi_{af} = \epsilon(f)\Psi_{af}$ [4] and remembering that $\omega^{-1} = \omega^3$

$$\varphi\omega^{-1} = \omega^2 \sum_a \xi_{af}\Psi_{af}\omega = \omega^2 \sum_a \xi_{af}\epsilon(f)\Psi_{af} = \omega^2\epsilon(f)\varphi = \pm\varphi$$

and so both $\mathbb{1}_V$ and PT behave as expected also on generic spinors of any S_f. The effect of reversion automorphism (10) on a generic spinor is

$$\tau\varphi\tau^{-1} = \sum_a \xi_{af}\Psi_{-a-f}$$

and if φ has a defined chirality, $\omega\varphi = \eta\,\varphi$, then

$$\omega\tau\varphi\tau^{-1} = \sum_a \epsilon(-a)\xi_{af}\Psi_{-a-f} = (-1)^m\eta\tau\varphi\tau^{-1}\ .$$

We consider now the solutions of equations like $v\varphi = 0$, where $v \in V$ and $\varphi \in S$. We observe that they must remain the same under any injective map and thus $xv\varphi x^{-1} = 0$ if and only if $v\varphi = 0$ and thus $\varphi x^{-1} = 0$ only if $\varphi = 0$. As it was intuitive, inner automorphisms do not change solutions of $v\varphi = 0$; in particular the solutions of $xv\varphi = 0$ are identical to those of $xv\varphi x^{-1} = 0$.

We conclude with a first characterization of transformations that stabilize spinor spaces:

Proposition 3. *The automorphims such that $xS_fx^{-1} = S_f$ form a subgroup of $C_g^\star$ (6).*

Proof. Let $x \in C_f := \{x \in C_g^\star : xS_fx^{-1} = S_f\}$, by previous proposition we have $S_fx^{-1} = S_f$ and right multiplying by x we get $S_f = S_fx$ with which we prove that also $x^{-1}S_fx = S_f$ i.e. that also $x^{-1} \in C_f$. Let $x, y \in C_f$ then $xyS_fy^{-1}x^{-1} = S_f$ thus also $xy \in C_f$. □

We remark that in general if x leaves invariant spinor space S_f nothing can be said on its properties on a different $S_{f'}$, moreover this subgroup is not normal as it is simple to see.

In turn $C_g^\star$ contains another subgroup of transformations that stabilizes *any* spinor space S_f; in this case one can prove that its generic element has the form

$$x = \sum_a \alpha_a\Psi_{aa} \qquad \alpha_a \in \mathbb{F} - \{0\}$$

where Ψ_{aa} are the primitive idempotents; these elements form an abelian subalgebra and a normal subgroup of $C_g^\star$. In a certain sense this group is the equivalent of the Clifford Lipschitz group in the sense that whereas the former is a subgroup of $C_g^\star$ that stabilizes vectors the latter stabilizes spinor spaces. The study of these groups will be pursued in a forthcoming paper.

5. Conclusions

We have proved that all orthogonal transformations of an even dimensional vector space V can be seen as the restrictions of inner automorphisms of $\mathcal{C}\ell(g)$. We are thus allowed to assume that also a spinor φ must transform as $x\varphi x^{-1}$ and that, in some cases like e.g. P and T, this has the effect of moving spinor $\varphi \in S_f$ to $x\varphi x^{-1} \in S_{-f}$. This has no influence on the solutions of equations like $v\varphi = 0$ but the moved spinor $x\varphi x^{-1}$ may have opposite chirality.

The perspectives appear interesting but many things remain to be done to complete this study, one for all the classification of continuous transformations of V since it is a simple exercise to verify that whereas all automorphisms generated by an odd number of generators, like e.g. $\varphi \to e_i\varphi e_i^{-1}$, move the spinor space of φ, automorphisms where the generators appear in couples, like e.g. $\varphi \to (e_{2i-1}e_{2i})\varphi(e_{2i-1}e_{2i})^{-1}$, do not move the spinor space of φ. This and other issues on the group of trasformtions that stabilize spinors will be tackled in ongoing work.

IOP Conf. Series: Journal of Physics: Conf. Series **845** (2017) 012031 doi:10.1088/1742-6596/845/1/012031

References

[1] Ian M. Benn and Robin W. Tucker. *An Introduction to Spinors and Geometry with Applications in Physics.* Adam Hilger, Bristol and Philadelphia, 1987.

[2] Marco Budinich. On Computational Complexity of Clifford Algebra. *Journal of Mathematical Physics*, 50(5):053514, May 2009. arXiv:0904.0417 [math-ph] 2 April 2009.

[3] Marco Budinich. The Extended Fock Basis of Clifford Algebra. *Advances in Applied Clifford Algebras*, 22(2):283–296, 2012.

[4] Marco Budinich. On spinors transformations. *Journal of Mathematical Physics*, 57(7):071703–1–11, July 2016. arXiv:1603.02181 [math-ph] 7 Mar 2016.

[5] Paolo Budinich and Andrzej Mariusz Trautman. *The Spinorial Chessboard.* Trieste Notes in Physics. Springer-Verlag, Berlin Heidelberg, 1988.

[6] Paolo Budinich and Andrzej Mariusz Trautman. Fock Space Description of Simple Spinors. *Journal of Mathematical Physics*, 30(9):2125–2131, September 1989.

[7] Élie Cartan. Les groupes projectifs qui ne laissent invariante aucune multiplicité plane. *Bulletin de la Société Mathématique de France*, 41:53–96, 1913.

[8] Élie Cartan. *The Theory of Spinors.* Hermann, Paris, 1966. first edition: 1938 in French.

[9] Kenneth M. Case. Biquadratic spinor identities. *Physical Review*, 97:810–823, Feb 1955.

[10] Claude C. Chevalley. *Algebraic Theory of Spinors.* Columbia University Press, New York, 1954.

[11] Ludwik Dabrowski. *Group Actions on Spinors.* Lecture Notes. Bibliopolis, Napoli, 1988.

[12] Matej Pavšič. Space Inversion of Spinors Revisited: A Possible Explanation of Chiral Behavior in Weak Interactions. *Physics Letters B*, 692(3):212–217, May 2010.

[13] Roger Penrose and Wolfgang Rindler. *Spinors and Space-Time: Volume 2, Spinor and Twistor Methods in Space-Time Geometry.* Cambridge monographs on mathematical physics. Cambridge University Press, 1988.

[14] Ian Robertson Porteous. *Clifford Algebras and the Classical Groups.* Cambridge Studies in Advanced Mathematics 50. Cambridge University Press, 1995.

[15] Andrzej Mariusz Trautman. *Clifford Algebras and Their Representations*, volume v. 1–5 of *Encyclopedia of Mathematical Physics*, pages 518–530. Elsevier, Oxford, UK, 2006.

[16] Vadim Valentinovich Varlamov. Discrete symmetries and clifford algebras. *International Journal of Theoretical Physics*, 40(4):769–805, April 2001.

[17] Vadim Valentinovich Varlamov. CPT Groups of Spinor Fields in de Sitter and Anti-de Sitter Spaces. *Advances in Applied Clifford Algebras*, 25(2):487–516, June 2015.